4th International Conference on Smart and Innovative Agriculture (ICoSIA 2023)

BIO Web of Conferences Volume 80 (2023)

Yogyakarta, Indonesia
10-11 October 2023

Editors:

M.Z. Abidin
R. Shamsudin
P. Tongdeenok
S.T. Sobue
A. Younf
A. Hoque

ISBN: 978-1-7138-8653-2

Printed from e-media with permission by:

Curran Associates, Inc.
57 Morehouse Lane
Red Hook, NY 12571

Some format issues inherent in the e-media version may also appear in this print version.

For additional information, please contact EDP Sciences – Web of Conferences at the address below.

EDP Sciences – Web of Conferences
17, Avenue du Hoggar
Parc d'Activité de Courtabœuf
BP 112
F-91944 Les Ulis Cedex A
France

Phone: +33 (0) 1 69 18 75 75
Fax: +33 (0) 1 69 28 84 91

contact-edps@webofconferences.org

Additional copies of this publication are available from:

Curran Associates, Inc.
57 Morehouse Lane
Red Hook, NY 12571 USA
Phone: 845-758-0400
Fax: 845-758-2633
Email: curran@proceedings.com
Web: www.proceedings.com

TABLE OF CONTENTS

AGRICULTURAL BIG DATA ANALYSIS

Application of Hyperspectral Imaging for Rapid and Nondestructive Detection of Paraffine-Contaminated Rice 01001 1
Mohammad Akbar Faqeerzada, Tanjima Akter, Umuhoza Aline, Muhammad Fahri Reza Pahlawan, Byoung-Kwan Cho

Implementation of Real-Time Image Processing on Bacterial Cellulose Formation Using Soybean-Boiled Wastewater with the Variation of Carbon Sources During Fermentation 01002 9
Ahmad Fatih Marzuki, Darmawan Ari Nugroho, Tyasto Prima Ahmadi, Atris Suyantohadi

AGRICULTURAL GEOGRAPHY

Analysis of Agricultural Land Carrying Capacity in North Sumatra Province in 2020 02001 17
Muhammad Bagus Fitriawan, Nabila Husna Risada, Yosi Yaseva, Ratih Fitria Putri

Agriculture Carrying Capacity Evaluation of Land Subsidence Area in Yogyakarta Special Province 02002 22
Alfindy Juliansari Alam, Azizah Nurkhalifah, Muhammad Zayyin Asrofi, Muhammad Raidan Azani, Ratih Fitria Putri

Analysis of the Carrying Capacity Agricultural Land of Jawa Tengah Province in 2022 Using the Soemarwoto Method 02003 27
Arianti Ade Maya Dian, Wicaksono Satrio Aji, Muhammad Zayyin Asrofi, Ratih Fitria Putri

Analysis of the Carrying Capacity of Agricultural Land in South Sulawesi Province 02004 33
Favian Daniswara Nugraha, Syifa Ayulia Anggraini, Muhammad Zayyin Asrofi, Ratih Fitria Putri

Agricultural Land Monitoring of Land Subsidence Area Using Spatio-Temporal Approach in East Java Indonesia 2018-2022 02005 37
Dyfany Aurariel Syahda, Melta Meliano Putri, Yosi Yaseva, Putri Yuko Nawang Wulan, Ratih Fitria Putri

Distribution of Oil Palm Plantations and Their Suitability with Spatial Planning: An Assessment in Aceh, Indonesia 02006 42
Ikhwan Amri, Elsa Widura, Muhammad Iqbal, Yan Andika, Asalia Raudhati Izzatillah

GIS-Based Assessment of Land Suitability for Rubber Cultivation in Chittagong District, Bangladesh 02007 49
Md. Ahasanul Hoque, Rezwan Ahmed

The Importance of Livelihood Diversification on Agroforestry Farmers in the Landslide Prone Area: A Case Study in Menoreh Hills Kulon Progo, Indonesia 02008 59
Arum Mutiarra Ridyo, Utami Arini Wahyu, Irham Irham

Women's Roles in Beef Cattle Farming Households in Nagari Lakitan Utara, Pesisir Selatan Regency, Sumatra Barat Province 02009 66
Siti Andarwati, Budi Guntoro, Muhammad Fajri

LAND AND ENVIRONMENTAL MANAGEMENT

Amelioration of Dry Land Suboptimal Using Biochar and Compost to Improve Soil Physical Properties and Soybean Yield 03001 72
Endriani, Diah Listyarini

Biodiversity in Lower Montane Forest of Petungkriyono: A Review 03002 79
Dian Pratiwi, Tri Wira Yuwati, Sri Lestari, Yonky Indrajaya, Astarina Eka Dewi

Can iPhone/IPad LiDAR Data Improve Canopy Height Model Derived from UAV? 03003 98
Deha Agus Umarhadi, Senawi, Wahyu Wardhana, Emma Soraya, Aqmal Nur Jihad, Fiqri Ardiansyah

Comparative Effect of Polymer Coating Materials on Physical Properties of NPK Fertilizer Pellet 03004 103
Gatot Trimulyadi Rekso

Crop Pattern Suitability Based on Water Availability in Dry-Land Agriculture Nawungan, Selopamioro, Imogiri, Bantul 03005 108
Khoirida Dian Putranti, Nur Ainun H. J. Pulungan, Makruf Nurudin

Cross-Sectional Study on Land and Seed of Oil Palm Smallholders in the Western Part of Aceh: Challenge for Sustainability Agenda 03006 114
Ahmad Humam Hamid, Agus Nugroho, Fajri, Agussabti, Litna Nurjannah Ginting, Muhammad Rusdi, Muyyasir, Akhmad Baihaqi

Effectiveness Combination of Leucaena Compost and Coconut Shell Biochar on Reduce Soil Compaction and Corn Growth in Ultisol 03007 121
Endriani, Diah Listyarini

Enhancing Food Operator Intention on Specialty Café Using 3R Waste Management Approach for Eco-Enzyme Production as an Implementation of SDGs 12 (Study on Akkar Specialty Cafés) 03008 128
Wahyu Putri Andhini, Darmawan Ari Nugroho, M. Prasetya Kurniawan

Erosion Control in the Tropical Rainforest Catchments Using Modified Intensive Rehabilitation 03009 136
Hatma Suryatmojo, Muhammad Chrisna Satriagasa, Widiyatno, Alifia Dyah Ayu

Formulating the Direction of Community Land Use to Support the Conservation of Javanese Gibbon (Hylobates Moloch) Habitat at Petungkriyono, Central Java 03010 141
Tri Wira Yuwati, Astarina Eka Dewi, Ahmad Arief Fahmi, Muhammad Ali Imron, Muhammad Tafrichan, Yudistira Soeherman, Sri Lestari, Dian Pratiwi

Assessment of Surface Runoff Potency Under Tropical Environment for Soil and Water Conservation Planning 03011 146
Safira Defiani Khoirotunnisa, Gina Isna Nafisa, Chandra Setyawan, Muchammad Nu'Man, Ngadisih

Impact of Bamboo-Based Agroforestry Demonstration Plots Based on Farmers' Perceptions 03012 151
Eva Fauziyah, Sanudin, Aditya Hani

Impact of Glyphosate Contamination on Chemical Properties of Inceptisols Amelioration with Biochar from Rice Husks, Young Coconut Waste, and Bamboo 03013 160
Herviyanti Herviyanti, Amsar Maulana, Arestha Leo Lita, Mimien Harianti, Teguh Budi Prasetyo, Rezha Tri Khurnianto, Pitri Juwita, Ridho Ryswaldi, Syafrimen Yasin

Plants Suitability for Landslides Mitigation in the Giritengah Catchment Area, Indonesia 03014 169
Satwika Indri Masrianti, Ambar Kusumandari, Hatma Suryatmojo

Relationship Between Tree Height-Diameter at Breast Height (DBH) and Crown Diameter-DBH of Fruit Trees in Bogor Botanical Gardens 03015 176
Indriani Ekasari, Rifaangga Kurnia

Retention and Availability of Andisol Soil Phosphorus in Monocultural and Policultural Crop Patterns in Ngablak and Getasan Subdistrict, Magelang and Semarang, Indonesia 03016 183
Elia Laila Rizqiyah, Sri Nuryani Hidayah Utami, Benito Heru Purwanto

Soil Erodibility: Influencing Factors and Its Relation to Soil Fertility in Nawungan, Selopamioro, Bantul Regency 03017 193
Umi Mar Atus Sholihah, Nur Ainun H. J. Pulungan, Fathi Alfinur Rizqi

Spatial Distributional of Irrigation Water Quality in Peaty Paddy Field of Paser Regency, East Kalimantan Province, Indonesia 03018 200
Moch Reza Wahyu Abdilah, Sahid Susanto, Chandra Setyawan, Muhamad Khoiru Zaki, Dehafiyyan Ramadhani

The Critical Level Mapping of the Infiltration Area in the Northern Part of Banyumanik District, Semarang City, Central Java, Indonesia 03019 207
Ahmad Arfiandi, Wawan Budianta, Esti Handini

The Effect of Land Cover Changes on Landslides in Grindulu Watershed, Pacitan Regency, East Java 03020 221
Raka Achmad Fauzi, Rika Harini, R. Rijanta

The Growth and Yield of Upland Rice (Oryza Sativa L.) Appliced with Rice Husk Zeolite and N Fertilizer 03021 227
Nelvia, Zulfatri, Idwar, Nursiani Lubis, Ferdi Saputra

The Impact of Water Management Systems and Biofertilizers on Soybean Production in Type B Tidal Swamp Lands in Kapuas District, Central Kalimantan, Indonesia 03022 233
Kartika Ayu Kinanti, Sri Nuryani Hidayah Utami, Jaka Widada, Hendri Sosiawan, Wahina Anisa Yusuf, Arif Mulyawan

The Positive Effect of the Forest Environment in the Ecotherapy Healing Forest Program on the Physical and Mental Health After COVID-19 03023 243
Vincencia Septaviani Issera Sulistya Putri, Lies Rahayu Wijayanti, Iswari Nur Hidayati

Watershed Management and Classification in Thailand 03024 252
Piyapong Tongdeenok

Yield Gap in Cayenne Pepper Production in South Kalimantan 03025 264
Lelya Pramudyani, Sri N. H. Utami, Taryono, Eka T. S. Putra, Anna Hairani

PRECISION NUTRITION TECHNOLOGY

Analysis of the Effect of Conveyor Speed in Cryo-Grinding System on Matcha Product Quality 04001 270
Febriana Rias Putri, Anggoro Cahyo Sukartiko, Wagiman, Wahyu Supartono

Changes of Dark Couverture Chocolate Hardness During Storage Tempered Using Automatic Tempering Machine with Tank and Tempering Temperature as Variables 04002 276
Shiammita Rosmindari, Arifin Dwi Saputro, Nursigit Bintoro

Extraction and Isolation of Cricket Protein Isolate with Ammonium Sulfate Addition Method and Its Effect on the Functional Properties of the Proteins 04003 284
Edward, Thanakorn Wongprasert, Thasorn Bunyakanchana, Wasin Pattaraprachyakul, Panattida Siripitakpong, Inthawoot Suppavorasatit

Gas Absorbent Based Modified Atmosphere Packaging to Optimize Dragon Fruit Shelf Life 04004 292
Satria Primatama, Fahrizal Yusuf Affandi, Anisa Kistanti, Musyrifah Kurniawati, Bayu Nugraha

Palm Oils Consumption Modulates Serum SGPT and SGOT in Rats Fed a High-Fat Diet 04005 303
Zaki Utama, Sri Raharjo, Agnes Murdiati, Andriati Ningrum

Physical Characteristics of Extrudate from Mixed Corn Gritsoybean Flour with Treatments of Moisture Content and Extruder Barrel Temperature 04006 307
Riski Sri Mahanani, Joko Nugroho Wahyu Karyadi, Devi Yuni Susanti, Ailsa Saraswati Samodra, Vincentius Ferry Surya Dwinata, Dian Kharisma Rahmawati, Prasetya Febriyansyech Personanta Sinuhaji

The Importance of LCA in the Production of Turmeric Powder and Chilli Powder at PT X 04007 313
Dyah Ayu Kurniawati, Wahyu Supartono, Pujo Saroyo

X-Ray CT Applications to Reveal Physical Defects of Onion During Storage 04008 318
Devi Priyanti, Ira Ayuningsih, An-Nidaa' Fatkhur Rahmah, Sri Rahayoe, Bayu Nugraha

SMART GENETICS RESOURCE MANAGEMENT AND UTILIZATION

Continental Vs. Tropical Breed: Immunity Comparison Under Heat Stress Conditions Utilizing qRT-PCR Technique 05001 323
Roland Fajardo, Putri Kusuma Astuti, Zoltán Bagi, Lilla Bodrogi, Tímea Pintér, Gabriella Skoda, Husein Ohran, George Wanjala, Szilvia Kusza

Morphology, Biomass Production, and Nutrient Analysis Four Napier Grass Cultivar (Pennisetum Purpureum Schumach.) in Teak Tree Forest Area Megeri Village, Blora, Central Java 05002 331
Yogi Sidik Prasojo, Mustafa Kamal, Arkan Roja Sukmana, Bowo Dwi Siswoko, Nafiatul Umami

Tiller Number and Flowering Ability on BC1F1 Progeny of Interspecific Hybridization in Ruzigrass (Urochloa Ruziziensis) 05003 335
Yogi Sidik Prasojo, Pattama Nitthaisong, Melody Muguerza, Sukanya Kamphayae, Nafiatul Umami, Genki Ishigaki

SMART AND PRECISION FARMING

Design and Development Smart Aquaculture in Freshwater Pond Based on Fuzzy Logic 06001 339
Endra Pitowarno, Bagas Oktavianto Leksono, Eko Budi Utomo, Muhammad Muamar

Development of Automatic Dual Sequence Control Temporary Immersion Bioreactor Systems for Micropropagation Coelogyne Pandurata Lindl. 06002 346
Anisa Dewi Rahayu, Lathief Al Umami, Shiddharta Arya Anggoro Cen, Bimo Rizki Abdusshamad, Leonard Sean, Kresna Budiman, Labriyantoko Kurniawan, Muhammad Akhsin Muflikhun, I. Wayan Mustika, Endang Semiarti

Identification of Maturity Stage of Cacao Using Visible Near Infrared (Vis-NIR) and Shortwave Near Infrared (SW-NIR) Reflectance Spectroscopy 06003 351
Riana Listanti, Rudiati Evi Masithoh, Arifin Dwi Saputro, Hanim Zuhrotul Amanah

Improved Image Recognition Via Synthetic Plants Using 3D Modelling with Stochastic Variations 06004 361
Chris C. Napier, David M. Cook, Leisa Armstrong, Dean Diepeveen

Innovation Characteristics in Configuring Farmers' Digital Literacy on E-Reporting Polygon in Wukirsari Indonesia 06005 368
Hisna Shania Alya, Widhiningsih Diah Fitria, Subejo

Multi-Agro: Design and Development of a Cartesian-Based SelfPlantation Robot for Home-Scale Gardens with Mobile App 06006 375
Endra Pitowarno, Adytia Darmawan, Fatih Ali Fikri El Munba'Its, Bagus Pramuditya Affandy

Performance of a Portable NIR Spectrometer to Distinguish Coffee Species Based on Qualitative Chemometric and Artificial Neural Network (ANN) Models 06007 381
Agus Dharmawan, Rudiati Evi Masithoh, Hanim Zuhrotul Amanah

Predicting Physicochemical Properties of Melon (Cucumis Melo L.) Using Ultrasonic Technology and Artificial Neural Network 06008 389
Fahar Yazid Izdihar, Nafis Khuriyati, Wagiman

Space Based Data Usage for Smart Farming 06009 396
Shin-Ichi Sobue

SUSTAINABLE FOOD PRODUCTION

Analysis of the Effect of Brand Equity on Purchasing Decision Towards Organic Rice in Special Region of Yogyakarta 07001 403
Nisa Puspita Mawadati, Novita Erma Kristanti, Nafis Khuriyati

Assessing the Viability of a Small Micro-Enterprise of an Urban Farming Group in Bausasran Village, Yogyakarta, in Developing a Vegetable-Based Noodle Products 07002 409
Habib Luthfi Ash Shiddiqie, Ryan Anugrah Putra, Ardi Wiranata, Teguh Ariyanto

Effect of Varying Color LED Lights on Porang (Amorphophallus Muelleri) Bulbil Seed Germination and Porang Plant Vegetative Growth Phase 07003 415
Aryanis Mutia Zahra, Esty Indrayanti, Bayu Dwi Apri Nugroho, Rudiati Evi Masithoh

Identification of Sustainable Supply Chain Performance in Primary Cocoa Processing (Case Study in Patuk, Yogyakarta) 07004 423
Tasyarafa Shelia Khairani, Muhammad Prasetya Kurniawan, Kuncoro Harto Widodo, Tri Marwati, Titiek Farianti Djaafar

Nuisance Insects of Rearing Mass Asian Citrus Psyllid (ACP) Diaphorina Citri Under Controlled Conditions 07005 430
Haris Setyaningrum, Edhi Martono, Alan Soffan, Jianhua Mo, Siti Subandiyah

The Quality Analysis of "JOSS" (Jatisarono Organik Sehat Sejahtera) Organic Rice 07006 434
Hasna Difa Salisa, Novita Erma Kristanti, Anggoro Cahyo Sukartiko

Using Superior Variety and Application of CI200 to Support Food Security in South Sumatra Swamp Land 07007 439
Waluyo, Agus Suprihatin, Suparwoto, Jumakir

Author Index

https://doi.org/10.1051/bioconf/20238000001

Preface

ICoSIA 2023 - 4th International Conference on Smart and Innovative Agriculture is an annual meeting organized by Directorate of Research Universitas Gadjah Mada (UGM), Yogyakarta, Indonesia. This year's conference was held on 10–11 October 2023. The scientific symposium program consisted of plenary sessions and parallel sessions and included the following topics: Agricultural Big Data Analysis; Agricultural Geography; Land and Environmental Management; Precision Nutrition Technology; Smart and Precision Farming; Smart Genetics Resource Management and Utilization; and Sustainable Food Production.

We are grateful for the support of the faculties in the Universitas Gadjah Mada which scientifically supported the conference from the reviewers until the editing of the papers published in the BIO Web on Conferences. The ICoSIA 2023 had about 223 attendees including the presenters and participants. Among those, the conference received 95 submitted papers and all papers were subjected to at least 2–3 reviews by experts on the related topics. For this, only 63 papers were selected for publication from eight countries: Australia, Bangladesh, Hungary, Indonesia, Japan, Philippines, South Korea, and Thailand.

We believed that during the two-day conference, we had enabled to facilitate plenty of networking opportunities and provided the participants with the opportunity to meet and interact with the leading scientists and researchers, friends as well as colleagues. We hope that this proceeding will be beneficial for the development of life science in general and biological sciences and public health in particular.

Editorial boards

https://doi.org/10.1051/bioconf/20238000002

Welcoming Remarks from the Chair

Assalamualaikum. Wr. Wb.

Distinguished Speakers, Guests, and all participants of ICoSIA 2023.

Good morning, welcome to the 4th International Conference on Smart and Innovative Agriculture (ICoSIA 2023) in Eastparc Hotel Yogyakarta, Indonesia. As the Chairman of ICoSIA 2023, I am deeply honoured to stand before you, to extend a warm welcome to all of you to Indonesia, especially to Yogyakarta with its warmth and culture. "*Sugeng rawuh wonten Yogyakarta*".

ICoSIA 2023 holds a special place in our hearts to explore the theme of "Food Security and Resilience in Multidimensional Crises." In a world where challenges are multifaceted and ever-evolving, the pursuit of food security is not only a noble endeavor but also a vital one for humanity. This theme underscores the urgency of our discussions during the next two days.

I would like to take this moment to express my sincere gratitude to our distinguished speakers, whose unwavering commitment to agricultural innovation and food security has brought us here today. We are privileged to have with us an exceptional lineup of experts:

1. Prof. Ts. Dr. Rosnah Shamsudin from Universiti Putra Malaysia, Malaysia
2. Dr. Glenn Gregorio from Southeast Asian Regional Center for Graduate Study and Research in Agriculture, Philippines
3. Dr. Sinichi Takashima Sobue from Japan Aerospace Exploration Agency, Japan
4. Dr. Anthony Young from The University of Queensland, Australia
5. Ahasanul Hoque from the World Bank, Bangladesh
6. Asst. Prof. Piyapong Tongdeenok from Kasetsart University, Thailand

This year, ICoSIA 2023 has brought together a wealth of knowledge under the umbrella of seven distinct symposia, including Agricultural Big Data Analysis Symposium; Agricultural Geography Symposium; Land and Environmental Management Symposium; Precision Nutrition Technology Symposium; Smart and Precision Farming Symposium; Smart Genetics Resource Management and Utilization Symposium and Sustainable Food Production Symposium.

With a total of 95 submissions from several countries around the world such as Indonesia, South Korea, Pakistan, Bangladesh, France, Thailand, Ethiopia, India, Australia, Japan, and Hungary, ICoSIA 2023 has indeed become a global forum for agricultural excellence. The diversity of perspectives and experiences represented here today is a testament to the conference's international reach and significance.

In conclusion, I want to express my sincere appreciation to the organizing committee, reviewers, and all those who have worked hard to make ICoSIA 2023 happen. Your dedication and hard work have not gone unnoticed, and we are grateful for your contributions.

Thank you very much for your presence and participation in ICoSIA 2023. I am confident that our collective efforts will lead to transformative outcomes that will benefit our world for generations to come.

Let us embark on this enriching journey together. Thank you.
Wassalamualaikum. Wr. Wb.

Yogyakarta, October 10, 2023

Chair of ICoSIA 2023
Dr. Ir. Mohammad Zainal Abidin, S.Pt., M.Biotech., IPM.

Welcoming Remarks
from the Rector of Universitas Gadjah Mada

Dear distinguished keynote speakers,
Distinguished invited speakers and participants,
Conference's committee,
Ladies and gentlemen.

On behalf of Universitas Gadjah Mada, it is my pleasure to welcome all of you today at the 4th International Conference on Smart and Innovative Agriculture (ICoSIA 2023), hosted by the Directorate of Research Universitas Gadjah Mada. First of all, I would like to express my sincere appreciation for our speakers, Prof. Ts. Dr. Rosnah Shamsudin, Asst. Prof. Piyapong Tongdeenok, Dr. Anthony Young, Dr. Sinichi Takashima Sobue, Dr. Glenn Gregorio, and Mr. Ahasanul Hoque.

As a pioneering university, the history of UGM's education has opened the boundaries between academics and professionals across the world, tso discover critically scientific inventions as the precious roots of knowledge for the benefit of humankind. The 4th International Conference on Smart and Innovative Agriculture (ICoSIA 2023) also provides an ideal academic platform for researchers to present the latest research findings and describe emerging technologies and directions in agriculture.

This year, the selected theme for the conference will focus on "**Food security and resilience in multidimensional crises**". Food security has been a longstanding area of emphasis in research and policy concerned with development. Food security is accomplished when individuals, regardless of their circumstances, have consistent and equitable access to an adequate quantity of safe and nutritious food that meets their dietary requirements and preferences for an active and healthy life. The challenge of food security is situated at the intersection of the fundamental rights to food and health, affecting not only developing nations but also prosperous, industrialized countries. On the other hand, resilience has emerged as a more recent field of study. Considering the intricate nature of both topics, the endeavor of integrating the inherently multidimensional concepts of resilience and food security is a challenging undertaking.

The 4th ICoSIA plays a crucial role preserving our commitment to nurturing an intellectual culture.It is our sincere hope that this year's conference will serve as a powerful source of inspiration for academicians and researchers, motivating them to achieve even greater impact in the field of life sciences. Furthermore, we aspire to encourage students to actively engage in scholarly writing and strive towards publishing their work in esteemed journals.

We are deeply honored and humbled by the presence of numerous esteemed experts who attend this year's conference. We extend our sincere gratitude to the speakers for their exceptional expertise and knowledge, which will enrich the discussions during the conference. Additionally, we would like to express our special thanks to the hardworking organizing committee members of ICoSIA 2023 for their dedication in preparing the conference, and also to the staff of the Directorate of Research for their valuable contributions. Lastly, we would like to express our gratitude to all the conference participants whose valuable contributions have made ICoSIA 2023 a truly remarkable event. We sincerely hope that the outcomes of this conference will have a lasting impact on the fields of science, agriculture, and humanity for many years to come. Moreover, we anticipate that this conference will foster stronger connections and friendships among the attendees. We are confident that you will engage in fruitful and rewarding exchanges, leading to valuable insights and collaborations.

Thank you.

Rector of Universitas Gadjah Mada,
Prof. dr. Ova Emilia, M.Med.Ed., Sp.OG(K)., Ph.D.

UNIVERSITAS GADJAH MADA

UGM ICoSIA 2023

The 4th International Conference on Smart and Innovative Agriculture

Food Security and Resilience in Multidimensional Crises

Conference Day : 10–11 October 2023 | Yogyakarta, Indonesia

Prof. **ROSNAH SHAMSUDIN**
Universiti Putra Malaysia, Malaysia

Asst. Prof. **PIYAPONG TONGDEENOK**
Kasetsart University, Thailand

Dr. **ANTHONY YOUNG**
The University of Queensland, Australia

Dr. **SHINICHI TAKASHIMA SOBUE**
Japan Aerospace Exploration Agency, Japan

Dr. **GLENN GREGORIO**
Southeast Asian Regional Center for Graduate Study and Research in Agriculture, Philippines

AHASANUL HOQUE, M.Sc.
GeoHabitatLab, Bangladesh

Overview

Held as part of the **Universitas Gadjah Mada Annual Scientific Conferences (UASC 2023)** series, the **4th International Conference on Smart and Innovative Agriculture (ICoSIA 2023)** provides an ideal academic platform for researchers to present the latest research findings and describe emerging technologies and directions in agriculture.

Symposia

- Agricultural Big Data Analysis
- Agricultural Geography
- Land and Environmental Management
- Precision Nutrition Technology
- Smart And Precision Farming
- Smart Genetics Resource Management and Utilization
- Sustainable Food Production

Important Dates

- **8 September 2023**
  Full paper submission deadline
- **20 September 2023**
  Notification of acceptance
- **25 September 2023**
  Early bird payment deadline
- **27 September 2023**
  Registration deadline
- **27 September 2023**
  Camera ready and presentation submission deadline
- **10-11 October 2023**
  THE CONFERENCE

The accepted and presented papers will be submitted in the BIO Web of Conferences that are indexed by Scopus

icosia.ugm.ac.id icosia@ugm.ac.id icosiaugm icosiaugm s.id/icosiasupport edp sciences

Web of Conferences

Statement of Peer review

In submitting conference proceedings to *Web of Conferences*, the editors of the proceedings certify to the Publisher that

1. They adhere to its **Policy on Publishing Integrity** in order to safeguard good scientific practice in publishing.
2. All articles have been subjected to peer review administered by the proceedings editors.
3. Reviews have been conducted by expert referees, who have been requested to provide unbiased and constructive comments aimed, whenever possible, at improving the work.
4. Proceedings editors have taken all reasonable steps to ensure the quality of the materials they publish and their decision to accept or reject a paper for publication has been based only on the merits of the work and the relevance to the journal.

Title, date and place of the conference

ICoSIA 2023 - 4th International Conference on Smart and Innovative Agriculture

Yogyakarta, Indonesia at 10–11 October 2023

Proceedings editor(s):

1. Dr. Mohammad Zainal Abidin, S.Pt., M.Biotech.; Universitas Gadjah Mada, Indonesia (m.zainal.abidin@ugm.ac.id)
2. Prof. Ts. Dr. Rosnah Shamsudin; Universiti Putra Malaysia, Malaysia (rosnahs@upm.edu.my)
3. Asst.Prof. Dr. Piyapong Tongdeenok; Kasetsart University, Thailand (piyapong.t@ku.th)
4. Dr. Shinichi Takashima Sobue; Japan Aerospace Exploration Agency, Japan (sobue.shinichi@jaxa.jp)
5. Dr. Anthony Young; The University of Queensland, Australia (anthony.young@uq.edu.au)
6. Ahasanul Hoque, MD.; The World Bank, Bangladesh (hoque.ahasan@gmail.com)

Date and editor's signature

November 9th, 2023

Dr. Mohammad Zainal Abidin, S.Pt., M.Biotech.

Application of Hyperspectral Imaging for Rapid and Nondestructive Detection of Paraffine-Contaminated Rice

Mohammad Akbar Faqeerzada[1], Tanjima Akter[2], Umuhoza Aline[1], Muhammad Fahri Reza Pahlawan[2], Byoung-Kwan Cho[1,2*]

[1]Department of Biosystems Machinery Engineering, Chungnam National University, Daejeon, Korea
[2]Department of Smart Agricultural Systems, Chungnam National University, Daejeon, Korea

Abstract. The emergence of paraffin-coated rice in China, aimed at enhancing its market appeal and achieving a translucent appearance, has given rise to a significant global food safety concern. This situation poses substantial health risks to consumers. Hyperspectral analysis, recognized as a powerful and nondestructive technique for assessing food quality and safety, offers a potential solution. This study conducted a comprehensive investigation using Visible-Near Infrared (VIS-NIR) hyperspectral imaging systems operating within the 400-1000 nm range to identify paraffin-contaminated rice. Various rice varieties from diverse regions were obtained and intentionally tainted with varying levels of paraffin. Imaged samples were further preprocessed for spectral data extraction from individual rice seeds' regions of interest (ROI). The dataset encompassed 3000 spectral records obtained from both non-contaminated and contaminated samples. The obtained spectral data were employed to develop partial least squares discriminant analysis (PLS-DA) and principal component linear discriminant analysis. The primary goal was to discriminate between contaminated and non-contaminated rice samples effectively. Notably, the results indicated that PLS-DA consistently achieved an accuracy exceeding 94% across various preprocessing techniques. Overall, this study showcased the potential of combining hyperspectral imaging with chemometrics to detect paraffin-contaminated rice seeds, providing a valuable contribution to food safety assessment in the industry.
Keywords: hyperspectral imaging, paraffine contaminated, PLS-DA, PCA-LDA, beta coefficient.

1 Introduction

Rice (*Oryza sativa L.)* has stood as a cornerstone in feeding a substantial fraction of the world's inhabitants. About half of the global populace depends on rice as a central source of sustenance. Across numerous nations, especially in Asia, rice holds a pivotal role in diets, serving as a fundamental nutritional element [1]. The adulteration of food products is a deceptive practice involving substituting high-cost ingredients with lower-grade and cheaper substitutes [2]. In recent years, rice adulteration has emerged as a significant concern, drawing attention from regulatory agencies, suppliers, and consumers alike. This adulteration trend is particularly prevalent in Chinese rice production and consumption, which was recently reported [3]. Notably, paraffin, an industrial wax, has become a central element in adulterating rice, leading to the alarming emergence of "Toxic rice." This practice involves treating rice with chemicals and applying a paraffin coating to enhance its visual appeal and marketability, resulting in a desirable translucent appearance. However, the consequences of consuming this adulterated rice are far from desirable, as it contains carcinogens [4] that pose serious health risks. Detecting adulterated rice contaminated with industrial paraffin is challenging due to the deceptive nature of its appearance. Current identification methods often rely on subjective factors, such as visual cues like color and texture, or more advanced physical and chemical tests like GC-MS [5], [6]. However, these methods offer higher precision; they are labor-intensive and require extensive separation and extraction of paraffins before qualitative and quantitative analysis can be performed.

In response to the challenges posed by traditional detection methods, emerging technologies such as hyperspectral imaging have gained attention. Hyperspectral imaging has emerged as a powerful and innovative tool for agricultural product quality inspection, which combines traditional imaging with spectral analysis, enabling the acquisition of extensive data on the surface and internal characteristics of measured samples. This technology has been widely utilized in nondestructive testing for adulterated food products, including grains, fruits, and vegetables [7]. For example, color adulterant in red chili [8], fraud detection in meat [9], and adulterated almond powder with apricots and peanuts [10]. Further, hyperspectral imaging demonstrated high potential for qualitative and quantitative analysis of agriculture crops, for instance, the determination of chemical contents of water and

*Corresponding author: Byoung-Kwan Cho (chobk@cnu.ac.kr)

tuber flour [11], prediction of anthocyanins in black rice [12], assessments of internal defects in macadamia [13] and quality analysis of stored bell peppers [14].

In this study, we capitalize on the capabilities of hyperspectral imaging technology to address the challenges posed by conventional detection methods in identifying adulterated rice. The utilization of hyperspectral imaging in detecting adulterated rice can enhance the efficiency of detection processes and ensure consumer protection from the health risks associated with consuming toxic rice. This study's Specific goal was to develop a system for nondestructive quality inspection of contaminated rice. The obtained results from this research could have broader implications for the application of hyperspectral imaging in ensuring the authenticity and quality of various food products.

2 Materials and Methods

2.1 Sample preparation

Three varieties of uncontaminated rice samples were prepared from the local market, while the paraffin liquid was purchased from authentic companies in South Korea. Contaminated rice was made deliberately by adding paraffin to the uncontaminated rice. Six contamination concentration were used 0, 1, 2, 3, 4, and 5 %w/w. A total of 50 g contaminated rice were made for each concentration. After the paraffin liquid were added to the rice, the mixture were shaken using vortex mixer for around 5 minutes each. Every samples were kept in the room temperature to equilibrate before spectra measurement.

2.2 System development

1. Camera lens	5. Line bar Light	9. Camera and light slider
2. Spectrograph	6. Optic fibers	
3. EMCCD	7. Sample holder	10. Sample
4. Light sources	8. DC motor	11. Translation stage

Fig. 1. Schematic representation of imaging system and components.

A line-scan hyperspectral imaging (HSI) technique was employed to assess contaminated rice non-destructively. During measurement, rice sample were placed on a 10x10 cm black plate. The system ranged the visible to near-infrared range from 400 to 1000 nm. The critical elements of the imaging setup, as depicted in Figure 1, consisted of a line-scan spectrograph provided by Headwall Photonics (Fitchburg, MA, U.S.A.), a c-mount objective lens from Schneider Optics (Hauppauge, NY, U.S.A.) with specifications of F/1.9 and 35mm, and an electron-multiplying charge-coupled device (EMCCD) camera, model MegaLuca R, manufactured by ANDOR Technology (South Windsor, CT, USA). The camera featured 1,004 spatial × 1,002 spectral pixels and operated at a maximum pixel-readout rate of 12.5 MHz. The entire system was equipped with a thermoelectric cooling mechanism, maintaining a temperature of -20 °C. The lighting setup comprised six 100 W halogen light sources (LS-F100HS) connected via optical fiber, and a stepper motor was coordinated with the camera's movements using custom software.

3 Image processing and analysis

Samples were imaged using hyperspectral imaging, and the initial hyperspectral images were converted to corresponding reflectance images to eliminate undesired noise stemming from both instruments and external sources of light; equation (1) was employed. Mainly, the correction procedure aimed to standardize the light distribution and generate images reflecting actual reflectance levels. A Teflon tile with an approximate reflectance of 99% was utilized to capture the white reference under the same lighting arrangement as the sample images. On the other hand, the dark reference was obtained by covering the lens with an opaque cap, resulting in a reflectance of approximately 0%, while the illumination unit was turned off. The refined images, denoted as Xcal, were computed by applying the formula to the measured raw hyperspectral image (Xraw), the white reference (Xwhite), and the dark reference (Xdark).

$$X_{cal} = \frac{X_{raw}-X_{dark}}{X_{ref}-X_{dark}} \qquad (1)$$

The background was removed from the rectified hyperspectral images by utilizing the mean reflectance of both the background pixels and the pixels corresponding to plant rice samples. This procedure aimed to eliminate irregular intensity and fluctuations in the background. Subsequently, the adjusted hyperspectral image was employed to extract spectral data from the ROI of the individual seed samples.

4 Data analysis

4.1 Spectral extraction and correction

After removing the background, calibrated hyperspectral images were utilized to extract spectral insights within specific areas of interest (ROIs) across the entire range of the spectra (VIS-NIR: 400-1000 nm) for rice seed samples. The average spectrum for each seed from the ROIs of 1,500 non-contaminated and 1,500 contaminated seeds and further analysis dataset was divided into 70% for calibration and 30% for validation. The acquired raw spectral data contained random noise that resulted from variations in light direction and scattered light produced by the equipment. Consequently, preprocessing the spectral data was necessary to eliminate unwanted spectral fluctuations and unveil the valuable information within the plants. This study employed multiple preprocessing techniques, including standard normal variate (SNV), multiplicative scatter correction (MSC), smoothing, and the Savitzky–Golay first and second derivatives approach. Detailed information regarding each preprocessing method can be found in published reports [15]. Figure 2 (a) displays the initial preprocessing of raw spectral data, whereas Figure 2 (b) depicts the data after undergoing raw and SNV preprocessing and shows apparent differences between preprocessed and raw spectral data. Notably, the higher performed preprocessing was further highlighted for further analysis.

4.2 Model Development

Processed spectral data were utilized to develop distinct models of principle component analysis combined with linear discrimination analysis (PCA-LDA) and partial least square discrimination analysis (PLS-DA). The spectral data were divided into 2000 (1000 contaminated and 1000 non-contaminated) spectral data for calibration and 1000 (500 contaminated and 500 non-contaminated) for validation, all derived from raw preprocessed spectral data.

PCA-LDA analysis, a statistical technique, synergistically merges principal component analysis (PCA) and linear discriminant analysis (LDA) to analyze intricate datasets. PCA effectively diminishes the dimensionality of high-dimensional data by identifying inherent patterns and correlations. Data is repositioned Through this transformation into a new coordinate system where the first principal component captures the most variance, and the subsequent components decrease variance. This dimensionality reduction serves to simplify data analysis. Conversely, LDA pinpoints the linear amalgamation of features that distinctly classify predefined groups or classes within the dataset. It achieves this by optimizing between-class variance while minimizing within-class variance. In the PCA-LDA analysis process, PCA takes precedence by initially reducing data dimensionality by identifying critical patterns and correlations. The altered data then undergoes LDA to determine the optimal linear amalgamation of features for effectively segregating the data into predefined groups or classes. Combining these two techniques in PCA-LDA enhances the precision of data classification by reducing dimensionality and enhancing data distinctiveness.

Furthermore, we have developed a PLS-DA model, a modified version of partial least square regression (PLS-R). PLS-DA is specifically designed for situations where the goal is to classify data points into different classes or categories based on their features. In PLS-DA, the method adapts the principles of PLS to classification problems. It constructs a linear relationship between the predictor variables (independent variables) and a set of dummy variables representing the class labels (dependent variables). The goal is to find latent variables that maximize the covariance between predictor variables and class labels and optimize class separation. Further information and implementation on the NIR spectral data can be found in previously published reports [16].

4.3 Models performance evaluation

The analysis was adapted to encompass two distinct groups of rice seeds: non-contaminated and contaminated samples. The evaluation process focused on the performance assessment of the PCA-LDA and PLS-DA models. Essential metrics, including sensitivity, specificity, precision, accuracy, and error rate, were computed using equations (2 to 6) derived from the confusion matrix.

$$Sensitivity = \left(\frac{TP}{TP+FN}\right) * 100 \quad (2)$$

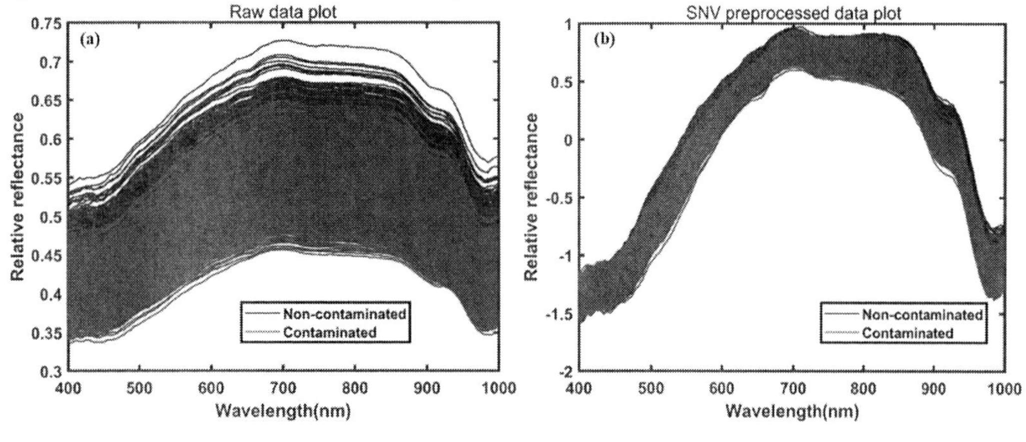

Fig. 2. Plots show (a) raw spectral data and (b) preprocessed spectral data.

Sensitivity, as expressed in Equation (2), delineated the models' efficacy in correctly identifying samples belonging to the contaminated group. This metric gauged the models' ability to discern actual positive cases while mitigating false negatives. Sensitivity served as a vital indicator of the models' capacity to accurately spot actual instances of contamination within the rice seed samples.

$$Specificity = \left(\frac{TN}{FP+TN}\right) * 100 \quad (3)$$

Equation (3) specificity measures parameter highlighted the models' adeptness in correctly recognizing true negatives while minimizing instances of false positives. Specificity provided insights into the models' capability to label negative cases accurately.

$$Precision = \left(\frac{TP}{TP+FP}\right) * 100 \quad (4)$$

Precision, outlined in Equation (4), assessed the models' precision in accurately predicting the number of accurate optimistic predictions within the contaminated group. It illuminated the models' accuracy in identifying contaminated instances among the predicted positives.

$$Accuracy = \left(\frac{FP+FN}{TP+FP+FN+TN}\right) * 100 \quad (5)$$

$$Error\ rate = \left(\frac{TP+TN}{TP+TN+FP+FN}\right) * 100 \quad (6)$$

Furthermore, the overall effectiveness of the models' classification was encapsulated by accuracy, denoted by Equation (5). Accuracy presented a comprehensive view of how well the models correctly categorized non-contaminated and contaminated rice seed samples. Conversely, the error rate, calculated using Equation (6), examined the models' classification performance by quantifying the proportion of inaccurately assigned samples concerning the total number of samples.

The terms are defined as follows: True Positive (TP) represented the instances where contaminated rice seeds were correctly identified as such. It corresponded to a positive response for an actual positive instance. False Positive (FP) indicated cases where non-contaminated rice seeds were wrongly classified as contaminated. It corresponded to a positive response for a negative instance. True Negative (TN) denoted the accurate recognition of non-contaminated rice seeds. It corresponded to a negative response for an actual negative instance. False Negative (FN) encompassed instances where contaminated rice seeds were erroneously identified as non-contaminated. It corresponded to a negative response for a favorable instance. The adapted analysis delved into the performance evaluation of the PCA-LDA and PLS-DA models, employing these metrics to comprehensively assess their accuracy in classifying non-contaminated and contaminated rice seed samples.

5 Results and Discussions

5.1 Spectral information of rice samples

Figure 3 (a) presents the raw mean spectral data and (b) demonstrates the SNV preprocessed of contaminated rice samples. The mean spectral shows an apparent difference between contaminated and non-contaminated spectra, in which the contaminated samples show lower intensity due to absorbing more light, and the non-contaminated spectra show higher intensity. Contaminated rice absorbs more light might due to presence of hydrocarbon bonds in parrafin [17]. However, both contaminated and non-contaminated spectral patterns demonstrated a similar pattern, and slight changes appear between 448-507 nm and 903-932 nm.

5.2 PCA-LDA and PLS-DA results

The PCA-LDA and PLS-DA models' effectiveness in classifying contaminated and non-contaminated rice samples into two classes was appraised using different preprocessing methods on spectral data. Nevertheless, the accuracy of the PCA-LDA and PLS-DA models varied based on the chosen preprocessing approach. The PCA-LDA and PLS-DA methods consistently demonstrated the lowest root mean square error of cross-validation (RMECV) for each preprocessing technique. To identify the optimal preprocessing methods, the ones resulting in the minimum RMSECV and the most appropriate number of latent variables (LVs) were chosen. The number of LVs was determined through a 10-fold cross-validation process. The results from the models are summarized in Table 1, depicting the calibration and prediction outcomes. The raw and range normalization (RN) preprocessing yielded enhanced accuracy across all models, including PCA-LDA and PLS-DA.

Comparing the performance of PCA-LDA and PLS-DA models, it becomes evident that they both achieved commendable levels of accuracy. However, the outcomes presented in Table 1, where these analysis methods were compared against highly effective preprocessing techniques, reveal distinct patterns. PCA-LDA displayed varying prediction set accuracies, ranging from 90% to 77%, and an elevated error rate surpassing 22%. In contrast, leveraging range normalization, PLS-DA showcased the highest accuracy, reaching 94.10% in the prediction set and a meager error rate of 5.9%. The classification outcomes of PLSDA for contaminated and non-contaminated rice samples, based on the application of range normalization, are visually depicted in Figure (4).

https://doi.org/10.1051/bioconf/20238001001

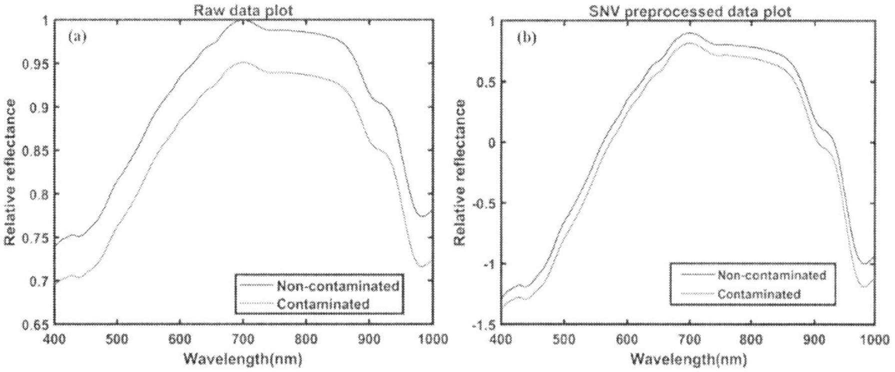

Fig. 3. Raw and preprocessed spectral data of contaminated and non-contaminated samples.

Models		Calibration set (1500)					Prediction set (1000)				
		Sen	Spec	Pre	Acc	E-rate	Sen	Spec	Pre	Acc	E-rate
PCA-LDA	Raw	92.40	92.40	92.42	92.40	7.60	90.10	90.10	90.24	90.10	9.90
	SNV	79.45	79.45	79.62	79.45	20.55	77.80	77.80	77.80	77.80	22.20
	RN	86.60	86.60	86.62	86.60	13.40	83.60	83.60	83.78	83.60	16.40
PLS-DA	Raw	94.90	94.90	94.90	94.90	5.10	93.50	93.50	93.54	93.50	6.50
	SNV	88.80	88.80	88.82	88.80	11.20	88.00	88.00	88.00	88.00	12.00
	RN	94.95	94.95	94.95	94.95	5.05	94.10	94.10	94.11	94.10	5.90

Table 1 summarizes the analysis results of the PCA-LDA and PLS-DA models.

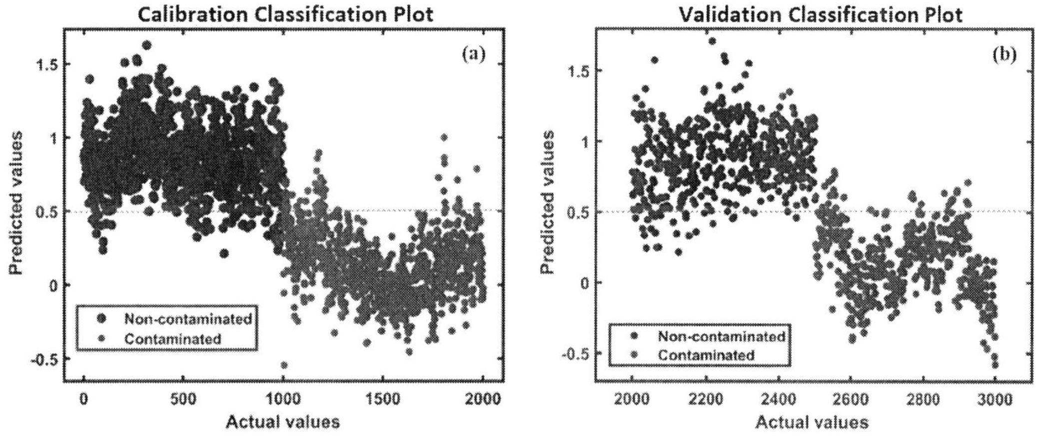

Fig. 4. PLS-DA classification analysis results (a) calibration and (b) validation, respectively.

Furthermore, to present a comprehensive overview of the outcomes for each class, confusion matrices were generated for the most effective models, as illustrated in Figure 5. The two axes of the matrices encompassed two categories: contaminated and non-contaminated samples. The labels assigned by the model were represented on the horizontal axis, while the vertical axis depicted the actual data labels. In particular, Figure 5 encompassed three subfigures: (a) showcasing the calibration results and (b) displaying the validation outcomes. These matrices pertained to the performance of the PLS-DA model, which employed range-normalization preprocessing, highlighting its remarkable achievements.

The confusion matrices illustrated a situation where a slightly higher number of contaminated samples were inaccurately classified as non-contaminated, and a small number of non-contaminated samples were misclassified as contaminated and showed the model error rate. Additionally, in multivariate analysis, the resulated beta coefficient from the models reveals the connection between each composition's absorption and specific spectral bands. Its importance stems from assisting in selecting informative wavelengths and simplifying the interpretation of outcomes. Figure (6) displays the derived PLS-DA beta coefficient using the raw spectral data, showcasing the contaminated and non-contaminated spectral insights linked to rice samples.

 https://doi.org/10.1051/bioconf/20238001001

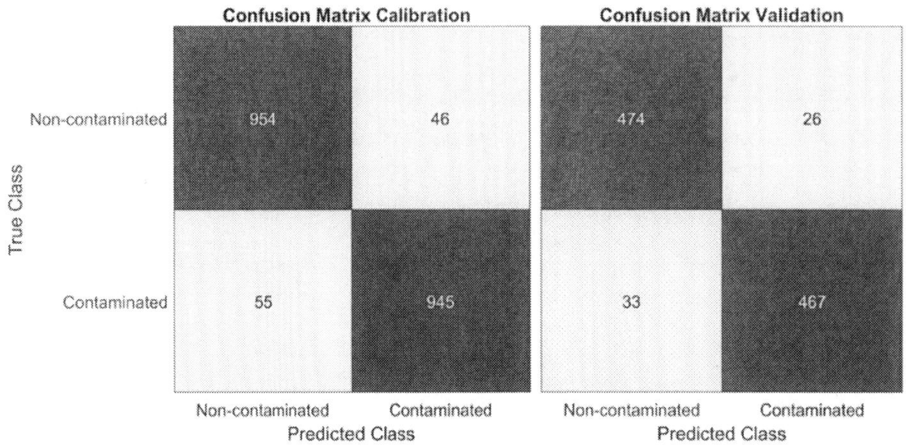

Fig. 5. Confusion matrices analysis of calibration and validation set.

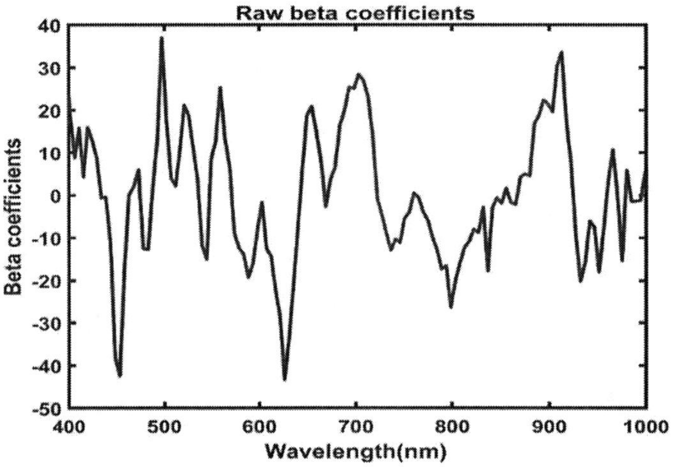

Fig. 6. Beta coefficient of the calibration model

The absorbed energy causes bonds to vibrate and stretch, leading to characteristic absorption bands in the VNIR spectrum. The energy levels associated with C-H, O-H, and N-H bonds align well with the energy range of VNIR light, making these regions particularly informative for analysis. The C-H bonds are prevalent in organic compounds, including fats, oils, carbohydrates, and proteins [18]. The O-H bonds are commonly found in water and various functional groups like alcohols and phenols [19]. Similarly, N-H bonds are present in molecules such as amino acids and proteins. As a result, VIS-NIR spectroscopy can be used to detect, quantify, and characterize a wide range of organic compounds based on their unique spectral signatures in the VNIR region. Within the spectral range encompassing VNIR wavelengths (400 to 1000 nm), we encountered spectra containing vital chemical details of rice constituents. These constituents include carbohydrates, proteins, amylose, and moisture, all of which correspond to the third overtone absorption of C–H bonds at 880 nm, as well as the second overtone absorption of O–H bonds in the range of 750 to 900 nm and N–H bonds spanning from 962 to 1000 nm [20], [21].

Furthermore, the amide bonds in proteins can cause absorption bands in the 600-900 nm range (Nirmal Thirunavookarasu, 2022). Amino acids have specific absorption features around 950 nm [22]. Starch and carbohydrates may contribute to absorption bands in the 800-1000 nm region due to their molecular vibrations [23]. Water absorption is significant in the 900-1000 nm range [24].

6 Conclusion

In conclusion, this study focused on applying hyperspectral imaging combined with chemometric analysis to detect paraffin-contaminated rice seeds, addressing a pressing food safety concern. The emergence of paraffin-coated rice in the market has raised alarms due to potential health risks, making the need for reliable detection methods crucial. Moreover, employed Visible-Near Infrared (VIS-NIR) hyperspectral imaging and advanced analytical techniques showcased promising results. Through a meticulous investigation involving different rice varieties from diverse regions contaminated with varying levels of paraffin, the study employed Partial Least Squares Discriminant Analysis (PLS-DA) and Principal Component Analysis Linear Discriminant Analysis (PCA-LDA) models for classification. The effectiveness of these models was evaluated using different preprocessing methods on the spectral data, including raw data, standard normal variate (SNV) transformation, and range normalization (RN). The results highlighted the potential of both PCA-LDA and PLS-DA models in accurately classifying contaminated and non-contaminated rice samples. However, a comparative analysis demonstrated that PLS-DA exhibited superior performance, especially with range normalization preprocessing. This preprocessing approach consistently yielded high accuracy rates, reaching 94.10% in the prediction set, with a minimal % error rate of 5.9%. At the same time, while generally accurate, PCA-LDA showed more variability in accuracy, ranging from 90% to 77%, and a higher error rate exceeding 22% based on different preprocessing techniques. This research underscores the potential of hyperspectral imaging and chemometric approaches, particularly PLS-DA with range normalization preprocessing, as a viable solution for detecting paraffin-contaminated rice. Furthermore, the study's findings contribute significantly to the field of food safety assessment, providing a valuable tool for the industry to ensure the quality and safety of rice products in the market. With the potential for further advancement, refinements in spectral data extraction and classification methodologies could enhance detection accuracy, making such techniques even more robust and reliable for real-world applications.

References

[1] FAO, "Rice Market Monitor, Food Agric. Organ. United Nations," *Food Agric. Organ. United Nations*, vol. XXI, no. 1, pp. 1–38, 2018.

[2] R. Johnson, "Food fraud and 'Economically motivated adulteration' of food and food ingredients," *Food Fraud Adulterated Ingredients Background, Issues, Fed. Action*, pp. 1–56, 2014.

[3] X. Feng, Q. Zhang, P. Cong, and Z. Zhu, "Preliminary study on classification of rice and detection of paraffin in the adulterated samples by Raman spectroscopy combined with multivariate analysis," *Talanta*, vol. 115, pp. 548–555, 2013, doi: 10.1016/j.talanta.2013.05.072.

[4] K. Renu *et al.*, "Molecular mechanism of heavy metals (Lead, Chromium, Arsenic, Mercury, Nickel and Cadmium) - induced hepatotoxicity – A review," *Chemosphere*, vol. 271, p. 129735, 2021, doi: 10.1016/j.chemosphere.2021.129735.

[5] J. F. X. Fen, L. Shiyao, C. Hui, and A. Di, "Determination of illegally added paraffin in rice by GC-MS/MS," *Hubei Prov. Eng. Technol. Res. Cent. Food Qual. Saf. Test, Hubei Prov. Inst. Food Superv. Test, Wuhan 430075, China)*, 2019, doi: 10.11882/j.issn.0254-5071.2019.03.038.

[6] D. Huang *et al.*, "Concentrations of and risks posed by short-chain and medium-chain chlorinated paraffins in soil at a chemical industrial park on the southeast coast of China," *Environ. Pollut.*, vol. 258, p. 113704, 2020, doi: 10.1016/j.envpol.2019.113704.

[7] U. Aline, T. Bhattacharya, M. A. Faqeerzada, M. S. Kim, I. Baek, and B. Cho, "Advancement of non-destructive spectral measurements for the quality of major tropical fruits and vegetables : a review," *Front. Plant Sci.*, no. August, pp. 1–18, 2023, doi: 10.3389/fpls.2023.1240361.

[8] M. Hussain Khan, Z. Saleem, M. Ahmad, A. Sohaib, H. Ayaz, and M. Mazzara, "Hyperspectral imaging for color adulteration detection in red chili," *Appl. Sci.*, vol. 10, no. 17, 2020, doi: 10.3390/app10175955.

[9] M. Kamruzzaman, "Fraud Detection in Meat Using Hyperspectral Imaging," *Meat Muscle Biol.*, vol. 5, no. 3, 2021, doi: 10.22175/mmb.12946.

[10] M. A. Faqeerzada *et al.*, "Hyperspectral shortwave infrared image analysis for detection of adulterants in almond powder with one-class classification method," *Sensors (Switzerland)*, vol. 20, no. 20, p. 5855, 2020, doi: 10.3390/s20205855.

[11] R. E. Masithoh, L. M. Kandpal, S. Lohumi, W. S. Yoon, H. Z. Amanah, and B. K. Cho, "Shortwave Infrared Hyperspectral Imaging for the Determination and Visualization of Chemical Contents of Wheat and Tuber Flour," *Int. J. Adv. Sci. Eng. Inf. Technol.*, vol. 12, no. 4, pp. 1574–1579, 2022, doi: 10.18517/ijaseit.12.4.14266.

[12] H. Z. Amanah *et al.*, "Near-infrared hyperspectral imaging (Nir-hsi) for nondestructive prediction of anthocyanins content in black rice seeds," *Appl. Sci.*, vol. 11, no. 11, 2021, doi: 10.3390/app11114841.

[13] A. Rahman, Q. Wu, H. Chang, S. Wang, J. Yan, and H. Xu, "Assessment of Intact Macadamia Nut Internal Defects Using Near-Infrared Spectroscopy," *ASABE Annu. Int. Meet.*, pp. 3–8, 2020.

[14] A. Rahman *et al.*, "Quality analysis of stored bell peppers using near-infrared hyperspectral imaging," *Trans. ASABE*, vol. 61, no. 4, pp. 1199–1207, 2018, doi: 10.13031/trans.12482.

[15] Å. Rinnan, F. van den Berg, and S. B. Engelsen, "Review of the most common pre-processing techniques for near-infrared spectra," *TrAC - Trends in Analytical Chemistry*. 2009, doi: 10.1016/j.trac.2009.07.007.

[16] J. Yasmin, M. R. Ahmed, S. Lohumi, C. Wakholi, M. S. Kim, and B. K. Cho, "Classification method for viability screening of naturally aged watermelon seeds using FT-NIR spectroscopy," *Sensors (Switzerland)*, vol. 19, no. 5, 2019, doi: 10.3390/s19051190.

[17] H. Chung, "Applications of near-infrared spectroscopy in refineries and important issues to address," *Applied Spectroscopy Reviews*, vol. 42, no. 3. pp. 251–285, 2007, doi: 10.1080/05704920701293778.

[18] S. Serranti, D. Cesare, and G. Bonifazi, "The development of a hyperspectral imaging method for the detection of Fusarium-damaged, yellow berry and vitreous Italian durum wheat kernels," *Biosyst. Eng.*, vol. 115, no. 1, pp. 20–30, 2013, doi: 10.1016/j.biosystemseng.2013.01.011.

[19] Z. Wu, G. Ouyang, X. Shi, Q. Ma, G. Wan, and Y. Qiao, "Absorption and quantitative characteristics of C-H bond and O-H bond of NIR," *Opt. Spectrosc. (English Transl. Opt. i Spektrosk.*, vol. 117, no. 5, pp. 703–709, 2014, doi: 10.1134/S0030400X1411023X.

[20] S. Weng *et al.*, "Hyperspectral imaging for accurate determination of rice variety using a deep learning network with multi-feature fusion," *Spectrochim. Acta - Part A Mol. Biomol. Spectrosc.*, vol. 234, p. 118237, 2020, doi: 10.1016/j.saa.2020.118237.

[21] L. Ma *et al.*, "Systematic discovery about NIR spectral assignment from chemical structural property to natural chemical compounds," *Sci. Rep.*, vol. 9, no. 1, pp. 1–17, 2019, doi: 10.1038/s41598-019-45945-y.

[22] J. Hang, D. Shi, J. Neufeld, K. E. Bett, and J. D. House, "Prediction of protein and amino acid contents in whole and ground lentils using near-infrared reflectance spectroscopy," *Lwt*, vol. 165, no. June, p. 113669, 2022, doi: 10.1016/j.lwt.2022.113669.

[23] R. Pourdarbani, S. Sabzi, S. Jarolmasjed, and T. Panagopoulos, "Determination of the most effective wavelengths for prediction of fuji apple starch and total soluble solids properties," *Appl. Sci.*, vol. 10, no. 22, pp. 1–16, 2020, doi: 10.3390/app10228145.

[24] J. H. Lam, K. J. Tu, and S. Kim, "Narrowband diffuse reflectance spectroscopy in the 900–1000 nm wavelength region to quantify water and lipid content of turbid media," *Biomed. Opt. Express*, vol. 12, no. 6, p. 3091, 2021, doi: 10.1364/boe.425451.

Implementation of real-time image processing on bacterial cellulose formation using soybean-boiled wastewater with the variation of carbon sources during fermentation

Ahmad Fatih Marzuki[1], *Darmawan Ari* Nugroho[1*], *Tyasto Prima* Ahmadi[1] and *Atris* Suyantohadi[1]

[1]Department of Agroindustrial Technology, Faculty of Agricultural Technology, Universitas Gadjah Mada, Flora Street Number 1, Yogyakarta 55281, Indonesia

Abstract. Bacterial cellulose (BC) is produced by aerobic bacteria through oxidative fermentation in synthetic and non-synthetic mediums. Several mediums reported to be used as BC formation mediums are coconut water and soybean-boiled wastewater. Carbon sources are needed to optimize the BC formation process. Recent study has implemented a real-time image processing approach for monitoring BC formation. This study aimed to investigate the correlation between variables that influence the fermentation and to determine the kinetic model of BC formation using an image processing approach with the variation of carbon sources during the fermentation. The results showed that the correlation between fermentation time and thickness had the highest percentage for glucose, sucrose, and mannitol mediums. The kinetic observation of BC formation in the medium using glucose, sucrose, and mannitol followed the Gompertz model equation, with the medium using sucrose having the fastest rate of increase at the 44th hour, followed by the medium using mannitol at the 112th hour, and the medium using glucose at the 149th hour.

1 Introduction

Bacterial cellulose (BC) is produced by aerobic bacteria such as *Acetobacter xylinum* through oxidative fermentation in synthetic and non-synthetic mediums [1]. *Acetobacter xylinum* is a Gram-negative bacterium that ferments actively at a pH range from 3 to 7 and a temperature range from 25°C to 30°C [2]. BC retains the fundamental structure of a fibril composed of β-1→4 glucan chains with the formula $(C_6H_{10}O_5)_n$ held together by intra- and inter-hydrogen bonds [3, 4]. Chen et al. [5] stated that through biosynthesis, bacteria secrete glucan chains through pores arranged in regular lines along the long axes of the cell's cytomembrane.

The most commonly used BC formation medium is coconut water [6, 7, 8]. Furthermore, wastewater can also be used as a BC formation medium, such as liquid waste from sugarcane processing during the jaggery [9], juice from Japanese pear, orange, apple, grape, and pineapple waste [10], waste beer yeast [11], rice wine distillation wastewater [12], liquid waste from tofu production [13], and soybean-boiled wastewater from tempeh production [14]. In addition, Maloringan et al. [14] used soybean-boiled wastewater as a BC formation medium because it contains nitrogen used by bacteria as a source of nutrients during fermentation.

Wang and Zhong [15] stated that a substrate in the form of carbon sources is needed to optimize the BC formation process. There are many carbon sources reported to be used, such as sucrose [2, 6-8, 10, 16-19], glucose [2, 10-12, 16-19], fructose [2, 10, 16, 19], mannitol [2, 16-19], galactose [2, 19], maltose [2], glycerol [16, 18, 19], date syrup [17], and lactose [18]. Ruka et al. [16] stated that of the various types of carbon sources, sucrose, mannitol, and glucose are carbon sources that produce consistently high yields of cellulose, regardless of the medium composition used. Other variables affecting BC formation besides carbon and nitrogen sources include BC layer thickness [7], fermentation time [19, 20], pH [20, 21], temperature [21], and turbidity conditions [22].

The fermentation commonly takes 6-8 days [7]. Traditionally, the observation of the formation of the BC during the fermentation is carried out after the fermentation has been completed by opening the lid of the tray and lifting a layer of BC to determine the thickness. The observation determines the formation amount that occurred during the fermentation. This step aims to ascertain how much the BC has expanded as a result of the fermentation [6]. Nugroho et al. [7] revealed a new observation method using real-time image processing (RTIP). In addition, Poka et al. [23] stated that the observation using RTIP was determined through edge detection. As a result, the edge detection algorithm has successfully separated BC layer object from other objects such as medium, BC non-layer, and air space between the fermenter cover and the medium [8].

In agricultural fields, RTIP that uses large amounts of data itself has been widely applied, including being used for early detection systems on cocoa pest attacks

* Corresponding author: ari-nugroho@ugm.ac.id

[24], classification of the quality of whole plant corn silage (WPCS) for cattle feed [25], and the creation of an automatic weed control system that can distinguish weeds and plants correctly in tomatoes [26], celeries [27], potatoes [28], and lettuces [29]. There are similarities in the functions of these studies, i.e. RTIP is used as an early warning system to identify anomalies that occur in the study objects. Furthermore, RTIP can become a decision support system based on the data that has been captured.

In microbiological fields, particularly in the production of BC, the mathematical model developed from RTIP has potential applications in industry. It can serve as an early warning system for contamination during BC fermentation through the detection of anomalies in BC formation patterns and a decision support tool for determining the fermentation time required for achieving certain BC layer thickness. However, the application of RTIP in observing BC formation is still limited to coconut water medium and sucrose carbon source. Therefore, this study aimed to investigate the correlation between variables that influence the fermentation and to determine the kinetic model of BC formation using a RTIP approach with the variation of carbon sources, i.e. sucrose, mannitol, and glucose during the fermentation using soybean-boiled wastewater.

2 Material and Method

2.1 Starter Production

The organism used as a starter was *Acetobacter xylinum*. The starter created using 1.5 L coconut water, 5% glucose, 0.2% ammonium sulfate, 0.2% acetic acid, and 20% pure starter. The fermentation was carried out for 7-13 days so that the starter is ready to use.

2.2 Medium Production

The BC medium was composed of coconut water and soybean-boiled wastewater in a ratio of 70:30 with a total volume of 2.5 L [14]. The other constituent components were carbon sources (10% glucose [30], 5% sucrose [6], or 0.83% mannitol [31]), 0.2% ammonium sulfate [30], 0.2% acetic acid [30], and 20% starter [32]. All components except the starter were heated (T = 100°C; t = 30 s) and then heated at a lower temperature (T = 60°C; t = 10 m). After that, the BC medium was left until it reached room temperature. Subsequently, the starter was added to the medium. Last, it was poured into the fermenter.

2.3 Fermentation Process

The fermentation method used in this study was static and batch fermentation [6]. No additional treatment was applied during fermentation. Fermentation was carried out until the stationary phase of bacterial growth was reached. After reaching this phase, the fermentation was stopped. Later, the thickness of the BC layer was measured manually with a caliper. The purpose of manual measurement is to be used as a reference in converting pixel units into cm units.

2.4 Fermenter

The fermenter used [8] composed of acrylic material painted with a non-reflective black color on the external surface and measuring 35 cm in length, 25 cm in width, and 6 cm in height. The transparent field was created facing the USB camera with a width of 6 cm and a height of 4 cm. The fermenter's lid is made of paper that allows oxygen to pass through. An LED light is positioned in front of the fermenter to increase the contrast of the object with the camera by providing lighting assistance. The USB camera with a resolution of 640 x 480 pixels and manual focus features is positioned directly in front of the transparent field. Figure 1. shows the fermenter model used.

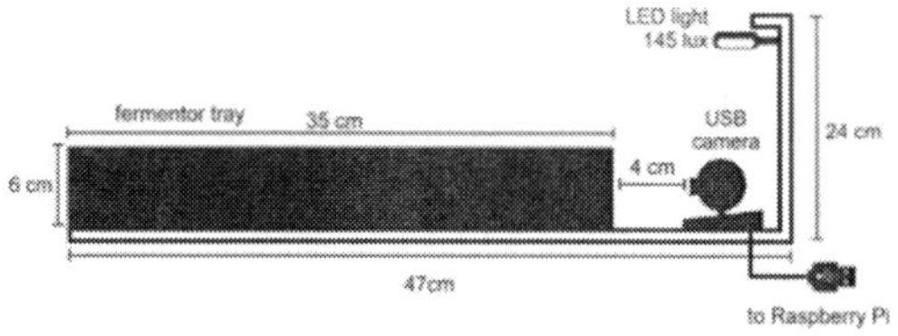

Fig. 1. Fermenter model [8].

2.5 Equipment Used

The equipment used refers to Nugroho et al. [8], which consists of Raspberry Pi 4, Arduino UNO, and four sensors, i.e. USB camera, temperature, pH, and turbidity sensor. The selection of four sensors used in this study is based on what variables are influential during fermentation, i.e. fermentation time, BC layer thickness, temperature, pH, and turbidity conditions. The USB camera sensor is used to capture BC layer thickness data which is directly connected to the Raspberry Pi 4. The temperature, pH, and turbidity sensor are positioned inside the fermentation medium at a depth of 1 cm from the bottom of the fermenter. The data obtained from temperature, pH, and turbidity sensor will be sent to the Arduino UNO, where it will be converted from analog-type data into digital-type data that the Raspberry Pi 4 can read.

Raspberry Pi 4 is a low-priced and pocket-sized computer. This board includes two USB 3.0 and two USB 2.0 ports, 40 pins, a Micro-SD card slot, two micro-HDMI ports, and WiFi and Ethernet support. The operating system is open-source and free software based on the Linux kernel called Raspbian Pi OS. The MySQL database's BC thickness calculation method and command insertion mechanism were developed using Python programming [8].

2.6 Telegram Bot

Telegram Bot can send messages automatically within a specific time range. Rianto et al. [33] reported that Telegram Bot could be used to send academic information services automatically. Trirahma [34] used

Telegram Bot as a real-time data-collecting tool for area mapping activity progress reporting.

In this study, Telegram Bot was used to monitor the progress of BC formation by sending messages containing information such as data collection time, fermentation code, BC layer thickness, temperature, pH, and BC medium turbidity. This was possible because the Telegram Bot was connected by PHP Handler to a Python script on the Raspberry Pi 4. Furthermore, the Raspberry Pi 4 was connected to the internet, and automatically the script was run every 15 minutes.

2.7 Data Collection

Nugroho et al. [8] explained that the data collection scheme starts with data transmission. The image the USB camera captures is sent to the Raspberry Pi 4 for BC thickness calculation. Simultaneously, the captured temperature, pH, and turbidity data will be sent to the Raspberry Pi 4 through the Arduino UNO. After that, all this data is entered and stored in the MySQL database installed on the Raspberry Pi 4. Concomitantly, the data is sent to the user's Telegram account. Data were collected every 15 minutes from the beginning until the stationary phase of fermentation. Figure 2. shows the data collection scheme.

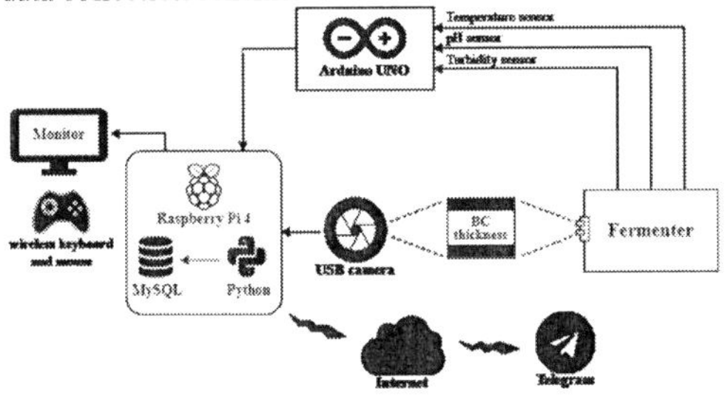

Fig. 2. Data collection scheme [8].

2.8 Statistical Analysis using R and Rstudio

R and Rstudio are used for statistical analysis [5]. The statistical analysis includes correlation, principal component, clustering, and kinetic model analysis. The R version used is 4.3.1, which can be downloaded from https://cran.r-project.org/bin/windows/base/. The GUI application uses the RStudio 2023.06.0+421 version, which can be downloaded from https://download1.rstudio.org/electron/windows/RStudio-2023.06.0-421.exe

3 Result and Discussion

3.1 Data Collection Results during BC Fermentation

The four sensors used successfully captured data on thickness, temperature, pH, and turbidity during fermentation with the medium using three different types of carbon sources. The data was then plotted into the graphs shown in Figure 3, 4, and 5.

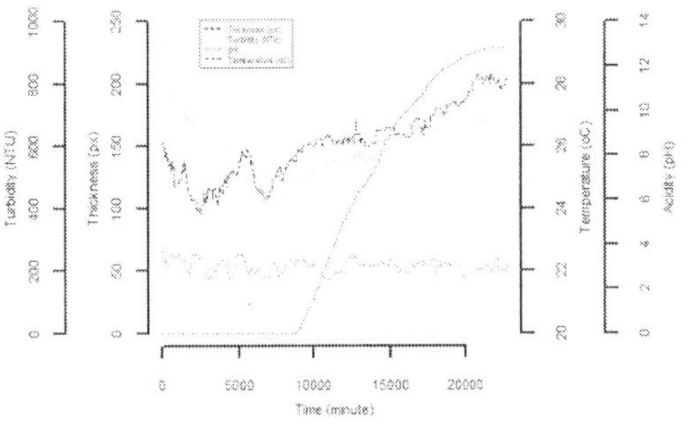

Fig. 3. Data collection result in the medium using glucose.

In the medium using glucose, as shown in Figure 3, the BC layer is visually visible in the 149th hour, and after that hour, there is an exponential increase in BC thickness until fermentation ends. The initial temperature condition before the BC layer appeared was at 24-26°C. When the BC layer became visually visible until the end of fermentation, the temperature was at 26-28°C. This increase occurs due to heat release by the inoculum, which indicates an increase in metabolic rate during the formation of the BC layer. The temperature condition is aligned with Nugroho et al. [8], although the medium used differs. Subsequently, the turbidity condition showed a significant decrease until the initial phase of the BC layer formation caused by biomass deposition in the fermenter. After that, the turbidity became stable until the end of fermentation because the inoculum increasingly secreted BC fibers, forming the BC layer. Furthermore, the pH condition was at 2-4 from the beginning of fermentation to the end. In this medium, fermentation lasted for 16 days with the final BC layer thickness of 1.08 cm.

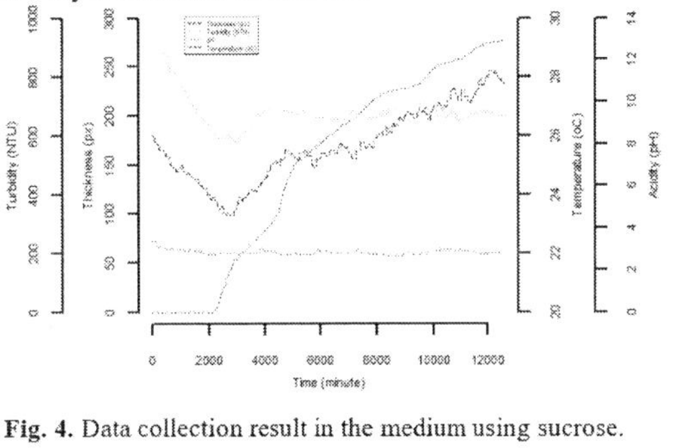

Fig. 4. Data collection result in the medium using sucrose.

In the medium using sucrose, as shown in Figure 4, the BC layer can be seen visually starting from the 37th hour, increasing exponentially until the end of fermentation. In the initial phase of pouring until before the formation of the BC layer, the temperature was at 24-26°C, and after the BC layer was formed until fermentation ended, the temperature was at 24-28°C. Hereafter, turbidity conditions in this medium have a similar pattern to those in the medium using glucose. From the initial phase to the BC formation phase, there was a decrease in turbidity, and then increased until fermentation ended. Moreover, the pH condition did not change significantly during fermentation. Nugroho et al. [8] reported similar pH results when using sucrose as a carbon source. In this medium, fermentation lasted for 9 days with the final BC layer thickness of 1.31 cm.

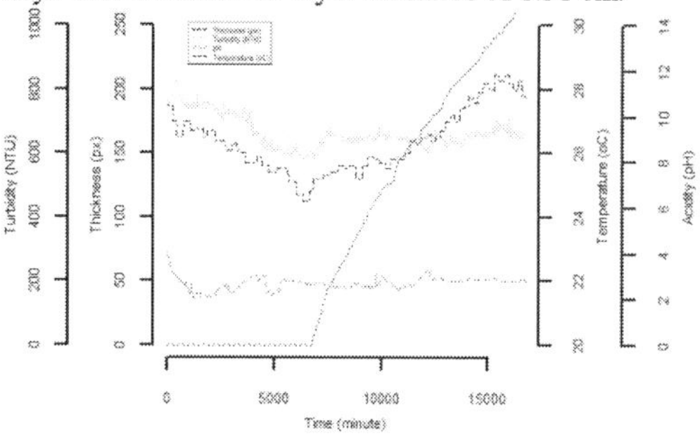

Fig. 5. Data collection result in the medium using mannitol.

In the medium using mannitol, as shown in Figure 5, the BC layer can be visually seen from the 112th hour, and the exponential increase in thickness occurs after that hour. In the phase after biomass decanting until before the formation of the BC layer, the temperature was 24-27°C, and after the formation of the BC layer until the end of fermentation, the temperature was 25-29°C. Furthermore, turbidity in the initial phase after biomass pouring decreased until the initial phase of BC layer formation and then became more stable until the end of fermentation. In addition, the pH condition was at 2-4 during fermentation. In this medium, fermentation lasted for 11 days with the final BC layer thickness of 1.27 cm.

The similarity of the pattern of temperature and turbidity changes from the use of three different carbon sources shows that the different types of carbon sources do not significantly impact the pattern of temperature and turbidity changes during fermentation. Furthermore, differences in carbon sources affect the pattern of pH changes. The medium using sucrose has the most stable pH pattern compared to the other two mediums. The medium using sucrose also produced the thickest BC layer with the fastest fermentation time compared to the other two mediums.

3.2 Correlation Analysis

Correlation analysis measures the level of relationship between a variable and another variable [35]. Analysis was performed on thickness, temperature, pH, turbidity, and fermentation time to determine which variable had the highest percentage level of relationship. Figure 6, 7, and 8 shows the results of the correlation analysis.

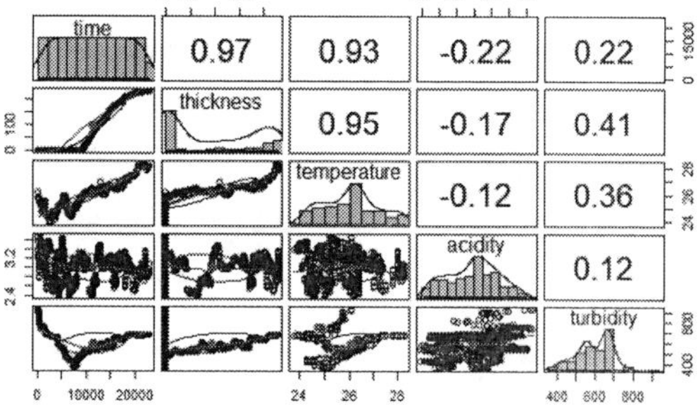

Fig. 6. Correlation analysis result in the medium using glucose.

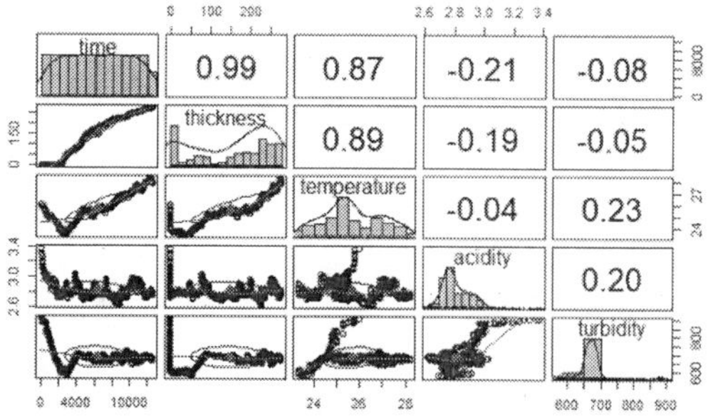

Fig. 7. Correlation analysis result in the medium using sucrose.

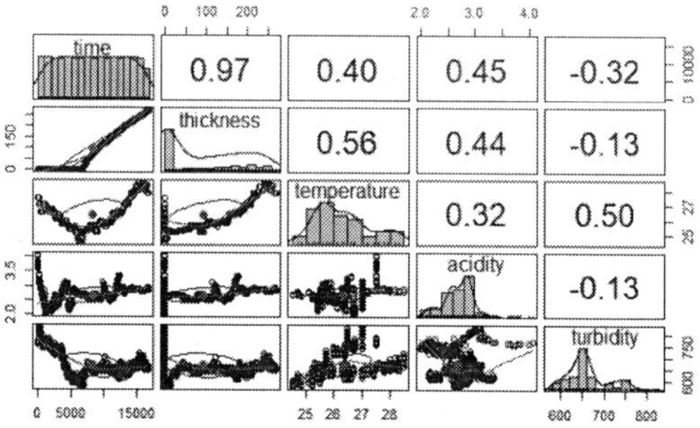

Fig. 8. Correlation analysis result in the medium using mannitol.

In the medium using glucose, as shown in Figure 6, fermentation time and thickness relationship had the highest correlation of 97%, followed by the correlation between temperature and thickness of 95%. In the medium using sucrose, fermentation time and thickness relationship had the highest correlation of 99%, followed by the relationship between temperature and thickness of 83%. In the medium using mannitol, fermentation time and thickness relationship had the highest correlation of 97%, followed by the relationship between temperature and thickness of 56%.

The results of the correlation analysis of the three mediums showed that the relationship with the highest correlation was between fermentation time and thickness. Mikkelsen et al. [19] explained that the longer the fermentation time, the more thickness of the BC layer will increase. Ruka et al. [16] also stated the same thing, regardless of the different types of carbon sources used. As a result, the length of fermentation time is highly correlated with the increase in BC layer thickness despite the different types of carbon sources used.

3.3 Principal Component Analysis

Principal component analysis (PCA) aims to reduce the variables in data to a smaller number of variables without losing the information contained in the original data [36]. Although there are fewer variables, only using the PCA data will produce the same value as using the original data. To determine it is based on the percentage of the cumulative proportion of variance. According to Jolliffe & Cadima [37], if the percentage of the cumulative proportion of variance has reached 70%, it is considered sufficient to represent the total diversity of the data. Therefore, the PCA result data will be used for the clustering analysis.

Table 1. Principal component analysis result.

Principal Component	Cummulative Proportion Values		
	Glucose	Sucrose	Mannitol
PC1	0.59790	0.58870	0.55100
PC2	0.84500	0.90770	0.82800
PC3	0.97680	0.98693	0.97490
PC4	0.99827	0.99850	0.99865
PC5	1.00000	1.00000	1.00000

Table 1 shows that PCA resulted in five principal components, namely PC1, PC2, PC3, PC4, and PC5. The cumulative proportion value of PC1 and PC2 in the medium using glucose is 84.5%. This result means that PC1 and PC2 were able 84.5% of the data diversity. The cumulative proportion value of PC1 and PC2 in the medium using sucrose is 86.07%. This result means that PC1 and PC2 were able 86.07% of the data diversity. The cumulative proportion value of PC1 and PC2 in the medium using mannitol is 82.8%. This result means that PC1 and PC2 were able 82.8% of the data diversity. Based on the results of the three mediums, it is known that data containing only PC1 and PC2 can be used to replace the original data because that data had cumulative proportion values more than 70%. The information included by PC1 and PC2 is proven to represent the information owned by the actual data [8].

3.4 Clustering Analysis

Clustering analysis aims to divide data into groups where data in the same group have similar characteristics compared to other groups [38]. Based on the relationship between fermentation time and thickness in Figure 6, 7, and 8, clustering analysis was performed on the data during fermentation. Analysis begins with determining the optimal number of clusters.

The methods used for the determination were the within-cluster-sum of square (wss) and silhouette [39]. The purpose of using two different methods is to compare the results of both so that the optimal number of clusters obtained is more accurate. Graphs of the clustering analysis results are shown in Figures 9, 10, and 11.

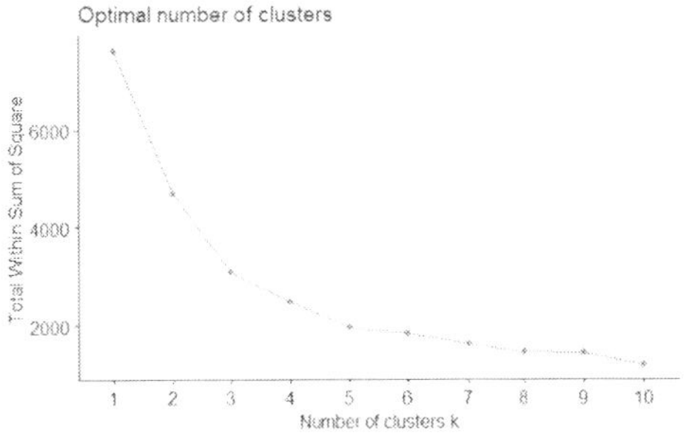

(a)

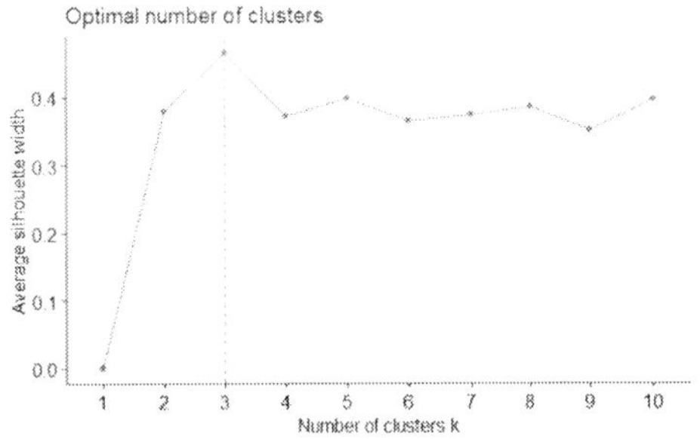

(b)

Fig. 9. The optimal cluster number determination result using the (a) wss and (b) silhouette methods in the medium using glucose.

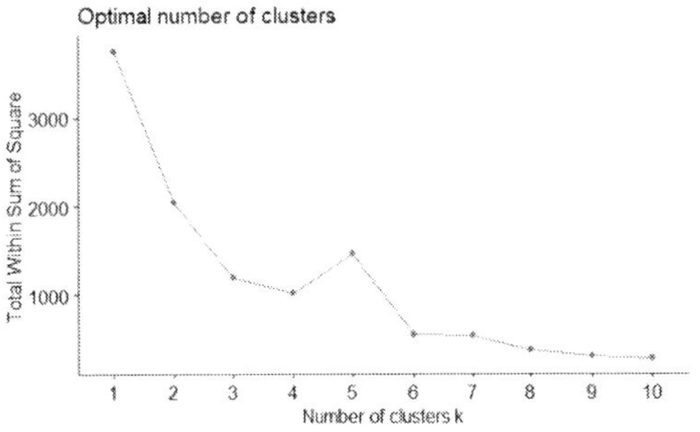

(a)

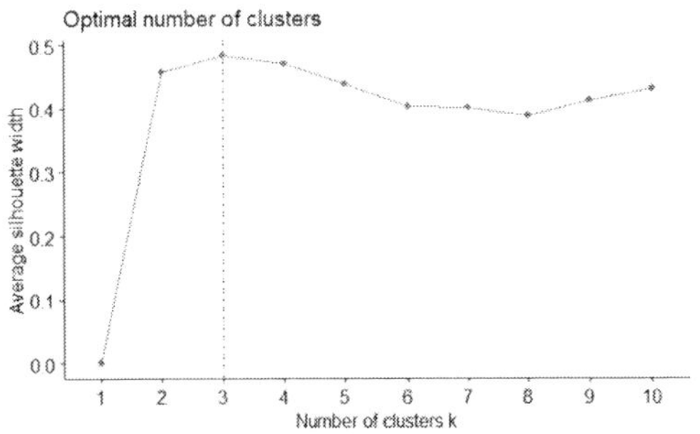

(b)

Fig. 10. The optimal cluster number determination result using the (a) wss and (b) silhouette methods in the medium using sucrose.

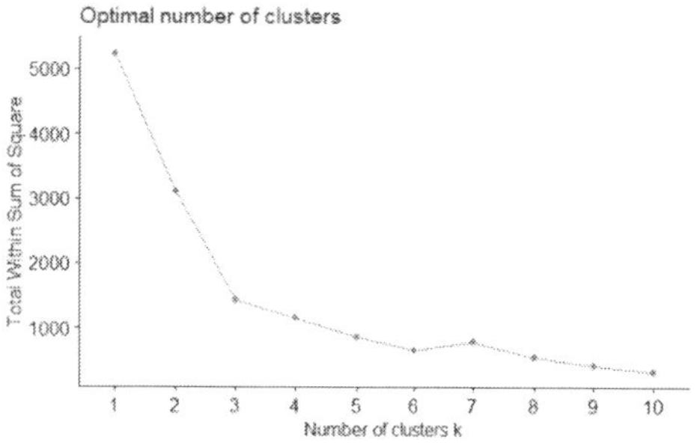

(a)

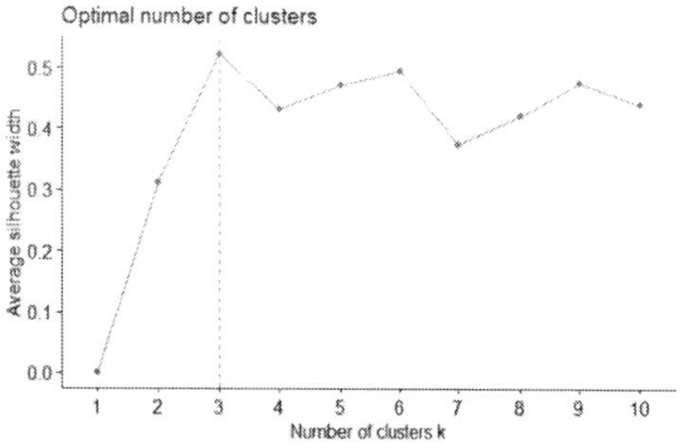

(b)

Fig. 11. The optimal cluster number determination result using the (a) wss and (b) silhouette methods in the medium using mannitol.

As a result, in the medium using glucose, shown in Figure 9, the optimal number of clusters produced by both methods is three. In the medium using sucrose, shown in Figure 10, the optimal number of clusters produced by both methods is three. In the medium using mannitol, shown in Figure 11, the optimal number of clusters produced by both methods is three. These results indicate that the different methods of the optimal number of clusters determination do not affect the results of the optimal number of clusters.

After the optimal number of clusters is determined, the data groups are mapped into the graph as shown in Figure 12, 13, and 14. Data group mapping is done to determine the position of each data group.

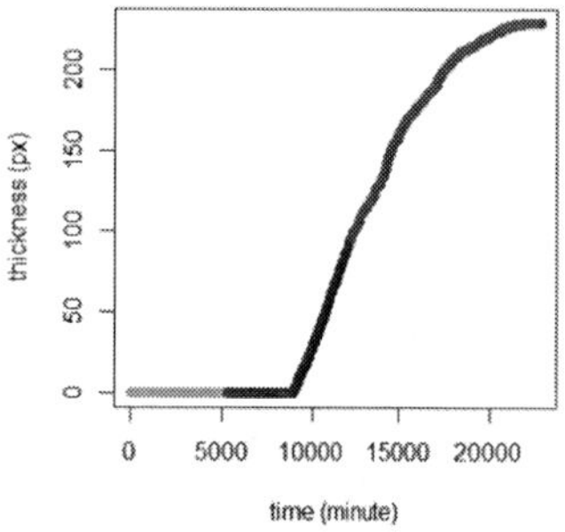

Fig. 12. Data group mapping in the medium using glucose.

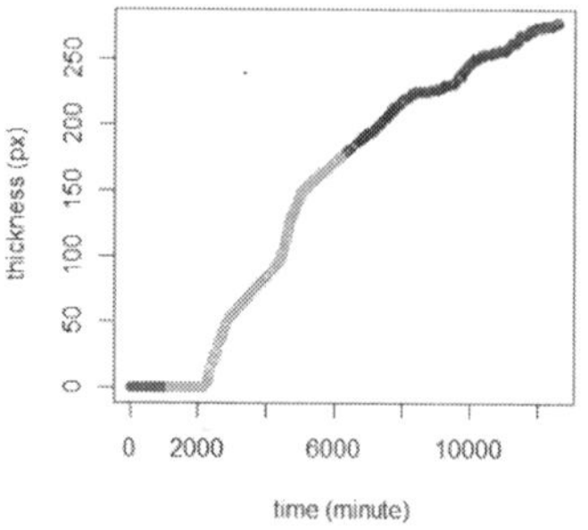

Fig. 13. Data group mapping in the medium using sucrose.

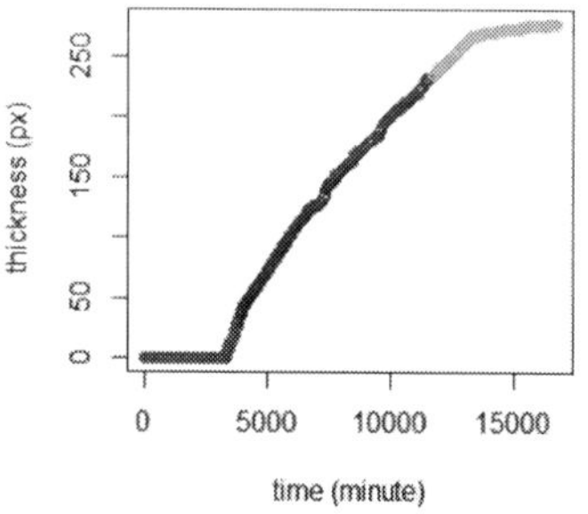

Fig. 14. Data group mapping in the medium using mannitol.

In the medium using glucose, as shown in Figure 12, the first cluster lasted until the 84th hour. The second cluster took place from the 85th until the 202nd hour. The third cluster took place at the 203rd hour until the end of fermentation. In the medium using sucrose, as shown in Figure 13, the first cluster lasted until the 18th hour. The second cluster took place from the 19th until the 115th hour. The third cluster took place at the 116th hour until the fermentation ended. In the medium using mannitol, as shown in Figure 14, the first cluster lasted until the 67th hour. The second cluster lasted from the 68th to the 204th hour. The third cluster lasted from the 205th hour until the end of fermentation.

3.5 Kinetic Model Determination of the BC Formation

After the optimal number of clusters in the medium using glucose, sucrose, and mannitol was obtained, an equation model was determined to define the BC layer's formation rate. The selection of the equation model is based on the highest R^2 value [8]. Therefore, the Gompertz model was chosen because it has been commonly used to describe the growth of animals, plants, and bacteria [40]. According to Oliviera Zardin et al. [41], this model is able to provide different biological interpretations relating to initial conditions or growth rates.

Table 2. Iteration, coefficients, and R^2 values as a result of the Gompertz equation.

Results		Carbon Sources		
		Glucose	Sucrose	Mannitol
Start Iteration Values	a	600	600	600
	b	1	1.2	1
	c	0.05	0.05	0.05
Coefficient Values	a	2.372e+02	0.027e+04	2.853e+02
	b	5.779e+01	6.492e+00	2.017e+01
	c	3.330e-04	4.359e-04	3.092e-04
R^2 Value		0.9987	0.9928	0.9948

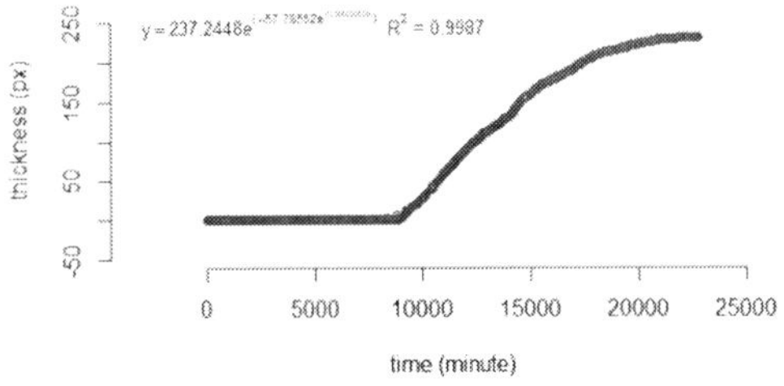

Fig. 15. Gompertz model plot of the medium thickness using glucose.

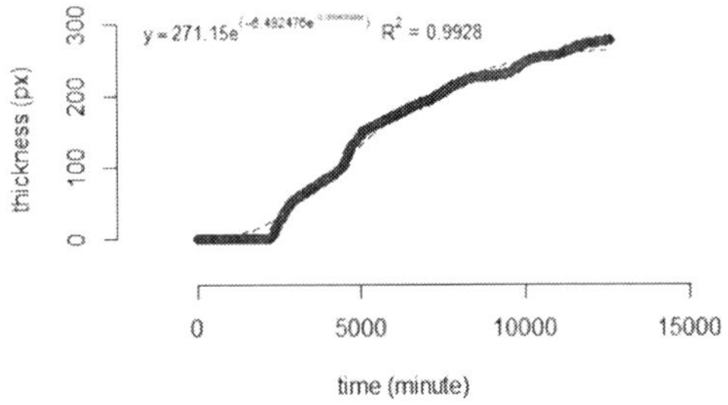

Fig. 16. Gompertz model plot of the medium thickness using sucrose.

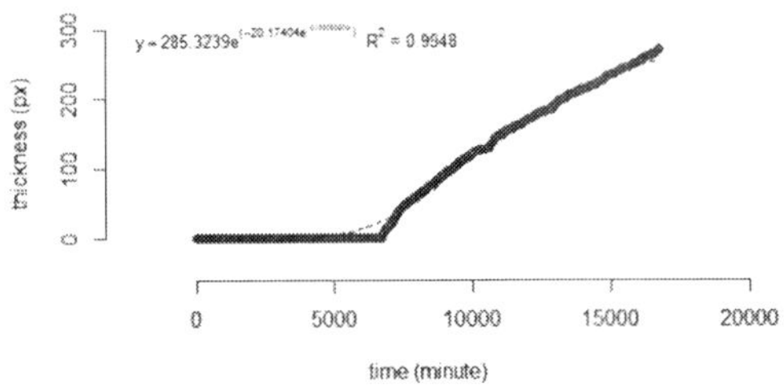

Fig. 17. Gompertz model plot of the medium thickness using mannitol.

Table 2. shows the start iteration values, the coefficient values, and the R^2 value of the Gompertz model of the medium using glucose, sucrose, and mannitol. Figure 15, 16, and 17 shows the plot of the Gompertz model used to measure BC layer thickness during fermentation. Based on the calculations that had been done, the Gompertz model is appropriate to define the BC layer's formation rate.

4 Conclusion

The correlation between fermentation time and thickness had the highest percentage in the medium using glucose, sucrose, and mannitol. The rate of BC layer formation with medium using glucose, sucrose, and mannitol followed the Gompertz model, with the medium using sucrose having the fastest rate of increase at the 44th hour, followed by the medium using mannitol at the 112th hour, and the medium using glucose at the 149th hour. Therefore, the mathematical model can serve as a reference for the future development of an early warning system and a decision support tool in the BC production industry.

Acknowledgment. The authors are grateful to Universitas Gadjah Mada for providing the research grant for this study through the Program of Final Assignment Recognition (Rekognisi Tugas Akhir/RTA) Batch 1 2022 based on letter No. 1525/UN1/DITLIT/Dit-Lit/PT.01.05/2022.

References

1. F. Esa, S.M. Tasirin, N.A. Rahman, Agriculture and Agricultural Science Procedia, **2**, 113–119 (2014)
2. C. Castro, R. Zuluaga, C. Álvarez, J. Putaux, G. Caro, O.J. Rojas, I. Mondragon, P. Ganán, Carbohydrate Polymers, **89**, 1033-1037 (2012)
3. Z. Ahmad, F. Qayyum, S.A., Shah, Y.O. Al-Ghamdi, S.A. Khan, Bacterial Cellulose: Synthesis, Production, and Applications, 201-212, CRC Press (2022)
4. A. Khalil, S.B. Khan, T. Kamal, Bacterial Cellulose: Synthesis, Production, and Applications, 79-98, CRC Press (2022)
5. C. Chen, W. Ding, H. Zhang, L. Zhang, Y. Huang, M. Fan, J. Yang, D. Sun, Carbohydrate Polymers, **278**, 118995 (2022)
6. D.A. Nugroho, L. Sutiarso, E.S. Rahayu, R.E. Masithoh, agriTECH, **40**, 118-123 (2020)
7. D.A. Nugroho, L. Sutiarso, E.S. Rahayu, R.E. Masithoh, IOP Conf. Series: Earth and Environmental Science, **752**, 012014 (2021)
8. D.A. Nugroho, L. Sutiarso, E.S. Rahayu, R.E. Masithoh, Food Research, **6**, 210-214 (2022)
9. W.A. Khattak, T. Khan, M. Ul-Islam, M.W. Ullah, S. Khan, F. Wahid, J.K. Park, Journal of Food Science and Technology, **52**, 8343-8349 (2015)
10. A. Kurosumi, C. Sasaki, Y. Yamashita, Y. Nakamura, Carbohydrate Polymers, **76**, 333–335 (2009)
11. D. Lin, P. Lopez-Sanchez, R. Li, Z. Li, Bioresource Technology, **151**, 113-119 (2014)
12. J.M. Wu, R.H. Liu, Journal of Bioscience and Bioengineering, **115**, 284-290 (2013)
13. C. Nisaa', D.A. Nugroho, G.T. Mulyati, Undergraduate Thesis, Universitas Gadjah Mada (2018)

https://doi.org/10.1051/bioconf/20238001002

14. Y.G. Maloringan, D.A. Nugroho, Wagiman, Undergraduate Thesis, Universitas Gadjah Mada (2016)
15. X. Wang, J. Zhong, Process Biochemistry, **122**, 95-102 (2022)
16. D.R. Ruka, G.P. Simon, K.M. Dean, Carbohydrate Polymers, **89**, 613-622 (2012)
17. F. Mohammadkazemi, M. Azinb, A. Ashori, Carbohydrate Polymers, **117**, 517-523 (2015)
18. N. Yodsuwan, O. Amorn, N. Atitaya, T. Nattaya, N. Soykeabkaew, December, 2005–2010 (2012)
19. D. Mikkelsen, B.M. Flanagan, G.A. Dykes, M.J. Gidley, Journal of Applied Microbiology, **107**, 576–583 (2009)
20. J.T. Hsieh, M.J. Wang, J.T. Lai, H.S. Liu, Journal of the Taiwan Institute of Chemical Engineers, **63**, 46-51 (2016)
21. S.M. Mohammad, N.A. Rahman, M.S. Khalil, S.R. Abdullah, Advances in Biological Research, **8**, 307-313 (2014)
22. P.D. Kasi, E.P. Tenriawaru, R. Ridwan, ICONSS Proceeding Series (2019)
23. G. Poka, B.Z. Balazs, Journal of Manufacturing Processes, **80**, 706-717 (2022)
24. Basri, Harli, Indrabayu, I.S. Areni, IOP Conf. Series: Journal of Physics: Conf. Series, **1244**, 012003 (2019)
25. E.M.C. Rocha, J.L. Drewry, R.M. Willett, B.D. Luck, Computers and Electronics in Agriculture, **203**, 107415 (2022)
26. H.J. Kennedy, S.A. Fennimore, D.C. Slaughter, T.T. Nguyen, V.L. Vuong, R. Raja, R.F. Smith, Weed Technology, **34**, 342-350 (2019)
27. W-H. Su, D.C. Slaughter, S.A. Fennimore, Computers and Electronics in Agriculture, **168**, 105155 (2020)
28. S.B. Rigoulot, T.M. Schimel, J.H. Lee, H. Brabazon, K.A. Meier, M.J. Schmid, E.M. Seaberry, M.R. Poindexter, J.S. Layton, J.W. Brabazon, J.A. Madajian, M.J. Finander, J. DiBenedetto, A. Occhialini, S.C. Lenaghan, C.N. Stewart Jr., bioRxiv, **865428** (2019)
29. R. Raja, T.T. Nguyen, D.C. Slaughter, S.A. Fennimore, Biosystems Engineering, **192**, 257-274 (2020)
30. D.A. Nugroho, P. Aji, Agriculture and Agricultural Science Procedia, **3**, 278-282 (2015)
31. S. Gea, K.M. Pasaribu, K. Sebayang, E. Julianti, S.C. Amaturahim, S.U. Rahayu, Y.A. Hutapea, AIP Conference Proceedings, **2049**, 020064 (2018)
32. Lusi, Periadnadi, Nurmiati, Jurnal Metamorfosa, **4**, 126-131 (2017)
33. Rianto, A. Rahmatulloh, T.A. Firmansah, Journal Publications & Informatics Engineering Research, **3**, 73-78 (2019)
34. A. Trirahma, Jurnal RESTI, **5**, 1182-1192 (2021)
35. N. Al-Rousan, N.A. Mat Isa, M.K. Mat Desa, International Journal of Energy Research, **45**, 453–477 (2021)
36. R.A. Johnson, D.W. Wichern, Pearson Prentice Hall (2007)
37. I. Jolliffe, J. Cadima, Phil. Trans. R. Soc. A, **374**, 20150202 (2016)
38. R.N. Pohan, A.M. Asih, Undergraduate Thesis, Universitas Gadjah Mada (2022)
39. C. Yuan, H. Yang, Multidisciplinary Scientific Journal, **2**, 226-235 (2019)
40. K. Tjørve, E. Tjørve, PLoS ONE, **12**, e0178691 (2017)
41. A. Oliveira Zardin, C. de Oliveira, S. de Oliveira, G. Yoshida, D. de Albuquerque, C. de Campos, R. Ribeiro, Aquaculture, **511**, 734243 (2019)

Analysis of agricultural land carrying capacity in North Sumatra Province in 2020

Muhammad Bagus Fitriawan, *Nabila* Husna Risada, *Yosi* Yaseva and *Ratih* Fitria Putri*

Department of Environmental Geography, Faculty of Geography, Universitas Gadjah Mada, Indonesia

Abstract. North Sumatra Province is an area with considerable agricultural potential in the Sumatra region. Analysis of the carrying capacity of agricultural land in North Sumatra in 2020 is very important to do in order to determine the carrying capacity of agricultural land in North Sumatra in meeting the food needs of its people. The purpose of this paper is to calculate and present the results of the analysis of the carrying capacity of agricultural land in North Sumatra Province. The data needed in this study is secondary data obtained from several journals and publications from related agencies, such as the Central Statistics Agency (BPS) of North Sumatra Province. The results showed that overall the carrying capacity of North Sumatra's land was classified as low. The low carrying capacity of rice farming land in North Sumatra is caused by increasing population growth and the quality of agricultural land which continues to decline so that the production of rice plants cannot meet the maximum food needs of the community. Therefore, it is necessary to make efforts to increase the carrying capacity of rice farming land in this area, for example by reducing the population and improving the quality of agricultural land resources.

1 Introduction

Indonesia is an agrarian country where the majority of its population relies on agriculture for their livelihoods. However, over time, agricultural land has been experiencing a decline in both quantity and quality. This is due to the increasing pressure of population on agricultural land [1]. The continuously growing population and the rampant development activities have led to the conversion of agricultural land, which should have been used for food production. Consequently, this has resulted in a diminishing availability of agricultural land to meet the food needs of the population.

The carrying capacity of agricultural land is defined as the ability of agricultural land to support the livelihoods of the community in a certain area, particularly in terms of food security [2], [3], [4]. The carrying capacity of agricultural land involves the calculation of the population that can be supported or sustained by a certain area of land resources in a prosperous state [5]. The analysis of agricultural land carrying capacity needs to be conducted to ensure adequate food availability for the population. Understanding the potential production of agricultural land becomes crucial in taking necessary measures to enhance agricultural productivity, optimize land use, and meet the increasing food demands in the future.

*Corresponding author:ratihfitria.putri@ugm.ac.id

The province of North Sumatra is an area with significant agricultural potential and serves as a food barn in the Sumatra region. This is due to its favorable physical geographical conditions, such as agroclimatic conditions and abundant natural resources, as well as the cultural practices of its population, a large portion of whom work in the agricultural sector, particularly in food crops. Therefore, North Sumatra has become one of the prime markets for the best agricultural products, with some commodities even being exported abroad.

The agricultural sector, along with forestry and fisheries, plays a vital role in the economy of North Sumatra. In 2020, the economic structure of North Sumatra was still dominated by the agricultural, forestry, and fisheries sectors. These sectors contributed 21.34 percent to the total Gross Regional Domestic Product (GRDP) of North Sumatra [6]. This makes them one of the largest contributors to the formation of GRDP in the province of North Sumatra [7].

The food crops produced in North Sumatra consist of rice, corn, soybeans, peanuts, mung beans, sweet potatoes, and cassava. Among these food crops, rice had the highest production in 2020, reaching 2,040,500.19 tons with a productivity of 5,251 kg/ha. Based on productivity, there were three regencies/cities with the highest productivity values, namely Karo Regency, Deli Serdang Regency, and Pematang Siantar City [8].

The agricultural sector, which is essentially the economic backbone of North Sumatra province, is increasingly threatened by the growing population, which leads to an increased demand for residential land and infrastructure. According to data from the BPS (2020), the population of North Sumatra province reached 14,799,361 individuals, consisting of 7,422,046 males and 7,377,315 females. The population growth rate in North Sumatra, which was 1.22% during the period of 2000-2010, increased to 1.28% during the period of 2010-2020. This increase has led to changes in land use for housing and development facilities, which on the other hand, can threaten the availability of sustainable agricultural land.

Based on the above problems, the research entitled Analysis of the carrying capacity of agricultural land in North Sumatra in 2020 is very important to do in order to determine the carrying capacity of agricultural land in North Sumatra in meeting the food needs of its people. In addition, the purpose of this writing is to calculate and present the results of the analysis of the carrying capacity of agricultural land in North Sumatra Province. Knowing the capacity of agricultural land in each district/city in North Sumatra Province can provide an overview to the government and the local community regarding the availability of paddy agricultural land for food production. The pressure on agricultural land in North Sumatra is a concern for farming communities, so that if this situation is not controlled it will result in a decline in the demand for agricultural land for food production in the future.

2 Methodology

This research is a quantitative descriptive research with a spatial approach. Quantitative descriptive research is research that aims to describe, record, analyze, interpret, and classify a condition that occurs. While the spatial approach is an approach that puts forward the principles of distribution, interrelation, and description [9].

2.1 Location determination

This research was conducted in North Sumatra Province which has a total area of approximately 182,414.25 km² consisting of a land area of approximately 72,981.23 km² and an ocean area of approximately 109,433.02 km². Administratively, the province of North Sumatra consists of 25 regencies and 8 cities with the capital city of Medan.

2.2 Data collection method and calculation

The data required for this research is secondary data obtained from several journals and publications of relevant institutions, such as the Central Statistics Agency (BPS). Agricultural land carrying capacity is one of the environmental carrying capacity analyses, which serves to calculate the availability of paddy agricultural land in supporting the population's needs. To determine the Land Carrying Capacity Status (DDL), it will be done by comparing the land availability (SL) with the land requirement (DL). The following is the interpretation of the analysis results, namely:

If SL > DL, it means that the land carrying capacity is surplus.
If SL < DL, it means that the land carrying capacity is deficit or exceeded.

The calculation is conducted to determine the region's capacity to achieve food self-sufficiency by comparing the optimum population with the food production output. The concept used to understand the critical threshold of carrying capacity is the existence of a limited population that can be supported without compromising the natural environment and ensuring the preservation of the ecosystem [10]. The agricultural land carrying capacity can be calculated using a formula based on a combination of theories from Odum, Chirtaler, Ebenezer Howard, and Issard (Suhardjo & Tukiran, 1990).

$$\sigma = \frac{Lp/Pd}{KFM/Pr} \quad (1)$$

Explanation:
σ = Agricultural land carrying capacity
Lp = Harvested land area (Ha)
Pd = Population
Pr = Average rice production per hectare (Kg/Ha)
KFM= Minimum Physical Needs (Kg/capita/year)

Regions capable of achieving food self-sufficiency are those that can meet the minimum physical needs of the population, which are 1600 calories/person/day or equivalent to 265 kg of rice/person/year. Regions capable of providing a decent standard of living for the population depend on the area of food crops that can meet the population's needs at an acceptable level, equivalent to 650 kg of rice/person/year or 2.46 times the minimum physical needs (KFM). Based on these values, the agricultural land carrying capacity is classified into three categories:

Class I ($\sigma > 2.46$):
Regions capable of achieving food self-sufficiency and meeting the food needs of the population adequately.

Class II ($1 \leq \sigma \leq 2.46$):
Regions capable of achieving food self-sufficiency but not yet able to meet the food needs of the population adequately.

Class III ($\sigma < 1$):
Regions that are not able to achieve food self-sufficiency and cannot meet the food needs of the population adequately.

3 Results and discussion

Carrying Capacity research location Land (DDL) agriculture This conducted in North Sumatra Province. kindly geographical North Sumatra Province is located at 1° - 4° North Latitude and 98° - 100° East Longitude, with a land area of 72981.23 km². Most of North Sumatra province is on the mainland Sumatra Island and parts small are on the Island Nias, Batu Islands, as well a number of island small, both in the west and part east beach Sumatra Island. North Sumatra Province consists of 25 districts and 8 municipalities of area the biggest is Regency Langkat with area of 6,262 km², Regency Mandailing Natal with area of 6,134 km², and Regency South Tapanuli with area of 6030.47 km², meanwhile wide area smallest is Tebing Tinggi City with area of 31 km² [6].

Land area harvest paddy on North Sumatra Province in 2020 general is whole existing paddy fields, that is wide 388591,22 hectare. If seen from years previously wide land paddy the experience change decline wide harvest, so matter the influence to results production paddy from in 2019 amounted to 2,078,901.59 tons become of 2,040,500.19 tons with an average of 5,251 kg/ha. Production rice produced in North Sumatra obtained from production paddy and upland rice.

Need to food will Keep going experience enhancement along with amount residents who will the more increase. North Sumatra is Province fourth with amount resident the biggest in Indonesia after West Java, East Java and Central Java. So that growth population in North Sumatra included high. In 2020 the population of North Sumatra Province is 14.703.532 people. Condition increase amount resident tend far more high, whereas for availability wide paddy fields as production food tend decreased. This is a factor triggering the low availability of land. The carrying capacity of the land in general can be influenced by the behavior of the community in an area itself, both rice fields that are converted to function, as well as an increasing population.

Table 1. The results of the carrying capacity of rice farming land in North Sumatra Province in 2020

No	Regency/City	Total population in 2020	Harvested area (Ha) in 2020	Average productivity (kg/Ha)	KFM	Carrying Capacity of Agricultural Land	Classification
1	Nias	143983	9513.49	3720	265	0.9275251913	Kelas 3
2	Mandailing Natal	451028	18198.88	4086	265	0.622147909	Kelas 3
3	Tapanuli Selatan	283389	17677.65	5140	265	1.209925939	Kelas 2
4	Tapanuli Tengah	382917	11903.68	3523	265	0.4132790257	Kelas 3
5	Tapanuli Utara	303688	21508.92	5126	265	1.370009885	Kelas 2
6	Toba Samosir	184493	17574.08	6041	265	2.171479171	Kelas 2
7	Labuhan Batu	501596	11594.12	5019	265	0.4377791244	Kelas 3
8	Asahan	735026	10737.39	5714	265	0.3149853739	Kelas 3
9	Simalungun	871678	33172.77	5270	265	0.7568161547	Kelas 3
10	Dairi	285481	6546.43	5394	265	0.4667586631	Kelas 3
11	Karo	421997	8601.24	6725	265	0.5172471628	Kelas 3
12	Deli Serdang	2234320	49658.5	6346	265	0.5322337589	Kelas 3
13	Langkat	1048100	27742.99	5040	265	0.5034254963	Kelas 3
14	Nias Selatan	322520	10803.5	4277	265	0.5406313196	Kelas 3
15	Humbang Hasundutan	191776	11968.69	4711	265	1.109480294	Kelas 2
16	Pakpak Bharat	49688	1064.93	3497	265	0.2828259821	Kelas 3
17	Samosir	126710	7927.89	4680	265	1.104960375	Kelas 2
18	Serdang Bedagai	617772	48862.29	6085	265	1.816185923	Kelas 2
19	Batu Bara	420103	12988.09	5693	265	0.6641785051	Kelas 3
20	Padang Lawas Utara	277423	8583.94	4018	265	0.4691462703	Kelas 3
21	Padang Lawas	286627	8374.84	3489	265	0.3846932057	Kelas 3
22	Labuhanbatu Selatan	344819	164.48	3796	265	0.006832855542	Kelas 3
23	Labuanbatu Utara	366603	12268.17	4575	265	0.5777354214	Kelas 3
24	Nias Utara	138800	7584.08	4183	265	0.862492704	Kelas 3
25	Nias Barat	82425	2691.48	3920	265	0.4830280976	Kelas 3
26	Sibolga	87791	0	0	265	0	0
27	Tanjungbalai	177005	75.45	5658	265	0.009101034882	Kelas 3
28	Pematangsiantar	257110	2055.44	6157	265	0.1857415713	Kelas 3
29	Tebing Tinggi	166100	484.97	5517	265	0.06078583009	Kelas 3
30	Medan	2295003	924.83	5435	265	0.008264799572	Kelas 3
31	Binjai	279302	1456.08	5405	265	0.1063312735	Kelas 3

32	Padangsidimpuan	224483	3532.72	5551	265	0.3296485067	Kelas 3
33	Gunungsitoli	143776	2349.21	5684	265	0.3504641822	Kelas 3
Total	Sumatera Utara	14703532	388591.22	5251	265	0.5236817714	Kelas 3

Source: Department of Agriculture and Food Security, 2021 (processed)

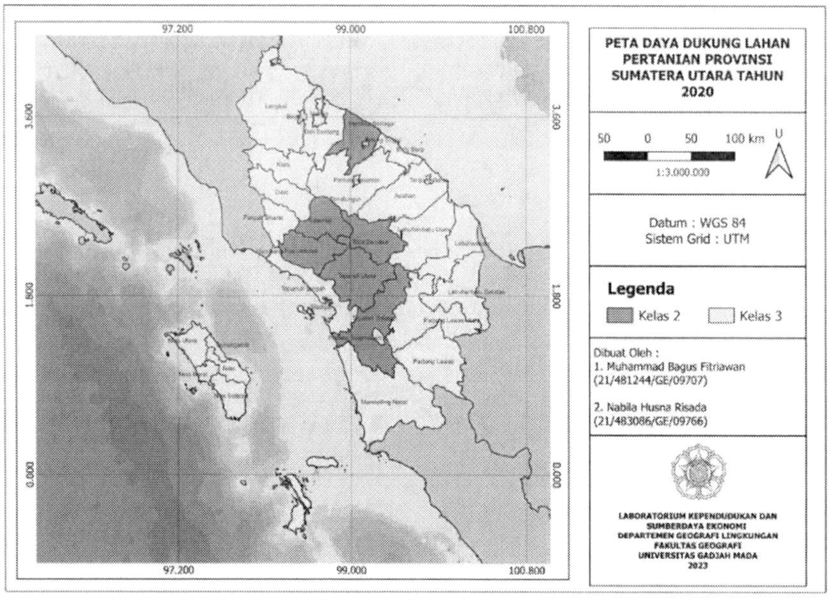

Fig. 1. Map of the carrying capacity of rice farming land in North Sumatra in 2020

The carrying capacity of agricultural land in general can change, it depends on the factors that can influence it. Population pressure is the main factor in exceeding the carrying capacity of agriculture. The need for land for rice production is indeed important, because it supports the realization of regional independence in achieving self-sufficiency. By Odum et al., in [11], in general the minimum annual food requirement is 265 kg/person/year. This basic need is the minimum size a person can live normally with the fulfillment of the necessities of life. Based on Table 1, the results of the analysis of the carrying capacity of paddy fields in North Sumatra Province in 2020 were obtained. Districts that are capable of self-sufficiency in food are only found in South Tapanuli, North Tapanuli, Toba Samosir, Samosir, and Serdang Bedagai Regencies. Meanwhile, other Regencies/Cities have not been able to be self-sufficient in food. This district is capable of self-sufficiency in food, but has not been able to properly meet the food needs of its people.

The results of the analysis of the carrying capacity of paddy fields in North Sumatra Province in 2020 showed that the carrying capacity of agricultural land was low, so that it had not been able to achieve self-sufficiency in food and had not been able to meet the food needs of its people properly. The low carrying capacity of paddy fields in North Sumatra Province is due to the increase in population from year to year, while the area of agricultural land tends to decrease with decreasing land quality. The low carrying capacity of agricultural land is also caused by agricultural production management factors that are not in accordance with conservation principles, resulting in low food production and crop failure.

The low carrying capacity of land in each of these regency indicates that the carrying capacity of agricultural land in North Sumatra needs to be increased either by reducing the population or by increasing the ability of agricultural land to provide maximum yields. Activities that can be carried out to increase the carrying capacity of land in order to fulfill self-sufficiency in food and provide a decent life is to increase agricultural yields by implementing agricultural intensification and extensification. Intensification activities that can be carried out are by selecting and using superior rice seeds, cultivating land in the right way, setting irrigation for rice fields, applying appropriate fertilizers and controlling pests. Meanwhile, an example of extensification that can be done is by opening or expanding new agricultural land.

4 Conclusion

The level of carrying capacity of agricultural land in North Sumatra in general is still relatively low, because the available agricultural land is lower than the available agricultural land needed. This shows that North Sumatra has not been able to achieve self-sufficiency in food and has not been able to meet the food needs of its people properly. The low carrying capacity of rice farming land in North Sumatra is due to the increasing population and the quality of the land agriculture continues to decline so that the production of rice plants can not meet the food needs of the community to the maximum. Therefore, it is necessary to suppress or reduce the population by reducing the birth rate. Apart from that, the most important thing is that there is a need to improve the management of agricultural land so that it is hoped that later yields can increase both in quantity and quality.

References

[1] Moniaga, V. (2011). Analasis Daya Dukung Lahan Pertanian. *Jurnal ASE*, 7(2), 61-68.
[2] Mubarokah, N. (2020). Analisis Daya Dukung Lahan Pertanian Tanaman Pangan Daerah Aliran Sungai Cibaliung, Provinsi Banten. *Jurnal JIPI*, 25(1), 73-80.
[3] Kafafa, U., Nadia, H., Fadilah, G.O., Abadi, A. W., & Putri R. F. Carrying capacity trend and projection analysis for Sumatra Selatan agricultural land in 2030. *IOP Conf. Ser.: Earth Environ. Sci*. 451 0120. (2020).
[4] Salma, N., Askarina, M., Romadhoniastri, S., Azahra, A. F., Karim, D. Isnain, M. N., & Putri, R. F. Study of agricultural economic potential in West Kalimantan using Regional Analysis Techniques. E3S Web of Conferences 325, 07008 (2021).
[5] Jakapratama, I. D. (2021). Analisis Daya Dukung Dan Kebutuhan Lahan Pertanian Di Kabupaten Purwakarta Tahun 2028. *Jurnal Samudra Geografi*, 4(1), 16-29.
[6] BPS. (2020). *Indikator Pertanian Provinsi Sumatera Utara Tahun 2020*. Medan: Badan Pusat Statistik Provinsi Sumatera Utara.
[7] Novita, D. (2010). Analisis Keterkaitan Sektor Pertanian Di Sumatera Utara. *Jurnal Agrica*, 3(1), 42-54.
[8] BPS. (2021). *Provinsi Sumatera Utara Dalam Angka 2021*. Medan: Badan Pusat Statistik Provinsi Sumatera Utara.
[9] Hilmanto, R. (2010). *Etnoekologi*. Bandar Lampung: Universitas Lampung.
[10] Muta'Ali, L. (2012). *Daya Dukung Lingkungan Untuk Perencanaan Pengembangan Wilayah*. Yogyakarta: Universitas Gadjah Mada.
[11] Suhardjo, & Tukiran. (1990). *Studi Literatur Konsep yang Sudah ada Mengenai Daya Tampung Wilayah*. Jakarta: Menteri Negara Kependudukan Dan Lingkungan Hidup.

Agriculture Carrying Capacity Evaluation of Land Subsidence Area in Yogyakarta Special Province

Alfindy Juliansari Alam, *Azizah* Nurkhalifah, *Muhammad* Zayyin Asrofi, *Muhammad* Raidan Azani and *Ratih* Fitria Putri*

Department of Environmental Geography, Faculty of Geography, Universitas Gadjah Mada, Indonesia

Abstract. The Special Region of Yogyakarta Province is one of the regions where the agricultural sector serves as a significant source of income. Due to the development of the region in the Special Region of Yogyakarta Province, several districts have experienced a decline in agricultural land due to land conversion for residential purposes. The ability of an area to achieve food self-sufficiency or meet its own food needs can be assessed through Land Carrying Capacity (DDL) and food availability. This analysis of Land Carrying Capacity for Agriculture and Food Availability employs a quantitative descriptive method. The land carrying capacity in the Special Region of Yogyakarta Province in 2022 indicates that all districts/cities have a DDL value of <1. This implies that the districts/cities in the Special Region of Yogyakarta Province cannot achieve food self-sufficiency and cannot adequately meet the population's food needs.

1. Introduction

Indonesia, as an agrarian country with fertile agricultural land, has led to most of its population working in the agricultural sector. The favorable agricultural conditions in Indonesia are influenced by its geographical location, resulting in a tropical climate conducive to significant rock weathering processes. The pivotal role of the Agricultural Sector in Indonesia's economy includes engaging a substantial portion of the population and contributing significantly to domestic food production. In other words, the agricultural sector also plays a role in enhancing the well-being of society. Furthermore, this sector can contribute to sustainable development through environmentally friendly farming practices and sustainable agricultural development.

As time progresses, population pressures have increased, leading to a decline in agricultural land. The growing population has led to increased development activities, resulting in the conversion of agricultural land into residential, industrial, and office areas, among others. Land conversion from agriculture has been experienced in various regions across Indonesia due to the expansion of built-up land for other sectors [1]. There are areas in the Special Region of Yogyakarta Province that have experienced a reduction in agricultural land due to conversion for residential use.

The Special Region of Yogyakarta Province is one of the regions where the agricultural sector is a significant source of income. [12] asserts that development orientation has favored urban areas. This is evident from the decrease in agricultural land area in recent years in various parts of the Special Region of Yogyakarta Province. The stagnant agricultural land area is not commensurate with the growing population. An increase in population translates to an increased demand for food, necessitating the enhancement of agricultural land resources in a region. To prevent a decline in food crop productivity and establish equilibrium, an analysis to enhance the carrying capacity of agricultural land is essential.

The problem of subsidence in agricultural land get serious attention along with the growing intensity of the issue of climate change and global warming. Therefore, it is necessary to revitalize agricultural lands that have been damaged due to over drying. Drainage canals must be immediately blocked or even (if possible) completely hoarded, so as not to cause increasingly serious damage to agricultural land environment. To support this, then regulate the ground water level with building blocks and forestry efforts back to reduce the rate of land subsidence the farm. The carrying capacity of agricultural land illustrates the ability of such land to support the food needs of the population [9]. The analysis of agricultural land carrying capacity serves as a valuable tool for sustainable development planning, especially within the agricultural sector [7], [14]. Determining agricultural land carrying capacity involves various parameters, including population size, harvest area, food requirements, and food productivity.

2. Methodology

2.1 Location Determination

The Carrying Capacity of Agricultural Land research was conducted in the Special Region of Yogyakarta Province. Considerations for determining locations in Special Region of Yogyakarta Province

*Corresponding author: ratihfitria.putri@ugm.ac.id

include the area of paddy fields which tends to increase year by year during the period 2015-2019 (Table 1). With the potential availability of these lands, it is hoped that the Special Region of Yogyakarta Province can not only be self-sufficient in food but also properly meet the community's food needs.

Table 1. Area of Rice Fields by City in Special Region of Yogyakarta Province during the period 2015—2019 (Ha)

No	Regency/ City	Year (ha)				
		2015	**2016**	**2017**	**2018**	**2019**
1	Kulon Progo	9,806	10,164	10,038	11,053	11,008
2	Bantul	14,116	14,067	14,857	14,857	14,945
3	Gunungkidul	7,718	7,862	7,861	31,896	31,973
4	Sleman	21,856	21,834	19,083	18,129	18,295
5	Yogyakarta	57	58	58	55	51
Special Region of Yogyakarta Province		53,553	53,985	51,897	75,990	76,272

Source: Ministry of Agriculture, 2020 processed.

2.2 Carrying Capacity of Agricultural Land

The ability of a region to achieve food self-sufficiency can be determined by comparing the optimum population size with food production output [6]. The concept used to understand the critical threshold of carrying capacity involves a limited and sustainable population that does not degrade the natural environment, thereby maintaining the ecosystem [10]. The carrying capacity of a region for agriculture can be calculated using a formula based on a combination of theories by Odum, Christeiler, and Ebenezer Howard, as follows [9]:

$$DDL = \frac{(^{Lp}/_{Pd})}{(^{KFM}/_{Pr})}$$

Information:
DDL = Carrying capacity of agricultural land
Lp = Area harvested (ha)
Pd = Total population (people)
KFM = Minimum physical requirements (kg/inhabitant/year) = 2,65 kw/inhabitant/year
Pr = Land productivity (kg/ha)

A region capable of meeting the minimum physical needs of its population, equivalent to 1600 calories/person/day or 265 kilograms of rice/person/year (KFM 2.65 kg rice/person/year). Meanwhile, a region capable of adequately meeting the dietary needs of its population is equivalent to 650 kilograms of rice/person/year or 2.47 times KFM. Based on these values, the established classification is as follows:

1) Class I (DDL > 2,47)
A region capable of food self-sufficiency and capable of offering a decent life to its inhabitants.
2) Class II (1 ≤ DDL ≤ 2,47)
Regions capable of food self-sufficiency but which have not been able to ensure a decent life for their inhabitants.
3) Class III (DDL < 1)
Regions that are not yet self-sufficient in food.

2.3 Food Security

Food security represents a condition where the food needs of a region are fulfilled, characterized by sufficient, safe, equitable, and accessible food supply (Government Regulation No. 68 of 2002 on Food Security). The concept used to assess food security involves the variables of Food Availability and Food Needs. Food availability and needs can be calculated using the following formulas:

Food availability = (Pr×Lp) × Paddy to rice conversion index

$$\text{Food demand} = Pd \times KFM$$

Information:
Pr = Land productivity (kg/ha)
Pd = Total population (people)
Lp = Area harvested (ha)
Paddy to rice conversion index = 64,02%
KFM = Minimum physical requirements (kg/inhabitant/year) = 2,65 kw/inhabitant/year

3. Results and Discussion

3.1 Carrying Capacity of Agricultural Land in Special Region of Yogyakarta Province

Table 2. Carrying Capacity of Agricultural Land in Special Region of Yogyakarta Province in 2022

No	Regency/ City	Harvested area of rice per capita	Area of land in food self-sufficiency	DDL Rate	Classification
1	Kulon Progo	0.043	0.044	0.97	Class III
2	Bantul	0.024	0.049	0.49	Class III
3	Gunungkidul	0.057	0.058	0.97	Class III
4	Sleman	0.021	0.054	0.38	Class III
5	Yogyakarta	0.000	0.049	0.00	Class III
Special Region of Yogyakarta Province		0.029	0.052	0.56	Class III

Source: Department of Agriculture and Food Security, 2023 processed.

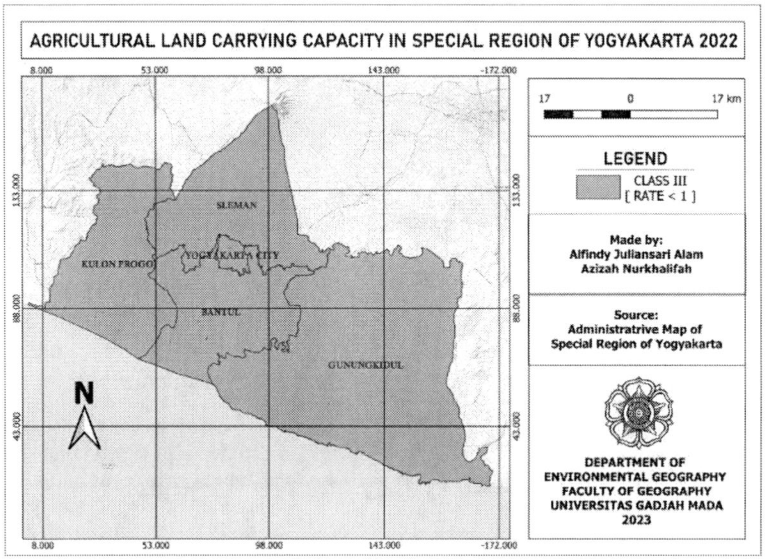

Fig. 1. Agricultural Lan-Carrying Capacity in Special Region of Yogyakarta 2022 Map

The analysis of land carrying capacity is linked to sustainable agricultural development, as understanding the land carrying capacity allows us to predict the future agricultural capabilities [4]. The results of the Land Carrying Capacity (DDL) calculation for Special Region of Yogyakarta Province in 2022 indicate that all districts/cities have DDL values < 1. The districts/cities in the Special Region of Yogyakarta Province cannot achieve food self-sufficiency and adequately meet the population's food needs.

A balanced land carrying capacity is determined when the agricultural land area in a region can meet the minimum physical needs of its population. The condition for a chieving food self-sufficiency (DDL = 1) in Special Region of Yogyakarta Province occurs when the agricultural land productivity reaches 89.86 kw/ha. Apart from productivity, achieving food self-sufficiency in Special Region of Yogyakarta Province also requires a population size of 2.119.619 individuals. This population size indicates that the Special Region of Yogyakarta Province currently has a population above the optimum level, necessitating increased food production and maintaining agricultural land availability.

The districts of Kulon Progo and Gunungkidul are targeted by one of the programs of the Special Region of Yogyakarta Agriculture and Food Security Office, the "Lumbung Mataram" program. According to jogjaprov.go.id, the Special Region of Yogyakarta Regional Government states that this program aims to fulfill the food needs of the community. The program is rooted in traditional community farming practices and serves to educate farmers and encourage them to develop better farming practices. In addition to education, the funds allocated are used for equipment procurement, dry land cultivation, and various activities to enhance agricultural productivity. The program's development can support the population's food needs and contribute to enhancing the carrying capacity of existing agricultural land.

Bantul District is known as the rice granary of Special Region of Yogyakarta. However, in 2022, it shows a lower DDL value compared to Kulon Progo and Gunungkidul Districts. Bantul District has been experiencing population pressure on agricultural land (paddy fields) since 2015 [11]. This indicates that the available agricultural land can no longer support the food needs or livelihoods of the existing farmers. To address this issue, the farming population in Bantul needs to consider sectors outside of agriculture or expand their farming activities. This is consistent with the physical conditions of Bantul District, which has sub-districts with potential for Sustainable Food Crop Agricultural Land (LP2B) in wetland areas.

The agricultural land carrying capacity of Sleman District shows the lowest value after Yogyakarta City. Based on its physical condition, Sleman District, which spans slopes, foothills, and the base of Mount Merapi, should ideally have a high DDL value. However, the agricultural land carrying capacity of Sleman District decreased from 2016 to 2019. Land use change is driven by population density, social infrastructure, industrial activities, Gross Regional Domestic Product (PDRB), and productivity [13]. The decrease in agricultural conditions in Sleman District is due to development trends directed towards Sleman District, which shares a direct border with Yogyakarta City, a high-growth center in the region.

3.2 Availability of Food in the Special Region of Yogyakarta Province

Table 3. Food demand and availability in the Special Region of Yogyakarta Province in 2022

No.	Regency/ City	Total population (people)	KFM	Food demand	Rice Production (kw GKG)	Food availability (kw Beras)	Availability-Demand	Information
1	Kulon Progo	451.342	2,65	1.196.056	1.160.560	742.991	-453.066	Deficit
2	Bantul	1.013.170	2,65	2.684.901	1.303.390	834.430	-1.850.470	Deficit
3	Gunungkidul	770.883	2,65	2.042.840	1.983.100	1.269.581	-773.259	Deficit
4	Sleman	1.147.562	2,65	3.041.039	1.169.510	748.720	-2.292.319	Deficit
5	Yogyakarta	378.913	2,65	1.004.120	430	275	-1.003.844	Deficit
Special Region of Yogyakarta Province		**3.761.870**	**2,65**	**9.968.956**	**5.616.990**	**3.595.997**	**-6.372.959**	**Deficit**

Source: Badan Pusat Statistik, 2023 processed.

The food availability in Special Region of Yogyakarta Province is at 3.595.997 kw/year during 2022, while the food requirement reaches 9.968.956 kw/year. When the available food is lower than the required amount, a deficit condition occurs. All districts/cities in the Special Region of Yogyakarta Province show deficit conditions, with the highest figure reaching 2.292.319 kw of rice in Sleman District. According to the Special Region of Yogyakarta Agriculture and Food Security Office, rice consumption in 2022 is only 0,8 kw/capita/year. This has a significant difference from the Minimum Physical Requirement (KFM) used as the reference, which is 2,65 kw/capita/year. The larger the KFM value in the calculation, the greater the difference between Food Availability and Food Requirement becomes.

Districts with deficit analysis results between Food Availability and Food Requirement do not necessarily show complex food scarcity issues. The availability of rice can still be found in various markets at prices that are still affordable for the local population. However, the rice production in the Special Region of Yogyakarta Province is not yet capable of achieving food self-sufficiency for its population. Consequently, it can be concluded that the food needs of the Special Region of Yogyakarta Province population are met in areas outside the province.

3.3 The Relationship between Agricultural Land Carrying Capacity and Food Availability

Variables affecting the value of Agricultural Land Carrying Capacity (DDL) include population, harvested land area, farmland productivity and KFM. Meanwhile, variables affecting food security in quantity include population, KFM, agricultural production, and the rate of paddy-to-rice conversion. Agricultural Land Carrying Capacity is linked to the extent of land, both for harvest and decent living, so a decrease in DDL can be addressed through land conservation, agricultural intensification, and land conversion.

Population growth also affects market demand and needs, which in turn affect market selling prices. A region's ability to achieve food self-sufficiency or meet its own food needs can be assessed through Agricultural Land Carrying Capacity and Food Availability. Agricultural Land Carrying Capacity (DDL) analysis focuses on rice productivity and the land's ability to achieve optimal conditions. Meanwhile, food availability analysis focuses on the population's consumption and the conversion rate of paddy to rice, which can depict the quality of rice, whether it is for consumption (GKG) or as seed (GKP). Thus, the relationship between paddy agricultural land carrying capacity and rice food availability has a strong correlation, where higher paddy agricultural land carrying capacity values correspond to better food security [4].

4. Conclusion

The agricultural land carrying capacity of Special Region of Yogyakarta Province in 2022 indicates that all districts/cities have DDL values < 1. The districts/cities in the Special Region of Yogyakarta Province cannot achieve food self-sufficiency and adequately meet the population's food needs. Furthermore, all districts/cities in the Special Region of Yogyakarta Province also exhibit deficit food availability conditions and still receive external food assistance to meet its food requirements. The relationship between paddy agricultural land carrying capacity and rice food availability reveals that as the value of paddy agricultural land carrying capacity decreases, food security worsens.

Acknowledgement

The writers would like to express our deep gratitude towards Directorate of Research Universitas Gadjah Mada Number 5075/UN1.P.II/Dit-Lit/PT.01.01/2023.

References

1. Ayun, Q., Kurniawan, S., & Saputro, W. A. (2020). Perkembangan Konversi Lahan Pertanian di Bagian Negara Agraris. *Vigor: Jurnal Ilmu Pertanian Tropika Dan Subtropika*, 5(2), 38-44.
2. BPS. (2023). Luas Panen dan Produksi Padi di DIY 2022 (Angka Tetap). Berita Resmi Statistik Mo. 29/04/34/Th.XXV
3. BPS. (2023). DIY Dalam Angka 2023.
4. Daniati, A., Sudrajat. (2019). Kajian Daya Dukung Lahan Pertanian dan Ketahanan Pangan Rumahtangga Tani di Desa Banjarharjo Kecamatan Kalibawang Kabupaten Kulonprogo (Thesis). Fakultas Geografi Universitas Gadjah Mada.
5. Humas DIY. (2022). "Wujudkan DIY Mandiri Pangan, Sri Sultan Segera Resmikan Lumbung Mataram di Gunungkidul." Accessed from https://jogjaprov.go.id/berita
6. Imansyah, Harisandi, D., Tamia, N., & Rahmawati, D. (2020). Analisis Daya Dukung Lahan Pertanian Terhadap Tekanan Penduduk Di Desa Sandik. *Media Komunikasi Geografi*, ***20***(2).
7. Kafafa, U., Nadia, H., Fadilah, G.O., Abadi, A. W., Putri R. F. Carrying capacity trend and projection analysis for Sumatra Selatan agricultural land in 2030. *IOP Conf. Ser.: Earth Environ. Sci.* 451 0120. (2020).
8. Kunu, P. J. (2020). Analisis Daya Dukung Lahan Pertanian untuk Menjamin Keamanan Pangan di Kepulauan Kei Besar Kabupaten Maluku Tenggara. *AGROLOGIA*, ***9***(2), 71—80.
9. Mubarokah, N., Rachman, L. M., & Tarigan, S. D. (2020). Analisis Daya Dukung Lahan Pertanian Tanaman Pangan Daerah Aliran Sungai Cibaliung, Provinsi Banten. *Jurnal Ilmu Pertanian Indonesia*, ***25***(1), 73-80.
10. Muta'ali, L. (2012). Daya Dukung Lahan Lingkungan untuk Perencanaan Pengembangan Wilayah. Yogyakarta: Badan Penerbit Fakultas Geografis (BPFG) Universitas Gadjah Mada.
11. Pridarsari, S. A. (2018). Daya Dukung Lahan Pertanian Dan Penentuan Lahan Pertanian Pangan Berkelanjutan Di Kabupaten Bantul. *Jurnal Bumi Indonesia*, *7*(1).
12. Prihatin, R. B., & Mutaali, L. (2015). Alih Fungsi Lahan di Perkotaan (Studi kasus di Kota Bandung dan Yogyakarta. *Jurnal Aspirasi*, ***6***(2), 105-118.
13. Rizal, M., Sudrajat. (2022). Pengaruh Laju Alih Fungsi Lahan Terhadap Daya Dukung Lahan Pertanian Di Kabupaten Sleman Tahun 2010 – 2019 (Thesis). Fakultas Geografi Universitas Gadjah Mada.
14. Salma, N., Askarina, M., Romadhoniastri, S., Azahra, A. F., Karim, D. Isnain, M. N., & Putri, R. F. Study of agricultural economic potential in West Kalimantan using Regional Analysis Techniques. E3S Web of Conferences 325, 07008 (2021).

Analysis of the Carrying Capacity Agricultural Land of Jawa Tengah Province in 2022 using the Soemarwoto Method

Arianti Ade Maya Dian, *Wicaksono* Satrio Aji, *Muhammad* Zayyin Asrofi, and *Ratih* Fitria Putri*
Department of Environmental Geography, Universitas Gadjah Mada, Indonesia

Abstract. Jawa Tengah is one of the provinces in Indonesia which has high agricultural productivity. That way, it is necessary to carry out an analysis of the carrying capacity of agricultural land in Jawa Tengah Province in 2022. The method used to calculate population pressure and the carrying capacity of agricultural land is the Soemarwoto method which is suitable for local conditions and can assess the factors that affect the level of carrying capacity of land in an area. Jawa Tengah Province has a carrying capacity of agricultural land of 1.43. The value of population pressure in Jawa Tengah Province with a dominance of less than 1 indicates that districts/cities in Jawa Tengah Province are dominated by the carrying capacity of high agricultural land. The high rate of population growth and not accompanied by an area of land capable of supporting the needs of the population can be a threat to the availability of sustainable agricultural land.

Keyword: *Agricultural, Capacity, Carrying, Method, Soemarwoto*

1 Introduction

An agrarian country is a country where most of the people work in the agricultural sector. Indonesia is an agricultural country because around 38 million Indonesians work in the agricultural sector [1]. That way, the need for adequate carrying capacity of agricultural land for the sustainability of the agricultural sector. The carrying capacity of agricultural land can be affected by population pressure and high population pressure can reduce the carrying capacity of agricultural land.

The need to know the role of the carrying capacity of agricultural land to analyse land capability. Land capability can be interpreted as land quality that can be assessed inclusively and differently between uses [2,3,4]. In analysing it is also necessary to use the theories that have been used before. There are many theories to analyse the level of carrying capacity of agricultural land such as the Combined Theory and Soemarwoto Theory.

Jawa Tengah is one of the provinces in Indonesia which has high agricultural productivity. One example is the rice plant. The productivity of rice plants in Jawa Tengah Province in 2022 will reach 55.41 ku/ha and production will reach 9 million tonnes [5]. That way, it is necessary to carry out an analysis of the carrying capacity of agricultural land in Jawa Tengah Province in 2022.

2 Methodology

Population pressure is a symptom of overpopulation in an area, given the availability of available resources for the needs of the population, in accordance with the desired standard of living in the area concerned [6]. An increase in population that continues continuously with the availability of land that remains in an area can cause population pressure on land to increase. Increasing population pressure on land can accelerate this occurrence, influenced by population growth factors, and shifts in livelihoods in the dynamics of urban growth [7]. There are 3 calculation models to calculate population pressure on agricultural land. The first model assumes that people only depend on the land they cultivate themselves. The second model assumes that the higher the income from the non-agricultural sector, the less population pressure. The third calculation model assumes that the greater the land productivity, the lower the population pressure [8]. Population pressure The Soemarwoto method is classified as supportive (if the population pressure value is >1), present (if the

*Corresponding author : ratihfitria.putri@ugm.ac.id

population pressure value is 1), and unsupported if the population pressure value is <1).

For this analysis using Model 2, the following formula [8]:

$$Tp = Z\,(1-\alpha)\,\frac{fPo\,(1+r)t}{L}$$

information:
TP : Population pressure on agricultural land
Z : Minimum land area for decent living (ha/person/year)
F : Percentage of farmers in the population (%)
Po : Total population in the first year (people)
r : Average annual population growth rate (% year)
t : Calculation time period
α : Percentage of non-agricultural income
L : Agricultural land area (ha)
Pt : Number of population at the end of the year (people) or Po(1+r)

The carrying capacity of agricultural land is closely related to the balance of supply and demand. Supply is generally limited, while demand is unlimited [9]. The supply of agricultural land is very limited due to the continuous demand for land for residential needs. Population pressure is inversely proportional to carrying capacity. The higher the population pressure, the lower the carrying capacity [10]. The carrying capacity of agricultural land can be classified as high (if the carrying capacity of agricultural land is> 1), optimal (if the carrying capacity of agricultural land is 1), and low (if the carrying capacity of agricultural land is <1) [11]. The following is the formula for knowing and analyzing the carrying capacity of agricultural land using the Soemarwoto method:

$$iDDL = \frac{1}{TP}$$

information:

iDDL: Carrying Capacity Land Agricultural Index

TP: Population Pressure

The reason for using the Soemarwoto method in calculating population pressure and the carrying capacity of agricultural land is because the method is in accordance with local conditions. The Soemarwoto method was developed based on studies and observations made in Indonesia, making it suitable for this analysis. The Soemarwoto method can also assess the factors that affect the level of land carrying capacity in an area. This theory is approached by the carrying capacity of the land which is the ratio between the available land and the number of farmers. The use of this analysis is based on land area standards with the availability of land according to its designation [12].

3 Discussion

Table 1. Population Pressure Value of Jawa Tengah Province in 2022 using Soemarwoto Method

Regrency/City	Population Pressure
Cilacap	0,56
Banyumas	0,70
Purbalingga	0,64
Banjarnegara	0,49
Kebumen	0,85
Purworejo	0,45
Wonosobo	0,65
Magelang	0,95
Boyolali	0,67
Klaten	0,80
Sukoharjo	0,60
Wonogiri	0,37
Karanganyar	0,76
Sragen	0,85
Grobogan	1,08
Blora	1,03
Rembang	0,44
Pati	0,72
Kudus	0,44
Jepara	0,51
Demak	0,68
Semarang	0,59
Temanggung	0,85
Kendal	0,60
Batang	0,47
Pekalongan	0,80
Pemalang	0,97
Tegal	0,90
Brebes	1,16
Magelang	1,65
Surakarta	3,08
Salatiga	0,23
Semarang	0,21
Pekalongan	0,54
Tegal	4,44
JAWA TENGAH	0,70

Table 2. The value of the carrying capacity of agricultural land in Jawa Tengah province in 2022 using Soemarwoto Method

Regency/City	Carrying capacity
Cilacap	1,77
Banyumas	1,43
Purbalingga	1,56
Banjarnegara	2,06
Kebumen	1,18
Purworejo	2,23
Wonosobo	1,55
Magelang	1,06
Boyolali	1,50
Klaten	1,25
Sukoharjo	1,68
Wonogiri	2,68
Karanganyar	1,32
Sragen	1,17
Grobogan	0,93
Blora	0,97
Rembang	2,27
Pati	1,39
Kudus	2,28
Jepara	1,98
Demak	1,47
Semarang	1,68
Temanggung	1,18
Kendal	1,66
Batang	2,11
Pekalongan	1,26
Pemalang	1,03
Tegal	1,11
Brebes	0,86
Magelang	0,60
Surakarta	0,32
Salatiga	4,33
Semarang	4,69
Pekalongan	1,85
Tegal	0,23
JAWA TENGAH	1,43

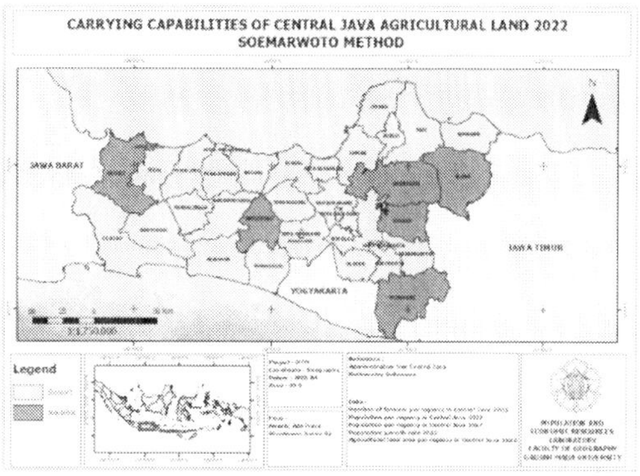

Fig 1. Map of Population Pressure on Agricultural Land in Jawa Tengah Province in 2022

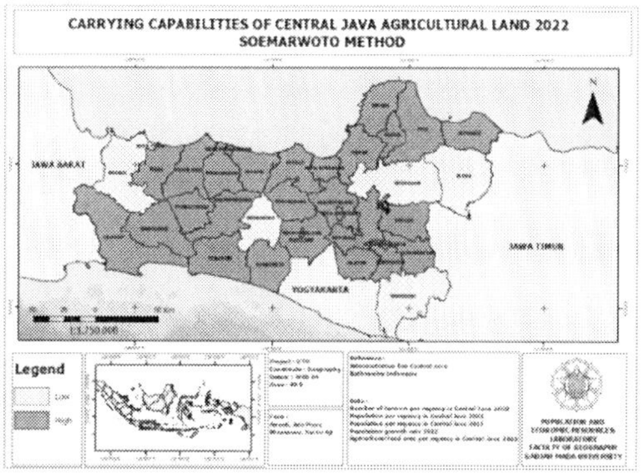

Fig 2. Map of the The carrying capacity of agricultural land of Jawa Tengah Province Soemarwoto Method in 2022

Administratively, Jawa Tengah Province consists of 35 regencies/cities. Regencies/cities in Jawa Tengah Province have various population pressures on agricultural land, this population pressure will affect the carrying capacity of agricultural land. Jawa Tengah Province has population pressure in the model 2 ITP calculation of 0.7, the ITP value is less than 1, indicating that there has not been population pressure on the land. A low ITP value indicates that there are a large number of people working in non-agricultural fields, thereby contributing to reducing population pressure on land. The calculation results also show that Jawa Tengah Province is dominated by agricultural land which is still able to support the needs of the population. Overall, Jawa Tengah Province has a carrying capacity for agricultural land of 1.43. The value of population pressure in Jawa Tengah Province with a dominance of less than 1 indicates that districts/cities in Jawa Tengah Province are dominated by the carrying capacity of high agricultural land. Meanwhile, different conditions occurred in several areas of Jawa Tengah Province, namely there were 6 regencies/cities that experienced high population pressure and low land carrying capacity, namely Grobogan Regency, Blora Regency, Brebes Regency, Magelang City, Surakarta City, and Tegal City.

Administratively, Tegal City is part of Jawa Tengah Province. This city has the highest population pressure compared to other regions. Population pressure in Tegal City was analyzed using model 2 to get a value of 4.44. Agricultural land in this region has the third smallest area after Magelang City and Surakarta City, which is only 652 ha. However, this condition is not supported by a high population parameter compared to Magelang City so that population pressure on land is relatively high. Food sustainability can support community welfare which can be pursued through increasing the carrying capacity of agricultural land by increasing the area of agricultural land.

Model 2 population pressure is low in Semarang City, which is equal to 0.21. This condition can occur marked by a large population of farmers as a livelihood, namely 14,517 people and a declining population from 2017 to 2022, the pattern of population decline only occurs in several districts/cities in Jawa Tengah Province. The low population pressure value indicates the effect of population income in the non-agricultural sector being able to support the welfare of farmers. The condition of low population pressure on agricultural land is also influenced by the area of agricultural land (irrigated and dry rice fields) occupying an area of 0.57 of the total area of Jawa Tengah Province with a balanced number of farmers and residents. The distribution pattern of population density in Semarang City, which has a high optimal population level, is generally classified as a rural area where many rice fields are still found.

Semarang City is an area with the highest carrying capacity of agricultural land in Jawa Tengah Province, namely 4.69 and is classified into the class of carrying capacity of high agricultural land. This happens because the irrigation conditions in Semarang City are supported by the West River Flood Canal [13]. The soil conditions in this area are also dominated by alluvial soil types which are fertile and have high levels of nutrients. The existence of an agricultural development policy regarding the development of this area which is oriented towards maintaining food security in Semarang City is also the main cause of this area having a high carrying capacity of rice plants. This phenomenon is different from Tegal City which only has a small land area and a large population. As a result, the City of Tegal has a carrying capacity of 0.23 agricultural land which is classified as a carrying capacity of low agricultural land. This means that this district has not been able to carry out food self-sufficiency and has not been able to meet the food needs of the community properly. The decline in food crop production was due to the decreased productivity of agricultural land.

Semarang city is a metropolitan city with a high population accompanied by food needs in the city of Semarang which continues to increase which is also followed by the need for space to accommodate the activities of its population. So that there is a struggle for land between residents who want to live with agricultural areas where the availability of land in the city of Semarang is increasingly limited. To control land conversion, the Semarang City government regulates land use regulations listed in Semarang City Regional Regulation No. 14 of 2011. The city of Semarang has agricultural potential to be developed in several areas to be able to support efforts to achieve food security, namely Banyumanik, Genuk, Gunungpati, Mijen, Ngaliyan, Pedurungan, Tembalang, and Tugu Districts for the development of the agricultural sector by empowering communities and introducing methods and the latest technology. The city of Semarang also has a farming community and farmer women's groups of around 10,285 farmers who are members of 389 farmer groups.

The population in each district varies so that it experiences different population growth. The smallest percentage of population growth rates occurred in Magelang City and Surakarta City with a population growth rate of 0.07%, while the highest growth rate occurred in Cilacap Regency. Population growth causes land use problems to become more complex and competitive [14]. However, the population pressure that occurred in Cilacap Regency of 0.56 was classified as being able to support the population because the large population was accompanied by the number of farmers with a percentage of the population working in this sector of 0.11%. Therefore, the carrying capacity of agricultural land is also high at 1.77. This condition also occurs in Wonogiri Regency with the percentage of farmers of 0.11% with low population pressure of 0.37 and wide agricultural land available, so the carrying capacity of the land is classified as high at 2.68. Different conditions occurred in the City of Magelang, Surakarta City, Semarang City, and Pekalongan City with a low percentage of farmers to the population of 0.01%.

The population pressure figure reflects the ratio between population parameters (population size and population growth rate) and the area of agricultural land. Along with the increase in population, simultaneously the need for agricultural land is also increasing [15]. The high rate of population growth and not accompanied by an area of land capable of supporting the needs of the population can be a threat to the availability of sustainable agricultural land. The population continues to increase and is not accompanied by a balanced distribution pattern causing the need for land to increase resulting in land conversion. Under the same conditions, food consumption also increases, while food production remains relatively constant or decreases. The population that continues to increase can lead to a number of food crop production that is not balanced with the needs of the population. This condition shows that the carrying capacity of agricultural land will be increasingly deficit. Indications in the field are known that the need for land is influenced by an increase in population, while the availability and area of land are fixed [16].

Wonogiri Regency has the largest number of farmers and agricultural land compared to regencies/cities in Jawa Tengah Province. The total population with people working in the agricultural sector is 270,879 people. However, the area of this area with land use in the form of agricultural land is 107,163 ha. This condition causes the minimum area of land to be able to live properly at 0.23 ha/person/year. High population growth and a large part of the population living as farmers can increase population pressure on agricultural land. If this condition continues, it can reduce the ability of agricultural land both in quality and quantity [17]. Areas with a population density that works in the agricultural sector will experience greater population pressure than areas with a population density where the population works in the non-agricultural sector [18]. This value indicates that there is an inverse relationship between the area of agricultural land which tends to decrease, while the population working in the agricultural sector increases. This condition causes people who work as farmers to lose agricultural land. [19] stated that population pressure can occur if population growth causes a decrease in the ratio of land to population. This condition causes farmers to expand agricultural land or work outside the agricultural sector. Based on the fact, the area of paddy fields in Jawa Tengah Province has decreased every year, including being converted as built-up land. This land conversion has not been accompanied by efforts to open new agricultural land. Reduced agricultural land resulted in reduced productivity of agricultural crops. Vigilance against increasing population pressure is an urgency for the government of Jawa Tengah Province in implementing a strategic program for the development of the agricultural sector in the future.

The highest minimum land area for decent living is owned by the City of Magelang with an area of 0.5 ha/person/year. This value is supported by the number of farmers of 1,049 people against a population that occupies this region in 2022 of 121,675 people. Meanwhile, the total agricultural land area is 206 ha. So that this Regency is classified on the existence of population pressure on land. The high population pressure on land through the ITP parameter illustrates that agricultural products in Magelang City are unable to meet the food needs of the population so that there is population pressure on land. This condition is evidenced by the low availability of agricultural land and low land productivity. There are many land uses other than agriculture so that the conversion of existing land functions is relatively high [20]. The land in this area is less productive but can be utilized based on the appropriate land capability.

The stability of agricultural land so that further land conversion does not occur, it is necessary to optimize agricultural land management efforts and reduce land conversion and divert the livelihoods of the farming population to the non-agricultural sector in districts with high population pressure on agriculture. One of the solutions offered in reducing population pressure is to increase the benefits of land for sharecroppers. High population pressure on agricultural land can result in a decrease in land resources and damage to ecosystems. This reflects the imbalance between the population and its growth rate with the availability of agricultural land.

Agricultural land has the potential to support the food security of a region. However, it cannot be denied that agricultural land has limitations in its processing. Limited land use needs to be identified or analyzed for land suitability to determine land capability in agricultural activities, especially in processing sustainable crop production. Sustainable crop production can meet food needs in a region. The increased demand for food is due to an increase in population accompanied by an increase in the use of agricultural land.

4 Conclusion

Jawa Tengah Province has a carrying capacity of agricultural land of 1.43. The value of population pressure in Jawa Tengah Province with a dominance of less than 1 indicates that districts/cities in Jawa Tengah Province are dominated by the carrying capacity of high agricultural land. Meanwhile, different conditions occurred in several areas of Jawa Tengah Province, namely there were 6 regencies/cities that experienced high population pressure and low land carrying capacity, namely Grobogan Regency, Blora Regency, Brebes Regency, Magelang City, Surakarta City, and Tegal City. Semarang City is an area with the highest carrying capacity of agricultural land in Jawa Tengah Province because the irrigation conditions in Semarang City are supported by the West Flood Canal River and the soil conditions of this area are also dominated by alluvial soil types which are fertile and have sufficient nutrients and regional policies regarding Food security makes Semarang City have a high carrying capacity of agricultural land.

The high rate of population growth and not accompanied by an area of land capable of supporting the needs of the population can be a threat to the availability of sustainable agricultural land. Areas with a population density that works in the agricultural sector will experience greater population pressure than areas with a population density where the population works in the non-agricultural sector. This value indicates that there is an inverse relationship between the area of agricultural land which tends to decrease, while the population working in the agricultural sector increases. This condition causes people who work as farmers to lose agricultural land. One of the solutions offered in reducing population pressure is to increase the benefits of land for sharecroppers. High population pressure on agricultural land can result in a decrease in land resources and damage to ecosystems.

Reference

1. Moniaga, V. R. (2011). Analisis daya dukung lahan pertanian. *Agri-Sosioekonomi*, *7*(2), 61-68
2. Notohadiprawiro. (1987). *Tanah, Tata Guna Lahan dan Tata Ruang Dalam Analisis Dampak Lingkungan*. Universitas Gadjah Mada Press. Penerbit: Yogyakarta.
3. U. Kafafa, H. Nadia, G. O. Fadilah, A. W. Abadi, R. F. Putri, *Carrying capacity trend and projection analysis for Sumatra Selatan agricultural land in 2030,* in IOP Conf. Ser.: Earth Environ. Sci. 451 0120. (2020).
4. N. Salma, M. Askarina, S. Romadhoniastri, A. F. Azahra, D. Karim, M. N. Isnain, R. F. Putri, *Study of agricultural economic potential in West Kalimantan using Regional Analysis Techniques* in E3S Web of Conferences 325, 07008 (2021)
5. Central Berau of Statistic. (2023). *Provinsi Jawa Tengah dalam Angka*. Semarang: Central Berau of Statistic.
6. Rusli, S., Widono, S., dan Indriana, H. (2009), Tekanan penduduk dan, overshoot ekologi pulau jawa, dan masa pemulihannya. *Jurnal Sodalit*, 3(1)
7. Esti S dan Suwarno. (2004). Pola spasial konversi lahan pertanian dan faktor-faktor yang berpengaruh di Purwokerto. *Jurnal Penelitian dan Pengabdian Masyarakat Saintek*. No.1 2004.
8. Soemarwoto, O. (1985). A Qualitative of Population Pressure and It's Potential Use in Development Planning. *Majalah Demografi Indonesia*, Vol.12 (24). Jakarta
9. Muta'ali, L. (2015). *Teknik Analisis Regional untuk Perencanaan Wilayah, Tata Ruang dan Lingkungan*. Badan Penerbit Fakultas Geografi (BPPFG) Universitas Gadjah Mada. Yogyakarta.
10. Muta'ali, L. (2012). *Daya Dukung Lingkungan untuk Perencanaan Pengembangan Wilayah*. Badan Penerbit Fakultas Geografi (BPFG) UGM. Yogyakarta
11. Soemarwoto, O. (1994). *Ekologi, Lingkungan Hidup dan Pembangunan*. Djambatan. Bandung
12. Apriyanto, M. (2022). Lahan Pertanian Panagan Berkelanjutan. Mulono Apriyanto. ISBN: 9786233489874.
13. Prasetyo, D., & Dermawan, V. (2015). Kajian Penanganan Sedimentasi Sungai Banjir Kanal Barat Kota Semarang (Study of Sedimentation Mitigation West Floodway Semarang City). *Jurnal Teknik Pengairan: Journal of Water Resources Engineering*, *6*(1), 76-87.
14. Jayadi, I. M. Y., Christiawan, P. I., & Sarmita, I. M. (2017). Dampak Pertumbuhan Penduduk Terhadap Daya Dukung Lahan Pertanian Di Desa Sambangan. *Jurnal Pendidikan Geografi Undiksha*, *5*(2).
15. Imansyah, I., Harisandi, D., Tamia, N., & Rahmawati, D. (2020). Analisis Daya Dukung Lahan Pertanian Terhadap Tekanan Penduduk Di Desa Sandik. *Media Komunikasi Geografi*, *21*(2), 120-129.
16. Cahyanto, A., & Fikriyah, V. N. (2022). *Analisis Daya Dukung Lahan dan Proyeksi Kebutuhan Lahan Pertanian di Kabupaten Klaten* (Doctoral dissertation, Universitas Muhammadiyah Surakarta).
17. Mubarokah, N., Rachman, L. M., & Tarigan, S. D. (2020). Analisis daya dukung lahan pertanian tanaman pangan daerah aliran Sungai Cibaliung, Provinsi Banten. *Jurnal Ilmu Pertanian Indonesia*, *25*(1), 73-80.
18. Herlindawati, A., Trimo, L., & Noor, T. I. (2018). Analisis Tekanan Penduduk Terhadap Petani Padi Sawah. *Mimbar Agribisnis: Jurnal Pemikiran Masyarakat Ilmiah Berwawasan Agribisnis*, *4*(1), 12-24.
19. Mardhatil, M., Purwati, I., & Iskarni, P. (2020). Analisis Tekanan Penduduk Terhadap Lahan Pertanian Sawah Kabupaten Pemalang, Jawa Tengah. *Jurnal Kependudukan dan Pembangunan Lingkungan*, *1*(3), 1-9.
20. Umar, S. (2020). *Perspektif Ekonomi Kesatuan Pengelolaan Hutan (Kph)*. Deepublish.

Analysis of The Carrying Capacity of Agricultural Land in South Sulawesi Province

Favian Daniswara Nugraha, *Syifa* Ayulia Anggraini, *Muhammad* Zayyin Asrofi and *Ratih* Fitria Putri*

Department of Environmental Geography, Faculty of Geography, Universitas Gadjah Mada, Indonesia

Abstract. *The agricultural sector is the field sector that contributes the largest Gross Regional Domestic Product (GRDP) in South Sulawesi, which has a valueof more than 109 billion. This study aims to analyze the carrying capacity offood agriculture land to be useful for evaluation and ongoing planning management of food agricultural land in South Sulawesi. The data obtained is secondary data originating from South Sulawesi in Figures 2021 published by the Central Statistic Agency (BPS). The method used is adescriptive quantitative approach to describe relationships between variables. The calculation of Carrying Capacity of Agricultural Land in this study used a combination of theories from Odum, Christeller, Ebenezer Howard, and Isard. The variables used in the calculation included the harvested land area (Ha), population size, average rice production per hectare (Kg/Ha), and the Minimum Physical Requirement with apredetermined value of 265. The result showed that there are 7 class I where Barru, Bone, Soppeng, Waju, Sidenreng Rappang, Pinrang, and East Luwu Regencies are included. While for class two classification consists of 10 districts and the remaining seven are included in class 3 indicating the lowestcarrying capacity among all urban districts in South Sulawesi.*

1 Introduction

Indonesia is a predominantly agricultural nation with a large agricultural workforce. Agriculture is the most significant economic activity, so agricultural land is crucial for their ability to survive. Agricultural land where farmers conduct their operations has been steadily declining recently. The cause of this condition is the increasing demand for agriculturalland due to population growth. As a result, the agricultural land's ability to hold the world's food supply is declining [1].

Environmental carrying capacity refers to an ecosystem's capacity to sustain healthy organism life while maintaining productivity, adaptability, and self-renewal. Because each region has unique geographical features and engages in human activity for a variety of reasons, the environmental carrying capacity varies greatly. The carrying capacity of an agricultural land can affect the carrying capacity of land in an area, determining whether or not agriculture can still be sustained there. The carrying capacity of agricultural land refers to the amount of agricultural land that can sustain the welfare of people in a region, particularly in terms of meeting food needs [2][3][4]. Therefore, it is crucial to conduct an analysis of the carrying capacity of agricultural land because it ensures the long- term viability of the industry. The harvested area and agricultural productivity are linked by analysis of the carrying capacity of agricultural land, and this relationship suggests that these two factors may have an impact on the rise in carrying capacity of agricultural land [5]

South Sulawesi has a vast area in Sulawesi Island. South Sulawesi has an area of 46.717,48 kilometers square. This province has 21 regencies and 3 city with Makassar becoming its capital. The agricultural sector is the field sector that contributes the largest Gross Regional Domestic Product (GRDP) in South Sulawesi, which has a value of more than 109 billion. South Sulawesi produced 5,054,166.96 tons of rice in 2020 from 978,192.54 ha of harvested rice plantation land. The district with the highest yields of rice, corn, and soybean sis Bone Regency. The district of Bone produces the most rice, 772,874.27 tons [6]. Sulawesi Selatan also produces horticultural products in addition to food crops. According to [6] South Sulawesi is the top five provinces with the largest rice production in Indonesia with 4 708 464,97 ton of rice production in 2020. However, the increasing population in South Sulawesi will impact the agricultural land and its carrying capacity.

The amount of food that is available in a region will depend on the agricultural land that is constantly changing. The state of the carrying capacity of the land will be affected by the declining food supply, whether it gets better or worse. More food resources are required as there is a greater increase in population. As based more than 150 years ago on the Malthusian theory of

*Corresponding author: ratihfitria.putri@ugm.ac.id

population and the environment. Malthus discusses the tension between population growth that is too rapid and the population's capacity to feed itself. In contrast to the finite supply of natural resources, population growth requires more land, water, energy, fertilizer, and other resources.

The use of agricultural food land including the size and productivity of rice fields owned as well as the population have an impact on the availability of food in a region. Therefore, it is important to take intoaccount the area of rice fields, the productivity of rice fields, and the population in order to calculate the carrying capacity of a region's food agriculture. This study aims to determine the carrying capacity of food agriculture in South Sulawesi Province. In order to determine sustainable agriculture policy strategies, it is expected to give direction and guidance based on knowledge of the effects of land use change andpopulation growth in this province.

2. Method

The research location covers the area of South Sulawesi Province which is geographically located at 116° 48'-122°36' East Longitude and 0° 12' - 8° South Latitude. South Sulawesi Province has 3 cities and 21 districts. South Sulawesi was chosen as the object of study because it is one of the highest rice producers in Indonesia. The data obtained is secondary data originating from South Sulawesi in Figures 2021 published by the Central Statistic Agency (BPS) [7]. The data obtained are the population in 2020, harvested area (ha) in 2020, and production (kg). The level of carrying capacity of agricultural land can be calculated based on the combined theory of Odum, et al. The theory is used as a calculation because it uses data that is easy to obtain compared to Soemarwoto's theory. The calculation of the carrying capacity of agricultural land can be seen in the following formula:

$$\sigma = X/K$$

Where :

σ = Level of Land Carrying Capacity

X = Total of Plant Per Capita

K = Total Area of Land for Self-Suffiiency

With :

X = Total Harvest Area (Ha) / Amount of Population (People)

K = Minimum Physical Need (MPN) / Food Productivity Plant(Ha/Year)

Regions with self-sufficiency capability are area that is able to meet the minimum physical needs (NPM) of 2600/calorie/person/day or equivalent 265 kg of rice/person/year. Areas that can deliver decent livelihood for the population, where the area depend on capable food crops meet the standard of living of a decent population that is equal to 650 kg of rice/person/year 2,466 NPM. Class I where $\sigma > 2.47$ area who are able to be self-sufficient in food and at the same time provide decent livelihood for the population, class II where $1 < \sigma < 2, 47$ regions are capable of self-sufficiency food but unable to provide a stable livelihood feasible for the population, class III where $\sigma <$ 1 area who are unable to be self-sufficient in food [8].

3. Result And Discussion

The results of the calculation of the carrying capacity of agricultural land in South Sulawesi Province in 2020 from all Regencies/Cities show various levels of carrying capacity of the land. The distribution of land carrying capacity in SouthSulawesi Province can be classified into three classes as can be seen in table 1 and 2 below.

Table 1. Carrying Capacity Index of agricultural land for regencies/cities in South Sulawesi

No	Regency/City	Index	No	Regency/City	Index
1	Kepulauan Selayar	0,21	13	Wajo	5,78
2	Bulukumba	1,57	14	Sidenreng Rappang	5,39
3	Bantaeng	0,97	15	Pinrang	4,85
4	Jeneponto	1,07	16	Enrekang	0,75
5	Takalar	1,31	17	Luwu	1,68
6	Gowa	1,19	18	Tana Toraja	0,75
7	Sinjai	1,31	89	Luwu Utara	1,90
8	Maros	1,78	20	Luwu Timur	2,89
9	Pangkajene Kepulauan	1,42	21	Toraja Utara	1,17
10	Barru	2,70	22	Kota Makassar	0,03
11	Bone	3,55	23	Kota Parepare	0,11
12	Soppeng	4,23	24	Kota Palopo	0,35

Table 2. Classification of Carrying Capacity of Agricultural Land Index for regencies/cities in South Sulawesi.

No	Regency/City	Class	No	Regency/City	Class
1	Kepulauan Selayar	III	13	Wajo	I
2	Bulukumba	II	14	Sidenreng Rappang	I
3	Bantaeng	III	15	Pinrang	I
4	Jeneponto	II	16	Enrekang	III
5	Takalar	II	17	Luwu	II
6	Gowa	II	18	Tana Toraja	III
7	Sinjai	II	89	Luwu Utara	II
8	Maros	II	20	Luwu Timur	I
9	Pangkajene Kepulauan	II	21	Toraja Utara	II
10	Barru	I	22	Kota Makassar	III
11	Bone	I	23	Kota Parepare	III
12	Soppeng	I	24	Kota Palopo	III

There are 7 districts with class I of land carrying capacity, which illustrates that these districts will be able to be self- sufficient in food in 2020 and their residents can live a decent life. Another classification result shows that 10 districts are included in the category of carrying capacity of class II agricultural land which indicates that the area can be self-sufficientin food for the next twenty years, but its population has not been able to make a decent living from the existing agricultural sector. There are 7 districts classified into the carrying capacity of class III agricultural land with the most deficit or the lowest level indicating not being able to be self-sufficient in food. For more details, the distribution of the carrying capacity of the South Sulawesi Province's landcarrying capacity in 2020 is presented in the following map.

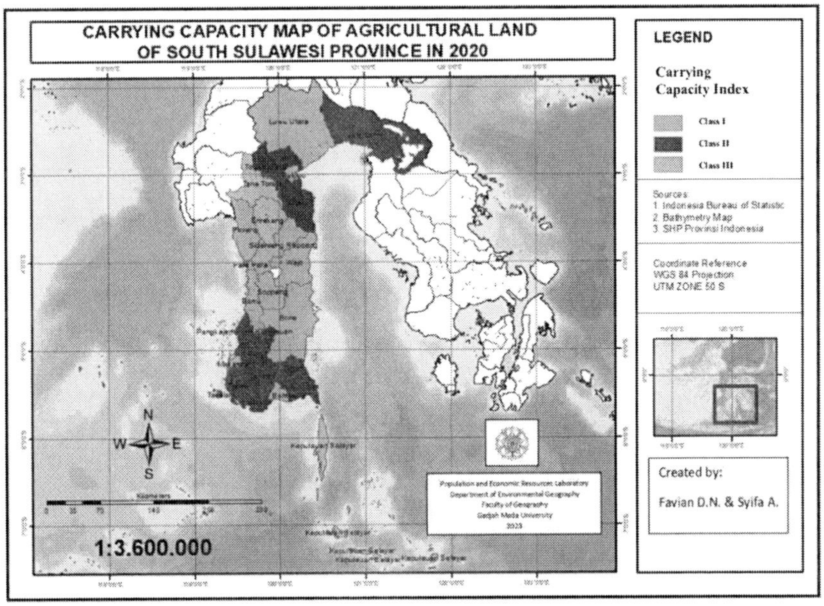

Fig.1. Carrying Capacity Map of Agricultural Land of South Sulawesi Province in 2020

Based on Odum et al's calculations, South Sulawesi Province is divided into 3 classes of carrying capacity of agricultural land. Selayar Regency included has a DDL value of 0.21 which belongs to class III. Bulukumba has a DDL value of 1.57 with class II. Bantaeng Regency has a DDL value of 0.97 with class III. Jeneponto Regency has a DDL value of 1.07 with class II. Takalar Regency has a DDL value of 1.31 which belongs to class II. Gowa Regency has a DDL value of 1.19 which belongs toclass II. Sinjai Regency has a DDL value of 1.31 which belongs to class II. Maros Regency has a DDL value of 1.78 with class II. Pangkajene and the archipelago districts have a DDL value of 1.42 with class II. Barru Regency has a DDL value of 2.7 with class I. Bone Regency has a DDL value of 3.5 with class I. Soppeng Regency has a DDL value of 4.23 which is in class I. Wajo Regency is in class I with a DDL of 5.78. Sidenreng Rappang Regency has a DDL value of 5.39 with class I. Pinrang Regency is included in class I with a DLL value of 4.85. Enrekang Regency has a DDL value of 0.75 with class II. Luwu Regency has class II with a DDL value of 1.68. Tana Toraja Regency which has a DDL of 0.75 with class III with class II. North Luwu Regency has a DDL value of 1.9 with class II. East Luwu Regency hasa score of 2.89 with class I. North Toraja Regency has a DDL of 1.17 with class II. Makassar City has a DDL value of 0.03 which belongs to class III. Parepare City has a DDL of 0.1 with class III. Palopo City has a DDL value of 0.35 with class III. From the all regencies and cities in South Sulawesi, there are 7 class I where Barru, Bone, Soppeng, Waju, Sidenreng Rappang, Pinrang, and East Luwu Regencies are included. This means that the district can fulfill self-sufficiency in food in its area and is able to meet food needs properly. Meanwhile, 10 districts namely Bulukumba, Jeneponto, Takalar, Gowa, Sinjai, Maros, Pangkajene and Islands, Luwu, North Luwu and North Toraja fall into class II. This indicates that the region is capable of self- sufficiency in food but has not been able to meet food needs properly. Meanwhile, 7 other regencies/cities are included in class III where there are Selayar Islands, Bantaeng, Enrekang, Tana Toraja, Makassar City, Parepare City, and Palopo City. This states that the regions have not been able to achieve self-sufficiency in food and have not been able to meet food needs properly in their regions.

There are still many areas that have not been able to meet food needs properly in their regions indicating that there are still many areas with low land carrying capacity which can be caused by population, rice field harvest area, average rice production which has an impact on high and low agricultural land support. in South Sulawesi. The high population but not in line with the high average rice production can result in low carrying capacity of agricultural land. This happens because rice production cannot meet the food needs of a very high population so that the carrying capacity of the land is low.

4. Conclusion

It was found that the carrying capacity of agricultural land in South Sulawesi is 7 districts that are included in class I, namely Barru, Bone, Soppeng, Waju, Sidenreng Rappang, Pinrang, and East Luwu Regencies. Class II consists of 10 districts namely Bulukumba, Jeneponto, Takalar, Gowa, Sinjai, Maros, Pangkajene and Islands, Luwu, North Luwu, and North Toraja. Class III has 7 regencies/cities namely Selayar Islands, Bantaeng, Enrekang, Tana Toraja, Makassar City, Parepare City, and Palopo City. The carrying capacity of agricultural land in South Sulawesi is still classified as being able to carry out food self-sufficiency but has not been able to meet food needs properly. This is caused by factors such as population, rice field harvest area, and average rice production. This shows that agriculture in South Sulawesi has not been able to properly meet the food needs of the region. In the future it is hoped that agriculture in South Sulawesi can meet their food needs properly.

References

1. S.I. Wayan, A.M. Ruslin, Soemarno, Jurnal Bumi Lestari 13, 115-123 (2011).
2. U. Kafafa, H. Nadia, G. O. Fadilah, A. W. Abadi, R. F. Putri, *Carrying capacity trend and projection analysis for Sumatra Selatan agricultural land in 2030,* in IOP Conf. Ser.: Earth Environ. Sci. 451 0120. (2020).
3. M.Nurlaila, R.L. Mahir, T.S. Darma, Jurnal Ilmu Pertanian Indonesia 1 (2020)
4. N. Salma, M. Askarina, S. Romadhoniastri, A. F. Azahra, D. Karim, M. N. Isnain, R. F. Putri, *Study of agricultural economic potential in West Kalimantan using Regional Analysis Techniques* in E3S Web of Conferences 325, 07008 (2021)
5. M.V.R.Bernhard, *Analisis Daya Dukung Lahan Pertanian ASE.* Manado: Jurusan Sosial Ekonomi Fakultas Pertanian Universitas Sam Ratulangi (2011).
6. BPS, Luas Panen, Produksi, dan Produktivitas Padi Menurut Provinsi 2020-2022. https://www.bps.go.id/indicator/53/1498/1/luas- panen-produksi-dan-produktivitas-padimenurutprovinsi.html.
7. BPS Sulawesi Selatan, Sulawesi Selatan Dalam Angka Tahun 2021. Makassar: Badan Pusat Statistik, 2022.
8. S. Barita, I. Dewata, Journal of Disaster, Geography, and Geography Education 4, 68-71 (2020).

Agricultural Land Monitoring of Land Subsidence Area Using Spatio-Temporal Approach in East Java Indonesia 2018-2022

Dyfany Aurariel Syahda, *Melta* Meliano Putri, *Yosi* Yaseva, *Putri* Yuko Nawang Wulan and *Ratih* Fitria Putri*

Department of Environmental Geography, Faculty of Geography, University of Gadjah Mada, Indonesia

Abstract. Java Island is the center of economy and government so that infrastructure and economic development are evenly distributed and the accessibility is high, including in East Java. As a result, East Java has become a migration destination which causes its population to continue to increase from 2018 to 2022 as it increases the chances of land conversion. As a province with high rice production, this is a challenge for East Java to maintain its rice production. A calculation was carried out to determine the carrying capacity class of East Java agricultural land in 2018-2022 using a combined method of Odum, Christaller, Howard, and Issard. The analytical method used in this study is quantitative descriptive analysis with a spatial-temporal approach. The results showed that the carrying capacity class of East Java's agricultural land from 2018 to 2022 continued to decline. Since 2021 there are no more areas in class I of the agricultural land carrying capacity. The lowest carrying capacity of agricultural land is in Surabaya Municipality, while the highest is Ngawi Regencies. East Java experienced a decrease in the carrying capacity of agricultural land from class II in 2018 to class III from 2019 to 2022.

1 Introduction

Referring to the results of the 2020 population census, Indonesia is the country with the fourth largest population in the world with a population of 270.20 million people [1]. It is stated that 56.10% of Indonesia's total population was concentrated on the island of Java. In fact, the area of Java Island is only 7% of the total area in Indonesia [2]. Java Island is the center of government and economic in Indonesia. It has high economic activities and has the largest contribution value to GDP in Indonesia [3]. As the center of government and economy, this causes Java Island to have a relatively large population. Equal infrastructure development, equitable distribution of economic centers, and ease of accessibility of service centers used by the community are strong pull factors for residents outside Java Island to urbanize [2]. The concentration of population in Java Island according to [4] Java Island almost has everything needed by the community, so it becomes an attractive gem for residents outside the island to try their luck there.

Java Island is administratively divided into several provinces, one of them is East Java. As the center of economic activity, East Java has the highest GDP of Constant Prices in Indonesia from 2012-2016. [5] stated that the GDP growth rate of East Java Constant Prices was higher than Special Capital Regencies (DKI) of Jakarta as the center of the capital of Indonesia.

The population of East Java is increasing from 2018-2020. This trend is visualized in the following graph.

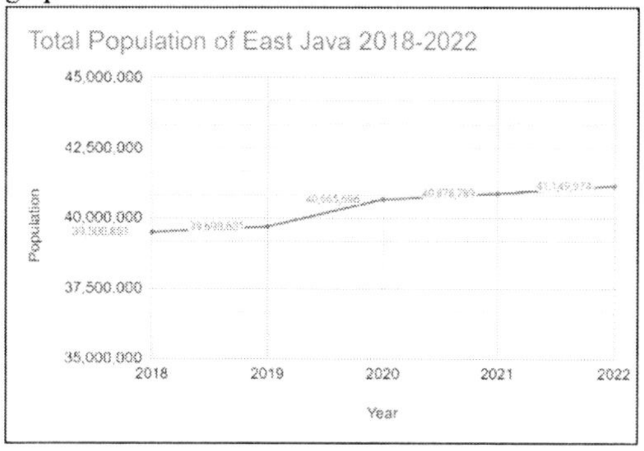

Fig. 1. Total Population of East Java 2018-2022

A fixed land area accompanied by greater population growth has implications for the threat of narrower agricultural land. [6] argues that increasing population will demand development activities that cause agricultural land conversion. According to [7], the increasing population, the area of agricultural land that continues to decrease, and the reduction of land area to live properly are factors that affect the carrying capacity of agricultural land.

The most critical problem related to environmental carrying capacity is population pressure on land, especially agricultural land [8]. Population pressure is inversely proportional to environmental carrying capacity. The higher the population density, the lower

*Corresponding author: ratihfitria.putri@ugm.ac.id

the carrying capacity [9]. Agricultural land carrying capacity is the ability of a agricultural land to support the lives of people in an area, especially related to meeting food needs [10]. [11] argues that the high population pressure on agricultural land causes the threat of a land to be able to meet food needs.

Land subsidence caused by fluid extraction (groundwater, oil, and gas) can reach 10 cm per year, the faster the fluid is extracted the faster the land subsidence. This land subsidence has a spatial pattern varying from patchy to regional. Land conversion such as reclamation, construction, and agriculture activities can cause subsidence by the same mechanism as subsidence land caused by natural phenomena. The study of the analysis of the carrying capacity of agricultural land needs to be carried out as material in development planning because it can provide an overview of the relationship between population, land use, and the environment [12, 13]. The calculation of the carrying capacity of agricultural land for East Java Province in 2018-2022 was carried out to determine the trend of changing carrying capacity classes according to the combined method of Odum, Christaller, Howard, and Issard which is based on production, harvested area, and average rice production per hectare [9], and its implications for agricultural activities in East Java Province.

2 Methodology

The analytical method in this study is quantitative descriptive analysis with spatial and temporal approaches.

2.1 East Java

East Java is located in Java Island, between 111.0′—114.4′ East Longitude and 7.12”—8.48” South Latitude. East Java Province is directly adjacent to Central Java Province, Java Sea, and Bali Strait. The plains of East Java are divided into highlands (>100 meter above sea level), medium (45-100 masl), and low (<45 masl). In addition, East Java is also watered by two main rivers, namely the Brantas River and Bengawan Solo River. The tropical wet climate makes East Java have less rainfall than western Java. East Java has 29 regencies and 9 municipalities with Surabaya as the provincial capital. With an area of 47,803.49 km2, East Java has a population of 39,500,851 people in 2018 and has increased to 41,149,974 people in 2022. BPS recorded an increase in East Java's population density from 2020 to 2021, from 851 to 857.

2.2 Data

This research requires data such as total population of each regencies/municipality, paddy harvested area (Ha), average rice production per hectare (kg/Ha), and KFM or Minimum Physical Needs (kg/capita/year). All data used are data from 2018 to 2022 because the trend of carrying capacity class of agricultural land in East Java Province is being studied. All data is obtained from the Central Bureau of Statistics (BPS) Indonesia.

2.3 Agricultural Land Carrying Capacity

Agricultural Land carrying Capacity is the ability of a agricultural land to support the lives of people in an area, especially related to meeting food needs [14]. The analysis serves to calculate the availability of paddy agricultural land in supporting population needs [11]. Land carrying capacity atatus will be carried out by comparing land availability (SL) with land demand (DL). The following is an interpretation of the results of the analysis, namely: If SL > DL means the carrying capacity of surplus land. If SL < DL, it means that the carrying capacity of the land is deficit or exceeded.

The following formula can be used to calculate agricultural land carrying capacity.

$$\ell = \frac{Lp/Pd}{KFM/Pr} \quad (1)$$

ℓ = agricultural land carrying capacity
Lp = harvested area (ha)
Pd = people population
KFM = Minimum Physical Needs (kg/capita/year) 265 kg/capita/year
Pr = average rice production per hectare (kg/ha)

Based on the results of the agricultiral land carrying capacity calculations, a classification is obtained into three classes, namely:

- Class III ($\ell < 1$) means that the region is unable to implement food self-sufficiency, or the population exceeds the optimal population.
- Class II ($1 \leq \ell \leq 2{,}46$) means that the region has an optimal carrying capacity.
- Class I ($\ell > 2{,}46$) means that the region is able to carry out food self-sufficiency or the population is below the optimal population.

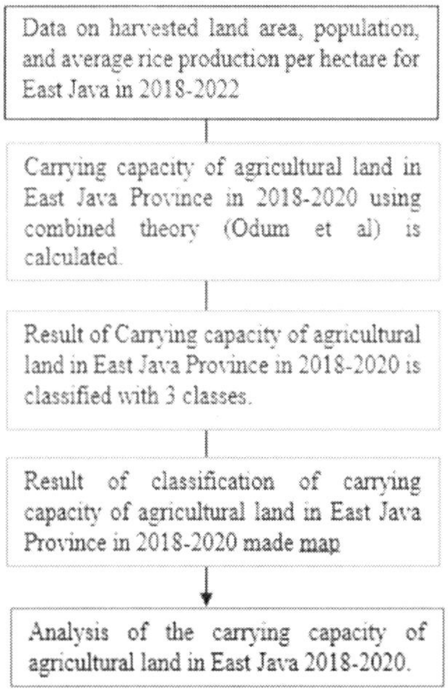

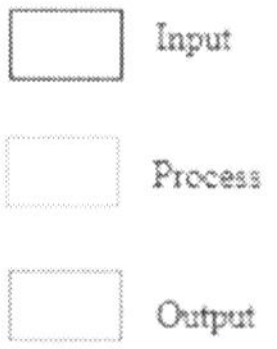

Fig. 2. Flow chart of the method used in this paper

3 Results and Discussion

The carrying capacity of agricultural land is the ability of agricultural land to support the lives of people in an area, especially related to meeting food needs [14]. Rice is one of the staple foods of the world community where Indonesia is one of the largest producing countries [15]. East Java is a region that has very high rice production. At the same time, East Java has challenges in maintaining rice production due to massive changes in rice field land cover. [16] argues that the growth of infrastructure, industry, and population growth is a trigger for the conversion of agricultural land into developed land that will threaten rice production.

Table 1. Agricultural Land Carrying Capacity Class of East Java 2018-2022

Year	Population	Harvested Area (Ha)	Productivity (Kg/Ha)	KFM	Carrying Capacity	Class
2018	39,500,851	1,828,700	5,763	265	1.01	II
2019	39,698,631	1,702,426	5,628		0.91	III
2020	40,665,696	1,754,380	5,713		0.93	III
2021	40,878,789	1,747,481	3,235		0.52	III
2022	41,149,974	1,693,210	3,303		0.51	III

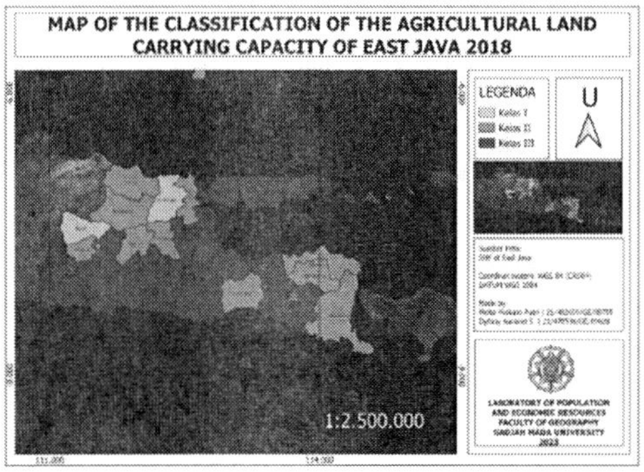

Fig. 3. Map of the classification of the agricultural land carrying capacity of East Java 2018

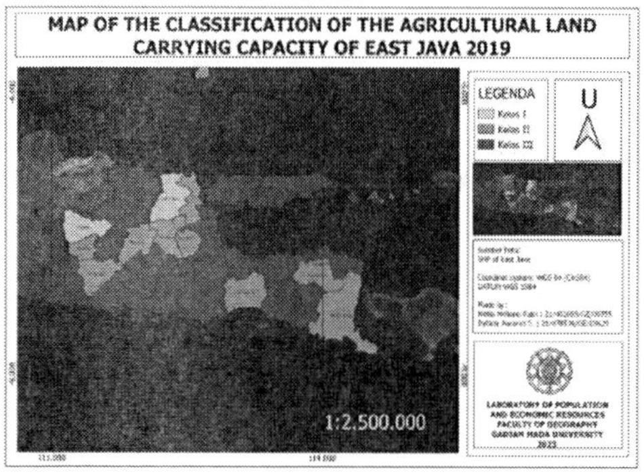

Fig. 4. Map of the classification of the agricultural land carrying capacity of East Java 2019

https://doi.org/10.1051/bioconf/20238002005

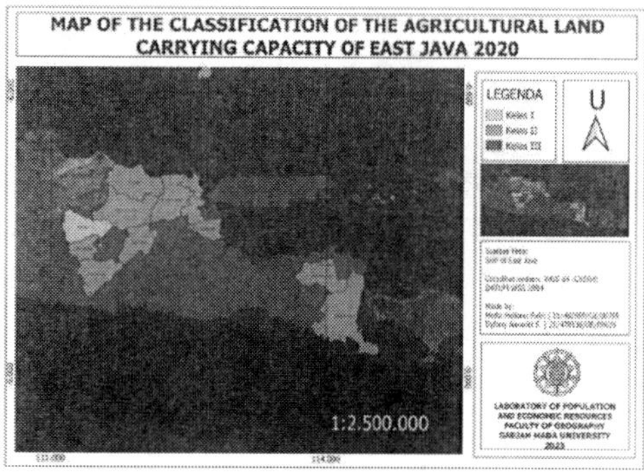

Fig. 5. Map of the classification of the agricultural land carrying capacity of East Java 2020

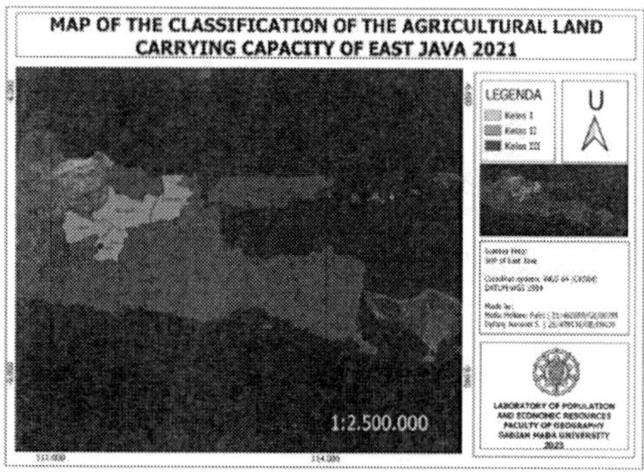

Fig. 6. Map of the classification of the agricultural land carrying capacity of East Java 2021

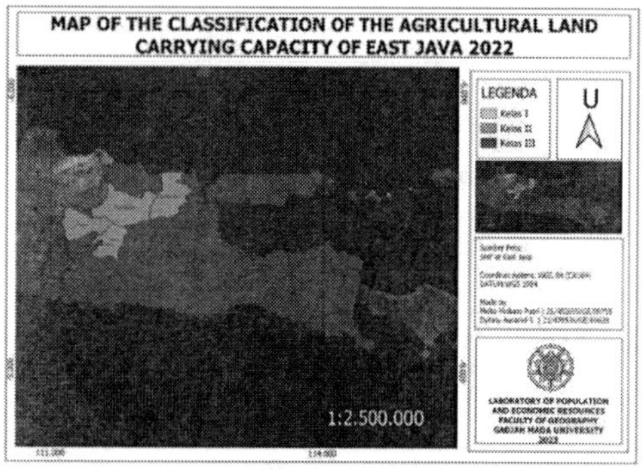

Fig. 7. Map of the classification of the agricultural land carrying capacity of East Java 2022

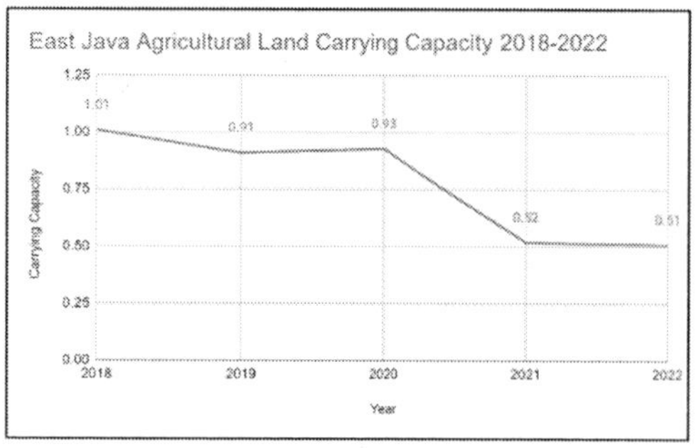

Fig 8. Graph of the agricultural land carrying capacity of East Java 2018-2022

The calculation results show that overall, the carrying capacity of East Java agricultural land in 2018 is in class II. However, from 2019 to 2022, the carrying capacity class of East Java agricultural land continues to decrease to class III. A drastic decrease occurred from 2020 to 2021, from 0.93 to 0.52. This happened because of a decrease in rice harvest area and considerable land productivity. In fact, the population of East Java itself has increased.

In general, the number of regencies or municipalities that have class I carrying capacity also continues to decrease. In 2018 and 2019, there are 2 regions with the carrying capacity of class I agricultural land which are Ngawi Regencies and Lamongan Regencies. Lamongan Regencies dropped to class II in 2020, then there are no more regencies or municipalities with the carrying capacity of class I agricultural land in 2021 and 2022. The same thing also happened to the carrying capacity of class II agricultural land, which was originally 13 in 2018, then became 12 in 2019, dropped to 11 in 2020, and finally only 4 regions in 2021 and 2022.

Table 2. Total Regencies/Municipalities in Each Carrying Capacity Class

Year	Carrying Capacity Class		
	I	II	III
2018	2	13	23
2019	2	12	24
2020	1	11	26
2021	0	4	34
2022	0	4	34

Over time, the carrying capacity of agricultural land in East Java Province is decreasing. This happens not only in general, but also in each regencies or municipality. The area that has the lowest carrying capacity of agricultural land is Surabaya Municipality, while the highest is Ngawi Regencies. The development of Surabaya Municipality causes developed land in this region to tend to increase. Land use change in East Java was recorded in 2019 at 9,597 hectares. Agricultural land is converted into warehousing areas, industrial areas, to property areas [17]. This makes massive land shrinkage. In fact, Sidoarjo Regencies, which used to have characteristics or local wisdom from agricultural products and ponds that are sufficient to meet independent food needs [18], has now gone through a rice deficit since 2018. One of the factors is because Sidoarjo Regencies became a supporting area for

Surabaya which later became the core area of development, resulting in land conversion.

Based on research conducted by [6] on the carrying capacity of agricultural land, it was found that the cause of the low carrying capacity of agricultural land of food crop agriculture in the Cibaliung watershed area was caused by an increase in population from year to year, while the area of agricultural land tended to be static and even decreased with poor land quality drop. In addition, according to [6] the low quality of agricultural land is caused by land management that is not in accordance with conservation principles, which results in low food production and crop failure. This inappropriate management of agricultural land is caused by the low knowledge of the community regarding conservation and conservation techniques in general, farming communities manage agricultural land only based on hereditary knowledge obtained from their ancestors.

Low carrying capacity of agricultural land has implications for meeting low food needs as well. Low food needs according to [10] indicates that people's welfare in terms of food is low. Therefore, the carrying capacity of agricultural land of agriculture, especially food crops, must be maintained so that the fulfilment of food needs can be fulfilled. Efforts that can be made to improve the quality of agricultural land according to [10] are suppressing or reducing the population and improving the management of agricultural land so that the yields produced increase both in quality and quantity.

4 Conclusion

The trend of the carrying capacity of agricultural land in East Java changes every year. In 2018 and 2019 there are still 2 regencies/municipalities that fall into the class I classification. In 2020 there is still 1 regencies/municipality that falls into the class I classification. In 2021 and 2022 there are no regencies/municipality that fall into the in class I classification. The number of regencies/municipalities that fall into class III always increases from year to year. The cause of the lower carrying capacity of agricultural land in East Java, namely due to the increase in population each year and due to land management that is not in accordance with conservation principles. Efforts are needed to improve the quality of agricultural land by suppressing or reducing populations and improving agricultural land management so that the yields produced increase both in quality and quantity.

Acknowledgement

The writers would like to express our deep gratitude towards Directorate of Research Universitas Gadjah Mada Number 5075/UN1.P.II/Dit-Lit/PT.01.01/2023.

References

1. BPS, Hasil Sensus Penduduk 2020, Jakarta: Badan Pusat Statistik (2020)
2. A. F. N. Arif, N. Nurwati, Jurnal Ilmu Kesejahteraan Sosial HUMANITAS, **4**(I), 54-70. (2022).
3. D. H. Prastiwi, L. T. Laut, R. Destiningsih, *DINAMIC:* Directory Journal of Economic, **2**(1), 33-50. (2020).
4. S. Adi, N. Istiyani, A. Widjajanti, e-Journal Ekonomi Bisnis dan Akuntansi, **4**(1), 79-82. (2017).
5. A. Darmawan, Pengaruh kemiskinan, pendidikan dan kesehatan terhadap perfumbuhan ekonomi di Jawa Timur. Skripsi, UIN Syarif Hidayatullah Jakarta. (2020).
6. V. R. Moniaga, *Agri-Sosioekonomi*, **7**(2), 61-68. (2011).
7. I. B. Mantra, Pengantar Studi Demografi, Yogyakarta: Nur Cahaya. (1986).
8. O. Soemarwoto, Ekologi, Lingkungan Hidup, dan Pembangunan, Bandung: Djambatan. (1994).
9. L. Muta'ali, Daya Dukung Lingkungan untuk Perencanaan Pengembangan Wilayah. Yogyakarta: Badan Penerbit Fakultas Geografi (BPPFG) Universitas Gadjah Mada. (2012).
10. N. Mubarokah, L. M. Rachman, S. D. Jurnal Ilmu Pertanian Indonesia, **25**(1), 73-80. (2020).
11. Imansyah, D. Harisandi, N. Tamia, D. Rahmawati, MKG, **2**(2). (2020).
12. U. Kafafa, H. Nadia, G. O. Fadilah, A. W. Abadi, R. F. Putri, *Carrying capacity trend and projection analysis for Sumatra Selatan agricultural land in 2030,* in IOP Conf. Ser.: Earth Environ. Sci. 451 0120. (2020).
13. N. Salma, M. Askarina, S. Romadhoniastri, A. F. Azahra, D. Karim, M. N. Isnain, R. F. Putri, *Study of agricultural economic potential in West Kalimantan using Regional Analysis Techniques* in E3S Web of Conferences 325, 07008 (2021)
14. M. N. Dwirani, T. R. Fariz, F. E. Riani, N. Safitri, A. F. Umam, A. Jabbar, F. Lutfiananda, Jurnal Teknologi Lingkungan Lahan Basah, **10**(2), 139-145. (2022)
15. L. Gandharum, M. E. Mulyani, D. M. Hartono, A. Karsidi, M. Ahmad, International Journal of Remote Sensing, **42**(5), 1738-1767. (2021).
16. I. Indarto, F. L Hakim, Socio-economic Series, **52**(52), 7-24. (2021).
17. F. Firmansyah, M. Yusuf, T. O. Argarini, D. Perencanaan, F. T. Sipil, P. Kebumian, Jurnal Penataan Ruang, **16**(1), 47-53.M (2021).
18. Najib, Komersialisasi Lahan Sawah dan Tambak, Studi Alih Fungsi Lahan di Kabupaten Sidoarjo. *Doctoral dissertation,* Universitas Airlangga (2018).

Distribution of oil palm plantations and their suitability with spatial planning: An assessment in Aceh, Indonesia

Ikhwan Amri[1*], *Elsa* Widura[2], *Muhammad* Iqbal[3], *Yan* Andika[4], and *Asalia Raudhati* Izzatillah[2]

[1]The Graduate School of Universitas Gadjah Mada, Yogyakarta 55281, Indonesia
[2]Ministry of Agrarian Affairs and Spatial Planning/National Land Agency, Jakarta 12110, Indonesia
[3]Faculty of Geography, Universitas Gadjah Mada, Yogyakarta 55281, Indonesia
[4]Regional Office of the National Land Agency of Aceh Province, Banda Aceh 23115, Indonesia

Abstract. Oil palm plantations are undoubtedly recognized as a significant contributor to economic development in Indonesia. However, uncontrolled expansion of oil palm cultivation presents challenges due to its potential negative consequences on the environment and society. This study aims to assess the distribution of oil palm plantations and their suitability with spatial planning in the Aceh Province, Indonesia. We obtained spatial data on plantations from official authorities, which underwent an extensive inventory process during the year 2021. The total area of oil palm plantations in Aceh was reported at 706,943.87 ha, primarily concentrated in the eastern and western lowland areas of Aceh mainland. Through spatial analysis, we identified that approximately 40% of the oil palm plantations are located in areas where plantation development should not be permitted according to existing spatial planning regulations. This study provides important information for land use management in Aceh and offers policy recommendations to promote sustainable oil palm cultivation practices.

1 Introduction

Oil palm industry holds significant importance for Indonesia's economy, considering its contribution to regional income generation and job opportunities. Oil palm is favored over other plantation commodities due to its advantages in terms of cultivation system efficiency, productivity, return on investment, and versatility [1]. Globally, Indonesia plays a major role as the leading producer of crude oil palm (CPO), which is further processed into various food, cosmetic, hygiene, and bioenergy products. The Ministry of Agriculture of the Republic of Indonesia reported that the volume of CPO exports reached 7,170,956 tons in 2020, valued at 4.7 billion USD [2]. Additionally, rural communities greatly benefit from the existence of oil palm plantations, providing livelihoods, accelerating development, and offering various socioeconomic advantages.

Despite its positive influences, oil palm production is frequently criticized for its detrimental effects on the environment and society. For instance, the expansion of oil palm plantations is deemed responsible for the loss of forest cover in recent decades [3]. Improper agricultural practices contribute to increased risks of forest fires and the destruction of critical lands, particularly tropical peat swamp ecosystems [4]. These issues become more complex when combined with escalating tenure conflicts among various actors, including local communities, private companies, and state [5].

Sustainable oil palm certification is regarded as a key solution to ensure the governance of plantations that adhere to environmental, social, and economic sustainability principles. At the global level, the concept of sustainable oil palm is promoted by the Roundtable on Sustainable Oil palm (RSPO). In Indonesia, this certification scheme is known as the Indonesian Sustainable Oil palm (ISPO), mandated by the government for both plantation companies and smallholder farmers. However, research indicates that the implementation of ISPO certification still faces structural and sociocultural challenges, leading to suboptimal control over sustainability standards [1].

In Indonesia, the largest oil palm production centers are found in Sumatra and Kalimantan Islands. Aceh is one of the provinces in Sumatra where oil palm dominates the plantation commodities, thanks to its ideal geographical settings. As shown in Figure 1, the availability of lowland areas and its proximity to the equator (which has implications for climate) are crucial indications of the suitability of oil palm production. Statistics Indonesia noted oil palm as the commodity with the largest plantation area in Aceh [6]. Its average productivity is 2745 kg/ha per year in the form of CPO. The production has shown an increasing trend in the past five years.

The aim of this study is to describe the spatial distribution of oil palm plantations throughout the Aceh

* Corresponding author: ikhwan.amri@mail.ugm.ac.id

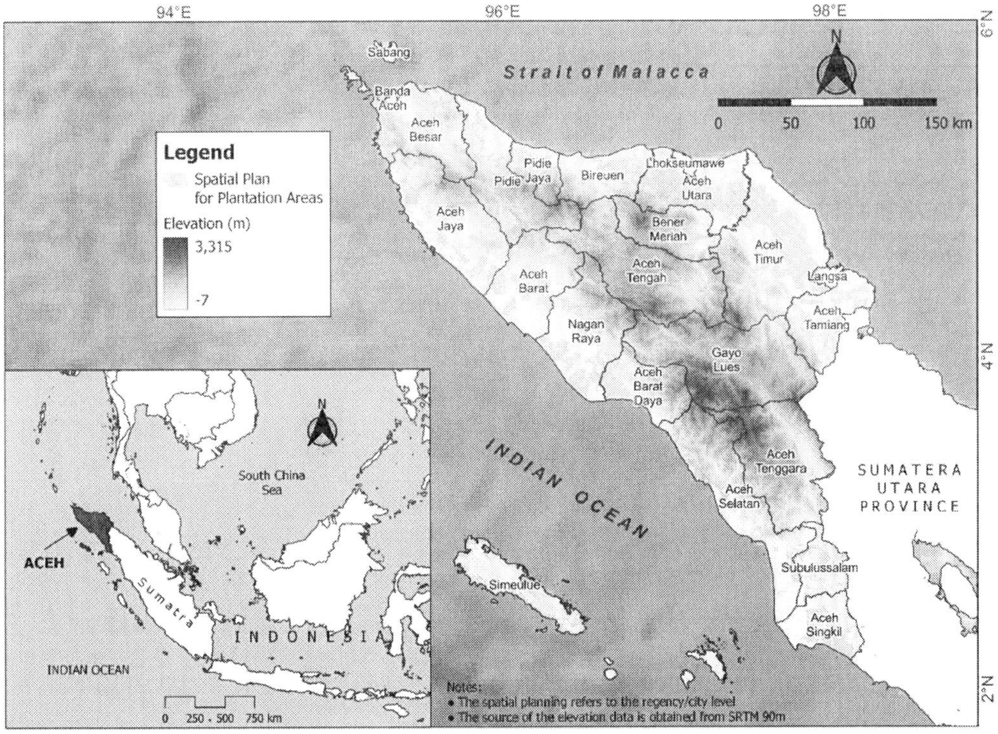

Fig. 1. Study area.

Province and identify their compliance with spatial planning policies. We utilized spatial data on oil palm plantations that were inventoried by official authorities in 2021. Evaluating the land use of plantations is an essential step in reviewing and reinforcing commitments to sustainable oil palm initiatives. For decision-makers, this study indicates the effectiveness of implementing land use planning regulations and serves as a consideration for policy improvements in the future.

2 Research method

The distribution of oil palm plantations and spatial pattern plans are two key data in this research. The plantation map was acquired from the National Land Agency through extensive inventory activities covering all commodities in 2021 [7]. The data collection process involved multiple stages over the course of one year, including on-screen digitization of high-resolution satellite imageries, field surveys, and several coordination meetings with key stakeholders. The data extracted from the plantation map focused on oil palm, considering it as the largest plantation commodity in Aceh based on recent statistics. While the inventory of plantation areas is updated annually by the authorities responsible for the agricultural sector, we argue that this data is the most comprehensive (including in tabular and spatial formats) and accurate for the current period.

Meanwhile, the spatial pattern plan maps were compiled from various regency/city-level regulations. Generally, the division of spatial patterns is agreed upon as protective and cultivation functions, but the specific classifications of each function vary across regions. For analysis purposes, we simplified and modified the classes into two broad categories: plantations and non-plantations.

Data processing and analysis were facilitated using QGIS software. The area of oil palm plantations in each regency/city was calculated, along with its proportion to the total area of all plantation commodities. Further analysis involved overlaying the land use layer with the spatial planning layer (Figure 2). With this technique, the suitability of land use with spatial planning could be estimated in detail for each region.

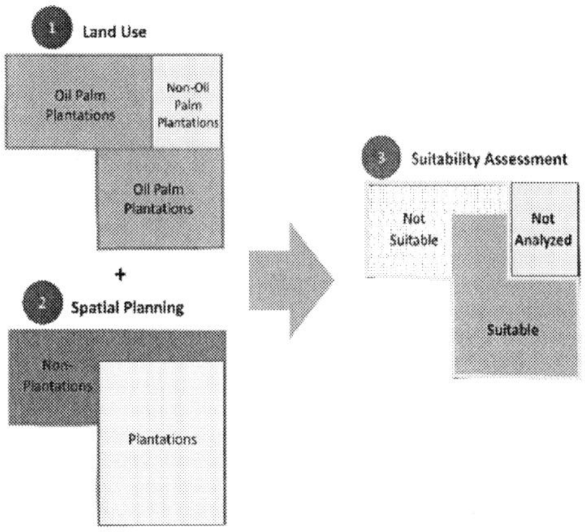

Fig. 2. Suitability assessment workflow.

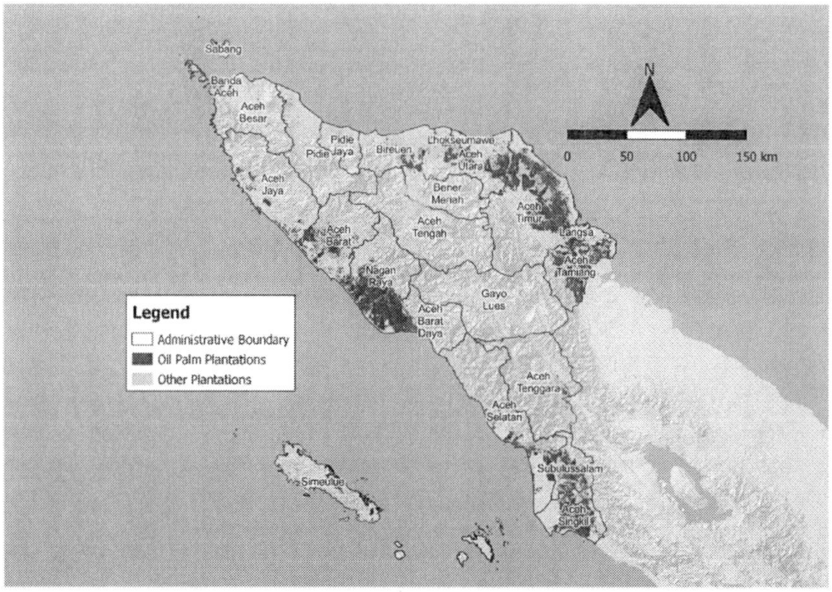

Fig. 3. Distribution of oil palm plantations in Aceh.

3 Results and discussion

3.1 Distribution of oil palm plantations

Based on the processing of spatial data, the total area of plantations in Aceh Province is 1,185,666.66 ha, which accounts for one-fifth of the province's total area. In this study, plantations were classified into two broad categories: oil palm and other plantations. The area of oil palm plantations reached 706,943.87 ha, representing 60% of the total plantation area. It is important to note that the scope of oil palm plantations in this study is not limited to monoculture and productive plantations. It also includes mixed-crop plantations, newly planted areas, and even newly cleared land for oil palm cultivation. Such criteria may differ from previous research that produced regional and global-scale oil palm map datasets [8, 9].

Oil palm plantations are distributed throughout the administrative regions of Aceh Province, with a significant concentration in the eastern and western part of the Aceh mainland (Figure 3). Environmental factors such as water availability, air temperature, land topography, and soil types significantly influence the suitability of oil palm cultivation. Nineteen out of the 23 regencies/cities in Aceh Province have oil palm plantations. The four regions that do not have oil palm plantations include Banda Aceh, Sabang, Gayo Lues, and Aceh Tengah. Due to its role as the administrative and urban center of Aceh Province, it is understandable that Banda Aceh does not have plantation areas. Sabang, a port city on a small island in Aceh waters, has limited land use for plantations, but oil palm plantations have not been identified in that region. Aceh Tengah and Gayo Lues, on the other hand, have geographical characteristics of highland areas that are not suitable for oil palm trees.

The calculation of oil palm plantation areas in each regency/city in Aceh is shown in Figure 4. The largest distribution of oil palm plantations is found in Aceh Timur (147,099.89 ha), followed by Nagan Raya (123,414.78 ha) and Aceh Tamiang (84,761.77 ha). Meanwhile, the three areas with the smallest oil palm plantations, in consecutive order, are Aceh Besar (693.81 ha), Lhokseumawe (761.00 ha), and Pidie (908.49 ha).

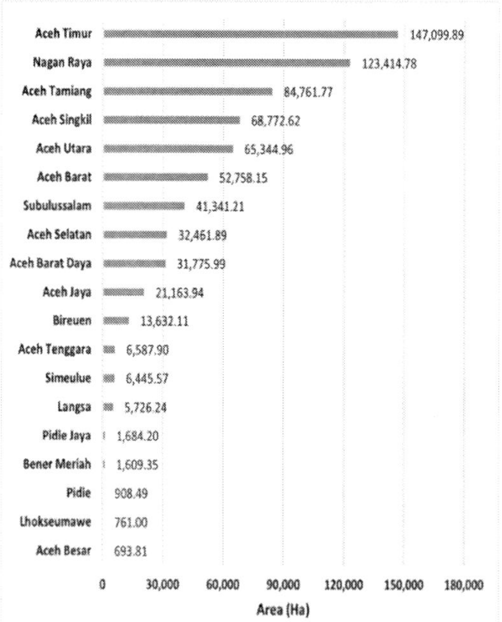

Fig. 4. Area of oil palm plantation by regency/city.

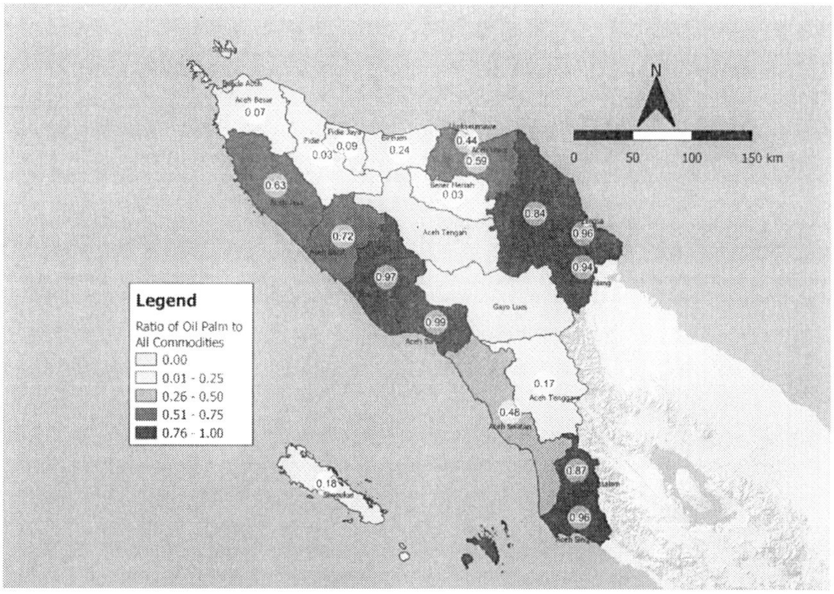

Fig. 5. Ratio of oil palm to all commodities.

Furthermore, this study calculates the proportion between the land area used for oil palm plantations and the total land area used for various types of plantations (Figure 5). This metric provides an overview of the contribution of oil palm plantations in the context of the total plantation sector in a region. The results show that there are seven regions where more than three-quarters of their plantations consist of oil palm as the main commodity. These regions include Aceh Barat Daya (99%), Nagan Raya (97%), Aceh Singkil (96%), Langsa (96%), Aceh Tamiang (94%), Subulussalam (87%), and Aceh Timur (86%). Aceh Utara, Aceh Jaya, and Aceh Barat can also be considered to have a significant dominance of oil palm plantations, with percentages ranging from 51% to 75%. On the other hand, three regions with the lowest ratios (excluding regions without oil palm plantations) are identified in Pidie (3%), Bener Meriah (3%), and Aceh Besar (7%).

3.2 Evaluation of spatial planning

The mismatch between existing land use and spatial planning can be accurately determined by overlaying these two thematic layers. Suitable conditions occur when oil palm plantations are located in areas designated for plantation purposes. Conversely, unsuitable conditions occur when oil palm plantations are situated in areas not designated for plantations (such as forests, food agriculture, horticulture, riverbanks, settlements, industries, etc.).

Table 1 shows the area of oil palm plantations based on their compatibility with spatial planning in each regency/city. The area of plantations that align with their designated land use is 421,572.75 ha, while the area that does not align is 285,371.12 ha. In essence, this means that 60% of the plantation land use is suitable, while the remaining 40% is unsuitable.

Table 1. Area of oil palm plantations based on their suitability with spatial planning.

Region	Area (Ha)	
	Suitable	Not Suitable
Aceh Timur	105,352.51	41,747.38
Nagan Raya	52,804.13	70,610.66
Aceh Tamiang	42,307.98	42,453.80
Aceh Singkil	33,816.56	34,956.06
Aceh Utara	60,831.14	4,513.82
Aceh Barat	26,947.03	25,811.12
Subulussalam	22,658.31	18,682.90
Aceh Selatan	17,005.12	15,456.77
Aceh Barat Daya	24,614.07	7,161.92
Aceh Jaya	11,954.96	9,208.98
Bireuen	8,159.56	5,472.55
Aceh Tenggara	2,142.04	4,445.87
Simeulue	5,481.41	964.16
Langsa	4,008.10	1,718.13
Pidie Jaya	962.49	721.71
Bener Meriah	1,554.49	54.86
Pidie	290.74	617.75
Lhokseumawe	354.30	406.70
Aceh Besar	327.82	365.99
Total	**421,572.75**	**285,371.12**

The proportion of spatial planning compatibility varies greatly in each region (Figure 6). While in absolute terms, the largest area of plantations that do not comply with spatial planning is identified in Nagan Raya (70,610.66 ha), the highest percentage of

 https://doi.org/10.1051/bioconf/20238002006

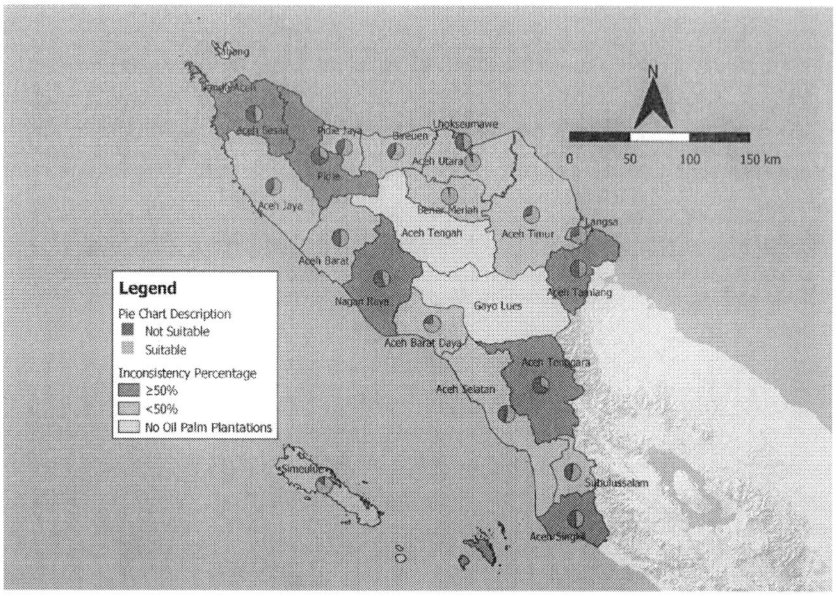

Fig. 6. Percentage of consistency of oil palm plantations with spatial planning.

inconsistency is found in Pidie (68%), despite its relatively small area. Apart from Pidie, there are six other regions where half or more of their oil palm plantations do not align with the spatial pattern plans, namely Aceh Tenggara (67%), Nagan Raya (57%), Aceh Besar (53%), Lhokseumawe (53%), Aceh Singkil (51%), and Aceh Tamiang (50%). The distribution of incompatible oil palm plantations in Aceh Timur and Aceh Tamiang is almost identical and quite substantial as well (following Nagan Raya), at around 40 thousand hectares. Meanwhile, both in terms of absolute and relative values, the extent of spatial planning incompatibility in Bener Meriah is the smallest in the entire Aceh Province, with only 54.86 ha or representing 3% of the total oil palm plantation area.

3.3 Issues and recommendations

Official data indicates that oil palm is a strategic plantation commodity in Aceh, as evidenced by its extensive coverage, surpassing other plantations in terms of land area. However, our findings show that approximately two-fifths of oil palm plantation areas (covering 285,371.12 ha) are incompatible with spatial planning. Six specific regions exhibit a more dominant percentage of inconsistency, namely Pidie, Aceh Tenggara, Nagan Raya, Aceh Besar, Lhokseumawe, Aceh Singkil, and Aceh Tamiang. Meanwhile, in terms of absolute area, the major contributors to spatial planning inconsistency are identified in Nagan Raya, Aceh Tamiang, and Aceh Timur. Therefore, it is crucial for the Aceh government to promptly implement corrective strategies and promote better governance of the oil palm industry.

Oil palm industry is widely recognized for its significant contribution to economic development through multiplier effects. However, the neglect of essential land use guidelines is counterproductive to the principles of sustainable development. Considering the promising benefits, the expansion of oil palm not only impacts existing non-oil palm plantations but frequently leads to uncontrolled transformations of land previously used for rice farming and forested areas [10, 11]. This conversion can be initiated by both small-scale farmers and plantation companies.

Oil palm expansion, especially involving vital ecosystems in tropical regions such as primary forests and peatlands, tends to attract special attention from experts and environmental activists. The establishment of unplanned oil palm plantations is widely acknowledged for its detrimental effects on the crucial functions of these environments, such as carbon storage, hydrological cycles, biodiversity conservation, and the provision of other essential ecosystem services. Despite regulations prohibiting such activities in Aceh, previous research revealed oil palm expansion is continuing in protected areas [11, 12]. The lack of effective law enforcement by local authorities has led to rampant illegal logging. Additionally, corrupt practices in granting plantation permits contribute to facilitating deforestation activities. The destruction of primary forests and peatlands can lead to multidimensional impacts such as greenhouse gas emissions, declining water and soil quality, disruption of biodiversity, conflicts between protected wildlife and humans, increased disaster risks, and social tensions [11-14]

Furthermore, the issue of overlapping spatial planning has long been a complex challenge in Indonesia. It arises from policy conflicts across various sectors, including agriculture, industry, trade, land administration, and forestry. Different sectors often prioritize their own interests, resulting in inconsistent land allocations. For example, agricultural policies tend to encourage the expansion of commercial crops such as oil palm, while forestry policies insist on protecting and preserving forest areas. The lack of transparency and accountability complicates the harmonization of natural resource management policies. In addition, the governance challenges of oil palm plantation permits in Aceh are found to be interconnected, demanding fundamental improvements in problematic licensing processes [15].

Improving land use planning regulations is the main pathway to reduce policy conflicts. This requires policy alignment related to natural resource management and spatial planning [16]. Such harmonization is crucial for providing legal certainty for investors. Another benefit is related to preventing agrarian conflicts [17]. To achieve these goals, increased transparency and active engagement of stakeholders, both governmental and non-governmental actors, are needed in the planning and decision-making processes.

Strengthening spatial utilization control as an instrument for spatial planning implementation needs to be reinforced, such as imposing sanctions on illegal oil palm plantation operators causing socio-environmental harm. Assessments of spatial planning implementation, monitoring, and evaluation can be conducted across sectors, accompanied by the establishment of complaint systems from the regional to the central levels. In addition to strengthening spatial utilization control, the government can consider these findings in spatial planning revisions, especially since most regencies/cities in Aceh are entering a period of spatial planning revision, ensuring consideration of environmental carrying capacity and land status (smallholder or large-scale plantation). However, spatial planning revisions should not merely align spatial patterns with existing permits or land use. For example, monoculture oil palm cultivation in protected areas with high conservation value cannot be justified or maintained for any reason [18, 19]. If possible, converted forest areas should be restored by incorporating strategies for improving community welfare based on non-timber forest products (such as the social forestry scheme).

For oil palm plantations that are in line with spatial planning, sustainable farming practices must still be evaluated. Oil palm production should ideally not cause negative consequences for the surrounding environment (e.g., good agricultural practices). Plantation locations adjacent to protected areas should be more closely monitored to prevent illegal encroachment. Relevant stakeholders need to evaluate oil palm plantation lands that do not achieve maximum production. These efforts can be integrated with ISPO or RSPO certification for communities or companies. Also, oil palm producers should shift their paradigm from an extensification-oriented approach to adopting environmentally friendly intensification strategies. Meanwhile, the negative impacts that have already occurred need to be addressed by the government through the implementation of incentive and disincentive mechanisms [20].

4 Conclusion

In tropical areas like Indonesia, oil palm plays a strategic role in the agricultural sector. Focusing on Aceh Province, our research aims to describe the spatial distribution of oil palm plantations and identify their compatibility with spatial planning. Official data reveals that oil palm is predominantly concentrated in the eastern and western lowland areas of Aceh mainland, covering a total area of 706,943.87 hectares, which accounts for nearly 60% of the total plantation area for all crops. This commodity is found in all regencies/cities except for Banda Aceh, Sabang, Aceh Tengah, and Gayo Lues. The three regencies with the largest oil palm plantations are Aceh Timur, Nagan Raya, and Aceh Tamiang. These regions also have a very high proportion of oil palm plantations (>75% of the total plantation area), along with four other regions: Aceh Barat Daya, Aceh Singkil, Subulussalam, and Langsa.

Appropriate plantation practices are an integral part of land management based on sustainability principles. To achieve this goal, it seems that the Aceh government needs to make more efforts to improve the governance of the oil palm industry. We found that approximately 40% of oil palm plantations are located in areas designated for non-plantation purposes. Areas with substantial spatial deviations should certainly receive special attention.

Our analysis contributes to laying the groundwork for future research and policy initiatives aimed at creating a balance between socio-economic development and environmental conservation. Regular monitoring of oil palm plantations is crucial in the context of land use supervision. In this regard, remote sensing technology offers concrete solutions to support multi-temporal observations of oil palm expansion. In addition to spatial planning, it is advisable to investigate the legality of plantation operations, taking into account location permits, forest area status, and peatland. Finally, further research could consider building upon this work to explore the determinants of spatial misuse in detail and evaluate the potential impacts on various dimensions.

The author would like to express gratitude to the staff and individual consultants at the Regional Office of the National Land Agency of Aceh Province who have assisted in gathering spatial data of oil palm plantations.

References

1. A.H. Dharmawan, D.I. Mardiyaningsih, F. Rahmadian, B.E. Yulian, H. Komarudin, P. Pacheco, J. Ghazoul, R. Amalia, Sustainability **13**, 2611 (2021), doi:10.3390/su13052611
2. Ministry of Agriculture of the Republic of Indonesia, *Statistical of National Leading Estate Crops Commodity 2020-2022* (Jakarta, Ministry of Agriculture of the Republic of Indonesia, 2021)
3. K.G. Austin, A. Mosnier, J. Pirker, I. McCallum, S. Fritz, P.S. Kasibhatla, Land Use Policy **69**, 41-48 (2017), doi:10.1016/j.landusepol.2017.08.036
4. H. Purnomo, B. Okarda, A.A. Dewayani, M. Ali, R. Achdiawan, H. Kartodihardjo, P. Pacheco, K.S. Juniwaty, For. Policy Econ. **91**, 94-106 (2018), doi:10.1016/j.forpol.2017.12.014

5. W. Herningtyas, Bhumi: Jurnal Agraria dan Pertanahan **7**, 199-209 (2021), doi:10.31292/bhumi.v7i2.504
6. BPS, *Aceh Province in Figures 2021* (Banda Aceh, BPS, 2021)
7. Kanwil BPN Provinsi Aceh, *Neraca Penatagunaan Tanah Sektoral Perkebunan Provinsi Aceh* (Banda Aceh, Kanwil BPN Provinsi Aceh, 2021)
8. Y. Xu, L. Yu, W. Li, P. Ciais, Y. Cheng, P. Gong, Earth Syst. Sci. Data **12**, 847-867 (2020), doi:10.5194/essd-12-847-2020
9. A. Descals, S. Wich, E. Meijaard, D.L.A. Gaveau, S. Peedell, Z. Szantoi, Earth Syst. Sci. Data **13**, 1211-1231 (2021), doi:10.5194/essd-13-1211-2021
10. S. Fitri, Faktor-faktor yang mempengaruhi alih fungsi lahan dari tanaman karet rakyat menjadi tanaman kelapa sawit (Studi kasus: Desa Selamat, Kecamatan Tenggulun, Kabupaten Aceh Tamiang), Undergraduate Thesis (Medan, Universitas Muhammadiyah Sumatera Utara, 2019)
11. D. Ruysschaert, M. Hufty, Land Use Policy **99**, 103359 (2020), doi:10.1016/j.landusepol.2018.04.034
12. Sufardi, Manfarizah, Khairullah, Jurnal Pertanian Tropik **3**, 267-277 (2016), doi:10.32734/jpt.v3i3.2987
13. J. Putri, Penguasaan tanah untuk kegiatan perkebunan kelapa sawit di kawasan hutan (Studi kasus di Kecamatan Tenggulun Kabupaten Aceh Tamiang), Undergraduate Thesis (Yogyakarta, Universitas Gadjah Mada, 2017)
14. N.L. Najmi, A.J. Al-Jaktsa, S. Suharno, A. Fariyanti, Agribusiness Forum **9**, 53-68 (2019), doi:10.29244/fagb.9.1.53-68
15. Syahrul, Kanun: Jurnal Ilmu Hukum **20**, 337-354 (2018), doi:10.24815/kanun.v20i2.10603
16. A.E. Pravitasari, F.P. Yudja, S.P. Mulya, Y.A. Stanny, IOP Conf. Ser.: Earth Environ. Sci. **694**, 012065 (2021), doi:10.1088/1755-1315/694/1/012065
17. A. Tarfi, I. Amri, Bhumi: Jurnal Agraria dan Pertanahan **7**, 210–225 (2021), doi:10.31292/bhumi.v7i2.509
18. V. Vijay, C.D. Reid, M. Finer, C.N. Jenkins, S.L. Pimm, Environ. Res. Lett. **13**, 114010 (2018), doi:10.1088/1748-9326/aae540
19. E.M. Ordway, R.L. Naylor, R.N. Nkongho, E.F. Lambin, Nat. Commun. **10**, 114 (2019), doi: 10.1038/s41467-018-07915-2
20. I. Sandi, T.C. Dawood, Jurnal Ilmiah Mahasiswa Ekonomi Pembangunan **4**, 375-382 (2019)

 https://doi.org/10.1051/bioconf/20238002007

GIS-Based Assessment of Land Suitability for Rubber Cultivation in Chittagong District, Bangladesh

Md. Ahasanul Hoque[1*], *Rezwan* Ahmed[2]

[1]Geospatial and Data Management Expert, The World Bank, Dhaka – 1207, Bangladesh
[2]Department of Geography and Environment, University of Dhaka, Dhaka - 1000, Bangladesh

Abstract. This research examines the viability of rubber cultivation in Chittagong, Bangladesh, employing Geographic Information System (GIS) methodologies and a detailed multi-criteria model. Given Bangladesh's conducive agroclimatic conditions, rubber cultivation offers a promising avenue for sustainable economic expansion. This study evaluates essential parameters such as temperature, elevation, slope, soil texture, pH, and depth to gauge the aptness of land for rubber plantations. The research's main objective is to identify the potential of rubber cultivation in Chittagong, given its climatic advantages and economic incentives. This analysis aids stakeholders, including policymakers, farmers, and investors, in aligning with larger goals of economic development and agricultural diversification. Key findings suggest that Chittagong's attributes, including optimal temperature, suitable elevations, moderate slopes, and fitting soil properties, make it ideal for rubber cultivation. Further analysis incorporates factors like land use and river proximity. The central region emerges as particularly promising, while urban and aquatic regions are omitted, highlighting conscientious land use. These insights can guide stakeholders toward sustainable rubber cultivation, fostering rural development, job creation, and foreign exchange benefits in Bangladesh. This research is pivotal for the economic diversification strategy and the sustainable growth of rubber cultivation in the area.

1 Introduction

Bangladesh, a country known for its agriculture, tea plantations, and fruit orchards, could thrive in the rubber (Hevea brasiliensis) plantation business, which is still relatively young in Bangladesh compared to other nations because the country's environment is ideal for the production of rubber. Due to these advantageous conditions, the government has encouraged plantings in hilly areas since 1980; however, only roughly 45,000 acres of land have been allocated to the BFIDC and 32,500 acres to private landowners for rubber plantations as of yet. In Bangladesh, rubber plantations are primarily located in areas such as Madhupur, Sherpur, Rangamati, Bandarban, Chittagong, Khagrachari, Cox's Bazar, Sylhet, and Mymensingh. The country features 1,304 private and 29 government-managed plantations, in addition to various community and privately-owned gardens. Altogether, these plantations cover an area of 140,000 hectares and produce an annual yield of 67,939 metric tons of latex. The ability of rubber trees to absorb carbon three times more efficiently than any other tree makes them a valuable resource for the Global Carbon Trading and Environmental Fund, which offers millions of dollars in foreign currency.[1] The development of this industry is also influenced by a wide range of additional factors. There is a significant gap left as more nations switch from producing rubber to producing palm oil, which might be filled by potential Bangladeshi producers. [2]

Bangladesh's rubber production is expanding as a result of rising internal demand and booming exports, mainly to India's neighbor. In the current fiscal year, rubber export revenues increased dramatically by 57 percent year over year to $28 million from $18.3 million in the prior year's comparable period. The most recent export statistics outperformed earnings data from previous years, with the exception of the most recent fiscal year, when overall earnings reached $34 million, the most in ten years. [3]

In summary, a well-executed land suitability analysis for rubber production in Bangladesh can not only boost agricultural and industrial output but also promote rural development, employment, and foreign exchange earnings. It plays a vital role in diversifying the economy, reducing import dependency, and contributing to sustainable economic growth. However, it's important to ensure that this expansion is carried out with responsible and sustainable practices to maximize long-term benefits.

The suitability of the land to provide the best prospects for a specific land use form directly and significantly affects the appropriateness of that land use form. Thus, land suitability is a crucial fundamental element in the management of

* Corresponding author: hoque.ahasan@gmail.com

the environment. The optimum approach to use a certain piece of land is determined by the results of a land suitability evaluation. In other words, it alludes to a certain category of land use. Geographic Information System (GIS) techniques have evolved into essential tools for land use planning, fostering sustainable development across the globe. In agriculture, they have played a pivotal role in determining optimal locations for different types of cultivation based on a variety of environmental, climatic, and socio-economic factors. One such promising venture is the cultivation of rubber especially in Bangladesh, where the agro-climatic conditions are highly favorable for this type of plantation.

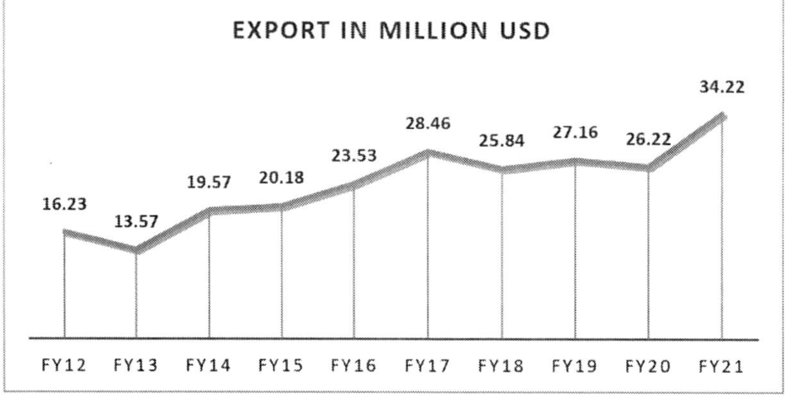

Fig. 1. Rubber export scenario in Bangladesh.

Implementing GIS techniques allows for a systematic analysis of the land characteristics, rainfall, soil type, precipitation patterns, and other critical factors to identify potential zones suitable for rubber cultivation in Bangladesh. By merging this technological advancement with the country's agricultural practices, Bangladesh can maximize the yield and economic benefits, while also contributing positively to global climate change mitigation efforts.

1.1 Objectives

- To systematically assess and determine the potential areas or parcels of land in Chittagong district are most suitable for cultivating rubber trees.
- To build a GIS based multi-criteria model for land suitability analysis for finding potential areas of land for rubber production in Bangladesh.

1.2 Scope of the Study

The study's primary focus lies within the Chittagong district of Bangladesh, where it endeavours to ascertain the suitability of land for rubber tree cultivation by conducting a comprehensive analysis of factors encompassing soil quality, climate, topography, and socio-economic conditions. The study's intended beneficiaries encompass policymakers, farmers, and prospective investors who harbour an interest in advancing rubber cultivation within the Chittagong district, with the overarching objective of harnessing rubber's economic potential for the development and diversification of Chittagong's agricultural landscape. In future use, this study provides an invaluable resource for those seeking to make informed decisions regarding rubber cultivation in this region, promoting sustainable economic growth and land use planning.

2 Materials and Method

2.1 Study Area

Chittagong is one of the oldest and biggest districts in Bangladesh located, in the south-eastern region of Bangladesh with an area of 5283 sq km^2. Chittagong District (Figure 2) is characterized by its diverse topography, encompassing both low-lying coastal areas and the Chittagong Hill Tracts in the east. The district is intersected by several rivers, including the Karnaphuli, Sangu, and Matamuhuri, providing a network of water resources. The hills and valleys offer varying altitudes and microclimates, which are essential considerations for rubber cultivation. Chittagong is known for its strategic location as the country's primary seaport. This district is not only economically significant but also ecologically diverse, making it an ideal candidate for site suitability analysis for rubber cultivation.

2.1.1 Climatic Conditions

Chittagong District experiences a tropical monsoon climate. It is influenced by the Bay of Bengal, leading to high humidity levels and substantial rainfall during the monsoon season. The climatic conditions of the district manifest as discernible wet and dry phases, exerting influence on the growth and productivity of rubber trees.

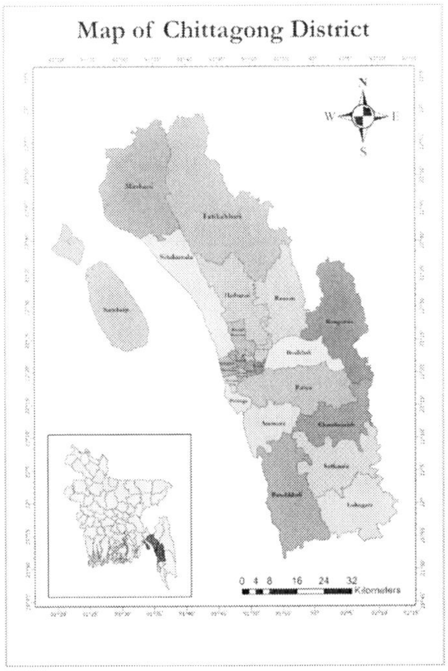

Fig. 2. Map of the study area.

2.1.2 Rainfall

In the climatic context of Chittagong, significant seasonal fluctuations in monthly precipitation levels are observed. According to data sourced from the Bangladesh Meteorological Department, the region undergoes a prolonged wet phase that spans approximately 9.8 months, commencing on February 16 and culminating on December 9. Throughout this interval, the mean cumulative rainfall is reported to be 2,700 millimetres. Of particular note is the month of July, which emerges as the wettest month, characterized by an average precipitation level of 443 millimetres. In contrast, the region witnesses a dry phase lasting for 2.2 months, which extends from December 9 to February 16. This study underscores the acute variability in precipitation patterns in Chittagong, which holds implications for water resource management and ecological sustainability in the area.

2.1.3 Vegetation and Land Use

The natural vegetation in Chittagong District includes evergreen forests, mangroves along the coast, and various types of grasslands. Agriculture is a significant contributor to the district's economy, with rice, jute, tea, and betel leaf being some of the primary crops. However, there is untapped potential for rubber cultivation, which can diversify the agricultural landscape.

2.1.4 Socio-Economic Context

Chittagong District is home to a diverse population, including indigenous communities in the Chittagong Hill Tracts. The socio-economic status of the inhabitants varies, with agriculture being a primary source of livelihood for many. Introducing rubber cultivation could contribute to increased income and employment opportunities.

2.2 Methodology

The methodology for site selection is predicated on a defined set of local criteria, encompassing factors such as current land use, slope gradient, soil depth, soil texture, precipitation levels, and proximity to water bodies.[4] To quantitatively evaluate site suitability, a scoring and weighting system was implemented across these multiple dimensions. A comprehensive schematic of the methodology is illustrated in Figure 3.

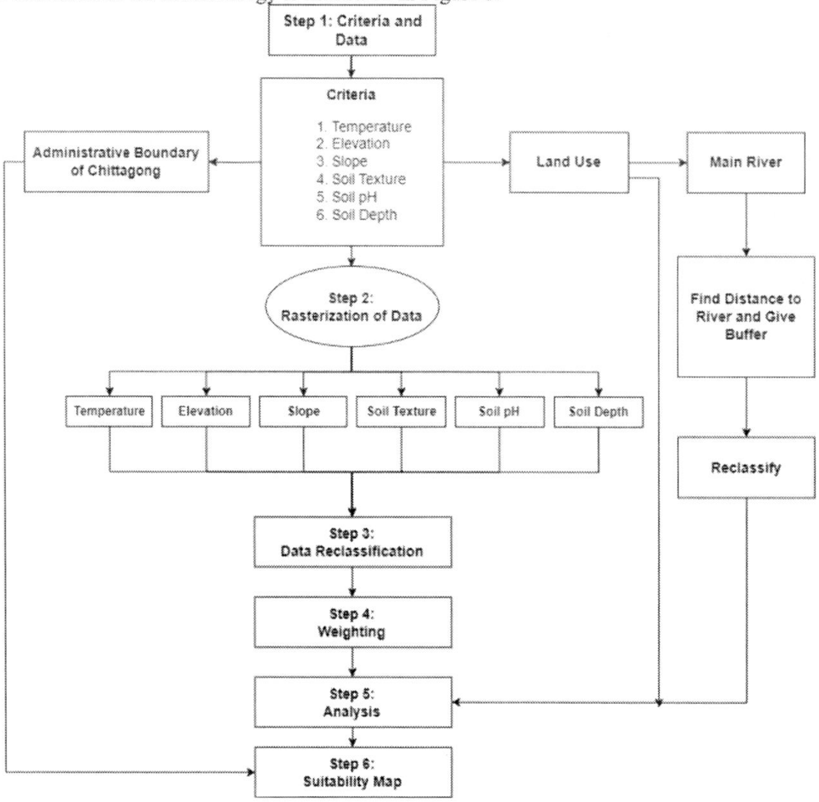

Fig. 3. Flow diagram of overall methodology of land suitability analysis.

2.2.1 GIS Based Land Suitability Analysis

In contemporary land use planning and environmental science, Geographical Information Systems (GIS) have emerged as indispensable tools for site suitability analysis. The multifaceted capabilities of GIS offer more than just spatial visualization; they provide a platform for the integration, manipulation, and interpretation of multifarious datasets that can depict land characteristics and potential impacts of various land uses.[5] Critical to any suitability analysis is the foundation of base data. Land qualities relevant to this study included existing land use, slope, water bodies, flood hazards, soil attributes, annual rainfall, and more. [6] A composite score is then generated, culminating from the weighted attributes affecting suitability. For this assessment, the parameters delineated for suitability analysis encompassed are: (i) Temperature; (ii) Elevation; (iii) Slope; (iv) Soil Texture; (v) Soil pH; (vi) Soil Depth; (vii) Existing Land use; (viii) Rainfall; (iv) Rivers/streamlines.

In the context of Chittagong District, Bangladesh, where diverse landscapes from hills to coastlines intersect, the nuanced approach offered by GIS is particularly crucial. Ensuring sustainable and beneficial land-use outcomes necessitates a delicate balance between environmental conservation, social imperatives, and economic development, a balance that can be efficiently negotiated with the analytical depth of GIS.

2.2.2 Data Collection

Data forms the cornerstone of any analysis. GIS amalgamates spatial and attribute data, creating potent insights by relating them. [7] The study required a comprehensive amount of spatial and attribute data, focusing on variables pertinent to site selection. The data were collected (Table 1), such as mean surface temperature, annual rainfall, soil texture, elevation, and existing land use patterns, from several GIS databases, tailored explicitly to Chittagong District. The following table catalogs the data types and their sources:

Table 1. Data, data type and data sources.

SL No.	Data Name	Data Type	Sources
1.	Temperature	Raster	BMD
2.	Elevation	Raster	NASA 30m SRTM DEM
3.	Slope	Raster	NASA 30m SRTM DEM
4.	Soil Texture	Vector	Bangladesh Agricultural Research Council
5.	Soil pH	Vector	Bangladesh Agricultural Research Council
6.	Soil Depth	Vector	Bangladesh Agricultural Research Council
7.	Land use	Vector	Supervised classification of Landsat 9 Imagery
8.	Rainfall	Raster	World Clim Historical Data
9.	Rivers	Vector	Survey of Bangladesh

2.2.3 Data Processing

For the digital manipulation of the data, the assessment leaned on the functionalities of the ArcGIS Pro. Through its myriad extensions, including spatial analysis and Geo-processing, the data were proficiently processed and generated the desired outputs.

2.2.4 Spatial Analysis

In the realm of geo-processing, satellite imagery played a pivotal role, enabling the identification of regions with the optimal vegetation index, thereby indicating their suitability for rubber cultivation. Advanced GIS tools such as weighted overlay, Euclidean distance, buffer, intersect etc. were employed to overlay multifaceted data layers, encompassing aspects such as soil texture, pH, slope, elevation, prevalent rainfall patterns, and current land use. This layered approach provided a comprehensive view of the land's potential for rubber cultivation. According to Bangladesh Meteorological Department, on average, the district receives around 2500 mm to 3000 mm of rainfall annually, with the majority of it occurring during the monsoon season, which typically spans from June to September. Hence, the entire Chittagong district falls under a highly suitable region for rubber growth and due to this important criterion (rainfall) for rubber cultivation and growth, has been ignored in the spatial analysis portion.

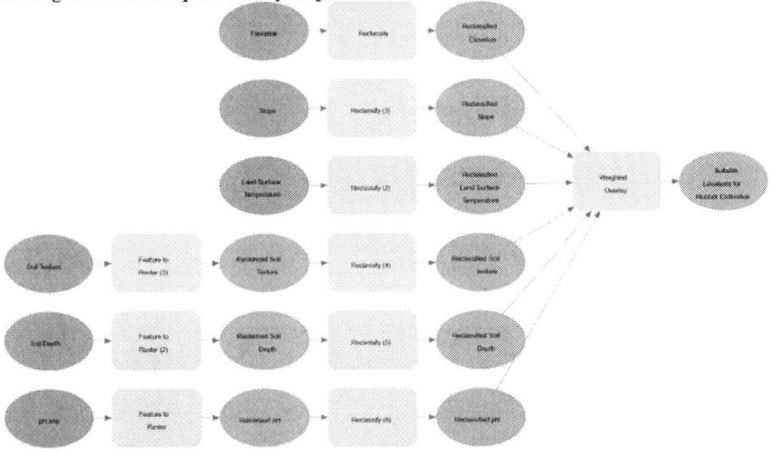

Fig. 4. Flow diagram of overall methodology of land suitability analysis.

2.2.5 Geo-processing

Geo-processing played a pivotal role in this assessment, acting as the bridge between raw data and actionable insights. Satellite imagery was employed as the primary tool for remote sensing, enabling the identification of regions with the optimal vegetation index, urban areas, waterbodies etc. These images provided a high-resolution view of the land's topography, vegetation density, and other critical factors that could influence rubber cultivation. The GIS mapping also facilitated the visualization of areas already under cultivation, urbanized zones, and protected regions, thereby aiding in pinpointing regions that could be converted to rubber plantations without causing significant environmental or socio-economic disruptions.

2.2.6 Weighted Overlay Based Suitability Assessment

This assessment was not a mere overlay of data but a deep dive into several critical ecological parameters using weighted overlay technique that play a pivotal role in land health and productivity. Weighted overlay represents a method employed in suitability modeling, and ArcGIS employs the subsequent procedure for conducting this analysis. In this assessment, within the framework of suitability analysis, each individual raster layer of criteria was allocated a specific weight. For the suitability assessment analysis, thorough literature review and sectoral expert opinions were conducted on the suitable conditions for rubber growth. The values contained within these raster layers are subsequently transformed and reclassified onto a uniform suitability scale (high, moderate, low) and used in a geo-processing model (Figure 4). The output of the review is noted in table 2.

Table 2. Criteria parameters, the level of suitability and scores for assessment. [8][9][10]

Factor	Class	Description	Weight
Temperature	25 - 34^0 C	High Suitable	25
	17 -25^0 C	Moderate Suitable	
	>34^0 C	Low Suitable	
Elevation	40 - 250 m (Low to Mid Elevations)	High Suitable	10
	250 - 450m (Mid to High Elevations)	Moderate Suitable	
	>450m (High Elevations)	Low Suitable	
Slope	0-8% (Gentle Slope)	High Suitable	15
	8-20% (Moderate Slope)	Moderate Suitable	
	>20% (Steep Slope)	Low Suitable	
Soil Texture	sandy clay loam	High Suitable	25
	sandy loam, loam	Moderate Suitable	
	Clay, clay loam	Low Suitable	
Soil pH	pH 6.0 – 7.0 (Slightly Acidic to Neutral pH)	High Suitable	10
	pH 5.5 – 6.5 Moderately Acidic to Slightly Alkaline pH	Moderate Suitable	
	pH < 5.5 or > 7.5 (Highly Acidic or Highly Alkaline pH)	Low Suitable	
Soil Depth	>1meter (Deep)	High Suitable	15
	0.6 -1meter (Medium)	Moderate Suitable	
	< 0.6 meter (Shallow)	Low Suitable	
Total			100

3 Result and Discussion

3.1 Criteria for rubber cultivation

3.1.1 Temperature

Temperature is of paramount importance in rubber cultivation because it directly influences the growth, development, and overall productivity of rubber trees (Hevea brasiliensis). The highly suitable range of temperature is 24°C to 35°C (75°F to 95°F) that provides the most favorable conditions for rubber cultivation. Temperatures falling between 18°C to 24°C (64°F to 75°F) are considered moderately suitable for rubber growth. [11] In this range the growth may be slower, and latex production may not be as high, but it's still feasible. Temperatures below 18°C (64°F) are generally considered low suitable for rubber cultivation due to slower growth and reduced latex production [8][9][10]. The temperature profile [Figure 5 a(i)] of Chittagong district mostly favorable for rubber cultivation because most of the area fall under high suitable in terms of temperature followed by moderate and low suitable areas after the raster reclassification [Figure 5 a(ii)].

3.1.2 Elevation

Rubber cultivation is highly suitable in regions with low to mid elevations typically ranging from sea level up to around 500 meters above sea level. These elevations provide favorable temperature and climatic conditions for the healthy growth of rubber trees. Elevations between 200 meters to 500 meters above sea level can be considered moderately suitable for rubber cultivation. Elevations above 500 meters are generally considered less suitable for rubber cultivation. At higher elevations, temperatures tend to be lower, which may negatively impact the growth and latex production of rubber trees.[10][13][14] The distribution of land elevation of study area illustrated that the overall land elevation is maximum 330 meter in Chittagong district [Figure 5 b(i)]. After reclassification, elevation in the study area is divided into three classes: 40-250 meter, 250-450 meters and greater than 500 meters, which is illustrated in [Figure 6 b(ii)]. It can be stated that overall study area is highly suitable for rubber cultivation considering land elevation.

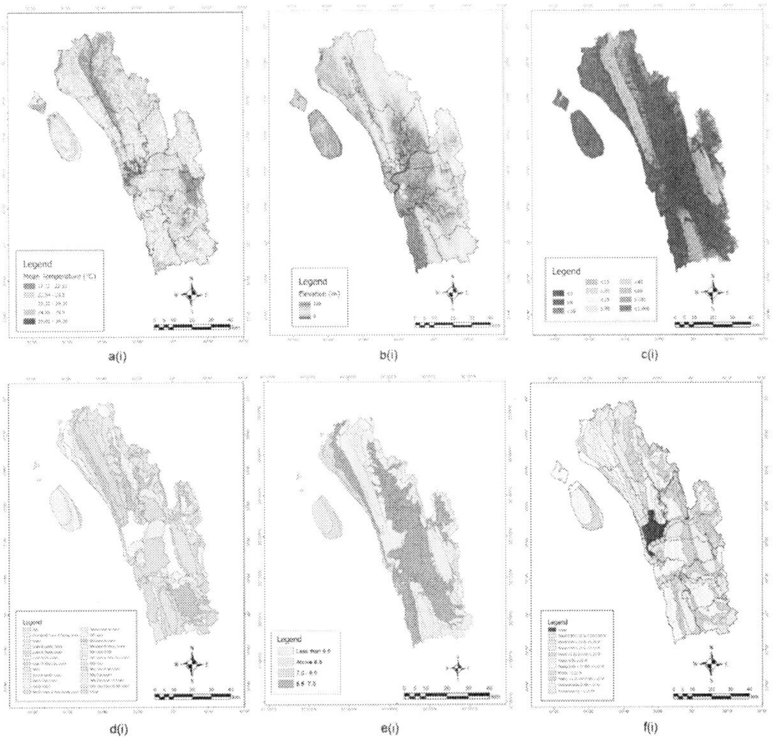

Fig. 5. Criteria for rubber cultivation.

3.1.3 Slope

Rubber cultivation is highly suitable on gentle slopes, typically ranging from 0% to 8%. These slopes provide adequate water drainage and reduce the risk of erosion, which is essential for the successful establishment and growth of rubber trees. Slopes between 8% to 20% can be considered moderately suitable for rubber cultivation[8][9][10] [Figure 5 c(i)] shows the distribution of land slope of study area. After reclassification, slope in the study area is divided into three classes, high suitable (0-8%), moderately suitable (8-20%) and low suitable (steeper slope, >20%), which is illustrated in [Figure 6 c(ii)]. It is found that considering land slope, Chittagong district is highly suitable for rubber cultivation.

3.1.4 Soil Texture

For optimal root penetration, nutrient absorption, and water drainage, rubber trees necessitate loamy soils. Such soils balance moisture retention and drainage, ensuring aeration and preventing waterlogged conditions detrimental to the roots. Sandy Clay Loams are best to grow rubbers in Bangladesh. Chittagong, blessed with a deltaic and fluvial landscape, boasts significant areas of loamy soil composition.[15] From the spatial data the distribution of soil texture portrays a vivid range of soil classes [Figure 5 d(i)]. The raster reclassified map of soil texture indicates that most of the areas of Chittagong district comprises sandy loam and loam soil, which is moderate suitable for rubber cultivation followed by high (sandy clay loam) and low (clay, clay loam) suitable areas [Figure 6 d(ii)]. Most of the higher suitable soil texture for rubber cultivation available in the central and southeastern part of the district.

3.1.5 Soil pH

Rubber trees can tolerate a slightly wider pH range; however, careful soil management practices may be necessary to ensure optimal nutrient availability in the soil. Rubber cultivation is highly suitable in soils with a slightly acidic to neutral pH ranging from 6.0 to 7.0. This pH range provides an optimal environment for the availability of essential nutrients required for the healthy growth and latex production of rubber trees. Soil pH ranging from 5.5 to 6.5 can be considered moderately suitable for rubber cultivation. Soils with highly acidic or highly alkaline pH levels, i.e., below 5.5 or above 7.5, are generally considered less suitable for rubber cultivation. Extreme pH levels can adversely affect nutrient availability and uptake by rubber trees, leading to nutrient deficiencies and poor growth [8][9][10] The soil pH of the Chittagong district is distributed from 6.5 to more than 8.5 [Figure 5 e(i)]. and after reclassification, the soil pH is divided into three classes high suitable (pH 6.0 – 7.0); moderate (pH 5.5 – 6.5; low (< 5.5 or > 7.5) suitable, which are illustrated in [Figure 6 e(ii)]. It also stated that the overall study area is highly and moderately suitable for rubber cultivation.

3.1.6 Soil Depth

Rubber cultivation is highly suitable in areas with deep soil profiles exceeding 1 meter. Deep soil provides ample room for root development, ensuring access to nutrients, water, and anchorage for the rubber trees. Soil depth ranging from 0.6 to 1 meter are considered moderately suitable for rubber cultivation. Soils with shallow depths of less than 0.6 meter are generally considered less suitable for rubber cultivation. Surveys of Chittagong's soil depth reveal profiles that are predominantly deeper than 1.5 meters, ensuring ample space for root expansion [15]. [Figure 5 f(i)]. shows the distribution of soil depth of Chittagong district and after reclassification, soil depth of Chittagong is divided into three classes: high, moderate and low in terms of suitability for rubber cultivation [Figure 6 f(ii)]. The analysis suggests that the majority of regions within the Chittagong district exhibit moderate suitability for rubber cultivation, followed by areas of high and low suitability.

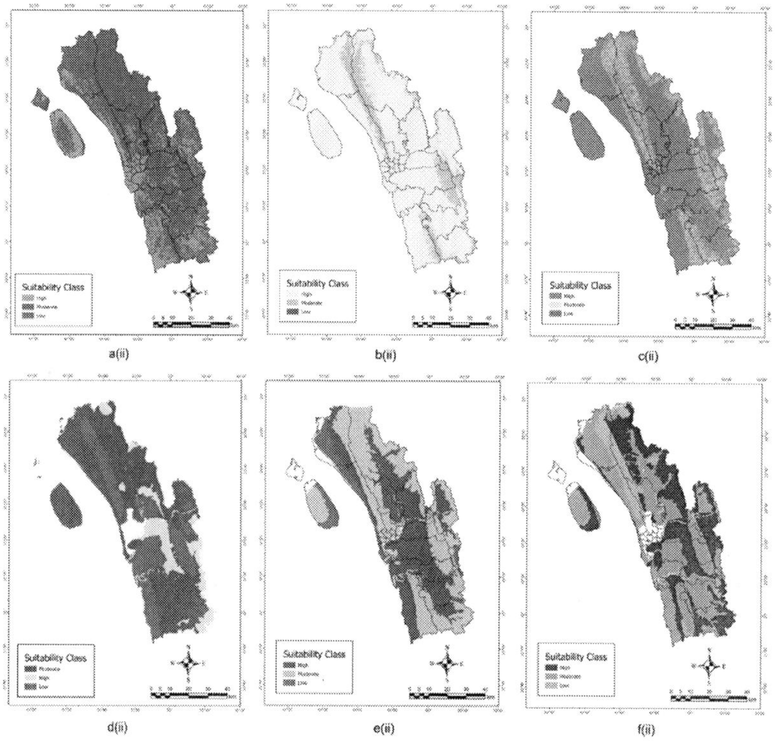

Fig. 6. Reclassified to suitability of criteria for rubber cultivation.

3.2 Suitability Map of Chittagong District

3.2.1 Weighted Suitability Map of Chittagong District

According to the weighted suitability map, which employs specified criteria parameters for rubber cultivation, three distinct suitability classes are identified: highly suitable, moderately suitable, and low suitable. The data reveals that the Chittagong district, in general, ranges from moderate to high suitability for rubber cultivation. Specifically, regions in the Southern, Eastern, lower Northern, and Western parts of the district are identified as highly suitable for this agricultural practice.

3.2.2 Sensitivity Analysis

In determining viable regions for rubber cultivation, a comprehensive evaluation of current land use is paramount. Parcels dedicated to water bodies or agricultural activities should be sidelined from consideration due to their inherent incompatibility with rubber farming conversion. Equally, the ecological and socio-cultural value of forested regions necessitates their preservation and exclusion from potential cultivation sites.[16] A weighted suitability map, pivoted on physical parameters, demarcates optimum zones predominantly in the southern and mid-eastern parts of the district. The natural proclivities of rubber trees, which display an aversion to inundation and water stagnation, mandate the incorporation of proximity to rivers as a pivotal criterion [Figure 7 a]. Urban landscapes and forested regions[Figure 7 c] should also be omitted. For enhanced clarity, suitability was categorized based on river proximities: 0-5 km (low suitability), 5-25 km (high suitability), and >25 km (moderate suitability) [Figure 7 b]. Land use data from the Chittagong district elucidate the prevalence of croplands, vegetative, and built-up sectors. Water bodies are discernible, particularly centrally [Figure 7 c]. The definitive rubber cultivation expanse is computed by excising these unsuitable tracts from the initially earmarked suitable terrain, grounded on stipulated criteria.

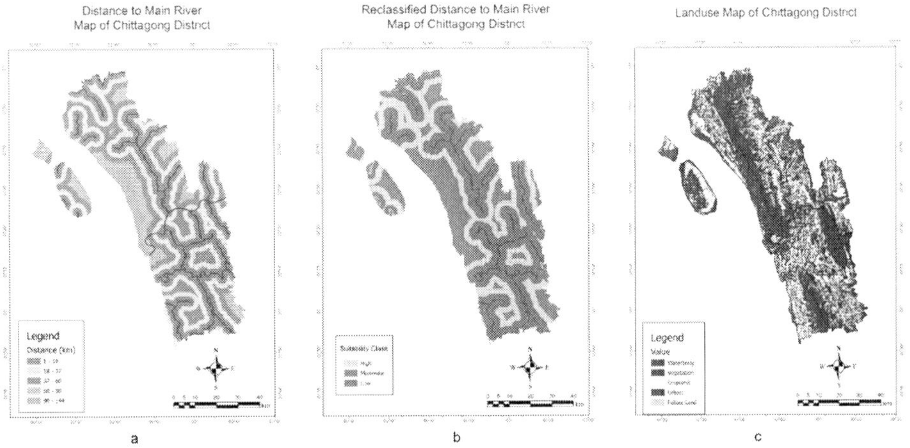

Fig. 7. Sensitivity criteria for suitability analysis for rubber cultivation in Chittagong district.

3.2.3 Final Suitability Map for Rubber Cultivation

The final suitability map involves the sensitivity analysis using distance to river [Figure 8 a] and land use profile [Figure 8 b] of Chittagong district to achieve the potential suitable area for rubber cultivation. From this figure it can be illustrated that the area which remains after subtracting the urban land use and considering the buffer of more than 5 kilometers, is suitable for rubber cultivation. Therefore, some regions, especially the central region becomes suitable for rubber cultivation [Figure 8 c].

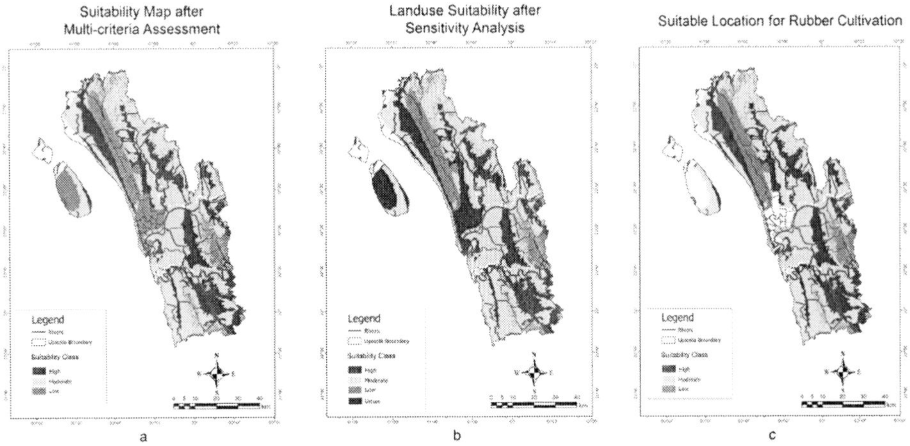

Fig. 8. Land suitability map for rubber cultivation in Chittagong district.

The final output reveals that there are 609.86 km^2 areas are found highly suitable, 2970.33 km^2 is moderately suitable, 1429.3 km^2 for rubber cultivation and 273.61 km^2 is urban.

4 Conclusion

This research delineates a comprehensive land suitability assessment for rubber cultivation in Bangladesh's Chittagong district, employing a Geographic Information System (GIS)-based multi-criteria decision analysis (MCDA). The study systematically scrutinizes an array of agro-climatic and edaphic parameters, including temperature, elevation, slope gradient, soil texture, soil pH, and soil depth, with the aim of providing actionable insights to an interdisciplinary audience of policymakers, agronomists, and economic investors.[17] Our analyses reveal that the Chittagong district manifests a confluence of favorable agro-climatic attributes conducive to rubber cultivation. Temperature indices align well within highly suitable climatic bounds, while elevational gradients predominantly favor optimal rubber tree growth. Additionally, the topographical undulations, characterized by gentle slopes, are conducive to efficient water drainage—an elemental factor in rubber cultivation. Furthermore, edaphic properties such as soil texture, pH, and depth are found to be commensurate with, or surpassing, the requisite conditions for rubber tree agronomy. In conclusion, the findings elucidate significant untapped potential for rubber cultivation in the Chittagong district, in alignment with broader sustainability goals and economic diversification targets. The strategic utilization of these inherent agro-climatic advantages, undergirded by ecologically responsible agricultural practices, has the potential to catalyze socio-economic transformation in rural communities, including job creation and foreign exchange revenue enhancement.[18] Therefore, this study should be regarded as an essential empirical foundation for stakeholders interested in optimizing the latent potential for rubber cultivation in the Chittagong district, offering a sustainable development trajectory that could benefit both local communities and the national economy.

5 Limitation of the study

The study's limitations include its restricted geographical focus on the Chittagong district, making it less directly applicable to other regions in Bangladesh or different countries with distinct environmental and socio-economic conditions. The study's findings are temporally constrained, based on conditions during the research period, which may not account for future changes in environmental and socio-economic factors. Furthermore, environmental impacts and mitigation measures, regulatory frameworks, alternative land uses, stakeholder engagement, economic viability assessment, and ethical considerations related to rubber cultivation are not explicitly addressed.

6 References

1. S.R. Maggiotto, D. de Oliveira, C.J. Marur, S.M.S. Stivari, M. Leclerc, C. Wagner-Riddle, Acta Scientiarum. Agronomy 36, 239-245 (2014).
2. S. Mostafiz, Business Post BD. (2022, September 12).
3. J. Chakma, The daily observer. (February 18, 2022).
4. S. Roy, S. Hazra, A. Chanda, S. Das, J. Earth Syst. Sci. 131, 201 (2022).
5. M. Giaoutzi, C. Papadopoulou, (2021).
6. S.W. Kamau, D. Kuria, M.K. Gachari, 5(6), 121–132 (2015).
7. G. Andrienko, N. Andrienko, P. Jankowski, D. Keim, M.-J. Kraak, A. MacEachren, S. Wrobel, Int. J. Geogr. Inf. Sci. 21(8), 839-857 (2007).
8. A.K. Azad, N.S. Pitol, Y. Hara, Int. J. Environ. Agric. Biotechnol. 6.3 (2021).
9. J. Waewsak, S. Ali, Y. Gagnon, Biomass Bioenergy 137 (2020).
10. F.U. Ahmed, N. Khan, Farming practices and sustainable development in the Chittagong Hill Tracts. CHTDB and VEFP-IC. pp105-112 (2002).
11. B. Kositsup, P. Montpied, P. Kasemsap, P. Thaler, T. Améglio, E. Dreyer, Trees 23, 357-365 (2009).
12. L. Samarappuli, W. Wijesuriya, D. Dissanayake, S. Karunaratne, K. Herath, J. Environ. Prof. Sri Lanka. 3 (2014).
13. M. Chandrakala, S. Ramasamy, A. Kumar, S. Karumathil, R. Hegde, S. Singh, H. Nirmala, Curr. J. Appl. Sci. Technol. 35, 1-9 (2019).
14. L. Agbogun, A. Umweni, H. Kadiri, F. Okunsebor, Ilmu Pertanian (Agricultural Science) 6, 156 (2021).
15. M. Alamgir, M. Islam, N. Hossain, M. Kibria, M. Rahman, Int. J. Plant Soil Sci. 7, 362-372 (2015).
16. M. Rendana, A. Rahim, S. Lihan, I. Tukimat, A. Rahman, Zulfahmi, 14, 1019-1025 (2014).
17. I. Hassan, M. Javed, M. Asif, M. Luqman, S. Ahmad, A. Ahmad, S. Akhtar, B. Hussain, Pak. J. Agric. Sci. 57, 1509-1519 (2020).
18. K.N. Ahmed, S.K. Khisa, Sustainable Development 8.1, 1-10 (2000).

The Importance of Livelihood Diversification on Agroforestry Farmers in The Landslide Prone Area: A Case Study in Menoreh Hills Kulon Progo, Indonesia

Arum Mutiarra Ridyo*, *Utami* Arini Wahyu, and *Irham* Irham[1]

[1]Department of Agriculture Socioeconomics, Faculty of Agriculture, Universitas Gadjah Mada, Yogyakarta, Indonesia

Abstract. Menoreh Hill in Kulon Progo is one of the most prone and vulnerable areas to natural disasters due to its location in an old volcanic complex, which can severely affect the livelihoods of agroforestry farmers. Diversification has become a livelihood improvement strategy in the landslide-prone area. This study specifically examines how and to what extent farmers diversify their livelihoods by comparing agroforestry farmers in high-risk areas (Samigaluh) and low-risk areas (Kalibawang). Primary data were collected from 54 farm households in Samigaluh and 51 farm households in Kalibawang districts using the census technique. The livelihood diversification index was calculated using the Simpson Diversification Index (SDI) and used an independent sample t-test to test whether groups of farmers would differ regarding livelihood diversification. The results showed that farmers in Samigaluh have a relatively higher diversification index (0.3994) than farmers in Kalibawang (0.3864). The results also show that diversification doesn't always bring higher income to farming households, but it is essential to survive in the landslide-prone area.

1 Introduction

Kulon Progo Regency is one of the regencies included in the category of high-risk areas affected by natural disasters, this is indicated by the disaster risk index score of Kulon Progo Regency, which is ranked the highest in DIY Province [1] and is one of the regencies in Java Island, which is included in the super priority category as an area vulnerable to natural disasters [2]. Looking at the data on disaster events, Kulon Progo Regency experienced an increase from 2018 to 2021 [3]. The highest total number of disasters occurred in 2021 with 297 disasters [4].

The dominat disaster in Kulon Progo Regency is a landslide. The results of [5] concluded that Kulon Progo Regency has the highest number of landslide risk points compared to other regencies located on the Menorah Hills. Menoreh Hills is one of the areas in Java Island that has a high frequency of land mass movements. This is triggered by the diverse constituent content of the relief, which is an old volcanic mountainous area [6]. Based on the disaster data report of BPBD Kulon Progo Regency, the losses of households affected by landslides were estimated at 9,000,000 idr.

Considering the risk of landslides in Kulon Progo Regency, encourag agroforestry farmer households to pursue livelihood strategies through diversification. Income diversification refers to a continuous adaptive process whereby households add new activities, maintain existing ones, or drop others, thereby maintaining diverse and changing livelihood portfolios. People may diversify their livelihood patterns by adopting a range of activities and income sources. Rural livelihood and diversification can be defined as the process by which rural households construct an increasingly diverse portfolio of activities and assets to survive and improve their standard of living [6] [7].

Livelihood strategies are a combination of activities and choices made to make ends meet. According to [9], livelihood strategies are the process by which farm households carry out various activities and obtain social support to fulfill and improve living standards. Livelihood strategies are created by two motivations, namely necessity and choice [9]. Livelihood diversification is the process of rural households undertaking diverse activities to survive and improve living standards. According to [10] there are two interpretations of livelihood diversification, the first being that diversification is a coping strategy or an involuntary response to a crisis. The second interpretation of diversification is that it is considered a decision by proactive households to reduce their risk by spreading their livelihoods. In the first sense, diversification is a survival strategy practiced by vulnerable households, while in the second sense, it is a accumulation strategy practiced by richer households. Diversification helps households deal with income instability, increases resilience, and reduces

* Corresponding author: mutiarra.ridyo.arum@mail.ugm.ac.id

vulnerability. Broadly speaking, households will benefit from diversification.

Research by [11] shows that livelihood diversification is a strategy used by more than 62% of farmers in Burat, Kenya, in dealing with drought. The same thing was conveyed by [12] to overcome drought, farmers carry out various strategies, one of which is the diversification of livelihood sources. Implementing livelihood diversification allows farmers to have a lower risk of loss from the effects of natural disasters. Research by [13] stated that farmers have the ability to overcome the impact of losses and accelerate the recovery time in the face of disasters by diversifying. Diversification means that farmers get income from various sources so that they have higher adaptive capabilities than farmers with fewer sources of income. However, research by [14] explains that the implementation of farmer diversification is faced with several obstacles. The constraints faced are climate risk and uncertainty (23%), inadequate natural resources (17%), limited level of training (15%), lack of institutional support (12%), lack of credit facilities (11%), poor infrastructure (markets and roads) (16%), and lack of labor availability (4%).

Considering the constraints faced by farmers in implementing livelihood diversification [15], [16]. The studies compared farmers' livelihood diversification in several disaster-prone locations. The results showed that locations that are most prone to natural disasters tend to have low diversification levels. Research by [15] explains that there is a strong negative relationship between the shock effect of natural disasters and livelihood diversification. Locations in the post-disaster period will experience a decline in welfare due to the loss of income from livelihood diversification. Research by [16] explains that the majority of households in disaster-prone locations depend on the utilization of natural resources, so their ability to diversify their livelihoods is low because their income comes from agricultural activities.

This study assesses the pattern of income sources and measures diversification by comparing agroforestry farming households at different levels of landslide-prone areas. Overall, this study will give significant attention to the income diversification portfolio by finding the conditions of appropriate livelihood strategies undertaken by farmer households. Thus, this study aims to analyse income diversification and the decisions taken by farmer households in landslide-prone areas.

2 Material and Method

2.1 Study Location

The study was carried out in the Samigaluh district and Kalibawang district, which were purposely chosen because they are the highest and lowest landslide-prone areas in the Menoreh Hills, Kulon Progo regency. The research was conducted in two locations. Samigaluh is located between latitudes 7039'50.18"-7039'01.17" S and longitude 110011'36.50"-110011'47.45" E, which are at the top of Menoreh hills. Based on data from the Regional Disaster Management Agency (Badan Penangulangan Bencana Daerah, BPBD), there was an increase in disaster events from 2018-2022, with the peak in 2022 six times more than 2021. Then the second location in Kalibawang was located between longitude 7°42'53.4"-7°43'20.9"S and latitude 110°11'17.9"-110°12'00.3"E, which is at the foothills. The village of Sidoharjo represents the location with the highest landslides prone-areas, while the village of Banjararum represents the location with the lowest prone areas. These two areas have implemented an agroforestry system in Kulon Progo regency.

2.2 Data Collection

The data were collected through a census of farmer households, with a total of 105 respondents consisting of farmer households in Samigaluh (n=54) and Kalibawang (n=51). The census method and questionnaire were used to explore the demographic characteristics of the village and the respondents. Meanwhile, observations and in-depth interviews were conducted to obtain detailed socio-economic dynamics and household demographic information. Each interview was preceded by an explanation of the research objectives to obtain informed consent. Data were collected from January to February 2023.

2.3 Data Analysis

Livelihood diversification was measured using the Simpson Diversification Index (SDI). The Simpson index is most widely used because of its simple calculation method and broad adequacy [17]. The equation of the index is shown below:

$$\mathrm{SDI} = 1 - \sum_{i=1}^{n} P_i^2$$

where n is the total number of livelihood income sources and Pi is the proportion of income from the i-th livelihood source. To measure livelihood diversification, it can be done through employment diversification and income diversification [18]. In this study, livelihood diversification is calculated based on income sources. Income sources come from different activities such as agroforestry income (AI), non-agroforestry income (NAI), livestock income (LI), off-farm income (OI), and non-farm income (NI), so the model is SDI:

$$\mathrm{SDI} = 1 - \left\{\left(\frac{\mathrm{AI}}{\mathrm{THI}}\right)^2 + \left(\frac{\mathrm{NAI}}{\mathrm{THI}}\right)^2 + \left(\frac{\mathrm{LI}}{\mathrm{THI}}\right)^2 + \left(\frac{\mathrm{FI}}{\mathrm{THI}}\right)^2 + \left(\frac{\mathrm{OI}}{\mathrm{THI}}\right)^2 + \left(\frac{\mathrm{NI}}{\mathrm{THI}}\right)^2\right\}$$

The SDI value is between 0 and 1. Households with the most diversified income sources have the highest SDI values and are close to 1. Meanwhile, households that do not diversify or specialize have an SDI value close to 0. The number of income sources and the distribution of income from various sources of activity both affect the value of the Smpson index. Basically, based on the SDI value, we also tested the difference between two locations : high-risk areas (Samigaluh) and low-risk areas (Kalibawang) using the Mann-Whitney statistical test [17]. In addition to analysing the SDI, the level of livelihood diversification has been defined as [2]: limited ($\leq$ 0.01), low (0.01-0.25), medium (0.26-0.50), high (0.51-0.75), and very high ($\geq$ 0.75) (Khatun and Roy 2012; Datta and Sing 2011). To assess the relationship between income and livelihood diversification, we relied more on descriptive analysis.

3 Result and Discussion

3.1 Source of Household Income and Percentage Shares

The finding in this study is consistent with the research conducted by [19] that households located in natural disaster-prone areas have low incomes and are vulnerable to poverty. Natural disasters are the main driving factor in the risk of unpredictable uncertainty and their impact on economic losses. Table 1 shows that farmer households in the highest landslide-prone areas have lower incomes than households in the lowest landslide prone areas. This encourage farming households to combine livelihood sources from on-farm, off-farm, and non-farm activities.

Table 1 shows the contribution and distribution of each source of income for the respondents. Households in the highest landslide-prone areas (Sidoharjo Village) have the largest contribution from on-farm income (69%), followed by non-farm activities (27%). When compared to farmer households in the lowest landslide-prone area (Banjararum Village), the largest contribution comes from non-farm income (48%), and the second from on-farm income (44%). From both research locations, the contribution of off-farm income does not dominate. This is because the majority of agricultural activities are carried out by labor within the family, so off-farm income in both locations is low. According to [20] the low contribution of off-farm income is due to the view that off-farm activities can reduce the availability of family labor.

The difference in income from on-farm activities in the two research locations is quite significantly different, with the income of farmer households in the highest landslide prone areas being greater. Farmer households located in the highest landslide prone areas are identified to manage their natural resources better. This is identified by the land management system and the types of crops cultivated. Farmers in landslide-prone areas are highly engaged in agroforestry land management through mixed farming systems with more economically valuable crop commodities, including chili, cardamom, coffee, and sengon. In addition, farmers integrate resources from agroforestry land with livestock businesses as a new source of livelihood. The results of interviews found that agroforestry land management activities are a form of adaptive capacity for farmers in the face of landslides [21] .Meanwhile, farmers in the lowest landslide-prone areas are not managing agroforestry land properly, this is indicated by the outpouring of labor in agroforestry land management carried out only during planting and harvesting.

Non-farm livelihoods contribute significantly to the income of farm households in the lowest landslide prone area. This is due to the existence of a location that supports farmer households to run livelihood sources outside of agroforestry land. This is in line with research [22] that low or high intensity of land management at the foot of the slope has no impact on landslides. In addition, the availability of access and facilities, as well as the nature of non-farm income that can be accessed throughout the year, support farmer households in landslide-prone areas to source livelihoods from non-farm activities. According to [23] [24], non-farm income sources can improve the livelihood of farmer households.

Table 1. Source of Household Income and Percentage Shares

https://doi.org/10.1051/bioconf/20238002008

Sources of Income	Kalibawang		Samigaluh	
	Amount	Percentage (%)	Amouth	Percentage (%)
		On-Farm		
Agroforestry	4,706,016	13	11,019,855	34
Paddy field	2,675,604	7	26,574	0.08
Livestock	8,927,902	24	11,307,259	35
Fisheries	97,451	0.26	2,890	0.01
		Off-Farm		
Agriculture labour	3,042,549	8	1,055,926	3
Post-harvest	-	0	133,333	0.41
		Non-Farm		
Non-agriculture wage labour	1,031,373	3	2,256,481	7
Service and salaried job	7,759,175	21	5,412,963	17
Small business	9,403,922	25	1,038,889	3
Total	37,643,991	100	32,254,171	100

Sources : Primary data analysis (2023)

3.2 Livelihood Diversification

This study measured diversification using the Simpson diversification index method. The results showed that agroforestry farmer households in the highest landslide-prone areas practiced higher diversification than farmers in the lowest landslide-prone areas. Livelihood diversification activities are carried out by farmer households with a drive to fulfill their needs and survive the risk of landslides (diversification for survival). This study underscores that diversification is not limited to undertaking several livelihood activities but is a combination of several activities with equal income from each livelihood source.

Farmer households in landslide-prone areas are highly engaged in several activities to earn a higher income in even or balanced proportions. Farmer households practice diversification with no tendency to focus on one source of livelihood. This indicates that farmer households try to spread the risk of disaster by diversifying their livelihoods. Diversifying livelihoods allows farmers to have a lower risk of losses caused by natural disasters. This research is in line with [13] that by diversifying farmer household, they have the ability to cope with the impact of losses and accelerate recovery time when facing disasters. Diversification is a risk minimization strategy that is effectively applied by rural farmer households to meet their needs [25].

Farmer households in the lowest landslide-prone areas have lower diversification scores or tend to practice livelihood specialization. Being located in a low landslide risk area causes farmers to focus on one source of livelihood that contributes significantly to household income. Table 1 shows that the household income of farmers in the very low landslide risk area is dominated by income from non-farm activities. According to [26] that non-farm income allows households to do work and earn income throughout the year.

Based on the results of the Mann-Whitney test, it shows that the diversification values in the two research locations are not statistically different. However, this does not mean that there is no difference between the two locations. Table 2 shows that households in the highest landslide-prone areas have a higher diversification value but a lower average income than households in the lowest landslide prone areas. This shows that households in the very high landslide risk area carry out livelihood diversification to avoid the risk of failure from the threat of landslides. Meanwhile, households in low-risk areas tend not to need livelihood diversification (specialization) but have a higher average income. Thus, this study emphasizes that the livelihood strategies of agroforestry farming households are influenced by the level of disaster risk.

3.3 The Importance of Livelihood Diversification on Agroforestry Farmers

Table 3 shows the distribution of diversification index values in both research locations. The results of the categorization of farmer households in the highest landslide-prone area are very high, which is classified as high diversification (22%), moderate (52%), and low (26%). Meanwhile, farmer households in lowest

Table 2 Simphson Diversification Index

SDI	Kalibawang	Samigaluh	Mann-Whitney Test
Mean	0.3864	0.3994	Z= -0.141, p>.05
Min	0.0000	0.0678	
Max	0.7197	0.6838	

Sources : Primary data analysis (2023)

landslide prone areas, which are classified as high (33%), medium (37%), low (27%), and very low (2%) diversification. Both research locations have an even distribution in the low, medium, and high diversification categories.

The research found that the higher value of diversification is not consistent with an increase in household income (Table 3). This is evidenced by the value of SDI and the average household income in both research locations. The higher the value of SDI, it does not show an increase in the household income of agroforestry farmers in the lowest landslide prone areas. However, agroforestry farmers in very high landslide disaster risk areas showed higher SDI values and higher income. This is in line with the results of research conducted [27] that diversification cannot be concluded to be consistent with increased income.

The results of the study prove that farmers who are in a very high landslide-prone area need to carry out livelihood diversification to meet their needs, or diversification for survival. This is indicated by the average household income of farmers, whose income has increased by diversifying their livelihoods. Meanwhile, for farmers in the lowest landslide-prone area, livelihood specialization is a good strategy to increase income. The low level of disaster vulnerability makes it easier for farmers to focus on and maximize one source of livelihood.

4 Conclusion

This study found that agroforestry farmers in very high landslide-prone areas practiced higher diversification than farmers in very low landslide-prone areas, but the higher value of livelihood diversification was not consistent with higher household income. This study found that livelihood diversification is necessary for agroforestry farmers in very high landslide-prone areas to spread the risk of failure and make ends meet. Meanwhile, in very low landslide prone areas specialization or not diversifying is a good strategy to increase income. The findings of this study suggest that the livelihood strategies of farmer households are strongly related to the level of disaster vulnerability.

Acknowledgment

We are grateful for the research funding from Badan Pengelolaan Dana Lingkungan Hidup (BPDLH), Ministry of Finance, Indonesia. We would like to also express our gratitute to community leaders and members of Banjararum and Sidoharjo Villages.

Table 3 The Interaction between Livelihood Diversification and Income Household

SDI	Category SDI	Kalibawang			Samigaluh		
		Amount	Percentage (%)	Avg. Income	Amount	Percentage (%)	Avg. Income
≤ 0.01	Limited	1	2	3,501,000	0	0	-
0.011-0.250	Low	14	27	43,142,452	14	26	18,804,507
0.251-0.50	Medium	19	37	41,730,200	28	52	34,186,775
0.51 – 0.75	High	17	33	30,557,269	12	22	43,436,208
≥ 0.751	Very high	0	0	-	0	0	-

Source: Primary data analysis (2023)

References

[1] A. W. Adi *et al.*, "Indeks risiko bencana Indonesia tahun 2021," pp. 11–3, 2022.

[2] P. Mahyastuti *et al.*, "Daftar lokasi & aksi ketahanan iklim," 2021.

[3] Badan Pusat Statistik, "Banyaknya desa/kelurahan yang mengalami bencana alam dalam tiga tahun terakhir," 2022. https://yogyakarta.bps.go.id/indicator/152/244/1/banyaknya-desa-kelurahan-yang- mengalami-bencana-alam-dalam-tiga-tahun-terakhir.html

[4] Badan Penangulangan Bencana Daerah, "Rekapitulasi Data Kejadian Bencana Kab. Kulon Progo," 2022. https://bpbd.kulonprogokab.go.id/detil/497/rekapitulasi-data-kejadian-bencana-alam-bpbd-kulon-progo-tahun-2021

[5] K. R. Fairuuz, "Pemetaan Risiko Bencana Longsor Perbukitan Menoreh," Universitas Gadjah Mada, 2017.

[6] R. Purwaningsih, J. Sartohadi, and M. Anggri, "Trees and crops arrangement in the agroforestry system based on slope units to control landslide reactivation on volcanic foot slopes in Java, Indonesia," *Land*, vol. 9, no. 9, 2020, doi: 10.3390/LAND9090327.

[7] D. Narayan and P. Petesch, "Crying Out," no. January 2000, 2015.

[8] S. Schwarze and M. Zeller, "Income diversification of rural households in Central Sulawesi, Indonesia," *Q. J. Int. Agric.*, vol. 44, no. 1, pp. 61–73, 2005.

[9] F. Ellis, "Rural Livelihood Diversity in Developing Countries," *Oxford Univ. Press*, no. 40, pp. 1–10, 2000.

[10] C. Be´ne´, K. Mindjimba, E. Belal, T. Jolley, and A. Neiland, "Inland fisheries, tenure systems and livelihood diversification in Africa: The case of the Yae´re´ floodplains in Lake Chad Basin," *Afr. Stud.*, vol. 62, no. 2, pp. 187–212, 2003, doi: 10.1080/0002018032000148759.

[11] A. Quandt, "Coping with drought: Narratives from smallholder farmers in semi-arid Kenya," *Int. J. Disaster Risk Reduct.*, vol. 57, p. 102168, 2021, doi: 10.1016/j.ijdrr.2021.102168.

[12] F. Opiyo, O. Wasonga, M. Nyangito, J. Schilling, and R. Munang, "Drought Adaptation and Coping Strategies Among the Turkana Pastoralists of Northern Kenya," *Int. J. Disaster Risk Sci.*, vol. 6, no. 3, pp. 295–309, 2015, doi: 10.1007/s13753-015-0063-4.

[13] N. T. T. Pham, D. Nong, A. Raghavan Sathyan, and M. Garschagen, "Vulnerability assessment of households to flash floods and landslides in the poor upland regions of Vietnam," *Clim. Risk Manag.*, vol. 28, no. January, p. 100215, 2020, doi: 10.1016/j.crm.2020.100215.

[14] A. A. Shah, Z. Gong, and N. A. Khan, "Livelihood diversification in managing catastrophic risks : evidence from flood-disaster regions of Khyber Pakhtunkhwa Province of Pakistan," pp. 40844–40857, 2021.

[15] S. S. Ibrahim, H. Ozdeser, and B. Cavusoglu, "Vulnerability to recurrent shocks and disparities in gendered livelihood diversification in remote areas of Nigeria," *Environ. Sci. Pollut. Res.*, vol. 26, no. 3, pp. 2939–2949, 2019, doi: 10.1007/s11356-018-3854-5.

[16] N. T. T. Thao, D. N. Khoi, T. T. Xuan, and B. Tychon, "Assessment of Livelihood Vulnerability to Drought: A Case Study in Dak Nong Province, Vietnam," *Int. J. Disaster Risk Sci.*, vol. 10, no. 4, pp. 604–615, 2019, doi: 10.1007/s13753-019-00230-4.

[17] N. Habib, P. Rankin, M. Alauddin, and R. Cramb, "Determinants of livelihood diversification in rural rain - fed region of Pakistan : evidence from fractional multinomial logit (FMLOGIT) estimation," *Environ. Sci. Pollut. Res.*, no. 0123456789, 2022, doi: 10.1007/s11356-022-23040-6.

[18] R. O. Babatunde and M. Qaim, "Patterns of income diversification in rural Nigeria: Determinants and impacts," *Q. J. Int. Agric.*, vol. 48, no. 4, pp. 305–320, 2009.

[19] R. B. Bista, "Does Disaster Change Income and Wealth Distribution Toward Extremity of Inequality and Poverty? Analysis of Flood and Landslides in the Vulnerable Locations of Nepal," *Forum Soc. Econ.*, vol. 51, no. 4, pp. 1–15, 2020, doi: 10.1080/07360932.2020.1715810.

[20] L. Pfeiffer, A. López-Feldman, and J. E. Taylor, "Is off-farm income reforming the farm? Evidence from Mexico," *Agric. Econ.*, vol. 40, no. 2, pp. 125–138, 2009, doi: 10.1111/j.1574-0862.2009.00365.x.

[21] J. Tan, L. Peng, and S. Guo, "Measuring Household Resilience in Hazard-Prone Mountain Areas: A Capacity-Based Approach," *Soc. Indic. Res.*, vol. 152, no. 3, pp. 1153–1176, 2020, doi: 10.1007/s11205-020-02479-5.

[22] N. Muddarisna, E. Yuniwati, H. Masruroh, and A. O, "Local Agroforestry as Landslide Mitigation in the Gede Catchment in Malang Regency," 2020, doi: 10.4108/eai.18-7-2019.2290363.

[23] G. W. Gebru, H. E. Ichoku, and P. O. Phil-Eze, "Determinants of livelihood diversification strategies in Eastern Tigray Region of Ethiopia," *Agric. Food Secur.*, vol. 7, no. 1, pp. 1–9, 2018, doi: 10.1186/s40066-018-0214-0.

[24] M. A. Iqbal *et al.*, "A quest for livelihood sustainability? Patterns, motives and determinants of non-farm income diversification among agricultural households in punjab, pakistan," *Sustain.*, vol. 13, no. 16, pp. 1–14, 2021, doi: 10.3390/su13169084.

[25] T. B. Shan and J. U. Ahmed, "Determinants of Livelihood Diversification of Rural Households in Sylhet," vol. 38, no. 11, pp. 97–104, 2020, doi: 10.9734/AJAEES/2020/v38i1130457.

[26] G. Danso-Abbeam, G. Dagunga, and D. S.

Ehiakpor, "Rural non-farm income diversification: implications on smallholder farmers' welfare and agricultural technology adoption in Ghana," *Heliyon*, vol. 6, no. 11, p. e05393, 2020, doi: 10.1016/j.heliyon.2020.e05393.

[27] K. Hussein and J. Nelson, "Sustainable Livelihoods and Livelihood Diversification," 1998. [Online]. Available: http://www.ids.ac.uk/ids/research/env/index.html

Women's Roles in Beef Cattle Farming Households in Nagari Lakitan Utara, Pesisir Selatan Regency, Sumatra Barat Province

Siti Andarwati*, *Budi* Guntoro, and *Muhammad* Fajri

Department of Livestock Social Economics, Faculty of Animal Science, Universitas Gadjah Mada, Yogyakarta 55281, Indonesia.
*Correspondence:

Abstract. Gender equality is one of the main aspects in the sustainable development goals. This study aimed to analyze the roles of women in the households of beef cattle raising farmers in Nagari Lakitan Utara, Pesisir Regency, Sumatra Barat Province. This research involved 100 farmer households and collaborating partners from related institutions, employing the Harvard gender model of analysis and using the approach of Level of Effort (LoE) values. Results showed that women had significantly important roles in the management of beef cattle raising, starting from providing the logistical needs to managing the finances from cattle sales. In particular, women showed greater roles in herding as well as contacting inseminators and veterinarians in the case of ill animals.

1 INTRODUCTION

One of the main goals in sustainable developments is gender equality. The Indonesian government showed the commitment to support that goal in the presidential decree *Perpres No. 59 tahun 2017* on the Implementations to Achieve the Goals of Sustainable Development. These implementations include preparing the National Action Plans for Family Farming consisting of sustainable family farming, gender equality, capacity advancement of farmer organizations, and farmers' welfare improvement, which is stated in the strategic planning of the Ministry of Agriculture for 2020 - 2024. This action is a manifestation of Indonesia in support of the United Nations' declaration of 2019 - 2028 as the decade for family farming. Additionally, *Rencana Pembangunan Jangka Menengah Nasional* (RPJMN – Medium-term National Development Plan) 2020-2024 states that the national development strategy must include gender perspective to achieve the advancement that is more just and equally distributed for all Indonesians, both men and women. This strategy's goal is to have an increased equality in all fields by the year 2025.

The score of 0.70 in the Gender Gap index for Indonesia places the country at the 85th place among the evaluated 153 countries [1]. In the Southeast Asian region, Singapore, Laos, and the Phillipines show better score and are ranked at 54th, 43th, and 16th, respectively. The Gender Gap score for Indonesia indicates that there is still a large gap between men and women in the country. The government issued a presidential instruction in response to the report to put gender in the mainstream of national development. This instruction defines that gender equality is a condition in which women and men have the same opportunity as well as equal human rights in order to be able to have the same roles in all fields, such as politics, social, culture, and so on. Since the subsector of animal husbandry has important roles in the national development, therefore, it needs the implementations of sustainable development.

The understanding of gender mainstreaming and equality between men and women is required to achieve a sustainable development in animal husbandry [2]. But in reality, women have not fully had the control of the operation [3]. Men are still the dominant force in accessing and controlling the resources and profit utilization. Women's roles in the various daily activities, whether planned or unplanned, essentially carry economic values, especially in relation to household income. The roles of women in improving the household income are aimed at increasing the family welfare. Therefore, public support must be increased for widening the vocational opportunity for women especially in the rural areas [4].

Pesisir Selatan is one of the regencies in Indonesia known for beef cattle raising and is located in the *agropolitant* of Sumatra Barat province [5]. Pesisir Selatan regency had the highest concentration (20.47%) of beef cattle in the province with a total number of 85,361 head in 2021. Women in this area are involved in the family economy, including in caring for the livestock. However, their participation is significantly small in improving the farmers' capacity in support of animal husbandry practices, such as being involved in guidance sessions. This leaves women with less knowledge and skill in caring for their animals. Studies on women's roles in these beef cattle raising families must be conducted to measure gender equality in an economic sector in order to achieve sustainable developments, especially in Pesisir Selatan regency.

2 Research Objectives

This research aimed to analyze the roles of women in the households of beef cattle raising farmers in Nagari

* Corresponding author: andar_siti@ugm.ac.id

Lakitan Utara, Pesisir Selatan Regency, Sumatra Barat province.

3 Methodology

This study involved 100 households of beef cattle raising farmers as well as collaborating partners from related institutions, including the Livestock and Animal Health Bureau of Pesisir Selatan regency Sumatra Barat Province, civil servants of Nagari Lakitan Utara, and Veterinary Center of Sutera and Lengayang areas. Collaborating partners played a role in providing information regarding the general description of beef cattle farming in Nagari Lakitan Utara, Lengayang District, Pesisir Selatan Regency, Sumatra Barat Province.

There were several criteria in determining the sample for this research, including female breeders who are married, livestock businesses that have been running for more than 1 year, patterns of raising beef cattle using semi-intensive and intensive system. Then interviews with selected respondents and data analyses were carried out based on the completed questionnaires.

The variables measured consisted of feeding and watering management, shelter management, livestock reproductive management, the livestock health management, the livestock procurement management, livestock sales and monetary management. The Harvard gender model of analysis was employed to analyze the data because it is a simple analytical technique for a set data in a micro scale that can be used to determine the different roles in a household. In this study, the household data observed were data on women's productive role in implementing the management of beef-cattle farming business.

The analyses of women's roles in the livestock subsector were conducted with the approach of Level of Effort (LoE) between husband (man) and wife (woman) in a household. In this study, women referred to wives and LoE was defined as degree of efforts shown by husband or wife or both in relation to the management of livestock. The parameters for LoE were presented as indices with the range of 0 – 4 [6]. An index value of 0 meant no effort and an index value of 4 meant a full effort.

4 Results and Discussion

4.1 Characteristics of respondents (women)

Out of 100 participants, 99% of the women in the beef-cattle raising families were in the productive age category, with only one older participant. Age determines the firmness in carrying out business [7]. At a productive age, farmers are able to coordinate and take effective steps in advancing their business [8]. The largest number of participants came with high school education at 42%, with only two women had higher education experience. Education level in farmers influences their learning process, in which the higher the education, the easier they understand the information [9].

In terms of vocation, the majority participants (80%) declared beef cattle raising as their main occupation. The amount of time allocated in a work can impact performance [10]. Meanwhile, more than half of the participants (54%) had more than 15 years of experience in raising beef cattle. [11]. Longer experience could lead to a better management of cow-calf operations. Many of these households (64%) had 1-3 family dependents.

4.2 Value of Level of Effort (LoE)

4.2.1 Values of LoE on the feeding and watering management

The values of LoE on the feeding and watering management are described in detail in Table 1.

Table 1. Values of LoE on the feeding and watering management.

No	Activity		LoE	
		Husband	Wife	Both
1	Gathering forage	3.00	0.25	0.75
2	Feeding forage	0.00	0.64	3.36
3	Procuring concentrate	1.74	0.47	1.79
4	Feeding concentrate	0.05	0.58	3.37
5	Supplying water	0.00	1.08	2.92
6	Herding	0.14	2.14	1.72
	Average	0.82	0.86	2.32

Source : primary data analysis (2022)

Husbands dominated forage gathering with the LoE value of 3.00. Meanwhile, herding is mostly conducted by wives (LoE = 2.14). Both spouses carried out the activities at about equal efforts on feeding forage, feeding concentrate, and supplying water with LoE values of 3.36, 3.37, and 2.92, respectively.

The time required to gather forage is about two hours, depending on the location, forage quantity, and precipitation amount. Feed is provided two to three times a day. The concentrate generally consists of rice bran, banana trunk, and sagu. The majority of farmers do not give concentrate routinely to their cattle. The amount of time to provide the concentrate ranges from 10 to 60 minutes, depending on the types of concentrate and the number of cattle. Water provision is done once or twice a day and it takes about ten minutes.

To add the amount of feed, farmers usually herd their cattle once a day. This activity can take place in farms, orchards, or road sides. In general, farmers take out their cattle at 16.00 to 17.45 WIB (Indonesian west standard time) or before the dusk prayer. Distance influences the amount of herding time also. Herding is mainly performed by women, which is in contrast to a previous report by [12] who observed herding to be mostly conducted by men.

4.2.2 Values of LoE on the shelter management

The values of LoE on the shelter management are described in detail in Table 2.

Table 2. Values of LoE on the shelter management

No	Activity		LoE	
		Husband	Wife	Both
1	Building	3.38	0.00	0.63
2	Cleaning	0.64	0.96	2.40
3	Maintaining	3.28	0.12	0.60
4	Taking out livestock	0.20	0.82	2.98
5	Putting in livestock	0.20	0.82	2.98
	Average	1.54	0.54	1.92

Source : primary data analysis (2022)

Husbands generally conduct activities that require more strength. Therefore building (LoE=3.38) and maintaining (LoE=3.28) shelters are carried out mostly by men. Meanwhile cleaning shelter (LoE=2.40) and bringing cattle in (LoE=2.98) and out (LoE=2.98) of shelter are performed by both spouses. To support the animal husbandry activities, farmers furnish supporting tools around animal shelters, such as feed and water troughs, hoe, rake, machete, etc [13]. In addition to feed and water troughs, waste holders and drainage could be found in a cow-calf operation.

Cattle pens in Nagari Lakitan Utara are close to the houses of the farmers. This arrangement facilitates the management of the cattle as well as prevents animal thefts [14]. Shelters function as a facilty to assist in managing livestock production, such as giving provisions, cleaning, breeding, preventing theft, and increasing manpower efficiency. Farmers clean the pens twice a day, in the morning and in the afternoon. The cattle are taken out of the shelters during the cleaning that takes about 10 to 20 minutes. Pen cleaning is performed to increase comfort for the animals and reduce the incidence as well as spread of diseases. Additionally, farmers burn some leftover forage near the pens in the early evening to repel insects and provide warmth to animals.

4.2.3 Values of LoE on the livestock reproductive management

The values of LoE on the livestock reproductive management are described in detail in Table 3.

Table 3. Values of LoE on the livestock reproductive management

No	Activity		LoE	
		Husband	Wife	Both
1	Monitoring estrus	0.00	0.24	3.76
2	Contacting inseminator	1.35	1.68	0.97
3	Overseeing AI	1.52	0.51	1.98
4	Monitoring pregnancy	0.24	0.20	3.56
5	Monitoring calving	0.12	0.20	3.67
6	Post-calving care	0.08	0.16	3.76
	Average	0.55	0.50	2.95

Source : primary data analysis (2022)

The management of reproduction is one of the important aspects in the sustainability of a cow-calf operation. Farmers' good knowledge of reproduction can increase the efficiency of their livestock business by accurate estrus detections and thus reduce repeat breeding [15]. Cows in estrous can be detected from vulvar swelling accompanied with increased pink coloration and clear mucous secretion from the vulva [16].

Contacting inseminators is mostly done by the wives with an LoE value of 1.68. This is most likely a result of women spending more time at home than men, to monitor the condition of their cattle. These women immediately call inseminators upon observing estrus for prompt AI procedures. Meanwhile both spouses contribute equally in estrus monitoring, AI supervision, pregnancy monitoring, calving monitoring, and post-partum calf care, with respective LoE of 3.76, 1.98, 3.56, 3.67, and 3.76.

The majority of study participants employ artificial insemination (AI) for their cow breeding needs. Natural breeding has become rare in Indonesia because of the significant increased efficiency of AI [17]. AI is performed by a technician (inseminator) depositing bull semen in a cow uterus using a metal rod, which is often called AI gun or rod. [18] The advantages of AI that include genetic improvement, cost efficiency, and disease prevention. AI improves the genetic makeup of a local herd because the semen is usually collected from superior bulls. By not having to raise breeding bulls, farmers can reduce expenditure in their operation. Hygienic practices can be conducted more easily in AI and therefore it can reduce the spread of sexually transmitted infections among cattle.

4.2.4 Values of LoE on the livestock health management

The values of LoE on the livestock health management are described in detail in Table 4.

Table 4. Values of LoE on the livestock health management

No	Activity		LoE	
		Husband	Wife	Both
1	Monitoring livestock health	0.00	0.12	3.88
2	Contacting Veterinarian	1.38	1.50	1.13
3	Applying	1.90	0.38	1.71

No	Activity		LoE	
		Husband	Wife	Both
	medication			
4	Caring for sick livestock	0.06	0.24	3.70
5	Applying vitamin	2.57	0.38	1.05
6	Bathing livestock	2.49	0.20	1.31
	Average	1.40	0.47	2.13

Source : primary data analysis (2022)

Men do the majority of applying medications and vitamins, as well as bathing the cattle with respective LoE values of 1.90, 2.57, and 2.49. Meanwhile, women do the majority of contacting the veterinarians (LoE=1.50) for illness in livestock. However, health monitoring is mostly conducted by both spouses (LoE=3.88).

Livestock health monitoring is conducted to reduce the incidence of diseases, that can reduce the efficiency of and even destroy the operation. The more common maladies in beef cattle include tick infestation, bovine ephemeral fever (locally known as *Jumalang*), and others. As previously mentioned in the shelter management, farmers clean their cow pens twice a day [19]. The 2 biosecurity aspects, which are hygiene and sanitation, must be performed for a good livestock operation. There are two types of livestock farmers in terms of health management, which are those who do not employ biosecurity measures at present nor in the future and those who briefly perform biosecurity measures [20].

4.2.5 Values of LoE on the livestock procurement management

The values of LoE on the livestock procurement management are described in detail in Table 5.

Table 5. Values of LoE on the livestock procurement management

No	Activity		LoE	
		Husband	Wife	Both
1	Purchasing	1.33	0.00	2.67
2	Selecting	1.33	0.00	2.67
	Average	1.33	0.00	2.67

Source : primary data analysis (2022)

The activities of purchasing and choosing cattle are in majority carried out by both spouses with identical LoE values of 2.67. Cattle procurement can be funded with self-financing, livestock share, or endowment. Self financing often comes from the joint savings of the spouses. Meanwhile, livestock share may come from other family members, friends, or even other community members. This option is popular to those who do not have sufficient fund to buy cattle on their own [21]. The cattle selection process is mostly based on visual examinations and farmers must choose their calves usually based on the available capital.

4.2.6 Values of LoE on the livestock sales management

The values of LoE on the livestock sales management are described in detail in Table 6.

Table 6. Values of LoE on the livestock sales management

No	Activity		LoE	
		Husband	Wife	Both
1	Selling	0.00	0.04	3.96
2	Determining sale price	0.00	0.00	4.00
	Average	0.00	0.02	3.98

Source : primary data analysis (2022)

The activities in selling (LoE=3.96) and determining sale price (LoE=4.00) involved both spouses the majority of time. These results contradict a report from [22], in which women had limited roles in the commercial side of livestock raising. There were frequent market imbalances caused by this gender gap. However, there seems to have been gender advancements in Nagari Lakitan Utara where women actively engage in this aspect of animal husbandry.

Beef cattle raising in Nagari Lakitan Utara can also be a social safety net, in which farmers only sell their cattle when they need some fund, instead of being based on market forces. There were 8 families among 100 participants who have yet to sell their charges. They consider their livestock as savings and only sell them for emergency.

The farmers have not optimally utilized the manure produced by their livestock. This waste is usually collected around the pens and used when needed in the farms without any special treatment, such as composting. This biowaste has the potential to increase family income and reduce pollution if treated properly [23].

4.2.7 Values of LoE on the monetary management

The values of LoE on the monetary management are described in detail in Table 7.

Table 7. Values of LoE on the monetary management

No	Activity		LoE	
		Husband	Wife	Both
1	Managing funds	0.00	0.61	3.39

Source : primary data analysis (2022)

In general, the money from cattle sale is held by women. However, when it comes to managing the finances, the majory effort and control are performed by both spouses, as indicated by LoE value of 3.39.

4.2.8 Average values of LoE on each management activity

The average values of LoE on each management activity are described in detail in Table 8.

 https://doi.org/10.1051/bioconf/20238002009

Table 8. Average values of LoE on each management activity

No	Management Activity	LoE		
		Husband	Wife	Both
1	Feed and water provision	0.82	0.86	2.32
2	Shelter	1.54	0.54	1.92
3	Reproduction	0.55	0.50	2.95
4	Health	1.40	0.47	2.13
5	Livestock procurement	1.33	0.00	2.67
6	Livestock sales	0.00	0.02	3.98
7	Monetary management	0.00	0.61	3.39

Source : primary data analysis (2022)

The LoE values in Table 8 show that women in general have a large role in beef cattle raising operations. They are closely involved in managements of feed and water provision, shelter, reproduction, health, sales, and finances. The roles of women are needed in the management of family-scale livestock business. Women provide significant resources for the continuity of the enterprise [24]. Family farms need women to help increase cost efficiency, especially in terms of manpower availability. When husbands assume dual roles as the head of the family and breadwinner, they cannot effectively carry out the management of animal husbandry. Women can enter the scene by assisting the management of the families' livestock business, which in turn can enhance the welfare of the families [25]. Increased roles of women, as well as acknowledgment of their rights of access and control, in livestock operations can increase family income and improve family welfare.

5 Conclusion

Gender equality is one of the main goals in sustainable developments. A gender perspective is needed in national development to achieve more justice and equality for both women and men. Women play a role in various fields including on livestock sector.

Women (wives) in beef cattle raising households in Nagari Lakitan Utara, Lengayang subdistrict, Pesisir Selatan regency, Sumatra Barat province, have significant roles in their livestock operations. In general, women's roles can be observed through the husbandry activities starting from managing the provisions to managing the finances resulted from livestock sales. The specific activities that the majorities are conducted by women include herding, and contacting inseminators and veterinarians. Meanwhile, men perform the majority of forage gathering, shelter building and maintenance, medicine applications, vitamin applications, and cattle bathing. Both spouses have the same efforts and controls on forage feeding, concentrate preparation and feeding, water provision, pen cleaning, moving cattle in and out of pens, estrus monitoring, AI supervision, pregnancy monitoring, calving monitoring, post-calving care, health monitoring, caring ill livestock, purchasing and selecting calves, and selling and managing the finances from cattle sales.

Acknowledgements

We express gratitude to Universitas Gadjah Mada that has provided funding through "Rekognisi Tugas Akhir". We also would like to thank the personnel of the following institutions: Dinas Peternakan, Dinas Pertanian Kabupaten Pesisir Selatan Provinsi Sumatera Barat, Dinas Kesehatan Hewan Wilayah Sutera dan Lengayang, and Kenagarian Lakitan Utara.

Funding

This study is funded by Universitas Gadjah Mada through *Rekognisi Tugas Akhir program*.

References

1. World Economic Forum. Global Gender Gap Report 2020. Retrieved from: http://www3.weforum.org/docs/WEF_GGGR_2020.pdf. (2019)
2. H CIM. Gender in Agriculture. International Fund for Agricultural Development. Wahington DC (US). (2009).
3. T Sumarti, A M Fuah. Women, Gender Equality in Livestock Development : Case Study from Papuan and Central Java. Sustainable Animal Production for Better Human Welfare and Environment in International Seminar on Animal Industry; (2015 Sep 17-18); Bogor (ID): Faculty of Animal Science IPB. **396- 399.** (2015).
4. E. Saleh dan Yunilas. Perbandingan Alokasi Waktu Tenaga Kerja Perempuan dan Pria dalam Usaha Penggemukan Sapi di Kecamatan Hamparan Perak Kabupaten Deli Serdang. Jurnal Komunikasi Penelitian, Vol **16 (6).** Jurusan Peternakan, Fakultas Pertanian, Universitas Sumatera Utara. Medan. (2004).
5. Badan Pusat Statistik Provinsi Sumatera Barat. Sumatera Barat Dalam Angka. Padang. (2021).
6. AIP-PRISMA. Gender Mainstreaming Guide for AIP-PRISMA. AIP PRISMA. (2016).
7. R.S. Prayitno. Analisis usaha ternak indukan sapi peranakan Simental di Kecamatan Patean Kabupaten Kendal. Agromedia **36(1): 97-105**. (2018).
8. H.Y Prawira, Muhtarudin, dan R. Sutrisna. Potensi pengembangan peternakan sapi potong di kecamatan tanjung bintang kabupaten lampung selatan. Jurnal ilmiah peternakan terpadu vol. **3(4): 250-255**, November 2015. (2015).
9. N.B. Tarmizi, Dasrul dan G. Riady. Keberhasilan Inseminasi Buatan (IB) pada Sapi Aceh menggunakan Semen Beku Sapi

Bali, Simental, dan Limosin di Kecamatan Mesjid Raya Kabupaten Aceh Besar. JIMVET, **2(3): 318 – 328.** (2018).
10. C.I. Novita, M.A.N. Abdullah, E.M. Sari dan Z. Zulfian. Evaluasi Program Inseminasi Buatan pada Sapi Lokal Betina di Kecamatan Juli, Kabupaten Bireuen, Provinsi Aceh. Jurnal Agripet **19(1): 31 – 39.** (2019).
11. K. Kastalani, H. Herlinae dan A. Kurniawan. Tingkat Keberhasilan Inseminasi Buatan (IB) pada Peternakan Sapi Potong di Kelurahan Kalampangan Kecamatan Sabangau Kota Palangka Raya. Jurnal Ilmu Hewani Tropika **8(2) : 82 – 88.** (2019).
12. U. Santoso dan Kususiyah. Kontribusi dan Status Perempuan dalam Usaha Peternakan Sapi Potong. Jurnal Sain Peternakan Indonesia. Vol. 10 No **1** Januari – Juni 2015. Hal **32 – 43.** (2015).
13. A. Rasyid & Hartati. Perkandangan Sapi Potong. Pusat Penelitian dan Pengembangan Peternakan. Badan Penelitian dan Pengembangan Pertanian. Departemen Pertanian. (2007).
14. F. Sukmawati & M. Kaharudin. Perkandangan Sapi Potong. Pusat Penelitian dan pengembangan Peternakan. Badan Penelitian dan Pengembangan Pertanian. Departemen Pertanian. (2010).
15. S. Suharyati dan M. Hartono. Pengaruh Manajemen Peternak Terhadap Efesiensi Reproduksi Sapi Bali Di Kabupaten Pringsewu Provinsi Lampung. Jurnal Penelitian Pertanian Terapan **16 (1): 61-67**. (2016)
16. MH. Richard. Bovine Reproduction. Iowa (United States): John Wiley & Sons, Inc. (2015).
17. Ismaya. Bioteknologi Inseminasi Buatan pada Sapi dan Kerbau. Gadjah Mada University Press. Yogyakarta. (2014)
18. D. Setiawan. Artificial Insemination of Beef Cattle UPSUS SIWAB Program Based on the Calculation of Non-Return Rate, Service Per Conception and Calving Rate In The North Kayong Regency. Int. J. Trop. Vet. Biomed. Res. 3(1): 7-11. (2018).
19. V.S. Lestari, S.N. Sirajuddin, I.M. Saleh & K.P. Indah. Perilaku Peternak Sapi Potong terhadap Pelaksanaan Biosekuriti. Prosiding Seminar Nasional Teknologi Peternakan dan Veteriner, pp. **263-71.** http://dx.doi.org/10.14334/Pros.Semnas. TPV-2019-p.251-259. (2020).
20. M. L. Brennan, N. Wright, W. Wapenaar, S. Jarratt, P. Hobson-West, I. F. Richens, J. Kaler, H. Buchanan, J. N. Huxley, and H. M. O'Connor. Exploring attitudes and beliefs towards implementing cattle disease prevention and control measures: A qualitative study with dairy farmers in Great Britain. Animals **6(10):61.** https://doi.org/10.3390/ani6100061. (2016).
21. A. Amam, H. B. Setyawan, M. W. Jadmiko, P.A Harsita, S. Rusdiana & M. Luthfi. Pengaruh Sumber Daya Manusia Terhadap Aksesibilitas Sumber Daya Usaha Ternak Sapi Potong Rakyat. Jurnal Ilmu dan Teknologi Peternakan Tropis. **8 (1): 57-65.** https://doi.org/10.33772/ jitro.v8i1.14118. (2021).
22. USAID (United States Agency for International Development). Enhancing Women's Access to Markets : an Overview of Donor Programs and Best Practices http://pdf.usaid.gov/pdf_docs/PNADH668.pdf . (2005).
23. I.N.G. Ustriyana. Analisis Nilai Tambah Dan Pendapatan Usaha Pengolahan Limbah Ternak: Studi Kasus Di Desa Babahan Kecamatan Penebel Kabupaten Tabanan. Dwijenagro Vol. 1 **No. 2** Issn : 1979-3901. (2011).
24. H. Hayati, S. Amanah, A.V Hubeis dan Tjitropanoto. Kemampuan Perempuan Tani Dalam Mendukung Ketahanan Pangan Rumah Tangga. Jurnal Sosiohumaniora Volume 18, No **03**, hlm. 229 – 235. (2015).
25. W. Bayer and B. Letty. The role of livestock in Developing Communities :Enhancing Multifuctionallity. Bloemfontein (ZA) : University of the Free State. (2010).

Amelioration of dry land suboptimal using biochar and compost to improve soil physical properties and soybean yield

Endriani [1]* and *Diah* Listyarini [1]

[1] Agroecotechnology Department, Faculty of Agriculture, Jambi University

Abstract. The primary constraint of suboptimal soils under wet tropical regions, besides their low pH, is their low soil organic matter (SOM) content. This research aimed to improve the soil properties of dry land suboptimal and the yield of soybeans by applying biochar and compost. This research was carried out to investigate the effects of coconut shell biochar (B) and Leucaena compost (C) applied alone (at a rate of 10 ton.ha^{-1}) or in combination (5 B: 10 C; 10 B: 15 C each. thus 10 [1] B: 10 C) on soil physicochemical properties, growth, and yield of Soybean on Ultisols in Jambi Province. Biochar and compost applied alone or in combination significantly increased soil organic matter (SOM), total porosity (TP), and Aeration Pore (AP); decreased bulk density (BD) and low drainage pore (LDP). Also, combined application and single application biochar or compost to impact beneficially on soil water retention and available water (AW). Additionally, combined application and single application biochar or compost additions increased components of yield as a number of pods of soybean. The study showed that biochar applied alone or in combination with compost offers the potential to enhance soil quality and improve soybean yield.

1. Introduction

The dryland areas with less-than-ideal conditions in Jambi are primarily characterized by Ultisols, which originate from acidic sedimentary rock parent materials. As a result, the soil has an acidic pH, and the base saturation levels are below 35% [1]. Ultisols exhibit poor aeration and soil stability indexes, making the soil prone to compaction and erosion susceptibility. Consequently, the permeability of roots into the soil is hindered, leading to inhibited plant root growth [2]. This mineral soil has been used intensively for agriculture. However. the big problem found in the use of Ultisols for soybean farming is the high clay content in subsoil, low stability aggregate, and relatively low organic matter. High clay content and low soil organic matter cause soil micropores to dominate and cause an imbalance between micro and macro pores [3]. This condition is unfavourable for soybean growth which requires porous soil with sufficient available air and water in the soil.

An approach to enhance the fertility and porosity of clayey soils like Ultisols in Tangkit Jambi involves the incorporation of biochar and compost. Compost has a long history of being utilized by farmers to enhance both the physical and chemical characteristics of soil, making it an effective method for increasing soil fertility.

Right now, using biochar and compost together is seen as a highly effective and sustainable agricultural approach to reduce environmental risks and boost soil fertility and crop yields [4,5]. Recent research suggests that biochar is most beneficial in enhancing soil qualities when it is used alongside or processed together with other soil amendments like biosolids, paper mill sludge, manure, and compost [6-9].

Biochar is a carbon-rich, solid, porous substance created through pyrolysis in an oxygen-deprived setting, known for its extensive surface area and cation exchange capacity (CEC) [10,11]. Many studies have shown that using biochar in soil can have positive impacts on crops and soil properties, such as improved increased hydraulic conductivity, reduced bulk density, infiltration rate, porosity, water retention, and soil structure [12-14], which may lead to an increase in crop yield [15-19]. Mensah and Frimpong [20] found that using biochar could enhance soil health by raising pH, increasing water retention capacity, cation exchange capacity (CEC), and influencing microbial communities.

Incorporating compost can improve soil structure by reducing compaction and increasing water content [21]. Plants that have a high biomass that can be used as compost is leucaena Plants that have a high biomass that can be used as compost is leucaena. According to Agbede [22] Leucaena leaves have a chemical composition are 39% OC; 3.23 % N; 12.1 C: N; 0.23 % P; 1.68 % K; 1.24 % Ca; dan 0.35 % Mg.

Adding organic manures can enhance root growth by altering soil physical properties [23]. Biochar, while carbon-rich, lacks sufficient nutrients for demanding crops. Combining biochar with chemical fertilizers and compost enhances soil water retention [24], aggregate stability [25], more efficient use of nutrients [26,27], enhance in microbial biomass and activities which lead plant growth [11,28,29], over the sole application [30,31]. In addition, the application of biochar with

*Corresponding author: endriani@unja.ac.id

compost or lime to its impact on soil parameters and plant growth decreases the leaching of nutrients and improves the efficiency of compost [26,27].

Structure stability, bulk density, hydraulic conductivity, and water retention are common indicators for studying soil physical health and hydraulic properties [32,33]. Limited research has examined the impact of compost and biochar on soil hydraulic properties [34].

In Jambi Province's Ultisols dry land, poor soil properties hinder plant growth. This study aimed to improve soybean yield and soil properties using biochar and compost. The hypothesis was that combining biochar and compost would enhance soil physical properties better than using them separately in sub-tropical conditions.

2. Material and Methods

2.1 Biochar and Soil Preparation

Coconut shell biochar was produced through low-temperature pyrolysis at (300–400 °C) using drum oil which on the bottom has been fitted with cavities. Coconut shells were burned separately for about 6-8 hours. According to some Authors, pyrolysis can be used in producing coconut shell biochar widely [35-43].

For composting, firstly Leucaena biomass was chopped and mixed with poultry manure with a balance of 3:1 and incubated with *Trichoderma sp* for one month. Chemical properties of coconut shell biochar and Leucaena compost were analyzed for pH (pH meter). Organic-C (Walkley and Black).

2.2. Experimental Design

The field experiment has been conducted at Tangkit Village. Sungai Gelam sub district. Muara Jambi Regency. Jambi Province in April – August 2023. The experiment used a randomized block design with four replicates to test various types of amendments, whereas treatments tested were some kind of amendment. that were A0: control; A1: Coconut shell biochar 10 kg ha^{-1} ; A2: Leucaena compost 10 kg ha^{-1}; A3: Coconut shell biochar 10 ton ha^{-1} + Leucaena compost 5 ton ha^{-1}; A4: Coconut shell biochar 5 ton ha^{-1} + Leucaena compost 10 ton ha^{-1}; A5: Coconut shell biochar 10 ton ha^{-1} +Leucaena compost 10 ton ha^{-1}. After one week of soil tillage, Soybean Variety Anjasmoro was planted with a spacing of 25 × 30 cm in plots measuring 2 × 3 m. Inorganic fertilizers (urea, SP-36, and KCl) were applied at planting time at rates of 50, 100, and 100 kg ha-^{1}, respectively. Plant parameter measurements, including pod count per plant, were taken at harvest after 93 days.

2.3. Soil Sampling Analyses.

Soil samples were collected before and after treatments, with undisturbed samples taken using a 5 cm diameter, 4 cm high ring at a 20 cm depth, and composite samples collected from six points at the same depth using a one-inch diameter soil auger. Soil analysis included bulk density (BD), water content at pF 1, 2, 2.54, and 4.2, as well as analysis of soil chemical properties such as organic carbon (C-organic). Yield components were also analyzed.

2.4. Data Analysis

Data were statistically analyzed using ANOVA or diversity tests at a 95% significance level. To identify significant differences among treatment variables, the Duncan Multiple Range Test (DMRT) was conducted at a 5% significance level.

3. Result and Discussion

3.1. Initial characteristics of soil.

According to Table 1, the soil had a sandy clay loam texture with low soil organic matter (SOM) content, poor hydraulic conductivity and infiltration rate, limited available water content, and high bulk and particle density. These characteristics rendered the soil unsuitable for soybean growth, as it lacked the necessary porosity and crumb structure essential for soybean root development. However, through soil amendment with organic matter, it has the potential to transform into a suitable medium for soybean cultivation.

According to Table 1, the soil had a sandy clay loam texture with low organic matter content, high bulk density, low total porosity, low permeability, low aggregate stability, and poor soil aggregation. Due to these characteristics, the soil was unsuitable for soybean growth as it requires a more porous and crumbly texture to support root development.

The soil had low pH, which made essential nutrients, particularly phosphorus, unavailable due to binding with aluminum and iron. To enhance its physical and chemical properties, soil amendment with organic matter is needed, potentially making it a suitable medium for soybean cultivation.

Table 1. Physicochemical attributes of initial soil

Properties	Initial Soil	Criteria
C-organic (%)	1.12	Low *
pH (H2O)	4.78	Acid*
Organic Matter (%)	1.93	Low*
Bulk Density (g cm^{-3})	1.57	High*
Total Porosity (%)	41.88	Low*
Particle Density (g cm^{-3})	2.84	High*
Texture: Sand %)	48.40	Sandy clay loam
Silt (%)	18.40	
Clay (%)	33.20	
Water content (%)	23.12	Low*
Field Capacity (%)	28.36	
Available Water (%)	8.88	Low**
Wilting Point WC (%)	19.48	
Permeability (cm jam^{-1})	6.03	Slow**
Infiltrasion Rate (cm jam^{-1})	14.2	Medium**
% Aggregate (%)	43.85	Low*
Stability Agregat (%)	52.41	Low*

* *Pusat Penelitian Tanah Bogor (1994);* ** *Sarief (1989).*

3.2. Characteristics of coconut shell biochar and Leucaena compost

The characteristics of coconut shell biochar and Leucaena compost are shown in Table 2. Both biochar and compost have pH>7 and other chemical properties are quite different. Some of the previous studies are also showing the pH of coconut shell biochar > 7 [35, 44].

Biochar as an ameliorant had very low mass and very high pores. so, it reduced soil BD and improved porosity. The biochar had 33.02% C and 4.68% ash. This amendment was very suitable to make soil porous and to provide a good medium for soybean growth.

This amendment was highly effective in improving soil porosity and creating an optimal environment for soybean root development.

Table 2. Characteristics of coconut shell biochar and Leucaena compost

Parameters of Analyzed	Coconut shell Biochar	Leucaena Compost
Water content (%)	9.40	34.28
Organic Carbon	33.02	26.97
pH (H_2O)	8.90	7.22
Ash content (%)	4.68	-
Volatile matter (%)	66.98	-
N total (%)	-	2.19
C/N	-	14.34

Coconut shell biochar

The coconut shell biochar exhibited a very high CEC (Cation Exchange Capacity) and a high specific surface area [45], indicating its strong reactivity in binding soil particles into aggregates. On the other hand, the Leucaena compost had a notably high organic carbon (OC) content, measuring at 26.97%, which was 3.3 times greater than that of the soil used. This suggests that using compost as an organic matter source can significantly increase soil organic matter, thereby improving both the physical and chemical properties of the soil.

3.3. Soil Physical Properties after application biochar and compost

Effect of biochar and compost application on SOM

The effect of biochar and compost on SOM was significant. However. addition of biochar and compost increased the SOM over the control (Figure 1). SOM was highest in A5 (5.83%), followed by A4 dan A2 (5.27% and 5.10 %) compared with control. The highest SOM content was found under treatment at 10:10. It was significantly different to treatment A0, A1, A2, A3, and A4.

A5 (BC 10: 10) has the highest percentage of soil organic matter, which could be attributed to the incorporation of biochar and compost. both of which are organic amendments high in organic matter. Because of its high stability towards degradation in the soil. the biochar application will make a lifelong increase in soil organic matter that will give benefit to all the main functions exhibited by the organic matter that is present in the soil [46].

Cox et al. [47], found that applying biochar to the soil surface increased soil organic matter (SOM) not only in the 0–10 cm depth but also in the 10–20 cm depth. They concluded that using biochar and compost boosted SOM in the 10–20 cm soil layer due to the movement and leaching of these amendments. Biochar, rich in carbon content, plays a role in humus production and enhances soil fertility [48,49].

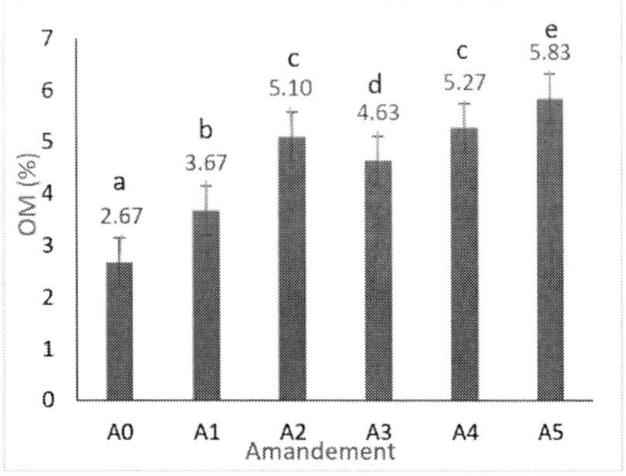

Fig 1. Effect of biochar and compost on SOM

The effect of biochar and compost application on BD, TP, AP, and LDP

Based on Table 3, the soil had a sandy clay loam texture, application of biochar and compost decreased the bulk density (BD) of the soil. This condition can occur because biochar is porous and compost-improved soil structure has a low BD from 0.08 to 0.17 g cm^{-3}. thus, contributable to the decrease of soil BD. Many studies reported that the application of biochar and organic fertilizer increased soil organic matter and decreased BD [46,47,50], found that using biochar, with or without compost, reduced bulk density and increased total porosity in the soil.

Table 3. Effect of biochar and compost on some soil physical properties

Amd	BD (g cm^{-3})		TP (%vol)		AP (%vol)		LDP (%vol)	
A0	1.28	a	47.27	b	16.23	a	11.53	a
A1	1.22	b	51.37	b	14.80	ab	9.47	b
A2	1.17	c	55.91	ab	12.31	b	5.27	d
A3	1.17	c	53.93	b	16.66	a	6.97	c
A4	1.16	c	58.28	ab	16.58	a	6.53	cd
A5	1.09	d	65.80	a	14.20	ab	5.70	cd

Notes: Amd=amandement; A0 = control; A1 = biochar 10; A2 = compost 10; A3 = biochar + compost 10:5; A4 = biochar + compost 5:10; A5 = biochar + compost 10:10

The results indicated that all treatments significantly boosted average total porosity from 11.43% to 39.20%, demonstrating increased water-holding capacity compared to the control. Biochar enhances porosity and pore spaces, leading to improved water retention and sandy soil stability. It also enhances soil structure, including bulk density (BD), porosity, water-holding capacity (WHC), available water for crops, and specific surface area [52].

The result also showed application of biochar and compost alone or combined significantly increased the average aeration pore (AP) and low drainage pore (LDP). Fast drainage and aeration pores (AP) were higher on average in biochar combined with compost compared to other treatments, although this trend wasn't consistent across all treatments.

The results showed that the treatments of sandy clay loam soil with compost 10-ton ha^{-1}. significantly decreased the average aeration pore (AP) from 31.84%, but non significantly with combined biochar and compost (10B:10C).

Aeration pores in all treatments remained relatively high (>10% by volume), falling within the moderate range (10 to 15% by volume). Conversely, the average low drainage pore content in all treatments remained very low.

The results indicated that the treatments of sandy clay loam soil with biochar and compost significantly reduced the average soil LDP from 21.75–111.79% with the following order A1 (9.47) < A3 (6.97) < A4(6.53) < A5 (5.70) <A2 (5.27). Decreasing the percentage of slow drainage pores will increase the arase pores. indicating more porous soil and reduced soil compaction.

Effect of biochar and compost on soil retention

Based on Figure 2, combined application and single application biochar or compost additions increased water retention. The retention curve changes dynamically following the soil pore balance. Biochar consistently impacts water retention more predictably than other soil hydraulic properties.

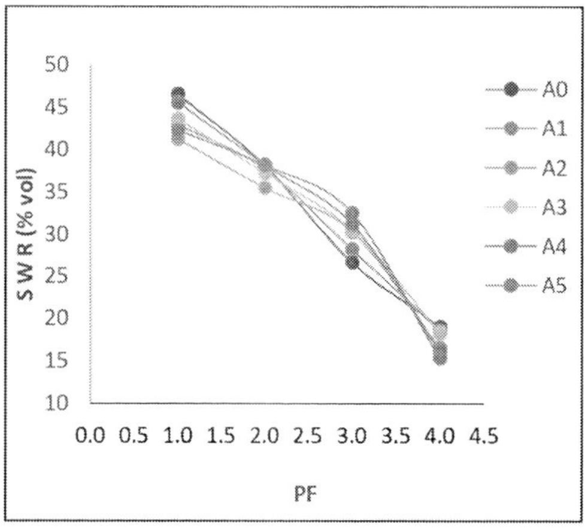

Fig 2. Effect of biochar and compost on Soil water retention

Blanco-Canqui [53] found that recent studies suggest biochar enhances water retention in 90% of cases, with 17 out of 19 soils showing improved retention. Biochar's water-absorbing capacity is highly effective. Notably, some studies reported increased water retention at low biochar application rates, while others observed improvements only at 15 tons per hectare [54,55]. Applying compost enhanced soil water retention properties [52,51].

Seyedsadr et al. [51] found that using biochar, with or without compost, can improve water retention in low-organic soils while maintaining or even enhancing nutrient retention.

Effect of biochar and compost on available water

Based on, Figure 3. shows that the treatments of sandy clay loam soil with biochar and compost alone or combined significantly increased available water (AW). The results showed that the treatments of sandy clay loam soil with combined biochar and compost. and compos alone, or biochar alone significantly increased the average soil AW from 32.91.75–137.92 % with the following order A1 (9.57) < A3 (12.23) < A2(13.87) < A4 (15.07) <A5 (17.13). The greater combination of biochar and compost (10:10) contributed the most to available water.

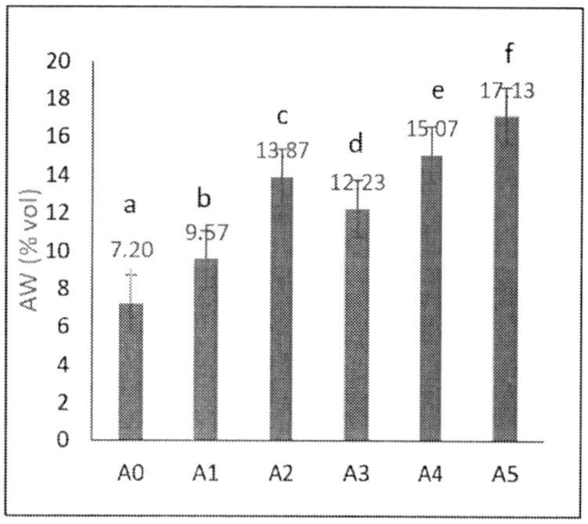

Fig 3. Effect of biochar and compost on available water

Biochar's enhanced water availability is likely due to its increased specific surface area and porosity, which can reach up to 3000 m^2 g^{-1}.

Biochar's porous nature retains water within its particles due to its high specific surface area. Micropores hold water more effectively than macropores and mesopores, thanks to capillary and adhesive forces. Adding biochar to soil can change total porosity, pore sizes, water transmission, and retention properties [53].

Ghorbani et al. [56] found that biochar and compost significantly increased available water capacity (AWC) in all conditions over two years. Biochar improved soil aggregation and porosity, enhancing water retention, particularly in low-water and polluted environments.

Panagea et al. [57] noted that the direct impact of soil organic carbon (SOC) on soil water retention (SWR) is minor. The indirect impact is more significant at higher matric potentials and varies with SOC composition. Different responses under various conditions suggest that combining practices, such as reducing soil disturbances and increasing organic matter inputs, is needed to enhance soil water-holding capacity.

Effect of biochar and compost on component yield

Based on, Figure 4. shows that the treatments of sandy clay loam soil with biochar and compost alone or combined significantly increased the number of pods. The results showed that the treatments of sandy clay loam soil with combined biochar and compost. and compos alone. or biochar alone significantly increased the average number of pods from 21.42–57.14% with the following order A2 (149) < A1 (153) < A3(171) < A4(176) < A5(197). The greater combination of biochar and compost (10:10) contributed the most to the number of pods.

Biochar and compost, alone or together, enhance hydro-physical properties of sandy clay loam soil. Improvement of soil properties (Table 3) also increased the effectiveness of soybean growth, as presented in Figure 4.

Currently, combining biochar and compost is a highly effective and sustainable agricultural practice for reducing environmental risks and enhancing soil fertility and crop yields (Izilan et al., 2022; Sharma et al., 2021; Ariani et al..2021; Sekaran et al., 2020). According to Nurida and Rahman [50], Soybeans gave better yield when the soil was treated with Biochar SP50 (50% biochar and 50% organic fertilizer). Improved soybean yields on acidic soil likely resulted from reduced soil acidity and increased availability of water pores.

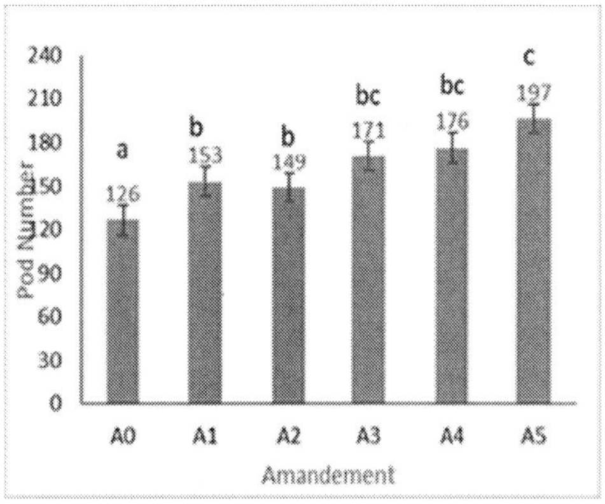

Fig 4. Effect of biochar and compost on the number of pods

4. Conclusion

The study showed that biochar applied alone or in combination with compost in combination significantly increased soil organic matter (SOM), total porosity (TP), and Aeration Pore (AP); decreased bulk density (BD) and low drainage pore (LDP). Both combined and individual applications of biochar or compost positively impacted soil water retention, available water (AW), and soybean yield components, such as pod numbers.

Acknowledgments This study was partially funded by Competitive Research Indonesia. The authors thank the Directorate General of Higher Education of Indonesia for superior university-applied research in 2023. Also, thanks to the Rector of Universitas Jambi who has supported this research.

Reference

1. Prasetyo and Suriadikarta. *Characteristic, potential, and management of Ultisols for agricultural upland in Indonesia.* J Litbang Pertanian **25**: 39-46. (2006).
2. Subowo, G. *Empowerment of Soil Biological Resources for Rehabilitation of Degraded Ultisols.* Jurnal Sumberdaya Lahan. **6** (2): 79-88 (2012).
3. Endriani, Sunarti dan Ajidirman. 2013. *Pemanfaatan Biochar Cangkang Kelapa Sawit Sebagai Soil Amandement Ultisol Sungai Bahar Jambi*. Jurnal Penelitian Universitas Jambi Seri Sains. **15** (1): 39-46 (2013).
4. N I S Izilan et al. *The effects of biochar-compost on soil properties and plant growth performance grown in sandy loam soil.* IOP Conf. Ser.: Earth Environ. Sci. **1059** 012021 (2022).
5. P Sharma, Vikas A, Vikas S, Shubham C, Ch. Srinivasa R, A.Q. Ganie, Daniel Ingo Hefft, Mohamed A. El-Sheikh, Sheikh M. *Effectiveness of biochar and compost on improving soil hydro-physical properties, crop yield and monetary returns in inceptisol subtropics*. Saudi Journal of Biological Sciences. Vol **28**, (12): December 2021, 7539-7549 (2021).
6. Manirakiza. E., Ziadi. N., Luce. M.S.. Hamel. C.. Antoun. H.. and Karam. A. 2022. *Changes in soil pH and nutrient extractability after co-applying biochar and paper mill biosolids.* Can. J. Soil Sci.. **102**(1). (2022). doi:10.1139/cjss-2019-0138.
7. Ziadi, N., Zhang, X., Gagnon, B., and Manirakiza, E. *Soil phosphorus fractionation after co-applying biochar and paper mill biosolids.* Can. J. Soil Sci. **102**(1). doi:10.1139/cjss2020-0098.(2022).
8. Hangs. R.D., Schoenau. J.J... and Knight. J.D . *Impact of manure and biochar additions on annual crop growth. nutrient uptake. and the fate of 15 N-labelled fertilizers in two contrasting temperate prairie soils after four years.* Can. J. Soil Sci. **102**(1).
9. T T Doan. Thierry Henry-des-Tureaux. Cornelia Rumpel, Jean-Louis Juneau. Pascal Jouquet. 2015. *Impact of compost. vermicompost and biochar on soil fertility. Maize yield and soil erosion in Northern Vietnam: A three-year mesocosm experiment.* Science of the Total Environment **514** (2015) 147–154 (2015).
10. Hussain. M.. Farooq. M., Nawaz. A., Al-Sadi. A.M., Solaiman. Z.M., Alghamdi. S.S., Yong. U.A., Siddique. K.H.M. *Biochar for crop*

production: potential benefits and risks. J. Soils Sediments **17** (3). 685–716 (2017).
11. Mansoor. S., Kour. N., Manhas. S.. Zahid. S., Wani. O.A., Sharma. V., Wijaya. L., Alyemeni. M.N., Alsahli. A.A. El-Serehy. H.A. Paray. B.A. *Biochar as a tool for effective management of drought and heavy metal toxicity*. Chemosphere. 129458. (2020).
12. Laird. D.A., Fleming. P., Davis. D.D., Horton. R., Vang. B., Karlen. D.L., 2010. *Impact of biochar amendments on the quality of a typical midwestern agricultural soil*. Geoderma **158** (3&4). 443–449 (2010).
13. Herath. H.M.S.K., Camps Arbestain. M., Hedley. M., 2013. *Effect of biochar on soil physical properties in two contrasting soils: an Alfisol and an Andisol*. Geoderma **209–210**. 188–197 (2013).
14. Abrol. V., Ben-Hur. M., Verheijen. F.G., Keizer. J.J., Martins. M.A., Tenaw. H., Tchehansky. L., Graber. E.R., 2016. *Biochar effects on soil water infiltration and erosion under seal formation conditions: rainfall simulation experiment*. J. Soils Sediments **16** (12). 2709–2719 (2016).
15. Basso. A.S., Ferrando. E.M., Laird. D.A., Horton. R., Westgate. M. *Assessing the potential of biochar for increasing the water-holding capacity of sandy soils*. Bioenergy **5** (2). 132–143 (2012).
16. Carlsson. M.. Andren. O.. Stenstrom. J.. Kirchmann. H.. Katterer. T.. 2012. *Charcoal application to arable soil: effects on CO2 emissions Commun*. Soil Sci. Plant Anal. **43**. 2262–2273 (2012).
17. H.M.S.K. Herath, Marta Camps-Arbestain, Mike H. *Effect of biochar on soil physical properties in two contrasting soils: An Alfisol and an Andisol*. Geoderma **209-210** (2013) 188–197.
18. Mukherjee, A., & Lal, R. *The biochar dilemma*. Soil research, **52**(3), 217-230 (2014).
19. Martinsen. V., Alling. V., Nurida. N.L., Mulder. J., Hale. S.E., Ritz. C., Rutherford. D.W., Heikens. A., Breedveld. G.D. Cornelissen. G. *pH effects of the addition of three biochars to acidic Indonesian mineral soils*. Soil Sci. Plant Nutrition. **61** (5). 821–834 (2015).
20. Hansen. V., Nielsen. H.H., Petersen. C.T., Mikkelsen. T.N., Stöver. D.M., *Effects of gasification biochar on plant-available water capacity and plant growth in two contrasting soil types*. Soil Tillage Res. **161**. 1–9 (2016).
21. Blanco-Canqui. H., Lal. R. *Crop residue removal impacts on soil productivity and environmental quality*. Crit. Rev. Plant Sci. **28**. 139–163 (2017).
22. Mensah. A.K., Frimpong. K.A. *Biochar and/or compost applications improve soil properties. growth. and yield of maize grown in acidic rainforest and coastal savannah soils in Ghana*. Int. J. Agron Vol **2018** (2018).
23. Evanylo. G.. Sherony. C.. Spargo. J.. Starner. D.. Brosius. M.. Hearing. K.. 2008. *Soil and water environmental effects of fertilizer. manure. and compost-based fertility practices in an organic vegetable cropping system*. Agric. Eco. Environ. **127**. 50–58. (2008).
24. Agbede. T. M. *Effect of Green Manure Application on Cassava (Manihot esculenta Crantz) Growth. Yield Quantity and Quality in Degraded Alfisols*. Pertanika J. Trop. Agric. Sc. **41** (4): 1757-1777 (2018).
25. Zhang P, Wei T, Jia Z, Han Q, Ren X, et al. (2014) *Effects of Straw Incorporation on Soil Organic Matter and Soil Water-Stable Aggregates Content in Semiarid Regions of Northwest China*. PLoS ONE **9**(3): e92839 (2014).
26. Abel. S., Peters. A., Trinks. S., Schonsky. H.. Facklam. M., Wessolek. G. *Impact of biochar and hydro-char addition on water retention and water repellency of sandy soil. Geoderma* 202–203. 83–191 (2013).
27. Sun. F., Lu. S.. 2014. Biochar improves aggregate stability. water retention. and porespace properties of clayey soil. J. Pl. Nutri. Soil Sci. **177** (1). 26–33 (2014).
28. Major. J. *A guide to conducting biochar trials*. Version 1.1. Int. Biochar Initiative. (2009)
29. Novak. J.M., Busscher. W.J., Laird. D.L., Ahmedna. M., Watts. D.W., Niandou. M.A.S. 2009. *Impact of biochar amendment on fertility of a southeastern coastal plain soil*. Soil Sci. **174** (2). 105–112 (2009)
30. Adekiya. A.O., Agbede. T.M., Aboyeji. C.M., Dunsin. O., Simevon. V.T.. 2018. Effects of biochar and poultry manure on soil characteristics and the yield of radish. Sci. Horti. **243** (3). 457–463.(2018)
31. Sadegh-Zadeh. F.. Tolekolai. S.F. Bahmanyar. M.A.. Emadi. M.. 2018. *Application of biochar and compost for enhancement of rice (OryzaSativa L.) grain yield in calcareous sandy soi*l. Commun. Soil Sci. Plant Anal. **49** (5). 552–566. (2018)
32. Ippolito. J.A., Stromberger. M.E., Lentz. R.D., Dungan. R.S., 2016. *Hardwood biochar and manure co-application to a calcareous soi*l. Chemosphere **142**: 84–91 (2016).
33. Hagemann. N., Harter. J., Kaldamukova. R.. Guzman-Bustamante. I., Ruser. R., Graeff. S. 2017. *Does soil aging affect the N2O mitigation potential of biochar? A combined microcosm and field study*. GCB Bioenergy. (2017).
34. Reynolds. W.D., Drury. C.F., Tan. C.S., Fox. C.A., Yang. X.M. 2009. *Use of indicators and pore volume-function characteristics to quantify soil physical quality*. Geoderma **152** : 252–263.
35. Castellini. M., Stellacci. A.M., Tomaiuolo. M. 2019. *Spatial variability of soil physical and hydraulic properties in a durum wheat field*: an assessment by BESTproducer. Water **11**: 1434–1437 (2019).
36. Kammann. C.,Glaser. B., Schmidt. H.P., 2016. *Combining biochar and organic amendments*. In: Shackley. S. Ruysschaert. G., Zwart. K., Glaser. B. (Eds.). Biochar in European Soils and Agriculture: Science and Practice. Routledge. New York. pp. 136–164 (2016).
37. A Ajien, Juferi I, Nurzawani Md S, Rafidah H and Hazman S. 2023. *Coconut shell and husk biochar:*

A review of production and activation technology. economic. financial aspect and application. Waste Management & Research **41**(1). 37-51 (2023).
38. Adorna J. Borines M. Dang VD. et al. *Coconut shell-derived activated biochar – manganese dioxide nanocomposites for high-performance capacitive deionization.* Desalination **492**: 114602 (2020)
39. Baharum NA. Nasir HM. Ishak MY. et al. (2020) *Highly efficient removal of diazinon pesticide from aqueous solutions by using coconut shell modified biochar.* Arabian Journal of Chemistry **13**: 6106–6121 (2020).
40. Behera B. Selvam MS. Dey B. et al. (2020) *Algal biodiesel production with engineered biochar as a heterogeneous solid acid catalyst.* Bioresource Technology **310**: 123392 (2020).
41. Khuenkaeo N and Tippayawong N. (2020) *Production and characterization of bio-oil and biochar from ablative pyrolysis of lignocellulosic biomass residues.* Chemical Engineering Communications **207**: 153–160 (2020).
42. Nuryana D. Alima MFR. Yahayu M. et al. (2020) *Methylene blue removal using coconut shell biochar synthesized through microwave-assisted pyrolysis.* Jurnal Teknologi **82**: 31–41 (2020).
43. Pituya P. Sriburi T and Wijitkosum S. (2016) *Properties of biochar prepared from acacia wood and coconut shell for soil amendment.* Engineering Journal **21**: 63–76 (2016).
44. Sari RM. Gea S. Wirjosentono B. et al. (2020). *Improving quality and yield production of coconut shell charcoal through a modified pyrolysis reactor with tar scrubber to reduce smoke pollution.* Journal of Environmental Study **29**: 1815–1824 (2020).
45. Windeatt JH. Ross AB. Williams PT. et al. (2014) *Characteristics of biochars from crop residues: Potential for carbon sequestration and soil amendment.* Journal of Environmental Management **146**: 189–197 (2014).
46. S. Abrishamkesh, M. Gorji, H. Asadi, G.H. Bagheri-Marandi, A.A. Pourbabaee. *Effects of rice husk biochar application on the properties of alkaline soil and lentil growth.* Plant, Soil and Environment · Vol. 61, 2015, No. **11**: 475–482 (2015).
47. Severo FF. da Silva LS. Moscôso JSC. Sarfaraz Q. Júnior LFR. Lopes AF. Marzari LB. and Molin GD. 2020. *Chemical and physical characterization of rice husk biochar and ashes and their iron adsorption capacity.* SN Appl. Sci. **2**. 1286 (2020).
48. Sánchez-Monedero. M. A.. Cayuela. M. L., Sánchez-García. M, Vandecasteele. B, D'Hose. T, López. G, Mondini. C. *Agronomic evaluation of biochar. compost and biochar-blended compost across different cropping systems: Perspective from the European project FERTIPLUS.* Agronomy. **9**(5). 225. (2019).
49. J. Cox. N. V. Hue. A. Ahmad. and K. D. Kobayashi. *Surface-applied or incorporated biochar and compost combination improves soil fertility. Chinese cabbage and papaya biomass.* Biochar. **3**(2): 213–227 (2021).
50. S. Yang. X. Chen. Z. Jiang. J. Ding. X. Sun. and J. Xu. *Effects of biochar application on soil organic carbon composition and enzyme activity in paddy soil under water-saving irrigation.* International Journal of Environmental Research and Public Health. vol. **17** (1): 333 (2020).
51. M. Ayaz. D. Feizien˙e. V. Tilvikien˙e. K. Akhtar. U. Stulpinait˙e and R. Iqbal. *Biochar's role in the sustainability of agriculture and environment.* Sustainability. vol. **13** (3): 1330 (2021).
52. Nurida, N.L., & Rachman, A. (2020). *Amelioration of acid upland to increase soil productivity and soybean yield.* AGRIVITA Journal of Agricultural Science, **42**(2), 350–359.
53. Seyedsadr S, Václav Š. Lukáš J, Michal S, Luke B, Michael P, Martin K, Lukáš T. 2022. *Biochar considerably increases the easily available water and nutrient content in low-organic soils amended with compost and manure.* J Chemosphere. Volume **293**. April 2022. 133586. (2022).
54. Głąb Andrzej Ż. Urszula S. Krzysztof G. Michał. M Mierzwa-Hersztek. Sylwester T. J Stanek-Tarkowska. *Fertilization effects of compost produced from maize. sewage sludge and biochar on soil water retention and chemical properties.* Soil and Tillage Research. Vol **197**. March 2020. 104493.
55. Blanco-Canqui H. *Biochar and Soil Physical Properties.* Soil Sci. Soc. Am. J. **81**:687–711 (2017).
56. Paneque. M., J.M. Rosa. J.D. Franco-Navarro. J.M. Colmenero-Flores. and H. Knicker. 2016. *Effect of biochar amendment on morphology. productivity and water relations of sunflower plants under non-irrigation conditions.* Catena **147** : 280–287 (2016).
57. Xiao. Q.. L. Zhu. H. Zhang. X. Li. Y. Shen. and S. Li. 2016. *Soil amendment with biochar increases maize yields in a semi-arid region by improving soil quality and root growth.* Crop Pasture Sci. **67**: 495–507 (2016).
58. M Ghorbani, Reinhard W. Neugschwandtner, Petr K, Hossein A, Marek K., Elnaz A. (2023) *Comparative effects of biochar and compost applications on water holding capacity and crop yield of rice under evaporation stress*: a two-year field study. Paddy and Water Environment. 23 January 2023. (2023).
59. Panagea. I.S., Berti. A., Cermak. P., Diels. J., Elsen. A., Kusá.ˇ H., Piccoli. I., Poesen. J., Stoate. C., Tits. M. et al. 2021. *Soil Water Retention as Affected by Management Induced Changes of Soil Organic Carbon*: Analysis of Long-Term Experiments in Europe. Land 2021. **10**. 1362. (2021)

Biodiversity in Lower Montane Forest of Petungkriyono : A Review

Dian Pratiwi[1*], *Tri Wira Yuwati*[1,2], *Sri Lestari*[1], *Yonky Indrajaya*[1], *Astarina Eka Dewi*[3]

[1]National Research and Innovation Agency, Republic of Indonesia, Jl. Jakarta-Bogor Km 46, Cibinong, Bogor, West Java, 16911
[2]Faculty of Forestry, Gadjah Mada University
[3]Relung Foundation
*Email: dian.pratiwi@brin.go.id

Abstract. One of the symbols of biodiversity in Indonesia's lower montane forest is Petungkriyono. Due to its location adjacent to populated areas, it has an impact on the survival of the local flora and animals in the Petungkriyono forest area. This study aims to provide a preliminary description of the plant and animal species found in the Petungkriyono forest area. A synoptic review method to Petungkriyono biodiversity publications was used in the writing of the article. The study's findings revealed that the Petungkriyono forest region consisted of 168 species of wildlife and 204 species of plants. Certain species are endemic, while others fall into the IUCN conservation status categories of Vulnerable (VU), Endangered (EN), and Critically Endangered (CR). It is also a certain that a number of species will be added to the CITES Appendix list in 2023 (Appendices I and II). The demands of population expansion and poaching are the current issues, burdening the forest areas that are already there. High biodiversity, stunning scenery, and the presence of indigenous knowledge can all present opportunities for ecotourism. Immediate implementation of conservative measures, such as maximizing current opportunities to preserve biodiversity, is essential.

Keywords: flora, fauna, endemic, IUCN, Appendix CITES, challenges, opportunities

1 Introduction

Indonesia has the most abundant species richness in the world. But unfortunately, currently the rate of species extinction is increasing rapidly, including in Indonesia (A. Setiawan, 2022), which is known as "biological failure". A growing number of human activities, including deforestation, changing land uses, and climate change, are endangering our biodiversity. This affects both our natural ecosystems and our capacity to provide for our fundamental necessities, including food, clean water, and medication (Desiana et al., 2019; Rasyid, 2014).

In this setting, it is crucial to carry out biodiversity research to enhance our comprehension of the susceptibility and flexibility of our ecosystems to contemporary pressures. It offers the knowledge required to create effective and long-lasting conservation plans.

According to (Damayanti et al., 2018), Petungkriyono Forest is one of the last remaining natural tropical forests in Java, home to a diverse range of flora and wildlife. Sixty percent of the Petungkriyono District's land is covered by secondary dryland forest (Abdurrahman et al., 2021; Relung Indonesia, 2023). The natural state of Petungkriyono features a comparatively high humidity and rainfall. The objectives of this study are to ascertain the biodiversity, conservation status, and interactions between the forest and the local community in the Petungkriyono forest environment. The goal of this research is to improve our knowledge of biodiversity and the difficulties it encounters, as well as to serve as a guide for stakeholders in managing the Petungkriyono ecosystem sustainably.

2 Methods

This research was conducted in the Petungkriyono forest area, Pekalongan Regency, Central Java Province. The research was conducted in desk study/desk review to achieve research objectives using thematic analysis based on the literature review method (Randolph, 2009; Cooper, 1998). The stages carried out in the desk study are: framing review question, searching relevant literature, assessing quality of literature,

https://doi.org/10.1051/bioconf/20238003002

summarizing evidence, and interpreting findings.

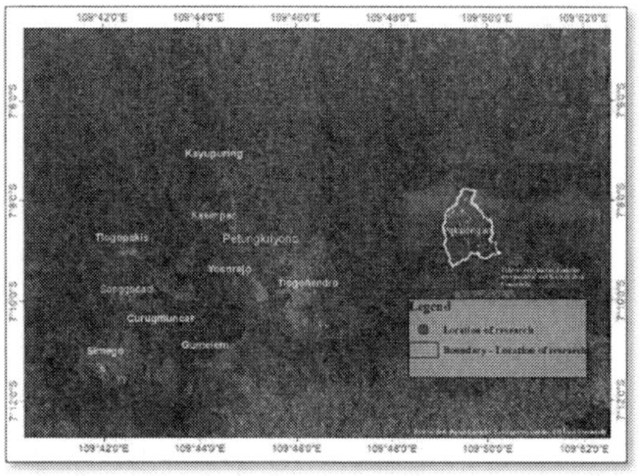

Fig. 1. Reseach location

3 Results and Discussions

3.1 The Sosiogeography of Petungkriyono

One of the subdistricts in the Pekalongan Regency is Petungkriyono. With an area of 7,359 hectares, it is situated in the southeast of the district, around 34 kilometers from Pekalongan, the district capital. Compared to other sub-districts in the Pekalongan district, this one has the lowest temperature (18–30 °C) (Damayanti et al., 2021a; Pradana et al., 2020). This is because the area is mostly covered in forests and has a high altitude of approximately 1,294 meters above sea level (BPS Pekalongan Regency, 2023). Its elevation ranges from 500 to 2,100 meters above sea level (Tanguay, 2020). Most of the Petungkriyono area's slopes are moderate (20–40°) slopes (Arrisaldi, 2018).

The Petungkriyono Forest itself has an area of 5,189,507 ha, which is officially managed by KPH East Pekalongan-BKPH Doro. It is divided into restricted production forest and mixed natural forest, both of which serve as protected forests (Ahmadin, 2016). The community's existence and its relationship with the forest are inextricably linked, given the state of the land cover in the Petungkriyono sub-district, which is dominated by forest regions. Petungkriyono's neighboring communities and the forest are both impacted by these intricate relationships.

Yosorejo village, the subdistrict capital, is the first of nine villages in Petungkriyono. The other villages are Simego, Songgodadi, Curugmuncar, Gumelem, Tlogohendro, Tlogopakis, Kasimpar, and Kayupuring. The community's daily requirements, which are primarily met by farmers, rely heavily on agricultural goods, particularly the resources found in the nearby forests. Within the context of the economy, people in Petungkriyono villages gather non-timber forest products; for instance, residents of Kayupuring village tap sugar palm in natural forests (Fikri & Adinugraha, 2021); residents of several nearby villages, including Kasimpar village, Kayupuring (Wibowo et al., 2022; Widodo, 2021; Tanguay, 2020), tap pine resin from forests with limited production; and people harvest food items such as bamboo, fruit, cloves, coconut, and other vegetables (BPS Pekalongan Regency, 2023; Damayanti et al., 2018). The aforementioned activities are all direct advantages that the Petungkriyono people can obtain from the forests to meet their financial demands.

The Petungkriyono people have very intimate ties to environment in their traditional traditions. The villages surrounding the Petungkriyono forest have local wisdom known as nyadran Gebyog (Ahmad & Soeprabowati, 2022). This is a synergistic combination of the community's efforts to preserve the forest and the surrounding natural environment while upholding traditions passed down from generation to generation. An earth-friendly charity event called Nyadran Gebyog is a way for people to give back to the natural world that has given them resources to meet their needs.

At the nyadran event, participants carried their harvested produce and traveled down a trail into the heart of the forest. In the jungle, before the Nyadran event starts in a location that has been chosen for generations and is revered by the locals, participants must bathe in a crystal-clear river as a sign of self-purification. People are instinctively conscious of the need to preserve these natural resources because of the tradition of taking a river bath as part of a sequence of nyadran activities in the forest, which are seen to have significant spiritual value. One significant example of how cultural and environmental preservation are integrated into the daily lives of the inhabitants surrounding the Petungkriyono forest is the persistence of clear rivers (Ahmad & Soeprabowati, 2022).

The woodland scenery of Petungkriyono offers great potential for use in ecotourism and education for the local population. Petungkriyono is a vital water supply for the life of villages downstream since it is traversed by eight rivers, six of which are tributaries (Banteng, Pakis, Totogan, Sawangan, Welo, and Mangunan rivers) and two of which are major rivers (Kupang and Sengkarang Rivers). The Petungkriyono forest is still intact, as seen by the crystal-clear water, and several of its locations—Curug Muncar, Welo Asri, Curug Bajing, Embung Rawa, Telaga Mangunan, and Telaga Sigebyar—are visited by tourists. Other natural tourist attractions include Puncak Tugu Jalak, Karang Srity, Black Canyon, Lingga Yoni Site, and Puncak Kendalisodo, in addition to water-based tourism (Andriyani et al., 2022; M. Anam Rahardyan et al., 2021; T. A. Setiawan et al., 2020; Yusron et al., 2020). These numerous tourist attractions make use of Petungkriyono's stunning natural resources. Furthermore, the process of education happens simultaneously with the interaction of humans and environment.

It is very important to maintain the environment while making various enhancements to educational buildings. Talk about creating butterfly gardens and captive breeding programs as a development strategy (Peggie, 2018). Apart from that, the existence of primate watching and bird watching spots also adds to the natural knowledge of the surrounding community and people who visit Petungkriyono. Meanwhile, from a medical perspective, the diversity of germplasm sources in the Petungkriyono forest also has the potential to become medicinal plants for the surrounding community. The assumption that herbal medicines are believed to carry minimum risk has made herbal medicines increasingly mushroom in Indonesian society, and Petungkriyono is no exception.

From the various explanations that have been given, a common thread can be drawn that Petungkriyono has an interesting sociogeography. The forest area around this community indirectly influences the community's life patterns in various aspects, especially regarding how to fulfill basic human living needs. Fulfillment of food, economic, cultural spirituality, ecotourism, education and forest firewood needs has been provided.

3.2 The biodiversity of Petungkriyono forest

3.2.1 The flora of Petungkriyono forest : IUCN conservation status

Petungkriyono Forest is a lower montane forest, or mountain tropical rain forest, with typical monsoon temperatures ranging from 18 to 300 degrees Celsius. It is situated between 700 and 1500 meters above sea level (masl).(Damayanti et al., 2021a). According to some (Damayanti et al., 2021b, 2018), Petungkriyono is the last remaining high biodiversity tropical forest in Java. There is a whole range of vegetation in Petungkriyono, from bushes to trees (Ahmadin, 2016). Petungkriyono forest region has 204 known species of live plants that are dispersed throughout the area. Magnoliopsida, liliopsida, polypodiopsida, lycopodiopsida and pinopsida are the five classes into which they are divided.

First off, the Magnoliopsida class, which is dominant in the Petungkriyono region and comprises 117 species, is distinguished by its two components. With fourteen species, the genus *Ficus spp.* is the most prevalent type within the magnoliopsida class in tropical forest (Prapitasari et al., 2020). The markers of this type's proliferation include tolerance to light, liking for warm (tropical) conditions, high humidity (Yulia & Yanti, 2010) in the lower montane forest of Petungkriyono, and the existence of specialized pollinators that coexist harmoniously with Ficus spp.

Second, there are 52 species in the Liliopsida class, which is distinguished by one-piece plants. In the Petungkriyono forest area, the Liliopsida class includes the orchid family (Orchidaceae), which is the most frequently seen family, including in Indonesia (Kusmana & Hikmat, 2015). *Dendrobium spp.* (7 species), *Bulbophyllumn spp.* (4 species), *Coelogyne spp.* (4 species), *Flickingeria spp.* (4 species), Liparis (4 species), and Asplenium (4 species) are the most abundant genera of Orchidaceae. Orchids love tropical rain forests, especially those that live on the surface of trees without harming their hosts (epiphytic species). These regions include low-lying mountain tropical rain forests. The tropics provide year-round uniform illumination- shady lighting and enough humidity, which makes them the perfect environment for a variety of orchid species (Agustin & Widowati, 2015).

Third, there are 29 species in the well-known Polypodiopsida class, which is also known as true ferns. The majority of fern species are members of the Polyopsida class. The majority of this class are epiphytes that grow on rocks or trees in humid environments, and they are distinguished by the spores on the underside of the leaves, or sori. This class is most frequently inhabited by three species of the genus Selliguea spp., followed by two species each of Microsorum spp. and Pyrrosia.

Fourth, only three species from the genus Huperzia spp. were discovered in the Lycopodiopsida class. These, sometimes known as wire nails or kumpai, are typically found growing on rocks or humus soil, which is wet, shaded, and receives little sunshine. Pinopsida, the final class, is comprised of two species: Pinus merkusii and Agathis dammara..

Table 1. Flora of Petungkriyono forest.

№	Nama lokal	Kelas	Spesies	Status	Sumber
1	Keruing bunga	Magnoliopsida	*Dipterocarpus hasseltii*	EN	(Damayanti et al., 2021b)
2	Keruing keladan	Magnoliopsida	*Dipterocarpus gracilis*	VU	(Damayanti et al., 2021b)
3	Damar mata-kucing	Magnoliopsida	*Shorea javanica*	EN	(Damayanti et al., 2021b)
4	Ki riung anak	Magnoliopsida	*Castanopsis acuminatissima*	LC	(Damayanti et al., 2021b, 2020)
5	Saninten	Magnoliopsida	*Castanopsis argente*	EN	(Damayanti et al., 2021b)
6	Akalipa	Magnoliopsida	*Acalypha lanceolata*	NE	(Damayanti et al., 2021b)
7	Huru bodas	Magnoliopsida	*Acer laurinum*	LC	(Damayanti et al., 2021b)
8	Ki jeruk	Magnoliopsida	*Acronychia trifoliolata*	LC	(Damayanti et al., 2021b)
9	-	Magnoliopsida	*Actinodaphne glabra*	LC	(Damayanti et al., 2021b)
10	Ki Careuh	Magnoliopsida	*Alangium rotundifolium*	NE	(Damayanti et al., 2021b)
11	Jabon putih	Magnoliopsida	*Anthocephalus cadamba*	NE	(Damayanti et al., 2021b)
12	Ipuh	Magnoliopsida	*Antiaris toxicaria*	LC	(Damayanti et al., 2021b)
13	Ande-ande	Magnoliopsida	*Antidesma montanus*	NE	(Damayanti et al., 2021b)

14	Jengkol	Magnoliopsida	*Archidendron pauciflorum*	NE	(Damayanti et al., 2021b)
15	Bendo	Magnoliopsida	*Artocarpus elasticus*	LC	(Damayanti et al., 2020, 2021b; Wulandari et al., 2016)
16	Nangka	Magnoliopsida	*Artocarpus heterophyllus*	NE	(Damayanti et al., 2021b)
17	Gembirung	Magnoliopsida	*Astronia spectabilis*	NE	(Damayanti et al., 2021b)
18	Lembayung malabar	Magnoliopsida	*Basella alba*	NE	(Damayanti et al., 2021b)
19	Gintung	Magnoliopsida	*Bischofia javanica*	LC	(Damayanti et al., 2021b, 2020)
20	Sembung	Magnoliopsida	*Blumea balsamifera*	LC	(Damayanti et al., 2021b)
21	Kanyere	Magnoliopsida	*Bridelia monoica*	NE	(Damayanti et al., 2021b)
22	-	Magnoliopsida	*Chisocheton macrophyllus*	LC	(Damayanti et al., 2021b)
23	Kina bulu	Magnoliopsida	*Cinchona pubescens*	LC	(Damayanti et al., 2021b)
24	Kayu manis	Magnoliopsida	*Cinnamomum iners*	LC	(Damayanti et al., 2021b)
25	Sintok	Magnoliopsida	*Cinnamomum sintoc*	LC	(Damayanti et al., 2021b)
26	Kopi robusta	Magnoliopsida	*Coffea canephora*	LC	(Damayanti et al., 2021b)
27	Ki banen	Magnoliopsida	*Crypteronia peniculata*	NE	(Damayanti et al., 2021b)
28	Ipis kulit	Magnoliopsida	*Decaspermum fruticosum*	NE	(Damayanti et al., 2021b)
29	Dahu	Magnoliopsida	*Dracontomelon mangiferum*	NE	(Widodo, 2021)
30	Sengkuang	Magnoliopsida	*Dracontomelon dao*	LC	(Damayanti et al., 2021b, 2020)
31	Durian	Magnoliopsida	*Durio zibethinus*	NE	(Damayanti et al., 2021b, 2020)
32	Bawangan	Magnoliopsida	*Dysoxylum blumei*	NE	(Damayanti et al., 2021b, 2020)
33	Kedoya	Magnoliopsida	*Dysoxylum gaudichaudianum*	LC	(Damayanti et al., 2021b, 2020)
34	-	Magnoliopsida	*Dysoxylum nutans*	LC	(Damayanti et al., 2021b, 2020)
35	Antap	Magnoliopsida	*Elaeocarpus obtusus*	LC	(Damayanti et al., 2021b; Widodo, 2021)
36	Ganitri	Magnoliopsida	*Elaeocarpus sphaericus*	NE	(Damayanti et al., 2021b)
37	Kilalayu	Magnoliopsida	*Erioglossum rubiginosum*	NE	(Damayanti et al., 2021b)
38	Dadap serep	Magnoliopsida	*Erythrina lithosperma*	NE	(Damayanti et al., 2021b)
39	Ki sapu	Magnoliopsida	*Eurya acuminata*	NE	(Damayanti et al., 2021b)
40	Beringin	Magnoliopsida	*Ficus benjamina*	LC	(Damayanti et al., 2020, 2021b; Widodo, 2021; Wulandari et al., 2016)
41	Tin	Magnoliopsida	*Ficus carica*	LC	(Damayanti et al., 2021b)
42	Terep areuy	Magnoliopsida	*Ficus globosa*	LC	(Damayanti et al., 2021b)
43	Luwing	Magnoliopsida	*Ficus lepicarpa*	LC	(Damayanti et al., 2021b)
44	Hamerang putih	Magnoliopsida	*Ficus padana*	LC	(Damayanti et al., 2021b)
45	Hamerang putih	Magnoliopsida	*Ficus toxicaria*	NE	(Damayanti et al., 2021b)
46	Loa	Magnoliopsida	*Ficus racemosa*	LC	(Damayanti et al., 2021b)
47	Darangdan	Magnoliopsida	*Ficus sinuata*	NE	(Damayanti et al., 2021b)
48	Ki ara jingkang	Magnoliopsida	*Ficus stupenda*	NE	(Damayanti et al., 2021b)
49	Kondang	Magnoliopsida	*Ficus variegata*	LC	(Damayanti et al., 2021b; Widodo, 2021)
50	Kebak	Magnoliopsida	*Ficus alba*	NE	(Widodo, 2021)
51	Kiara bulu	Magnoliopsida	*Ficus procera*	NE	(Widodo, 2021)
52	Spri	Magnoliopsida	*Ficus rostrata*	NE	(Widodo, 2021)
53	Ipik	Magnoliopsida	*Ficus superba*	NE	(Widodo, 2021)

54	Pari	Magnoliopsida	*Glochidion lutescens*	LC	(Damayanti et al., 2021b)
55	Ki mareme	Magnoliopsida	*Glochidion rubrum*	LC	(Damayanti et al., 2021b)
56	Kendung	Magnoliopsida	*Helicia javanica*	LC	(Damayanti et al., 2021b)
57	Waru	Magnoliopsida	*Hibiscus tiliaceus L*	LC	(Damayanti et al., 2021b)
58	Kareumbi	Magnoliopsida	*Homalanthus populneus*	LC	(Damayanti et al., 2021b)
59	Kemadu	Magnoliopsida	*Laportea sinuata*	NE	(Damayanti et al., 2021b)
60	Girang merah	Magnoliopsida	*Leea indica*	LC	(Damayanti et al., 2021b)
61	Pasang bodas	Magnoliopsida	*Lithocarpus elegans*	NE	(Damayanti et al., 2021b)
62	Wuru banyu	Magnoliopsida	*Litsea umbellate*	NE	(Damayanti et al., 2021b)
63	Wuru kunyit	Magnoliopsida	*Litsea angulata*	NE	(Damayanti et al., 2021b)
64	Mara	Magnoliopsida	*Macaranga tanarius*	LC	(Damayanti et al., 2021b)
65	Huru Leuleus	Magnoliopsida	*Machilus rimosa*	NE	(Damayanti et al., 2021b)
66	Pampung	Magnoliopsida	*Macropanax dispermus*	LC	(Damayanti et al., 2021b)
67	Pohon payung	Magnoliopsida	*Maesopsis eminii*	LC	(Damayanti et al., 2021b)
68	Cempaka hutan	Magnoliopsida	*Magnolia montana*	DD	(Damayanti et al., 2021b)
69	Sampang	Magnoliopsida	*Melicope latifolia*	NE	(Damayanti et al., 2021b)
70	Bintinu	Magnoliopsida	*Melochia umbellata*	LC	(Damayanti et al., 2021b)
71	Angrit	Magnoliopsida	*Nauclea lanceolata*	LC	(Damayanti et al., 2021b)
72	Bengang	Magnoliopsida	*Neesia altissima*	NE	(Damayanti et al., 2021b)
73	Rambutan	Magnoliopsida	*Nephelium lappaceum*	LC	(Damayanti et al., 2021b)
74	Mumuncan gan	Magnoliopsida	*Ostodes paniculata*	LC	(Damayanti et al., 2021b)
75	Nyatuh pucung	Magnoliopsida	*Palaquium rostratum*	LC	(Damayanti et al., 2021b)
76	Picung	Magnoliopsida	*Pangium edule*	LC	(Damayanti et al., 2021b)
77	-	Magnoliopsida	*Pittosporum ramiflorum*	LC	(Damayanti et al., 2021b)
78	Sengiran	Magnoliopsida	*Pittosporum ferrugineum*	EN	(Damayanti et al., 2021b)
79	Putat	Magnoliopsida	*Planchonia valida Blume*	NE	(Damayanti et al., 2021b)
80	Matoa	Magnoliopsida	*Pometia pinnata*	LC	(Damayanti et al., 2021b)
81	Kawoyang	Magnoliopsida	*Prunus arborea*	LC	(Damayanti et al., 2021b)
82	Bayur	Magnoliopsida	*Pterospermum javanicum*	LC	(Damayanti et al., 2021b)
83	Bedali	Magnoliopsida	*Radermachera gigantea*	LC	(Damayanti et al., 2021b)
84	Kecapi	Magnoliopsida	*Sandoricum koetjape*	LC	(Damayanti et al., 2021b, 2020)
85	Lotrok	Magnoliopsida	*Saurauia bracteosa*	VU	(Damayanti et al., 2021b)
86	Puspa	Magnoliopsida	*Schima wallichii*	LC	(Damayanti et al., 2021b, 2020)
87	Sindur	Magnoliopsida	*Sindora javanica*	VU	(Damayanti et al., 2021b)
88	Ki Somang	Magnoliopsida	*Sloanea sigun*	NE	(Damayanti et al., 2021b)
89	Hantap	Magnoliopsida	*Sterculia oblongata*	LC	(Damayanti et al., 2021b)
90	Kemenyan	Magnoliopsida	*Styrax benzoin*	NE	(Damayanti et al., 2021b)
91	Mahoni	Magnoliopsida	*Swietenia mahagoni*	NT	(Damayanti et al., 2021b)
92	Jirak	Magnoliopsida	*Symplocos fasciculata*	NE	(Damayanti et al., 2021b)
93	-	Magnoliopsida	*Syzygium densiflorum*	VU	(Damayanti et al., 2021b)
94	Salam	Magnoliopsida	*Syzygium polyanthum*	NE	(Damayanti et al., 2021b)
95	Baheda	Magnoliopsida	*Terminalia bellirica*	LC	(Damayanti et al., 2021b)
96	Suren	Magnoliopsida	*Toona sureni*	LC	(Damayanti et al., 2021b; Wulandari et al., 2016)

97	Mengkirai	Magnoliopsida	*Trema orientalis*	LC	(Damayanti et al., 2021b)
98	Kiramo giling	Magnoliopsida	*Trevesia sundaica*	NE	(Damayanti et al., 2021b)
99	Gletang	Magnoliopsida	*Tridax procumbens*	NE	(Damayanti et al., 2021b)
100	Ki Bangkong	Magnoliopsida	*Turpinia pomifera*	LC	(Damayanti et al., 2021b)
101	Ki Bancet	Magnoliopsida	*Turpinia sphaerocarpa*	NE	(Wulandari et al., 2016)
102	Kluwak	Magnoliopsida	*Pangium edule*	LC	(Wulandari et al., 2016)
103	Kantung semar	Magnoliopsida	*Nepenthes adrianii*	NE	(Widhiono, 2009)
104	Sengon	Magnoliopsida	*Albizia chinensis*	NE	(Damayanti et al., 2021b)
105	Sengon laut	Magnoliopsida	*Albizzia falcata*	NE	(Widodo, 2021)
106	Pulai peucang	Magnoliopsida	*Alstonia spectabilis*	LC	(Damayanti et al., 2021b)
107	Pule	Magnoliopsida	*Alstonia scholaris*	LC	(Widodo, 2021)
108	Jabon	Magnoliopsida	*Anthocephalus cadamba*	NE	(Widodo, 2021)
109	Senduduk bulu	Magnoliopsida	*Clidemia hirta*	NE	(Widodo, 2021)
110	Sempayang	Magnoliopsida	*Evodia aromatica*	NE	(Widodo, 2021)
111	Punyam	Magnoliopsida	*Mallotus sp.*	NE	(Widodo, 2021)
112	Muncang cina	Magnoliopsida	*Mallotus paniculatus*	LC	(Damayanti et al., 2021b; Widodo, 2021)
113	Gembirit	Magnoliopsida	*Tabernaemontana sphaerocarpa*	LC	(Widodo, 2021)
114	-	Magnoliopsida	*Eugenia sp*	NE	(Nuraini, 2018)
115	-	Magnoliopsida	*Myristica spp*	NE	(Nuraini, 2018)
116	Miniran	Magnoliopsida	*Myrica javanica*	NE	(Damayanti et al., 2021b)
117	Ki Mangkok	Magnoliopsida	*Micromelum minutum*	LC	(Damayanti et al., 2021b)
118	Lumbah	Liliopsida	*Curculigo capitulata*	NE	(Damayanti et al., 2021b)
119	Palem merah	Liliopsida	*Cyrtostachys lakka*	NE	(Damayanti et al., 2021b)
120	Palem jingga	Liliopsida	*Cyrtostachys renda*	NE	(Damayanti et al., 2021b)
121	Pinang jawa	Liliopsida	*Pinanga javana*	NE	(Damayanti et al., 2021b)
122	-	Liliopsida	*Malaxis junghuhnii*	NE	(Mardiyana et al., 2019)
123	Anggrek peri	Liliopsida	*Oberonia similis*	NE	(Mardiyana et al., 2019)
124	-	Liliopsida	*Acriopsis sp1*	NE	(Mardiyana et al., 2019)
125	-	Liliopsida	*Agrostophyllum laxum*	NE	(Mardiyana et al., 2019)
126	-	Liliopsida	*Agrostophyllum sp1*	NE	(Mardiyana et al., 2019)
127	-	Liliopsida	*Appendicula reflexa*	NE	(Mardiyana et al., 2019)
128	-	Liliopsida	*Appendicula undulata*	NE	(Mardiyana et al., 2019)
129	-	Liliopsida	*Bulbophyllum blumei*	NE	(Mardiyana et al., 2019)
130	-	Liliopsida	*Bulbophyllum sp1*	NE	(Mardiyana et al., 2019)
131	-	Liliopsida	*Bulbophyllum sp2*	NE	(Mardiyana et al., 2019)
132	-	Liliopsida	*Bulbophyllum sp3*	NE	(Mardiyana et al., 2019)
133	-	Liliopsida	*Ceratostylis subulata*	NE	(Mardiyana et al., 2019)
134	-	Liliopsida	*Coelogyne sp1*	NE	(Mardiyana et al., 2019)
135	-	Liliopsida	*Coelogyne sp2*	NE	(Mardiyana et al., 2019)
136	-	Liliopsida	*Coelogyne sp3*	NE	(Mardiyana et al., 2019)
137	-	Liliopsida	*Coelogyne sp4*	NE	(Mardiyana et al., 2019)
138	-	Liliopsida	*Dendrobium lobullatum*	NE	(Mardiyana et al., 2019)
139	-	Liliopsida	*Dendrobium mutabile*	NE	(Mardiyana et al., 2019)
140	-	Liliopsida	*Dendrobium sp1*	NE	(Mardiyana et al., 2019)
141	-	Liliopsida	*Dendrobium sp2*	NE	(Mardiyana et al., 2019)

142	-	Liliopsida	*Dendrobium sp3*	NE	(Mardiyana et al., 2019)
143	-	Liliopsida	*Dendrobium sp4*	NE	(Mardiyana et al., 2019)
144	-	Liliopsida	*Dendrobium sp5*	NE	(Mardiyana et al., 2019)
145	-	Liliopsida	*Eria multiflora*	NE	(Mardiyana et al., 2019)
146	-	Liliopsida	*Eria sp1*	NE	(Mardiyana et al., 2019)
147	-	Liliopsida	*Eria sp2*	NE	(Mardiyana et al., 2019)
148	-	Liliopsida	*Flickingeria sp1*	NE	(Mardiyana et al., 2019)
149	-	Liliopsida	*Flickingeria sp2*	NE	(Mardiyana et al., 2019)
150	-	Liliopsida	*Flickingeria sp3*	NE	(Mardiyana et al., 2019)
151	-	Liliopsida	*Flickingeria sp4*	NE	(Mardiyana et al., 2019)
152	-	Liliopsida	*Liparis rhombea*	NE	(Mardiyana et al., 2019)
153	-	Liliopsida	*Liparis sp1*	NE	(Mardiyana et al., 2019)
154	-	Liliopsida	*Liparis sp2*	NE	(Mardiyana et al., 2019)
155	-	Liliopsida	*Liparis sp3*	NE	(Mardiyana et al., 2019)
156	-	Liliopsida	*Luisia Javanica*	NE	(Mardiyana et al., 2019)
157	Anggrek Bulan	Liliopsida	*Phalaenopsis amabilis*	NE	(Mardiyana et al., 2019)
158	-	Liliopsida	*Podochilus sp1*	NE	(Mardiyana et al., 2019)
159	-	Liliopsida	*Pholidota imbricata*	NE	(Mardiyana et al., 2019)
160	-	Liliopsida	*Polystacyha concentra*	NE	(Mardiyana et al., 2019)
161	-	Liliopsida	*Pteroceras teysmannii*	NE	(Mardiyana et al., 2019)
162	-	Liliopsida	*Robiquetia spatulata*	NE	(Mardiyana et al., 2019)
163	-	Liliopsida	*sp1*	NE	(Mardiyana et al., 2019)
164	-	Liliopsida	*sp2*	NE	(Mardiyana et al., 2019)
165	Anggrek bulu	Liliopsida	*Trichotosia ferox*	NE	(Mardiyana et al., 2019)
166	-	Liliopsida	*Trichotosia pauciflora*	NE	(Mardiyana et al., 2019)
167	-	Liliopsida	*Trichotosia sp1*	NE	(Mardiyana et al., 2019)
168	Anggrek permata	Liliopsida	*Macodes petola*	NE	(Widhiono, 2009)
169	Aren	Liliopsida	*Arenga pinnata*	NE	(Fikri & Adinugraha, 2021)
170	Penawar jambi	Polypodiopsida	*Cibotium barometz*	NE	(Damayanti et al., 2021b)
171	-	Polypodiopsida	*Asplenium caudatum*	NE	(Lestari, 2019)
172	-	Polypodiopsida	*Asplenium cuneatum*	NE	(Lestari, 2019)
173	Paku sarang burung	Polypodiopsida	*Asplenium nidus*	NE	(Lestari, 2019)
174		Polypodiopsida	*Asplenium longissimum*	NE	(Lestari, 2019)
175	Paku tertutup	Polypodiopsida	*Davallia denticulata*	NE	(Lestari, 2019)
176	-	Polypodiopsida	*Davallia sp*	NE	(Lestari, 2019)
177	-	Polypodiopsida	*Crepidomanes minutum*	NE	(Lestari, 2019)
178	-	Polypodiopsida	*Hymenophyllum sp*	NE	(Lestari, 2019)
179	-	Polypodiopsida	*Oleandra neriiformis*	NE	(Lestari, 2019)
180	Paku ular	Polypodiopsida	*Aglaomorpha heraclea*	NE	(Lestari, 2019)
181	-	Polypodiopsida	*Belvisia sp*	NE	(Lestari, 2019)
182	Paku daun kepala tupai	Polypodiopsida	*Drynaria quercifolia*	NE	(Lestari, 2019)
183	-	Polypodiopsida	*Goniophlebium percussum*	NE	(Lestari, 2019)
184	-	Polypodiopsida	*Lepisorus longifolius*	NE	(Lestari, 2019)
185	-	Polypodiopsida	*Leptochilus sp*	NE	(Lestari, 2019)
186	-	Polypodiopsida	*Microsorum heterocarpum*	NE	(Lestari, 2019)

87	-	Polypodiopsida	*Microsorum membranifolium*	NE	(Lestari, 2019)
88	Paku daun picisan	Polypodiopsida	*Pyrrosia nummularifolia*	NE	(Lestari, 2019)
89	Paku sisik naga	Polypodiopsida	*Pyrrosia piloselloides*	NE	(Lestari, 2019)
90	-	Polypodiopsida	*Selliguea enervis*	NE	(Lestari, 2019)
91	-	Polypodiopsida	*Selliguea albidosquamata*	NE	(Lestari, 2019)
92	-	Polypodiopsida	*Selliguea stenophylla*	NE	(Lestari, 2019)
93	-	Polypodiopsida	*Lemmaphilum accendens*	NE	(Lestari, 2019)
94	-	Polypodiopsida	*Antrophyum parvulum*	NE	(Lestari, 2019)
95	-	Polypodiopsida	*Antrophyum semicostatum*	NE	(Lestari, 2019)
96	-	Polypodiopsida	*Vittaria sp*	NE	(Lestari, 2019)
97	-	Polypodiopsida	*Sp 13*	NE	(Lestari, 2019)
98	-	Polypodiopsida	*Sp 2*	NE	(Lestari, 2019)
99	Kumpai ekor tikus	Lycopodiopsida	*Huperzia carinata*	NE	(Lestari, 2019)
100	-	Lycopodiopsida	*Huperzia serrata*	NE	(Lestari, 2019)
101	Kumpai ekor monyet	Lycopodiopsida	*Huperzia squarrosa*	NE	(Lestari, 2019)
102	Damar	Pinopsida	*Agathis dammara*	VU	(Damayanti et al., 2021b)
103	Pinus	Pinopsida	*Pinus merkusii*	VU	(Damayanti et al., 2021b)
104	Wuru kangkang	-	-	NE	(Damayanti et al., 2021b)

IUCN red list which is a list of the conservation status of species based on their risk of extinction. The track record of research related to Petungkriyono biodiversity has been carried out for many years and provides information about the natural riches and uniqueness of this forest area. This area is home to hundreds of species of flora and fauna and some are categorized as endangered according to the IUICN Red List of Threatened Species. From the results of data analysis, 135 species were in the Not Evaluated (NE) category, 1 species was in the Data Deficient (DD) category, 57 species were in the Least Concern (LC) category, 1 species was in the Near Threatened (NT) category, 6 species were in the Vulnerable (VU) category, and 4 species categorized as Endangered (EN). Requiring special attention here are species that are categorized as Vulnerable and Endangered. Species with Vulnerable status include: from the magnoliopsida class (*Dipterocarpus gracilis, Saurauia bracteosa, Sindora javanica,* and *Syzygium densiflorum*) and the pinopsida class (*Agathis dammara* and *Pinus merkusii*). These six species are experiencing significant reductions in natural numbers or threats, making them vulnerable to extinction. Meanwhile, species in the endangered category include: *Dipterocarpus hasseltii, Shorea javanica, Castanopsis argente,* and *Pittosporum ferrugineum* from the magnoliopsida class. These four species are very at risk of extinction, so urgent conservation action is needed to restore their populations.

3.2.2 Fauna of Petungkriyono forest : IUCN conservation status

Petungkriyono has an abundant diversity of fauna. The results of data processing show that there are 168 species of fauna that can be classified into 5 classes. The class consists of: aves, insects, reptiles, amphibians, and mammals.

The aves class is the majority fauna found in the Petungkriyono forest area. About 91 species of aves with various important roles in a forest ecosystem. Genus *Aethopyga spp.* (4 species) and the genus *Megalaima spp.* (4 species) is the most bird species found. *Aethopyga spp.* is a class of pollinator agents, helping the pollination of plant flowers. As for *Megalaima spp.* are insectivorous and fruit-eating birds. Aktivitas memangsanya membantu ekosistem dalam menjaga keseimbangan populasi serangga mangsa,

On the other hand, eating fruit contributes to the spatial dispersal of forest plant seeds. Three more genera with a sizable number of species are the genus Rhipidura spp. (3 species), the genus *Arachnothera spp.* (3 species), and the genus *Zosterops spp.* (3 species), which are primarily insectivores.

Second, there are 47 species in the Insecta class, which includes dragonfly and butterfly members. The genus *Mycalesis spp.* (3 species) and the genus *Melanitis spp.* (3 species) are the most prevalent butterfly genera. As pollinator agents, these two genera play a major role in the Petungkriyono forest area. Butterflies move pollen from one flower to another when they are seeking nectar. Smaller insect predators include the three species of dragonflies in the genus Orthetrum and the three species of Drepanosticta.

Third, a class of 15 species of reptiles that are present in the Petungkriyono forest. The class of reptiles in this region is divided into three major genera: *Eutropis spp.* (2 species), *Bronchocela spp.* (2 species), and Draco spp. (2 species). A flying lizard known as a Draco species feeds effectively on little insects found in forest tree habitats. Reptiles in the genus Eutropis are omnivores, while those in the genus Bronchocela are mostly herbivorous.

Fourth, the Petungkriyono forest region is home to the twelve species of amphibians. Only the genus *Limnonectes spp.* was determined to have more than one species in this class of amphibians (two species). This genus consumes a variety of small insects, small aquatic animals like larvae of water insects, and other small amphibians.

Fifth, the class Mammal comprises three species. *Presbytis comata, Hylobates moloch*, and *Trachypithecus auratus* are the mammals found in the Petungkriyono forest area. Being primates, all three have a propensity for herbivory.

Table 2. Fauna of Petungkriyono forest.

No	Local name	Class	Spesies	Status	Source
1	Lutung Hitam	Mammalia	*Trachypithecus auratus*	VU	(Damayanti et al., 2018; Fatmasari et al., 2017; Nijman & Setiawan, 2000)
2	Surili	Mammalia	*Presbytis comata*	VU	(Damayanti et al., 2018; Nijman & Setiawan, 2000)
3	Owa Jawa	Mammalia	*Hylobates moloch*	EN	(Damayanti et al., 2018; Nijman & Setiawan, 2000; Nofiyani & Nugroho, 2016; A. Setiawan et al., 2012)
4	Burung pekakak biru	Aves	*Alcedo euryzona*	CE	(Rachmawati et al., 2022)
5	Elang Hitam	Aves	*Ictinaetus malaiensis*	LC	(Damayanti et al., 2018; Apriliyani, 2017; Nijman & Setiawan, 2000)
6	Elang Jawa	Aves	*Nisaetus bartelsi*	EN	(Widodo, 2021; Damayanti et al., 2018; Nijman & Setiawan, 2000)
7	Julang Mas	Aves	*Aceros undulatus*	VU	(Damayanti et al., 2018; Nijman & Setiawan, 2000)
8	Seriwang sangihe	Aves	*Eutrichomyias rowleyi*	CE	(Hendriati et al., 2019)
9	Butbut besar	Aves	*Centropus sinensis*	LC	(Hendriati et al., 2019; Nijman & Setiawan, 2000)
10	Burung kopi	Aves	*Pomatorhinus montanus*	LC	(Hendriati et al., 2019)
11	Tepus leher putih	Aves	*Stachyris thoracica*	LC	(Hendriati et al., 2019; Nijman & Setiawan, 2000)
12	Tepus dada putih	Aves	*Stachyris grammiceps*	NT	(Ahmadin, 2016; Hendriati et al., 2019)
13	Cica matahari	Aves	*Laniellus albonotatus*	NT	(Hendriati et al., 2019)
14	Brencet wergan	Aves	*Alcippe pyrrhoptera*	LC	(Hendriati et al., 2019)
15	Bintayung	Aves	*Fregata andrewsi*	VU	(Hendriati et al., 2019)
16	Kasumba	Aves	*Apalharpactes reinwardtii*	VU	(Hendriati et al., 2019)
17	Piong	Aves	*Lophozosterops javanica*	LC	(Hendriati et al., 2019; Widodo, 2021)
18	Elang Perut-karat	Aves	*Lophotriorchis kienerii*	NT	(Apriliyani, 2017; Nijman & Setiawan, 2000)
19	Pleci	Aves	*Zosterops flavus*	EN	(Hendriati et al., 2019)
20	Kacamata gunung	Aves	*Zosterops montanus*	LC	(Hendriati et al., 2019)
21	Kacamata biasa	Aves	*Zosterops palpebrosus*	LC	(Hendriati et al., 2019; Nijman & Setiawan, 2000)
22	Burung madu pipi merah	Aves	*Anthreptes singalensis*	LC	(Hendriati et al., 2019)
23	Burung Madu Kelapa	Aves	*Anthreptes malacensis*	LC	(Hendriati et al., 2019)
24	Burung madu merah ekor panjang	Aves	*Aethopyga mystacalis*	LC	(Hendriati et al., 2019)
25	Burung madu merah	Aves	*Aethopyga siparaja*	LC	(Hendriati et al., 2019)
26	Burung madu gunung	Aves	*Aethopyga eximia*	LC	(Hendriati et al., 2019; Nijman & Setiawan, 2000)
27	Burung madu Sepah Raja	Aves	*Aethopyga siparaja*	LC	(Nijman & Setiawan, 2000)
28	Jantung kecil	Aves	*Arachnothera longirostra*	LC	(Widodo, 2021; Hendriati et al., 2019; Nijman & Setiawan, 2000)
29	Burung jantung besar	Aves	*Arachnothera robusta*	LC	(Hendriati et al., 2019)
30	Burung jantung gunung	Aves	*Arachnothera affinis*	LC	(Hendriati et al., 2019; Nijman & Setiawan, 2000)
31	Glatik gunung	Aves	*Psaltria exilis*	LC	(Hendriati et al., 2019)
32	Celepuk jawa	Aves	*Otus angelinae*	VU	(Hendriati et al., 2019)

33	Burung kipas	Aves	*Rhipidura javanica*	LC	(Widodo, 2021; Hendriati et al., 2019)
34	Kipasan mutiara	Aves	*Rhipidura euryura*	LC	(Widodo, 2021; Hendriati et al., 2019)
35	Kipasan merah	Aves	*Rhipidura phoenicura*	LC	(Widodo, 2021; Hendriati et al., 2019)
36	Sikatan dada merah	Aves	*Philentoma velatum*	NT	(Hendriati et al., 2019)
37	Burung kipas biru	Aves	*Muscicapa ruecki*	NE	(Hendriati et al., 2019)
38	sikatan bubik	Aves	*Muscicapa dauurica*	LC	(Widodo, 2021)
39	Cekakak jawa	Aves	*Halcyon cyanoventris*	LC	(Widodo, 2021; Hendriati et al., 2019)
40	Pelatuk bawang	Aves	*Dinopium javanense*	LC	(Widodo, 2021; Hendriati et al., 2019)
41	Kacer	Aves	*Copsychus saularis*	LC	(Hendriati et al., 2019)
42	Cucak hijau	Aves	*Chloropsis sonnerati*	EN	(Hendriati et al., 2019)
43	Anis kembang	Aves	*Geokichla interpres*	EN	(Hendriati et al., 2019)
44	Takur gunung	Aves	*Psilopogon monticola*	LC	(Hendriati et al., 2019)
45	Gelatik jawa	Aves	*Lonchura oryzivora*	EN	(Hendriati et al., 2019)
46	Bondol Jawa	Aves	*Lonchura leucogastroides*	LC	Nijman et al., 2000
47	Jalak kebo	Aves	*Acridotheres javanicus*	VU	(Hendriati et al., 2019)
48	Ciung mungkal jawa	Aves	*Cochoa azurea*	VU	(Hendriati et al., 2019)
49	Opior jawa	Aves	*Heleia javanica*	LC	(Hendriati et al., 2019)
50	Serindit jawa	Aves	*Loriculus pusillus*	NT	(Widodo, 2021)
51	Kepudang hutan	Aves	*Oriolus xanthonotus*	NT	(Widodo, 2021)
52	Sikatan Dada-merah	Aves	*Ficedula dumetoria*	LC	(Widodo, 2021)
53	Puyuh gonggong	Aves	*Arborophila javanica*	LC	(Nijman & Setiawan, 2000)
54	Sikep madu Asia	Aves	*Pernis ptilorhynchus*	LC	(Nijman & Setiawan, 2000)
55	Elang brontok	Aves	*Spizaetus cirrhatus*	NE	(Nijman & Setiawan, 2000)
56	Ayam Hutan Hijau	Aves	*Gallus varius*	LC	(Nijman & Setiawan, 2000)
57	Tekukur Biasa	Aves	*Spilopelia chinensis*	LC	(Nijman & Setiawan, 2000)
58	Elang ular bido	Aves	*Spilornis cheela*	LC	(Apriliyani, 2017; Nijman & Setiawan, 2000)
59	Kangkok Ranting	Aves	*Cuculus saturatus*	LC	(Nijman & Setiawan, 2000)
60	Wiwik Uncuing	Aves	*Cacomantis sepulcralis*	LC	(Nijman & Setiawan, 2000)
61	Kadalan Kembang	Aves	*Phaenicophaeus javanicus*	LC	(Nijman & Setiawan, 2000)
62	Kadalan Birah	Aves	*Phaenicophaeus curvirostris*	LC	(Nijman & Setiawan, 2000)
63	Walet Sarang Putih	Aves	*Collocalia fuciphaga*	NE	(Nijman & Setiawan, 2000)
64	Walet sapi	Aves	*Collocalia linchi*	LC	(Nijman & Setiawan, 2000)
65	Walet Palem-Asia	Aves	*Cypsiurus balasiensis*	LC	(Nijman & Setiawan, 2000)
66	Tepekong jambul	Aves	*Hemiprocne longipennis*	LC	(Nijman & Setiawan, 2000)
67	Luntur Harimau	Aves	*Harpactes oreskios*	LC	(Nijman & Setiawan, 2000)
68	Udang Punggung Merah	Aves	*Ceyx rufidorsa*	LC	(Nijman & Setiawan, 2000)
69	Cekakak Batu	Aves	*Lacedo pulchella*	LC	(Nijman & Setiawan, 2000)
70	Takur Bututut	Aves	*Megalaima corvina*	LC	(Nijman & Setiawan, 2000)
71	Takur Tulungtumpuk	Aves	*Megalaima javensis*	NT	(Hendriati et al., 2019; Nijman & Setiawan, 2000)
72	Takur Tohtor	Aves	*Megalaima armillaris*	LC	(Widodo, 2021; Hendriati et al., 2019; Nijman & Setiawan, 2000)
73	Takur Tengeret	Aves	*Megalaima australis*	NE	(Hendriati et al., 2019; Nijman & Setiawan, 2000)
74	Sempur Hujan Rimba	Aves	*Eurylaimus javanicus*	NT	(Nijman & Setiawan, 2000)
75	Paok Pancawarna	Aves	*Pitta guajana*	LC	(Nijman & Setiawan, 2000)

76	Jingjing Batu	Aves	*Hemipus hirundinaceus*	LC	(Nijman & Setiawan, 2000)
77	Sepah Hutan	Aves	*Pericrocotus flammeus*	LC	(Nijman & Setiawan, 2000)
78	Gagak Hutan	Aves	*Corvus enca*	LC	(Nijman & Setiawan, 2000)
79	Gelatik batu Kelabu	Aves	*Parus major*	LC	(Nijman & Setiawan, 2000)
80	Munguk Beledu	Aves	*Sitta frontalis*	LC	(Nijman & Setiawan, 2000)
81	Pelanduk Semak	Aves	*Malacocincla sepiaria*	LC	(Nijman & Setiawan, 2000)
82	Berencet Berkening	Aves	*Napothera epilepidota*	LC	(Nijman & Setiawan, 2000)
83	Meninting Kecil	Aves	*Enicurus velatus*	LC	(Nijman & Setiawan, 2000)
84	Ciung Batu Siul	Aves	*Myophonus caeruleus*	LC	(Nijman & Setiawan, 2000)
85	Remetuk Laut	Aves	*Gerygone sulphurea*	LC	(Nijman & Setiawan, 2000)
86	Cinenen Jawa	Aves	*Orthotomus sepium*	LC	(Nijman & Setiawan, 2000)
87	Perenjak Jawa	Aves	*Prinia familiaris*	NT	(Nijman & Setiawan, 2000)
88	Sikatan Biru Muda	Aves	*Cyornis unicolor*	LC	(Nijman & Setiawan, 2000)
89	Sikatan Cacing	Aves	*Cyornis banyumas*	CR	(Nijman & Setiawan, 2000)
90	Pentis Pelangi	Aves	*Prionochilus percussus*	LC	(Nijman & Setiawan, 2000)
91	Cabai Bunga Api	Aves	*Dicaeum trigonostigma*	LC	(Nijman & Setiawan, 2000)
92	Cabai Gunung	Aves	*Dicaeum sanguinolentum*	LC	(Nijman & Setiawan, 2000)
93	Cabai jawa	Aves	*Dicaeum trochileum*	LC	(Nijman & Setiawan, 2000)
94	Bondol Hijau Binglis	Aves	*Erythrura prasina*	LC	(Nijman & Setiawan, 2000)
95	Bangkong tuli	Amphibia	*Limnonectes kuhlii*	LC	(Eprilurahman et al., 2010)
96	-	Amphibia	*Limnonectes sp*	NE	(Eprilurahman et al., 2010)
97	Bancet rawa	Amphibia	*Occidozyga sumatrana*	LC	(Eprilurahman et al., 2010)
98	Kongkang kolam	Amphibia	*Hylarana chalconota*	LC	(Eprilurahman et al., 2010)
99	Kongkang jeram	Amphibia	*Huia masonii*	LC	(Eprilurahman et al., 2010)
100	Katak-tanduk Jawa	Amphibia	*Megophrys montana*	LC	(Eprilurahman et al., 2010)
101	Bangkong Serasah	Amphibia	*Leptobrachium hasseltii*	LC	(Eprilurahman et al., 2010)
102	Katak-pohon bergaris	Amphibia	*Polypedates leucomystax*	LC	(Eprilurahman et al., 2010)
103	Katak terbang Jawa	Amphibia	*Rhacophorus margaritifer*	LC	(Eprilurahman et al., 2010)
104	Katak-pohon Emas	Amphibia	*Philautus aurifasciatus*	LC	(Eprilurahman et al., 2010)
105	-	Amphibia	*Ichthyophis sp*	NE	(Eprilurahman et al., 2010)
106	Cecak Batu	Amphibia	*Cyrtodactylus marmoratus*	LC	(Eprilurahman et al., 2010)
107	Cecak terbang	Reptilia	*Draco volans*	LC	(Eprilurahman et al., 2010)
108	Naga terbang berjanggut merah	Reptilia	*Draco haematopogon*	LC	(Eprilurahman et al., 2010)
109	Bunglon jambul hijau	Reptilia	*Bronchocela cristatella*	LC	(Eprilurahman et al., 2010)
110	Bunglon surai	Reptilia	*Bronchocela jubata*	LC	(Eprilurahman et al., 2010)
111	Bunglon Hutan	Reptilia	*Gonocephalus chamaeleontinus*	LC	(Eprilurahman et al., 2010)
112	Kadal mimikri	Reptilia	*Pseudocalotes tympanistriga*	LC	(Eprilurahman et al., 2010)
113	Kadal pohon hijau	Reptilia	*Dasia olivacea*	LC	(Eprilurahman et al., 2010)
114	Kadal kebun	Reptilia	*Eutropis multifasciata*	LC	(Eprilurahman et al., 2010)
115	Kadal serasah cokelat	Reptilia	*Eutropis rudis*	LC	(Eprilurahman et al., 2010)
116	Kadal hutan	Reptilia	*Sphenomorphus puncticentralis*	LC	(Eprilurahman et al., 2010)
117	Ular gadung	Reptilia	*Ahaetulla prasina*	LC	(Eprilurahman et al., 2010)
118	Ular viper palsu	Reptilia	*Psammodynastes pulverulentus*	LC	(Eprilurahman et al., 2010)
119	Ular alang-alang	Reptilia	*Calamaria linnaei*	LC	(Eprilurahman et al., 2010)
120	Sanca kembang	Reptilia	*Python reticulatus*	LC	(Eprilurahman et al., 2010)

121	Viper Pohon Hijau	Reptilia	*Cryptelytrops albolabris*	LC	(Eprilurahman et al., 2010)
122	Kupu male	Insecta	*Terinos clarissa*	NE	(Nuraini, 2018; Peggie, 2018)
123	Kupu pangeran hitam	Insecta	*Rohana parisatis*	NE	(Nuraini, 2018; Peggie, 2018)
124	-	Insecta	*Chersonesia peraka*	NE	(Peggie, 2018)
125	-	Insecta	*Chersonesia rahria*	NE	(Nuraini, 2018)
126	Kupu baron	Insecta	*Euthalia aconthea*	NE	(Nuraini, 2018; Peggie, 2018)
127	-	Insecta	*Discophora necho*	NE	(Nuraini, 2018; Peggie, 2018)
128	-	Insecta	*Discophora sondaica*	NE	(Nuraini, 2018; Peggie, 2018)
129	-	Insecta	*Ramelana jangala*	NE	(Nuraini, 2018)
130	-	Insecta	*Amathusia phidippus*	NE	(Nuraini, 2018)
131	-	Insecta	*Faunis canens*	NE	(Nuraini, 2018)
132	Kupu Merak	Insecta	*Junonia almana*	LC	(Nuraini, 2018)
133	-	Insecta	*Lethe confusa*	NE	(Nuraini, 2018)
134	Kupu Ilalang	Insecta	*Melanitis leda*	LC	(Nuraini, 2018)
135	-	Insecta	*Melanitis phedima*	NE	(Nuraini, 2018)
136	-	Insecta	*Melanitis zitenius*	NE	(Nuraini, 2018)
137	-	Insecta	*Mycalesis horsfieldii*	NE	(Nuraini, 2018)
138	-	Insecta	*Mycalesis mineus*	NE	(Nuraini, 2018)
139	-	Insecta	*Mycalesis moorei*	NE	(Nuraini, 2018)
140	-	Insecta	*Neorina crishna*	NE	(Nuraini, 2018)
141	Kupu Rumput Hitam	Insecta	*Orsotriaena medus*	NE	(Nuraini, 2018)
142	Kupu Ekor Gunting	Insecta	*Polyura athamas*	NE	(Nuraini, 2018)
143	-	Insecta	*Ragadia makuta*	NE	(Nuraini, 2018)
144	-	Insecta	*Stibochiona coresia*	NE	(Nuraini, 2018)
145	-	Insecta	*Tanaecia palguna*	NE	(Nuraini, 2018)
146	Kupu sayap biru	Insecta	*Graphium sarpedon*	LC	(Nuraini, 2018)
147	Kupu Pastur	Insecta	*Papilio memnon*	NE	(Nuraini, 2018)
148	Kupu belerang	Insecta	*Eurema blanda*	NE	(Nuraini, 2018)
149	Kupu rumput biasa	Insecta	*Eurema hecabe*	LC	(Nuraini, 2018)
150	-	Insecta	*Heliogomphus drescheri*	NT	(Nafisah & Soesilohadi, 2021)
151	-	Insecta	*Heliocypha fenestrata*	LC	(Nafisah & Soesilohadi, 2021)
152	-	Insecta	*Leptogomphus lansbergei*	LC	(Nafisah & Soesilohadi, 2021)
153	Capung tengget jala tunggal	Insecta	*Neurothemis ramburii*	LC	(Nafisah & Soesilohadi, 2021)
154	Capung jala lurus	Insecta	*Neurothemis terminata*	LC	(Nafisah & Soesilohadi, 2021)
155	-	Insecta	*Orthetrum glaucum*	LC	(Nafisah & Soesilohadi, 2021)
156	Capung sambar merah	Insecta	*Orthetrum pruinosum*	LC	(Nafisah & Soesilohadi, 2021)
157	Capung sambar hijau	Insecta	*Orthetrum sabina*	LC	(Nafisah & Soesilohadi, 2021)
158	Capung ciwet	Insecta	*Pantala flavescens*	LC	(Nafisah & Soesilohadi, 2021)
159	Capung Jemur Tarum	Insecta	*Potamarcha congener*	LC	(Nafisah & Soesilohadi, 2021)
160	-	Insecta	*Zygonix ida*	NE	(Nafisah & Soesilohadi, 2021)
161	-	Insecta	*Vestalis luctuosa*	LC	(Nafisah & Soesilohadi, 2021)
162	-	Insecta	*Rhinocypha heterostigma*	NT	(Nafisah & Soesilohadi, 2021)
163	-	Insecta	*Pericnemis stictica*	LC	(Nafisah & Soesilohadi, 2021)
164	-	Insecta	*Coeliccia membranipes*	LC	(Nafisah & Soesilohadi, 2021)
165	-	Insecta	*Drepanosticta gazella*	LC	(Nafisah & Soesilohadi, 2021)

166	-	Insecta	*Drepanosticta spatulifera*	DD	(Nafisah & Soesilohadi, 2021; Nugrahaningrum & Soesilohadi, 2021)
167	-	Insecta	*Drepanosticta sundana*	DD	(Nafisah & Soesilohadi, 2021)
168	Capung beludru sunda	Insecta	*Euphaea variegata*	LC	(Nafisah & Soesilohadi, 2021)

The data analysis results showed that 31 species were classified as Not Evaluated (NE), 2 species as Data Deficient (DD), 107 species as Least Concern (LC), 11 species as Near Threatened (NT), 8 species as Vulnerable (VU), 6 species as Endangered (EN), and 3 species as Critically Endangered (CR).

Two mammalian species (*Trachypithecus auratus* and *Presbytis comata*) and six avian species (*Aceros undulatus, Fregata andrewsi, Apalharpactes reinwardtii, Otus angelinae, Acridotheres javanicus*, and *Cochoa azurea*) are included in the Vulnerable group. Because of a decline in their geographic habitat or population size, these eight species face extinction.

Of the species classified as Endangered, five belong to the avian class (*Nisaetus bartelsi, Zosterops flavus, Chloropsis sonnerati, Geokichla interpres*, and *Lonchura oryzivora*), and one belongs to the mammals class (*Hylobates moloch*). The conservation and preservation of nature are necessary for species that fall into the endangered category because they are very vulnerable to extinction. The aves class includes the species *Alcedo euryzona, Eutrichomyias rowleyi*, and *Cyornis banyumas* that are classified as Critically Endangered. This species faces immediate extinction if appropriate conservation measures are not implemented, as it is critically endangered.

3.2.3 Endemic species in CITES appendix

There are many endemic species of plants and animals in the Petungkriyono Forest. Because they have evolved and adapted to a specific geographic environment, endemic species are limited to that geographic area. since of the geographic restrictions on their range, endemic species are more susceptible to the threat of extinction since they will go extinct in the wild if they disappear from their endemic locations. The Petungkriyono forest is reported to have a large number of endemic plants and animals, according to tables 3 and 4. There are 56 endemic species of animals in Java, Bali, Lombok, Kalimantan, Sumatra, and the Bangka Belitung Islands, and eight endemic plants that are unique to Java, Sumatra, Kalimantan, or Indonesia.

Table 3. Endemic flora categorized in CITES appendix

No	Local name	Class	Species	Endemic	CITES 2023
1	Damar mata-kucing	Magnoliopsida	*Shorea javanica*	Sumatera, jawa	-
2	Saninten	Magnoliopsida	*Castanopsis argente*	Sumatera, jawa, kalimantan	-
3	-	Magnoliopsida	*Dysoxylum nutans*	Indonesia	-
4	Hamerang putih	Magnoliopsida	*Ficus padana*	Indonesia	-
5	Kecapi	Magnoliopsida	*Sandoricum koetjape*	-	Appendix II
6	Mahoni	Magnoliopsida	*Swietenia mahagoni*	-	Appendix II
7	Kantung semar	Magnoliopsida	*Nepenthes adrianii*	Jawa	Appendix II
8	Gembirit	Magnoliopsida	*Tabernaemontana sphaerocarpa*	Indonesia	-
9	-	Liliopsida	*Malaxis junghuhnii*	Jawa	-
10	Anggrek peri	Liliopsida	*Oberonia similis*	Jawa	-
11	-	Liliopsida	*Dendrobium lobullatum*	-	Appendix II
12	-	Liliopsida	*Dendrobium mutabile*	-	Appendix II
13	-	Liliopsida	*Dendrobium sp1*	-	Appendix II
14	-	Liliopsida	*Dendrobium sp2*	-	Appendix II
15	-	Liliopsida	*Dendrobium sp3*	-	Appendix II
16	-	Liliopsida	*Dendrobium sp4*	-	Appendix II
17	-	Liliopsida	*Dendrobium sp5*	-	Appendix II
18	-	Liliopsida	*Eria multiflora*	-	Appendix II
19	-	Liliopsida	*Eria sp1*	-	Appendix II
20	-	Liliopsida	*Eria sp2*	-	Appendix II
21	-	Liliopsida	*Flickingeria sp1*	-	Appendix II
22	-	Liliopsida	*Flickingeria sp2*	-	Appendix II
23	-	Liliopsida	*Flickingeria sp3*	-	Appendix II
24	-	Liliopsida	*Flickingeria sp4*	-	Appendix II
25	-	Liliopsida	*Liparis rhombea*	-	Appendix II
26	-	Liliopsida	*Liparis sp1*	-	Appendix II
27	-	Liliopsida	*Liparis sp2*	-	Appendix II

 https://doi.org/10.1051/bioconf/20238003002

28	-	Liliopsida	*Liparis sp3*	-	Appendix II
29	-	Liliopsida	*Luisia Javanica*	-	Appendix II
30	Anggrek Bulan	Liliopsida	*Phalaenopsis amabilis*	-	Appendix II
31	-	Liliopsida	*Podochilus sp1*	-	Appendix II
32	-	Liliopsida	*Pholidota imbricata*	-	Appendix II
33	-	Liliopsida	*Polystacyha concentra*	-	Appendix II
34	-	Liliopsida	*Pteroceras teysmannii*	-	Appendix II
35	-	Liliopsida	*Robiquetia spatulata*	-	Appendix II
36	-	Liliopsida	*sp1*	-	Appendix II
37	-	Liliopsida	*sp2*	-	Appendix II
38	Anggrek bulu	Liliopsida	*Trichotosia ferox*	-	Appendix II
39	-	Liliopsida	*Trichotosia pauciflora*	-	Appendix II
40	-	Liliopsida	*Trichotosia sp1*	-	Appendix II
41	Anggrek permata	Liliopsida	*Macodes petola*	-	Appendix II
42	Penawar jambi	Polypodiopsida	*Cibotium barometz*	-	Appendix II

A number of species found in the Petungkriyono forest region are included in Appendices I, II, and III of the CITES, which will become effective on May 21, 2023. A list of international agreements that control trade in specific animal and plant species worldwide can be found in Appendix CITES (Convention on International Trade in Endangered Species of Wild Fauna and Flora). Appendices I, II, and III of this list each contain three sets of regulations pertaining to the control of trade in that species. The appendix (appendix II) had 35 species of Petungkriyono flora, and the appendix (4 appendix II, 2 appendix I) contained 6 species of Petungkriyono fauna. These findings were derived from the data processing results. The species included in appendix I—Hylobates moloch and Fregata andrewsi—should be taken very seriously since they face the greatest threat to extinction. Trade in these two species is virtually usually forbidden, with the exception of very particular uses and controlled environments, such as scientific study.

Table 4. Endemic fauna included in CITES Appendices

No	Local name	Class	Species	Endemic	CITES 2023
1	Lutung Hitam	Mammalia	*Trachypithecus auratus*	Jawa, bali, lombok	-
2	Surili	Mammalia	*Presbytis comata*	Jawa	-
3	Owa Jawa	Mammalia	*Hylobates moloch*	Jawa	Appendix I
4	Burung pekakak biru	Aves	*Alcedo euryzona*	Jawa	-
5	Elang Jawa	Aves	*Nisaetus bartelsi*	Jawa	-
6	Julang Mas	Aves	*Aceros undulatus*		Appendix II
7	Tepus leher putih	Aves	*Stachyris thoracica*	Jawa, bali	-
8	Tepus dada putih	Aves	*Stachyris grammiceps*	Jawa	-
9	Cica matahari	Aves	*Laniellus albonotatus*	Jawa	-
10	Brencet wergan	Aves	*Alcippe pyrrhoptera*	Jawa	-
11	Bintayung	Aves	*Fregata andrewsi*		Appendix I
12	Kasumba	Aves	*Apalharpactes reinwardtii*	Jawa	-
13	Piong	Aves	*Lophozosterops javanica*	Jawa, bali	-
14	Burung madu merah ekor panjang	Aves	*Aethopyga mystacalis*	Jawa	-
15	Burung madu gunung	Aves	*Aethopyga eximia*	Jawa	-
16	Glatik gunung	Aves	*Psaltria exilis*	Jawa	-
17	Celepuk jawa	Aves	*Otus angelinae*	Jawa	-
18	Kipasan mutiara	Aves	*Rhipidura euryura*	Jawa	-
19	Kipasan merah	Aves	*Rhipidura phoenicura*	Jawa	-
20	Cekakak jawa	Aves	*Halcyon cyanoventris*	Jawa	-
21	Gelatik jawa	Aves	*Lonchura oryzivora*		Appendix II
22	Bondol Jawa	Aves	*Lonchura leucogastroides*	Jawa, bali, lombok, sumatera	-

23	Ciung mungkal jawa	Aves	*Cochoa azurea*	Jawa	-
24	Opior jawa	Aves	*Heleia javanica*	Jawa	-
25	Serindit jawa	Aves	*Loriculus pusillus*	Jawa	-
26	Puyuh gonggong	Aves	*Arborophila javanica*	Jawa	-
27	Takur Bututut	Aves	*Megalaima corvina*	Jawa	-
28	Takur Tulungtumpuk	Aves	*Megalaima javensis*	Jawa	-
29	Takur Tohtor	Aves	*Megalaima armillaris*	Jawa	-
30	Sempur Hujan Rimba	Aves	*Eurylaimus javanicus*	Jawa	-
31	Paok Pancawarna	Aves	*Pitta guajana*	Jawa, bali	Appendix II
32	Meninting Kecil	Aves	*Enicurus velatus*	Jawa, sumatera	-
33	Cinenen Jawa	Aves	*Orthotomus sepium*	Jawa, bali, lombok	-
34	Perenjak Jawa	Aves	*Prinia familiaris*	Jawa, sumatera, bali	-
35	Cabai Gunung	Aves	*Dicaeum sanguinolentum*	Jawa, bali	-
36	Cabai jawa	Aves	*Dicaeum trochileum*	Jawa, bali, lombok, kalimantan, sumatera, kepulauan bangka belitung	-
37	Bancet rawa	Amphibia	*Occidozyga sumatrana*	Jawa, sumatera	-
38	Kongkang kolam	Amphibia	*Hylarana chalconota*	Jawa, bali, sumatera	-
39	Kongkang jeram	Amphibia	*Huia masonii*	Jawa	-
40	Katak-tanduk Jawa	Amphibia	*Megophrys montana*	Jawa	-
41	Bangkong Serasah	Amphibia	*Leptobrachium hasseltii*	Jawa, bali, sumatera	-
42	Katak terbang Jawa	Amphibia	*Rhacophorus margaritifer*	Jawa	-
43	Katak-pohon Emas	Amphibia	*Philautus aurifasciatus*	Jawa, sumatera	-
44	Cecak Batu	Amphibia	*Cyrtodactylus marmoratus*	Jawa, bali	-
45	Cecak terbang	Reptilia	*Draco volans*	Jawa, bali	-
46	Bunglon surai	Reptilia	*Bronchocela jubata*	Jawa, bali, kalimantan	-
47	Kadal mimikri	Reptilia	*Pseudocalotes tympanistriga*	Jawa, sumatera	-
48	Kadal hutan	Reptilia	*Sphenomorphus puncticentralis*	Jawa	-
49	Ular alang-alang	Reptilia	*Calamaria linnaei*	Jawa	-
50	Sanca kembang	Reptilia	*Python reticulatus*		Appendix II
51	-	Insecta	*Heliogomphus drescheri*	Jawa, sumatera	-
52	-	Insecta	*Heliocypha fenestrata*	Jawa, bali	-
53	-	Insecta	*Leptogomphus lansbergei*	Jawa, sumatera	-
54	-	Insecta	*Vestalis luctuosa*	Jawa, bali, sumatera	-
55	-	Insecta	*Rhinocypha heterostigma*	Jawa	-
56	-	Insecta	*Coeliccia membranipes*	Jawa, bali, sumatera	-
57	-	Insecta	*Drepanosticta gazella*	Jawa	-
58	-	Insecta	*Drepanosticta spatulifera*	Jawa	-
59	-	Insecta	*Drepanosticta sundana*	Jawa	-
60	Capung beludru sunda	Insecta	*Euphaea variegata*	Jawa, bali, sumatera	-

3.3 The urgency of conservative efforts: opportunities and challenges

Within the Petungkriyono forest, there is a tight link between the flora and wildlife. There are more various animal species in a region, or there may be more individual animals within a species, the more complicated the vertical structure is and the greater the diversity of the surrounding flora. A study conducted in 2016 on the white-breasted tepus bird (Strachyris grammiceps) found that the bird's population increased in a more complex microhabitat with higher levels of vegetation cover and vertical arrangement than in a smaller microhabitat with lower levels of flora biodiversity.

An additional example would be the research conducted on Lepidoptera butterflies in the Petungkriyono forest. The more varied flora that makes up the forest is consistent with the increased diversity of butterfly fauna seen in research conducted in the Sokokembang forest (Nuraini, 2018). The individual density of the Javanese gibbon in the Sokokembang forest was also found to be higher than that of the gibbon in other Petungkriyono woods, including Tlogohendro (Salmah Widyastuti & Siti Nurleily Marliana, 2016). This gibbon density was 5.15/km2. One possible explanation for this could be because the habitat structure of the Sokokembang forest in Kayupuring Village is more adapted to the requirements of Javan gibbons. Soko Kembang is a lowland tropical rain forest with a more complex forest composition than other villages in Petungkriyono. Its density is relatively high and continuous, and its altitude is appropriate for the environment (Nofiyani & Nugroho, 2016). The height of the trees that make up Sokokembang's vegetation is ideal for sleeping trees and cover trees for the Javanese Owa, as evidenced by the huge diameter trees that were discovered there (Damayanti et al., 2021a).

The intricate natural resources found in the Petungkriyono forest are a gift from Almighty God that need to be protected. Needless to say, achieving sustainability requires addressing a number of opportunities and difficulties. The following is a summary of these opportunities and challenges.

Poaching is a problem that the people in the Petungkriyono area face, particularly with regard to the aves class. This includes poaching of birds that are endangered, like Lonchura oryzivora and Fregata andrewsi, as well as several wild animal species that are protected under government regulations Number 106 of 2018 and Number 20 of 2018 and Number 7 of 1999 concerning the preservation of plant and animal species, like mongooses, pangolins, langurs, squirrels, and Javan gibbons (Hendriati et al., 2019). One of the challenges lies in the fact that Indonesia's rapid pyramidal population growth (BPS Pekalongan Regency, 2023) raises the likelihood of increased strain on forest regions due to land use changes, particularly in the Petungkriyono forest. This poses a threat to the sustainability of current species.

Two important opportunities that can be used to the fullest in attempts to protect endemic and endangered flora and wildlife are the possibility of ecotourism as a substitute for stopping unlawful hunting or poaching, and local wisdom. Planning for conservation can be made more sustainable and effective by incorporating indigenous knowledge, such as that of the nyadran gebyog culture (Ahmad & Soeprabowati, 2022). Local communities can have a significant impact on the preservation of nature and enabling coexistence with it by making use of their knowledge of the local environment.

An additional example would be the research conducted on Lepidoptera butterflies in the Petungkriyono forest. The more varied flora that makes up the forest is consistent with the increased diversity of butterfly fauna seen in research conducted in the Sokokembang forest (Nuraini, 2018). The individual density of the Javanese gibbon in the Sokokembang forest was also found to be higher than that of the gibbon in other Petungkriyono woods, including Tlogohendro (Salmah Widyastuti & Siti Nurleily Marliana, 2016). This gibbon density was 5.15/km2. One possible explanation for this could be because the habitat structure of the Sokokembang forest in Kayupuring Village is more adapted to the requirements of Javan gibbons. Soko Kembang is a lowland tropical rain forest with a more complex forest composition than other villages in Petungkriyono. Its density is relatively high and continuous, and its altitude is appropriate for the environment (Nofiyani & Nugroho, 2016). The height of the trees that make up Sokokembang's vegetation is ideal for sleeping trees and cover trees for the Javanese Owa, as evidenced by the huge diameter trees that were discovered there (Damayanti et al., 2021a).

The intricate natural resources found in the Petungkriyono forest are a gift from Almighty God that need to be protected. Needless to say, achieving sustainability requires addressing a number of opportunities and difficulties. The following is a summary of these opportunities and challenges.

Poaching is a problem that the people in the Petungkriyono area face, particularly with regard to the aves class. This includes poaching of birds that are endangered, like Lonchura oryzivora and Fregata andrewsi, as well as several wild animal species that are protected under government regulations Number 106 of 2018 and Number 20 of 2018 and Number 7 of 1999 concerning the preservation of plant and animal species, like mongooses, pangolins, langurs, squirrels, and Javan gibbons (Hendriati et al., 2019). One of the challenges lies in the fact that Indonesia's rapid pyramidal population growth (BPS Pekalongan Regency, 2023) raises the likelihood of increased strain on forest regions due to land use changes, particularly in the Petungkriyono forest. This poses a threat to the sustainability of current species.

Two important opportunities that can be used to the fullest in attempts to protect endemic and endangered flora and wildlife are the possibility of ecotourism as a substitute for stopping unlawful hunting or poaching, and local wisdom. Planning for conservation can be made more sustainable and effective by incorporating indigenous knowledge, such as that of the nyadran gebyog culture (Ahmad & Soeprabowati, 2022). Local communities can have a significant impact on the preservation of nature and enabling coexistence with it by making use of their knowledge of the local environment.

Reducing unlawful hunting and poaching activities is another benefit of ecotourism. An industry of tourism known as "ecotourism" is one that places an emphasis on pleasurable and environmentally friendly encounters with the natural world. When done responsibly, ecotourism may give local communities an additional source of income without harming natural environments or jeopardizing threatened plant and animal populations. While making money to help environmental protection, ecotourism offers visitors the chance to witness endangered creatures in their natural habitat. According to (Andriyani et al., 2022; M. Anam Rahardyan et al., 2021; T. A. Setiawan et al., 2020; Yusron et al., 2020), Petungkriyono offers a variety of ODTWA, including water-based tourism, butterfly gardens, and monkey and bird watching.

Two effective strategies for promoting conservation include making use of local knowledge and growing ecotourism that is sustainable. In addition to safeguarding native and threatened plant and animal species, this also offers prosperous business prospects to the surrounding community, which may inspire them to preserve the environment. This approach can preserve both human wellbeing and natural biodiversity at the same time.

4 Conclusion

Numerous species of Indonesian plants and animals can be found in the Petungkriyono forest area. Because of its close proximity to the village, it will always be impacted by the different ways that the locals engage with the Petungkriyono forest, which will have both positive and bad effects on the sustainability of the current flora and fauna.

It was noted that the Petungkriyono forest area was home to 204 species of plants and 168 species of animals. According to the IUCN, six of these flora are classified as vulnerable (VU) species, while four are classified as endangered (EN). A total of 35 species are listed in the CITES 2023 Appendix (Appendix II), of which 8 are endemic. Regarding the fauna, eight species are classified as Vulnerable (VU), six as Endangered (EN), and three as Critically Endangered (CR). Six of the 56 indigenous species in the fauna are included in the CITES Appendix (Appendice I, II) for 2023. It is imperative that conservative measures be taken right away in order to preserve the biodiversity that currently exists. This can be done by maximizing possibilities and overcoming obstacles.

Acknowledgements

We would like to convey our gratitude to Relung Foundation for providing us with the opportunity to join the research. This research is also part of research of RP 1 Tema 4 OR HLBRIN 2023.

References

1. Abdurrahman, I., Mujianto, M., Dewi, A. E., & Fahmi, A. A. (2021). M e l e s t a r i k a n Petungkriyono Reportase hasil studi eksploratif dan assessment awal Pengembangan ekonomi hijau. Relung Indonesia.Agustin, D., & Widowati, H. (2015). Inventarisasi Keanekaragaman Anggrek (Orchidaceae) Di Hutan Resort Way Kanan Balai Aman Nasional Way Kambas Sebagai Sumber Informasi Dalam Melestarikan Plasma Nutfah. BIOEDUKASI (Jurnal Pendidikan Biologi), 6(1). https://doi.org/10.24127/bioedukasi.v6i1.156
2. Ahmad, H., & Soeprabowati, T. R. (2022). Kearifan Tradisi Budaya Sebagai Sarana Pelestaraian Lingkungan Serta Potensi Ekowisata Pada Komunitas Masyarakat Di Sekitar Hutan Lindung Petungkriyono (Studi Kasus Nyadran Gebyog). 03(02).
3. Ahmadin, K. (2016). Petungkriyono Forest Sub-District Petungkriyono, Pekalongan Regency, Central Java.
4. Andriyani, S. D., Sifa, M., & Rif, R. A. (2022). Kajian Pengelolaan Potensi Ekowisata Petungkriyono, Kabupaten Pekalongan. Jounal of Enviromental Science Sustainable, 3(1), Article 1. https://doi.org/10.31331/envoist.v3i1.2250
5. Apriliyani, E. (2017). Ekologi Burung Julang Emas (Rhyticeros undulatus Shaw, 1881) di Hutan Sokokembang Pekalongan Jawa Tengah: Vol. Skripsi. Universitas Islam Negeri Syarif Hidayatullah.
6. Arrisaldi, T. (2018). Pemetaan Risiko Gerakan Tanah Di Kecamatan Petungkriyono. Kajen: Jurnal Penelitian Dan Pengembangan Pembangunan, 2. https://doi.org/10.54687/jurnalkajenv02i01.1
7. BPS Pekalongan Regency. (2023). BPS Pekalongan Regency. https://pekalongankab.bps.go.id/publication.html
8. Cooper, H. M. (1998). Synthesizing research: A guide for literature reviews: Vol. (Vol. 2). Sage Publications, Inc.
9. Damayanti, I., Bambang, A. N., & Soeprobowati, T. R. (2021a). Carbon stocks of Petungkriyono mixed forest as climate change mitigation. IOP Conference Series: Earth and Environmental Science, 724(1), 012064. https://doi.org/10.1088/1755-1315/724/1/012064
10. Damayanti, I., Bambang, A. N., & Soeprobowati, T. R. (2021b). Plant diversity of Petungkriyono Forest of Dieng Plateau, Central Java, Indonesia. Biodiversitas Journal of Biological Diversity, 22(8). https://doi.org/10.13057/biodiv/d220849
11. Damayanti, I., Nur Bambang, A., & Soeprobowati, T. (2020). The Analyze of Collaborative Management Perspective of Petungkriyono Forest. E3S Web of Conferences, 202, 05014. https://doi.org/10.1051/e3sconf/202020205014
12. Damayanti, I., Nur Bambang, A., & Soeprobowati, T. R. (2018). Extended benefit cost analysis as an instrument of economic valuated in Petungkriyono forest ecosystem services. Journal of Physics: Conference Series, 1025, 012034. https://doi.org/10.1088/1742-6596/1025/1/012034
13. Desiana, ., Roslinda, E., & Kartikawati, S. M. (2019). Jasa Lingkungan Taman Keanekaragaman Hayati (Kehati) Badan Lingkungan Hidup Di Kabupaten Sekadau Tahun 2017. Jurnal Hutan Lestari, 7(1). https://doi.org/10.26418/jhl.v7i1.32168
14. Eprilurahman, R., Qurniawan, T. F., & Kusuma, K. I. (2010). Studi Awal Keanekaragaman

Herpetofauna Dipetungkriyono, Kabupaten Pekalongan, Provinsi Jawa Tengah .,.
15. Fatmasari, D., Nugroho, A. S., & Kaswinarni, F. (2017). Struktur Umur dan Rasio Jenis Kelamin Lutung Budeng (Trachypithecus Auratus) di Hutan Lindung Sokokembang Petungkriyono Pekalongan Jawa Tengah.
16. Fikri, M. K., & Adinugraha, H. H. (2021). Pendampingan Regenerasi Penderes Nira Aren Desa Kayupuring Kecamatan Petungkriyono.
17. Hendriati, I. T., Abdoellah, O. S., & Parikesit, P. (2019). SEMINAR MENYONGSONG SATU ABAD INDONESIA.
18. Kusmana, C., & Hikmat, A. (2015). Keanekaragaman Hayati Flora Di Indonesia. Jurnal Pengelolaan Sumberdaya Alam Dan Lingkungan (Journal of Natural Resources and Environmental Management), 5(2), 187. https://doi.org/10.29244/jpsl.5.2.187
19. Lestari, I. (2019). Keanekaragaman jenis tumbuhan paku epifit di Hutan Petungkriyono Kabupaten Pekalongan, Jawa Tengah.
20. M. Anam Rahardyan, Loekman Mohamadi, & Wawan Destiawan. (2021). Perancangan Kawasan Wisata Telaga Mangunan di Petungkriyono Kabupaten Pekalongan. SARGA: Journal of Architecture and Urbanism, 15(2), 28–38. https://doi.org/10.56444/sarga.v15i2.151
21. Mardiyana, M., Murningsih, M., & Utami, S. (2019). Inventarisasi Anggrek (Orchidaceae) Epifit di Kawasan Hutan Petungkriyono Pekalongan Jawa Tengah. Jurnal Akademika Biologi, 8(2), Article 2.
22. Nafisah, N. A., & Soesilohadi, R. C. H. (2021). Community Structure of Dragonfly (Ordo: Odonata) in Natural Forest and Tourist Sites Petungkriyono Forest, Central Java, Indonesia. Journal of Tropical Biodiversity and Biotechnology, 6(3), 67328. https://doi.org/10.22146/jtbb.67328
23. Nijman, V. J., & Setiawan, I. (2000). Penilaian sepintas keragaman fauna di Pegunungan Dieng. Yayasan Pribumi Alam Lestari.
24. Nofiyani, T., & Nugroho, A. S. (2016). Aktivitas Harian Owa Jawa (Hylobates moloch) di Hutan Lindung Petungkriyono Kabupaten Pekalongan Jawa Tengah. Prosiding Seminar Nasional Sains dan Entrepreneurship III Tahun 2016.
25. Nugrahaningrum, A., & Soesilohadi, R. C. H. (2021). Variations of Movement, Dispersal, and Morphometrics among Subpopulations of Javan Endemic Damselfly, Drepanosticta spatulifera (Odonata: Platystictidae) in Petungkriyono Forest. Journal of Tropical Biodiversity and Biotechnology, 6(3), 65612. https://doi.org/10.22146/jtbb.65612
26. Nuraini, S. (2018). Komunitas Kupu-Kupu (Lepidoptera:Rhopalocera) Di Hutan Sokokembang Pekalongan, Jawa Tengah.
27. Peggie, D. (2018). KAJIAN DIVERSITAS KUPU-KUPU (Lepidoptera: Papilionoidea & Hesperioidea) DAN POTENSI PEMANFAATANNYA. 02(02).
28. Pradana, B., Ariani, N. M., & Pugara, A. (2020). Pengaruh Penggunaan Lahan terhadap Suhu Permukaan di Kabupaten Pekalongan Tahun 2020. Kajen: Jurnal Penelitian dan Pengembangan Pembangunan, 4(02), 92–100. https://doi.org/10.54687/jurnalkajenv4i02.1
29. Prapitasari, B., Kurniawan, A. P., & Muharam, D. H. (2020). Keanekaragaman dan Kemelimpahan Jenis Anggrek (Orchidaceae) di Resort Selabintana Taman Nasional Gunung Gede Pangrango (TNGGP) Jawa Barat. Biosfer : Jurnal Biologi Dan Pendidikan Biologi, 5(1), Article 1. https://doi.org/10.23969/biosfer.v5i1.2569
30. Rachmawati, Y., Pramudya, M., & Setiawan, A. (2022). Distribution of the critically endangered javan blue-banded kingfisher Alcedo euryzona along the Welo River flow in the Petungkriyono forest. IOP Conference Series: Earth and Environmental Science, 976(1), 012004. https://doi.org/10.1088/1755-1315/976/1/012004
31. Randolph, J. (2009). A guide to writing the dissertation literature review. Practical Assessment, Research, and Evaluation,: Vol. 14(1), 13.
32. Rasyid, F. (2014). Permasalahan dan Dampak Kebakaran Hutan. 4.
33. Relung Indonesia. (2023). Analisis Spasial Multi Kriteria (SCMA). Relung Indonesia.
34. Salmah Widyastuti, & Siti Nurleily Marliana. (2016). Densitas Dan Distribusi Populasi Owa Jawa (Hylobates Moloch Audebert, 1798) Di Hutan Tlogohendro, Petungkriyono, Pekalongan, Jawa Tengah. https://doi.org/10.13140/RG.2.2.28194.45760
35. Setiawan, A. (2022). Keanekaragaman Hayati Indonesia: Masalah dan Upaya Konservasinya. Indonesian Journal of Conservation, 11(1). https://doi.org/DOI 10.15294/ijc.v11i1.34532
36. Setiawan, A., Nugroho, T. S., Wibisono, Y., Ikawati, V., & Sugardjito, J. (2012). Population density and distribution of Javan gibbon (Hylobates moloch) in Central Java, Indonesia. Biodiversitas Journal of Biological Diversity, 13(1). https://doi.org/10.13057/biodiv/d130105
37. Setiawan, T. A., Ilyas, A., & Binabar, S. W. (2020). Pemetaan Potensi Pariwisata Petungkriyono Berbasis Web Geographycal Information System (Gis) Sebagai Cultural Techno Forestry Park Kabupaten Pekalongan. I.
38. Tanguay, L. (2020). How Preserving Biodiversity Mitigates the Impacts of Small-scale Land Grab on Livelihoods and Agricultural Production in Central Java. 35(1).
39. Wibowo, D. E., Jannah, K. D., & Permanasari, P. (2022). Pengembangan Rural Tourism Melalui Pemberdayaan Pokdarwis Menggunakan Life Skill di Kecamatan Petungkriyono Kabupaten Pekalongan. Ruang Cendekia : Jurnal Pengabdian Kepada Masyarakat, 1(2), 121–131. https://doi.org/10.55904/ruangcendekia.v1i2.68
40. Widhiono, I. (2009, October 6). " Konservasi Keanekaragaman Hayati Hutan Petungkriyono Melalui Ekoswisata" ("Biodiversity Conservation

Of Petungkriyono Forest Remnant With Ecotourism"). Dr. Rer.Nat. IMAM WIDHIONO MZ, MS. https://widhiono.wordpress.com/2009/10/06/%e2%80%9c-konservasi-keanekaragaman-hayati-hutan-petungkriyono-melalui-ekoswisata%e2%80%9d-%e2%80%9cbiodiversity-conservation-of-petungkriyono-forest-remnant-with-ecotourism%e2%80%9d/

41. Widodo, W. (2021). Perbandingan Komunitas Burung di Lahan Perkebunan Kopi dengan Naungan Pohon Alami. BIOEDUSAINS:Jurnal Pendidikan Biologi dan Sains, 4(2), 336–345. https://doi.org/10.31539/bioedusains.v4i2.2414
42. Wulandari, Y., Nugroho, A. S., & Ulfah, M. (2016). Karakteristik Pohon Tidur dan Pohon Cover Owa Jawa (Hylobates moloch) di Hutan Lindung Petungkriyono Jawa Tengah. SNSE III 2016.
43. Yulia, N. D., & Yanti, R. M. (2010). Anggrek Epifit Dan Pohon Inangnya Di Kawasan Gunung Penanggungan, Pasuruan – Jawa Timur. Berk. Penel. Hayati Edisi Khusus, 4A (37–40).
44. Yusron, A., Rosyadi, I., & Handayani, H. (2020). Sistem Informasi Wisata Petungkriyono Berbasis Android (Studi Kasus Telaga Mangunan). Vol ., 1.

Can iPhone/iPad LiDAR data improve canopy height model derived from UAV?

Deha Agus Umarhadi[1] , Senawi[1], *Wahyu* Wardhana[1], *Emma* Soraya[1*], *Aqmal Nur* Jihad[2], and *Fiqri* Ardiansyah[2]

[1]Department of Forest Management, Faculty of Forestry, Universitas Gadjah Mada, 55281 Yogyakarta, Indonesia
[2]Department of Silviculture, Faculty of Forestry, Universitas Gadjah Mada, 55281 Yogyakarta, Indonesia

Abstract. Aerial images resulting from unmanned aerial vehicle (UAV) are widely used to estimate tree height. The filtering method is required to distinguish between ground and off-ground point clouds to generate a canopy height model. However, the filtering method is not always perfect since UAV data cannot penetrate canopies into the forest floor. The release of iPhone/iPad devices with built-in LiDAR sensors enables the more affordable use of LiDAR for forestry study, including the measurement of local topography below forest stands. This study investigates to what extent iPhone/iPad LiDAR can improve the accuracy of canopy height model from the UAV. The integration of UAV and iPhone/iPad LiDAR data managed to increase the accuracy of tree height model with a mean absolute error (MAE) of 2.188 m, compared to UAV data (MAE = 2.446 m). This preliminary study showed the potential of combining UAV and iPhone/iPad LiDAR data for estimating tree height.

1 Introduction

Three-dimensional arrangement of individual trees within a forest has a significant impact on ecosystem functions as well as nutrients, water, and carbon cycles [1]. Tree height has been widely used to describe tree volume, biomass, and carbon stock, which are important for forest management [2,3].

Traditionally, tree height is measured using tangent method by means of survey tools that can quantity angles, such as clinometer [4]. The more recent terrestrial method employs laser rangefinders to directly measure tree height using laser technology [5]. However, this conventional approach takes intensive labour and time and is limited when implemented in dense and tall canopy forests [4,6,7].

The use of photogrammetry technique from aerial photos taken by unmanned aerial vehicles (UAVs) to calculate tree height has been commonly used these years [8,9]. UAV image captured the height of surface; thus, to obtain information on tree height commonly called canopy height model, the terrain conditions need to be eliminated using filtering method [10,11]. Despite filtering method implemented, in some cases, an inaccurate canopy height model may occur since aerial image cannot penetrate canopies.

Light detection and ranging (LiDAR) sensor has the ability to penetrate the gaps within tree crowns and capture the middle part of stands and even the forest floor [12,13]. Nonetheless, the use of LiDAR technology is costly in current. Alternatively, a low-cost Hand-held Laser Scanning (HLS) has been introduced by Apple in 2020 by releasing iPhone and iPad products with a built-in LiDAR sensor [14]. Although the measurement range is far below traditional LiDAR sensor (maximum distance of 5 m), iPhone/iPad LiDAR can precisely measure objects [15]. This device has been explored for forest inventory and reported to provide reliable information comparable to terrestrial and mobile laser scanner [16,17].

As iPhone/iPad LiDAR scans objects on the ground, it offers an opportunity to be integrated with UAV-based surface model to generate canopy height mode (CHM). Thus, this present study investigates the feasibility of apple lidar to improve CHM generated by UAV.

2 Methodology

2.1 Study area

Our study area is located in Wanagama I Education Forest or simply called Wanagama Forest. Wanagama Forest is situated in Yogyakarta, Indonesia with a total area of 622.25 ha. This study focused on the teak stands (*Tectona grandis*) in Plot 13 of Wanagama Forest with an extent between 110.5286°–110.5334° E and 7.8968°–7.9066° S or 448041–448565 mE and 9127080–9125999 mN (UTM Zone 49M) (Figure 1). The teak stands comprise two different types, i.e., Mega-clone type and conventional type. The Mega teak stands were planted in 2004 with a planting distance of 6 × 2 m [18]. Compared to conventional teak, Mega teak has superior growth characteristics in diameter, height, and stem alignment [19].

* Corresponding author: esoraya@ugm.ac.id

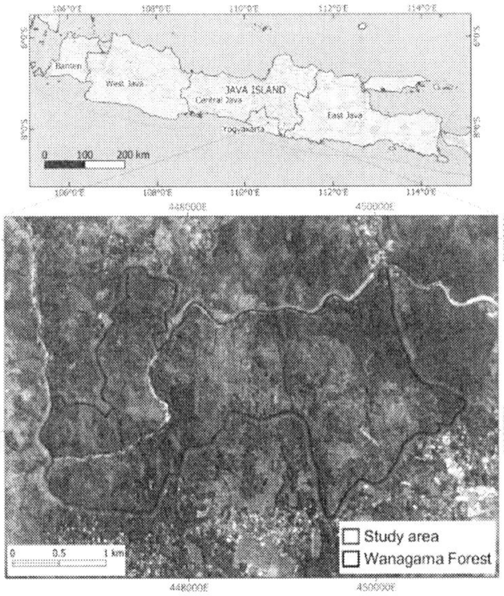

Fig. 1. Study area located in Wanagama I Education Forest, Yogyakarta, Indonesia.

2.2 Materials and methods

This study compared tree height estimates derived from UAV and the integration between UAV and LiDAR data. The flowchart is illustrated in Figure 2, while the detailed methods are provided in the following sub-sections.

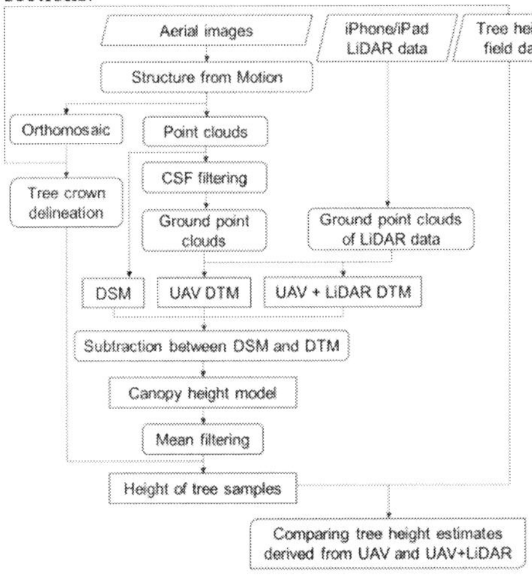

Fig. 2. Flowchart of the present study.

2.2.1 Aerial images

Aerial images were taken by deploying three quadcopter drones simultaneously to minimise the time difference in capturing the study area. We used DJI Mavic 2 Pro with an autopilot according to the planned flight paths programmed on DroneDeploy. The flight height was 150 m from the ground at the base station, while the overlap and sidelap were set to 80%. The aerial images were processed in Agisoft Metashape using Structure from Motion (SfM) algorithm to generate georeferenced orthomosaic and 3D data in the form of point clouds [20]. The resulted orthomosaic has a spatial resolution of 4 cm (Figure 3a).

2.2.2 iPhone/iPad LiDAR data

The Apple products equipped by a LiDAR sensor are iPhone 12 Pro, iPad Pro 2020, and the latest Pro version of those models. This study utilised iPhone and iPad mobile devices to record the point clouds of the ground surface using the LiDAR sensor operated with ForestScanner application [15].

The iPhone/iPad LiDAR scanning system was operated by holding the devices at the breast height (±1.3 m) while walking along the transects. The transects were designed to with a minimum distance of 40 m and considers the variability of topography. There were 11 transects within the teak plantation areas (Figure 3b).

2.2.3 Canopy height model

The point clouds generated from SfM processing contain X and Y coordinates, and Z value which represents a relative height to the ground of all surface objects. A filtering method was implemented to distinguish off-ground from ground objects. We used a cloth simulation filtering (CSF) approach by inverting the Z plane before a simulated cloth drops the surface to cover the inverted surface then performing a re-inversion [21].

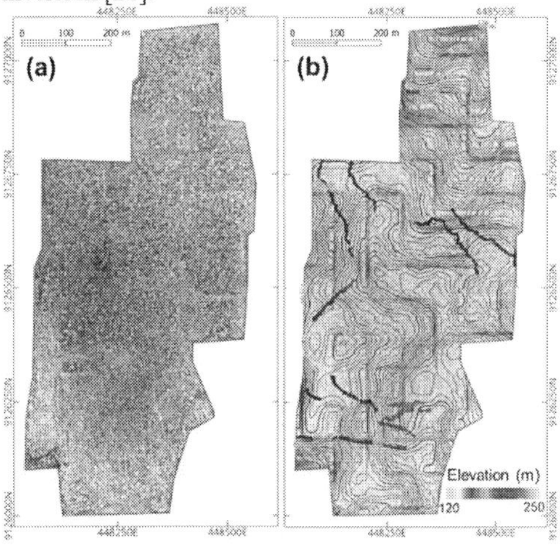

Fig. 3. (a) Aerial images captured from UAV and (b) visualization of filtered LiDAR point clouds measured showing the elevation of the terrain condition captured by iPhone/iPad.

The point cloud processing was conducted in CloudCompare software. CSF algorithm was applied on UAV point clouds with user-defined parameters. Cloth resolution, maximum iteration, and classification threshold were set to 1.2, 500, and 0.5, respectively, and a slope factor was set to "Relief" terrain condition. Subsequently, point cloud noise elimination was performed. CSF method was also implemented for the iPhone/iPad LiDAR data to eliminate tree stems and other non-surface objects which were recorded during the scanning. The parameters selected to filter iPhone/iPad LiDAR data are cloth resolution = 0.8, maximum iteration = 500, and classification threshold = 0.5.

Using the Rasterize tool in CloudCompare, the off-ground UAV point clouds were transformed into digital surface model (DSM) with a resolution of 4 cm. Similarly, the ground UAV point clouds were rasterized to produce digital terrain model (DTM). We also combined the ground UAV point clouds with filtered iPhone/iPad LiDAR data to generate DTM. Canopy height models (CHMs) were generated by a subtraction between DSM and DTMs with and without the integration with LiDAR data. A mean filtering was applied to the CHMs using a 3 × 3 window to minimise the extremely different values.

2.2.4 Tree height data

The ground truth data of tree height were measured using Nikon Forestry Pro II laser rangefinder with 2-points measurement mode. The tree height here refers to Larjavaara and Muller-Landau [4], who defined it as the vertical distance between the topmost of the tree (either living or dead part) and the base of the tree. We measured the tree height within approximately 20 × 20 m plots with a total of 6 sample plots. The measured tree locations were recorded using the iPhone/iPad LiDAR devices.

The tree height data were derived from the CHMs by calculating the maximum height value within the crown of the corresponding trees. The tree crowns were delineated visually from the orthomosaic. In total, there are 148 tree height data for the analysis (Figure 4).

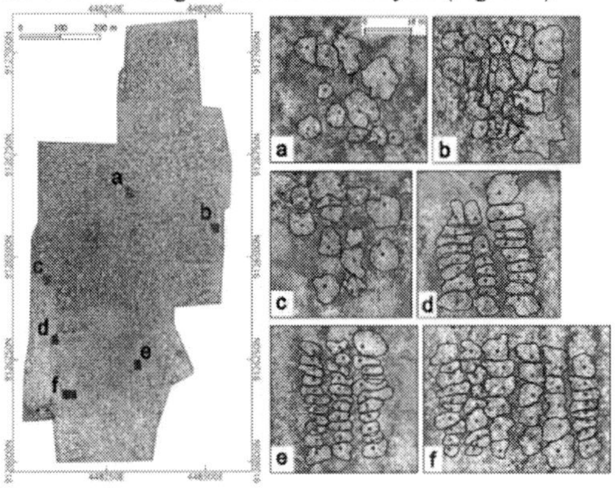

Fig. 4. Orthomosaic of the teak plantation area and the distribution of tree height samples collected in the field.

2.2.5 Data analysis

One-way analysis of variance (ANOVA) ($\alpha = 0.05$) was carried out to test the significant differences of tree height and CHMs generated from UAV and UAV + iPhone/iPad. In addition, Pearson's correlation analysis was conducted between tree height data and both CHMs, separately. Further, a linear regression analysis was performed with the ground truth data as the independent variable. We randomly divided the tree height data into two groups, i.e., training (104 samples) and validation (44 samples). The accuracy assessment was conducted by calculating the mean absolute error (MAE).

3 Results and discussion

CSF algorithm was intended to separate between the ground and off-ground surface of the point clouds generated from UAV data. The point clouds containing off-ground surface were used to produce digital surface model as shown in Figure 5a. The digital terrain models were derived from the UAV point clouds only and the combination of UAV point clouds and iPhone/iPad LiDAR. As depicted in Figure 5, visually, there are no significant differences between both resulted DTMs. While the profiles plotted as shown in Figure 6, the discrepancy is quite clear, where local terrain conditions below the teak stands varied as shown by the DTM from UAV + iPhone/iPad LiDAR. It can be seen that DTM derived from UAV + iPhone/iPad is smoother following the local topography compared to DTM from UAV that interpolated area between ground points indicated by the plain terrain as shown in Profile A. However, some errors were also recorded using iPhone/iPad, showing the rough topography and extreme slopes. This might be due to an imperfect filtering method in excluding the off-ground points.

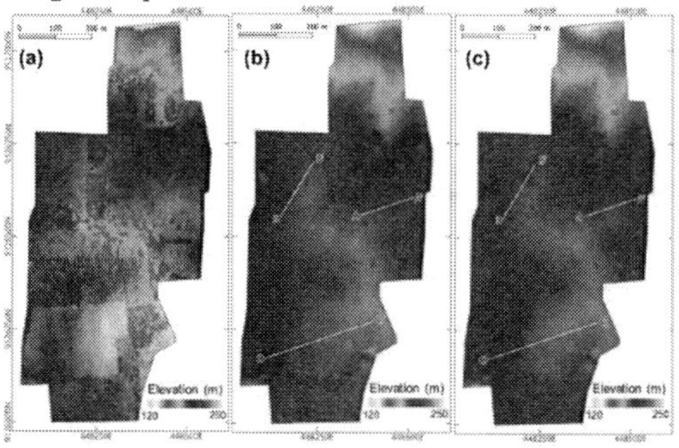

Fig. 5. Rasterization results of point clouds: (a) digital surface model, (b) digital terrain model derived from UAV, and (c) digital terrain model derived from the combination of UAV and iPhone/iPad LiDAR.

Canopy height models (CHMs) were generated from the subtraction results between DTMs and DSM, thus resulting in CHM from UAV and CHM from UAV + iPhone/iPad (Figure 7). At a glance, the UAV CHM shows higher values than UAV + iPhone/iPad CHM, particularly in the southern part of the study area.

The ANOVA test showed a significant difference between three populations (tree height, CHM from UAV, and CHM from UAV + iPhone/iPad) with a p-value smaller than 0.001. We plotted scatterplots to see the relationship between CHMs and tree height data. Figure 8 shows that both CHMs are overestimated, particularly for CHMs' values of more than 15 m. The MAE values are 5.24 and 4.05 m, while the correlation values are 0.542 and 0.601 for UAV and UAV + iPhone/iPad CHMs, respectively.

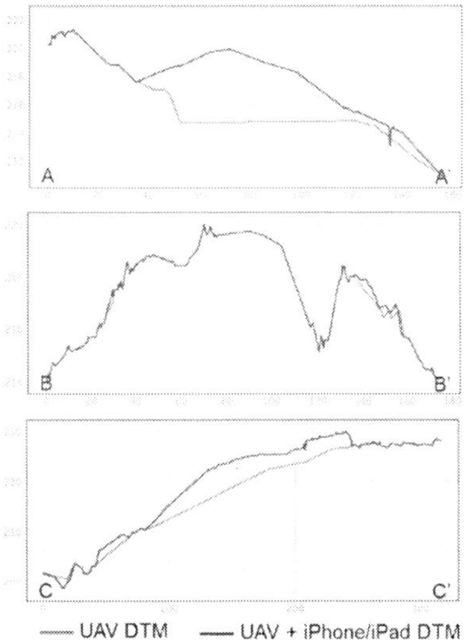

Fig. 6. Profiles showing the differences between DTMs from UAV and UAV + iPhone/iPad.

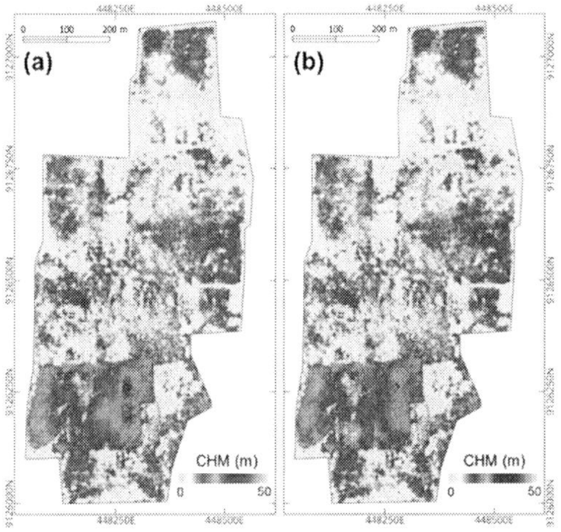

Fig. 7. Canopy height models derived from (a) UAV and (b) UAV + iPhone/iPad combined.

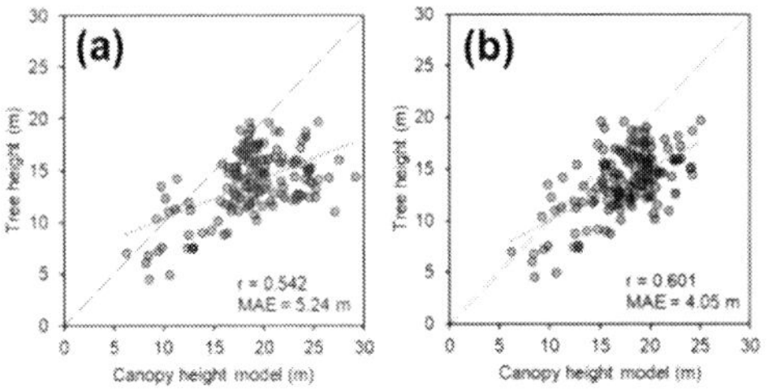

Fig. 8. Scatterplots between tree height and CHMs derived from (a) UAV and (b) UAV + iPhone/iPad.

This study extracted the maximum value of the CHMs within the corresponding tree crowns [9]. From above, UAV can capture the condition of the object's surface so that the maximum height can be easily detected. In contrast, the field measurement using laser rangefinders can lead to erroneous since the topmost of the tree might be obscured by the leaves and branches, especially in the dense coverage as reported by Larjavaara and Muller-Landau [4].

As the correlation values are significant (p-values < 0.05), we modelled the tree height based on the CHMs to correct the overestimation. The total samples are 148, of which 104 samples are for constructing linear regression model. The linear regression plots are illustrated in Figure 9. As observed, the coefficients of determination (R^2) are 0.3082 for UAV only and 0.3761 for UAV + iPhone/iPad CHMs.

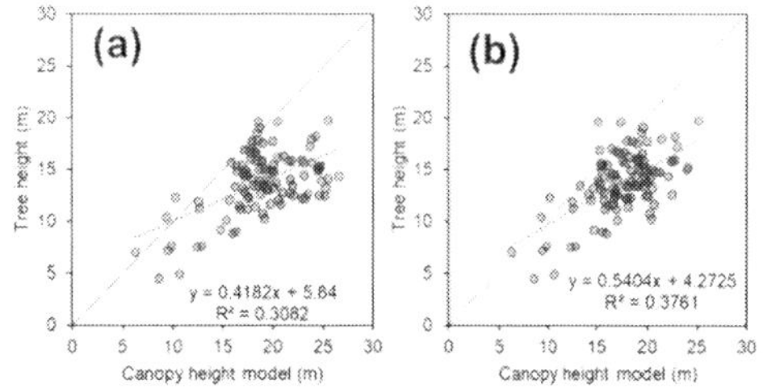

Fig. 9. Linear regression plots between tree height (y-axis) and CHMs (x-axis) derived from (a) UAV and (b) UAV + iPhone/iPad.

The regression equations were then applied to CHMs to produce tree height models from UAV and UAV + iPhone/iPad (Figure 10). Subsequently, we assessed the accuracy using the validation samples. Using the MAE metric, we found that the tree height estimated using the UAV + iPhone/iPad managed to produce a slightly higher accuracy (MAE = 2.188 m) compared to UAV only (MAE = 2.446 m). This improved accuracy is observed due to the local topography correction as the terrain conditions below teak stands were measured by iPhone/iPad.

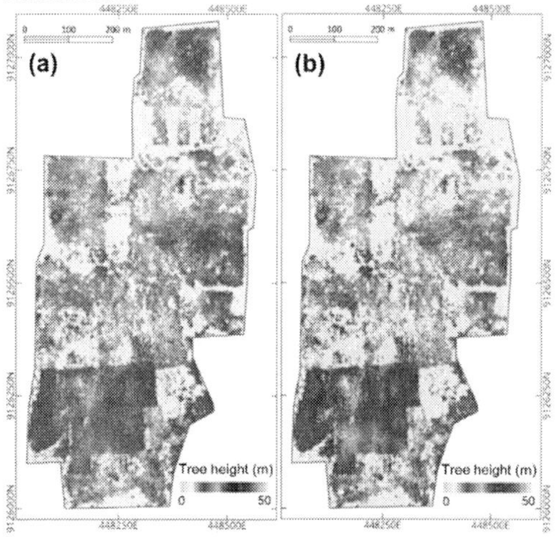

Fig. 10. Tree height modelled from canopy height models derived from (a) UAV and (b) UAV + iPhone/iPad combined.

Although the accuracy increased by 0.258 m, this preliminary study shows the potential of iPhone/iPad LiDAR to be integrated with UAV to generate a more accurate a tree height model. The slight increase is also due to the sparse iPhone/iPad LiDAR measurement transects. Future studies should consider denser and more systematic transects to cover the high variability of local topography.

4 Conclusion

This study demonstrates LiDAR data measured using the advanced iPhone/iPad devices to be integrated with UAV data to generate canopy height model (CHM). CHMs resulting from UAV and UAV + iPhone/iPad show overestimation compared to field data. Linear regression analysis was used to model tree height from CHMs. We found that the combination of UAV and iPhone/iPad can improve the accuracy of the tree height model, showing an MAE of 2.188 m, compared to the UAV-only result (MAE = 2.446 m).

Acknowledgements

This research is funded by the Young Lecturer Research Grant Universitas Gadjah Mada Number 5985/UN1.P.II/Dit-Lit/PT.01.03/2023. We are thankful to assistants of the Laboratory of Spatial Information System and Forest Mapping, Faculty of Forestry, Universitas Gadjah Mada (i.e., Vito, Rafindra, Megan, Farhan, Johan, Aufa, Opi, Arion, Wafiq, Gery, Lian, and Chica) for their help in data acquisition and pre-processing.

References

1. H. H. Shugart, S. Saatchi, and F. G. Hall, J. Geophys. Res. **115**, 1 (2010)
2. M. O. Hunter, M. Keller, D. Victoria, and D. C. Morton, Biogeosciences **10**, 8385 (2013)
3. D. A. King, S. J. Wright, and J. H. Connell, J. Trop. Ecol. **22**, 11 (2006)
4. M. Larjavaara and H. C. Muller-Landau, Methods Ecol Evol **4**, 793 (2013)
5. I. S. Saliu, B. Satyanarayana, M. A. Bin Fisol, G. Wolswijk, C. Decannière, R. Lucas, V. Otero, and F. Dahdouh-Guebas, Estuarine, Coastal and Shelf Science **248**, 106971 (2021)
6. C.-H. Chung, C.-H. Wang, H.-C. Hsieh, and C.-Y. Huang, GIScience & Remote Sensing **56**, 1289 (2019)
7. G. B. West, J. H. Brown, and B. J. Enquist, Nature **400**, 664 (1999)
8. R. Avtar, S. A. Suab, M. S. Syukur, A. Korom, D. A. Umarhadi, and A. P. Yunus, Remote Sensing **12**, 3030 (2020)
9. V. Nasiri, A. A. Darvishsefat, H. Arefi, M. Pierrot-Deseilligny, M. Namiranian, and A. Le Bris, Can. J. For. Res. **51**, 962 (2021)
10. C. Serifoglu Yilmaz and O. Gungor, Geocarto International **33**, 522 (2018)
11. M. Zeybek and İ. Şanlıoğlu, Measurement **133**, 99 (2019)
12. Q. Liu, L. Fu, Q. Chen, G. Wang, P. Luo, R. P. Sharma, P. He, M. Li, M. Wang, and G. Duan, Remote Sensing **12**, 2884 (2020)
13. F. Morsdorf, C. Eck, C. Zgraggen, B. Imbach, F. D. Schneider, and D. Kükenbrink, The Leading Edge **36**, 566 (2017)
14. M. Mokroš, T. Mikita, A. Singh, J. Tomaštík, J. Chudá, P. Wężyk, K. Kuželka, P. Surový, M. Klimánek, K. Zięba-Kulawik, R. Bobrowski, and X. Liang, International Journal of Applied Earth Observation and Geoinformation **104**, 102512 (2021)
15. S. Tatsumi, K. Yamaguchi, and N. Furuya, Methods Ecol Evol **14**, 1603 (2023)
16. R. Bobrowski, M. Winczek, K. Zięba-Kulawik, and P. Wężyk, Urban Forestry & Urban Greening **79**, 127815 (2023)
17. C. Gollob, T. Ritter, R. Kraßnitzer, A. Tockner, and A. Nothdurft, Remote Sensing **13**, 3129 (2021)
18. M. Na'iem, P. A. Rudian, S. M. Hasibuan, A. M. Idhom, A. Mustaqim, Sutriyati, and M. F. Cahyono, *Wanagama: Kisah Terciptanya Hutan Pendidikan, Konservasi Dan Kesejahteraan Sosial Ekonomi Bagi Rakyat Sekitar* (Samudra Biru, Yogyakarta, Indonesia, 2020)
19. F. Hidayati, I. T. Fajrin, M. R. Ridho, W. D. Nugroho, S. N. Marsoem, and M. Na'iem, Jurnal Ilmu Kehutanan **10**, (2016)
20. J. Iglhaut, C. Cabo, S. Puliti, L. Piermattei, J. O'Connor, and J. Rosette, Curr Forestry Rep **5**, 155 (2019)
21. W. Zhang, J. Qi, P. Wan, H. Wang, D. Xie, X. Wang, and G. Yan, Remote Sensing **8**, 501 (2016)

Comparative effect of polymer coating materials on physical properties of NPK fertilizer pellet

Gatot Trimulyadi Rekso*

Research and Technology Center for Application of Isotope and Radiation-BRIN, Jl. Lebak Bulus Raya No. 49, Jakarta 12440, Indonesia

Abstract. Slow-release fertilizer is one method to reduce the rate of loss of fertilizer from the soil due to leaching by rainwater or irrigation water. One of the agents that can be used for slow-release fertilizers is polymeric materials coated on NPK fertilizers. The NPK fertilizer coating polymer material used was a mixture of starch, acrylamide, PVA, and chitosan that was formulated and irradiated at a dose of 20 kGy. The purpose of this study was to obtain the best comparison between coating materials and NPK fertilizers made with a pelletizer machine. NPK fertilizer is mashed and mixed with polymer coating material with a weight ratio of NPK coating material/fertilizer; 1/9, 2/8, and 3/7. The results obtained showed the largest pellet yield capacity using a ratio of coating material and NPK fertilizer with a weight ratio of 1/9 of 4.28 g/minute and the greatest durability value at a ratio of 3/7 with a value of 87.1%. NPK released in NPK fertilizer coated with a polymer coating and fertilizer in a ratio of 1/9, 2/8, and 3/7 resulted in the release of NPK with not too big a difference. In the analysis of the functional group by FTIR between non-irradiated formulation and irradiated 20 kGy, there was no addition or change of molecular structure in the functional group. whereas thermal analysis with DSC displayed endotherm at temperature 50^0C to 120^0C and exotherm at temperature 250^0C until 320^0C for analysis of TGA degradation of the formulation is divided into 2 phases, in the first phase degradation begins at a temperature of 50^0C to 150^0C losing weight of 5% and in the second phase of degradation at 250^0C to 500^0C losing weighs 47%

1. Introduction

One of the efforts to increase the effectiveness and efficiency of fertilization is to modify it into a slow-release fertilizer (SRF) with supporting and environmentally friendly basic ingredients [1]. The main problem with the use of chemical fertilizers such as urea or NPK on agricultural land is their low efficiency due to their high solubility and loss due to the water-soluble, evaporation, and denitrification process of the fertilizer itself. Therefore, a formulation for the manufacture of coating materials with natural polymers based on natural polymers will be carried out which has three functions (multi-function) namely as a slow-release, plant growth inducer, and as a water-absorbent. Slow-release fertilizer is a fertilizer with a mechanism for releasing nutrients periodically following the pattern of nutrient absorption by plants. Several mechanisms that can be applied in the production of SRF are the mechanism of coating fertilizers with semi-permeable membranes and the mechanism of fusion of fertilizer nutrients in a matrix. The main principle of the two mechanisms is to create a barrier in the form of molecular interactions so that the nutrients in the fertilizer granules are not easily released into the environment. Controlled-release fertilizer (CRF) is one method to reduce the rate of loss of fertilizer from the soil due to leaching by rainwater or irrigation water and maintain water or mineral supplies for a long time until dry land recovery [2]. One of the agents that can be used for CRF is natural polymer materials. By using a polymer material that functions as a hydrogel, the release of water and nutrients can be slowed or even controlled so that plants can absorb more nutrients and water without wasting. Thus, hydrogel- controlled release fertilizer can increase the efficiency of nutrient application, and increase plant growth while reducing the impact on the ecological environment [3].

The research that will be carried out is to obtain the optimum ratio of coating material and urea fertilizer in the form of pellets. Parameters include the ratio of coating material and fertilizer weight and tool rotation speed. Characterization includes bulk density, durability, water absorption, and hardness.

* Corresponding author: gatot2811@yahoo.com

2. Material

The materials used are starch, polyvinyl alcohol, acrylamide, potassium dihydrogen phosphate, sodium hydroxide (NaOH), sulfuric acid, nitric acid, ammonium vanadate, and ammonium molybdate. While the solvent is aquadest. Then for testing the slow release of phosphate, NPK (10:55:10) phosphate fertilizer was used.

2.1 Preparation of copolymer coating material-starch-PVA-AAm-oligo-chitosan [4]

First, the starch is added to the acrylamide solution little by little while stirring. Then the mixture of these

https://doi.org/10.1051/bioconf/20238003004

solutions was added to a solution of polyvinyl alcohol (PVA), acrylamide, and chitosan, stirred, and heated at 90oC for 1 hour. The composition of the coating material is as follows: 5% starch, 50% PVA (10%), 40% acrylamide (3%), and 5% chitosan (2%).

2.2 Irradiation

The resulting polymer mixture was irradiated by gamma rays of 60Co in the air at room temperature to allow the desired polymerization to occur. In this irradiation process, the sample is irradiated with an irradiation dose of 20 kGy.

2.3 Characterization of the starch-acrylamide-oligochitosan copolymer

The copolymer of starch-polyvinyl alcohol- acrylamide obtained was analyzed for functional groups using Fourier Transform InfraRed (FTIR) at wave numbers from 4000 to 400 cm-1 using KBr powder. In addition, the thermal stability of the copolymers was also analyzed using a Themo Gravimetric Analysis (TGA) instrument, and to ensure the surface morphology of the formed copolymers, the copolymers were analyzed using a Scanning Electron Microscope (SEM) with a magnification of 100 times and 500 times.

2.4 Process of making pellet

The process of making pellets is the formation of a mixture in the form of pellets. The first time the NPK was ground and mixed with the copolymer material at the ratio of NPK coatings/fertilizers = 1/9, 2/8, and 3/7 until smooth. The tool used is a pelletizer machine with a maximum working capacity of 50 kg/hour and a working enginespeed of 60 rpm.

2.5 Evaluation properties of pellets

2.5.1. The yield capacity of the pellet material

The yield capacity of the pellet material is done by dividing the weight of the pellet that is printed against the time it takes to print the pellet.
$KH = BC/T$. Where : KH = Yield capacity (g/min), BC = Printed weight (g), T = Time (minutes)

2.5.2. Durability test

Durability test is used to determine the physical qualityof pellets, namely knowing the percentage of the numberof pellets that are still intact after going through physicaltreatment with mechanical devices. The steps in this test are to prepare 500 grams of coated NPK fertilizer samples. Then put the weighed sample into a vibrating machine tool (shaker) and turn it on for 10 minutes. After 10 minutes, then the machine was turned off and the samples were weighed which were still intact/not broken. The durability value is calculated by the equation:

% Durability = Mo/Mi x 100%

where: Mo=mass of an intact sample (g), Mi=mass of the sample before the test (g)

2.5.3. Determination of Slow Release

Determination of Slow Release of Phosphate-NPK Fertilizer [5].

The polymer that has been weighed as much as 0.1 g and is contacted with a fertilizer solution is measured forits slow-release properties in water. Every certain period, 10 mL of soaking water is pipetted and then 10 mL of distilled water is added back into the immersion container. To see the slow-release properties of the polymer, the immersion was carried out for a period of 10, 20, 30, 40, 50, 60, 120, 180, 240, 300, and 360 minutes. The immersion water was measured for its phosphate content using the molybdate-vanadate method on a UV-VIS spectrophotometer. Then the results of the measurement of phosphate levels in the water are compared with the immersion time.

3. Results and Discussion

3.1. Characterization of polymer NPK fertilizer coating materials with FTIR

Figure 1. Shows the difference in FTIR spectra between copolymers before and after radiation on starch-PVA-acrylamide-chitosan samples.

Fig. 1. IR Spectra Coating material formula at a) 0 kGy and

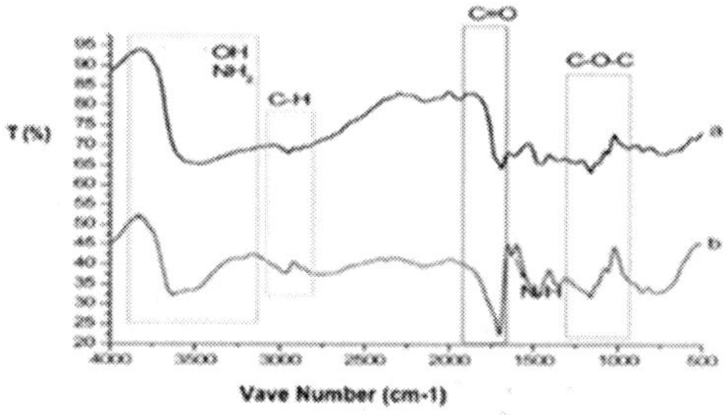

b) at 20 kGy

The presence of narrowed spectra from 3400 - 3654 cm^{-1} is thought to be due to the reduction of the free -OH functional group due to copolymerization. In this case, it shows the existence of hydrogen bonds that occur between N-H derived from acrylamide and –OH derived from starch and also PVA with each other. In addition, it can also show -OH stretching of the intra and intermolecular hydrogen bonds between starch and PVA or also of free -NH on acrylamide and free -OH on PVA and starch [4]. Then there was a shift in the wave number from 1611 cm^{-1} to 1626 cm^{-1} and an increase in the intensity at wave number 1626. cm-1. The shift that occurs may occur due to the presence of inter and intra-molecular between PVA and acrylamide through intermolecular hydrogen bonds [6]. With the increasing number of these interactions, it will be seen that the spectra are getting stronger and sharper [7]. At the wave number of 1699 cm-1, it is seen that the shift and peak

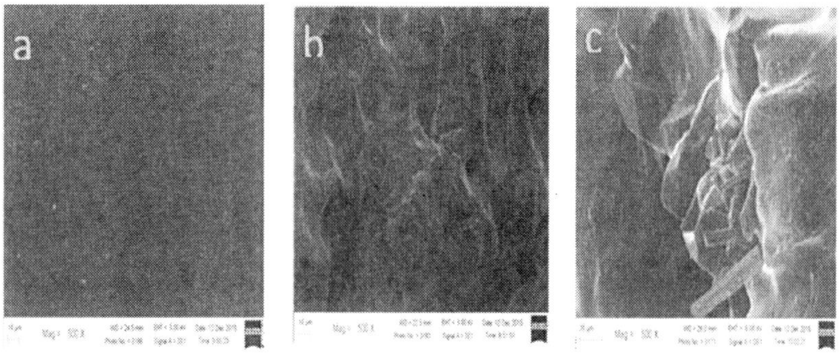

shape are getting stronger and sharper when compared to the wave number of 1688.75 cm^{-1}. The shift and sharper peak shape are thought to be due to the stretching of C=O bonds from starch that appears other than acrylamide in the copolymer. At wave numbers, 1699 cm-1, 1626 cm^{-1}, and 1354 cm^{-1} indicated the presence of C=O stretching, N-H bending, and C-N stretching, respectively, and indicated the –CONH2 group [4]. In the spectrum at wave numbers around 900 cm-1 and 1050 cm^{-1}, there is a reduction between before radiation and after radiation. This is presumably because at C-O-C and C-OH copolymerization occurs with monomers or polymers in the mixture.

3.2. Thermal properties

Characterization using TGA was carried out on 3% starch and acrylamide samples. This was done to see the thermal stability of the polymer synthesis material and the effect of adding acrylamide on polymer synthesis.

In Figure 2. the sample has three stages of decomposition which are indicated by the appearance of three peaks on the thermogram.

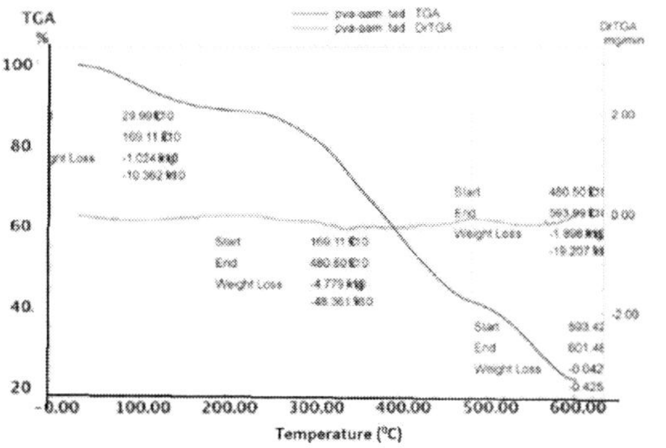

Fig. 2. Thermogram copolymer of starch and 5% acrylamide monomer

In the first stage, decomposition begins to occur up to a temperature of 176.78 °C. At this stage, it is suspected that there will be the decomposition of water in the copolymer and the homopolymer that may have been from the functional groups in the polyvinyl alcohol-acrylamide copolymer. In the last stage, decomposition begins to occur up to a temperature of 575.20 °C. Decomposition which occurs at this last stage is thought formed during the synthesis of the copolymer [7,8]. In the second stage, decomposition begins to occur up to atemperature of 487.72 °C. At this stage, it is suspected that decomposition occurs to come from the remains of components that have not been decomposed at a lower temperature so that in the end it produces residual ash of 0.81%.

3.3. SEM

Characterization using SEM was carried out on control samples of starch and acrylamide before irradiation (a) and samples irradiated at a dose of 20 kGy (b) with a magnification of 500 X (c). From the picture above, 5% starch-acrylamide irradiated at 20 kGy formed pores surface for slowrelease.

3.4. Pellet material yield capacity

To determine the capacity of the pellet material is calculated by the weight of the pellet produced in a certain time unit. In this study, it was carried out for 5 minutes,after 5 minutes the pellet material produced was weighed in grams so that pellet production was obtainedwithin 5 minutes.

The weight of the pellet material with variations of coating material (polymer) and NPK fertilizer for 5 minutes is shown in Figure 4.

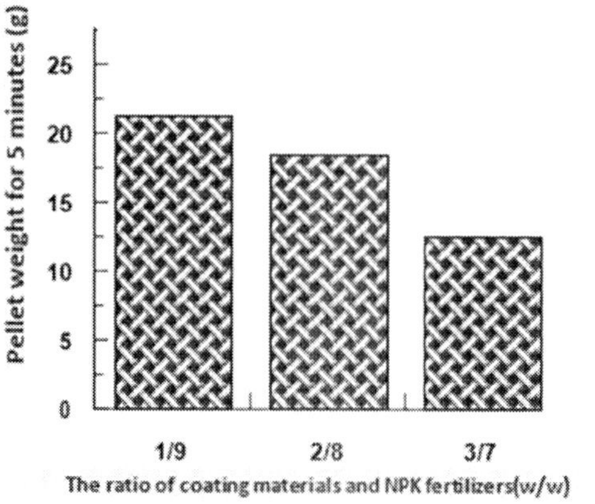

Fig. 4. The effect of variations in the ratio of the coating material (polymer) and NPK on the weight of the pellet material for 5 minutes

Figure 4 shows that the ratio of coating material and NPK fertilizer of 1/9 produces pellet material which is at most 21.4 g and the ratio of coating material on NPK fertilizer increases, namely at a ratio of 2/8, 18.6 g, and 3/7 are produced of 12.62 g. The cause of the decrease in the number of pellets produced is that the mixture of fertilizer with more coating materials becomes softer and sticky so that when it is made in the pelletizer machine there will be blockages in the discharge holes and sticking to the equipment [8].

Figure 5 shows that the ratio of coating material and NPK fertilizer of 1/9 produces pellet material which is at most 21.4 g and the ratio of coating material on NPK fertilizer increases, namely at a ratio of 2/8, 18.6 g, and 3/ 7 are produced. of 12.62 g. The cause of the decrease in the number of pellets produced is that the mixture of fertilizer with more coating materials becomes softer results in a higher yield capacity compared to a larger ratio (2/8 and 3/7). This is because, with the increase in the amount of coating material added to the fertilizer mixture, the pelleted fertilizer produced at a certain time and sticky so that when it is made in the pelletizer machine there will be blockages in the discharge holes and sticking to the equipment.

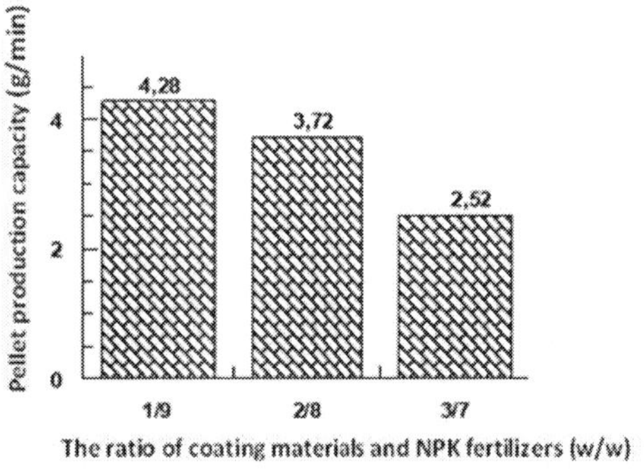

Fig. 5. Effect of coating material (polymer) ratio on pellet yield capacity (g/min).

A lower ratio of coating material and fertilizer is lower. This is because the dough is softer and stickier in the cavity of the pelletizer and clogs the outlet of the pelletizer so that the production capacity decreases [1].

3.5. Durability Test

Pellet durability is the resistance of pellet particles which is formulated as a percentage of the amount of whole pellet fertilizer after going through physical treatment in the test equipment. Durability is related to various processes in the use of pellets, such as the transportation process (transportation), as well as the distribution of the resulting pellets, therefore it is important to measure the durability of pellets [9]. The effect of coating material composition (polymer) and NPK on durability (%) is shown in Figure 6.

Good pellets are pellets that have a good durability so that in the process of handling and transporting pellets, they are not physically damaged, remain compact, sturdy, and not easily brittle. Dozier (2001) states that the minimum pellet durability specification standard is 80% [10]. Pellet durability is influenced by the chemical composition of the material, namely the coating polymer material and the composition of the polymer material and NPK fertilizer and durability can also be affected by the particle size of the pellet.

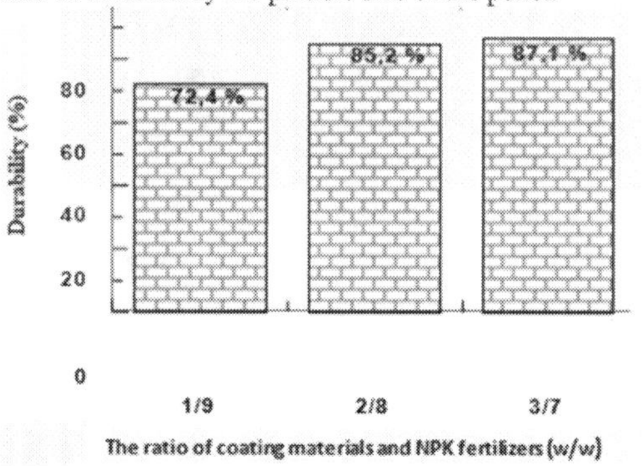

Fig. 6. Effect of coating material composition (polymer) on fertilizer on durability (%).

From the research conducted, the durability value on pellets at the ratio of coating material and NPK 1/9 was obtained at 72.4%, a comparison of 2/8 was obtained by 85.2%, and a ratio of 3/7 was obtained by 87.1%. The more coating materials added to the NPK fertilizer the more durability value increases. This is because the greater the coating material (polymer) given the bond between the fertilizers, the greater the coating material functions as an adhesive [11].

The results are shown in Figure 6. Composition of coating materials and NPK fertilizers with a ratio of 2/8 and 3/7. Durability values are above 80%, meeting the durability value standards for pellets. This shows that the pellets produced are not easily destroyed during handling and transportation.

3.6. Slow-Release properties of NPK coated with a polymer coating material

The test of slow-release properties on NPK coatedwith polymer with variations in the ratio of coating material and NPK is shown in Figure 7 which was carried out using the absorption-desorption method. The amount of phosphate absorbed and released in the copolymer was measured chemically using the molybdate-vanadate method [8].

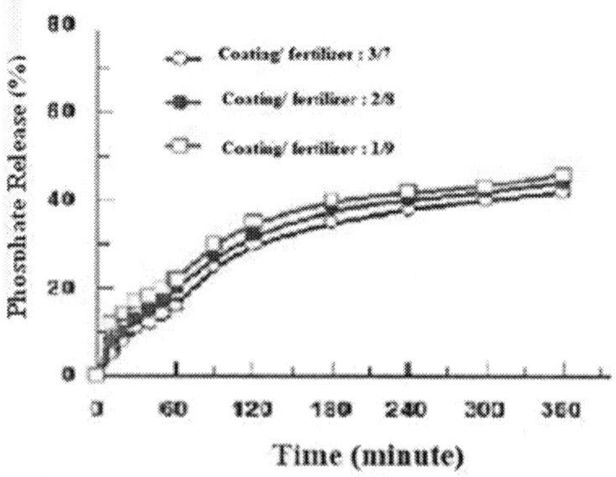

Fig. 7. Slow-release properties of NPK fertilizer coated with a polymer coating material

The results of the analysis are shown in Figure 7. The results show that the released NPK in NPK fertilizer coated with a polymeric coating and fertilizer with a ratio of 1/9, 2/8, and 3/7 resulted in the release of NPK with not too large a difference. At a ratio of 1/9, the slow-release time is slightly greater than that of fertilizers with a composition of coating material and NPK with a ratio of 2/8 and 1/9. This result is because more coating material will bind more NPK fertilizer [8]. After all, the coating material contains starch which functions as an adhesive and closes the NPK pores.

4. Conclusion

Based on the results of the research that has been carried out, it can be concluded: A copolymer between starch-PVA and acrylamide has been obtained which can be used as a coating material for NPK fertilizer and made into pellets. The largest pellet yield capacity uses a ratio of coating material and NPK fertilizer with a ratio of 1/9 of 4 .28 g/minute and the greatest durability value is at a ratio of 3/7 with a value of 87.1%. NPK released in NPK fertilizer coated with polymeric coating and fertilizer in a ratio of 1/9, 2/8, and 3/7 resulted in the release of NPK with not too big a difference.

References

[1] Quynh T M Binh N Van Thoa D K and Luong L T M, 2021 The growth promotion effect of microelement fertilizers containing low molecular weight chitosan and xanthan on radish (Raphanus sativus L.) *Nucl. Sci. Technol.* **10**, 3.

[2] Rekso G T, 2019 Kopolimerisasi Iradiasi Kitosan-Polivinyl Alkohol-Akrilamida sebagai Bahan Pelapis Pupuk Gatot Trimulyadi Rekso *Pros. Semin. Nasiona Tek. Kim. "Kejuangan"* April p. 1–7.

[3] Malerba M and Cerana R, 2019, Recent applications of chitin- and chitosan-based polymers in plants, *Polymers*. .

[4] Rekso G T and Rahmawati R, 2022 Structure- property relationships of Cassava starch- acrylamide gel copolymerized by Gamma irradiation *AIP Conf. Proc.* **2391**, March.

[5] Trimulyadi Rekso G, 2014 Chemical and physical properties of cassava starch-CM-chitosan-acrylic acid gel copolymerization by gamma irradiation *Indones. J. Chem.* **14**, 1 p. 37–42.

[6] Rekso G T and Sudradjat A, 2019 Synthesis of Ag nano-chitosan in lactic acid solvent by irradiation method in *IOP Conference Series: Materials Science and Engineering* **673**, 1.

[7] Rekso G T, 2022 Mass-production of oligo chitosan in the liquid phase by irradiation technique *IOP Conf. Ser. Earth Environ. Sci.* **1017**, 1 p. 0–8.

[8] Fortuna D *et al.*, 2020 Potensi pembuatan pupukdalam bentuk pelet dari bottom ash dan fly ash industri pulp dan kertas untuk hutan tanaman industri **7** p. 1–6.

[9] Bano I Ghauri M A Yasin T Huang Q and Palaparthi A D S, 2014 Characterization and potential applications of gamma irradiated chitosan and its blends with poly(vinyl alcohol)*Int. J. Biol. Macromol.*

[10] Sabharwal S Varshney L Chaudhari a D and Rammani S P, 2004 Radiation processing of polysaccharides November p. 29–38.

[11] Uthairatanakij A Teixeira da Silva J a and Obsuwan K, 2007 Chitosan for improving orchid production and quality *Orchid Sci. Biotechnol.* **1**, 1.

Crop Pattern Suitability Based on Water Availability in Dry-Land Agriculture Nawungan, Selopamioro, Imogiri, Bantul

Khoirida Dian Putranti[1] , *Nur Ainun H J Pulungan*[2], and *Makruf Nurudin*[2]
[1]Student of Soil Department, Faculty of Agriculture, Universitas Gadjah Mada, Indonesia
[2]Soil Department, Faculty of Agriculture, Universitas Gadjah Mada, Indonesia

Abstract. Climate change has changed the rainfall pattern. It causes a shifting in the beginning of the rainy and dry seasons as well as the planting season. This will affect changes in the calculation of the potential availability of agricultural water due to the distribution of water that is not in accordance with needs. Strategies and adaptation technology so that the agricultural sector is resilient to climate change are urgently needed. Crop patterns setting is considered an alternative to adapt to the limited water availability in Nawungan. The purpose of this research is to evaluate the suitability of cropping patterns in Nawungan agricultural land to find out whether the cropping patterns that have been or are being implemented are still appropriate or not with climate change. The research methods included interviews, surveys and field measurements, as well as laboratory analysis. Interviews were conducted with 30 respondents to collect information regarding variations in crop patterns applied. Field surveys were carried out to determine the zoning of crop patterns based on interview data. Field measurements were made of landscape analysis. Laboratory analysis was carried out on the measurement of soil parameters. The results showed that crop pattern that is suitable with the water availability at the research location is P3 (paddy-secondary crop-bero) zoning. Cropping period that has sufficient water availability is MT-2.

1. Introduction

In the past, farmers used pranoto mongso as a benchmark in farming, such as determining the start of the rice planting season. Farmers' understanding in reading phenomena or natural phenomena that are happening at this time tends to be difficult due to uncertain climate change [1]. As a result of climate change, the rainfall pattern has also changed, causing a shift in the beginning of the rainy and dry seasons as well as the planting season. This will affect changes in the calculation of the potential availability of agricultural water due to the distribution of water that is not in accordance with needs [2].

Changes in rain patterns due to climate change affect the amount of water availability. According to Ayu et al. [3], mastery of rainfall variations throughout the season can be used as an estimate of the amount of groundwater available in a rain-fed area for a certain period of time. Rainfall is one of the climatic factors that affect the success of plant productivity.

The impact of climate change, especially the agricultural sector, is feared to bring problems to the sustainability of agricultural production. For example, research by Runtunuwu and Syahbudin [4] states that there has been a shift in the beginning of the season and planting period as a result of a decrease in the amount of rainfall and changes in rain patterns in 1879 – 2006 in the Tasikmalaya area, West Java, thus reducing the potential for one rice planting period. . Another example of a plant that is vulnerable to changes in rainfall patterns is shallots. If the rainfall is high, the shallots will be more susceptible to fungal attack until they rot and if the drought conditions will cause the plants to dry [5].

Strategies and adaptation technology so that the agricultural sector is resilient to climate change are urgently needed. Adaptation technology related to excess water management during the rainy season by making ponds can be used to irrigate crops during the dry season where water availability is limited [6]. Apart from that, research by Banjarnahor and Simanjuntak [7] in Central Sumba regarding the regulation of plant types as well as research by Rusastra et. al. [8] who apply land use for lowland rice-rice-soybean crops can be used as a form of limited water management during the dry season so that production can be maintained.

Rainfall in the Selopamioro area tends to be low. According to Kurniawan [9] rainfall in the Selopamioro area is 1,564 mm/year with the peak of the rainy season in December, January and February. The amount of rainfall is very important because rainfall is a natural input for agricultural cultivation in Selopamioro. Apart from that, rainfall patterns are important so that planting periods and crops can be determined according to their potential water availability.

Crop patterns setting is considered an adaptation to the limited availability of water for agriculture on Nawungan. The planting pattern applied on Nawungan agricultural land refers to Bantul Regent Decree No. 434 in 2020 is rice-secondary crops-secondary crops based on rainfall which decreased slightly from the previous year. A shift in the planting period causes the cropping pattern to also shift, as a result the second planting season has a risk of crop failure or less than optimal yields due to the approach of the dry season. Therefore, it is necessary to evaluate the suitability of cropping patterns in Nawungan agricultural land to find out whether the cropping patterns that have been or are being implemented are still appropriate or not with climate change.

2. Methodology

This research was carried out in the Nawungan Imogiri River Basin, Bantul. The research location has an area of 551,483 m2. The research stages include interviews, surveys and field measurements, as well as laboratory analysis. Interviews were conducted with 30 respondents to collect information regarding variations in cropping patterns applied. Field surveys were carried out to determine the zoning of planting patterns at the research location based on interview data. Field measurements were made of the slope (using a clinometer) and infiltration rate (using a single ring infiltrometer). Laboratory analysis was carried out on the measurement of soil parameters such as texture (using the pipette method), bulk density (using a pycnometer), bulk volume (using the ring method), organic matter (using the Walkley-Black method), and water retention (using a pF device for pF 0 , 2, 2.54, 4.2). Secondary data collection was also carried out on rainfall data from BMKG for 2013 - 2022 and climatological data from AWS during the research implementation, from January to June 2023.

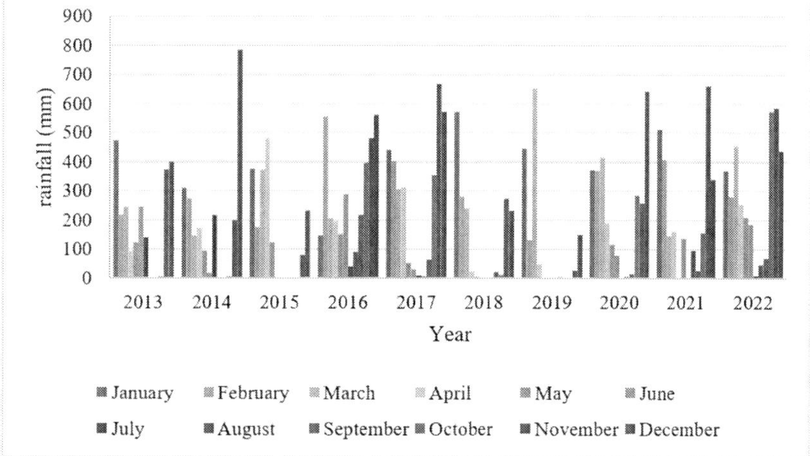

Fig. 1. Rainfall Pattern in Research Area

Data processing techniques are carried out quantitatively for the following parameters:

1. Availability of ground air
   Calculation of groundwater using effective rainfall and availability is directly intended to meet plant water needs for growth [10]. Rainfall data for 10 years is sought for 80% rainfall (R80), namely rainfall probability of 80% occurs or maybe greater than its rainfall.
2. Water requirements for crops
   The calculation of crop water needs uses the actual evapotranspiration calculation which is obtained from the potential evapotranspiration multiplied by the crop coefficient.

$$Eta = Eto \times KC$$

Where Eto is resulted from Penman-montheit equation:

$$ETo = \frac{0.408\,x\,\underline{\Delta}(Rn - G) + \frac{\gamma Cn}{T + 273} u2(es - ea)}{\underline{\Delta} + \gamma\,(1 + Cd)}$$

3. Evaluation of crop pattern suitability
   Evaluate the suitability of crop patterns using the water balance between soil water availability and plant water needs. Calculation of the water balance to determine the condition of groundwater deficit or surplus, then the calculation of alternative planting arrangements is carried out during the second planting period.

The data analysis technique was carried out using a descriptive-quantitative method based on the value of groundwater availability and plant water requirements at the study site. Analysis of the water balance, namely using the comparative method, compares the value of water availability with the water needs of plants during the second planting season by adjusting the climate and soil characteristics at the study site. Water balance analysis to determine which planting periods experience water shortages.

 https://doi.org/10.1051/bioconf/20238003005

3. Results

3.1 Cropping pattern zone

Based on the interview results, it was found that 3 zoning cropping patterns were applied by farmers at the study site. Planting pattern zone 1 has a plant pattern of rice-shallots-shallots-vegetables (mustard/chilli). Planting pattern zone 2 has a rice-shallot-vegetable (mustard/chilli). And planting pattern zone 3 has a rice-palawija crop pattern (peanuts/cassava).

The existing zoning of planting patterns is spread over land on various slopes. The slope of the slope in each zone includes flat, very gentle, gentle, slightly steep and steep slopes. Each zoning on a particular slope is the basis for determining soil sampling which were about 28 points.

3.2 Climate Characteristic

The climate characteristics at the study site based on the 10-year rainfall trend show a monsoonal pattern characterized by a type of rainfall with one peak rainy season (unimodial). In June, July and August there is a dry season while in December, January and February there is a rainy season.

3.3 Soil Characteristics

The clay content in each slope class and planting pattern zone of soil texture is dominated by the clay fraction. High clay content will cause the soil to be watertight [11].

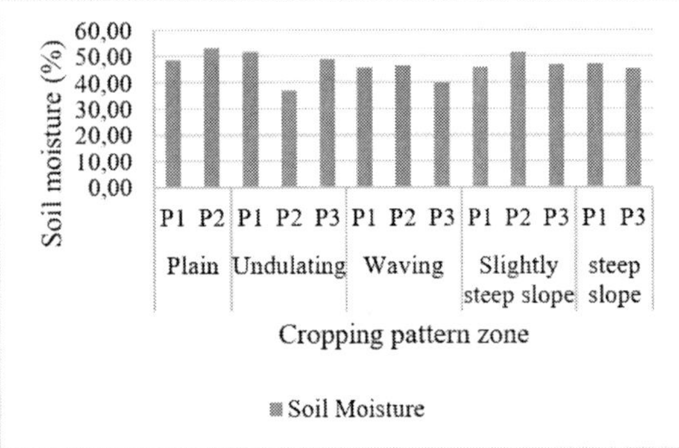

Fig. 2. Soil Moisture

Table 1. Soil Texture

Slope class	Cropping pattern zone	Sand (%)	Silt (%)	Clay (%)	Soil texture
Plain	P1	25	33	42	Clay
	P2	73	9	17	Sandy loam
Undulating	P1	14	16	70	Clay
	P2	20	35	44	Clay
	P3	54	21	24	Sandy clay loam
Waving	P1	18	17	65	Clay
	P2	9	9	82	Clay
	P3	34	31	35	Clay loam
Slightly steep slope	P1	25	23	52	Clay
	P2	26	30	44	Clay
	P3	44	29	27	Loam
Steep slope	P1	38	29	33	Clay loam
	P3	38	20	42	Clay

Moisture levels in each cropping pattern zone and slope class show values that do not vary too much. The moisture content value ranges from 36.87 – 53.63%. The lowest moisture content values are found in planting pattern zone 2 with a very gentle slope class. The low moisture content is caused by the very poor percentage of porosity so that the pores are small and the percentage of micro pores that store water is also small so that the soil's ability to bind water becomes very weak.

Fast drainage pores occupy 55.43% of the total pore space of the soil, while slow drainage pores only occupy 2.69% of the total pore space of the soil. Fast drainage pores are macro pores that will lose water at low matrix suction [12]. The high level of fast drainage pores indicates that the soil in the catchment area of the study allows water to pass through quite easily.

The organic carbon content shows that the soil organic matter content in the research catchment area ranges from 0.00% - 1.7% which is classified as very low to low with an average of 0.81%. Yang et al. [13] stated that the organic matter content of soil in agricultural land systems is influenced by the practice of returning harvest residues to the land. The low organic matter content in the research catchment area can be caused by harvest residues that are not returned to the land.

Table 2. Soil Organic Matters

Slope class	cropping pattern zone	C-Organic (%)	SOM (%)	Level
Plain	P1	0.73	1.25	Low
	P2	0.00	0.00	Very low
Undulating	P1	0.82	1.42	Low
	P2	0.98	1.70	Low
	P3	0.05	0.08	Very low
Waving	P1	0.62	1.07	Low
	P2	0.78	1.35	Low
	P3	0.32	0.54	Low
Slightly steep slope	P1	0.63	1.08	Low
	P2	0.37	0.63	Low
	P3	0.22	0.39	Very low
steep slope	P1	0.35	0.61	Low
	P3	0.22	0.39	Very low

The majority of land infiltration rates showed an insignificant decrease, except for land in cropping pattern zone 1 of the steep slope class showing a significant decrease in infiltration rates. Differences in infiltration rates in cropping pattern zone 1 land with steep slope classes are influenced by land processing. Add explanation. Tillage affects the volume of soil [12].

Table 3. Soil Physic Characteristics

Slope class	Copping pattern zone	BD (g/cm^3)	PD (g/cm^3)	Porosity (%)	Fast drainage pores (%)	Slow drainage pores (%)	Soil texture
Plain	P1	0.77	1.85	58	39.00	4.61	Clay
	P2	0.73	1.75	58	35.71	1.98	Sandy loam
Undulating	P1	0.81	1.85	56	35.32	4.94	Clay
	P2	1.07	1.98	46	28.67	2.42	Clay
	P3	0.79	1.82	56	30.10	2.20	Sandy clay loam
Waving	P1	0.86	1.97	56	32.83	5.93	Clay
	P2	0.87	2.04	57	32.41	0.22	Clay
	P3	0.98	1.91	49	36.20	5.60	Clay loam
Slightly steep slope	P1	0.89	1.81	51	32.27	4.46	Clay
	P2	0.79	1.84	57	35.82	2.97	Clay
	P3	0.86	1.88	54	30.31	1.17	Loam
Steep	P1	0.86	1.88	54	31.30	2.91	Clay loam
	P3	0.82	1.70	51	35.48	0.66	Clay

3.4 Water Availability

Based on the table of available water content, it shows that the value of available water content at the research location is quite varied. The highest average available water content was found in cropping pattern zone 3 on land with a rather steep slope. The differences in texture, density, porosity and organic material in each planting pattern zone cause quite significant differences in values. This is in accordance with Ayu et al. [3] stated that there are differences in soil surface conditions, organic matter, texture, structure and vegetation is a factor causing the difference in water holding capacity.

Table 4. Soil Water Availibility

Slope class	cropping pattern zone	Soil Water Availability (%)
Plain	P1	11.04
	P2	7.56
Undulating	P1	8.87
	P2	7.25
	P3	8.27
Waving	P1	5.33
	P2	10.88
	P3	8.40
Slightly steep slope	P1	6.92
	P2	9.56
	P3	13.07
steep slope	P1	10.00
	P3	12.25

3.5 Potential evapotranspiration

Based on the results of the analysis, the average amount of potential evapotranspiration (ETo) ranges from 2.9 – 3.49 mm/day. Based on the modified penman method, the amount of potential evapotranspiration (ETo) is influenced by the reflection coefficient. The reflection coefficient is the ratio of the amplitude of the reflected wave compared to the amplitude of the incoming wave. The reflection coefficient depends on the amount of vegetation distribution with land use.

3.6 Crop Water Needed

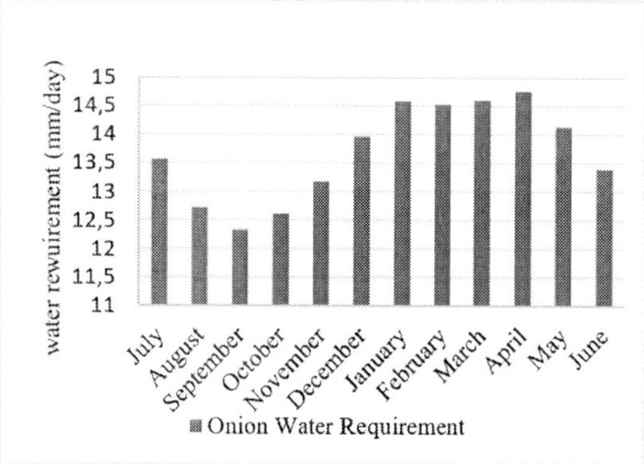

Fig. 3. Onion Water Requirement

The value of plant water needs was obtained in the range of 12.8 – 14.8 mm/day. The water requirement for shallots is lowest when planted in September and highest in April. The water needs of shallot plants depend on several factors, for example the age of the plant, temperature and humidity, rainfall.

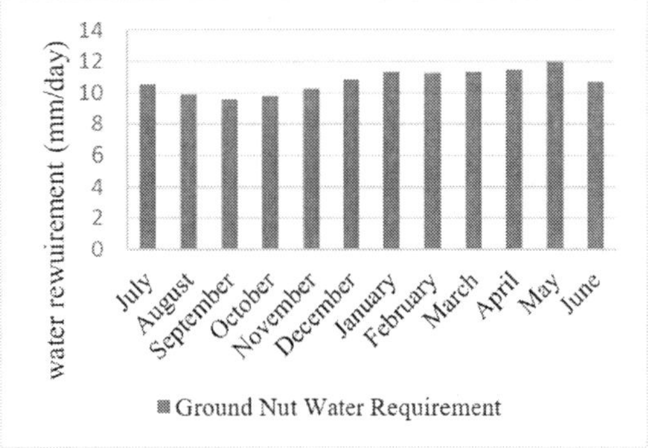

Fig. 4. Ground Nut Water Requirement

The value of water requirements for peanut plants obtained as shown in the picture below ranges from 8.8 – 12 mm/day. The water requirement value for peanuts is lowest if planted in September and highest in April.

3.7 Evaluation of Crop Pattern Suitability

Evaluation of the suitability of planting patterns is seen based on the balance of soil water availability and plant water needs. Groundwater availability is calculated from effective rainfall. Effective rainfall is rainfall that is effectively and directly used to meet plant water needs for growth [10].

Table 5. Crop Pattern

Slope Class	Crop Pattern Zone	Crop Pattern Term			
		I (Dec - Mar)	II (Apr - June)	III (July - Sep)	IV (Oct - Nov)
Plain	P1	Paddy	Onion	Onion	Caism
	P2	Paddy	Onion	Chilli	
Undulating	P1	Paddy	Onion	Onion	Chilli
	P2	Paddy	Onion	Chilli	
	P3	Paddy	Cabai		
Waving	P1	Paddy	Onion	Onion	Caism
	P2	Paddy	Onion	Chilli	
	P3	Paddy	Cassava		
Slightly steep slope	P1	Paddy	Onion	Onion	Chilli
	P2	Paddy	Onion	Chilli	
	P3	Paddy	Groundnut		
Steep slope	P1	Paddy	Onion	Onion	Caism
	P3	Paddy	Cassava		

Based on the figure below, through the calculation of an effective rainfall of 80% and surface runoff considered zero, it was found that each zoning pattern experienced a water deficit during the second planting season in May and June. During these months the availability of ground water is between 5 – 7 mm/day, while the water needs of shallots and peanuts are more than 7 mm/day and even reach 14 mm/day for shallots.

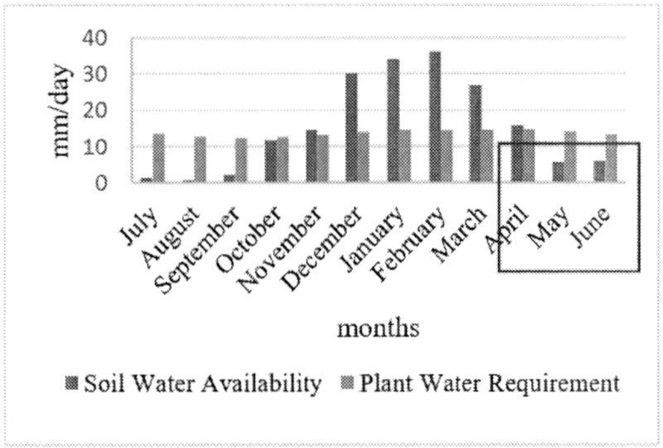

Fig. 5. Water Balance Cropping Pattern Zone 1 Plain Slope

The maximum deficit in Nawungan agricultural land during the second planting season occurred in zones 1 and 2 on cropping pattern zones 1 and 2 on flat to gentle slopes of 9 mm/day with shallots planted. The maximum surplus occurred during the second growing season in zone 3 of cropping pattern on land with rather steep and steep slopes of 5 mm/day with peanuts being planted. While the minimum deficit occurs in cropping pattern zone 3 of 3 mm/day and the minimum surplus occurs in cropping pattern zones 1 and 2 on a flat to gentle slope of 0.5 mm/day..

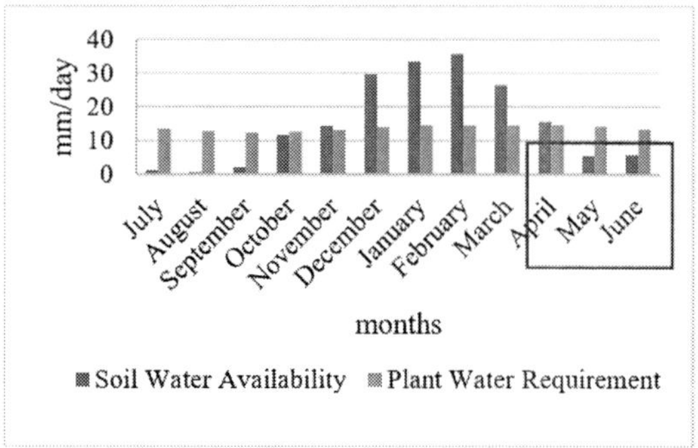

Fig. 6. Water Balance Cropping Pattern Zone 2 Waving Slope

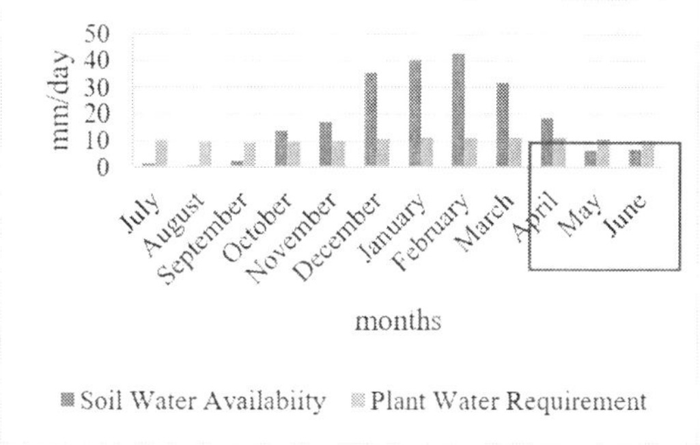

Fig. 7. Water Balance Cropping Pattern Zone 3 Slightly Steep Slope

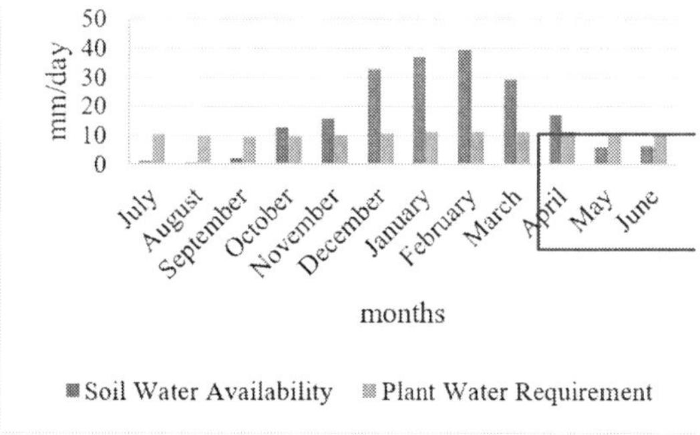

Fig. 8. Water Balance Cropping Pattern Zone 3 Steep Slope

4. Conclusion

1. The climate characteristics at the research location reflect low rainfall which also influences low water availability
2. Soil characteristics at the research location are dominated by fast drainage pores
3. The cropping pattern that is in accordance with water availability at the research location is P3 (paddy-palawija) zoning
4. Cropping period that has sufficient water availability is MT-2.

Acknowledgement

The author would like to thank to Nawungan-I research member team for their help and support during the research. This publication is a part of Bachelor thesis of first author, and was funded by *Final Project Recognition Grant* Universitas Gadjah Mada *Number* 5075/UN1.P.II/Dit-Lit/PT.01.01/2023.

References

1. Anazifa, R. D., Jurnal Pendidikan Biologi Program Pascasarjana Universitas Negeri Yogyakarta, **1**, 1 – 10 (2016).
2. Latifa, R., and Suprayogi, S., Jurnal Bumi Indonesia **1**, 1 – 15 (2018).
3. Ayu, I.W., S. Prijono, and Soemarno, J-PAL 4, 18 – 25 (2013).
4. Runtunuwu, E., and H., Syahbuddin. Jurnal Tanah dan Iklim, **26**, 1-12 (2007).
5. Ariska. N., Rahmawati, and Diah, Jurnal Agrotek Lestari, **4**, 42 – 50 (2017).
6. Surmaini, E., E. Runtunuwu, and I. Las, Jurnal Litbang Pertanian, **30**, 1 – 7 (2011).
7. Banjarnahor, D., and B. H., Simanjuntak. *Pola tanam Kabupaten Sumba Tengah yang sesuai dengan curah hujan setempat*. Prosiding Konser Karya Ilmiah, **1**, 97 – 107 (2015).
8. Rusastra, I.W., Saliem, H.P., Supriati, dan Saptana, Forum Penelitian Agroekonomi, **22**, 37 – 53 (2004).
9. Kurniawan, R. *Analisis Neraca Air Pada Lahan Sirsak (Annona Muricata L.) dengan Pemanenan Air Hujan di Kebun Buah Nawungan, Desa Selopamioro, Kecamatan Imogiri, Kabupaten Bantul*. Universitas Gadjah Mada. Skripsi (2020).
10. Jonizar dan Martini, Teknik, **4**, 131 – 137 (2016).
11. Widiyono, Penyakit Tropis: Epidemiologi, Penularan, Pencegahan dan Pemberantasannya (2008).
12. Haryati, U., Jurnal Sumberdaya Lahan, **8**, 125-138 (2014).
13. Yang, L., M. Song, A-X. Zhu, C. Qin, C. Zhou, F. Qi, X. Li, Z. Chen, and B. Gao, Geoderma, **340**, 289-302 (2019).

Cross-Sectional Study on Land and Seed of Oil Palm Smallholders in The Western Part of Aceh: Challenge for Sustainability Agenda

Ahmad Humam Hamid[1,2], *Agus Nugroho*[1], *Fajri*[1], *Agussabti*[1], *Litna Nurjannah Ginting*[2*], *Muhammad Rusdi*[3], *Muyyasir*[3], and *Akhmad Baihaqi*[1],

[1]Agribusiness Department, Faculty of Agriculture, Universitas Syiah Kuala, Banda Aceh 23111, Indonesia
[2]Center for Sustainable Agriculture and Rural Development, Universitas Syiah Kuala, Banda Aceh 23111, Indonesia
[3]Soil Science Department, Faculty of Agriculture, Universitas Syiah Kuala, Banda Aceh 23111, Indonesia

Abstract. Palm oil sectors in Indonesia have to faced critiques due to its compliance towards sustainability agenda. As the main palm oil producing country, Indonesia may lose its competitiveness since the global consumers demand more environmentally products. The two issues in terms of traceability are land legal clarity and the source of seeds. This study aims to examine the current stage of land legal clarity and the adoption of certified seeds of oil palm smallholders. A cross sectional study has been done across four palm oil producing regions in the western part of Aceh Province-Indonesia. The results show a significant consideration for future improvement as most of the smallholders' land and oil palm seeds were lack of traceability. These findings imply that the government's intervention is very urgent to accelerate the transformation of smallholders' oil palm practices towards a global sustainability agenda.

1 Introduction

The growth of oil palm estates has sparked considerable interest among socioeconomic audiences both at the domestic and international levels. Palm oil is considered to have a negative influence, particularly on deforestation and the rights of individuals violations. CPO exporting destination countries, particularly the European Union, require Indonesia to adopt sustainable palm oil management.

To support sustainable palm oil in the international market, the government began to make Sustainable Palm Oil certification enforceable. Based on this, Indonesian palm oil is required to be able to meet certification standards either ISPO or RSPO which are designed to keep palm oil expansion away from deforestation and to preserve human rights and fair prices for smallholders [1].

The plantation sector in Aceh Province- Indonesia, especially oil palm estates, is one of the drivers of economic growth in the province and is expected to increase the income and welfare of farmers. The increase in oil palm area in the last five years amounted to 5.38% compared to 2017. This increase in area is due to high public interest because it is considered to have a promising market. The largest oil palm areas in Aceh are in the districts of East Aceh (23,479 ha), West Aceh (23,479 ha), and Aceh Tamiang (16,138 ha) [2]. The average area of oil palm in Aceh Province is 1.8 Ha [2]. This small landholding will affect the income and livelihood of farmers [3].

The palm oil commodity is considered profitable and easy to market, encouraging farmers to switch their existing commodities to palm oil. However, the expansion of smallholder oil palm plantations has constraints related to low quality and productivity [3]. This is due to the inefficient use of inputs such as the use of poor seedlings and the absence of proper plant care and maintenance in accordance with Good Agriculture Practice (GAP). In addition, environmental issues related to land clearing for oil palm plantations may still be carried out irresponsibly. These things are still not compatible with the sustainability agenda required by certification regimes.

The growing interest in planting oil palm was not accompanied by an increase in farmers' knowledge of sustainable oil palm cultivation. This is due to the lack of assistance and understanding of sustainable palm oil by farmers. Two issues that become the root of the main problems in the implementation of sustainable palm oil certification are the clarity of land status and seeds. This study aims to describe and analyze the existing conditions of land and oil palm seedlings in smallholder plantations in Aceh. This research contributes to the selection of appropriate strategies to accelerate the transition of traditional oil palm cultivation to a sustainable scheme for the welfare of farmers and to maintain environmental quality.

Either ISPO or RSPO is a challenge for palm oil sustainability schemes in Aceh. Business actors consider economic aspects more than social and environmental aspects. Compliance of independent

*Corresponding author: litnanurjannah@usk.ac.id

smallholders with land status clarity remains a major obstacle. Accusations of land logging and deforestation are easily aimed at weakening the position of independent oil palm farmers. Land certificates, seeds, pesticide use, fertilization, and documentation are things that need to be supported in order to be included in the certification scheme [4]. Small-scale farmers also face the challenge of global market demand. In this case, the productivity and quality of smallholder production are sought to be stable and maximum to meet future demand [3]. One of the causes of low palm oil productivity is that the technology used is still very simple from nursery to harvesting. Therefore, it is necessary to identify the conditions of seedlings and land use related to the readiness of farmers to adopt sustainability certification.

2 Method

This research was conducted in the four largest palm oil producing districts in Aceh Province, namely Aceh Barat Daya, Nagan Raya, Subulussalam, and Aceh Singkil. These four districts are also the districts with the largest oil palm in the Western part of Aceh. Purposive sampling was applied with a minimum number of 30 palm oil farmers in each regency. In this study, the total sample size is 172 respondents spread across the four districts with the following details:

Table 1. Sampel Size.

Districts	Number of Farmers
Nagan Raya	41
Aceh Barat Daya	39
Subulussalam	34
Aceh Singkil	58
Total	**172**

The research used structured questions/questionnaires and direct interviews with sample oil palm farmers. GPS tagging was also carried out to see the location points of respondents and farmers' oil palm plantations. The research location can be seen in the following figures

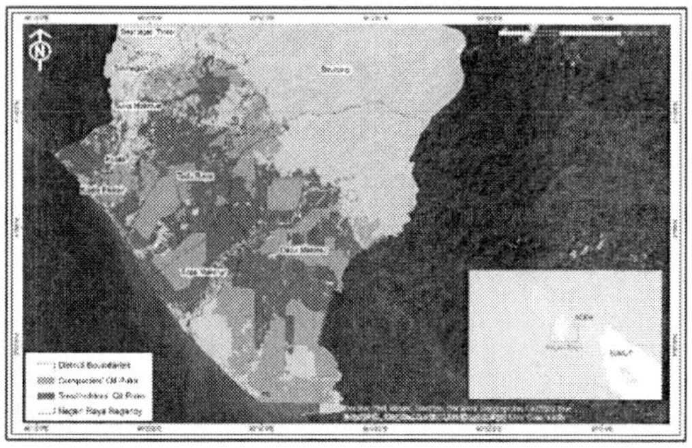

Fig. 1. Study Area in Nagan Raya

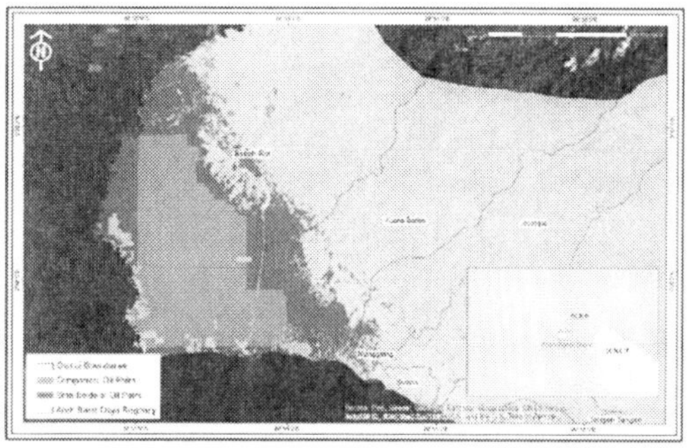

Fig. 2. Study Area in Aceh Barat Daya

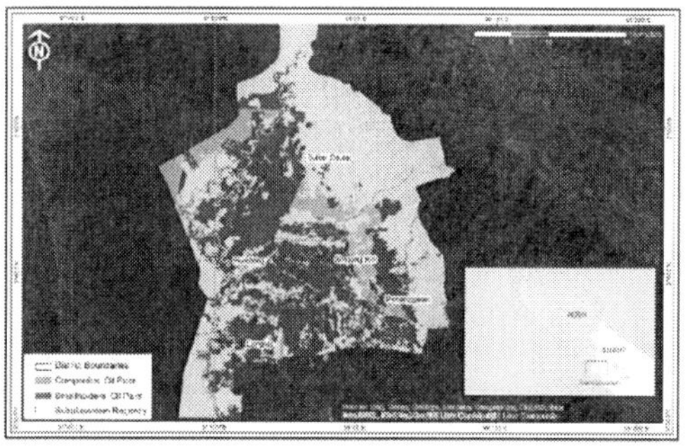

Fig. 3. Study Area in Subulussalam

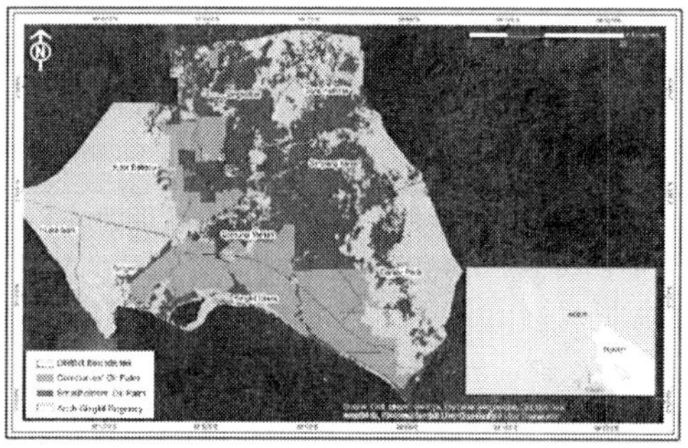

Fig. 4. Study Area in Aceh Singkil

According to the data, the majority of oil palm in Nagan Raya, Subulussalam, and Aceh Singkil is owned by smallholders. Meanwhile, estates in Aceh Barat Daya are dominated by private companies.

This research uses cross sectional studies to analyze data from a population at a single point in time. This method is useful for establishing preliminary evidence in planning a future advanced study. Descriptive cross-sectional studies simply characterize the condition of a specific population [5]. The descriptive data were analyzed by summarizing large amounts of information such as frequencies and percentages. The content of data is systematically analyzed, splitting it into meaningful pieces, and sorting those pieces in a way that allows the characteristics and meaning to be better understood [6].

3 Results and discussion

3.1. Farmers' characteristics

The descriptive statistics of the respondents are illustrated in Table 2.

Table 2. Characteristics of the Respondents

Districts	Gender (%)		Age	Experienc in Oil Paln
	Male	Female		
Aceh Barat Daya	100	0	47.59	11.36
Aceh Singkil	79.31	20.69	46.59	14.07
Nagan Raya	92.68	7.32	47.93	13.32
Subulussalam	88.24	11.76	45.32	14.94
Total	**88.95**	**11.05**	**46.88**	**13.45**

Most oil palm farmers are of productive age. At this age, farmers are still active in managing their oil palm estates both physically and mentally. Yet, there is a decreasing interest among young people to be involved in oil palm. Based on gender, oil palm farmers are dominated by males. Female oil palm farmers are normally inherited assets by their husbands and the management is handed over to other people. Women play a role in helping their husbands during the farming process such as fertilizing and harvesting. On average, farmers have been cultivating oil palms for 11-14 years, so they have sufficient experience in oil palm farming.

3.2 Seeds

The use of good-quality seeds is very important for farmers' productivity and welfare [3]. However, based on the field survey data, it was found that farmers still have very little knowledge regarding quality seeds. Most famers do not know the variety of seedlings used because many farmers still use random (*lelesan*) seedlings. Other farmers collected the fallen fruits from the company's plantation as seedlings.

The data in Figure 5 shows that 71 - 86% of farmers do not know for sure whether the varieties they use. A total of 4-8% of farmers use dura varieties while 5-10% use Tenera varieties. Only farmers in Aceh singkil use psifera. 5.17-10.26% of farmers use the *Marihat*, *Simalungun*, *Lonsum*, and Socfindo varieties.

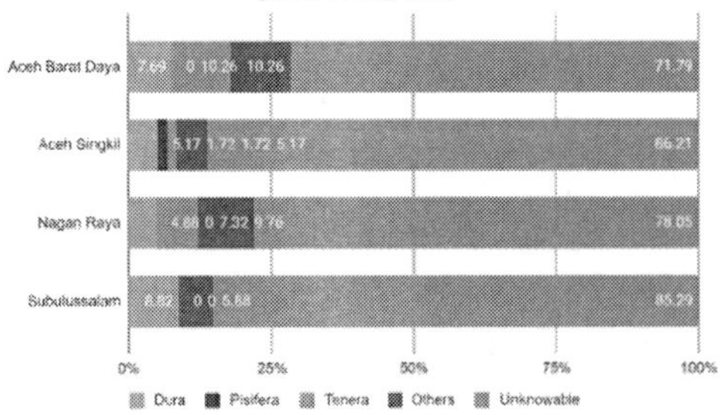

Fig. 5. Types of Seedlings

50% of farmers in Nagan Raya purchase their seed and Subulussalam farmers use the most random seeds among the four districts (45.68%). Asalan seedlings are seedlings that farmers make themselves and have no clear origin. A total of 6.96-15.05% of farmers received seeds from companies and assistance from the government. It is not uncommon for famers to buy seeds at low prices but of poor quality. Sometimes the farmers collect hybrid seeds from the ground, and turn out to be unfruitful after four years because the seeds are sterile [7].

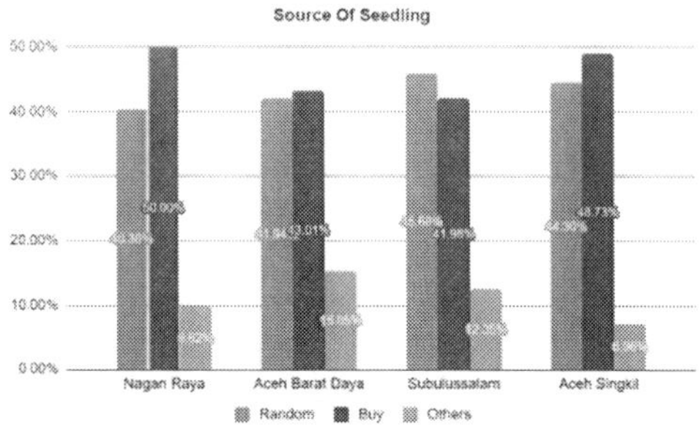

Fig. 6. Sources of Seedlings

The highest number of farmers using certified seeds was in Aceh Singkil District at 34.48% and Aceh Barat Daya at 33.33%. Farmers in Singkil and Aceh Barat Daya districts are starting to realize the importance of using certified seeds. Most farmers (23% - 50%) do not know whether the seeds they use are certified or not. This is due to a lack of understanding about certified seeds.

Farmers who use certified seed are mostly farmers who have large land areas and are aware of the importance of quality seedlings to produce high yields. The constraints in purchasing certified seed from PPKS felt by farmers are the long waiting period for orders. In addition, the number of seeds ordered must be in large quantities. The minimum seed order by farmers from PPKS Medan is 400 seeds and have to wait for several months. Therefore, an institution (local government) is needed that can provide seedlings according to the needs of farmers.

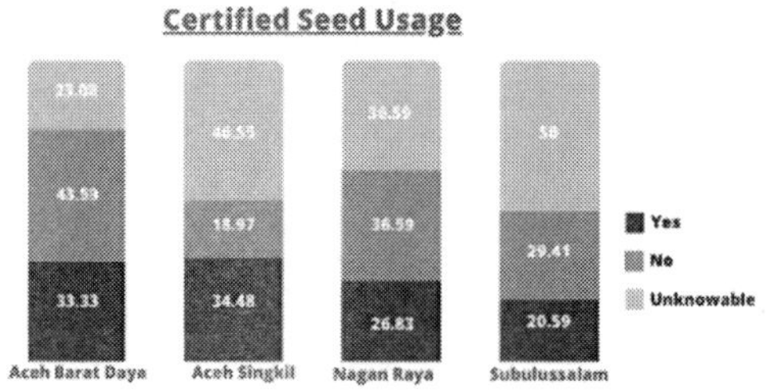

Fig. 7. Certified Seed Usage

The percentage of farmers using certified seeds was highest in Aceh Singkil (34.48) and Aceh Barat Daya (33.33%) districts, and lowest in Subulussalam district (20.59). Many farmers do not know whether the seeds they use are certified or not. This is due to farmers' low knowledge of certified seeds. Farmers who use certified seedlings are motivated to obtain good seedlings to obtain good Fresh Fruit Bunches (FFB) [8]. Some farmers who use certified seedlings are aware that non-certified seedlings will produce fruit with frequent and abundant growing seasons compared to certified ones.

Farmers' knowledge about certified seedlings will lead to positive perceptions. Farmers' motivation to use non-certified seedlings is due to their low price. Research conducted by [8] shows that price has a positive influence on farmers' decision to use certified seedlings. Infomation on the benefits of using certified seeds is more trusted by farmers if it comes from farmers and farmer groups that have used them. This is because farmers can obtain tangible evidence of the effect of using certified seeds on productivity. This is in accordance with [9] which states that the most effective information comes from personal sources. The standard of certified seedlings will be difficult for smallholders to meet shortly. It's because the age of the plants is currently at a productive age [10].

3.3 Land

The size of land owned by farmers varies between 0.5 to more than 20 Ha per farmer. On average, oil palm farmers in the four districts are entitle to less than 5 ha of land. Farmers with a land area of more than 20 ha are found in Aceh Singkil District at 5.17% and in Subulussalam at 2.94%. Oil palm farmers with access to capital can access more land while farmers with limited capital remain with limited land area and productivity [11].

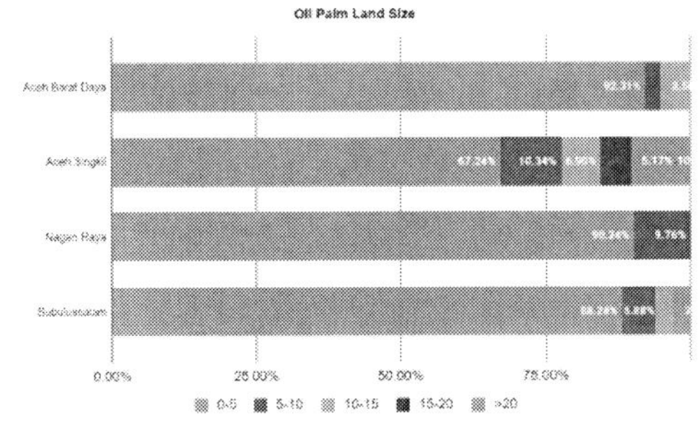

Fig. 8. Oil Palm Land Size

The promising market has led palm oil farmers to increase their profits by expanding their plantations and accessing more land both within and outside the village and even in other districts. Some farmers have plots in two locations and even in six different locations.

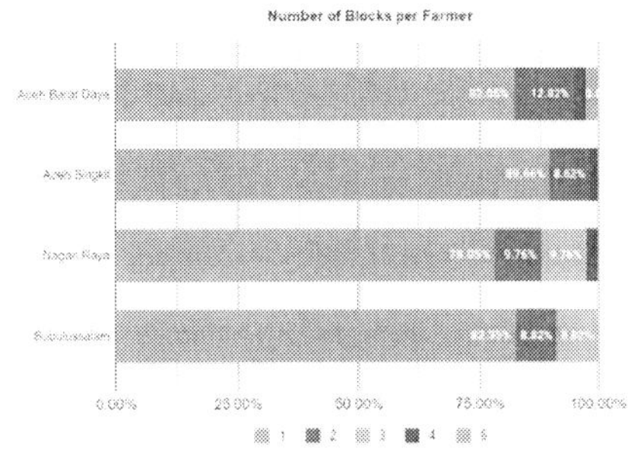

Fig. 9. Number of Blocks per Farmer

Oil palm plantation development is only permitted on other use areas (APL) and conversion forests that are carried out in a planned and systematic manner [3]. Most of the land planted by oil palm farmers today was converted from forests and other crops such as rubber or paddy fields. The great prospect of oil palm has made farmers interested in cultivating this crop. 7-10% of the farmers changed their paddy crops to oil palm because irrigation water began to dry up. This happened because many famers in the vicinity started planting oil palm, so the water supply began to be limited. In addition, the easier maintenance of oil palm plants compared to rice is also a factor in farmers converting to this crop. If this continues to happen, it will disrupt rice production. The government must implement policies so that the conversion from rice to palm oil can be controlled [12].

35-55% of the farmer convert their land from forest. In some cases, farmers convert production forests into oil palm plantations. Farmers do this because they do not know the status of the land beforehand. Spatial information is not accessible to farmers, so they do not know whether their farms are in accordance with the spatial plan[13]. Most smallholder oil palms were found in the production forest [14]. The percentage of production forest, limited production forest, and convertible production forest spatial patterns in the Existing Oil Palm Land can be seen in the table below.

Table 3. Production forest, limited production forest, and convertible production forest spatial patterns in the existing oil palm land

Districts	Pola Ruang		
	Production Forest (ha)	Limited Production Forest (ha)	Convertible Production Forest (ha)
Nagan Raya	152.64 (0.14%)	30.71 (0.03%)	-
Aceh Barat Daya	-	-	-
Subulussalam	29.34 (0.82 %)	-	452.45 (1.10%)
Aceh Singkil	296,31 (0.40%)	-	-

Source: Satellite image interpretation data

Data in the Table was the overlay of spacial planning policy (RTRW) and the smallholder oil palm map. From the data, Subulussalam has the highest percentage of oil palm land in production forest 0.82%, and conversion production forest 1.10% compared to the other three districts. Local community lands are often incorrectly categorized as part of the forest estate. Farmers whose land is located in an illegal area cannot obtain a cultivation registration letter. This will be an obstacle in the process of document legality [15].

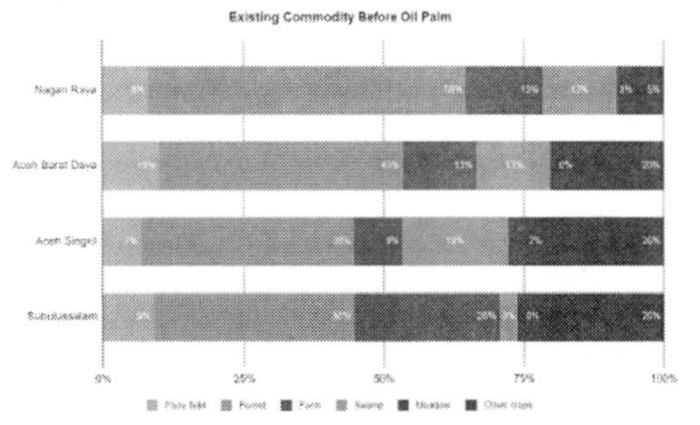

Fig. 10. Existing commodity before oil palm

Most of the land clearing for oil palm is done by manual logging (45-53%). Farmers use both non-family and family labor in the manual logging process. This land clearing normally takes longer periods due to limited resources and capital. 6-15% of farmers also still use burning when clearing land because it is cheaper and more effective [16], [17]. Land clearing by burning can be changed by using disincentives or coercion by the government [18]. Only 3-8% number of farmers use excavators in land clearing especially the one with sufficient capital and larger size of land.

Land clearing has not been implemented properly by farmers due to various factors, one of which is farmers' lack of awareness of the prohibition of burning, lack of information about land clearing regulations by the government, inadequate technology, and low economy [13], [19].

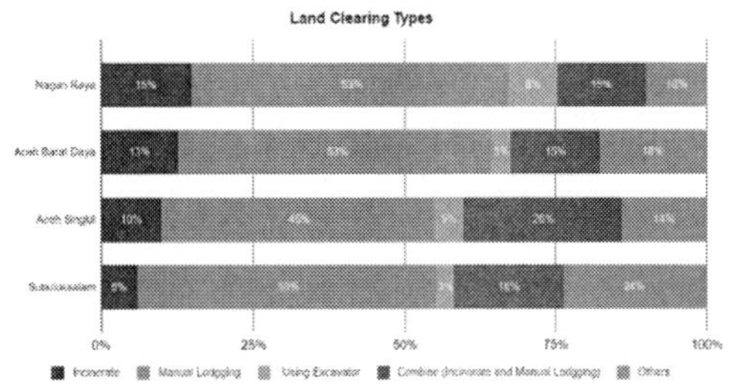

Fig. 11. Land Clearing Types

Conformity of land concessions with spatial plans will provide legal certainty in terms of ownership or control and management of areas used as plantation businesses [20][15]. Most farmers' land certificates are in the form of Certificates of Ownership (SHM) and SKT as well as other types of land certificates. SHM has the most benefits for the owner. SKT does not show strong ownership and is easy to cancel [21]. Farmers with the highest SHM ownership are in Nagan Raya district at 56.10% while the lowest is Southwest Aceh at 38.46%. Enhancing smallholder farmers' land certificates will be effective through institutional approaches (farmer groups).

Fig. 12. Type of Legal Certificate on Land

Independent smallholders have limited formal recognition of land tenure through land certificates and the ownership of the land is not harmonized with official spatial plans. This is the most difficult to comply without support [15][22]. So it is necessary to provide assistance to farmers in order to overcome problems and land legality.

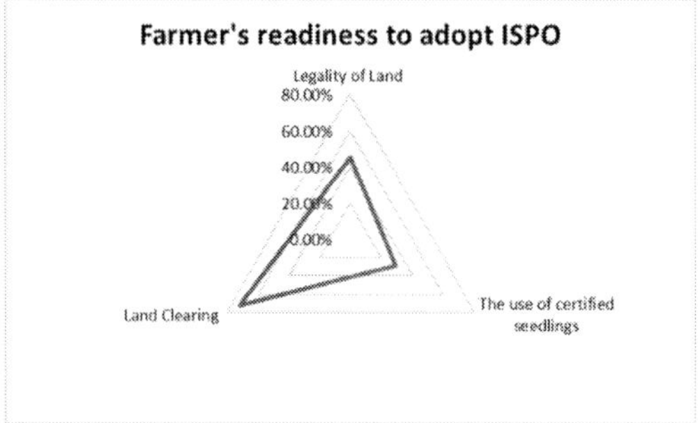

Fig. 13. Farmer's readiness to adopt ISPO

Fig. 13 shows the farmer's readiness to adopt ISPO based on land and seed criteria. The method of land clearing carried out by farmers is 72% compliant. Meanwhile, land legality is 45.35% and still needs to be improved through the Complete Systematic Land Registration (PTSL) program conducted by the land office in each district. The use of certified seeds is still very low at 28.81%. Therefore, there is a need for socialization to farmers regarding certified seeds. In addition, it is expected that through the People's Palm Oil Rejuvenation (PSR) program, the use of certified seeds will increase in the future.

The three indicators above are only a small part of the principles and criteria for sustainability certification. There are still many indicators that must receive attention so that the ISPO mandatory can be implemented in smallholders.

4 Conclusion and Recommendation

The number of farmers who use indigenous seeds in the four districts is still high. The highest number of farmers using certified seeds was in Aceh Singkil District, with 34.48%. When viewed from the land area, farmers in Aceh Singkil also have a larger land area. Farmers with access to capital can access land as well as quality seeds. Farmers with capital and knowledge will purchase certified seeds to ensure high FFB yields.

For land legality, Nagan Raya has the largest number of farmers with SHM ownership at 56.10% and the lowest is Aceh Barat Daya District at 38.46%. Therefore, the Land Office of each District needs to encourage the acceleration of land legality by involving farmer groups or working with plantation companies.

This study found that the clarity on land and seed still become the main problems in adopting sustainability schemes. One way to accelerate sustainable oil palm practices regarding land and seeds is to better understand the oil palm agroecosystem and test sustainability strategies. Land expansion and management can be achieved by encouraging the use of degraded and abandoned land for oil palm agriculture. To achieve sustainable land management methods, strong land-use planning and zoning restrictions should be implemented. Encouraging the use of high-yielding, disease-resistant oil palm seedlings that have been certified is another way to do this. Farmers should be given training and technical help in sustainable oil palm farming practices and financial incentives provision and support to farmers who adopt sustainable practices are important.

References

[1] M. Colchester, C. Sophie, and V. T. D. dan J. V. Jonas Dallinger, H.E.P. Sokhannaro, "Ekspansi Kelapa Sawit di Asia Tenggara : Sebuah Tinjauan," p. 292, (2011).

[2] Distanbun, "Statistik Perkebunan Aceh 2021," *2022*. (2021).

[3] H. Herdiansyah, H. A. Negoro, N. Rusdayanti, and S. Shara, "Palm oil plantation and cultivation: Prosperity and productivity of smallholders," *Open Agric.*, vol. 5, no. 1, pp. 617–630, (2020), doi: 10.1515/opag-2020-0063.

[4] C. Brandi, T. Cabani, C. Hosang, S. Schirmbeck, L. Westermann, and H. Wiese, "Sustainability Standards for Palm Oil: Challenges for Smallholder Certification Under the RSPO," *J. Environ. Dev.*, vol. 24, no. 3, pp. 292–314, (2015), doi: 10.1177/1070496515593775.

[5] X. Wang and Z. Cheng, "Cross-Sectional Studies: Strengths, Weaknesses, and Recommendations," *Chest*, vol. 158, no. 1, pp. S65–S71, (2020), doi: 10.1016/j.chest.2020.03.012.

[6] N. Saadun *et al.*, "Socio-ecological perspectives of engaging smallholders in environmental-friendly palm oil certification schemes," *Land use policy*, vol. 72, no. January, pp. 333–340, (2018), doi: 10.1016/j.landusepol.2017.12.057.

[7] E. Degli Innocenti and P. Oosterveer, "Opportunities and bottlenecks for upstream learning within RSPO certified palm oil value chains: A comparative analysis between Indonesia and Thailand," *J. Rural Stud.*, vol. 78, no. May, pp. 426–437, (2020), doi: 10.1016/j.jrurstud.2020.07.004.

[8] B. Utoyo and C. Yolandika, "Farmers' decision analysis to select certified palm oil seedlings in Lampung, Indonesia," *IOP Conf. Ser. Earth Environ. Sci.*, vol. 141, no. 1, (2018), doi: 10.1088/1755-1315/141/1/012034.

[9] N. J. Setiadi, *Perilaku Konsumen Konsep dan Implikasi Untuk Strategi dan Penelitian Pemasaran*. Jakarta: Prenada Media, (2003).

[10] S. Nuryanti, S. Hutabarat, and J. Yusri, "Analysis on Sustainability of Oil Palm Independent Smallholder (A Case Study of Oil Palm Independent Smallholder Kelompok Tani Petani Makmur at Lubuk Ogong Village, Bandar Sei Kijang Sub District, Pelalawan District)," *J. Sungkai*, vol. 7, no. Februari, pp. 61–78, (2019).

[11] G. Ramadhan, "Pengetahuan Sebagai Kuasa Eksklusi: Kisah Petani Kelapa Sawit Mandiri di Desa Harapan Jaya, Kalimantan Timur," Institut Pertanian Bogor, (2022).

[12] I. Zahri, E. Wildayana, A. T. Ak, D. Adriani, and M. U. Harun, "Impact of conversion from rice farms to oil palm plantations on socio-economic aspects of ex-migrants in Indonesia," *Agric. Econ. (Czech Republic)*, vol. 65, no. 12, pp. 579–586, (2019), doi: 10.17221/349/2018-AGRICECON.

[13] M. Rachmat and S. Nuryanti, "Daya Saing Produk Olahan Pertanian :," in *Memperkuat Daya Saing Produk Pertanian*, (2014), pp. 401–425.

[14] D. O. Pribadi *et al.*, "Mapping smallholder plantation as a key to sustainable oil palm: A deep learning approach to high-resolution satellite imagery," *Appl. Geogr.*, vol. 153, no. March, p. 102921, (2023), doi: 10.1016/j.apgeog.2023.102921.

[15] J. D. Watts *et al.*, "Challenges faced by smallholders in achieving sustainable palm oil certification in Indonesia," *World Dev.*, vol. 146, (2021), doi: 10.1016/j.worlddev.2021.105565.

[16] H. Purnomo *et al.*, "Fire economy and actor network of forest and land fires in Indonesia," *For. Policy Econ.*, vol. 78, pp. 21–31, (2017), doi: https://doi.org/10.1016/j.forpol.2017.01.001.

[17] E. P. Purnomo, R. Ramdani, Agustiyara, Q. P. V. Tomaro, and G. S. Samidjo, "Land ownership transformation before and after forest fires in Indonesian palm oil plantation areas," *J. Land Use Sci.*, vol. 14, no. 1, pp. 37–51, (2019), doi: 10.1080/1747423X.2019.1614686.

[18] H. Purnomo *et al.*, "Reducing forest and land fires through good palm oil value chain governance," *Forest Policy and Economics*, vol. 91. pp. 94–106, (2018), doi: 10.1016/j.forpol.2017.12.014.

[19] S. Hutabarat, M. Slingerland, P. Rietberg, and L. Dries, "Costs and benefits of certification of

independent oil palm smallholders in Indonesia," *Int. Food Agribus. Manag. Rev.*, vol. 21, no. 6, pp. 681–700, (2018), doi: 10.22434/IFAMR2016.0162.

[20] W. Widiati, A. Mulyadi, A. Syahza, and Mubarak, "Analysis of plantation management achievement based on sustainable development," *Int. J. Sustain. Dev. Plan.*, vol. 15, no. 4, pp. 575–584, (2020), doi: 10.18280/ijsdp.150418.

[21] L. Potter, "Who is ' land grabbing '? Who is deforesting ? Will certification help prevent bad practice ?," *L. grabbing, Confl. Agrar. Transform. Perspect. from East Southeast Asia*, no. 40, (2015), [Online]. Available: http://www.iss.nl/fileadmin/ASSETS/iss/Research_and_projects/Research_networks/LDPI/CMCP_40-_Potter.pdf.

[22] E. Apriani, Y. S. Kim, L. A. Fisher, and H. Baral, "Non-state certification of smallholders for sustainable palm oil in Sumatra, Indonesia," *Land use policy*, vol. 99, p. 105112, (2020), doi: 10.1016/j.landusepol.2020.105112.

Effectiveness combination of *leucaena* compost and coconut shell biochar on reduce soil compaction and corn growth in ultisol.

Endriani[1*], and *Diah Listyarini*[1]

[1]Department of Agroecotechnology, Faculty of Agriculture, Jambi University

Abstract. This study aims to identify optimal dose combinations that significantly impact reducing soil compaction in Ultisols and corn growth. The randomized block design (RBD) was utilized to conduct the study. The ameliorant materials consisted of leucaena compost and coconut shell biochar, which were divided into ten treatments and three treatments with 10 treatments & 3 group for each treatment. The treatments were combinations of coconut shell biochar leucaena compost and inorganic fertilizer. Soil physical parameters, such as soil organic matter, bulk density, porosity, and penetration resistance, were monitored during the study. The analysis was conducted using ANOVA at 95% confidence level (α = 5%). Subsequently, Duncan Multiple Range Test (DMRT) was conducted to evaluate the average effect of treatment. The result of the study indicated that Treatment of 5 tons/ha compost + 5 tons/ha biochar + ½ dose of inorganic fertilizer was able to reduce soil density compared to the control and showed a decrease in bulk density, as well as an increase in soil porosity, and a decrease in the value of soil penetration resistance. Treatment of 10 tons/ha of compost with 10 tons/ha of biochar + ½ dose of inorganic fertilizer is the best treatment for corn growth.

supports plant growth.

INTRODUCTION

Ultisol is a type of soil used for agricultural cultivation in Jambi Province. Continuous use of Ultisols for agriculture without efforts to prevent soil damage will decrease productivity. Ultisol has soil physical properties that do not support plant growth: less stable soil structure, slow infiltration and permeability, low soil organic matter (SOM), and low porosity, so the soil tends to be denser—the low aggregate stability results in increased erosion hazard and high bulk density [1]. High soil bulk density indicates that the soil has a high soil-specific density. High soil bulk density inhibits plant growth and development because it is difficult for plant roots to penetrate the soil to obtain nutrients.

Soil that is too compact, air exchange becomes slow, oxygen content in the soil is quite low, and permeability is hampered, so water will stagnate and inhibit plant growth. The effect is that the plants become stunted and thin, making it difficult to absorb nutrients and water optimally [2]. Control of soil density by increasing soil organic matter. Soil organic matter can increase soil aggregation, make the soil structure crumbly, and improve aeration and soil percolation. Soil organic matter creates balanced soil micro and macro pores for water transmission and retention [3]. Compost can improve soil that experiences high density and can increase plant growth. One of the raw materials for compost that is easy and widely available is lamtoro (Leucaena). The leucaena plant is widespread throughout rural areas and can easily grow in almost all areas that receive sufficient rainfall. Compost can improve the physical properties of soil, especially reducing soil density. This can happen because there is organic material in the compost. Composted organic material can provide micro nutrients for plants, loosen the soil, improve soil structure, increase porosity, improve soil aerase, increase soil binding capacity. to water and facilitate root growth [4].

On the one hand, adding organic fertilizers such as compost, manure, and plant residues in tropical areas' agricultural systems can immediately provide nutrients. However, C-Organic stability lasts relatively short, only a few seasons. This matter is due to the rapid decomposition of organic matter. So that Soil amendments are needed to maintain C-organic stability in the long term. One material that is difficult to decompose in soil is biochar. The use of soil improvement materials made from agricultural waste, such as coconut shell, which is difficult to decompose, is one alternative that can be taken to accelerate the improvement of the quality of soil physical properties [5].

* Corresponding author: endriani

Biochar is a soil amendment that can improve soil physical properties such as aggregate stability, C-organic content, water, and nutrient retention. Biochar can reduce volume weight by increasing soil porosity. A decrease in volume weight indicates a decrease in soil density [6]. Biochar of oil palm shells, coconut shells, and Sawdust given at 5 tons/ha and 10 tons/ha can improve soil density by reducing bulk density and increasing soil penetration resistance and porosity [7]. Rice husk biochar, sawdust, and coconut shells can reduce the soil volume weight value. Giving 5 tons/face can reduce soil density and increase soil porosity. However, the highest soybean yield was in plants given 10 tons/ha of rice husk biochar [8].

The weakness of biochar was the low content of organic matter,while improving the physical properties of the soil requires active organic matter. Therefore, it is necessary to combine biochar with other organic materials. Combining 5 tons/ha of biochar and 5 tons/ha of compost can increase available water content (8.39%), reduce soil penetration resistance so that sandy soil is more compact, and consistently increase aggregate stability. The combination of 10 tons/ha of compost and 40 tons/ha of corn stalk biochar can reduce bulk density (26.5%), increase porosity (9.2%), and reduce total plant water requirements (34.4%) as well as increase available soil water. (61.9%) [9]. Combining compost and biochar is expected to improve soil density and corn growth. This study aims to identify optimal dose combinations that significantly impact reducing soil compaction in Ultisols and corn growth.

MATERIALS AND METHODS

The study was conducted in Tangkit Village, Sungai Gelam Sub-District, Muaro Jambi District, Jambi Province. The soil type at the research site was Ultisol. The duration of the study was approximately six months. The research utilized various instruments, such as pyrolysis drums, weighing scales, soil ovens, sample rings, cone penetrometer. Research materials include fresh leucaena, coconut shell, corn seeds, inorganic fertilizer (Urea, KCl, TSP), chicken feces, rock phosphate, trichoderma, furan, and insecticides. This study was an experimental study using a randomized block design (RBD) with ten treatments with three repetitions for each treatment to obtain 30 experimental units. The treatments were combinations of oil palm shell biochar and leucaena compost as described below: **a0**: control (without Biochar and compost) + full dose inorganic fertilizer; **a1**: compost 5 tonnes/ha + biochar 10 tonnes/ha + 1/2 inorganic dose fertilizer; **a2**: compost 10 tonnes/ha + biochar 5 tonnes/ha + 1/2 dose inorganic fertilizer; **a3**: compost 10 tonnes/ha + without biochar + 1/2 dose inorganic fertilizer; **a4:** without compost + biochar 10 tonnes/ha + 1/2 dose inorganic fertilizer; **a5:** compost 5 tonnes/ha + biochar 5 tonnes/ha + 1/2 dose inorganic fertilizer; **a6:** compost 5 tonnes/ha + biochar 5 tonnes/ha + full dose inorganic fertilizer: **a7:** compost 10 tonnes/ha + without biochar + full dose inorganic fertilizer; **a8:** without compost + biochar 10 tonnes/ha + full dose inorganic fertilizer; and **a9:** compost 10 tonnes/ha + biochar 10 tonnes/ha + 1/2 dose inorganic fertilizer.

Research Stages

The biochar used in this study was derived from coconut shells that underwent pyrolysis and were subsequently sifted through 50 mesh (100% pass) and 80 mesh (50% pass) sieves. The compost used in the study was composed of fresh leucaena forage with Trichoderma decomposers and was subjected to a 4-week composting process. The experimental plots were sized at 3 x 4 meters and spaced at 75 x 40 cm. Treatment was carried out by broadcasting the materials onto the soil surface and incorporating them to a depth of approximately 20 cm. Treatment was administered based on the dry weight dose of organic matter and subsequently incubated for one week following even distribution.

Observational Variables

The plant variables observed in this study were: plant height. Soil variables observed include:

a. Soil organic matter (SOM)

The gravimetric method was utilized to measure bulk density. Intact soil samples were subjected to a 2 x 24-hour oven-drying process at 105°C, after which bulk density was determined using the equation below:

$$\%\,SOM = \left(\frac{sample\ dry\ weight - ash\ weight}{sample\ dry\ weight}\right) x100\%$$

b. Bulk density

The gravimetric method was utilized to measure bulk density. Intact soil samples were subjected to a 2 x 24-hour oven-drying process at 105°C, after which bulk density was determined using the equation below

$$Bulk\ density = \frac{soil\ dry\ weight\ (gr)}{soil\ volume\ (cm^3)}$$

c. Soil porosity

The gravimetric method was utilized to determine the soil porosity. Intact soil samples baked in sample rings for 2 x 24 hours at 105°C were used for the analysis. Based on the organic matter content value, which was greater than 1%, the total pore space was calculated using the equation provided below:

$$Total\ pore\ space = \left(1 - \frac{BD}{BJ - (0{,}02\ x\ \%\ BO)}\right) x\ 100\%$$

Description: BD = bulk density (gr/cm^3); BO = organic matter (%); BJ = weight of soil per unit volume of soil particles (gr/cm^3).

d. Soil penetration resistance

Soil penetration resistance measurements are carried out using a cone penetrometer by pushing the rod of the penetrometer cone into the ground with constant strength at a depth of 10 cm and 20 cm. The measurement value is read on a scale with 3 points on each plot. Measurement of penetration resistance begins two weeks after planting and is carried out once a week.

Data Analysis

In this study, a range of observational data was collected and analyzed. This included variables such as: soil organic matter (SOM), bulk density, soil porosity, and soil penetration resistance. The analysis was conducted using ANOVA at a 95% confidence level (α = 5%). Subsequently, the Duncan Multiple Range Test (DMRT) was conducted to evaluate the average effect of treatment.

RESULT AND DISCUSSION

Soil Physical Properties Before Treatment

The physical characteristics of the soil, as observed prior to treatment, indicated the presence of several physical properties that were not conducive to plant growth. The relevant physical properties of the soil prior to the completion of total treatment have been documented in Table 1.

Table 1. Physical soil properties before treatment.

Soil parameters	Measurement Results	Criteria
Organic matter (%)	2.12	low*
C-organik	1.23	low*
Bulk density (g/cm^3)	1.42	medium*
Porosity (%)	44.77	low*
Soil penetration resistance (KgF/cm):		
Depth 10 cm	18.36	high*
Depth 20 cm	31.61	high*

Note: * criteria based on Bogor Soil Research Center (1994); ** criteria based on National Soil Survey Center Natural Resource Conservation Service-USDA (1912).

Tabel 1 indicates that the soil in the low category contains 2.12% organic matter. This low organic matter leads to a medium bulk density value, which can reach 1.42 in the high category. Combination of low organic matter and medium bulk density results in soil compaction and a low porosity, which can be as low as 44.77% in the low category. Low soil organic matter is caused by organic matter. The soil has rotted, and there is no additional organic material, so it is necessary to add organic material. The organic matter content in the soil greatly influences other physical properties of the soil. Low soil organic matter content will result in high specific gravity and low porosity, as well as weak binding power between particles in the soil, so the impact of rainwater easily damages the soil. The crushed particles can cause the soil to clog the pores so that the soil becomes dense easily. Soil bulk density has a close relationship with soil porosity. Soil porosity can be obtained by determining soil bulk density, and the higher the bulk density of soil, the lower the porosity [10].

The research location, soil penetration resistance at a depth of 10 cm, was 18.36 KgFcm^{-2}, and at a depth of 20 cm, namely 31.61 KgFcm^{-2}, included in the high criteria. Soil with high density makes it difficult for corn seeds to grow. Soil compaction can reduce soil aeration, reduce water availability for plants, and inhibit root growth and plant germination [11].

Characteristics of Compost and Biochar.

Table 2 displays the complete analysis results of leucaena compost before its application. The analysis reveals that the compost adheres to the Indonesian National Standard (SNI) 19-7030-2005, indicating its high quality. Furthermore, the compost has undergone complete decomposition, making it appropriate for application with a C/N value of 15.41%. It is worth noting that the C/N ratio of the compost has also reached 12.56.

Tabel 2 Compost and biochar analysis

Parameters	Compost	Biochar
Water content (%)	38.00	9.40
C-organik (%)	28.97	32.82
N-total (%)	1.88	-
C/N	15.41	-
Ash content (%)	-	8.60

Results of analysis of leucaena compost and coconut shell biochar. Table 2 shows that the c-organic content in biochar is quite high, namely 32.82% and is also in accordance with the c-organic biochar quality standard of 9.8-32. Biochar has strong properties so it can provide a place for microorganisms to grow, because biochar contains high C-organic, so it can be used as an energy source for microorganisms in the soil. The process of making biochar shows that around 50% of the organic C present in the base material can be contained in biochar. Biological decomposition of biochar is usually less than 20% after 5-10 years [12].

The water and ash content of biochar have values of 9.4% and 8.6%, respectively, where the water content and ash content meet the standard criteria for biochar quality. Therefore, the ash content in biochar greatly affects the quality of the biochar because it can cause blockage of the pores in the biochar so that its surface area will be reduced [13]. The water content in the compost is 38.00%, c-organic 28,97%. It is in accordance with compost quality standards where the carbon content contained in the compost is a minimum of 9.8% and a maximum of 32.82%. Compost will increase the soil's ability to store water, besides that compost can loosen the soil so that it can become a medium for plant growth.

The nitrogen content in the compost is 1.88% and meets the criteria for compost quality standards, namely 0.4%. The C/N ratio of leucena compost is 15.41, where the maturity level of the compost is by the Compost Quality Standards of 10-20. This condition indicates that the compost is completely decomposed and ready to use.

Effects of treatment on soil organic matter

Tabel 3 displays effect of compost and biochar significantly increased soil organic matter (SOM). Treatment a2 and a9 significantly from control (a0) to increased soil organic matter. Treatment a1 not significant to increased soil organic matter.

Tabel 3 Effect of treatment doses on SOM

Treatments	SOM (%)
Control (without Biochar and compost) + full dose inorganic fertilizer (a0)	2.33 e
Compost 5 tonnes/ha + biochar 10 tonnes/ha + 1/2 inorganic dose fertilizer (a1)	5.67 ab
Compost 10 tonnes/ha + biochar 5 tonnes/ha + 1/2 dose inorganic fertilizer (a2)	6.00 a
Compost 10 tonnes/ha + without biochar + 1/2 dose inorganic fertilizer (a3)	5.33 abc
Without compost + biochar 10 tonnes/ha + 1/2 dose inorganic fertilizer (a4)	4.00 d
Compost 5 tonnes/ha + biochar 5 tonnes/ha + 1/2 dose inorganic fertilizer (a5)	5.33 abc
Compost 5 tonnes/ha + biochar 5 tonnes/ha + full dose inorganic fertilizer (a6)	4.33 cd
Compost 10 tonnes/ha + without biochar + full dose inorganic fertilizer (a7)	4.67 bcd
Without compost + biochar 10 tonnes/ha + full dose inorganic fertilizer (a8)	5.40 bcd
Compost 10 tonnes/ha + biochar 10 tonnes/ha + 1/2 dose inorganic fertilizer (a9)	6.33 a

According to Table 3, effect of compost and biochar on soil organic matter was significant. The treatments a2 and a9 are significantly different from treatment a0. Treatment a1 is not significantly different from treatments a3, a5, a7, and a8 but is significantly different from treatments a4 and a6 and without treatment. The analysis results showed that 5 tonnes/ha of compost and 5 tonnes/ha of biochar were able to increase the organic matter content in the soil compared to the control. Organic matter added to the soil can increase the SOM content, and increasing the SOM influences the physical, chemical, and biological properties of the soil for the better [14]. The analysis showed that the combination of compost and biochar was able to increase the SOM by 71.67% - 171.67%. Compost decomposes slowly, providing organic matter in the soil, and biochar is also biological charcoal that functions as a habitat for carbon-degrading microorganisms.

Effects of treatment on bulk density

Bulk density is an indicator of soil density. Table 4 displays effect of compost and biochar significantly reduce bulk density. Combination 10 tons/ha of compost and 10 tons/ha of biochar can reduce the largest bulk density value.

Tabel 4 Effect of treatment doses on bulk density

Treatments	Bulk density (g/cm^3)
Control (without Biochar and compost) + full dose inorganic fertilizer (a0)	1.58 a
Compost 5 tonnes/ha + biochar 10 tonnes/ha + 1/2 inorganic dose fertilizer (a1)	1.15 c
Compost 10 tonnes/ha + biochar 5 tonnes/ha + 1/2 dose inorganic fertilizer (a2)	1.16 c
Compost 10 tonnes/ha + without biochar + 1/2 dose inorganic fertilizer (a3)	1.25 bc
Without compost + biochar 10 tonnes/ha + 1/2 dose inorganic fertilizer (a4)	1.33 b
Compost 5 tonnes/ha + biochar 5 tonnes/ha + 1/2 dose inorganic fertilizer (a5)	1.23 bc
Compost 5 tonnes/ha + biochar 5 tonnes/ha + full dose inorganic fertilizer (a6)	1.24 bc
Compost 10 tonnes/ha + without biochar + full dose inorganic fertilizer (a7)	1,23 bc
Without compost + biochar 10 tonnes/ha + full dose inorganic fertilizer (a8)	1.29 bc
Compost 10 tonnes/ha + biochar 10 tonnes/ha + 1/2 dose inorganic fertilizer (a9)	1.14 c

Tabel 4 illustrates that treatments a9, a1, a2 were significantly different from treatments a4 and a0 but not significantly different from treatments a3, a5, a6, a7 and a8 in reducing bulk density. The combination of compost and biochar in a ratio of 1:1, 1:2, or 2:1 with the addition of half the recommended inorganic fertilizer can reduce soil bulk density by 15.92% - 27.35%. Sedangkan perlakuan 10ton/ha kompos dan 10 ton/ha biochar mampu menurunkan bulk density sebesar 28.05%. The decrease in bulk density occurred because an increase influenced it in SOM (Table 3), which caused a decrease in bulk density. The higher the SOM in the soil, the lower the bulk density value. The higher the som causes the bulk density to be lower, and the porosity value becomes high. Bulk density is inversely proportional to porosity. If the soil has a high bulk density, then the porosity is low, and if the lower the bulk density value, the higher the porosity value [15].

Effects of treatment on porosity

The combination of compost and biochar had the effect of increasing porosity compared to the control (without compost and biochar). However, the research results showed that the treatments were not significantly different and increased total porosity significantly.

Tabel 5 Effect of treatment doses on porosity

Treatments	Porosity (g/cm^3)
Control (without Biochar and compost) + full dose inorganic fertilizer (a0)	39.18 b
Compost 5 tonnes/ha + biochar 10 tonnes/ha + 1/2 inorganic dose fertilizer (a1)	54.68 a
Compost 10 tonnes/ha + biochar 5 tonnes/ha + 1/2 dose inorganic fertilizer (a2)	54,31 a
Compost 10 tonnes/ha + without biochar + 1/2 dose inorganic fertilizer (a3)	51.02 a
Without compost + biochar 10 tonnes/ha + 1/2 dôse inorganic fertilizer (a4)	48.21 a
Compost 5 tonnes/ha + biochar 5 tonnes/ha + 1/2 dose inorganic fertilizer (a5)	51.74 a
Compost 5 tonnes/ha + biochar 5 tonnes/ha + full dose inorganic fertilizer (a6)	51.74 a
Compost 10 tonnes/ha + without biochar + full dose inorganic fertilizer (a7)	51.80 a
Without compost + biochar 10 tonnes/ha + full dose inorganic fertilizer (a8)	49.44 a
Compost 10 tonnes/ha + biochar 10 tonnes/ha + 1/2 dose inorganic fertilizer (a9)	54.84 a

 https://doi.org/10.1051/bioconf/20238003007

Tabel 4 illustrates that Treatment of compost, biochar, and a combination of compost - biochar at doses of 1:1, 1:2, or 2:1 with the addition of ½ recommended inorganic fertilizer has been able to increase porosity by 22.84% - 39.38%. Organic materials derived from compost and biochar can improve soil structure. Soil with a good soil structure results in the soil becoming loose and crumbly and having high porosity.

The application of biochar causes increased soil porosity. Biochar can reduce soil density so that soil porosity increases. Biochar has nesting properties, so when mixed into the soil, it causes the soil to become loose. Biochar from various materials and various manufacturing temperatures, which, when added to the soil, increases soil porosity. When added to the soil, biochar from various materials can increase soil porosity. Biochar can also reduce soil density in the top layer of soil and grow organic carbon.l [16,17,18,19].

Effects of treatment on soil penetration resistance.

Soil penetration resistance shows the soil's ability to be penetrated by plant roots. Low soil penetration resistance causes roots to develop and penetrate the soil layers quickly. The effect of treatments on soil penetration resistance at depths of 10 cm and 20 cm was showed Tabel 6.

Tabel 6 Effect of treatment doses on soil penetration resistance

Treatments	Soil penetration resistance (KgF/cm)	
	10 cm	20 cm
Control (without Biochar and compost) + full dose inorganic fertilizer (a0)	29.12 a	35.92 a
Compost 5 tonnes/ha + biochar 10 tonnes/ha + 1/2 inorganic dose fertilizer (a1)	25.49 d	32.06 d
Compost 10 tonnes/ha + biochar 5 tonnes/ha + 1/2 dose inorganic fertilizer (a2)	26.28 cd	32.86 cd
Compost 10 tonnes/ha + without biochar + 1/2 dose inorganic fertilizer (a3)	27.76 bc	34.44 b
Without compost + biochar 10 tonnes/ha + 1/2 dose inorganic fertilizer (a4)	26.74 bcd	33.61 cd
Compost 5 tonnes/ha + biochar 5 tonnes/ha + 1/2 dose inorganic fertilizer (a5)	27.08 bcd	32.40 cd
Compost 5 tonnes/ha + biochar 5 tonnes/ha + full dose inorganic fertilizer (a6)	26.51 bcd	32.74 cd
Compost 10 tonnes/ha + without biochar + full dose inorganic fertilizer (a7)	28.21 bc	34.33 b
Without compost + biochar 10 tonnes/ha + full dose inorganic fertilizer (a8)	28.41 bc	32.63 cd
Compost 10 tonnes/ha + biochar 10 tonnes/ha + 1/2 dose inorganic fertilizer (a9)	25.04 d	32.06 d

Tabel 4 illustrates that penetration resistance to a depth of 10 cm indicates treatment a9 significantly different from treatments a0, a3, and a7. Treatments a2, a4, a5, a6 and a8 are not significantly different from treatments a1 and a9. However, regarding penetration resistance to a depth of 20 cm, treatment a9 significantly differed from treatments a3, a7, a8, and a0.

The treatment of a2, a4, a5, a6 is not significantly different from a1 and a9. Providing compost, biochar, or a combination of coconut shell compost-biochar in a ratio of 1:1, 1:2, or 2:1 with the addition of ½ recommended inorganic fertilizer has been able to reduce penetration resistance value by 9.57% - 17.22% at a depth of 10 cm 7.76 % - 16.22 % at a depth of 20 cm. This condition occurs because the higher organic material content affects bulk density and porosity. Increased SOM is able to bind soil grains, the soil structure becomes crumbly, reduces density, and can improve soil pores. Penetration resistance will be lower with increasing levels of organic material added to the soil [20].

The application of compost, biochar, and a combination of compost-biochar significantly affected soil penetration resistance compared to no treatment. Observation of penetration resistance at a depth of 10 cm, the lowest penetration resistance value was 15.00 (kgF/cm^2) in the combination treatment of 10 tons/ha of compost and 10 tons/ha of biochar. Meanwhile, the highest penetration resistance value was 18.29 (kgF/cm^2) in the control treatment (without compost and biochar). Penetration resistance at a depth of 20 cm: the lowest penetration resistance was 20.29 (kgF/cm^2) in the combination treatment of 10 tons/ha of compost and 10 tons/ha of biochar, while the highest penetration resistance was 24.22 (kgF/cm^2) in the control treatment (without compost and biochar). The dynamics of soil penetration in each treatment are presented in Figures 1 and 2.

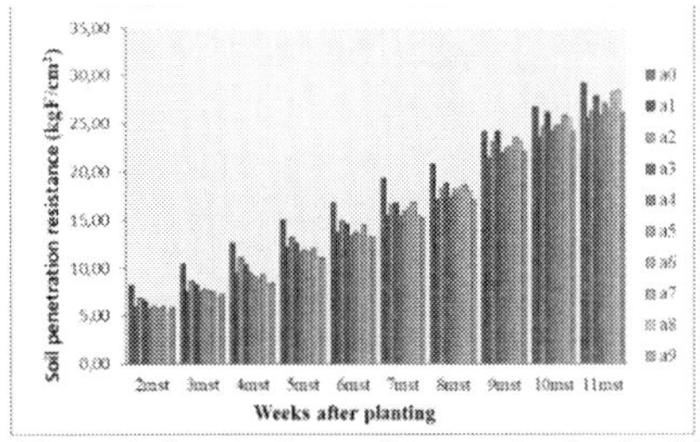

Figure 1. Penetration resistance at a depth of 10 cm

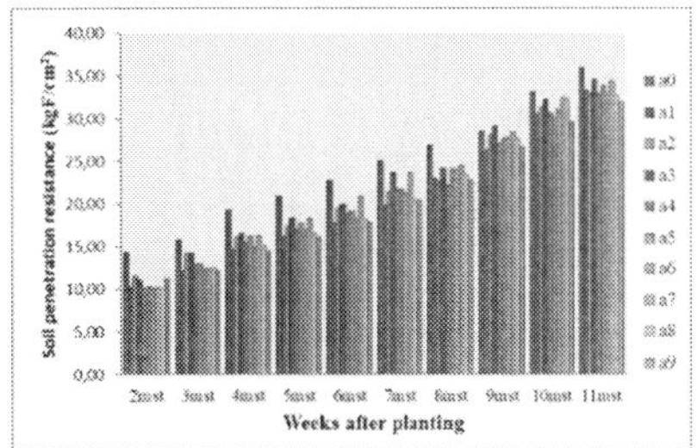

Figure 2. Penetration resistance at a depth of 20 cm

Two graphs illustrates that A0 treatment has the highest soil penetration resistance. Soil penetration at a depth of 10 cm and 20 cm has increased quite significant resistance. As the plant ages, soil penetration resistance will increase. This condition is caused by the impact of raindrops on the ground surface so that the soil becomes solid. In addition, plant roots with high vegetative growth will bind soil particles more tightly so that soil penetration resistance is higher. High penetration resistance indicates dense soil that has high bulk density and low porosity and organic matter.

Effects of treatment on corn growth.

The results of the research showed that the provision of compost and biochar had a significant effect on the growth of corn plants. The average values of plant height and corn yield as well as the results of further tests using the Duncan Multiple Range Test (DMRT) are presented in Table 6.

Table 7. Effects of treatment on corn growth

Treatments	height
Control (without Biochar and compost) + full dose inorganic fertilizer (a0)	168.42 c
Compost 5 tonnes/ha + biochar 10 tonnes/ha + 1/2 inorganic dose fertilizer (a1)	238.83 a
Compost 10 tonnes/ha + biochar 5 tonnes/ha + 1/2 dose inorganic fertilizer (a2)	239.42 a
Compost 10 tonnes/ha + without biochar + 1/2 dose inorganic fertilizer (a3)	198.25 bc
Without compost + biochar 10 tonnes/ha + 1/2 dose inorganic fertilizer (a4)	194.17 bc
Compost 5 tonnes/ha + biochar 5 tonnes/ha + 1/2 dose inorganic fertilizer (a5)	220.42 ab
Compost 5 tonnes/ha + biochar 5 tonnes/ha + full dose inorganic fertilizer (a6)	229.33 ab
Compost 10 tonnes/ha + without biochar + full dose inorganic fertilizer (a7)	228.58 ab
Without compost + biochar 10 tonnes/ha + full dose inorganic fertilizer (a8)	227.25 ab
Compost 10 tonnes/ha + biochar 10 tonnes/ha + 1/2 dose inorganic fertilizer (a9)	245.58 a

Tabel 7 illustrates that Plant height in treatments a1, a2, and a9 was significantly different from treatments a3, a4, and without but not substantially different from treatments a5, a6, a7, and a8. Providing compost and biochar and half the recommended inorganic fertilizer increased the percentage of corn plant height by 15.28% - 45.81%. The treatment contributes organic material to the soil as a plant nutrient supply. Organic materials play a role in absorbing nutrient elements, such as nitrogen element. The element nitrogen plays an important role in plant growth, especially in the vegetative phase such as root, stem and leaf growth, so that increasing nitrogen in the soil results in plant height increasing. Nitrogen is needed by plants in the vegetative and generative phases and is mobile in plants. The nitrogen nutrient content is required to stimulate stem growth, which stimulates plant height growth [21].

Treatment of 10 tons/ha of compost and 10 tons/ha of biochar contributed better to height growth. Ten tons/ha of compost and 10 tons/ha of biochar provide more organic material and nutrients than other treatments. Organic fertilizer in the form of compost can provide nutrients for plants. Organic fertilizer can also maintain and improve soil fertility so that plant growth is better.

CONCLUSION

Research shows that the application of 5 tons/ha of compost with 5 tons/ha of biochar with the addition of ½ recommended inorganic fertilizer can reduce soil density compared to without treatment. This condition is characterized by a decrease in bulk density, an increase in porosity, and a reduction in the value of soil penetration resistance. Ten tons/ha of compost with 10 tons/ha of coconut shell biochar and adding ½ recommended inorganic fertilizer is the best treatment for improving soil density.

ACKNOWLEDGEMENT

This study was partially funded by Competitive Research Indonesia. The authors thank the Directorate General of Higher Education of Indonesia for superior university-applied research in 2022. Also thanks to the Rector of Jambi University who has supported this research

REFERENCE

1. Utomo B. 2008. Improvement of the Physical Properties of Ultisols to Increase the Growth of Eucalyptus urophylla at an Altitude of 0-400 Meters. Faculty of Agriculture. University of North Sumatra. Medan.
2. Haridjaja. O. Hidayat. Y. & Maryamah. L.S. 2010. Effect of Soil Bulk Density on Soil Physical Properties and Seed Germinations of Peanut and Soybean. Indonesian Journal of Agricultural Sciences. **15**(3): 147-152.
3. Aqbar D. 2018. The Effect of Chicken Manure and EM4 on Growth and Yield of Water spinach (*Ipomoea reptans Poir.*) Thesis. Faculty of Agriculture Jambi University. Jambi.
4. Murbandono L. 2008. Make Compost. Niaga Swadaya, Jakarta
5. Sukartono. Utomo. W.H..Kusuma. Z. dan Nugroho. W.H. 2011. Soil fertility status. nutrient uptake. and maize (*Zea mays L.*) yield following biochar application on sandy soils of Lombok. Indonesia. Journal of Tropical Agriculure **49**:47-52.
6. Githinji L. 2014. Efect Biochar Aplication Rate on Soil Physical and Hydraulic Properties of A Sandy

Loam. Archives of Agronomy and Soil Science. **60**(4): 457-470

7. Putri. 2019. Improvement of Ultisol Density and Soybean Yield Through the Use of Several Types of Biochar. Thesis. Jambi University Faculty of Agriculture.
8. Endriani, Kurniawan, A. 2018. Soil and Carbon Conservation Through the Use of Biochar in Soybean Plantings. Jurnal Ilmu Terapan Uversitas Jambi. **2**(2):94-106
9. Syaikhu, A, H, F. Budi, H. Didik, S. 2016. Test of the Benefits of Biochar and Soil Improvement Materials to Improve Several Physical Properties of Sandy Soils and Their Impact on Sugarcane Growth and Production. Journal of Soil and Land Resources. **3**(2):345-357.
10. Sarief S. 1989. Physics and Chemistry of Agricultural Soils. Pustaka Buana. Bandung.
11. Wilson, E. 2006. Soil Density Due to Skidding by Forwarders and Its Effect on Seedling Growth. Thesis. Forest Products Department. Faculty of Forestry. Bogor Agricultural Institute.
12. Gani A. 2009. Biochar Saves the Environment. Warta Penelitian dan Pengembangan Pertanian. **31**(6): 15-16.
13. Scroder E. 2006. Experiment on the Generation of activated carbon from Biomassa. Institute for Nuclear and energy Technologies Forschungs Karlsruhe.Journal of Analytical and Applied Pyrolysis. **79**(1):106-111.
14. Utami, S. N. H, Handayani, S. 2003. Chemical Properties of Entisol in Organic Farming Systems. Ilmu Pertanian. **10**(2): 63-69
15. Endriani, Sunarti dan Ajidirman. 2013. Utilization of Palm Oil Shell Biochar as a Soil Amendment for Ultisol in the Bahar River, Jambi. Jurnal Penelitian Universitas Jambi Seri Sains. **15**(1): 39-46.
16. Doan TT, H des T Thierry, R Cornelia, LJ Jean and J Pascal. 2015. Impact of Compost, Vermicompost and Biochar on Soil Fertility, Maize Yield and Soil Erosion in Northern Vietnam: A Three Year Mesocosm Experiment. Journal Science of the Total Environment. **514**: 147–154.
17. Carvalho JLN, CEP Cerri, BJ Feigl, M de C Piccolo, V de P Godinho, U Herpin and CC Cerri. (2009). Conversion of Cerrado into Agricultural Land in The South-Western Amazon: Carbon Stocks and Soil Fertility. Scientia Agricola. **66**(2): 233-241.
18. Nurida, NL. 2017. Potential for Using Biochar for Dry Land Rehabilitation in Indonesia. **8**(3):57-68.
19. Masulili A. 2010. Rice Husk Biochar for Rice Based Cropping System in Acid Soil 1. The Characteristics of Rice Husk Biochar and its Influence on the Properties of Acid Sulfate Soils and Rice Growth in West Kalimantan, Indonesia. Journal of Agricultural Science **2**(1): 39-47.
20. Junedi H, IA Mahbub, dan Zurhalena. 2013. Utilization of Cow Manure Compost and Breech Figs to Reduce Ultisol Soil Density. Jurnal Penelitian Universitas Jambi. **15**(1): 47-52.
21. Setyamidjaja D. 2006. Palm Oil Cultivation Series, Cultivation Techniques, Harvesting, Processing. Yogyakarta.

 https://doi.org/10.1051/bioconf/20238003008

Enhancing food operator intention on specialty café using 3R waste management approach for eco-enzyme production as an implementation of SDGs 12 (Study on Akkar Specialty Cafés)

Wahyu Putri Andhini[1], *Darmawan Ari Nugroho*[1], and *M. Prasetya Kurniawan*[1*]
[1]Department of Agricultural Industrial Technology, Gadjah Mada University, Yogyakarta, Indonesia, 55281

Abstract
Handling café business waste is essential for environmental protection, especially for organic food waste. After all, organic food waste decomposes quickly and produces methane gas, which has a pungent odor if not appropriately managed. This research analyzes and describes the waste management activities of specialty café businesses, starting with showing the development potential for managing by-product waste to increase added value and increase customer awareness. The research method used is descriptive qualitative analysis based on literature study, observation, interviews, and analysis of other supporting documents. The practice of waste management can positively impact the amount of waste generated from the café business because the amount of waste that must be submitted to the landfill can be reduced and provide by-products to increase income and café aesthetics. Waste management can be implemented by collecting and sorting waste and then processing it into by-products, one of which is by making eco-enzymes. The impact of this activity is that café businesses can educate employees and consumers to care more about the environment through simple waste management. The limitation of this study is that the research was conducted without a waste treatment trial process. Although this research is locally focused and qualitative, the results and methods of this research can be used to guide the implementation of wiser and more sustainable café business waste management.

Keywords *Eco-enzyme, Reduce-Reuse-Recycle, SDGs 12, Specialty café*, Waste *management*

1. Introduction

Indonesia is one of the influential contributors to the world's food waste. Indonesia, in 2021, is ranked third as the country with the highest amount of food waste in the world after China and India, with a total of 20.9 million tons/year of food waste [1]. DI. Yogyakarta Province also experiences much waste and accumulation, precisely at the Piyungan Integrated Waste Management Site (in Bahasa: *Tempat Pembuangan Sampah Terpadu*/TPST), where in mid-2023 DI. Yogyakarta is experiencing a waste emergency because the amount of waste continues to increase until it exceeds the capacity of TPST. The amount of waste must be further reduced to prevent various environmental, social, and economic impacts that will arise. The accumulation of food waste, especially food waste, can impact the environment by producing methane gas from the decomposition of organic waste, as much as 8.2% of greenhouse gas emissions in the atmosphere and almost comparable to 10% of greenhouse gas emissions from the transportation sector [2]. Managing food waste has a positive social impact, allowing more people access to food [3]. Food waste has an energy content of 618-989 kcal/capita/day, equivalent to the needs of 29-47% of Indonesia's population that can be supplied from wasted food [4]. Food waste also has an economic impact; if consumers can reduce food waste, it can reduce the need for food, and the costs incurred to process the food are also wasted. The freed-up economic resources can be used for commodities other than food or indirectly have high economic growth effects [5]. Details of the effects of unmanaged food waste on environmental, social, and economic aspects are depicted in Fig 1.

The waste emergency problem experienced by DI. Yogyakarta requires the community to start processing waste independently. Therefore, it aligns with the Circular Letter of the Regional Secretariat of DI. Yogyakarta number 658/8312 related to the temporary closure of TPST Piyungan for 45 days due to the overcrowded condition of TPST, which requires each region in Bantul, Sleman, and Yogyakarta City to handle waste independently in each region. Efforts that the government can make to overcome the problem of waste accumulation have started since the beginning of 2023 by issuing Yogyakarta Mayor

*Corresponding author: kurniawan_prasetya@ugm.ac.id

Circular Letter number 660/6123/SE/2022 concerning the zero inorganic waste movement. This policy asks people to reduce, utilize, and eliminate inorganic waste in the city. This policy can increase public awareness about the need to process waste independently, including business actors, so that they play an active role in disseminating waste management and processing, one of which is for café businesses.

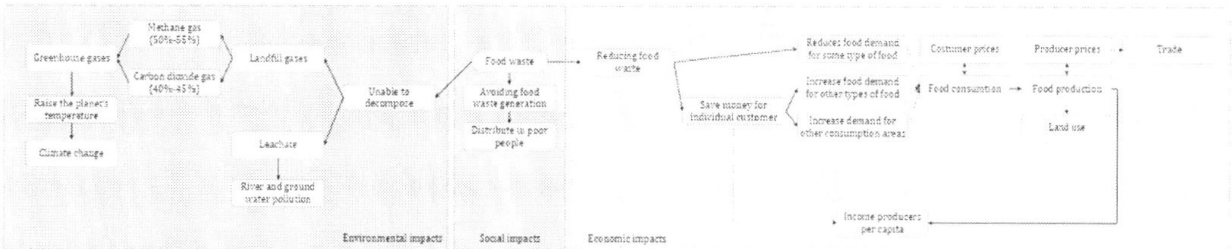

Fig 1. Schematic of the environmental, social, and economic impacts of food waste [6][7]

Café business waste is waste generated during daily operations. Of the various types of café businesses, there are juice café businesses that sell fruit and vegetable-based products. The types of waste generated consist of organic waste in the form of decaying raw materials to food waste, decaying waste of materials that have passed the date of use or are damaged, and inorganic waste in the form of packaging waste, waste of damaged production equipment, and waste due to excessive production. The percentage comparison of the amount of organic and inorganic waste of cafés in Indonesia and AKKAR's organic and inorganic waste is depicted in Fig 2.

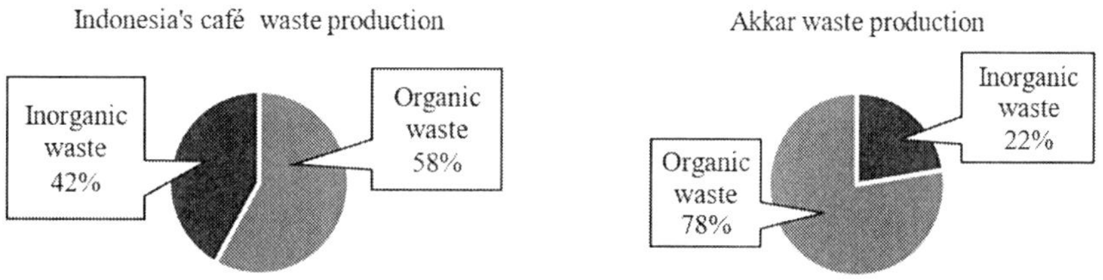

Fig 2. Percentage of café waste in Indonesia and Akkar juice bar [8]

The most straightforward juice café waste management is through the application of the 3R approach (Reduce, Reuse, and Recycle). The 3R approach is carried out by separating each type of waste based on the processing process that will then be carried out so that each waste can be managed appropriately and does not hinder the decomposition process. The advantages of the 3R approach over other waste management approaches are that it is more environmentally friendly because it helps reduce the creation of waste, saves resources, and is cost-effective [9][10]. Reduce action is an activity to minimize waste production by reducing or limiting the amount for the production process. Reuse is a management activity by reusing production equipment and product packaging. Reducing and reusing products can impact reducing new products, so resources and costs must be reduced [9][10]. Recycle activities are carried out by recycling organic waste into compost and inorganic waste handed over to waste processing companies for recycling. Thus, the 3R approach is a sustainable and low-cost waste management method compared to other methods.

Akkar juice bar is a café that provides processed fruits and vegetables from cold-pressed juice, smoothies, and salads produced with two production models. The made-for-stock production model is applied for cold-pressed juice products, while smoothies and salads are produced using the made-by-order model. Akkar Juice Bar is a café supporting healthy living by serving quality food and beverages with fresh and natural ingredients. Following its vision and mission, Akkar is present in the world of food and beverage in Yogyakarta with three values to create a healthy, belonging, and positive environment, namely:

a. Mindful, Akkar is aware that all production actions will impact all aspects of life, including human health and environmental responsibility.
b. Fruitful, Akkar always provides positive intentions for ourselves and others to grow together.
c. Impactful, Akkar ensures that its products come from the best ingredients to be easily digested, function optimally, and are sustainable.

To always realize these values, Akkar has made efforts to manage café waste by implementing the 5Rs, consisting of:

a. Refuse: AKKAR rejects using plastic packaging, replacing it with reusable bottles.
b. Reduce: AKKAR reduces plastic bag use and switches to reusable bags or containers.

 https://doi.org/10.1051/bioconf/20238003008

c. Reuse: AKKAR provides cashback for reusing every bottle exchange after sterilization.
d. Recycle: AKKAR works with local waste management services to sort inorganic waste.
e. Rot: AKKAR works with local farmers to process residual production waste into compost from organic waste.

Overall, Akkar already has a reasonably good waste management system through the application of the 3Rs to demonstrate an attitude of responsibility to the environment in sustainable production and consumption so that any residual production waste has been separated for processing and not directly disposed of in landfills by Sustainable Development Goals (SDGs) 12.5 even though the waste treatment still depends on third parties. Therefore, this research analyzes and describes the activities carried out by Akkar juice bar to increase the behavioral intention of business actors in waste management for sustainable businesses.

2. Impact of waste management on the sustainability of the café business

Waste management and treatment are crucial to achieving sustainable goals and actively contributing to the environment. Through waste management and treatment efforts by café business actors, café business actors become agents of change in achieving the 12th SDGs related to responsible consumption and production to encourage sustainable and efficient consumption and production patterns to reduce negative impacts on the environment. The impacts of handling the sustainability of café waste based on the application of the 3Rs include:

a. Reduce
   1) Waste accumulation: The disposal of food waste without processing leads to the accumulation of waste in landfills. The accumulation of food waste in landfills will increase the ambient temperature and cause an unpleasant odor because the bottom layer of the waste pile undergoes an anaerobic decay process [11].
   2) Ecosystem damage: Food waste deposited in landfills produces leachate from the decomposition of waste that can damage the soil. The accumulated waste also produces methane gas that contributes to greenhouse gas emissions, and food waste disposal is also the same as wasting 25% of the water needed in the manufacturing process [12]. Waste management activities to reduce potential damage to ecosystems are by SDGs 12.5 regarding reducing waste production by implementing the 3Rs.

b. Reuse
   1) Economic loss: Food waste behavior can cause economic losses for businesses because the costs incurred to produce food are wasted when the food is thrown away. The costs include raw material costs, labor costs, energy costs, transportation costs, processing costs, and the cost of disposing of food waste [13]. So, it is not in line with SDG 12.2, which targets sustainable management and efficient utilization of resources by 2030.
   2) Decreased human health: Hunger is still a global problem due to the lack of access to quality and nutritious food, which can impact health. Good food waste management can be done by distributing food that is almost expired but still suitable for consumption to people in need, or businesses can sell these products at a 50% discount.

c. Recycle

   Decrease in business sustainability: People realize the importance of protecting the environment. Over time, people have a preference for products and services that are environmentally friendly. Café business with the concept of green practice is a competitive advantage to improve customer relationships in the long term [14]. The closer relationship with customers makes it easier for cafés to involve customers in waste management so that awareness of a lifestyle that is in harmony with nature can be formed by SDG 12.8.

3. Methodology

The research method used is a descriptive qualitative analysis based on literature study, observation, interviews, and analysis of other supporting documents in the context of café waste management and treatment. The research steps are depicted in Fig 3.

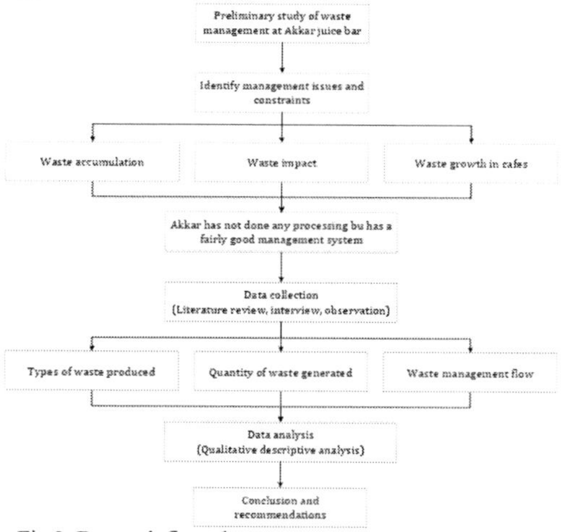

Fig 3. Research flow chart

4. Result and findings

4.1 Strategy for implementing the 3R concept of Akkar juice bar waste management

In order to support creating a healthy, sustainable environment and provide a sense of comfort for consumers. Akkar has a policy in managing café waste by implementing 3R (Reduce, Reuse, Recycle) as a realization of Akkar juice bar's commitment to grow together with the environment because Akkar realizes the importance of protecting the environment for natural balance. Therefore, Akkar implements a waste management system during café operations through activities described in Fig 4 with the following details:

a. Scheduling raw material orders. Akkar juice bar applies the FIFO (First In, First Out) system to all raw materials. According to the production schedule, the raw materials at Akkar are ordered from suppliers about two to three days before being processed. It is done to reduce the possibility of raw materials being damaged due to long storage.
b. Preparing raw materials. Newly arrived raw materials are prepared first by washing, peeling, and cutting some raw materials into smaller sizes, such as carrots, jicama, pineapple, and beets. The raw material preparation process is done carefully and thoroughly to minimize the amount of waste generated. Low-temperature storage is carried out using a chiller at 6°C and a freezer at -18°C for raw materials not in the production schedule. This activity is carried out to extend the shelf life of raw materials so that they do not quickly rot and are wasted.
c. Reusing packaging bottles. Akkar realizes that product packaging waste plays a role in the accumulation of waste. Therefore, Akkar applies the reusable concept to the packaging of its cold-pressed juice products to consumers who help return the bottle packaging to Akkar and are rewarded with cashback on their next purchase.
d. Separating organic and inorganic waste. Waste from production is sorted to produce two types of waste: organic and inorganic. Organic waste is derived from fruit peels, vegetable stalks, pulp from juicing, and rotting fruits and vegetables. In contrast, inorganic waste comprises clear glass packaging, colored glass, metal, tetrapack, cardboard, plastic bottles, duplex, white-blowing HDPE, PVC sheets, and non-aluminum foil plastic sheets. The two sources of waste are separated into different places to be processed through different processes. For inorganic waste, separate containers are provided to separate waste based on its type, while organic waste is placed in containers with a capacity of 20 kg per container.
e. Establishing cooperation with third parties for waste treatment. Organic waste collected is then handed over to local farmers to be processed into compost, with collection times every two times/week. At the same time, inorganic waste is channeled to a waste recycling company once, with a collection period in the first ten days of each month. These services can help reduce the volume of waste disposed to landfills and provide material benefits to business actors.

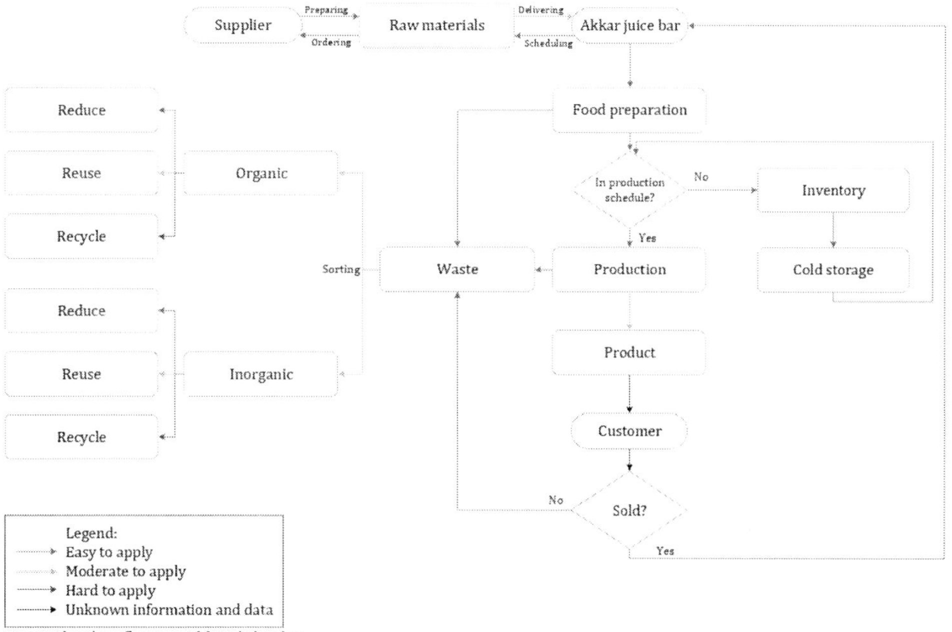

Fig 4. Production flow at Akkar juice bar

 https://doi.org/10.1051/bioconf/20238003008

Akkar has 3 product variants: cold-pressed juice, smoothies, and salads. Based on these three products, cold-pressed juice is Akkar's flagship product, produced daily, although only for some variants. The amount of cold-pressed juice production waste measurements were taken using a mass balance with repetition three times for seven observations so that the average organic waste result is 10.5 kg/day. These results can be influenced by the composition of the raw materials used, the number of production batches, and the production period with details of each variant, as in Fig 5. Of the seven observations, the average total waste was dominated by the red-colored juice variant consisting of the summer crush, hangover doctor, root boost, and sweet punch variants, with the most waste coming from the summer crush variant. The summer crush variant has ingredients from watermelon, pineapple, and lime. Based on the nature of the raw materials, the high water content causes the waste generated to be higher because the pulp from juicing still has a high water content. The organic waste is collected and then handed over to local farmers to be processed into compost.

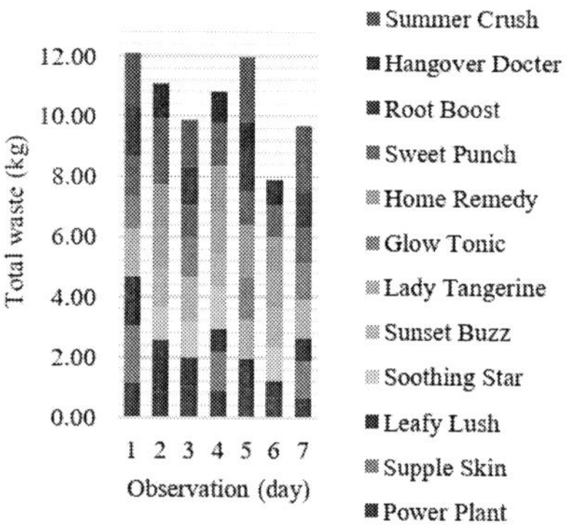

Fig 5. Amount of organic waste of *cold-pressed juice* Akkar juice bar

In addition to organic waste, the café business also produces inorganic waste derived from packaging raw materials for products in Akkar, as described in Fig 6. The inorganic waste produced consists of cardboard fruit packaging, metal on the lid of cold-pressed juice packaging, tetrapack for making smoothies, transparent polypropylene, white blowing HDPE, and colored glass from salad dressing bottle packaging, clear glass fragments from cold-pressed juice packaging bottles that cannot be reused, PVC sheets from berry packaging, plastic sheets from date packaging as an additional ingredient in making almond milk, and duplexes from take away packaging of salad and smoothies products.

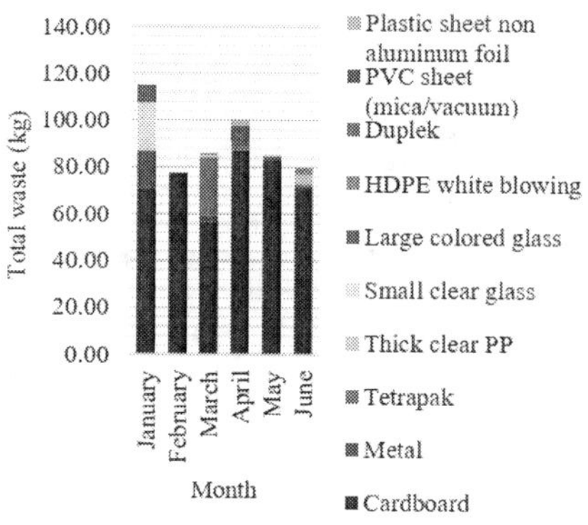

Fig 6. Amount of inorganic waste of Akkar juice bar

Based on the types of inorganic waste, the total waste is measured using a monthly mass balance; the calculated mass is the total mass of waste submitted to the recycling company. Through the results of inorganic waste measurements, it is known that the amount of inorganic waste is dominated by cardboard waste, with an average amount of waste of 74.5 kg/month. The large amount of cardboard is because cardboard is the packaging waste of raw materials supplied daily for the production of all Akkar products. All inorganic waste is recycled by a sustainable waste processing company that picks up waste at Akkar regularly once a month.

The processing of Akkar's production waste is mostly through waste recycling for both organic and inorganic waste, while in recycling waste, Akkar still relies on third parties for waste processing. This condition will be better if Akkar can independently process waste through simple and sustainable processing.

5. Discussion

5.1 Café business waste management-processing

Some of the steps that Akkar has implemented in waste management have shown that as a café business actor, Akkar is responsible for production and consumption by Sustainable Development Goals 12.3 regarding food waste handling. To increase the sense of environmental responsibility, Akkar can carry out the waste treatment process independently. One way is through composting. Akkar has managed the remaining organic waste from production into compost. However, the composting process is still carried out by a third party, with the composting results not returning to Akkar, and Akkar needs fees for the processing service. The improvement strategy can be applied is to process organic waste into eco-enzyme. Eco-enzyme is a versatile liquid from the synthesis of organisms by decomposing complex organic matter into simpler, environmentally friendly ones [15]. Making eco-enzymes requires organic waste such as fruit peels or vegetable scraps, molasses, and water. Eco-enzyme can be used as a cleaning liquid, fertilizer mixture, mosquito repellent, and recycled as raw material for soap making. Manufacturing eco-enzymes as an organic waste treatment can help reduce the amount of waste entering the landfill. Enzymes have unique characteristics that allow them to work efficiently and effectively on different types of agricultural waste. Developing environmentally friendly and efficient enzymatic treatment methods is a priority in agricultural waste management [16]. Therefore, eco-enzyme is one of the sustainable solutions that Akkar can apply to recycle organic waste because it is a simple practice but can positively impact the environment..

5.2Processing of citrus waste for eco-enzyme production

Akkar's cold-pressed juice products are juices with various fruits and vegetables. One of the raw materials used in all cold-pressed juice variants is citrus, which consists of limes, lemons, and oranges. Of the twelve cold-pressed juice variants, three do not use citrus in their ingredient composition: power plant, leafy lush, and sunset buzz. Using citrus plays a role in providing a refreshing sour taste to cold-pressed juice products. Its essential role in making cold-pressed juice is directly proportional to the amount of production and citrus waste produced. In one citrus fruit, only 45% can be consumed in the form of juice, and as much as 55% of the citrus must be discarded, consisting of skin (flavedo) as much as 27%, pulp (albedo and endocarp) as much as 26%, and seeds as much as 2% [17].

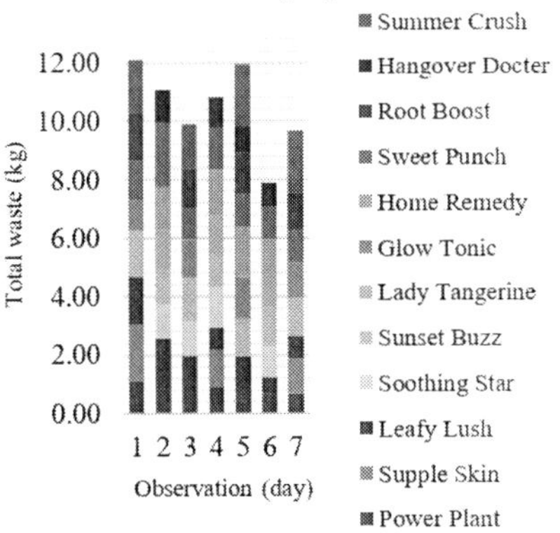

Fig 7. Amount of organic waste of cold-pressed juice Akkar juice bar without citrus waste

The practice of waste treatment by utilizing citrus waste as a base material for eco-enzyme is the most beneficial solution because it is the most economical and efficient method in terms of human and natural resources based on the short lead team and low operational costs required. Through citrus waste treatment, Akkar can reduce the amount of waste generated by about 9% or as much as 9.56 kg/day, with details in Fig 7. In addition to the impact on reducing the amount of waste, the manufacture of eco-enzymes using citrus has several advantages, among others:

1. Citrus peels contain vitamin C and a high level of acidity that makes the eco-enzyme results potent disinfectants and insecticides and can provide a sharp fragrance [18].
2. Manufacturing eco-enzymes requires low costs and can be done on a small or large scale. The tools and materials needed consist of citrus waste, sugar, water, and containers. Organic waste and water can be obtained for free, so the costs incurred are only for the purchase of sugar and containers. The results of the eco-enzyme were obtained to meet the need for environmentally friendly sanitation for daily use so that Akkar does not need to buy sanitation equipment [19].
3. Eco-enzyme is made from organic materials so that its use does not have a destructive impact that can pollute the environment because it does not contain chemicals that are toxic to humans [19]. Eco-enzyme is a creative way of treating waste to reduce, reuse, and recycle waste [18].

Making eco-enzyme does not have to be in large quantities but can also be done on a home scale. Using simple tools and materials makes it possible to make a versatile eco-enzyme.

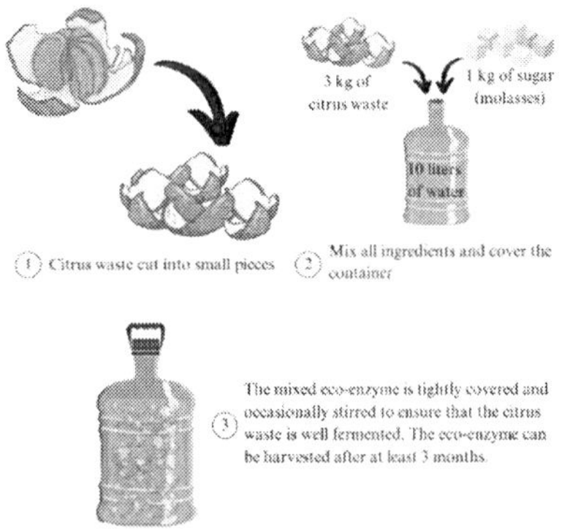

Fig 8. Step to make eco-enzyme

The steps that must be taken to make eco-enzyme are depicted in Fig 8 with the details of the activities, as follows:

1. Citrus waste from cold-pressed juice production is collected and cut into small pieces to expand the fermentation surface area [20][21].
2. Sugar, citrus waste, and water are mixed in a ratio of 1:3:10 into a tightly closed container to prevent air from entering [18].

3. The eco-enzyme mixture was stirred once a week to ensure that the citrus waste was evenly fermented, and then the mixture was allowed to stand for three months [18].
4. After three months, the eco-enzyme was filtered to remove solid impurities. The solid impurities can be used as compost, and the liquid eco-enzyme can be used for environmentally friendly sanitary products [20].

Through the simple steps of making eco-enzyme, organic waste treatment is accessible for Akkar to implement. Besides positively impacting the environment, eco-enzymes also have a social and economic impact. Through the manufacture of eco-enzyme, Akkar also obtains by-products from waste treatment so that it has a selling value and can make the waste treatment process by making eco-enzyme as a means of educating consumers to be applied in treating household waste, as illustrated in Fig 9. A good waste management and treatment process can make Akkar a juice cafe business actor who cares more about the environment and is an example for other cafe business actors. Increasing Akkar's participation in management practices and disseminating information about food waste treatment proves the seriousness of the cafe business that is displayed to customers [6].

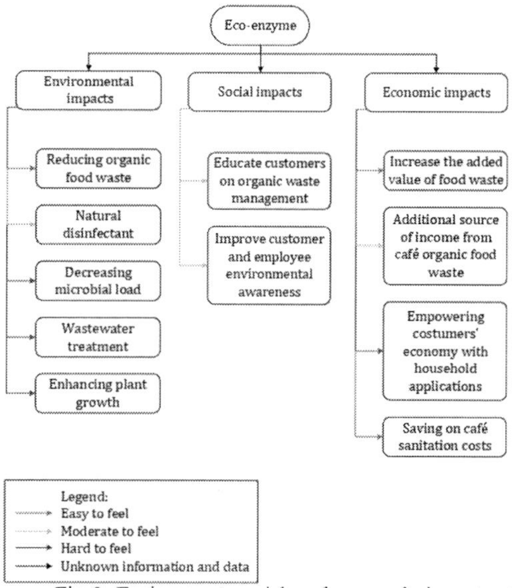

Fig 9. Environment, social, and economic impacts of *eco-enzyme* manufacturing [17]

6. Limitation and future research

In recent years, the development of the cafe business has continued to multiply. Not only coffee-based café businesses but juice café businesses have also begun to develop, which in production use fruits and vegetables as raw materials for the products sold. The remaining raw materials are a source of organic food waste that can pollute the environment if waste management and processing are not done. Therefore, Akkar Juice Bar, as a juice cafe business, has managed waste to realize its responsibility to the environment. However, because the business is still in the medium category, there is no proper waste treatment. Therefore, in this study, Akkar is advised to treat waste by making eco-enzymes. Manufacturing eco-enzymes can benefit Akkar both environmentally, socially and economically favorably. The limitations of this study are that the research was only conducted at the Akkar juice bar production house because it has the most dense production activities, and the waste treatment suggestions have not been tested. Thus, waste treatment trials can be carried out for further research, and the percentage of changes that occur from applying organic waste treatment in the juice cafe business can be measured.

7. Conclusion

Implementing waste management at Akkar Juice Bar has been running quite well because it has implemented reduction, reuse, and recycling. However, in the recycling process, Akkar still involves a third party for the waste treatment process. Increasing Akkar's role in waste management can be done by starting to process waste independently, namely by utilizing citrus waste to make eco-enzymes. Through the manufacture of eco-enzyme, Akkar not only treats waste at a low cost and is environmentally friendly but also gets by-products that have selling value and enhance consumer awareness and employees to care for the environment by carrying out sustainable waste treatment.

Acknowledgments

The authors thank Akkar Juice Bar for their support of the data collection, and all authors of this manuscript contributed equally to this research.

8. Reference

[1] United Nations Environment Programme. (2021). *Food Wasste Index Report 2021.* United Nations Environment Programme (UNEP).

[2] Ministry of Environment and Forestry of the Republic of Indonesia. (2017). Changing Lifestyle: Be Wise in Buying Food. Retrieved from http://ditjenppi.menlhk.go.id/kcpi/index.php/inovasi/352-perubahan-pola-hidup-bijak-dalam-membeli-makanan

[3] E. Sitorus, E. Mohamad, F. E. Srimulat, I. Marzuki, R. I. NNPS, G. Sugiyanto, . . . J. S. Adiansyah, (2021). *Environmental Knowledge.* Medan: Yayasan Kita Menulis.

[4] Ministry of National Development Planning/Bappenas. (2021). *Study Report: Food*

Loss and Waste in Indonesia. Retrieved from https://lcdi-indonesia.id/wp-content/uploads/2021/07/Report-Kajian-FLW-ENG.pdf

[5] P. F. Albizzati, P. Rocchi, M. Cai, D. Tonini, & T. F. Astrup (2022). Rebound effects of food waste prevention: Environmental impacts. *Waste Management, 153*, 138-146. doi: https://doi.org/10.1016/j.wasman.2022.08.020

[6] BIO Intelligence Service. (2013). *Modelling of Milestones for achieving Resource Efficiency, Turning Milestone into Quantified Objectives: Food waste.* European Commission.

[7] M. Bigdeloo, T. Teymourian, E. Kowsari, S. Ramakrishna, & A. Ehsani (2021). Sustainability and Circular Economy of Food Wastes: Waste Reduction Strategies, Higher Recycling Methods, and Improved Valorization. *Materials Circular Economy, 3*(3). doi: https://doi-org.ezproxy.ugm.ac.id/10.1007/s42824-021-00017-3

[8] R. Anasih, Y.Andriani, W.Lili, Iskandar, & I. Zidni (2021). The Restaurant Waste Fermentation as Feed. *World Scientific News: An International Scientific Journal, 158*, 285-298.

[9] United States Environmental Protection Agency. (2023, March 20). *Reducing and Reusing Basics.* Retrieved from United States Environmental Protection Agency: https://www.epa.gov/recycle/reducing-and-reusing-basics

[10] Sustainable San Antonio. (2018). *3R - Reduce, Reuse & Recycle.* Retrieved from SustainableSA.com: https://www.sustainablesanantonio.com/practices-technology/reduce-reuse-recycle/

[11] A. Lutviyani, F. F. Firdausi, & H. Hanim (2022). Review of Food Waste on the Environment in the Perspective of Islam and Science. *Proceedings of Conference in Integration of Interconnection of Islam and Sains, 4*, pp. 49-53.

[12] S. Chaturvedi, Vandana, S. Sharma, & Anshika (2021). Digital Knowledge Ecosystem: A New Weapon to Achieve Sustainable Food Waste Management. Dalam M. M. Thakur, *Sustainable Food Waste Management* (hal. 403-413). Springer Nature Singapore. doi: https://doi.org/10.1007/978-981-15-8967-6_22

[13] M. Lins, R. P.Zandonadi, A. Raposo, & V. C. Ginani, (2021). Food Waste on Foodservice: An Overview through the Perspective of Sustainable Dimensions. *Foods, 10*(1175), 1-14.

[14] M. R. Shalsabilla, S. Alam, Y. H. Candella, & R. Andari (2023). Analysis of Ecological Green Marketing Implementation at Work Coffee Indonesia. *Journal of Tourism Industry, 5*(2), 219-227. doi: https://doi.org/10.36441/pariwisata.v5i2.1427

[15] Waste 4 Change. (2018, December 31). *Eco-Enzyme: Multipurpose Liquid from Organic Waste.* Retrieved from Waste 4 change: https://waste4change.com/blog/eco-enzyme-multipurpose-liquid-from-organic-waste/

[16] T. Kuthiala, K.Thakur, D. Sharma, , G. Singh, M. Khatri, & S. K. Arya (2022). Review: The eco-friendly approach of cocktail enzyme in agricultural waste treatment: A comprehensive review. *International Journal of Biological Macromolecules, 209*, 1956-1974. doi:https://doi.org/10.1016/j.ijbiomac.2022.04.173

[17] N. Mahato, K. Sharma, M. Sinha, & M. H. Cho, (2018). Citrus waste derived nutra-/pharmaceuticals for health benefits: Current trends and future perspectives. *Journal of Functional Foods, 40*, 307-316. doi: https://doi.org/10.1016/j.jff.2017.11.015

[18] N. Benny, R. Shams, K. K. Dash, V. K. Pandey, & O. Bashir (2023). Recent trends in utilization of citrus fruits in production of eco-enzyme. *Journal of Agriculture and Food Research, 13*, 1-10. doi: https://doi.org/10.1016/j.jafr.2023.100657

[19] L. Vama, & M. N. Cherekar (2020). Production, extraction and uses of eco-enzyme using citrus fruit waste: wealth from waste. *Asian Journal of Microbiology, Biotechnology & Environmental Sciences, 22*(2), 346-351.

[20] C. Russo, A. Maugeri, G. E. Lombardo, L. Musumeci, D. Barreca, A. Rapisarda, . . . M. Navarra (2021). The Second Life of Citrus Fruit Waste: A Valuable Source of Bioactive Compounds. *Molecules, 26*(19), 2-20. doi:10.3390/molecules26195991

[21] M. Muktiarni, N. I. Rahayu, & R. Maryati, (2022). Orange and strawberry skins for eco-enzyme: experiment and bibliometric analysis. *Journal of Engineering Science and Technology*, 195-206.

Erosion Control in the Tropical Rainforest Catchments using Modified Intensive Rehabilitation

Hatma Suryatmojo[1*], *Muhammad* Chrisna Satriagasa[1], *Widiyatno*[2], *Alifia* Dyah Ayu[1]

[1]Laboratory of Watershed Management, Department of Forest Resource Conservation, Faculty of Forestry, Universitas Gadjah Mada, Jl. Agro No. 1, Bulaksumur, Yogyakarta 55281, Indonesia
[2]Laboratory of Tropical Silviculture, Department of Silviculture, Faculty of Forestry, Universitas Gadjah Mada, Jl. Agro No. 1, Bulaksumur, Yogyakarta 55281, Indonesia

Abstract. Tropical rainforests in Indonesia is currently managed by Intensive Forest Management System (IFMS), characterized by selective timber harvesting and intensive rehabilitation to enrich standing stock. This study conducted quantitative evaluations on catchment scale impacts of IFMS during each selective logging (SL) and intensive rehabilitation (IR) stage. SS concentrations were measured in three headwater catchments located in tropical rainforests of the Bukit Baka Experimental Catchments, Central Kalimantan, Indonesia. While no forestry operation was conducted in catchment A, operations based on IFMS were conducted in catchments B and C. A combination among the great surface disturbance by skid trails, large tree extraction ratio, and high slope angle of catchment B resulted in the greatest SS. During the post-IR period, SS concentration of catchment B did not show decreases in comparison with that during the post-SL period. In catchment C, increases in SS concentration were observed during the post-IR period. Thus, this study indicated that IR of IFMS was not effective to decrease SS concentration in the initial stage of the post operation period, but the modified intensive rehabilitation with considering the contour and topography potential*ly significant to control erosion.*

Keywords: tropical rainforest, intensive rehabilitation, paired catchment, suspended sediment, erosion control

1 Introduction

Tropical rainforests in Indonesia is currently managed by Intensive Forest Management System (IFMS), characterized by selective timber harvesting and intensive rehabilitation to enrich standing stock. In this operation, 15-20% of whole area is manually striped for intensive rehabilitation. Forest biomass and litter are left over the forest floor, and typically about 200 seedlings per hectare are planted. An expected standing stock at the end of a 30-year rotation is approximately 400 m3 per hectare [1-2]. Details on IFMS are described by [3-5].

IFMS potentially increases forest potency through the intensive rehabilitation, but on the other hand, it increases line clearing area (i.e., open area) for the intensive rehabilitation. Reductions in vegetation cover from forest clearing likely change sediment yields from catchments. Evaluations of the intensive rehabilitation method of IFMS have been rarely conducted whereas several previous studies have focused on impacts of the selective logging [6-11]. Therefore, a need exists to clarify responses of tropical rainforests managed under IFMS. This study employs the paired catchment method for investigating the effectiveness of modified intensive rehabilitation to control the impact on erosion.

2 Methods

2.1 IFMS conducted in paired catchment studies

No forestry operation was conducted in catchment A. Regular IFMS was conducted in catchments B, while IFMS with modified intensive rehabilitation based on contour topography was conducted in catchment C (Fig. 1). Data obtained in the pre-treatment period were used for calibrations to establish the hydrological relationships between the control and treated catchments. For each stage of IFMS, the data following a specific operational period were used in the analysis.

In the SL operational period, 4-m wide skid trails were constructed in catchment B and C (Fig. 1). In the IR operational period, a 3-m wide strip of rehabilitation was conducted in catchments B and C. In catchment B, intensive strip-rehabilitation was conducted, which is a standard rehabilitation method that uses a North–South direction to establish the planted lines as shown in Fig. 1a. Spacing between each planting line was 20 m. In catchment C, intensive contour-rehabilitation was conducted, which is a modified rehabilitation method that considers catchment topography (as shown in Fig.

* Corresponding author: hsuryatmojo@ugm.ac.id

1b) and was expected to reduce impacts on surface runoff and soil erosion.

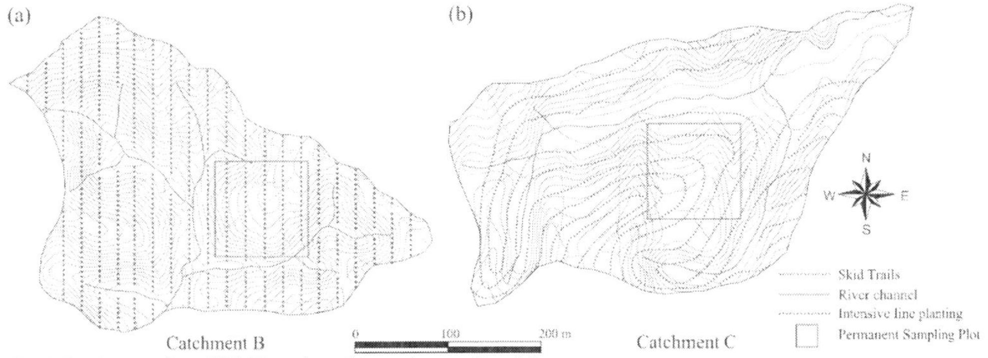

Fig. 1. Implementation of IFMS conducted in catchments B and C

Table 1. Number of trees and tree canopy cover in catchments A, B, and C based on PSP measurements.

Catchment	Number of trees*			Tree canopy cover (%)		
	Natural	Post-SL	Post-IR	Natural	Post-SL	Post-IR
A	206			95%		
B	218	135	80	88%	60%	55%
C	170	102	81	79%	56%	51%

* Diameter > 20 cm (at 1.3 m above land surface)

Changes in forest canopy cover were monitored in a permanent sampling plot (PSP) established in each catchment, as shown in Fig. 1. The PSP is used for the long-term observation of forest growth, including measurements of diameter increment, volume increment, and stand structure dynamics. The PSP consisted of one square hectare of forest area located in the middle slope of each catchment. Measurements were counted under the natural condition, and during the post-SL and -IR periods (Table 1).

Based on the PSP measurements, the SL operation greatly decreased forest canopy cover by reducing the number of trees. The IR operation also significantly reduced the number of trees and canopy cover. Under the natural condition, canopy cover in the three catchments ranged 79–95%. Catchment C had the smallest canopy cover of the three catchments. In the post-SL and post-IR periods, the difference in canopy cover between catchments B and C was reduced to 4%.

2.2 Paired catchment observation and analysis

Precipitation was monitored with an interval of 15 min using an automatic tipping-bucket rain gauge located near the three catchments. The discharge rates were measured using 90° V-notch weirs (0.9 × 1.4 × 3m) and water-level loggers (HOBO U-20) with a time interval of 15 min at each catchment outlet. The observed runoff hydrograph was divided into direct runoff and base flow using a straight-line method [12].

At each weir, sediment concentration was measured using a rising stage suspended sediment sampler [13]. Rising stage suspended samplers work on the siphon principle from the inlet-pipe to bottle samples. In this method, water samples including suspended sediment were collected at each water level. A rising stage was installed at a sidewall of the weir and attached to 60 inlet pipes (3.5 mm in diameter with a distance of 1 cm for each pipe). The inlet-pipe connected to a number of bottles (260 ml in volume) arranged on top of each other in frame, installed near the rising stage. Each bottle is fitted with a two-hole stopper, one hole for the sample inlet and one for the air exhaust. Suspended sediment samples were collected during storm rainfall occurrences.

3 Results and Discussions

3.1 Suspended sediment hysteresis loops

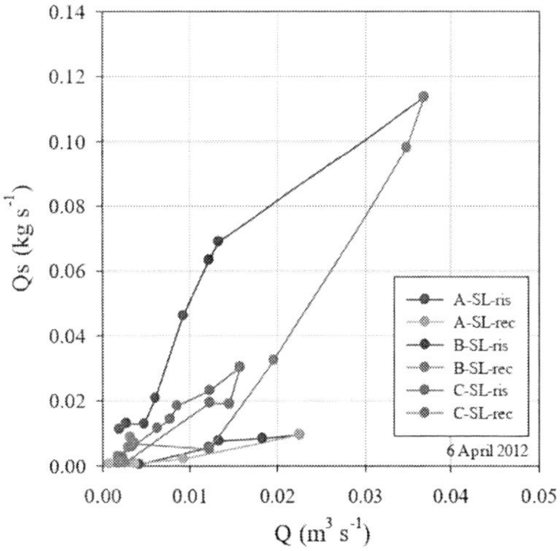

Fig. 2. Water discharge (Q) and suspended sediment discharge (Qs) relationship of catchments A, B, and C observed in the post-SL period. (ris=rising-limb; rec=recessing-limb)

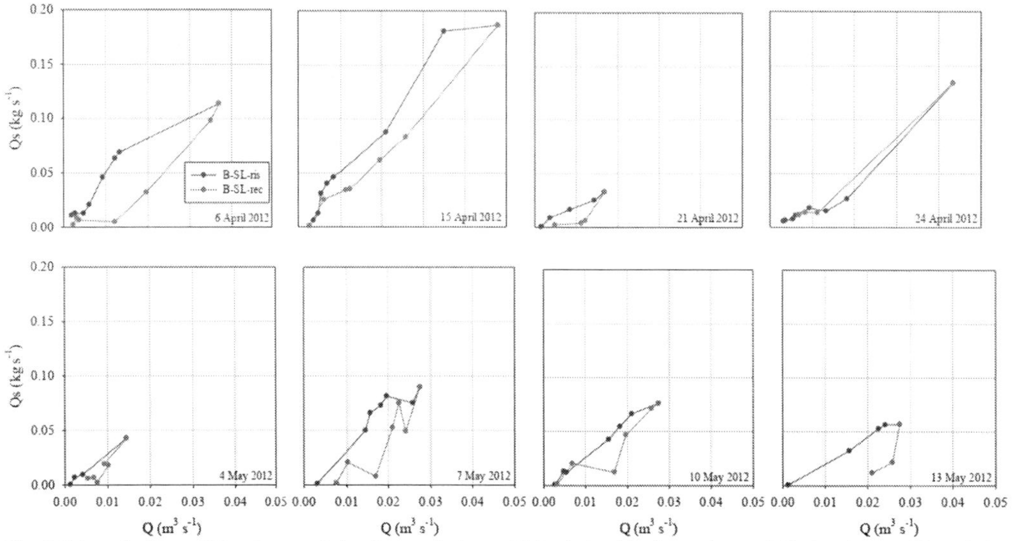

Fig. 3. Water discharge (Q) and suspended sediment discharge (Qs) relationship for catchment B during the post-SL period.

Fig. 2. shows the results of suspended sediment (SS) measurements conducted for a storm event (i.e., during the post-SL period). The figure indicates that catchment B produced a clear clockwise hysteresis loop in the water discharge (Q) and SS discharge (Qs) relationship. This means that SS concentration for the rising-limb (SL-ris) was greater than SS concentration for the recessing-limb (SL-rec). Hysteresis was not so clear for catchments A and C.

As shown in Fig. 3, during the post-SL period, catchment B produced clear hysteresis loops in Q and Qs relationships for most of the storm events. Again, no clear hysteresis was found for catchments A and C during the same period. Fig. 3c showed that catchment B had higher Qp than catchment C. It is likely that the greater amount of surface runoff as well as the higher slope gradient of catchment B produced larger ability for transporting SS. Therefore, in catchment B, the amount of sediment sources probably is the controlling factor for SS discharge; while large amount of SS material was washed away in rising-limbs of storm hydrographs, the SS concentration in the recessing-limb decreased because of lacks sediment sources. On the other hand, in catchment C, the ability of SS transporting by surface flow probably is the controlling factor. Consequently, the relationships between Q and Qs in catchment C were similar in both of rising- and recessing-limbs (Fig. 3).

During the post-IR period, hysteresis became unclear in catchment B. In catchment C, no hysteresis loop was observed during the post-IR period as well. The IR operation left the biomass litters in the forest floor, which leads to increases in the surface roughness and works as traps of the flowing SS material. Therefore, we consider that the IR operation reduces ability of SS transporting. Consequently, SS discharge during the post-IR period was not controlled by the amount of sediment source but by the transporting ability, which resulted in the diminishing of hysteresis in catchment B.

3.2 Suspended sediment concentration

Fig. 4 shows the relationship between Q and Qs for each catchment and each IFMS operation period. For catchment B, data obtained during the post-SL period were divided into the rising- and recessing-limbs, due to the clear hysteresis as shown in Figs. 2 and 3.

Figs. 4a, b and c show that the IFMS treatments resulted in higher suspended sediment concentrations in catchments B and C than in catchment A. In catchment B, Qs on the rising-limb during the post-SL period was greater than Qs during the post-IR period (Fig. 4a). On the recessing-limb during the post-SL period, catchment B produced lower Qs than that during the post-IR period (Fig. 4b). Overall, in catchment B, Qs during the post-SL period was similar to Qs during the post-IR period. Fig. 4c shows that during the post-SL period, catchment C exhibited slightly higher Qs than catchment A. Furthermore, the Qs of catchment C during the post-IR period was higher than that during the post-SL period.

Figs. 4d and 4e show that during the post-SL period, both the rising-limb and the recessing-limb of the Q-Qp relationship in catchment B produced higher Qs than the Q-Qp relationship in catchment C. This result indicates that the SS concentration in catchment B was greater than that in catchment C during the post-SL period. A combination of the larger skid trail area and steeper catchment topography in catchment B likely caused the greater magnitudes of SS discharge.

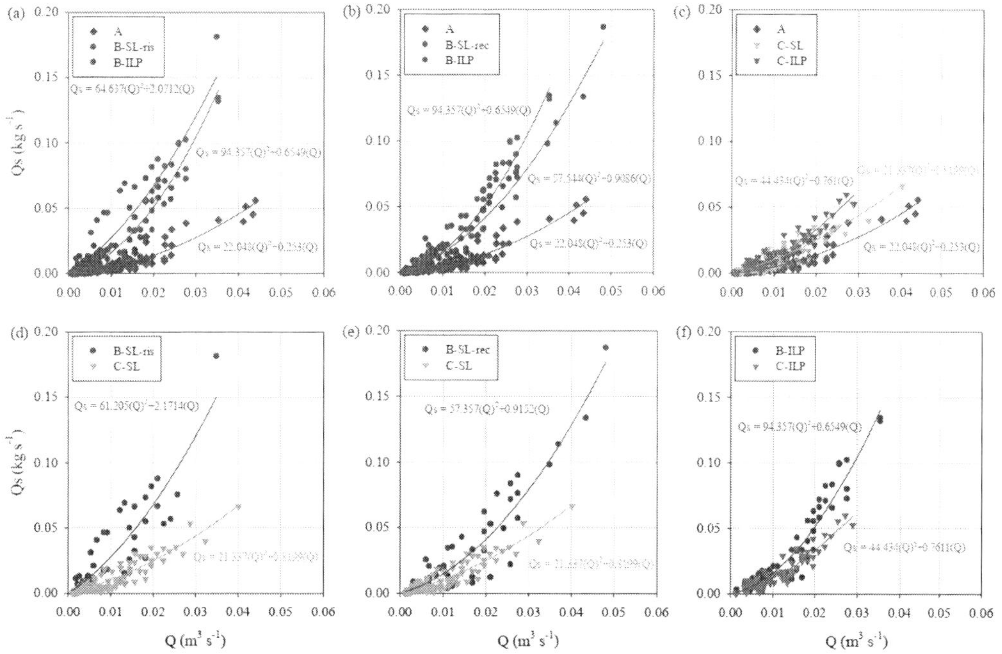

Fig. 4. Water discharge (Q) and suspended sediment discharge (Qs) relationships for the natural catchment (A), and treated catchments (B and C) during the post-SL (SL) and post-IR (IR) periods. (ris=rising-limb; rec=recessing-limb)

During the post-IR period, catchment B also produced greater Qs than catchment C (Fig. 4f). However, the differences between catchments B and C were smaller in comparison with those during the post-SL period (Figs. 4d and 4e). In catchment C, the sediment concentration during the post-IR period was slightly higher than that during the post-SL period (Fig. 4c). This increase was likely related to the addition of surface disturbance caused by IR operations. Although the manual line clearing operation left biomass litter on the forest floor, IR caused an increase in surface disturbance area and an overall decrease in canopy coverage (Table 1). Probably, effects of such disturbances outstrip effects of the biomass litter coverage in the forest floor, resulting in the increase of the SS concentration. The disturbances caused by the IR operations were pronounced in catchment C which suffered less disturbances than catchment B during the SL operation (Fig. 4d and e). Thus, our observations during the initial three months in the post-IR period suggested that the contour-rehabilitation conducted in catchment C was not effective to reduce the SS concentration of discharge water.

4 Conclusions

During the post-SL period, the greater amount of surface runoff as well as the higher slope gradient of catchment B attributed to high ability for transporting SS, which caused hysteresis in the relationship between water discharge and SS discharge and resulted in large amount of SS production. We concluded that a combination among the great surface disturbance by skid trails, large tree extraction ratio, and high slope angle brought the greatest SS yield in catchment B.

In catchment B, SS concentration during the post-IR period did not decrease in comparison with that during the post-SL period. In catchment C, IR operations increased SS concentration slightly higher than that during the post-SL period. Thus, both catchments B and C had high SS production rate in the initial stage of the post-IR period. This was attributed to additional surface disturbances by IR operations; although the manual line clearing left biomass litter on forest floor, IR caused increases in soil disturbance area and decreases in canopy coverage. Effects of such disturbances outstripped effects of the biomass litter coverage, resulting in increases in SS concentration.

Acknowledgements
This research was implemented by Laboratory of Watershed Management, Faculty of Forestry, Universitas Gadjah Mada, and Sari Bumi Kusuma forest concession company, Central Kalimantan, Indonesia for facilitating the field work activities. Thank you to Research Project Recognition (Rekognisi Tugas Akhir) under Directorate of Research Universitas Gadjah Mada for the dissemination support of this research.

References

1. Na'iem, M. and Faridah, E. 2006. Model of Intensive Enrichment Planting (TPTII). In: A. Rimbawanto (Eds.), Silviculture systems of indonesia's dipterocarps forest management a lesson learned. Yogyakarta: Faculty of Forestry

https://doi.org/10.1051/bioconf/20238003009

Gadjah Mada University and ITTO. Technical Report: ITTO Project PD 41/00 Rev. 3 (F,M), pp. 25–36.

2. E. Farida, S.P. Widiyatno, M. Naiem, *Close to nature technique: Applied studies on dipterocarps Indonesian tropical rain forest rehabilitation*, in Proceedings of the International Conference on Sustainable Forest Development in view of Climate Change, 8-11 August 2016, Putrajaya, Malaysia (2016)
3. H. Suryatmojo, M. Fujimoto, K. Kosugi, T. Mizuyama. *Impact of selective logging and intensive rehabilitation system on runoff and soil erosion in a tropical Indonesia rainforest*, in Proceedings of River Basin Management VI, 25-27 May 2011, California, USA (2011)
4. H. Suryatmojo, M. Fujimoto, K. Kosugi, T. Mizuyama, J-SustaiN **2**, 2 (2013)
5. H. Suryatmojo, K. Kosugi, Water **13**, 17 (2021)
6. I. Douglas, T. Spencer, T. Greer, K. Bidin, W. Sinun, W. W. Meng, Philos. Trans. Roy. Soc. B – Biol. Sci **335**, 1275 (1992)
7. K. Baharuddin, N. Abdul Rahim, J. Trop. For. Sci. **7**, 2 (1994)
8. J.N. Negishi, R.C. Sidle, A.D. Ziegler, S. Noguchi, N. Abdul Rahim, Earth Surf Process Landf **33**, 8 (2008)
9. R.P.D Walsh, K. Bidin, W.H. Blake, N.A. Chappell, M.A. Clarke, I. Douglas, R. Ghazali, A.M. Sayer, J. Suhaimi, W. Tych, K.V. Annammala, Philos. Trans. Roy. Soc. B – Biol. Sci **366**, 1582 (2011)
10. R.C. Gatti, S. Castaldi, J.A. Lindsell, D.A. Coomes, M. Marchetti, A.D. Paola, F. Paparella, R. Valentini, Ecol. Res. **30**, 1 (2015)
11. P.W. Ellis, T. Goapakhrisna, R.C. Goodman, F.E. Putz, A. Roopsind, P.M. Umunay, J. Zalman, E.A. Ellish, K. Mo, T.G. Gregoire, B.W. Griscom, For. Ecol. Manag. **438**, 255-266 (2019)
12. Black, P.E. Watershed Hydrology. New Jersey: Prentice-Hall, Inc., pp. 183-205. 1991.
13. Gordon, N.D., McMahon, T.A., Finlayson, B.L. 1992. Stream hydrology. An introduction for ecologists. University of Melbourne: John Wiley & Sons, pp.179–185

Formulating the Direction of Community Land Use to Support the Conservation of Javanese Gibbon (*Hylobates moloch*) Habitat at Petungkriyono, Central Java

Tri Wira Yuwati[1,2]*, *Astarina Eka* Dewi[3], *Ahmad Arief* Fahmi[3], *Muhammad Ali* Imron[2], *Muhammad* Tafrichan[4], *Yudistira* Soeherman[3], *Sri* Lestari[1] and *Dian* Pratiwi[1]

[1]Research Center for Ecology and Ethnobiology, BRIN , Jl. Raya Jakarta Bogor km 46, Cibinong, Indonesia
[2]Faculty of Forestry, Gadjah Mada University, Jl. Agro Bulaksumur No 1, Yogyakarta, Indonesia
[3]Relung Indonesia , Mangunan, Berbah Sleman, Yogyarta
[4]Javan Wildlife Institute (JAWI), Gedongkiwo Mantrijeron, Yogyakarta, Indonesia

Abstract. - The Petungkriyono forest is one of the few remnant tropical forests on the island of Java, particularly in Central Java. This forest is known for its high conservation value due to its abundance of biodiversity, particularly the endangered Javan gibbon or Owa Jawa (*Hylobates moloch*). The Petungkriyono people, on the other hand, rely on upland agriculture and cultivation on steep slopes as a means of subsistence, which poses a threat to the Petungkriyono forest. This paper aims to formulate the direction of the community land use at the habitat of Javanese gibbon at Petungkriyono. The method used the Spatial Multi-Criteria Analysis (SMCA). SMCA was applied to create the initial model of the zoning description of sustainable land use in Petungkriyono. The SMCA analysis process includes goal setting and conceptualization, data collection, analysis, field ground checks, and Focus Group Discussions (FGD) with the community and related parties. The result showed that considering the massive changes of land use at Petungkriyono almost 70% of the area is formulated for perservation and protection to support the conservation of Javan gibbon.

1 Introduction

Petungkriyono forest, located in the district of Pekalongan, Central Java province is the remaining tropical forest at Java island, especially in Central Java [1;2; 3]. It is said to own high conservation areas including 108 tree species from 88 genus and 43 families[1]. More than 30 species of epiphyte from 7 families are staying in this forest [2]. The most important thing is that this forest is the habitat of Owa Jawa or Javan gibbon / silvery gibbon (*Hylobates moloch*), an endemic primate of Java island [3; 4]. The Javan gibbon has been protected since 1931 through Wild Animal Protection Regulation No. 266 which was strengthened by Law No. 5 of 1990 and Minister of Forestry Decree 10 June 1991 No. 301/Kpts-II/1991, however the natural population is considered decreasing. Javan gibbon has most recently been assessed for the IUCN Red List of Threatened Species and listed as endangered under criteria A4cd.
In the perspective of State Forests, Petungkriyono Forest is included in the Doro Forest Management Unit (BKPH) Section, East Pekalongan Forest Management Unit (KPH) which has an area of 5,189,507 ha, consisting of Limited Production Forest with Pine Tree Plants and Other Natural Timber Forest or forest nature which functions as a Limited Protected Forest (HLT) for the Hydrological Protection function. The protected forest has an area of 1,931.90 ha (Decree of the Minister of Forestry Number: 359/Menhut.II/2004 dated 1 October 2004). The Petungkriyono protected forest functions as a Limited Protection Forest. The Petungkriyono protected forest is still a relatively well-maintained primary forest, with tropical rain forest vegetation types.
In the perspective of the administrative area, the area is within a certain administrative area. In this context, Petungkriyono Forest is a forest located in the administrative area of Petungkriyono District. Sixty-four percent of the Petungkriyono District area is in the form of State Forest which is currently managed by Perhutani. In fact, there are villages with very dominant forest areas such as in Curugmuncar Village where 80% of the area is forest area. Economically, the community has some dependence on the forest area in their area. Petungkriyono community utilizes the forest by cultivating grasses, harvesting coffee under forest stand, tapping pinus latex, coffee planting and vegetable planting [5]. One of the developing economic potentials in the Petungkriyono region is upland farming which is being developed in several villages [5]. If this intensive farming continues, there will be a risk of land expansion due to the growing needs of the community. In addition, the erosion and depletion of soil nutrients that will occur

*Corresponding author: tri.wira.yuwati@brin.go.id

are also higher on land with steep slopes. On the other hand, there is potential for production forests in the form of *Pinus merkusii*, Sengon and Aren (*Arenga pinnata*) which are cultivated on community-owned lands. Based on the study of [6], it is possible that the Petungkriyono mixed forest has a significant potential for reducing the effects of climate change by sequestering carbon due to its carbon stocks capacity up to 1,769,390.46 Mg. Moreover, [7] conducted a study to calculate the total economic value of Petungkriyono forest and found that the entire economic value of water value, landslide prevention structures, and carbon sequestration is IDR.163.065.858.080, and the Extended Benefit Cost Ratio in Petungkriyono is 281.35%. However, based on the outcome, it is believed that the Pekalongan local administration will be highly motivated to protect the survival of the Petungkriyono forest.

The problem that needs to be answered is how to maintain the biodiversity of the Petungkriyono forest especially in conserving the Javan gibbon, but the community also has a sustainable livelihood. To obtain that vision, the community should utilize it in a sustainable way. The aim of this paper is to formulate a direction of the community land use at Petungkriyono to support the habitat of Javan gibbon. A zoning map of community sustainable land use will be made for Petungkriyono.

2 Material and Methods

Spatial Multi-Criteria Analysis (SMCA) was used to make a basic model for a zoning description of sustainable land use in Petungkriyono District. The principle of this analysis is to combine several aspects into one with predetermined weighting between aspects [8]. The principle of this analysis is to use a limited rational model to produce an output in the form of a block direction map according to its function (functional area). The SMCA analysis process includes goal setting and conceptualization, data collection, SMCA analysis, field ground checks, and Focus Group Discussions (FGD) with the community and related parties. In the initial process, the first step is determining the focus or purpose of the analysis and to look at the potential for land cover management in Petungkriyono District. At this stage, the aspects and criteria that will be used in the SMCA analysis are also determined. Aspects to be used include ecological and socio-economic-cultural aspects. These two aspects were chosen because the direction map that will be made does not only concern ecological aspects, but will also take into account the socio-economic-cultural conditions in Petungkriyono District. Each aspect is given a value or score according to the magnitude of the influence on sustainable land use models including agroforestry, intensive farming, protected zones, and others. Ecological aspects as well as socio-economic and cultural aspects are prepared using several related criteria. For ecological aspects, the criteria used include land/area conditions, the presence of water sources, topographical conditions, and biodiversity conditions. For the socio-economic and cultural aspects, the criteria used include accessibility, housing, economy, and culture. Then, each of these criteria is implemented into one or more variables that can describe the conditions of the related criteria. The list of variables used in this activity can be seen in the Table 1. Each of these variables is deducted based on the boundaries of the Petungkriyono District. Each variable is classified with class weights one through five. The greater the weight value indicates the more important the area is to be protected. Determination of the weighting classification is carried out semi-experimentally adjusting the range of values obtained by each variable. However, for land cover and area function/status variables, the classification is based on the classification from the Ministry of Environment and Forestry. Each variable is then overlaid according to its aspect with equal weight considerations to obtain a map of aspects of ecological sensitivity and social, cultural, and cultural sensitivity. To make an indicative model of land use in Petungkriyono, the ecological sensitivity aspect is given a weight of 60%, while the social, cultural and cultural sensitivity aspect is given a weight of 40%. The weight determination is based on the assumption that ecological considerations have a larger portion in determining sustainable land use zoning. Verification and validation were carried out through ground check and focus group discussions. FGD was carried out to determine the aspirations of Petungkriyono community regarding their future land use. While ground check was carried out by stratified samplings and survey. Survey points were chosen randomly for each SMCA class namely at Tlogopakis, Tlogohendro, Yosorejo and Kayupuring. Vegetation condition, biodiversity, slopes and other ecological aspects were checked. As for survey, were conducted to give an overview of the actual socio economy and cultural condition.

Tabel 1. List of variables used for SMCA Analysis

Aspect	Criteria	Variable	Raw Data	Source
Ecology	Land condition	Land cover	Land cover and land use map 2019	MoEF
		Status/land function	Forest area map 2017	MoEF
		NDVI	Sentinel-2	European Space Agency
	Water source	Distance from river	Rupa Bumi Indonesia (RBI) map	Geospatial Information Agency, RI
	Biodiversity	Distance from important animal	Owa Jawa observation data	SwaraOwa
	Topography	Altitude	DEMNAS	Geospatial Information Agency, RI
		Slopes	DEMNAS	Geospatial Information Agency, RI

 https://doi.org/10.1051/bioconf/20238003010

Social, economy and culture	Accessibility	Distant from the road	RBI map	Geospatial Information Agency, RI
	Settlements	Distance from settlements	RBI map	Geospatial Information Agency, RI
	Economy	Distance from tourist spots	Field survey 2021, Google Maps	RELUNG
	Cultural	Distance from cultural sites	Field survey 2021, Google Maps	RELUNG

3 Results and Discussions

The ecological variables (land cover, status/land function, NDVI, distance from river, distance from important animal/key species/ altitude and slopes) were overlayed to create an ecological sensitivity map. The aspects of ecological sensitivity map resulted in the scores of 12-32 which were then classified into 5 classes. From the ecological sensitivity, more than 50% of Petungkriyono sub district with ecological score 5 were at Kayu Puring village. Most of Owa Jawa were seen at this village. The location of Kayu Puring is very close to forest proximity. It means that there is an urgent need for further arrangements of sustainable land use at Petungkriyono.

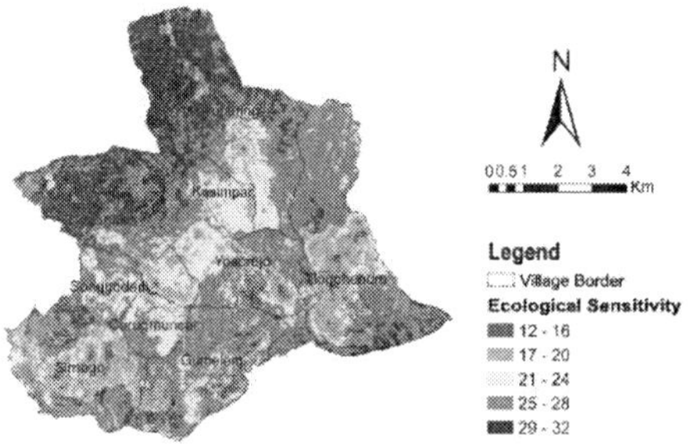

Figure 1. Ecological sensitivity map at Petungkriyono, the higher scores showed the higher urgency for protection

Four variables such as distance from the road, distance from settlements, distance from tourist spot and distance from cultural sites were overlayed to create socio-eco-culture sensitivity map. This map resulted in score 4-17 and those were classified into 5 classes. As for socio-eco-culture sensitivity, 44.1 % of Petungkriyono has score of 2 meaning that arrangement for sustainable land use is needed. The map of socio-eco-culture sensitivity of Petungkriyono is presented in Figure 2.

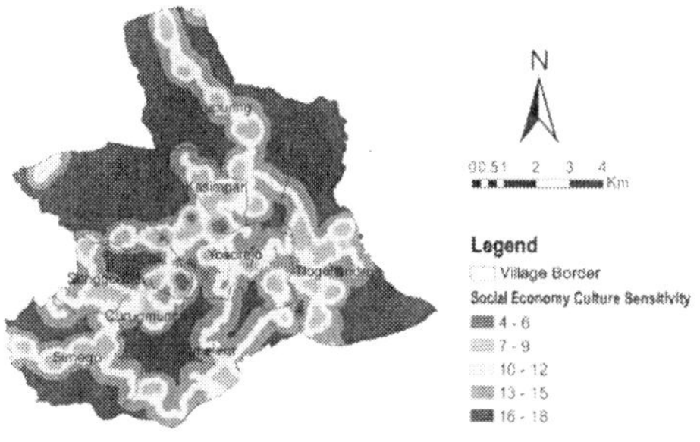

Figure 2. Socio-eco-culture sensitivity map at Petungkriyono

To create the initial model of land use zoning at Petungkriyono, we overlayed the two maps with consideration of 60% ecological sensitivity and 40% socio-eco-cultural map. Keeping the forest cover is a priority because the purpose of zoning map is to support the conservation of Owa Jawa habitat at Petungkriyono. Consequently, the focal point lies on the ecological dimension. Then, we classified 5 classes of land use including : business as usual (BAU), utilization, limited utilization, protection and preservation.

Business as usual showed vital locations for community with degraded ecological condition. In this class, people is free to utilize the land. Utilization class is directed for usage with no negative massive impact. Basically, utilization area consists of settlements, local government offices and roads. Limited utilization is directed for usage with small scale impact to environment. In this category, cultivation is authorized with certain limitations, such as soil and water conservation buildings. For example, for agriculture, it is mandatory to use terraces so that hydrological functions are maintained. Apart from that, this category is also urged to conduct planting with agroforestry method. Protection is directed for area with good vegetation cover condition. In this area, restoration and rehabilitation is needed when the vegetation is degrading. This category also emphasizes the cultivation of food trees for Javan Gibbons. If dryland farming continues to be practiced in this region, it is imperative that it be oriented towards the implementation of agroforestry planting techniques, with the ultimate goal of restoring the area to its original forested state in the future. Community are able to utilize the natural resources in this area without disturbing the biophysical and ecological function. Preservation areas were directed for limited utilization such as research, education and environmental services. From the initial indicative zoning map before groundceck, limited utilization area was the largest (36.7%) at Petungkriyono followed by protection (29.4%), utilization (24.2%) , BAU (6.5%) and preservation (3.2%).

https://doi.org/10.1051/bioconf/20238003010

The initial indicative zoning map before groundcheck is presented in Figure 3.

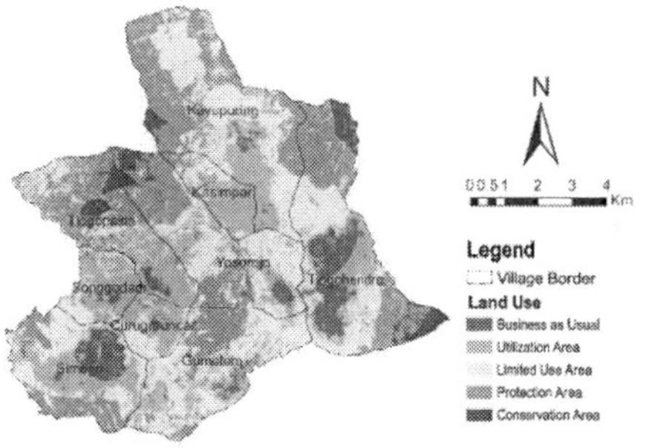

Figure 3. Initial indicative zoning map of Petungkriyono (before groundcheck)

After ground check and survey, it showed that agricultural expansion at the four villages were massive during 2019-2023. There were quite a lot the changes of land cover during 2019 and the actual condition. Land openings at steep slopes and high elevations are potential to cause natural disasters. In those areas, agroforestry implementation is needed. The presence of tree canopies and deep rooting systems can reduce the chances of soil erosion which is the main cause of land slides. Trees along border can be implemented for agricultural species which are not shading tolerant.

Furthermore, the income from non-timber forest products is not yet guaranteed which is the main cause of the clearing of forest cover for timber and the opening of new agricultural land. In fact, there are many potential non-timber forest product commodities in Petungkriyono namely pine resin, coffee, honey, durian, cloves, cardamon, palm sugar and ginseng. Other than pine resin, coffee, honey and ginseng, other non-timber forest product commodities have not been used and managed optimally by the community. The main problem related to the use of non-timber forest products is the absence of a market chain that benefits the community. Based on the results of discussions with several communities, they got very cheap prices from collectors/middleman. Meanwhile, prices in the downstream market can be worth several times the price they sell. So that in the future it is necessary to carry out special studies related to the market chain for non-timber forest product commodities in Petungkriyono District which are sustainable and benefit the local community.

Non-timber forest products in the form of environmental services have been used and managed by the local community such as water and natural attractions. However, there are still some problems related to these two things. The first is related to water, based on the results of field surveys, several places in Petungkriyono sub district do not have a good supply of water, especially in terms of quality. We predict the amount of erosion caused by clearing of forest cover at certain points as the main cause. Thus, it is very important to replant forest areas that are not forested, land at an altitude of more than 1500 meters above sea level, as well as steep land. Regarding the management of ecotourism, it is necessary to include cultural elements as one of the attractions apart from natural attractions. The most important thing was that during the ground check, several Owa Jawa were observed in Tlogopakis and Tlogohendro villages ; meaning that the ecological dimension needs to be more focussed.

Based on the survey and FGD, it was then decided to adjust the several variables according to the actual conditions in the field. The latest SMCA models were quite different with the initial zoning (Table 2). Table 2 showed that the area of preservation class is increasing 3 higher than the initial while for BAU area is decreasing half from the initial model. Figure 4 showed the final zoning map of Petungkriyono sub district with SMCA after ground check and field survey

Tabel 2. Changes of initial and final SMCA of Petungkriyono

Class	Recommendation	Initial (ha)	Final (ha)
1	Business as usual (BAU)	234.87	99.40
2	Utilization	1199.14	945.43
3	Limited utilization	3242.84	1483.34
4	Protection	2811.11	2472.37
5	Preservation	910.61	3398.03

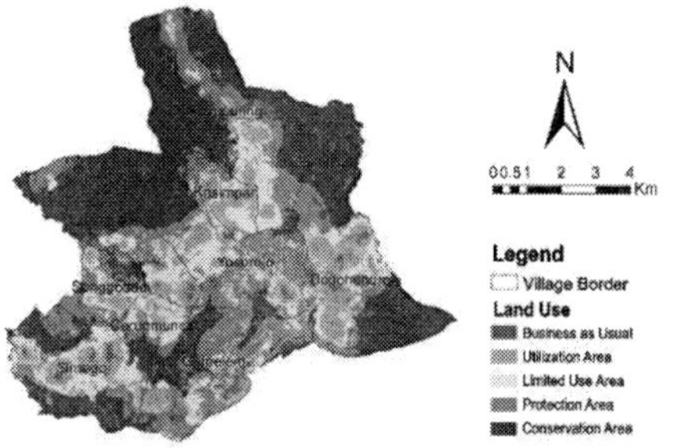

Figure 4. Final indicative zoning map of Petungkriyono with SMCA (after ground check)

Based on the study of [9], the people in the Petungkriyono area have had a very strong connection with the forest area since its historical history, long before the area was declared as a protected forest area. Indigenous people who adhere to customary rules and norms as agreements that control their daily patterns are one of two major community groups in the area. Traditional knowledge that controls how people interact with and use the environment and its resources [9] is just as important as knowledge about how to live with other people.As a result, it does not rule out the prospect of expanding the protective area receiving support from Petungkriyono community.

The Petungkriyono regional government has also formed a multi-stakeholder forum consisting of local government, companies, community and village representatives in Petungkriyono as well as NGOs that provide assistance there [10]. This forum has agreed that the Petungkriyono forest must be protected because it is the habitat of the endangered Javan Gibbon.

4 Conclusion

The initial model for zoning is divided into five classes, namely BAU, utilization, limited use, protection, and preservation. Each class has a proportion of area sequentially, namely limited use (36.7%), protection (29.4%), utilization (24.2%), BAU (6.5%), and preservation (3.2%)). The final SMCA models is generally more conservative. The preservation class of the new model has increased by more than three times the area of the first model. Meanwhile, the Business As Usual (BAU) class experienced a decrease in area to less than half of the initial model. The final zoning map of community land use in Petungkriyono including: Business as Usual (99.40 ha), utilization (945.43 ha), limited utilization (1483.34 ha), protection (2472.37 ha) and preservation (3398.03 ha). This zoning map is expected to support the conservation of Owa Jawa habitat.

Acknowledgements

The research is funded by BPDLH (Badan Pendanaan Lingkungan Hidup) under the project of " Pengembangan Agroforestry dan Pertanian Terpadu Untuk Perbaikan Kualitas Lingkungan Hidup di Hulu DAS Sengkarang, Jawa Tengah" implemented by Relung Indonesia. This research is also part of Rumah Program 1 Tema 4 "Pengungkapan dan Pemanfaatan Biodiversitas Nusantara", Organisasi Riset Hayati dan Lingkungan BRIN tahun 2023.

References

1. I. Damayanti, A. Nurbambang, and T. R. Soeprobowati, "Plant diversity of petungkriyono forest of dieng plateau, central java, indonesia," *Biodiversitas*, vol. 22, no. 8, pp. 3497–3507, doi: 10.13057/biodiv/d220849 (2021)

2. I. Lestari, Murningsih, and S. Utami, "Keanekaragaman jenis tumbuhan paku epifit di Hutan Petungkriyono Kabupaten Pekalongan, Jawa Tengah," *Niche J. Trop. Biol.*, vol. 2, no. 2, pp. 14–21, https://ejournal2.undip.ac.id/index.php/niche (2019)

3. Y. Rachmawati, M. Pramudya, and A. Setiawan, "Distribution of the critically endangered javan blue-banded kingfisher Alcedo euryzona along the Welo River flow in the Petungkriyono forest," *IOP Conf. Ser. Earth Environ. Sci.*, vol. 976, no. 1, doi: 10.1088/1755-1315/976/1/012004 (2022).

4. Y. Wulandari, A. S. Nugroho, and M. Ulfah, "Karakteristik Pohon Tidur dan Pohon Cover Owa Jawa (Hylobates moloch) di Hutan Lindung Petungkriyono Jawa Tengah," *Pros. Semin. Nas. Sains dan Entrep. III Tahun 2016 "Reorientasi Bioteknol. dan Pembelajarannya Untuk Menyiapkan Gener. Indones. Emas Berlandaskan Entrep.*, no. 266, pp. 303–307, (2016).

5. A. Abdurahman, I., Mujiyanto, M., EkaDewi, "M e l e s t a r i k a n Petungkriyono," Yayasan Relung Indonesia (2021).

6. I. Damayanti, A. N. Bambang, and T. R. Soeprobowati, "Carbon stocks of Petungkriyono mixed forest as climate change mitigation," *IOP Conf. Ser. Earth Environ. Sci.*, vol. 724, no. 1, doi: 10.1088/1755-1315/724/1/012064 (2021)

7. I. Damayanti, A. Nur Bambang, and T. R. Soeprobowati, "Extended benefit cost analysis as an instrument of economic valuated in Petungkriyono forest ecosystem services," *J. Phys. Conf. Ser.*, vol. 1025, no. 1, doi: 10.1088/1742-6596/1025/1/012034 (2018)

8. A. Zucca, A. M. Sharifi, and A. G. Fabbri, "Application of spatial multi-criteria analysis to site selection for a local park: A case study in the Bergamo Province, Italy," *J. Environ. Manage.*, vol. 88, no. 4, pp. 752–769, doi: 10.1016/j.jenvman.2007.04.026 (2008).

9. M. Woelansari, M. Maryono, F. Muhammad, M. A. R. Halim, and A. Setyawan, "Community participation in conservation of petungkriyono protected areas in pekalongan district," *E3S Web Conf.*, vol. 202, pp. 1–8, doi: 10.1051/e3sconf/202020206002 (2020)

10. I. Damayanti, A. Nur Bambang, and T. Soeprobowati, "The Analyze of Collaborative Management Perspective of Petungkriyono Forest," *E3S Web Conf.*, vol. 202, pp. 1–11, doi: 10.1051/e3sconf/202020205014 (2020)

Assessment of Surface Runoff Potency under Tropical Environment for Soil and Water Conservation Planning

Safira Defiani Khoirotunnisa[1], *Gina Isna Nafisa,*[1] *Chandra* Setyawan[1*], *Muchammad Nu'man*[1], *Ngadisih*[1]

[1]Department of Agricultural and Biosystems Engineering, Faculty of Agricultural Technology, Universitas Gadjah Mada, Yogyakarta 55281, Indonesia.

Abstract. Water is essential for living organisms, including crops. Its presence is a crucial factor for agriculture. Soil and water conservation is an effort to sustainably maintain the availability of water, thereby meeting the water needs of crops in the agricultural sector. This research aims to estimate the potency of surface runoff as a hydrological indicator of watershed critically for soil and water conservation purposes. A hydrological tank model was used to estimate surface runoff. The results showed a potential for surface runoff of approximately 133.82 mm/month, occurring primarily during the peak rainy season from December to April. Soil and water conservation (SWC) technology using water harvesting ponds (WHP) on farmland was proposed to store surface runoff. Data analysis indicates that the use of WHP provides significant benefits from environmental and economic aspects. Based on an average WHP storage capacity of 10 m^3, approximately 40% of the total watershed area is required for constructing water harvesting structures to accommodate all surface runoff. Harvesting all surface runoffs increases the base flow during the dry season by 225.14 mm. This study serves as a valuable reference for soil and water conservation planning, particularly in tropical watersheds.

1 Introduction

Water is an important factor for crops. Hence, water availability has become a crucial thing for food security [1]. As a tropical country, Indonesia has a reliable potency for water resources. However, the monthly water availability varies climatically every year. A good management of water resources is required to ensure water availability at all times. The primary source of water comes from rainfall that occurs during the wet season. Meanwhile, the water requirements remain consistent year-round, including during the dry season, so water reserves are needed to meet these needs.

Optimizing rainwater harvesting is crucial to ensure secure water availability. To optimize rain harvesting during the wet season, soil and water conservation technology should be applied to store water in the soil, groundwater systems, or on the earth's surface, including reservoirs in the rivers and water storage structures on farmland [2]. The natural water storage system in the soil, in the form of soil moisture and groundwater, can store significant amounts of water and channeling it as base flow during the dry season. The natural water storage function is suboptimal due to damage to the watershed. Many human activities, such as agriculture, mining, and others, can cause damage to watersheds when conservation principles are not implemented.

Landscape changes, particularly in the form of reduced vegetation and increased impervious surfaces, which occur in many watersheds in Indonesia [3], lead to decrease groundwater storage due to the disruption of the infiltration process. This condition increases surface runoff or overland flow and increases the potential for soil erosion and sedimentation in the rivers. Sediment deposition downstream leads to the shallowing of reservoirs and irrigation canals. The use of artificial soil and water conservation technology provides a solution for optimizing rainwater harvesting.

Soil and water conservation technology for rainwater harvesting is developed based on the potential water availability. The calculation of rainfall-runoff transformation is essential for estimating the potential for surface runoff harvesting. In this study, the potency of surface runoff harvesting was estimated as the basis for soil and water conservation development, especially through the construction of WHP on farmland. This engineering conservation building is cost-effective, environmentally friendly, easy to construct, and well-recognized by the community in the study area. From both social and economic perspectives, it meets the requirements.

Surface runoff is the focus of study in this study because it is one of the triggers for problems on land (soil erosion) and downstream areas (sediment deposit and flood). Surface runoff harvesting will reduce the potential for land damage, sedimentation and flooding and increase infiltration, soil moisture and groundwater reserves.

* Corresponding author: chandra.tep@ugm.ac.id

2 Method

2.1. Study Area

This study was performed in Kalirukem Watershed which occupies about 24 km^2 area (Figure 1). Administratively, the watershed is located in Wonosobo Regency, Central Java Province of Indonesia and a part of the Wadalintang Reservoir catchment area. Climatology Station of Wadaslintang near the study area recorded an average annual rainfall of 3680 mm from 2011-2020, while the average humidity and temperature were about 82% and 26°C respectively [4].

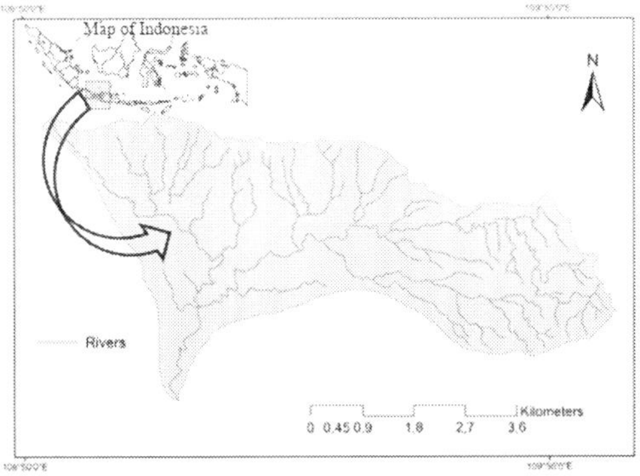

Fig. 1. Map of the study area

The watershed is covered by latosolic red-yellow soil, with an average land slope greater than 15%, and is predominantly dominated by farmland [5].

2.2. Hydrological Model

The tank hydrological model was used to identify the process of rainfall flow transformation and estimate surface runoff values. This model consists of three main components that describe the hydrological cycle in the atmosphere, land surface, and groundwater as shown in Figure 2 [6,7].

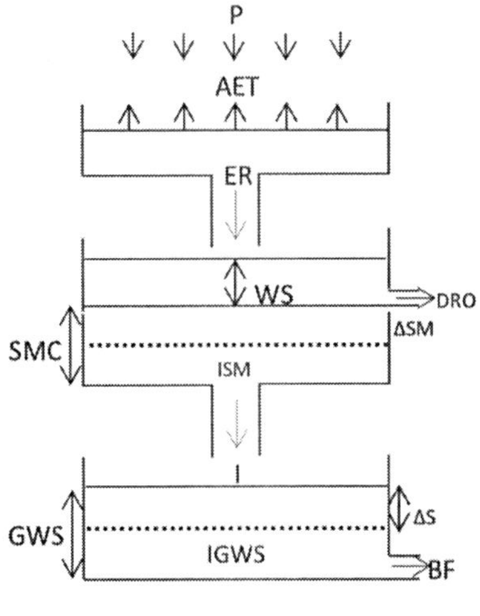

ER	=	P-AET	IGWS	=	GWS_{i-1}
WS	=	ER-ΔSM	DRO	=	WS-I
ΔSM	=	SMC-ISM	BF	=	I-ΔS
ISM	=	SM_{i-1}	ΔS	=	GWS – IGWS
I	=	Ci x WS	Q_{TOT}	=	DRO + BF
GWS	=	0.5 (1+K) I + K x IGWS			

P	=	precipitation	I	=	infiltration
AET	=	actual evapotranspiration	WS	=	water surplus
ER	=	excess rainfall	BF	=	base flow
DRO	=	direct runoff	Ci	=	infiltration coeficient
ΔSM	=	change of soil moisture	SMC	=	soil moisture capacity
ISM	=	initial soil moisture	K	=	recession constant
ΔS	=	change of ground water volume	Q_{TOT}	=	total discharge
GWS	=	ground water storage			
IGWS	=	initial ground water storage			

Fig. 2. Structure of hydrological model of Mock

This model has been validated through calibration and verification using observed discharge data from the Wadaslintang River in 1999-2001, downstream of the Kalirukem Watershed. Observed river discharge data was calculated using the calibration curve, which establishes the relationship between reservoir water level downstream of the watershed and water discharge. It's important to note that discharge data after 2001 is not recommended for model validation due to reservoir shallowing, which compromises the validity of the calibration curve. The result of the hydrological model calibration is presented in Table 1.

Table 1. Optimization result of model calibration

Parameter	Symbol	Unit	Value
1. Area of watershed	A	km^2	192.53
2. Wet Infiltr. Coeff.	WIC	-	0.50
3. Dry Infiltr. Coeff.	DIC	-	0.65
4. Initial soil moisture	ISM	(mm)	100
5. Soil moisture capacity	SMC	(mm)	400
6. Initial groundwater storage	IGWS	(mm)	1000
7. Groundwater reces.constant	K	-	0.85

The model calibration showed a correlation coefficient of 0.91, an error volume of 0.01, and an efficiency coefficient of 0.93 (see. Figure 3), indicating high accuracy of the model for runoff calculation [7,8].

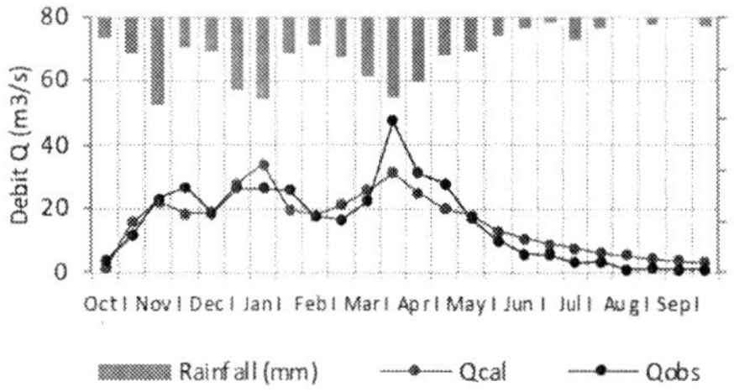

Fig. 3. The result of model calibration

The optimized model parameter values (Table 1) obtained through calibration were then used to calculate the value of direct runoff (DRO) or surface runoff, based flow (BF) and total runoff (TRO) using rainfall and climate data from 2011 to 2020.

The surface runoff value was obtained from the results of direct runoff (DRO) calculations, which is the difference between rainfall and excess rainfall (ER). TRO is the sum of DRO and BF. DRO only occurs during the rainy season with high monthly rainfall, so that during the dry season, river water discharge (TRO) only comes from BF [9]. The formula for calculating DRO, BF, and TRO are presented in Figure 2.

3 Result and Discussions

3.1. Rainfall-Runoff

The rain-runoff transformation reflects the distribution process of rainwater, transforming into evaporation, runoff (including surface runoff, interflow, and river discharge), and infiltration into the soil, ultimately becoming groundwater (base flow).

A watershed in good condition will be able to optimize groundwater recharge through infiltration during the rainy (wet) season and minimize surface runoff which causes soil erosion upstream, flooding and sedimentation downstream. Changes in the watershed landscape, especially the reduction in vegetation, have disrupted the hydrological function of the watersheds. An increase in surface runoff indicates disruption of the hydrological function of the watershed.

The rainfall-runoff modelling in this study shows a significant amount of surface runoff that occurs during the wet season as shown in Figure 4. Total surface runoff reached 802.95 mm or equal to 19,270,722.2 m³ at the study area (Kalirukem Watershed). The potency for surface runoff is high because the study area is located under a tropical climate region. Low vegetation cover and high farmland cover increase the potency of surface runoff [10].

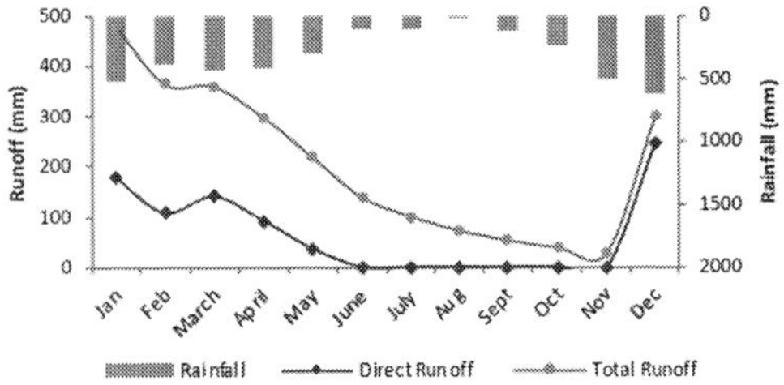

Fig. 4. Rainfall-Runoff Correlation in Study Area

The average annual rainfall of 3680 mm can sustain a consistent river water discharge throughout both the wet and dry seasons. The availability of river water discharge (TRO) follows the monthly rainfall patterns, with river water discharge tending to be high during the rainy season [11]. During the dry season, the water in the river comes from base flow which originates from the groundwater storage system. Groundwater recharge is very important to ensure the availability of river water discharge (based flow), especially during the dry season [9]. Soil and water conservation must be focused on efforts to optimize rain harvesting so that groundwater can be fully replenished.

This study also calculates the surface runoff coefficient (Cs), which represents the ratio between surface runoff or direct runoff (DRO) and rainfall (P) in monthly units. The value of Cs between 0-0.1 indicates good watershed conditions with low potential for surface runoff [12]. In this study, the Cs value during the wet season at the study location ranged from 0.12 to 0.39, with an average value of 0.28. The highest C value occurs in the month with the highest rainfall (P) reaching 625.0 mm/month as shown in Table 2. The surface runoff coefficient (Cs) value is smaller than the total runoff coefficient (C) which can reach 0.50-0.70, due to the additional base flow [13].

Table 2. Surface runoff coefficient (Cs) in the study area

Month	P (mm)	DRO (mm)	TRO (mm)	BF (mm)	Cs
Jan	510.9	178.9	483.2	304.3	0.35
Feb	378.6	108.3	365.9	257.6	0.29
March	424.1	140.6	360.8	220.2	0.33
April	417.3	92.4	298.9	206.5	0.22
May	295.1	36.7	218.5	181.8	0.12
June	108.0	0.0	138.0	138.0	-
July	102.9	0.0	99.7	99.7	-
Aug	12.3	0.0	72.0	72.0	-
Sept	119.7	0.0	52.0	52.0	-
Oct	230.9	0.0	37.6	37.6	-
Nov	505.5	0.0	27.2	27.2	-
Dec	625.0	246.0	299.6	53.6	0.39

The average value of Cs indicates that around 28% of rainfall that occurred in the study area is transformed into surface runoff. This value reveals a potential problem in the watershed [14]. Surface runoff usually occurs shortly after rain therefore, it is also called direct runoff (DRO). Generally, the DRO value shows a linear correlation with the rainfall value. However, the Cs value, representing the ratio of DRO to P, varies and does not always have a similar value. It indicates the influence of other factors in the rainfall-flow transformation, particularly related to environmental characteristics [15]. Table 2 also shows the value of base flow, which constitutes the primary water source in rivers during the dry season. Base flow comes from rainfall in previous months.

3.2. SWC Development

The potential for surface runoff in the study area is quite high. If not controlled, surface runoff can result in soil erosion, sedimentation, and floods. This condition also inhibits groundwater recharge, and triggers drought during the dry season. Appropriate strategies are needed to control surface runoff. Harvesting surface runoff by using soil and water conservation (SWC) technology is the most appropriate way to overcome this problem.

The technical, social, and economic aspects need to be considered in selecting the appropriate surface runoff harvesting technology, as most of the study area is dominated by farmland where local community activity in farmland is high. These three factors are important to ensure the sustainability of conservation based on community participation.

Water harvesting ponds (WHP) on agricultural land are the most suitable technology. This conservation structures collect surface runoff from farmland and optimize infiltration for groundwater recharge [16]. Additionally, these structures can be constructed permanently, allowing them to store water for irrigation or other uses during the dry season. This structure is a square, rectangular, or circular-shaped pool which is easy to construct at a low cost. From a social perspective, this structure is well-known among people in Indonesia and has been implemented in many dry farming areas [17].

WHP is relatively easy to maintain with small to medium scale. In this study, a WHP with dimensions of 2.5 meters in length, 2 meters in width, and 2 meters in depth is used to calculate surface runoff capture potential. This dimension of WHP allows it to hold 10 m^3 of water or surface runoff. This dimension is the most found in the study area. Figure 5 shows a typical example of a WHP commonly found around the study area.

Fig. 5. Example of a WHP applied in the farmland of Indonesia

Considering the potential surface runoff in the research area, which reaches 802.95 mm or 19,270,722.2 m^3, while the average Water Harvesting Pond (WHP) storage capacity is around 10 m^3, approximately 1.9 million WHPs are needed to accommodate all the runoff. This number of WHPs is equivalent to 40% of the study area required to accommodate all surface runoff. Another alternative type of structure that can be used is a long water storage system. Combining engineering and vegetative SWC methods (such as re-greening barren areas, contour strip cropping, and crop residue application), maximizes efforts to control surface runoff [2].

3.3. Economic Engineering Value of SWC

In addition to its environmental benefits, the SWC program must also be economically profitable. In this study, two simple economic analyses i.e. the Benefit-Cost (B/C) ratio and Net Present Value were carried out to assess the economic feasibility of a WHP construction (Table 3). The B/C ratio and NPV values must be more than 1 and 0 respectively, to indicate that the WHP is economically feasible to construct.

Table 3. Economic benefit and cost of a WHP construction

Component	Benefit	Cost
Construction Cost	-	1.500.000
Ground Water Increasing	1.690.000	-
Land Prod. Increasing	750.000	-
Soil Erosion Red.	900.000	-
Total	3.340.000	1.500.000

The calculation of revenue and costs in Table 3 shows that the B/C ratio and NPV value were 2.23 and IDR 1,840,000 respectively. These values indicate that WHP construction is economically profitable as reported by Singh in 2015 [18].

Construction costs are assumed to be approximately IDR 150,000 per m^3, following the applicable standard prices in the study area. This cost may be lower if the WHP construction is undertaken independently by the community, without labor assistance. The increase of groundwater was estimated based on the total surface runoff (direct runoff) that occurs in the study area as provided in Table 2 (802.95 mm), which is then multiplied by the base flow coefficient (Bc). Bc represents the ratio between baseflow during the dry season (April- September) and the total rainfall during the wet season (October- March).

The Bc value in this study was obtained at 0.28. This value was then multiplied by the total surface runoff that occurred during the six months of the wet season, with a total value of 802.95 mm (equal to 133.83 mm/month). By using the Bc value and the total surface runoff, we estimated the amount of runoff that was collected and became base flow. For one WHP with a dimension of 2 x 2.5 meters, the increase in base flow was calculated to be 1,125.68 liter, which was then multiplied by the current price of clean water, (approximately IDR 1,500 per liter in the study area, Central Java Province), resulting in a savings value of IDR 1,690,000.

The increase in land productivity was estimated for dry season farming in shallot crops. This plant has a high economic value and is widely planted during the dry season. The productivity of shallot planting land reaches 10 tons/ha, with water requirements around 400 ml (equal to 4000 m^3/ha) for a planting season [19]. Hence, a WHP can provide water for 0.0025 ha of farmland and produces around 25 kg of shallots. The selling price reaches IDR 30,000 per kilogram in 2021-2023 so that a WHP can provide water to produce around 25 kg of shallots with an economic value of approximately IDR 750,000.

Meanwhile, the economic value of soil erosion reduction was estimated based on the total protected area by a WHP. Based on an average surface runoff value of 133.82 mm per month, one WHP (with a capacity of 10 m^3) can protect about 74.72 m^2 (equal to 0.007472 ha) area from the risk of soil erosion caused by surface runoff. A study by Sutrisno et.al. (2012) [20] noted that economic loss due to soil erosion (loss of soil and nutrients) was about IDR 3,300,000 per ha per year.

Hence, the estimated economic saving value of soil erosion due to a WHP construction was about IDR 150,000 per month or equal to IDR 900,000 during the wet season (six months) when surface runoff is occurred.

The construction of a WHP must adhere to conservation principles to maximize economic benefits. Several important aspects that need to be considered in WHP construction include the topography of the area and the presence of water catchment areas in each location where the WHP is constructed.

4 Conclusions

Surface runoff in the study area occurred during the peak of the wet season, especially when monthly rainfall was high, typically from December to May, with an average value reaching 133.82 mm/month. Based on the potential value of surface runoff, soil and water conservation technologies using engineering methods with the construction of water harvesting ponds (WHP) on farmland offer a viable way to reduce surface runoff and increase groundwater recharge. Engineering economic analysis showed that WHP construction is economically profitable. By using WHPs with a capacity of 10 m^3 (large is 5 m^2) for one structure, approximately 40% of the total area was required to accommodate all surface runoff.

Acknowledgments

We would like to extend our heartfelt gratitude to the Directorate of Research Universitas Gadjah Mada for the support through *Final Project Recognition (RTA) Grant* Universitas Gadjah Mada *Number* 5075/UN1. P. II/Dit-Lit/PT.01.01/2023. This grant has played a pivotal role in the successful completion of our research.

References

1. C. Ingrao, R. Strippoli, G. Lagioia, D. Huisingh, Heliyon **9** (2023)
2. C. Setyawan, S. Susanto, C. Y. Lee, J. Deg. Min. Land and Man. **6 (2)** (2019).
3. A. Prasetyo, C. Setyawan, Ngadisih, R. Tirtalistyani, IOP Conf. Series: Earth and Enviro. Sci. **653** (2021)
4. Main Office of Serayu Opak River (BBWS.SO). Climatology Data of Wawar Watershed 2011-2020. The Ministry of Public Works, Indonesia (2023).
5. C. Setyawan, C.Y. Lee, M. Prawitasari, Nat. Hazard **98** (2019)
6. F. Nurrochmad, J. Sujono, D. Damanjaya, Media Tek. J. **2** (1998)
7. C. Setyawan, C.Y. Lee, M. Prawitasari, Ame. J. Eng. Res. **5 (11)** (2016)
8. L.A. Alvarenga, C.R. Mello, A. Colombo, L.A. Cuartas, L.C. Bowling, CATENA **143** (2016).
9. S.R. Oktavia, N.B. Rustiati, V.W. Andiese, T. Amaliah, M. Labombang, O.A. Mantika, IOP Conf. Series: Earth and Envir. Sci. **1075** (2022).
10. W. Anurogo, K. Pratiwi, M. Z. Lubis, M.K. Mufida, L.R. Sari, S.N. Chayati, J. Pend. Geografi **27** (2022).
11. N.B. Ansori, N. Anwar, Int. J. of Geomate **23** (2022).
12. Z. Zomlot, B. Verbeiren, M. Huysmans, O. Batelaan, J. of Hyd: Reg. Stud. **4** (2015).
13. D.S. Krisnayanti, W. Bunganaen, E. E. Hangge, F. Munaisyah, N. A. Nursyam, D.N. Khaerudin, Civil and Envir. Sci. J. **2 (1)** (2019)
14. G. Ghiglieri, A. Carletti, D. Pittalis, J. of Hyd. **519** (2014).
15. S. Wang, W. Yuan, J. Zhou, J. of Physics: Conf. Series **2333** (2022).
16. V.A. Swami, J. of Em. Tech. and Innov. Res. **5** (2018).
17. M.S. Dangda, J. Galung Trop. **8 (3)** (2019).
18. A.K. Singh, J. Res. and Dev. and Man. **8** (2015)
19. N.L.P.A. Ariastuti, I M. Suryana, C. Javandira, Agrimeta **7** (2017).
20. J. Sutrisno, B. Sanim, A. Saefuddin, S. R.P. Sitorus, Sepa **8** (2012).

Impact of Bamboo-Based Agroforestry Demonstration Plots Based on Farmers' Perceptions

Eva Fauziyah[12*], *Sanudin*[2], and *Aditya* Hani[13]

[1]Doctoral Program of Forestry Science, Faculty of Forestry, Universitas Gadjah Mada, Jl. Agro Bulaksumur No.1 Yogyakarta, Indonesia
[2]Research Center for Population, National Research and Innovation Agency, BRIN, Jl. Gatot Subroto No 10, Jakarta 12710, Indonesia
3Research Center for Ecology and Ethnobiology, National Research and Innovation Agency, BRIN, Jl. Raya Jakarta-Bogor KM 46, Cibinong 16911, Indonesia

Abstract. One of the efforts to develop bamboo is to apply agroforestry patterns, including the establishment bamboo-based agroforestry demonstration plots (BAF demplots). This study aims to determine the impact of BAF demplots based on farmers' perceptions. The research was conducted in Sukaharja Village, Rajadesa District, Ciamis Regency, West Java. The respondents in this study were 46 farmers who worked on BAF demplots and 31 farmers who did not work on BAF demplots, so the total number of respondents was 77. Primary data were collected through questionnaires, interviews, and direct observation. The data obtained were then processed and analyzed descriptively. Important Performance Analysis was used to analyse the impact of the BAF demplots based on the farmers' perceptions. The results showed that the BAF demplot had a positive impact on farmers involved in and outside the demplots, in term of social and ecological impacts. However, the economic impact of the BAF demplots is considered to be below farmers' expectations. Therefore, for the demplot to provide more economic benefits, it is necessary to develop potential bamboo-based businesses as part of the demplot development by involving various related stakeholders.

1 Introduction

The value of bamboo, popularly known as "green gold," is increasing in the global economy [1]. Given that it grows more quickly [1,2] and reaches maturity at the earliest [2]. It is a commonly exploited economic resource that can reduce poverty, especially for the tribal community [1]. Bamboo is a widespread plant in Indonesia [3] and is one of the plants essential to people's lives in rural areas [4].

Bamboo is an environmentally friendly, inexpensive, and renewable building material [5]. Bamboo is used to build conventional houses and as fuel in rural areas, which benefits the environment. Moreover, it shields dwellings from the wind [6]. Bamboo can be sustainable if it s harvested selectively rather than by clear cutting [7].

Bamboo-based agroforestry (BAF) hold great promise for the economies of developing countries by increasing productivity, sustainability, and resource conservation [8,9]. For example, the "talun-kebun" bamboo system in West Java combines fast food and timber production cycles and preserves the ecosystem [10].

The bamboo clumps consist of species that have long been cultivated in the community. The community has not yet to cultivated bamboo intensively from planting to harvesting [11]. However, the population of bamboo clumps tends to decrease as the area is used for settlement or replaced by other plant commodities that are considered more profitable [12]. Some of the reason for the the decrease in the number of bamboo clumps include very intensive logging that damage to the bamboo clumps, the conversion of land to other uses, and the conversion of bamboo plants to other plants that are considered more profitable [13]. The demand for bamboo raw materials continues to increase in line with population growth and the development of scientific progress. In the future, this increased demand can only be met by relying on something other than natural bamboo resources. It is therefore necessary to support efforts to plant or cultivate bamboo [12].

One way to support bamboo cultivation and development efforts is to establish a demonstration plot (demplot). Demplots can not only be used for community learning but can also have a social, economic, and ecological impact [14]. Assesing the impact felt by the community can be done through the perception of the community, both directly and indirectly involved. Perception is a process by which a person become aware of an object outside of him or herself [15]. As perception is the starting point for

* Corresponding author: evaf001@brin.go.id

https://doi.org/10.1051/bioconf/20238003012

development attitudes and behaviours, an accurate perception of an object is necessary. It is crucial to understand how the community feels in order apply community empowerment in good agroforestry forest management and to find solutions to any problems that may arise [16]. Research on the public perception of demplots has been done, such as research by [17] on sharecroppers on private land whose crops result of technological development by research institutes. However, research on smallholders' perceptions of BAF developed on village land has never been conducted. The development of bamboo BAF demplots is carried out to maintain the existence of bamboo, become a learning tool for the community and alternative community empowerment activities, and to motivate farmers to cultivate bamboo. This study aims to determine the impact of BAF demplots based on farmers' perceptions.

2 Method

2.1 Research Location

This research was conducted in 2021 in Sukaharja Village, Rajadesa Subdistrict, Ciamis Regency, West Java, Indonesia. This village is the location for a bamboo-based agroforestry (BAF) demplots built in 2015. A map of research locations is presented in Figure 1.

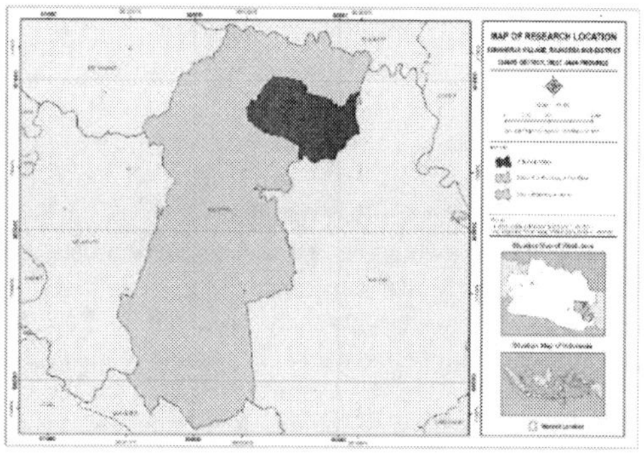

Fig. 1. A map of Research Location

2.2 Data Collection and Analysis

There were 77 respondents in this study, 46 farmers involved in the management of the demplot and 31 farmers not involved in the management of the bamboo-based agroforestry demplot. Primary data were collected through questionnaires, interviews, and direct observation. The questionnaire consisted of 30 questions on the impacts of BAF based on farmers' perceptions, with ten questions each on social, economic and ecological impacts. Measurements were scored using a Likert scale, i.e. 4 for strongly agree, 3 for agree, 2 for neutral, 1 for disagree and 0 for strongly disagree. Next the assessment compares the number of answer scores with the expected (highest) score multiplied by 100%. The results are a percentage of the expected [18]. Furthermore, the percentage value of answers is interpreted in qualitative sentences with references as in Table 2.

Tab 2. Percentage scores and interpretation criteria of perception analysis

Score (%)	Criteria of Interpretation
0 -33,32	Low (Not available, not enough, never, none, very often, hopeless, very influential, and unimportant)
33,33 – 66,66	Medium (Less available, less than enough, less effort, little, often, seldom, less hopeful, less threatened, influential, and less important)
66,67-100	High (Available, sufficient, effort enough, much, seldom, often, hopeful, threatened, less influential, significant, and important)

Source: [18]

Data analysis is presented qualitatively and descriptively through data tabulations and narrative explanations [19]. Data analysis used perception formation analysis [20] to determine farmers' perceptions of bamboo agroforestry adoption.

Importance-Performance Analysis (IPA) was used to determine the impact of the BAF demonstrations. The IPA approach, or quadrant analysis, seeks to quantify the relationship between consumer perceptions and goals for improving product or service quality [21]. Because of its ease of use and its ability to provide analysis results that facilitate suggestions for performance improvement [22], IPA has gained widespread acceptance and use in various academic disciplines. The first step in applying the IPA technique is to identify the attributes that are relevant to the observed decision situation [23]. The list of attributes can be generated by consulting written sources, conducting interviews, and exercising managerial discretion. The attributes compiled include the social, economic, and ecological impacts of the BAF.

3 Result and Discussion

3.1 Development of BAF Demplots

The 'bamboo talun' is one of the traditional agroforestry systems in West Java, Indonesia. *Gigantochloa ater*, *Gigantochloa verticillata*, *Gigantochloa apus*, and *Bambusa vulgaris* are among the dominant species in the 'bamboo talun'. The bamboo clumps are interspersed with *Albizia falcataria, Parkia speciosa, Arenga pinnata* (sugar palm), and a few fruit trees (mango, durian). Perennial trees and bamboo are combined to create a multi-layered canopy structure [24]. In Sukaharja Village in Ciamis Regency, the most common bamboo species were tali bamboo (*Gigantochloa apus*), betung bamboo (*Dendrocalamus asper*), ater bamboo (*Gigantochloa atter*), gombong bamboo (*Gigantochloa pseudoarundinaceae*), and haur bamboo (*Bambusa vulgaris*) [13].

The bamboo-based agroforestry (BAF) demplots in Sukaharja Village, Ciamis Regency, Indonesia, were established in 2015 and 2018 on six hectares and involved 64 smallholders. In 2015, three types of bamboo were planted, namely black bamboo (*Gigantochloa atroviolacea*), betung bamboo (*Dendrocalamus asper*), and green ampel bamboo (*Bambusa vulgaris* var vitata) were planted at 10 m x 10 m spacing, while in 2018, two types of bamboo were planted, namely yellow ampel bamboo (*Bambusa vulgaris*) and ater bamboo (*Gigantochloa atter*) with a spacing of 10 m x 15 m [25]. The fifth type of bamboo planted in the demplots can be used for house building, handicrafts, and as food ingredient (bamboo shoots).

The newly developed bamboo culms are called bamboo shoots. They are used to make a range of traditional and contemporary specialities that can be consumed fresh, dried, shredded, or pickled [26]. Bamboo shoots are also a excelent source of dietary fibre, phytosterols, low fat, high protein, amino acids, and carbohydrates [27,28]. Bamboo shoots are regularly harvested in form of arrangement so that the bamboo clumps are not too dense and the canopy is not too dense.

The arrangement of thebamboo clumps is also done so that farmers can still plant various types of seasonal plants between the bamboo plants. So far, the community's perception is that the presence of bamboo clumps makes it difficult for annual plants to growth. These demplots show how to properly manage bamboo clumps to allow annual crops to continue.

Bamboo is an important l part of agroforestry systems. Proper spacing of bamboo is essential to maximise the combinedperformance of bamboo and associated crops. Although intercrop performance was higher with reduced bamboo cover, the likely reduction in bamboo yield may not be economically beneficial. They developed a trade-off for bamboo-based agroforestry systems to maximise intercrop production without reducinglowering bamboo yields. It is best to use intermediate bamboo plantings, lsuch as eight by nine inches or ten by ten inches [8]. The best bamboo species for intercropping systems may be monopodial ones.

Depending on the species clump size, bamboo can be planted in such intercropping systems at different spacings ranging from 4 m 9 m to 8 m 9 m. Other suitable crops in India include finger millet, cowpea, bottle gourd, turmeric, sesame, and sweet potato [29]. According to [30], the success of any agroforestry systemsdepends on how the woody perennial and assciated elements interact. This interaction can take place above or below ground. Depending on how the components are arranged, how densely they are planted, and how much technical management is applied, competitive and complementary interactions can occur within a bamboo agroforestry system [31,32].

The introduction of agroforestry was facilitated by the diversity matching between species and demands [33]. Based on the suitability of the land and the wishes of the community, several types of annual or undergrowth plants were planted among the bamboo plants in the BAF in Sukaharja Village, including corn (*Zea mays*), peanuts (*Arachis hypogaea*), black potatoes (*Plectranthus rotundifolius*), grass animal feed, and cassava. Meanwhile, the multipurpose tree species (MPTs) developed are based on dog fruit (*Archidendron pauciflorum*), stink bean (*Parkia speciosa*), nutmeg (*Myristica fragrans*), and coffee (*Coffea* sp). Bamboo-peanut-corn agroforestry and bamboo-coffee agroforestry are shows in Figure 2.

(a) (b)

Fig. 2. (a) Bamboo-peanut -corn agroforestry and (b) Bamboo-coffee agroforestry

3.2 The Benefits of BAF Demplots

The adoption of new technology requires time and the appropriate medium. If the new technology increases the utility in terms of benefits and gains over current alternatives, it will adopted ; if not, it will be rejected. Innovations that offer more relative benefits (social, economic, and ecological), compatibility, simplicity, trialability, and observability will be adopted faster than other innovations [34]. Agricultural extension helps rural smallholders adopt new technologies more quickly by providing them with information and guidance [35,36].

Some of the factors that facilitate the adoption of innovations include: higher utility or profitability (usually multipurpose innovations), higher income, or better selling prices [37,38], the existence of related non-financial benefits (ecological conservation), complementarity with current practice (in term of knowledge requred, parctocess, and scale ofproduction), and livelihoods (contribution to risk reduction or food security) [39], a variety of households would be attractive, and was more intensive and appropriate for a wider range of environmental condition [40], matching species and needs [33,41]. However, increased costs of the innovation [38], reducing profitability [42], and the lack of any discernible revenues [37] have reduced the adoption of agroforestry innovations.

Farmers manage bamboo traditionally, such as with less optimal maintenance and no additions to other types of bamboo. This condition causes land productivity not to be optimal, making it less attractive to farmers, and some farmers replace bamboo with other types of plants of high economic value. Therefore, efforts are needed to introduce innovations in bamboo cultivation with agroforestry patterns through demplots. Demontrations, which are the most popular extension technique, allow for wider adoption by demonstrating viability under farmer-specific conditions [37].

The existence of BAF demplots is also expected to function as a provider of extension materials or information, play a role in the development of extension methods, and become a site for various types of training or application of various technological practices. Farmers often participate in demontrations of result and method offered by local extension and research groups to verify the effectiveness of new technologies [43]. Farmers' participation in training and demplots increased agroforestry adoption [33,44].

3.3 Social, Economic, and Ecology Impact of BAF Demplots

Adoption of agroforestry also involves socio-psychological elements, including the expectations and beliefs of family members' . Socio-psychological aspects (such as preferences) need to be incorporated into development programmes to increase the effectiveness of agroforestry interventions [45]. The development of bamboo-based agroforestry (BAF) demplots in the study involves farmer who joint in bamboo cultivation and farmers outside the demplots. The perceptions of farmers, both demplot farmers and non-demplot farmer, on the impact of the BAF demplot are as follows:

a) Social impact of BAF demplots

The social impact attributes of the BAF demplot consist of 1) increased activities/jobs in the garden, 2) increased knowledge of bamboo cultivation, 3) increased motivation to grow bamboo, 4) dynamic farmer groups, 5) many training opportunities related to BAF, 6) interaction between farmers, 7) related government attention, 8) infrastructure development, 9) farming activities in the garden, and 10) community activities in the village. The social impacts of BAF demplot based on farmers' perceptions are presented in Figure 3. Figure 3 shows that the social impacts of the BAF demplots for importance and performance levels are in the medium and high categories. Socially, the BAF demplots, has generally increased social interaction between fellow farmers and fellow communities in Sukaharja Village, as because the village governmentt often receives visits from various communities/related agencies to see the BAF demplots and learn obout BAF management.

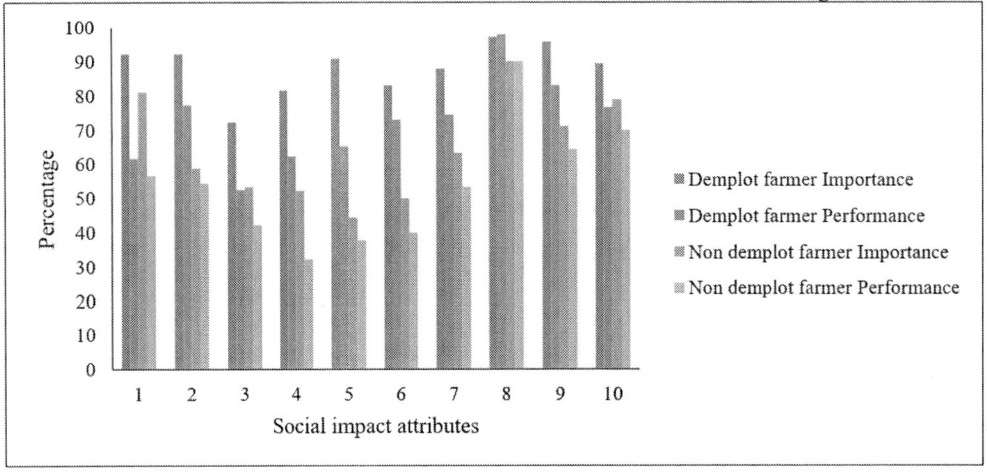

Fig. 3. Social impact of BAF demplot based on farmers' perception

Social factors play an important role in any tree planting programme. In short, the tree's survival rate shows that the community accepting the program well and that there is no conflict, and long-term (>5 years) community monitoring of the plants will affect the sustainability of the programme [46]. The existence of social demplots impacts the attention of various parties. The Sukaharja Village government responded to the existence of the BAF demplot by building several infrastructure facilities such as road repairs, mosque construction, meeting houses, and circuit tourist sites. This infrastructure will support the development of BAF, facilitate extension and training activities, hold smallholder farmer meetings;etc.. This condition is verydifferent from the land before the BAF programme.

It is important because the era of modernisation threatening the development of bamboo, where bamboo is beginning to be abandoned, especially by industry. This is because bamboo has an asymmetrical shape that most industry players find unattractive. There is a perception that bamboo material is only used for traditional communities or those from Indonesia, from low or poor economic background [47]. Socially, people have abandoned bamboo as a support in their daily lives. Plastic materials have now replaced bamboo due to the effects of modernisation [48].

Bamboo is an essential part of human life. In Indonesia, the cultivation of bamboo plantations still still largely a community effort. Thus, the availability of bamboo to meet current needs still depends on forest products or farms. Moreover, bamboo applications are mostly limited to traditional uses.

b) Economic impact of BAF demplot based on farmers' perception

The attributes of the economic impact of the BAF demplots consist of 1) land resources, 2) encouraging other economic businesses, 3) increasing farming, 4) increasing income, 5) increasing the economic value of bamboo, 6) increase farm productivity, 7) acces to agricultural capital, 8) the availability of wage labour, 9) the availability of family labour, and 10) the sustainability of farming.

The attributes of the economic impact of the BAF demplots consist of the economic impact of the BAF demplot based on farmers' perceptions is presented in Figure 4.

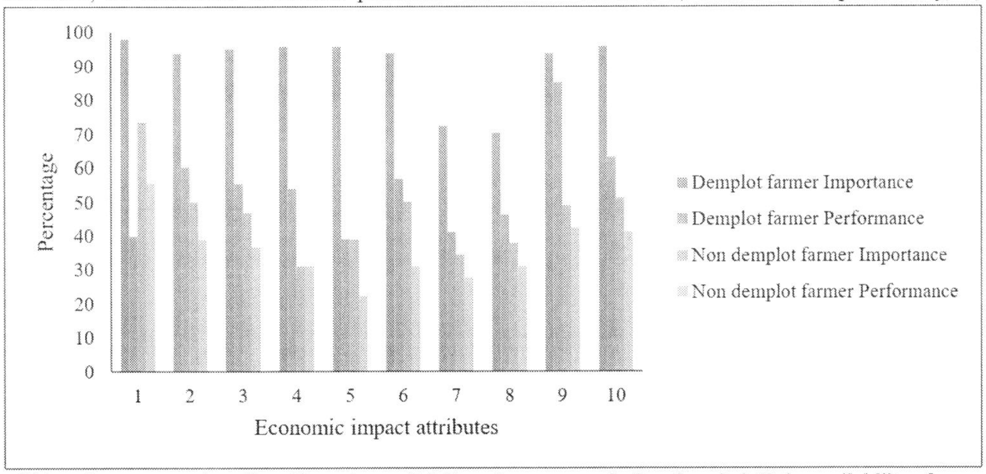

Fig. 4. The economic impact of BAF demplot based on farmers' perception

Figure 4 shows that the economic impact of the BAF demplots based on farmers' perceptions of importance and performance levels are in the medium and high categories. The demplot farmers feel that the BAF demplots have started to have a positive impact on increasing their income. However, this is not yet significant. [13] mentioned that agroforestry development this is not yet significantly improved their economy. Bamboo is a source of income for farmers, although the selling value still needs to be higher. The use of bamboo can generate and increase new income for the community, which will support the improvement of community welfare [46, 47]. However, the community is not yet fullfocused entirely on agroforestry, so there is a need to intensify agroforestry practices and improve agroforestry land management [51].

c) Ecology impact of BAF demplot based on farmers' perception

The ecological impact attributes of the agroforestry demplots consist of 1) reducing the potential for drought, 2) maintaining the sustainability of the planting period, 3) reducing the impact of drought, 4) improving land cover, 5) maintaining springs; 6) reducing forest pest disturbance, 7) improving air conditions, 8) increasing the growth of other plants, 9) increasing plant biodiversity, and 10) increasing animal biodiversity .

The ecological impacts of BAF demplot based on farmers' perceptions are shows in Figure 4. Figure 4 shows that the ecological impact of the BAF demplots for importance and performance levels are in the medium and high categories. People still believe that bamboo is good for protecting springs and farmers' land from erosion and landslides, so people still maintain the existence of bamboo plants [13]. Brazilian communities cultivate bamboo species such as *Bambusa blumeana* and *Phyllostachys pubescens* to control soil erosion, prevent loss of soil nutrients, and improve soil structure [52].

Bamboo has an extensive fibrous root system, systematically connected rhizome roots, and relatively thick bamboo leaves that can protect against rain. Bamboo produces new stems from underground rhizomes that can be harvested without disturbing the soil.

In addition, bamboo serves ecological purposes such as preventing soil erosion, conserving water, rehabilitating land, and sequestering carbon [53,54]. Other benefits to farmers include the availability of clean water and the appearance of several animals (monkeys, birds, and bees) that make community forests their habitat.

The important and the interests of the existence of demplots is medium and high categories on social, economic, and ecology. For demplots farmers, the existence of BAF demplots can fulfil some of the farmers' social, economic, and ecological expectations (their interests). The difference in perception between demplot farmers and non-demplot farmers is possible because the intensity of farmer involvement in demplots is greater than that of non-demplot farmers. The perception of non-demplot farmers is lower in term of importance and availability but is still in the moderate category. For non-demplot farmers, demplots strongly support efforts to increase motivation for farmers/communities to plant bamboo on their land and cultivated areas.

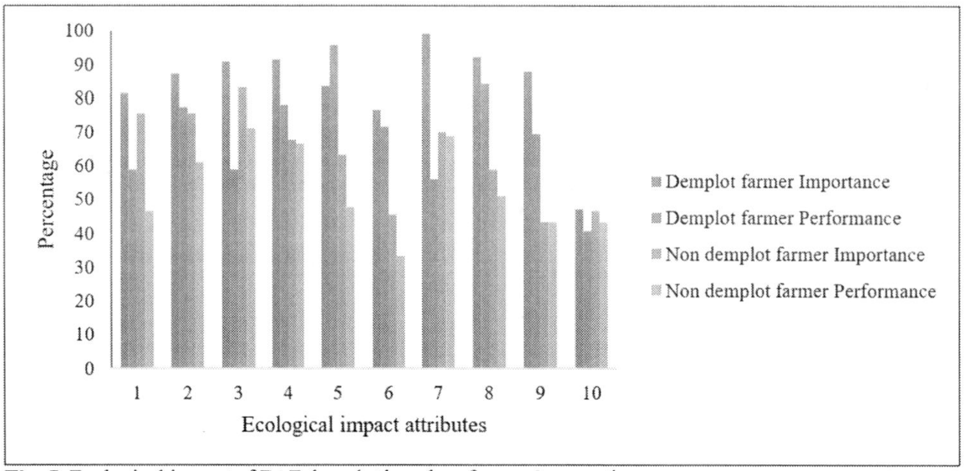

Fig. 5. Ecological impact of BAF demplot based on farmers' perception

d). Importance Performance Analysis (IPA) of BAF demplots impact

The Cartesian graph of the IPA of BAF demplots impact is shows in Figure 6. Figure 6 (a) shows that the IPA attributes are distributed in quadrants I, II, and III. Demplot farmers have a higher interest than non-demplot farmers. According to the perception of non-demplot farmers, socially, demplots provide something in the interests of farmers. However, there are two attributes whose availability has not been fulfilled but which are very important, so that the BAF demplot manager must pay priority attention to them, namely: for demplot farmers, how the BAF demplot can increase the opportunity to work on the plantation (attribute 1) and for non-demplot farmers, how the existence of the demplot motivates farmers to plant, maintain, and sustainably use bamboo (attribute 3).

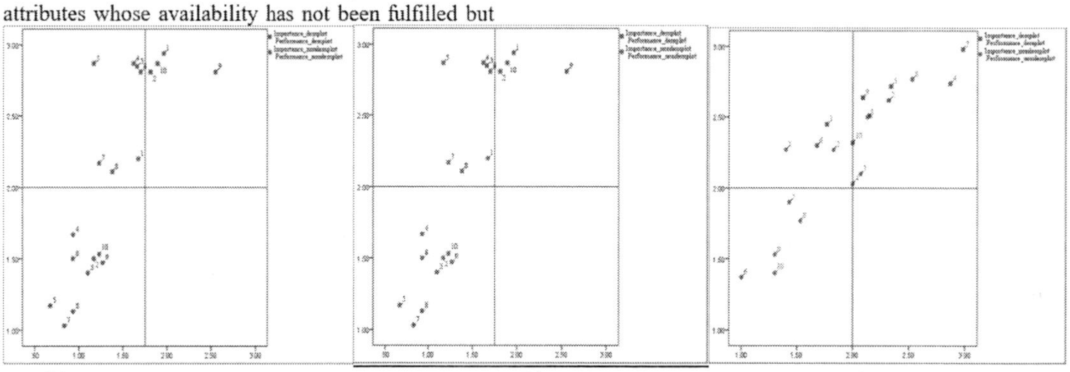

(a) (b) c)

Fig. 6. IPA of the social (a), economy (b) and ecology (c) impacts of BAF demplots

Figure 6 (b) shows that the demplots have yet to fully meet the farmers' economic expectations . The farmers' interest in developing bamboo through the production of BAF demplots is very high but has yet to be fulfilled by the demplots. In the economic aspect, many things need be considered, namely attributes 1, 3, 4, 5, 6, 7, and 8. In order to provide economic benefits to demplot farmers, it is necessary to strengthen the farming community in marketing bamboo and other agroforestry products because farmers are still face problems-barriers to marketing bamboo in small quantities. Planting trees can increase social cohesion [55] so that the community can share economic benefits and increase group cooperation.

Farmers around the demplot feel the ecological impact of the bamboo demplot (Figure 6 c). Non-demplot farmers also perceived some of the attributes. The attributes that need to be considered because they are very important but did not meet farmers' expectations are attributes 1, 2, and 6, namely optimising BAF to reduce the impact of drought in the dry season, keeping the planting period from decreasing, and how BAF reduces pests.

4 Conclusion

Bamboo-based agroforestry (BAF) is one of the interesting models in bamboo development efforts. The BAF demplot is an important tool that can be a medium for transferring cultivation bamboo knowledge and technology to farmers and the wider community. BAF demplots are also have an impact on farmers within the demplots and outside the demplot. The existence of demplots is more felt more by farmers who are directly involved in the development of demplots than by farmers who are not involved in demplots. The existence of BAFs has had a positive impact on farmers, particularly in term social and ecological impacts. However, the economic impact of the BAF demplots is considered to be below farmers' expectations. Farmers hope that in addition to bamboo development, they can also learn to process bamboo into products that have economic value and increase their income.

References

[1] Thakur A 2020 Green gold and the Himalayas. *Biot. Res. Today.* **2** 480–3

[2] Liese, Walter and M K 2015 *Bamboo: The plant and its uses. Switzerland* (Switzerland: Springer)

[3] Salim R, Cahyana B T, Putra Prabawa I D G and Hamdi S 2019 Potensi bambu untuk pemanfaatan sebagai bahan bakar arang dengan metode pengarangan retort tungku drum *J. Ris. Teknol. Ind.* **13** 230

[4] Arsad E 2015 Teknologi pengolahan dan pemanfaatan bambu *J. Ris. Ind.* **7** 45–52

[5] Muhtar 2020 Precast bridges of bamboo reinforced concrete in disadvantaged village areas in Indonesia *Appl. Sci.* **10** 25

[6] Nath A J, Das G and Das A K 2009 Above ground standing biomass and carbon storage in village bamboos in North East India *Biomass Bioenergy* **33** 1188–1196

[7] Irawan B, Partasasmita R, Rahayu N, Setiawati T I A and Iskandar J 2019 Indigenous knowledge of bamboos by Naga community , Tasikmalaya District , West Java , Indonesia **20** 1423–34

[8] Kittur B H, Sudhakara K, Mohan Kumar B, Kunhamu T K and Sureshkumar P 2016 Bamboo based agroforestry systems in Kerala, India: performance of turmeric (Curcuma longa L.) in the subcanopy of differentially spaced seven year-old bamboo stand *Agrofor. Syst.* **90** 237–50

[9] Tewari S, Banik R L, Kaushal R, Bhardwaj D R, Chaturvedi O P and Gupta A 2015 Bamboo based agroforestry systems *Bamboos in India* 261–84

[10] Christanty L 1997 "Without bamboo, the land dies": A conceptual model of the biogeochemical role of bamboo in an Indonesian agroforestry system *For. Ecol. Manage.* **91** 83–91

[11] Ekawati D, Karlinasari L, Soekmadi R and Machfud 2022 Drivers, barriers, and strategies in the community-based supply of bamboo for industrial-scale bamboo utilization in Ngada Regency, East Nusa Tenggara, Indonesia *Sustainability* **14** 1–23

[12] Khotimah H and Sutiono 2014 Analisis kelayakan finansial usaha budidaya bambu *J. Ilmu Kehutan.* **8** 14–24

[13] Hani A, Fauziyah E, Widyaningsih T and Kuswantoro D 2018 Potensi dan pola agroforestri yang mendukung kelestarian bambu di Desa Sukaharja Kabupaten Ciamis *J. Wasian* **5** 115–25

[14] Widyaningsih T S, Fauziyah E, Hani A and Kuswantoro D P 2016 Demplot agroforestry bambu: Media belajar bersama masyarakat *Prosiding Seminar Nasional dan Pertemuan Ilmiah Tahunan ke-2 KOMHINDO* ed Hafizianor, Y Nugroho and Susilawati (Banjarbaru: Lambung Mangkurat University Press)

[15] Wulandari, C., & Inoue M 2018 The importance of social learning for the development of community based forest management in Indonesia: the case of community forestry in Lampung Province *Small-Scale For.* **17** 361–76

[16] Masria M, Golar G and Ihsan M 2015 Persepsi dan sikap masyarakat lokal terhadap hutan di Desa Labuan Toposo Kecamatan Labuhan Kabupaten Donggala *War. Rimba* **3** 57–64

[17] Desmiwati, Yuniarti N, Zanzibar M, Bramasto Y, Sianturi R U D and Megawati 2021 Persepsi petani penggarap terhadap keberadaan demplot jati muna hasil mutation breeding di hutan rakyat Cariu *J. Perbenihan Hutan Tanam.* **9** 75–86

[18] Sugiyono 2008 *Metode Penelitian Kuantitatif Kualitatif dan R&D* (Bandung: Alfabeta)

[19] Yusuf M 2015 *Metode Penelitian Kuantitatif, Kualitatif & Penelitian Gabungan.* (Jakarta: Prenadamedia Group)

[20] Robbins S P and Judge T A 2013 *Organizational Behavior 15th Edition* (Upper Saddle River: Pearson Education Limited)

[21] Brandt D R 2000 *An "Outside-In" Approach to Determining Customer Driven Priorities for Improvement and Innovasion*

[22] Luo Y F, Yang S C and Kang S 2022 New media literacy and news trustworthiness: An application of importance–performance analysis *Comput. Educ.* **185**

[23] Aghajanzadeh M, Aghabayk K, Esmailpour J and De Gruyter C 2022 Importance – Performance Analysis (IPA) of metro service attributes during the COVID-19 pandemic *Case Stud. Transp. Policy* **10** 1661–72

[24] Christanty L, Kimmins J P and Mailly D 1997 "Without bamboo, the land dies": A conceptual

model of the biogeochemical role of bamboo in an Indonesian agroforestry system *For. Ecol. Manage.* **91** 83–91
[25] Fauziyah E, Widyaningsih T S, Kuswantoro D P and Sanudin 2021 Management of Village Treasury Land Based on Bamboo Agroforestry *IOP Conf. Ser. Earth Environ. Sci.* **940**
[26] Chongtham N, Bisht M S and Haorongbam S 2011 Nutritional properties of bamboo shoots: Potential and prospects for utilization as a health food *Compr. Rev. Food Sci. Food Saf.* **10** 153–68
[27] Wang Y, Chen J, Wang D, Ye F, He Y, Hu Z and Zhao G 2020 A systematic review on the composition, storage, processing of bamboo shoots: Focusing the nutritional and functional benefits *J. Funct. Foods* **71** 1–16
[28] Nongdam P and Tikendra L 2014 The nutritional facts of bamboo shoots and their usage as important traditional foods of Northeast India *Int. Sch. Res. Not.* **2014** 1–17
[29] Hiwale S 2015 *Sustainable horticulture in semiarid dry lands.* (India: Springer)
[30] Ong CK, Black C W J 2015 *Tree–crop Interactions: agroforestry in a changing climate* (UK: CABI, Wallingford)
[31] Atangana A, Khasa D, Degrande A and Chang S 2014 *Tropical Agroforestry* (New York: Springer)
[32] Partey S T, Sarfo D A, Frith O, Kwaku M and Thevathasan N V 2017 Potentials of Bamboo-Based Agroforestry for Sustainable Development in Sub-Saharan Africa: A Review *Agric. Res.* **6** 22–32
[33] Iiyama M, Derero A, Kelemu K, Muthuri C, Kinuthia R, Ayenkulu E, Kiptot E, Hadgu K, Mowo J and Sinclair F L 2017 Understanding patterns of tree adoption on farms in semi-arid and sub-humid Ethiopia *Agrofor. Syst.* **91** 271–93
[34] Espinosa A 2014 Involving technical advancement for agroforestry extension : an innovation-decision approach *Int. J. Agrofor. Silvic.* **1** 61–7
[35] Anderson J R and Feder G 2007 Chapter 44 Agricultural Extension *Handb. Agric. Econ.* **3** 2343–78
[36] Feder, G., Birner, R., Anderson J R 2011 The private sector's role in agricultural ex_tension systems: potential and limitations. *J. Agribus. Dev. Emerg. Econ.* **1** 31–54
[37] Conteh A M H, Moiwo J P and Yan X 2016 Using a logistic regression model to analyze alley farming adoption factors in Sierra Leone *Small-scale For.* **15** 109–25
[38] Kiptot E, Hebinck P, Franzel S and Richards P 2007 Adopters, testers or pseudo-adopters? Dynamics of the use of improved tree fallows by farmers in western Kenya *Agric. Syst.* **94** 509–19
[39] Gyau A, Chiatoh M, Franzel S, Asaah E and Donovan J 2012 Determinants of farmers' tree planting behaviour in the north west region of Cameroon: The case of prunus Africana *Int. For. Rev.* **14** 265–74
[40] Ajayi O C, Place F, Akinnifesi F K and Sileshi G W 2011 Agricultural success from Africa: The case of fertilizer tree systems in Southern Africa (Malawi, Tanzania, Mozambique, Zambia and Zimbabwe) *Int. J. Agric. Sustain.* **9** 129–36
[41] Rogers E M 2003 *Diffusion of innovations* (New York: Free Press)
[42] Sirrine D, Shennan C, Snapp S, Kanyama-Phiri G, Kamanga B and Sirrine J R 2010 Improving recommendations resulting from on-farm research: agroforestry, risk, profitability and vulnerability in Southern Malawi *Int. J. Agric. Sustain.* **8** 290–304
[43] Kabir K H, Sarker S, Uddin M N, Leggette H R, Schneider U A, Darr D and Knierim A 2022 Furthering climate-smart farming with the introduction of floating agriculture in Bangladeshi wetlands: Successes and limitations of an innovation transfer *J. Environ. Manage.* **323**
[44] Mgendi B G, Mao S and Qiao F 2022 Does agricultural training and demonstration matter in technology adoption? The empirical evidence from small rice farmers in Tanzania *Technol. Soc.* **70** 1–20
[45] Meijer SS 2014 The role of knowledge, attitudes and perceptions in the uptake of agricultural and agroforestry innovations among smallholder farmers in sub-Saharan Africa *Int. J. Agric. Sustain* 1–15
[46] Rana P and Miller D C 2021 Predicting the long-term social and ecological impacts of tree-planting programs : Evidence from northern India *World Dev.* **140** 105367
[47] Suriani E 2017 Bambu sebagai alternatif penerapan material ekologis: potensi dan tantangannya *EMARA Indones. J. Archit.* **3** 33–42
[48] Yu X 2007 *Bamboo: Structure and Culture: Utilizing Bamboo in the Industrial Context with Reference to Its Structure and Cultural Dimensions.* (VDM Publishing)
[49] Chen X 2003 Promotion of bamboo for poverty allevation and economic development. *J. Bamboo Ratt.* **2** 345–50
[50] Herliana S 2014 Model of household economic behavior bamboo creative industries: A preliminary study *Procedia - Soc. Behav. Sci.* **115** 1–11
[51] Suparwata D O 2018 Pandangan masyarakat pinggiran hutan terhadap program pengembangan agroforestri *J. Penelit. Sos. dan Ekon. Kehutan.* **15** 47–62
[52] Ben-Zhi, Z., Mao-Yi, F., Jin-Zhong, X., Xiao-Sheng Y and & Zheng-Cai L 2005 Ecological functions of bamboo forest: research and application. *J. For. Res.* **16** 143–7
[53] Iqbal M, Laratu N, Pitopang R and Suleman S M 2014 Keanekaragaman jenis tumbuhan herba pada dua tipe hutan di Desa Bobo

Kawasan Taman Nasional Lore Lindu Sulawesi Tengah *J. Biocelebes* **(8)2** 13–25

[54] Lobovikov et al. 2012 Bamboo in climate change and rural livelihoods. *Mitig. Adapt. Strateg. Glob. Chang.* **17** 261–76

[55] Donovan G H, Prestemon J P and Kaminski A R 2022 The natural environment and social cohesion : Tree planting is associated with increased voter turnout in Portland , Oregon *Trees, For. People* **7** 1–12

Impact of Glyphosate Contamination on Chemical Properties of Inceptisols Amelioration with Biochar from Rice Husks, Young Coconut Waste, and Bamboo

Herviyanti Herviyanti[1*], *Amsar* Maulana[2], *Arestha Leo* Lita[3], *Mimien* Harianti[1], *Teguh Budi* Prasetyo[1], *Rezha Tri* Khurnianto[4], *Pitri* Juwita[4], *Ridho* Ryswaldi[5] and *Syafrimen* Yasin[1]

[1]*Department of Soil Science and Land Resource, Agriculture Faculty, Andalas University, Limau Manis, Padang City, 25164, Indonesia*
[2]*Doctoral Program of Agricultural Science, Postgraduate of Andalas University, Limau Manis, Padang City, 25164, Indonesia*
[3]*Magister Program of Soil Science, Postgraduate of Andalas University, Limau Manis, Padang City, 25164, Indonesia*
[4]*Bachelor Program of Soil Science, Agriculture Faculty, Andalas University, Limau Manis, Padang City, 25164, Indonesia*
[5]*Department of Management, Economic Faculty, Padang State University, Air Tawar, Padang City, 25131, Indonesia*

Abstract. Growing concerns over glyphosate herbicides, if not applied carefully, can have unintended consequences on agroecosystems. The objective of this research was to study glyphosate contamination on the chemical properties of Inceptisols ameliorated with biochar of rice husk (B-RH), young coconut waste (B-YCW), and bamboo (B-B). This study used a completely randomized design (CRD) with five treatments and three replications, namely A = Control (without ameliorants and glyphosate); B = Soil + 100 mg l^{-1} glyphosate; and C = Soil + B-RH + 100 mg l^{-1} glyphosate; D = Soil + B-YCW + 100 mg l^{-1} glyphosate and E = Soil + B-B + 100 mg l^{-1} glyphosate. The results showed that contamination from glyphosate had a significant effect on surface changes (pH, EC, CEC, Mineral, and SOM) and nutrients (SOC, total N, and available P) of Inceptisols that had been improved with 40 t ha^{-1} biochar. The correlation and equation of linear regression between residues on glyphosate (RG) had a significant interaction with chemical properties of Inceptisols amended with biochar, namely exchangeable Ca [r=0.611* and RG=0.0232(Ca)-0.0079; R^2=0.3728]. CEC [r=0.593* and RG=0.0018 (CEC)-0.0312; R^2=0.3514]; available P [r=0.590* and RG=0.0061(P_2O_5)-0.0232; R^2=0.3472] and total N [r=0.570* and RG=0.257(N)-0.0621; R^2=0.4312].

Keyword. Bamboo Biochar, Glyphosate, Inceptisols, Rice husk, Young coconut waste

1 Introduction

Glyphosate, as the active component in the herbicide chemical Roundup, can have a profound contaminating effect on ecosystems. Glyphosate is an environmental pollutant used at a rate of 600,000 to 750,000 tons per year over an area of about 36 million square kilometers. Its use in agricultural and other contexts has steadily increased from 0.6 Mg in 1974 to 125.5 Mg in 2014, making it the most widely used herbicide in the world [1]. The presence of glyphosate in the soil is complex and related to mineralization, degradation, immobilization, and leaching. Its presence in the soil can affect adsorption, leaching, and mineralization [2]. Glyphosate is an amphoteric small molecule that has three polar functional groups. The groups are phosphonomethyl, amine, and carboxymethyl groups arranged linearly. Glyphosate is a highly polar ionic molecule (log KOW = -3.20) that is soluble in water (10.5 g L^{-1} at 20°C). Glyphosate is a polyprotic acid with pKa values ranging from 0.7 to 2.2, 5.9 to 10.6. The pH of the solution affects the speciation of the molecule. Monovalent and divalent anions are the dominant species at normal soil pH levels [3].

The impact of glyphosate contamination and its residues are determined by soil type, the suitability of the herbicide's biophysical-chemical properties, and the time gap between pesticide application and rainfall [4]. The quantity of glyphosate and its mineralization kinetics in the soil is influenced by temperature, pH, and total organic carbon (TOC) [5]. Glyphosate degrades rapidly in most soils, with half-lives ranging from 7 to 60 days [6]. The increase in pH, EC, CEC, and SOM at a glyphosate concentration of 100 ppm with a glyphosate solution pH of 5.20 indicates that Inceptisols ameliorated with 40 t ha^{-1} rice husk biochar as a soil improver can adsorb glyphosate in the soil [7]. This is also evident from the results of Fourier Transform Infra-Red (FT-IR) spectroscopy which shows a decrease in transmittance in O-H, C=C, C-O, C-H, and mineral groups indicating an increase in sorption capacity in the improved Inceptisols soil. Changes in functional groups also favor glyphosate adsorption. Interestingly, the physicochemical characteristics of Inceptisols have a significant role in regulating the adsorption capacity of glyphosate to bind to the soil. Therefore, there is a need

* Corresponding author: herviyanti@agr.unand.ac.id

to test with the comparison of potential biochar types as different ameliorant technologies to address glyphosate residues in Inceptisols.

Biochar is a porous carbonaceous substance made by pyrolyzing some forms of biomass [8][9]. Biochar has promising applications in the reduction of agricultural environmental pollutants such as pesticides, particularly herbicides [10]. Biochar has various benefits, including a large surface area with a porous structure, decreased toxicity, abundance of functional groups, and redox activity. The sorption and desorption behavior of glyphosate in soil has been widely researched and reviewed. pH, phosphate content, and the presence of iron and aluminum oxides are known to affect glyphosate sorption in soil. Glyphosate sorption increased in soils amended with biochar, indicating the role of soil physicochemical parameters in influencing glyphosate sorption capacity. In addition, biochar has been shown to sorb glyphosate from aqueous solution, and attempts to investigate the sorption process using isotherm models and kinetics data revealed some possible interactions, including - electron donor-acceptor interactions, H-bonding, and heterogeneous H-bonding [11]. Biochar has great potential with diverse materials such as rice husk, young coconut waste, and bamboo, and strategic comparisons among these materials may provide much-needed new insights into sorption mechanisms and identify biochar properties that affect glyphosate residues in soil.

2 Material and Methods

This study will be conducted in the Laboratory of Chemistry and Soil Fertility, Faculty of Agriculture, Andalas University (UNAND), Padang starting from January to May 2023.

2.1 *Experimental Design*

This research used a completely randomized design (CRD) consisting of five treatments with three replications. The experiment was conducted by observing the effect of 100 ppm glyphosate equivalent to 0.07 ml of glyphosate in 250 ml of water on the chemical properties of inceptisols with 40 t ha^{-1} biochar. The experiment design is presented in the following:
A = Control (without ameliorants and glyphosate);
B = Soil + 100 mg l^{-1} glyphosate;
C = Soil + rice husk biochar (B-RH) + 100 mg l^{-1} glyphosate;
D = Soil + young coconut waste biochar (B-YCWB) + 100 mg l^{-1} glyphosate and
E = Soil + Bamboo biochar (B-B) + 100 mg l^{-1} glyphosate.

2.2 *Ameliorant, Glyphosate and Soil Sampling*

Biochar Production [12] [13][14].

1. Rice Husk

Rice husk (RH) is a waste from the rice milling industry. Biochar from RH is produced using the Drum method through a pyrolysis process. Pyrolysis was performed in iron drums with a diameter of 58 cm, a height of 86 cm, and a capacity of 200 liters and combined with the Kon-Tiki method at temperatures up to 650^{0}C. The maximum pyrolysis temperature was kept constant for 45 min to ensure the completion of the final process.

2. Young Coconut Waste

Young coconut waste (YCW) from green coconut (Cocos nucifera L.) is cut to a length of about 10*5 cm and dried for one week in the Greenhouse of the Faculty of Agriculture, Andalas University until it achieves a moisture content of 18.20%. Furthermore, the manufacturing process is carried out using a method that has been tested three times. Figure 1 shows the process of producing YCW biochar with a weight of up to 10 kg and the following technique specifications, The Kon-Tiki method is made of conical steel which has a top diameter of 100 cm, a height of 90 cm, a wall slope of 63.50^0, and a capacity of 827 liters.

3. Bamboo

The preparation process for biochar production uses bamboo (B) biomass from the type of bamboo Betung (Dendrocalamus asper) and is cut to a length of approximately 30*6 cm and dried for one week in the Greenhouse of the Faculty of Agriculture, Andalas University until it reaches moisture of 20.48%. Furthermore, the production process is carried out based on the method that has been carried out with three repetitions. In the production process of bamboo biomass biochar used as much as 10 kg with the specifications of The Kon-Tiki method is made of conical steel which has a top diameter of 100 cm, a height of 90 cm, a wall slope of 63.50^0, and a capacity of 827 liters. The results of the production of biochar from each method are watered to stop the combustion process.

All biochar is dried in a 40-70^{0}C oven for 2*24 hours with the aim of homogeneous biochar water content and the next step is to analyze the characteristics of biochar in the laboratory.

Glyphosate Concentration from Round-Up Herbicide

Glyphosate used in soil contamination is a type of herbicide with the trademark Roundup Biosorb 486 SL (GRB-486 SL). Glyphosate used in soil contamination is a type of herbicide with the trademark Roundup Biosorb 486 SL (GRB-486 SL). The concentration of glyphosate contaminated in the soil was 100 mg l^{-1} glyphosate from Round-up Biosorb 486 SL herbicide or equivalent to 0.07 ml GRB-486 SL in 250 ml H_2O with 100g SC 5kg^{-1} soil based on recommendations [7].

Soil Sampling

Inceptisols was taken in Nagari Sariak, Sungai Pua District, Agam Regency in West Sumatera, Indonesia with GPS coordinates of -0021'56" LS and 100024'0" BT. Soil samples were taken at a depth of 0-20 cm from 5 collection points which were then composited. Then, the soil was air-dried for 2*24 hours. After air-drying, the samples were pulverized and sieved with a 2 mm sieve. After that, soil samples were taken for preliminary soil analysis. The soil samples were weighed as much as 5 kg equivalent absolute dry weight into 6 polybags. The soil in 3 polybags was watered with glyphosate solution until it reached field capacity and the other 3 polybags were added biochar based on recommendations, then mixed well and watered with glyphosate to field capacity. Furthermore, the soil was incubated for 2 weeks. After the incubation, soil samples were analyzed in the laboratory.

2.3 *Analysis of Soil, Residual of Glyphosate, and Statistical Analysis*

Soil analysis conducted such as pH (H_2O, KCl, and PZC by using formula 1), Mineral, SOM, CEC, SOC, total N, available P, K, exchangeable Ca, and Mg [15][16][17].

$$\text{pH PZC} = (2*\text{pH KCl}) - \text{pH } H_2O \qquad (F.1)$$

Analysis of glyphosate residues was conducted using the High-Performance Liquid Chromatography (HPLC) method [18][19][20]. The results of all analyses were statistically analyzed using SPSS 16.0, Statistic 8, and Excel 2016 software. R^2 coefficient and p-values [** = Significant correlation at the 0.01 (2-tailed) level and * = Significant correlation at the 0.05 (2-tailed) level] were calculated using bivariate correlation and two-sided significance tests.

3 Result and Discussion

3.1 The impact of glyphosate on surface change of Inceptisols ameliorated with biochar

The effect of glyphosate contamination on Inceptisols was substantial, with a decrease in pH, SOM, and CEC of Inceptisols by 0.16; 6.26; and 0.27 cmol(+) kg^{-1} at a concentration level of 100 mg l^{-1} glyphosate (Tables 1 and 2). Glyphosate is a pure chemical known as a highly polar molecule, a polyprotic acid with pKa values of 2; 2.3; and 5.6 [21]. The soil solution can lower the pH. Glyphosate contamination in the soil decreases the SOM composition, which is inversely related to the mineral composition of the soil. Increased accumulation of glyphosate residues in the soil will drastically reduce soil carbon [22]. This can also be seen in the CEC value of inceptisol soil which has decreased due to glyphosate contamination. The presence of glyphosate in the soil is influenced by the cation exchange capacity (CEC) of the soil [23]. The pH of Inceptisols PZC increased by 0.06 due to the effect of glyphosate contamination compared to the control (Table 1), indicating a decrease in the sorption capacity and negative charge of Inceptisols soil. The main site of glyphosate sorption is on the negatively charged surface of the soil mineral layer [24].

The considerable impact of glyphosate contamination necessitates the use of appropriate technologies, such as biochar-based amelioration technologies, to address glyphosate residues in Inceptisols. It has been demonstrated that the application of three forms of biochar at the rate of 40 t ha^{-1} can improve Inceptisols soil surface changes by increasing pH, CEC, SOM, and CEC. Young coconut waste biochar has the largest increase in pH and CEC values when compared to rice husk and bamboo biochar by 0.83; 0.99; and 0.28 dS m^{-1}, respectively, compared to the control and glyphosate-contaminated Inceptisols. This is because young coconut waste biochar has higher pH and CEC values compared to rice husk and bamboo biochar, which are 10 and > 2 dS m^{-1}, respectively compared to the control and glyphosate-contaminated Inceptisols, bamboo biochar increased SOM by 7.07 and 13.33%, respectively. The increase in pH and SOM in Inceptisols influenced the CEC, where the CEC in the application of young coconut waste biochar was greater than bamboo charcoal and rice husk by 20.78 and 21.05 cmol(+)kg^{-1} respectively compared to the control and glyphosate-contaminated Inceptisols.

Table 1. The impact of glyphosate on surface change of Inceptisols ameliorated with biochar

Treatment	pH H_2O	pH KCl	[a]pH PZC	EC
	unit			dS m^{-1}
Control	4.97 c	4.77 c	4.57 ab	0.08 d
Soil + 100 mg l^{-1} Glyphosate	4.81 d	4.72 c	4.63 a	0.08 d
Soil + B-RH + 100 mg l^{-1} Glyphosate	5.40 b	4.93 b	4.47 b	0.25 c
Soil + B-YCW + 100 mg l^{-1} Glyphosate	5.80 a	5.07 a	4.33 c	0.36 a
Soil + B-B + 100 mg l^{-1} Glyphosate	5.67 a	5.07 a	4.47 b	0.33 b
CV	**1.40**	**1.08**	**1.43**	**3.11**
Duncan's Test	**	**	**	**
SE	**0.06**	**0.04**	**0.05**	**0.006**

Remarks: EC = electrical conductivity; SE = standard error; n = 15; ** = significant at the 0.01 level (2-tailed).

Glyphosate uptake in Inceptisols can be improved by applying 40 t ha^{-1} biochar to the soil. Biochar contains OH functional groups, which can increase soil pH when added to it. Glyphosate belongs to anions in the relevant soil pH range and can only be adsorbed on variable charged surfaces, not permanently (negatively) charged surfaces, on the distribution of glyphosate species as a function of pH with acid dissociation constants (pKa1 = 2.22, pKa2 = 5.44, and pKa3 = 10.13) of carboxyl and amino groups [25].

Table 2. The impact of glyphosate on mineral, SOM, and CEC of Inceptisols ameliorated with biochar

Treatment	Mineral	SOM	CEC
	%		cmol(+)kg^{-1}
Control	74.47 b	25.53 c	30.25 c
Soil + 100 mg l^{-1} Glyphosate	80.73 a	19.27 d	29.98 c
Soil + B-RH + 100 mg l^{-1} Glyphosate	69.13 c	30.97 a	46.05 b
Soil + B-YCW + 100 mg l^{-1} Glyphosate	70.27 c	29.60 b	51.03 a
Soil + B-B + 100 mg l^{-1} Glyphosate	67.40 d	32.60 a	46.03 b
CV	**1.02**	**1.80**	**1.67**
Duncan's Test	**	**	**
SE	**0.60**	**0.39**	**0.55**

Remarks: SOM = soil organic matter by gravimetry methods; CEC = cations exchange capacity; SE = Standart error; n = 15; ** = significant at the 0.01 level (2-tailed).

The increase in CEC of Inceptisols due to the effect of biochar application is determined by a simple force

balance involving an electric field and solution viscosity resistance. The most important soil surface electrochemical parameters (E0) include exchangeable sodium percentage (ESP), cation exchange capacity (CEC), soil specific surface area (S), soil surface charge density (0), soil surface potential (0), and soil surface electric field strength [22]. while biochar application on Inceptisols increases the concentration of SOM, which helps to decrease the phytotoxicity of glyphosate. On the other hand, soil minerals play an important role in glyphosate uptake, although mineral composition was reduced after biochar application. The addition of SOM to the soil affects the amount of complex organic molecules present in the soil SOM components [26].
Biochar clay-carbon interactions with a strong affinity for hydroxyl groups. The amount of aluminum hydroxy complexes included in the clay, as well as its chemical composition and charge characteristics, all affect changes in its electrochemical properties. The point of zero charges (PZC) is affected by the number of interlayers on the cation exchange capacity of clays [27]. Cyclic organic compounds have high molecular weight, long chains, and active phenolics (-OH), which act as binders to bind cations, and anions (pH-dependent charge) at certain pH levels. Some carboxylates are released below pH 6, leaving negatively charged functional groups. The use of biochar can broaden the spectrum of negative charges on the surface charge of soil colloids. The pH PZC of Inceptisols on young coconut waste biochar decreased by 0.24 and 0.30, respectively compared to the control and glyphosate-contaminated Inceptisols. However, the biochar showed an increase in the negative charge of Inceptisols in preventing glyphosate contamination and sorption after treatment in general. On hydroxyl functional sites, protons and other inorganic and organic contaminants are adsorbed. PZC is defined as the presence of equal amounts of positive and negative charges. This indicator is crucial for determining good soil quality to combat soil contamination and pollution. Several parameters, including mineral composition, SOM, and pH, affect the presence and behavior of glyphosate [26].

3.2 The impact of glyphosate on the nutrient of Inceptisols ameliorated with biochar

Glyphosate contamination had a significant effect on the nutrient of Inceptisols, with reductions in SOC, available P, K, Ca, and exchangeable Mg at 100 mg l^{-1} glyphosate-containing 0.80% C; 0.12 ppm P_2O_5 and 0.03; 0.04 and 0.06 cmol(+)kg^{-1} compared to the control (Table 3 and Figure 1). The nutrient reduction of Inceptisol is affected by glyphosate contamination, and phosphate is taken up through ligand exchange at differently charged surface sites under strong bond formation. As a result, there is competition for absorption sites between glyphosate and nutrients, which can impact glyphosate binding and consequent leaching. Since prescriptions prevent glyphosate uptake and ingested glyphosate is mobilized (absorbed), soil mineral glyphosate and nutrient uptake show competition and uptake preference. Glyphosate and nutrients may compete for soil surface area, thereby reducing glyphosate uptake. In addition to competitive uptake, glyphosate, and nutrient uptake can be additive, meaning that certain surface areas can absorb both sorbates while others can only absorb glyphosate or nutrients. Glyphosate has a sorption capacity of 2.61 mol m^{-2}, phosphate has a sorption capacity of 2.85 mol m^{-2}, and glyphosate + phosphate has a sorption capacity of 4.17 mol m^{-2}. The presence of two types of sites, general and specialized, explains this difference. It can also be explained by sorption on different planes, which occurs when glyphosate is adsorbed onto phosphate via metal cations [28].

Table 3. The Impact of glyphosate on SOC, total N, and available P of Inceptisols ameliorated with biochar

Treatment	SOC	Total N	Available P
	%		ppm P_2O_5
Control	2.97 d	0.32 b	7.40 b
Soil + 100 mg l^{-1} Glyphosate	2.17 c	0.32 b	7.28 b
Soil + B-RH + 100 mg l^{-1} Glyphosate	4.10 a	0.46 a	12.35 a
Soil + B-YCW + 100 mg l^{-1} Glyphosate	4.22 a	0.46 a	11.99 a
Soil + B-B + 100 mg l^{-1} Glyphosate	3.83 b	0.44 a	12.54 a
CV	**2.29**	**3.96**	**7.16**
Duncan's Test	**	**	**
SE	**0.08**	**0.01**	**0.60**

Remarks: SOC = soil organic carbon by Walkley and Black methods; SE = Standart error; n = 15; ** = significant at the 0.01 level (2-tailed).

The interaction between glyphosate and nutrients is more than just a matter of fluctuating or permanently charged soil mineralogy. Soils have general and specialized sorption sites, but the sorption sites are highly regulated by the soil mineral type. After this herbicide was introduced in 1974, competition in the uptake of glyphosate and nutrients by plants proved to be a major problem. The most important element affecting glyphosate uptake or leaching is nutrient content. Environmental problems associated with glyphosate competition for nutrients caused by reduced glyphosate uptake in the soil can increase the danger of glyphosate leaching into the water environment. Soils have a large capacity for glyphosate uptake at equilibrium sorption, but soils are also affected by high hydraulic conductivity, which implies an increased risk of leaching if applied before rainfall. In soil solution, glyphosate can form chelates or complexes with micronutrient metal ions. Compared to Ca and Mg ions, Cu and Zn ions in solution can combine quite strongly with glyphosate at physiologically relevant pH values. Plant uptake of complex metal ions has been linked to glyphosate contamination [29].
Application, 40 t ha^{-1} of the three biochar can increase SOC, total N, available P, K, exchangeable Ca, and Mg. However, young coconut waste biochar can increase the proportion of C and N up to 4.22% and 0.46% respectively when compared with other biochar products, such as rice husk (4.10% and 0.46%) and bamboo (3.83% and 0.44%). Compared to rice husk biochar (12.35 ppm P_2O_5) and young coconut husk

(11.99), bamboo biochar was able to increase P availability up to 12.54 ppm P_2O_5 (Table 3). Biochar application has the potential to increase soil uptake of glyphosate at variably charged surface sites, while uptake by the permanently charged silicate layer is limited. Organic matter treatment increases specific surface area and sorption, which has an indirect influence on glyphosate sorption but may stabilize oxide minerals [30].

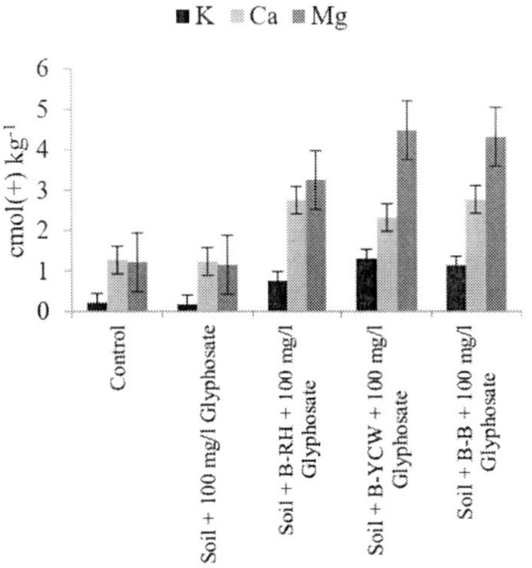

Fig. 1. Effect of glyphosate on exchangeable K, Ca, and Mg of Inceptisols ameliorated with biochar

Absorbed glyphosate generates mononuclear, monodentate, or binuclear, bidentate surface complexes with one coordinated surface site. For the available surface space, glyphosate and nutrients compete [31]. Glyphosate uptake in the soil is also time-dependent, doubling in soils with 3.1% organic carbon [32]. Biochar-based amelioration technology can reduce metal oxide phosphate sorption, increase organic C, total N (Table 3), and soil-accessible P, as well as exchangeable K, Ca, and Mg in Inceptisols soils. The largest increase in K and Mg values in young coconut waste biochar was 1.30 and 4.48 cmol(+)kg^{-1}, respectively, while the highest increase in Ca value in bamboo biochar was 2.77 cmol(+)kg^{-1} (Figure 1).

3.3 Correlation Between Residual of Glyphosate with Surface Change and Nutrient of Inceptisols ameliorated with biochar

The effect of glyphosate contamination on Inceptisols has an influence on the residual value of glyphosate in the soil. Inceptisols contaminated with glyphosate had a residue value of 0.04 mg kg. This shows that Inceptisols can naturally absorb glyphosate (Figure 2). The uptake of glyphosate in soil usually consists of a rapid initial uptake that constitutes the bulk of the uptake, followed by a delayed reaction that can last several days and absorbs only a small amount of glyphosate. Temperature impacts the kinetics of glyphosate uptake and degradation, but the rate of uptake with temperature is unknown. Every 10-degree Celsius increase in temperature is estimated to cause interactions between solutes and soil components [33]. Through the use of biochar amelioration technologies can be used to reduce the impact of glyphosate contamination and residues in the soil.

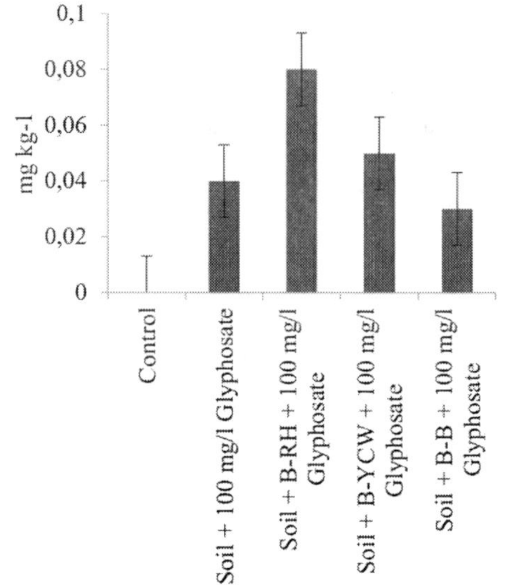

Fig. 2. Residual of glyphosate on Inceptisols ameliorated with biochar from young coconut waste, rice husks, and bamboo

Biochar has been shown to improve the chemical properties of Inceptisols and also acts as a glyphosate sink in Inceptisols, where the application of 40 t ha^{-1} biochar can reduce glyphosate residues in Inceptisols by 0.08 ppm (rice husk) > 0.05% (young coconut husk) > 0.03 ppm (bamboo), and respectively compared to the control and glyphosate-contaminated Inceptisols. Glyphosate absorption can be improved by increasing pH, EC, SOM, and CEC with biochar treatment. Effect of biochar on oxide mineral composition and glyphosate sorption on increasing pH and SOM. Glyphosate sorption by oxides is 1.45-1.85 mol m^{-2}. This is important in terms of the overall number of singly coordinated (possibly adsorbing) hydroxyl groups on the oxide surface, as iron oxide has an average of 3 OH groups nm^{-2}, which is equivalent to 5.0 mol m^{-2} [34]. Between glyphosate and the oxide surface, mononuclear and monodentate bonds occupy less than half of the singly coordinated OH groups, while binuclear and bidentate surface complexes develop, leaving some OH groups vacant. Mineral composition and SOM appear to have a function in glyphosate sorption in soil. This is due to the clay content interacting with the phenolic groups as well as the pores in the biochar. Thus, biochar has a very high absorption of glyphosate. This absorption is due to the chemical formation of hydrogen bonds with glyphosate, as well as the absorption of glyphosate within the pores of biochar.

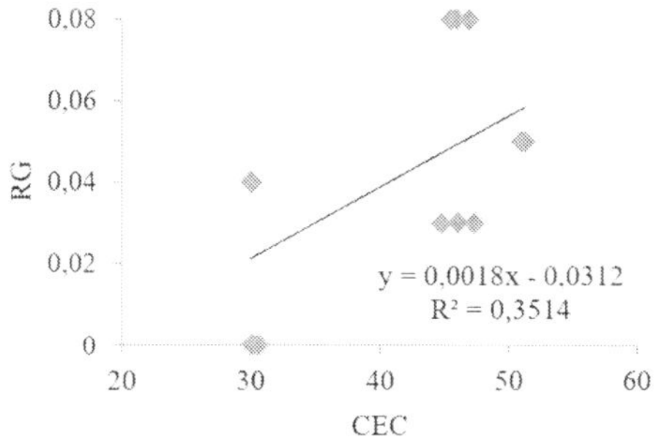

Fig. 3 Linear regression between residual of glyphosate (RG) with CEC on inceptisols meliorated with biochar

Table 4. Matrix correlation Pearson (r) of chemical properties of Inceptisols ameliorated with biochar

Correlation	pH H_2O	pH KCl	pH PZC	EC	CEC	Mineral	SOM	SOC	Total N	Available P	exch-K	exch-Ca	exch-Mg	Residual of Glyphosate (RG)
pH H_2O	1	0.974**	-0.820**	0.971**	-0.845**	0.837**	0.948**	0.894**	.888**	0.878**	0.981**	0.840**	0.977**	0.371
pH KCl		1	-0.672**	0.937**	-0.818**	0.810**	0.905**	0.841**	0.867**	0.866**	0.949**	0.840**	0.953**	0.336
pH PZC			1	-0.819**	0.725**	-0.718**	-0.832**	-0.830**	-0.735**	-0.702**	-0.825**	-0.640*	-0.802**	-0.359
EC (dS m^{-1})				1	-0.814**	0.809**	0.973**	0.887**	0.890**	0.922**	0.982**	0.868**	0.984**	0.481
Mineral (%)					1	-1.000**	-0.833**	-0.929**	-0.808**	-0.876**	-0.812**	-0.902**	-0.849**	-0.294
SOM (%)						1	0.829**	0.927**	0.808**	0.874**	0.804**	0.903**	0.844**	0.298
CEC [cmol(+)kg^{-1}]							1	0.933**	0.940**	0.930**	0.959**	0.901**	0.970**	**0.593***
SOC (%)								1	0.882**	0.901**	0.869**	0.877**	0.886**	0.481
Total N (%)									1	0.835**	0.900**	0.879**	0.917**	**0.570***
Available P (ppm)										1	.880**	0.957**	0.919**	**0.590***
K-exch [cmol(+)kg^{-1}]											1	0.833**	0.993**	0.387
Ca-exch [cmol(+)kg^{-1}]												1	0.891**	**0.611***
Mg-exch [cmol(÷)kg^{-1}]													1	0.439
Residual of Glyphosate (mg kg^{-1})														1

Remarks: PZC = Point of zero change; EC = electrical conductivity; CEC = cations exchange capacity; SOM = soil organic matter by gravimetry method; SOC = soil organic carbon by Walkley and Black methods; exch = exchange; n = 15; ** = Correlation is significant

at the 0.01 level (2-tailed); * = Correlation is significant at the 0.05 level (2-tailed).

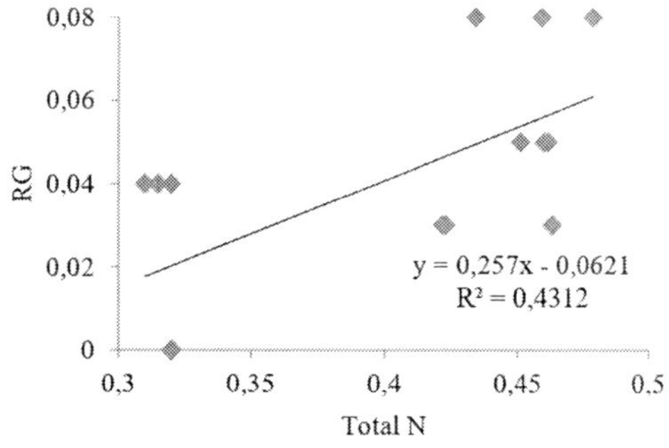

(A)

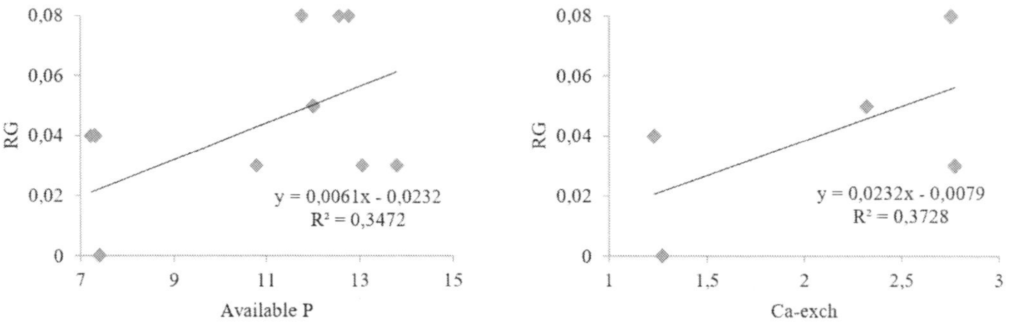

(B) **(C)**

Fig. 4 Linear regression between residual of glyphosate (RG) with total N (A), available P (B), and exchangeable Ca (C) on Inceptisols ameliorated with biochar

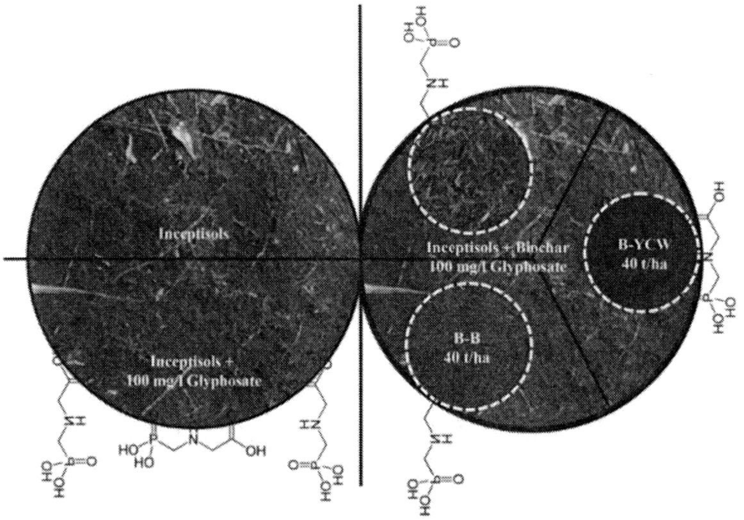

Fig. 5 Prediction of mechanism process of contamination from glyphosate on Inceptisols ameliorated with biochar

Determining the uptake and leaching of glyphosate in a particular soil is not easy, as it seems to depend on several soil characteristics such as pH, minerals, SOM, and CEC. There is an interaction between residues of glyphosate and CEC [r = 0.593* and RG = 0.0018 (CEC) - 0.0312; R^2 = 0.3514]. The increased negative charge caused by the chelation process has a major impact on glyphosate uptake on Inceptisols (Fig. 3). The potential of glyphosate as a chelating agent impacts the competition for binding sites in the soil, demobilizes particle-associated components into the soil solution, and increases their solubility [35]. While in nutrient composition there is also an interaction between residues with Ca [r = 0.611* and RG = 0.0232 (Ca) - 0.0079; R^2 = 0.3728] > available P [r = 0.590* and RG = 0.0061 (P2O5) -0.0232; R^2 = 0.3472] > total N [r = 0.570* and RG = 0.257 (N) - 0.0621; R^2 = 0.4312]. This also explains that glyphosate can affect nutrient composition in Inceptisols (Fig. 4). The presence of glyphosate affects the binding mechanism between glyphosate and phosphate due to P mobility in the soil through adsorption competition. This competition varies greatly among different soil types [36]. Consequently, the competition for adsorption sites in the soil between glyphosate and P appears to be quite important and has a considerable influence on the mobility and availability of P as a plant nutrient. The predicted mechanism is depicted in Figure 5, where biochar has hydroxyl groups as the basis of negative charge. Application of biochar on Inceptisols increased OH groups (pH H_2O) and the negative charge of the soil (CEC). This indicates that Inceptisols become more active in increasing surface charge supported by SOM content and inter-charge interaction (EC). Thus, soil absorption of glyphosate becomes greater with the decrease of glyphosate residue in Inceptisols after biochar treatment.

4 Conclusion

The Contamination from glyphosate had a significant effect on surface change (pH, EC, Mineral, SOM, and CEC) and nutrient (SOC, total N, and available P) of Inceptisols ameliorated with 40 t ha^{-1} biochar. The application of bamboo biochar had a significant and better effect than young coconut waste and rice husk biochar on Inceptisols on increasing pH (0.70); SOM (7.07%); 0.12% N; 5.14 ppm P_2O_5, compared to control.

Residual glyphosate concentrations on Inceptisol ameliorated with biochar were 0.08 (rice husk biochar) > 0.05 (young coconut waste biochar) > 0.04 (Inceptisol) > 0.03 mg kg^{-1} (bamboo biochar) compared to the initial glyphosate concentration of 100 mg l^{-1}, respectively. The correlation between residues on glyphosate (RG) had a significant interaction with chemical properties of Inceptisols amended with biochar (rice husk, young coconut waste, and bamboo) with a linear equation, namely exchangeable Ca (r = 0.611*); CEC (r = 0.593*); available P (r = 0.590*) and total N (r = 0.570*).

References

[1] C. Benbrook M, "Trends in glyphosate herbicide use in the United States and globally," *Environ. Sci. Eur.*, vol. 28, no. 3, p. 15, 2016, doi: 10.1186/s12302-016-0070-0.

[2] R. Mesnage, C. Benbrook, and M. N. Antoniou, "Insight into the confusion over surfactant co-formulants in glyphosate-based herbicides," *Food Chem. Toxicol.*, vol. 128, no. March, pp. 137–145, 2019, doi 10.1016/j.fct.2019.03.053.

[3] K. Gandhi *et al.*, "Exposure risk and environmental impacts of glyphosate : Highlights on the toxicity of herbicide co-formulants," *Environ. Challenges*, vol. 4, no. March, p. 100149, 2021, doi: 10.1016/j.envc.2021.100149.

[4] E. Tzanetou and H. Karasali, "Glyphosate Residues in Soil and Air: An Integrated Review," in *Pests, Weeds and Diseases in Agricultural Crop and Animal Husbandry Production, Intech Open*, 2020, p. 33.

[5] A. M. Muskus, M. Krauss, A. Miltner, U. Hamer, and K. M. Nowak, "Science of the Total Environment Effect of temperature, pH and total organic carbon variations on microbial turnover of 13 C 15 3 N-glyphosate in agricultural soil," *Sci. Total Environ.*, vol. 658, pp. 697–707, 2019, doi: 10.1016/j.scitotenv.2018.12.195.

[6] M. Yunus and N. Adeela, "Environmental Fate and Degradation of Glyphosate in Soil," *Pertanika J. Sch. Res. Rev.*, vol. 4, no. 1, pp. 102–116, 2018.

[7] Herviyanti, A. Maulana, A. L. Lita, T. B. Prasetyo, M. Monikasari, and R. Ryswaldi, "Characteristics of inceptisol ameliorated with rice husk biochar to glyphosate adsorption," *Sains Tanah*, vol. 19, no. 2, pp. 230–240, 2022, doi: 10.20961/stjssa.v19i2.61614.

[8] M. Sieradzka, C. Kirczuk, I. Kalemba-rec, A. Mlonka-mędrala, and A. Magdziarz, "Pyrolysis of Biomass Wastes into Carbon Materials," *Energies*, vol. 15, no. 5, pp. 1–12, 2022, doi: 10.3390/en15051941.

[9] A. Mutolib, A. Rahmat, E. Triwisesa, H. Hidayat, H. Hariadi, and K. Kurniawan, "Biochar from Agricultural Waste for Soil Amendment Candidate under Different Pyrolysis Temperatures," *Ejournal.Upi.Edu*, vol. 8, no. 2, pp. 243–258, 2023, [Online]. Available: https://ejournal.upi.edu/index.php/ijost/article/view/55193.

[10] K. E. Hall, K. A. Spokas, B. Gamiz, L. Cox, S. K. Papiernik, and W. C. Koskinen, "Glyphosate sorption/desorption on biochars – interactions of physical and chemical processes," *Pest Manag. Sci.*, vol. 74, no. 5, pp. 1206–1212, 2018, doi: 10.1002/ps.4530.

[11] A. Qureashi *et al.*, "Electrochemical analysis of glyphosate using porous biochar surface corrosive nZVI nanoparticles," *Nanoscale Adv.*, vol. 5, no. 3, pp. 742–755, 2023, doi: 10.1039/d2na00610c.

[12] Herviyanti, A. Maulana, A. L. Lita, T. B. Prasetyo, and R. Ryswaldi, "Characteristics of biochar methods from bamboo as ameliorant," in *IOP Conference Series: Earth and Environmental Science*, 2022, vol. 959, no. 1, p. 6, doi: 10.1088/1755-1315/959/1/012036.

[13] A. Maulana, Herviyanti, T. B. Prasetyo, M. Harianti, and A. L. Lita, "Effect of Pyrolysis Methods on Characteristics of Biochar from Young Coconut Waste as Ameliorant," in *IOP Conference Series: Earth and Environmental Science*, 2022, vol. 959, no. 1, p. 8, doi: 10.1088/1755-1315/959/1/012035.

[14] A. L. Lita, A. Maulana, Yulnafatmawita, Gusmini, Herviyanti, and R. Ryswaldi, "Characteristics Biochar from Young Coconut Waste based on Particle Size as Ameliorant," in *IOP Conference Series: Earth and Environmental Science*, 2022, vol. 959, no. 1, p. 6, doi: 10.1088/1755-1315/959/1/012034.

[15] Eviati and Sulaeman, *Petunjuk Teknis : Analisis Kimia Tanah, Tanaman, Air dan Pupuk*, 2nd ed., vol. 148. Bogor, Jawa Barat: BALAI PENELITIAN TANAH, 2012.

[16] G. Islas, J. A. Rodriguez, L. H. Mendoza-Huizar, F. Pérez-Moreno, and E. G. Carrillo, "Determination of glyphosate and aminomethylphosphonic acid in soils by HPLC with pre-column derivatization using 1,2-naphthoquinone-4-sulfonate," *J. Liq. Chromatogr. Relat. Technol.*, vol. 37, no. 9, pp. 1298–1309, 2014, doi: 10.1080/10826076.2013.789801.

[17] B. Singh, M. Camps-Arbestain, and J. Lehmann, *Biochar : A Guide To Analytical Methods*. Clayton South VIC 3169 Australia: CSIRO, 2017.

[18] R. L. Glass, "Liquid Chromatographic Determination of Glyphosate in Fortified Soil and Water Samples," *J. Agric. Food Chem.*, vol. 31, no. 2, pp. 280–282, 1983, doi: 10.1021/jf00116a023.

[19] M. I. G. Bi, A. J. Yapo, A. Dembele, A. S. Ello, and A. Trokourey, "Determination of glyphosate by high-performance liquid chromatography (HPLC) without prior extraction," *Int. J. Biol. Chem. Sci.*, vol. 5, no. 1, pp. 314–320, 2011.

[20] L. Sun *et al.*, "Determination of glyphosate in soil/sludge by high-performance liquid chromatography," *J. Chromatogr. A*, vol. 1502, pp. 8–13, 2017, doi: 10.1016/j.chroma.2017.04.018.

[21] M. Schweizer, K. Brilisauer, R. Triebskorn, K. Forchhammer, and H. R. Köhler, "How glyphosate and its associated acidity affect early development in zebrafish (Danio rerio)," *PeerJ*, vol. 2019, no. 6,

pp. 1–25, 2019, doi: 10.7717/peerj.7094.

[22] H. Liu *et al.*, "Interactive effects of microplastics and glyphosate on the dynamics of soil dissolved organic matter in a Chinese loess soil," *Catena*, vol. 182, no. May 2018, p. 104177, 2019, doi 10.1016/j.catena.2019.104177.

[23] S. Maccario, M. Lucotte, M. Moingt, É. Samson-Brais, É. Smedbol, and M. Labrecque, "Impact of Soil Characteristics and Weed Management Practices on Glyphosate and AMPA Persistence in Field Crops Soils from the St. Lawrence Lowlands (Quebec, Canada)," *MDPI - Agron.*, vol. 12, no. 5, p. 15, 2022, doi: 10.3390/agronomy12050992.

[24] M. Wang *et al.*, "Montmorillonites Can Tightly Bind Glyphosate and Paraquat Reducing Toxin Exposures and Toxicity," *ACS Omega*, vol. 4, no. 18, pp. 17702–17713, 2019, doi: 10.1021/acsomega.9b02051.

[25] Z. W. Windom, M. Datta, M. M. Huda, M. A. Sabuj, and N. Rai, "Understanding speciation and solvation of glyphosate from first principles simulations," *J. Mol. Liq.*, vol. 365, p. 120154, 2022, doi: 10.1016/j.molliq.2022.120154.

[26] C. Soares, P. Mateus, F. Fidalgo, and R. Pereira, "Modulation of the non-target phytotoxicity of glyphosate by soil organic matter in tomato (Solanum lycopersicum L.) plants," *Sci. Hortic. (Amsterdam).*, vol. 310, no. March 2022, p. 111773, 2023, doi: 10.1016/j.scienta.2022.111773.

[27] X. Wen *et al.*, "Surface charge properties of variable charge soils influenced by environmental factors," *Appl. Clay Sci.*, vol. 189, no. 808, p. 105522, 2020, doi 10.1016/j.clay.2020.105522.

[28] X. Zhang, S. Li, F. Guo, S. Chao, C. Liu, and G. Zhu, "Experimental Study on Friction Characteristics of Eccentric Cam–Tappet Pairs," *Neiranji Gongcheng/Chinese Intern. Combust. Engine Eng.*, vol. 43, no. 5, pp. 84–90, 2022, doi 10.13949/j.cnki.nrjgc.2022.05.012.

[29] S. O. Duke, J. Lydon, W. C. Koskinen, T. B. Moorman, R. L. Chaney, and R. Hammerschmidt, "Correction to Glyphosate Effects on Plant Mineral Nutrition, Crop Rhizosphere Microbiota, and Plant Disease in Glyphosate-Resistant Crops," *J. Agric. Food Chem.*, vol. 61, no. 51, pp. 12745–12745, 2012, doi 10.1021/jf405497g.

[30] N. Rampazzo, G. R. Todorovic, A. Mentler, and W. E. H. Blum, "Adsorption of glyphosate and aminomethylphosphonic acid in soils," *Int. Agrophys*, vol. 27, pp. 203–209, 2013, doi: 10.2478/v10247-012-0086-7.

[31] M. D. F. Souza *et al.*, "Adsorption mechanisms of atrazine isolated and mixed with glyphosate formulations in soil," *PLoS One*, p. 15, 2020, doi: 10.1371/journal.pone.0242350.

[32] R. Schmidt *et al.*, "Glyphosate Effects on Earthworms: Active Ingredients vs. Commercial Herbicides at Different Temperature and Soil Organic Matter Levels," *Agrochemicals*, vol. 2, no. 1, pp. 1–16, 2022, doi: 10.3390/agrochemicals2010001.

[33] S. Singh, V. Kumar, J. Pal, K. Gill, and S. Datta, "Herbicide Glyphosate: Toxicity and Microbial Degradation," *Int. J. Environ. Res. Public Health*, vol. 17, no. 7519, p. 18, 2020.

[34] R. C. Pereira, C. S. Antonio, F. Ivashita, A. Paesano, and D. A. M. Zaia, "Interaction between glyphosate and montmorillonite in the presence of artificial seawater," *Heliyon*, vol. 6, p. 10, 2020, doi: 10.1016/j.heliyon.2020.e03532.

[35] M. Mertens, S. Höss, G. Neumann, J. Afzal, and W. Reichenbecher, "Glyphosate, a chelating agent—relevant for ecological risk assessment," *Real-Time Imaging VII*, vol. 25, pp. 5298–5317, 2018, doi: 10.1117/12.477503.

[36] R. Kanssery, B. Gairhe, D. Kadyampakeni, O. Batuman, and F. Alferez, "Glyphosate: It's Environmental Persistence and Impact on Crop Health and Nutrition," *MDPI - Plant*, no. Table 1, pp. 1–11, 2019.

Acknowledgments

Thanks to Andalas University for funding this research with No. T/7/UN16.19/Pangan-PDU-KRP1GB-UNAND/2023.

Plants Suitability for Landslides Mitigation in the Giritengah Catchment Area, Indonesia

Satwika Indri Masrianti[1], *Ambar* Kusumandari[2*], *Hatma* Suryatmojo[2]

[1]Graduate student at the Faculty of Forestry, Universitas Gadjah Mada, Jalan Agro No. 1, Bulaksumur, Yogyakarta 55281
[2]Senior lecturer at the Faculty of Forestry Universitas Gadjah Mada, Jalan Agro No. 1, Bulaksumur, Yogyakarta 55281

Abstract. Changes in land use caused by human dependence on land to meet food, clothing, and settlement needs are in line with population growth. This triggers the destruction of watersheds and results in an increased of disasters such as landslides that lead to decrease land productivity and less water throughout the year. This study aims to determine the plant suitability for both annual plant species (forestry plants), and seasonal crops (agricultural crops) in the Giritengah catchment area. In addition, this study also aims to provide the proposed species that are suitable and at the same time allow for land improvement efforts. The results showed that the suitability classes for forestry plant species (teak, sengon, and mahogany) and annual plant species (chili, banana, and cassava) in Giritengah catchment were unsuitable/incompatible (N) with erosion hazard level as the heaviest limiting factor. There are two variable levels of erosion hazard, which are one of the reasons for the unsuitability of these plant species, namely the slope, which tends to be steep and the shallow soil solumn. To overcome this, it is necessary to apply soil and water conservation techniques both technically and vegetative.

* Corresponding author: ambar_kusumandari@ugm.ac.id

1 Introduction

In general, land use change usually is not followed by land degradation management techniques. Changes in land use are the result of human dependence on land to meet food, clothing, and shelter needs in line with population growth. The lack of balance between human activities in land use causes the watershed balance disrupted. This triggers the destruction of watersheds (DAS) and results in an increase in disasters such as landslides.

Land suitability is the suitability of a particular land type for a particular use. The process of land suitability classification is the assessment and grouping of certain areas of land in terms of their suitability for a specified use (FAO Framework, 1976). The prerequisite for land use planning is land suitability assessment (Akinci et al., 2013). Ahmadi et al. (2017) concluded that the findings of this study clearly show the suitability of land for industrial development, agriculture, tourism land use and can be the basis for planning the future development of the island based on ecological sustainability.

The topography of the Giritengah catchment is a hilly to mountainous area with relatively steep slopes. According to Paimin et al., (2009) landslides can occur under conditions of steep slopes, where there is a watertight sliding surface beneath the soil mass, and the soil mass became saturated with water. So that the condition of the land in the Giritengah catchment has the potential for landslides.

Tukiye, et.al, (2023) states that the impact of land use, especially agricultural productivity and climate change, requires a better solution for handling the movement of nutrients or nutrients and sediment in a land. Therefore, proper land use planning, such as afforestation, and water management strategies are climate resilience, such as water harvesting, will be very important to manage the risks associated with climate change and rapid land use change (Mengistu, et.al. 2023).

According to Nugroho et al. (2022) the landslide susceptibility factor can be analysed from the physical condition and management of the area concerned. Physical conditions include rainfall conditions, slopes, and geological conditions such as rocks or whether there are faults or faults. While management conditions include the type of land use or land use, settlements, and infrastructure. According to Aldrian et al., (2009) the form of handling landslides can be in the form of mitigation actions and early warning. One form of mitigation is by means of an ecological approach (vegetative/bioengineering). Mitigation by vegetative means is the use of plants to prevent landslides. Bioengineering has the same meaning but it combined with engineering techniques. The role of vegetation is to stabilize slopes (Riyanto, 2016). Plants with their typologies, namely crowns and roots, have an important role in preventing or eliminating landslides. We expect the vegetation with its wide distribution, diverse structure and composition to be able to provide great benefits and play a role in interception, evapotranspiration, infiltration, soil moisture, and others.

Selection of plant species is an important key in the success of vegetative control and handling of landslides. The selection of species based on the height where the tree grows. Selection of plant species needs to consider both the short and long term (Munawaroh, 2008). One of the weaknesses of agroforestry practices in various regions in Indonesia is that most of them are not land suitability based analysis, resulting in non-optimal production (Butarbutar, et al., 2017). So that the selection of species does not only pay attention to one ecological side, but also the economic side for the community.

Land evaluation is the process of estimating land suitability classes and land potency uses, both for agriculture and non-agriculture (Djaenudin et al., 2000 in Muhamad Yusuf Hidayat, 2006). Land evaluation is a process of assessing land resources to achieve for a specific purpose using a tested approach or method.

2 Method

2.1 Place and time of research

The location of this research was in the Giritengah Water Catchment Area (DTA), which is administratively located entirely in Giritengah Village, Borobudur District, and Magelang Regency in April - June 2022.

2.2 Tools and Research Material

The tools and materials used in the research include survey purposes, observation of land expanses and plant observation, and laboratory analysis. Tools and materials used for survey activities include Global Positioning System (GPS), Clinometer, Digital Camera, Pedological Hammer, Dodos, Pacul, Shovel, Boardlist, Plastic Clip, Meter, Hammer for Sample Ring, Sample Ring, Rope, Soil Drill, and Munsell Soil Color Chart (CC).

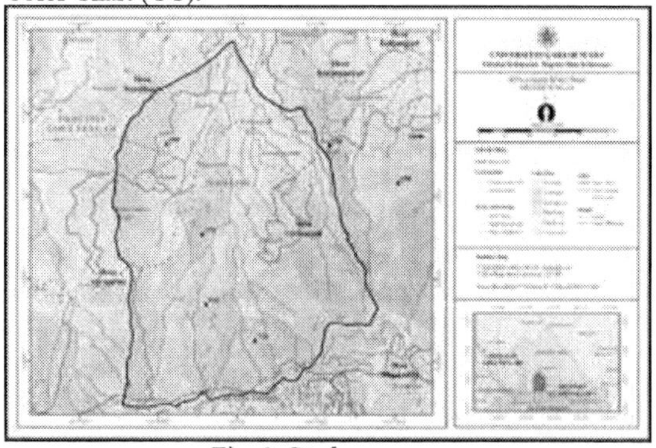

Fig. 1. Study area

2.3. Observation and Data Collection

Primary data obtained from survey data collection including soil sampling, data collection on plant species, interviews with local communities, actual land use, and laboratory results. Primary data collected during the survey included data and criteria for matching in land suitability assessments.

The secondary data collected is data on land maps, climate data and land use maps collected from the agencies holding each data guardian. We compiled the land suitability and table assessment for plants from various literatures.

2.4. Processing and analysis of data

The data analysis method used in this study utilizes the concept of evaluation and land suitability to assess the type of vegetation that is suitable for dealing with landslides that often occur in the study area. To analyse land suitability classification by referring to FAO framework (1976). The structure of land suitability classification distinguished according to its level, namely the level of Order, Class, Subclass, and Unit. The order itself is the state of global land suitability. At this level, land suitability is distinguished between land that is classified as suitable (S = Suitable) and land that is not suitable (N = Not Suitable).

The arrangement of land characteristics to evaluate suitability in this study grouped into three main factors to be able to present in detail the results of the formation of soil map units. The main factors in supporting the research process are topography, soil, and climate. After the data and information are obtained, the activities carried out are analysing and evaluating the land using different approaches such as multiplication of parameters, parameter sum systems, and matching systems between land quality and land characteristics with plant growth requirements.

Explanation Subclass is the state of the level in the land suitability class. Land suitability classes are divided into subclasses based on land quality and characteristics (soil properties and other physical environment) which are the heaviest limiting factors, for example Subclass S3rc, according to marginal suitability with limiting rooting conditions (rc = rooting condition). The level in the land suitability is in subclass and additional characteristics that affect its management based.

3 Result and discussion

3.1 Land characteristics

Land characteristics are a combination of several characteristics of the land and its environment. Land characteristics are described in each soil map unit (SPT) of the soil map, which includes the shape of the area or slope, soil drainage, soil depth, soil texture, soil pH, clay CEC, salinity, pyrite content, flood inundation and surface outcrops (rock outcrops)). The climate consists of rainfall and the number of dry months. The results of the arrangement of land characteristics are 27 land units Figure 3).

In accordance with the modification of Arsyad (2010) that the characteristics of the identifying land in the classification of land capability are inhibiting factors that are permanent or difficult to change. Soil texture in the Giritengah catchment area classified as fine with a slightly neutral pH, medium CEC and high KB in the karts hills and rather fine with slightly acidic pH, high CEC, and low KB in the volcanic hills. The fine-textured class consists of dusty clays and clays.

While the rather fine class includes clay loam, sandy clay loam. Some land units are also somewhat rough with a clay texture. In general, in all land units, the root media is well drained. An indication of good drainage is that the soil has good air circulation. The entire soil profile from the top to the bottom layer is a light color that has no manure.

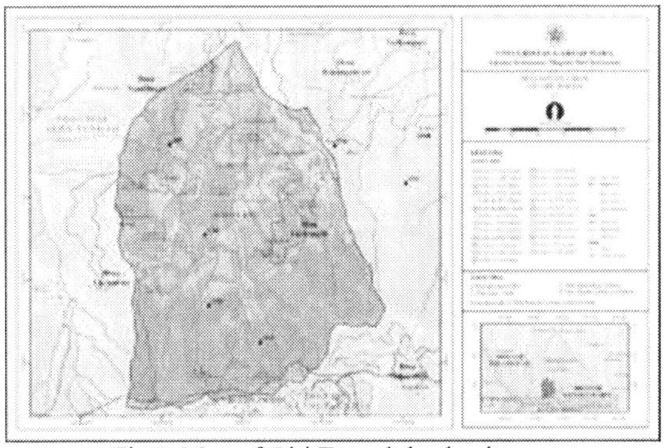

Fig 2. Map of Giri Tengah land unit.

3.2 Suitability of Forest Plant Type

The results of the assessment are in the form of land suitability classes and subclasses of the assessed plants determined by the heaviest limiting factor. These limiting factors may consist of one or more depending on the characteristics of the land. The types of forestry plants that dominate and spread over 27 land units are sengon, mahogany and teak. For all land units the three types are not suitable (N2) where for sengon and mahogany the limiting factors are rooting media, nutrient availability, and land slope. As for teak, the limiting factor is the availability of water.

The results of the matching analysis of sengon type plants in the Giritengah DTA show that for land units 1, 2, 3, 4, 9, 10, 11, 12, 13, 14, 15, 16, 17, 18, and 26 are N2 with a limiting factor heaviest, namely rooting media (r), nutrient availability (n) and erosion hazard level (e). Whereas for land units 5,6,7,8,19,20,21,22,23,24,25 and 27 the land suitability class is N2, namely the root medium (r), and nutrient availability (e).

The results of the suitability assessment of mahogany plants for all land units, the land suitability value is N2 with the limiting factors of root media (r), nutrient availability (n), and erosion hazard level (e). In terms of rooting media parameters, the main obstacle is the effective depth character (< 50 cm). In general, mahogany plants grow well and optimally in fertile soil. Based on the rainfall characteristics, the suitability class is S1. In a sense, even though the rainfall is only 2393 mm per year, it turns out that mahogany is able to survive in arid land. The soil texture dominated by clayey clay so that it is included in the suitability class S1 and S2. There has never been a pool of water, so actually mahogany is still suitable for living on the land.

Teak growing in the Giritengah catchment has land suitability classes S3 to N2 with the heaviest limiting factor being the availability of slope water with a very severe erosion hazard. However, sometimes the limitations on water availability can change because every year high or low rainfall can change according to climate phenomena.

3.3. Land Suitability Analysis for Seasonal Plants

Chili plants in the Giritengah catchment are in the land suitability class S3 to N2 with the heaviest limiting factors being root conditions and the level of erosion hazard. Of the 27 land units, only land unit number 24 fits the S3 class (according to marginal). This land has large boundaries, but conditions improved to reduce the risk of erosion that results in loss of top soil so that the nutrients in the top soil are depleted thereby reducing production or profits.

Suitability class of banana plant species is not suitable (N1 and N2) with the weighting factors being the effective depth/solumn (r), and the level of erosion hazard (e). The suitability of sweet potato species was also in the non-conforming suitability class (N1 and N2). The weighting factors in the suitability class for sweet potato species in the Giritengah watershed are the effective depth (solumn) and the level of erosion hazard.

3.4. Constraints and Efforts to Develop Plant Cultivation in DTA Giritengah

Land use planning requires evaluation of land suitability to reduce the risk of failure and increase the economic value of the land. Land suitability needs to pay attention to the characteristics of plants because plants have different growing requirements and different characteristics to produce optimally (Saputra et al., 2021). From the results of the matching analysis on annual plant species (sengon, mahogany, and teak) and seasonal crops (chili, banana, sweet potato) there are various obstacles. To overcome the various limiting factors, it is necessary to do handling efforts. For the chat for the development of forestry plant cultivation and food crops served in the following table.

Table 3.1 Constraints solution for plants

Nu	Plants	Constraints	Land unit	Proposed penanganan
1	Sengon	- Root media system, - Nutrient availability, - Slopes or erosion hazards	1,2,3,4,9,10,11,12,13,14,15,16,17,18,26	- Intercropping planting pattern with crops - Implement the use of fertilizers, pesticides, herbicides, fungicides in doses that are balanced with nutrient losses - Planting in strips according to contour lines and/or field strips if the slope is irregular.
	Sengon	- Root media system, - Nutrient availability	5,6,7,8,19,20,21,22,23,24,25,27	- Intercropping planting patterns with crops - Implement the use of fertilizers, pesticides, herbicides, fungicides in doses that are balanced with nutrient losses.
2	Mahagony	- Root media system, - Nutrient availability, - Slopes or erosion hazards	1-27	- cropping in buffer strips (buffer strip cropping) consisting of grasses or Leguminosae made between staple crops - ground cover planting - application of compost, manure, and other organic materials. - Inter cropping with short rotation crops that have similar growing characteristics.
3	Teak	- Water availability - Slopes - Erosion hazard	1-27	- Inter-cropping with short rotationof crops that have similar growing characteristics.
4	Chilli	- Water availability - Root media system - Erosion hazard	1-27	- Inter-cropping with short rotation of crops that have similar growing characteristics.
5	Banana	- Root media system - slopes - Erosion hazard		- Addition of compost, manure, and other organic materials. - Making terraces or planting parallel to the contour
6	Sweet potatoes	- Water availability - Root media system - Slopes - Erosion hazard		Making ditches or waterways and adding organic matter

The matching results for woody plants and annuals show that the dominant limiting factor for suitability is the root media, especially the thickness of the solumn or the effective depth of the soil and the level of erosion hazard. The soil solumn in the Giritengah catchment is thin, if the level of erosion hazard tends to be severe, the loss of fertile soil in the topmost layer will also be higher. Naturally, the displacement of surface material is a result of the physical forces of water or wind or due to agricultural activities (Nugroho, et.al, 2022). The suitability class of each plant is largely determined by the limiting factor of the level of erosion hazard. Based on the Universal Soil Loss Equation (USLE) formula (Wischmeier and Smith, 1978) the parameters used to obtain the amount of erosion per year are the rain erosivity factor (from rain intensity), soil erodibility factor, slope slope factor, crop factor (soil use) and soil conservation engineering factors. Land with a relatively flat level of erosion hazard is only moderate, so erosion reduced by constructing bench terraces to reduce the volume of surface runoff by up to 90% compared to land without bench terraces (Nugroho, et al, 2022).

3.5. Solutions

Landslide is a phenomenon of material movement down the slope due to the influence of slope destabilization due to the force of gravity and contact with water (Noviyanto, 2020). Landslides cause significant soil loss, deposition of landslide sedimentary material, changes in the morphological and physicochemical characteristics of the soil. Noviyanto also mentioned that landslides considered as one of the factors causing land damage, because they reduce soil fertility. Sparling et.al. (2003) explained that the decrease in c-organic soil and nutrients in landslide-affected material is due to mixing of soil material with parent material that is poor in c-organic and nutrients, presence of gravity and contact with water. At the research location in 2021 there are 9 location points (figure 3) of landslide events (BPBD, 2021).

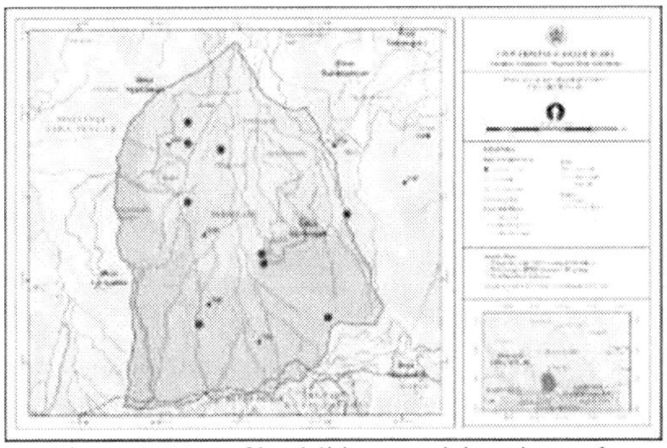

Figure 3. Map of landslide potential zoning and landslide occurrence points

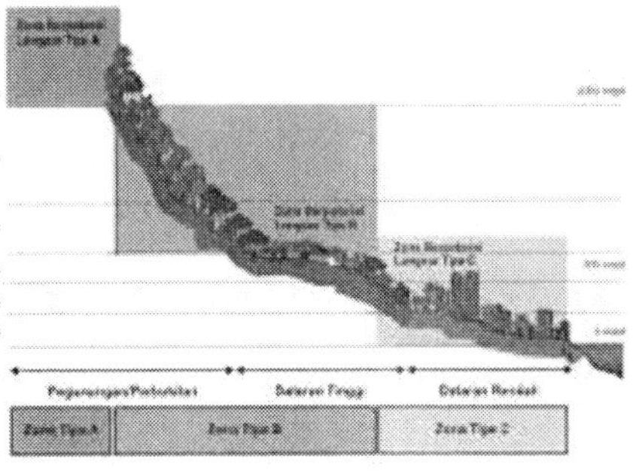

Figure 4. Zones of the watershed.

Based on Public Works Regulation No. 22 of 2007 there are 3 zones with the potential for landslides as shown in Figure 4. These zones are divided based on height above sea level and slope. Based on these zones, DTA Giritengah is in 2 (two zones), namely zones A and B. In selecting plant species based on Riyanto (2016), Zone A is at an altitude of 0-500 meters above sea level and Zone B at the research site is included in Zone B1. i.e. altitude 500-100 m above sea level. Then the proposed types of plants that matched are cloves, durian and *rambutan*.

Around the 90s the local government provided assistance with other plants, namely clove plants. For example, matching plants was carried out on one of the land units numbered 3, 10, 19, and 23. Apart from these types, the community also tried to plant durian plants in the hope of obtaining additional yields as a commodity. Types of durian and *rambutan* plants also have an unsuitable land suitability class (N). Of the three types, the slope zone is the heaviest limiting factor in developing plant species in the Giritengah catchment area.

The principle of landslide prevention is to prevent water from being concentrated in the sliding plane, binding the soil mass so it does not slide by seeping water into the deeper layers of the soil from the waterproof layer (gliding area) (Riyanto, 2016). Prevention of landslides and erosion done with plant-based interventions to strengthen slopes. Vegetation plays an important role in the prevention and mitigation of natural disasters and in the provision of various socio-ecological benefits (Olla uri, 2021).
Meanwhile, for mechanical handling on steep slopes, they plant clove, durian and *rambutan* planting with individual terraces as a solution. Individual terraces (figure 6) is a soil conservation technique that is suitable for staple crops planting.

4 Conclussion

1. In the Giritengah catchment area, Sengon, Mahogany and Teak plants, which are annual woody plants, are included in the land suitability class N (not suitable) with the heaviest limiting factor in the level of erosion hazard. Likewise for annual crops with the types of chili, banana and cassava the land suitability class is N (not suitable) with the heaviest limiting factor on the level of erosion hazard.
2. Variables that add to the level of erosion hazard are shallow solumn and steep slopes. To overcome this, several alternative and required soil and water conservation techniques that are appropriate to the field of tillage/agricultural land.
3. The types of plants proposed, namely clove, durian and *rambutan* according to land suitability class, are not suitable, but to anticipate that they can still grow these commodities, the efforts made are planting with agroforestry patterns accompanied by individual terrace techniques.

This research supported by the RTA UGM funding year 2022. Thank you very to UGM for the support and we hope this rearch can benefits in developing the vegetative methods for reduuucing the landslides.

References

1. Ahmadi, A., Mosammam, H.M. & Mirzael, N. 2017. Land suitability evaluation to determine the appropriate areas of development: A Case Study of Hormuz Island. Open Journal of Ecology, 7: 518-534.
2. Asdak, C., 2022. Hidrologi dan Pengelolaan Daerah Aliran Sungai. Gadjah Mada University Press, Jakarta.
3. Kusumandari, A. (2019). Crops and Trees for Function to The Existing Land Use of Blongkeng Sub Watershed, Java, Indonesia. Malaysian Applied Biology, 48(5), 113–119. Retrieved from https://j.mabjournal.com/ index.php/mab/article/view/1593.
4. Bailey, K.M., McCleery, R.A., Binford, M.W. & Zweig, C. 2016. Land-cover change within and around protected areas in a biodiversity hotspot. Journal of Land Use Science, 11(2): 154-176.
5. Bi, Yixian., et al., 2019. Interspecific interactions contribute to higher forage yield and are affected by phosphorus application in a fully-mixed perennial legume and grass intercropping system. Field Crops Research 244 (2019) 107636
6. Djaenuddin , D., Marwan, H., Subagjo, H., & A. Hidayat (2011). Petunjuk Teknis Evaluasi Lahan Untuk Komoditas Pertanian. Bogor: Balai Besar Litbang Sumberdaya Lahan Pertanian, Badan Litbang Pertanian.
7. FAO and Agricultural Organisation of the United Nations. 1976. Food and Agriculture Organi-sation of United Nations, Rome, Italy.
8. Harahap, S. H., et al. 2019. Evaluasi kesesuaian lahan tanaman pisang (musa acuminata colla.) di Kecamatan Salak Kabupaten Pakpak Bharat. jurnal.polbangtan-bogor.ac.id
9. Hardjowigeno, S. & Widiatmaka. (2011). Evaluasi Kesesuaian Lahan dan Perencanaan Tataguna Lahan. Yogyakarta: Gajah Mada University Press.
10. Hidayat, M.Y., Fauzi, R. and Siregar, C.A., 2021. Kesesuaian lahan beberapa jenis tanaman untuk perbaikan kualitas lahan di Hutan Lindung Sekaroh. Jurnal Penelitian Hutan dan Konservasi Alam, 18(1), pp.13-27.
11. Hopkins, L.D. 1977. Methods of generating land suitability maps: A comparative evaluation. Journal of American Institute of Planners, 43(4): 386-400.
12. Komaruddin, N. 2008. Penilaian tingkat bahaya erosi di sub DAS Cileunyi, Bogor (Evaluation of Erosion Hazard Index of Cileunyi Sub watershed, Bogor). Jurnal Agrikultura, 19(1): 173-178.
13. Kusumandari, A. 2016. Suitability of Land Area Function to the Existing Land Use of Blongkeng Sub watershed. Java. Indonesia. Proceeding of the International Seminar on Food Security. Vol 1: 230-235. Bandar Lampung.
14. Mendoza (2015). A GIS-based multi criteria approaches to land use suitability assessment and allocation. 7 pages. https://pdfs.semantic scholar.org
15. Mengistu, A.G., Woldesenbet, T.A., Dile, T.D., Bayabil, H.K., Tefera & Tefera, G.W. 2023. Modeling impacts of projected land use and climate changes on the water balance in the Baro basin, Ethiopia. Heliyon 9 (2023) e13965.
16. Nganji, M.U., Simanjuntak, B. H., & Suprihati. 2018. Evaluasi Kesesuaian Lahan Komoditas Pangan Utama di Kecamatan Umbu Ratu Nggay Barat Kabupaten Sumba Tengah. Agritech, 3 8 (2) 2018, 172-177.
17. Noviyanto, Amir,. 2020. Tesis : Karakteristik Morfologi Tanah dan Lapisan Klei Sensitif Pada Tiga Longsor Aktif di Sisi Selatan Gunung Api Sumbing, Jawa Tengah. Program Pascasarjana Fakultas Pertanian, Universitas Gadjah Mada Yogyakarta.
18. Nugroho, H.Y.S.H., Dewi, I. N., Sallata, M. K., 2022. Pengelolaan Daerah Aliran Sungai (DAS) & Konservasi Tanah dan Air. Penerbit Andi, Yogyakarta
19. Nugraha, S. 2009. Kesesuaian fungsi kawasan dengan pemanfaatan lahan di Daerah Aliran Sungai Samin (The suitability of land function to the usage of land in Samin watershed). MIIPS: 8(2): 67-76.
20. Krisanti, Oktaviana Karmia. "Evaluasi Kesesuaian Lahan untuk Tanaman Durian, Rambutan, dan Mangga di Desa Cukilan, Kecamatan Suruh, Kabupaten Semarang." PhD diss., 2023.
21. Ollauri, Alejandro Gonzalez, at al. The 'Rocket Framework': A Novel Framework to Define Key Performance Indicators for Nature-based Solutions Against Shallow Landslides and Erosion. Frontiers in Earth Science | www.frontiersin.org. Volume 9. 04 November 2021. doi: 10.3389/feart.2021.676059.
22. Pertami, R.R.D, et al. 2022. Optimasi Penggunaan Lahan Berdasarkan Kelas Kesesuaian Lahan Untuk Pengembangan Tanaman Cabai Merah di Kabupaten Jember. Jurnal Tanah dan Sumberdaya Lahan Vol 9 No 1: 163-170, 2022.
23. Ritung, S., Kusumo Nugroho, Anny Mulyani & Erna Suryani. (2011). Petunjuk Teknis Evaluasi Lahan Untuk Komoditas Pertanian (edisi Revisi). Bogor : Balai Besar Penelitian dan Pengembangan Sumberdaya Lahan Pertanian, Badan Penelitian dan Pengembangan Pertanian.
24. Riyanto, Heru Dwi., 2016. Rekayasa Vegetatif Untuk Mengurangi Risiko Longsor. Balai Penelitian dan Pengembangan Teknologi Pengelolaan Daerah Aliran Sungai.
25. Saputra, Muhammad & Adyatma, Sidharta & Arisanty, Deasy. (2021). Evaluasi Kesesuaian Lahan Untuk Tanaman Durian Menggunakan

Metode Matching. Jambura Geoscience Review. 3. 18-31. 10.34312/jgeosrev.v3i1.5652.

26. Sparling, G., D. Ross, N. Trustrum, G. Arnold, A. West, T. Speir, & L. Schipper. 2003. Recovery of topsoil characteristics after landslip erosion in dry hill country of New Zealand, and a test of the space-for-time hypothesis. Soil Bio. & Biochem. 35: 1575 – 1586. doi: 10.1016/j.soilbio.2003.08.002
27. Steiner, E., Tyler, M., Theilacker, J., Roe, D. & Hunter, L. 1984. Whitman County, Washington. Pages 37–44 in Adapting the agricultural land evaluation and site assessment (LESA) system in the Pacific Northwest. Publication No. 26. Western Rural Development Center, Oregon State University, Corvallis, Oregon.
28. Sudomo, et al. 2021. Intercropping Short Rotation Timber Species with Teak: Enabling Smallholder Silviculture Practices. Forests 2021, 12, 1761. https://doi.org/10.3390/f12121761
29. Supriyanto & Sayid., 2022. Kesesuaian Lahan Jenis Tanaman Kehutanan Pada Areal Lahan Terbuka di Kawasan Hutan Pendidikan Gunung Walat (HPGW), Sukabumi, Jawa Barat. Jurnal Silvikultur Tropika Vol. 13 No. 2, Agustus 2022, Hal 103-110.
30. Tan, K.H. 1991. Principles of Soil Chemistry. Dasar- dasar Kimia Tanah (Fundamentals of Soil Chemical) translated by Goenadi, D.H. Gadjah Mada University Press. Yogyakarta.
31. Tikuye, B.G., Gill. L., 2023. Modelling the impacts of changing land use and climate on sediment and nutrient retention in Lake Tana Basin, Upper Blue Nile River Basin, Ethiopia. Ecological Modelling 482 (2023) 110383
32. Veldkamp, E., et al. 2023. Multifunctionality of temperate alley-cropping agroforestry outperfor open cropland and grassland. Communications Earth & Environment | (2023) 4:20 | https://doi.org/10.1038/s43247-023-00680-1 |www.nature.com/comenvA. Mecke, I. Lee, J.R. Baker jr., M.M. Banaszak Holl, B.G. Orr, Eur. Phys. J. E **14**, 7 (2004)

Relationship Between Tree Height-Diameter at Breast Height (DBH) and Crown Diameter-DBH of Fruit Trees in Bogor Botanical Gardens

Indriani Ekasari[1]*and *Rifaangga* Kurnia[2]

[1]Research Center for Plant Conservation, Botanic Gardens, and Forestry, National Research and Innovation Agency (BRIN), Jl. Ir. H. Juanda No 13 Bogor, West Java, Indonesia
[2]Bachelor Degree Program, Agrotechnology Department, Faculty of Agricultural, Universitas Padjadjaran, Jalan Raya Bandung-Sumedang km 21, Sumedang,, West Java, Indonesia

Abstract. Ex-situ conservation of Bogor Botanical Gardens (BBG) represents a collection of predominantly Indonesian species and constitutes an exceptional tropical garden. The gardens' living collections, which include fruit trees, were mostly native to the forest and uncultivated, preserving their original characteristics. The study's aim was to examine the relationship between tree height and breast height diameter and breast height diameter and canopy diameter of fruit trees at BBG. Three individuals from each of the fifteen selected fruit species were measured for their diameter breast height (DBH), crown diameter, and tree height. The highest mean of all parameters was for *Durio zibethinus,* while the lowest mean of all parameters was for *Garcinia mangostana*. Mostly fruit tree species have a positive and strong correlation. The highest and the lowest relationship between tree height and DBH was for *C. burmanii* (R^2=99.88%) and for *C. nucifera* (R^2=1.21%). The highest and the lowest relationship between crown diameter and DBH was for *C. burmanii* with linear regression (R^2=99.94%) and for *A. catechu* with logarithmic regression (R^2=24.06%). There was a positive relationship between DBH and tree height and between crown diameter and DBH for the growth prediction of fifteen fruit trees in BBG.

Keywords: *Tree height, diameter breast height, crown diameter, Bogor Botanical Gardens, correlation.*

1 Introduction

Plants are fundamental for ecosystems and human well-being [1] however human activities cause habitat loss, emergence of new harmful pathogens, environmental pollution, climate crisis, etc. It is about 40% of the world's plants are confronted with the risk of extinction, including plant species in Indonesia. Indonesia, a country with rich biodiversity, has about 40,000 plant species or 15,5% of world plants [2], and needs a strategic method to conserve biodiversity. The loss of plant diversity will cause damage to biological groups including humans, so more effective protection of threatened plants is urgently needed.

In-situ conservation and ex-situ conservation are two essential means for plant diversity conservation. Ex-situ conservation, an important supplementary means of in-situ conservation, can make up for the limitations of in-situ conservation and also is a progressive direction of Indonesia's plant protection work [3]. Ex-situ conservation is to transfer living plants or seeds from the natural habitat to an artificial environment for preservation, so as to prevent plants from being damaged by natural or human factors. Botanical gardens protect plants by introduction and cultivation, and germplasm banks preserve plant seeds in vitro with a low-temperature technique [4].

One of the main tasks of Indonesian botanical gardens is the ex-situ conservation of plant species. Indonesian botanical garden has been conserving at least 24% of Indonesian native species listed in the Global IUCN Red List. Botanical gardens also grow many exotic species of plants, some of which are threatened in their native range [5]. Somehow, exotic species are of great benefit to humankind such as fruit tree species. For example Durian (*Durio zibethinus* L.), Rambutan (*Nephelium lappaceum* L.), Kelapa (*Cocos nucifera* L.), etc.

The Bogor Botanical Gardens (BBG) is not only the most prominent in Indonesia /or in Southeast Asia but represents one of the world's outstanding gardens and features a considerable plant collection of offers 17,000 species representing tropical trees, palm trees, orchids, etc [6]. This botanical garden is the oldest ex-situ conservation area in Southeast Asia, with a total collection of 14,057 specimens belonging to 213 families, 1248 genera, and 3404 species. The plant collections are arranged in a combination of taxonomic

* Corresponding author: indriani.ekasari@brin.go.id; rifaangga20001@mail.unpad.ac.id

and thematic classification patterns. This botanical garden is focused on conserving lowland humid and wet plant species. BBG collections originate from Indonesia and abroad [7]. The botanical garden's living collection is mostly of uncultivated origin (native to the forest) and keeps the original characteristics of the species. It is critical to conduct a study on the growth processes of living collections at the BBG, especially fruit trees. All living and herbarium collections are documented and arranged based on the taxonomic, bioregion, thematic or combination pattern of those patterns for conservation, research, education, tourism and environmental services [9]. The aim of this study was to investigate a relationship between tree height-diameter at breast height and diameter at breast height-crown diameter on fruit trees in the BBG.

2 Methods

2.1 Study Site

The study was conducted at the BBG located in Bogor, West Java, Indonesia (60 35'51.46" S, 1060 47'58.45" E) with an altitude of 230–270 m above sea level (Fig. 1). Bogor has a tropical climate with an average temperature of 25–27.4 °C, average humidity of 80%, and rainfall of 3658 mm/year.

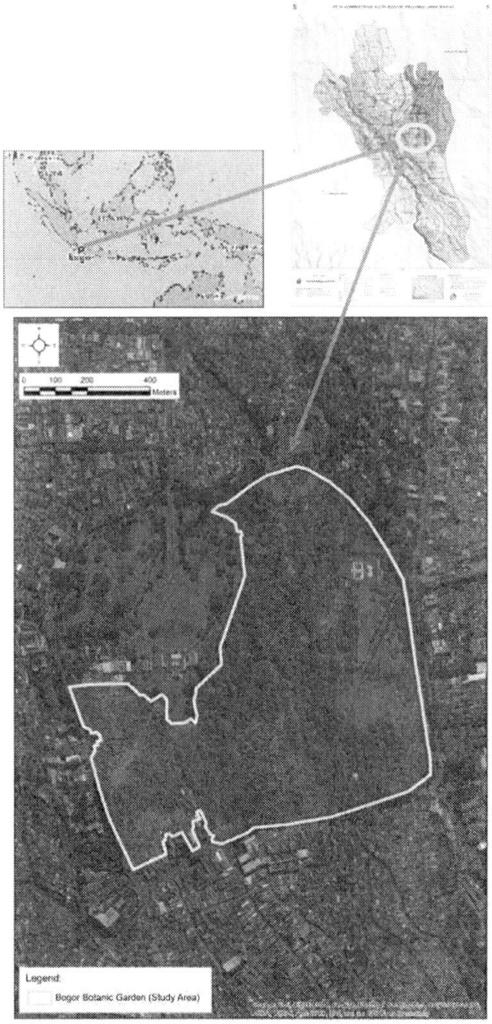

Fig. 1. Research site in Bogor Botanic Gardens. Source [8,9]

2.2 Data sampling

For data collection, the sampling method was used where three trees for each species had the possibility to be selected. After selecting trees, DBH, crown diameter, and total tree height were measured. A total of 45 trees were measured as the sample of the total population (three trees for fifteen fruit species). The DBH, tree height, and crown width were taken as independent and dependent variables for data analysis. The determination of the model of the architect's tree crown model was also observed by Halle and Oldeman [10].

2.3 Data collection

Data were collected from July to August 2023 in the gardens. All data needed for this study were collected by non-destructive measurements. Three persons were involved in height, DBH, and crown diameter measurements. Height (m) was measured by the ruler arch according to the formula for measuring tree height.

Tree diameter at breast height (DBH) was measured by diameter tape (1.3 m from the ground level) with intensive care. The crown diameter (m) was measured by taking the arithmetic average of the horizontal crown diameter on the north-south axis and on the east-west axis measured by measuring tape. The crown shape was also identified using manual instruction.

2.4 Data analysis

Data were compiled, and Pearson's correlations, as well as these regression models, were investigated using Microsoft Excel and SPSS of version 26. The complete dataset provided information for developing regression models between DBH and tree height as well as between DBH and crown diameter. Models were fitted using the enter method in SPSS of version 26.

Table 1. The equation used to develop the model of the correlation between height and crown diameter with diameter at breast height (dbh).

No	Equations	Form
1	$y = b_0 + b_1(dbh)$	Linear
2	$Y = b_0 + b_1(dbh) + b_2(dbh^2)$	Quadratic
3	$y = b_0 \ln(dbh) - b_1$	Logarithmic
4	$y = b_0(dbh)^{b1}$	Power
5	$y = b_0e^{b1(dbh)}$	Exponential

Note : y = tree height or crown diameter (m); x = diameter at breast height (cm); b_0, b_1 = constant. Source : [11].

3 Results and discussion

3.1 Distribution of tree height, diameter at breast height and crown diameter

As a tropical botanical garden, the existence of BBG remains important for its records of local floras. This work is the essential basis for all other studies of plants, whether scientific or technical. The planting management in BBG might have different methods for other commercial or private plantations. BBG's living collection is the result of field expeditions or other countries' donations. Every living collection has a record of the origin, planting location, date of planting in the garden, date of flowering and fruiting, and date of death (if applicable). Collected data of DBH, tree height, and crown diameter for forty-five individuals from fifteen species were analysed in this study. The fifteen species were Durian, Rambutan, Kelapa, Kakao (*Theobroma cacao* L.), Kopi (*Coffea arabica* L.), Pala (*Myristica fragrans* Houtt.), Manggis (*Garcinia mangostana* L.), Kayu Manis (*Cinnamomum burmanii* Ness & N. Tees) Blume, Cengkeh (*Syzygium acuminatisimum* (Blume) DC), Belimbing (*Averrhoa carambola* L.), Jamblang/ Duwet (*Syzygium cumini* (L.) Skeels), Nangka (*Artocarpus heterophyllus* Lam.), Mangga (*Mangifera indica* L.), Cempedak (*Artocarpus integer* (Thunb.) Merr.), and Aren (*Areca catechu* L.). Three individuals were selected for each species and they were in the same treatment management in the garden. The fifteen fruit species were chosen due to their fruit has high potential in the market but limited information about their performance using growth model prediction. Distribution tree height, DBH and crown diameter were varied but they were chosen according to the exact provenance for each species (Table 2).

Table 2. Distribution tree height, diameter at breast height and crown diameter of fifteen fruit species

No	Species	Parameters	Max	Min	Mean
1	*Durio zibethinus*	Tree Height DBH Crown	26 450 20.5	20 280 17.25	23.33 343.33 18.82
2	*Nephelium lappaceum*	Tree Height DBH Crown	13.1 74 10.05	10.04 54 8.1	11.7 66 9.21
3	*Cocos nucifera*	Tree Height DBH Crown	15.5 103 10.4	12.3 95 9.12	13.63 98.66 9.80
4	*Theobroma cacao*	Tree Height DBH Crown	10 24.18 10.4	4.2 13.17 4.75	7.3 17.86 6.83
5	*Cofeea arabica*	Tree Height DBH Crown	16.8 56.04 16.55	6.6 26.33 6.35	5.6 18.68 5.51
6	*Myristica fragrans*	Tree Height DBH Crown	8.1 22 3.89	5.8 12.75 3.59	6.96 17.25 3.76
7	*Garcinia mangostana*	Tree Height DBH Crown	6.6 22 4.75	2.54 10 1.97	4.17 15 3.16
8	*Cinamomum burmanii*	Tree Height DBH Crown	17 134 10.1	7.7 46 6.07	10.8 76 7.42
9	*Syzygium acuminatisimum*	Tree Height DBH Crown	20 400 14.8	5 12 3.14	14.33 237.33 9.86
10	*Averrhoa carambola*	Tree Height DBH Crown	12 128 12	10 61 6.75	11 89.33 8.78
11	*Syzygium cumini*	Tree Height DBH Crown	14.6 123 13.15	5 14.5 4.36	8.86 54.14 7.80
12	*Artocarpus heterophyllus*	Tree Height DBH Crown	11.2 60 7.75	6.6 25 4.23	8.76 46 6.39
13	*Mangifera indica*	Tree Height DBH Crown	11.2 130 12.25	6.2 46 5.62	8.63 84 8.27
14	*Artocarpus integer*	Tree Height DBH Crown	13 130 10.4	12 69 7.62	12.5 100 8.62
15	*Areca catechu*	Tree Height DBH Crown	19.3 60 3.31	4.35 30 2.15	10.71 45 2.71

The highest mean of tree height, DBH and crown diameter was for *D. zibethinus*, while the lowest mean of tree height, DBH and crown diameter was for *G. mangostana* (Table 2). Species *D. zibethinus* is a commercial fruit crop grown in several Southeast Asian countries, including Indonesia [12]. The minimum, maximum, and mean tree height for *D. zibethinus* were recorded at 20m, 26m, and 23.33m, respectively, in this study. In contrast, the minimum, maximum, and mean DBH for *D. zibethinus* recorded were 280cm, 450cm, and 343.33cm, respectively. Besides this, the minimum, maximum, and mean crown diameters of *D. zibethinus* evaluated were 17.25m, 20.50m, and 18.82m, respectively. Species *D. zibethinus* has a high ability to adapt to the BBG in this study. The optimal growing circumstances for *D. zibethinus*'s tree height, DBH, and crown diameter influence the amount of fruit produced [13]. The minimum, maximum, and mean tree heights for *G.mangostana* were recorded at 2.54m, 6.6m, and 4.17m, whereas the minimum, maximum, and mean tree DBH were 10cm, 22cm and 15cm.

The plantings of the BBG were fairly dense, hence some species were frequently shaded by tree species higher above them. Species *T. cacao* is one type of fruit tree that provides shade. In this study, the species *T. cacao* showed limited growth in mean height (7.3m), mean DBH (17.86cm), and mean crown diameter (6.83m). A previous study showed shade trees had a net negative effect on *T. cocoa* tree growth. However, *T. cocoa* yields were not significantly decreased under shade trees. Shade tree traits such as litter quality or tree morphology were also found to have significant effects on aggregate formation and the stabilization of nutrients in different aggregate-size classes, confirming the potential for soil improvements under shade trees [14].

Furthermore, shaded trees affect the photosynthetic process and are the main reason that influences diameter growth in plants. Photosynthetic results in carbohydrates that were stored as leaf replacement, root growth, and tree height. As a result, the difference in growth between these fifteen species of fruit trees can be influenced by both hereditary and environmental variables [15]. The environmental factors such as growth sites for soil mineral content, soil moisture, and sunlight. The internal factors such as the genetic factor for each species, as well as the balance of genetic traits between growth in tree height and DBH.

3.2 Relationship between tree height and diameter at breast height (DBH)

There were fifteen regression statistics in different models developed from tree height, DBH, and crown diameter. Tree height was used as a dependent variable and DBH as an independent variable. The highest relationship between tree height and DBH is for *C. burmanii* (R^2=99.88%), and the lowest was for *C. nucifera* (R^2=1.21%). Mostly fruit tree species have a positive and strong correlation (R^2 above 70%). Only two species (*C. nucifera* and *M. fragrans*) have weak correlation, they were 1.12% and 28.94%, respectively. Therefore, this study concluded that there was a probability of increasing height with increasing DBH but not for *C. nucifera* and *M. fragrans*. The size and growth of tree diameter were strongly correlated to DBH, which could be used for the growth model in trees. Growth equations of trees were indispensable for modelling tree functions as critical components of ecosystem services and directly depend on tree size and growth rate [16].

The plant growth model was commonly to illustrate the relationship between tree height and DBH. They come in various shapes and sizes depending on the species, research location, and other factors [17]. Besides shaded trees and photosynthesis having an effect on plant growth, the other abiotic factors, such as temperature and precipitation, are major climatic elements that generally influence tree height development. The results model of fifteen fruit species varied from linear regression to exponential regression. It illustrates that fifteen fruit species have their own growth characteristics in BBG.

Table 3. Relationship between tree height and DBH for fifteen fruit tree species in Bogor Botanical Gardens

No	Species	Equations	R^2
1	*Durio zibethinus*	$Y = 9.97\ln(X) - 24.69$ (Logarithmic)	70.08%
2	*Nephelium lappaceum*	$Y = 5.94^{e0.01X}$ (Exponential)	86.22%
3	*Cocos nucifera*	$Y = 4.48\ln(X) - 6.95$ (Logarithmic)	1.21%
4	*Theobroma cacao*	$Y = 9.02\ln(X) - 18.44$ (Logarithmic)	83.88%
5	*Cofeea arabica*	$Y = -0.14X + 8.32$ (Linear)	76.87%
6	*Myristica fragrans*	$Y = 5.12e^{0.017X}$ (Exponential)	28.94%
7	*Garcinia mangostana*	$Y = 1.18e^{0.078X}$ (Exponential)	99.87%
8	*Cinamomum burmanii*	$Y = 8.86\ln(X) - 26.42$ (Logarithmic)	99.88%
9	*Syzygium acuminatisimum*	$Y = 4.97e^{0.0038X}$ (Exponential)	91.52%
10	*Averrhoa carambola*	$Y = 8.75e^{0.0025X}$ (Exponential)	92.44%
11	*Syzygium cumini*	$Y = 4.92e^{0.0089X}$ (Exponential)	97.91%
12	*Artocarpus heterophyllus*	$Y = 4.63e^{0.01X}$ (Exponential)	85.00%
13	*Mangifera indica*	$Y = 3.59X - 2.50$ (Linear)	94.17%
14	*Artocarpus integer*	$Y = 1.55\ln(X) + 5.37$ (Logarithmic)	98.64%
15	*Areca catechu*	$Y = 1.88X + 24.82$ (Linear)	93.81%

All fifteen fruit species showed three regression models, they were linear, logarithmic, and exponential regression models. No quadratic and power regression model in this study. There were five fruit tree species *D. zibethinus, C. nucifera, T. cacao, C. burmanii,* and *A. integer* showed a logarithmic regression model in this study. Furthermore, species *N. lappaceum, M. fragrans, G. mangostana, S. acuminatisimum, A. carambola, S. cumini,* and *A. heterophyllus* showed an exponential regression model. The rest of the three fruit tree species illustrated a linear regression model (Table 3). All of these models showed that the growth of DBH as an independent variable affected the growth of tree height positively but for species *C. arabica*. A previous study found a very strong relationship between DBH and height in species *D. zibethinus and T. cacao* [14,18].

In this study, only fruit tree species *C. arabica* showed an inversely proportional relationship between height and DBH. This species is a living collection from the Netherlands and it might be hard to survive in a new ecosystem in BBG. The difference in climate and temperature from the origin site is the main reason affecting the growth of this species. A previous study found abiotic stress, such as heat temperature and drought, greatly inhibit the growth and photosynthetic activity of coffee plants [19,20].

Fig 2. Scatterplots showing trends in two of the highest species and two of the lowest species of the relationship of tree height and DBH.

No	Species	Scatterplots
1	*Cinamomum burmanii*	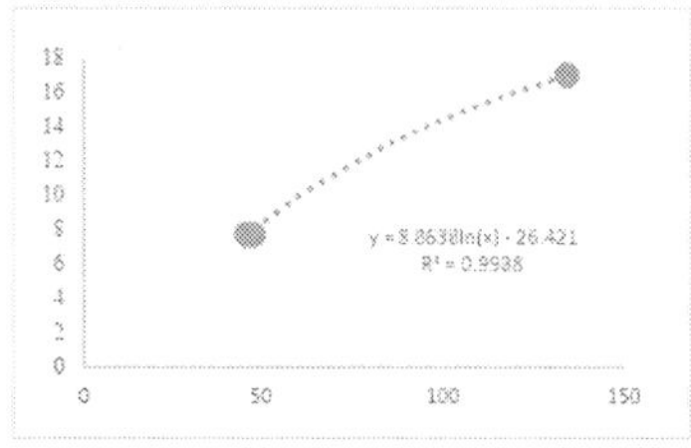
2	*Garcinia mangostana*	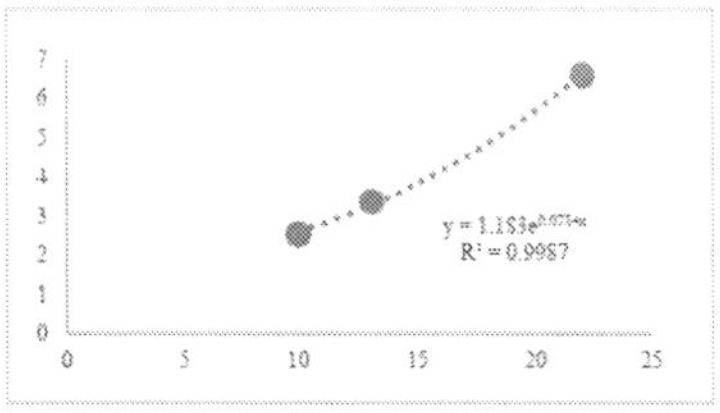
3	*Myristica fragrans*	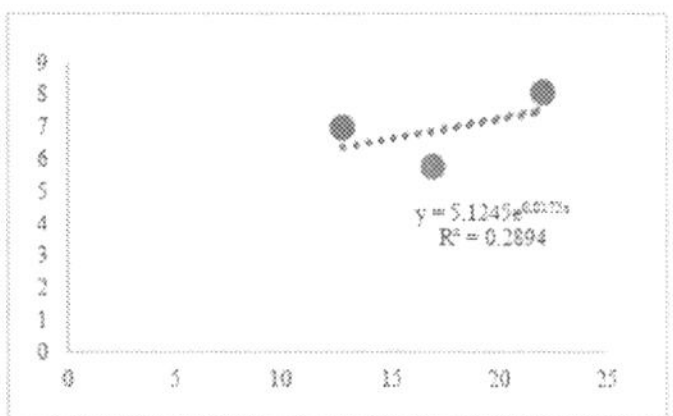
4	*Cocos nucifera*	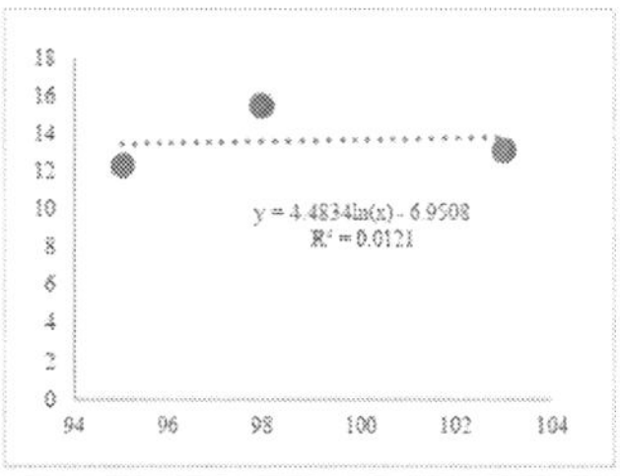

Species *C. nucifera* showed the lowest relationship between tree height and DBH in this study (Fig 2). This species was grown along the coasts and islands that are vulnerable to climate change-induced sea level rise. Growth, physiological processes, biochemical constituents and nutrient *C. nucifera* content were significantly affected by the increasing seawater treatment [21]. This case can be concluded that different ecosystems affect the growth of *C. nucifera* in BBG. Only 1.21% of the role of DBH influences the tree height of this species, the rest of *C. nucifera* growth was affected by other variables.

3.3 Relationship between crown diameter and DBH

Our study provides baseline information on the reliability of using equations established in other locations, as well as the possibility of examining how management practices impact the growth of fruit trees. Estimating accurate diameter growth rate and growth equations could improve the accuracy of evaluation on ecosystem services [16]. The majority of regression models for the relationship between crown diameter and DBH for eight fruit tree species was logarithmic. Only the species *Syzygium cumini* and *Mangifera indica* have exponential regression models in this study (Table 4). The best models of crown diameter and DBH both included predictor and their interactions. However, the contribution of the independent variable to deviance reduction differed significantly between the five models (linear, quadratic, logarithmic, exponential, and power). Here, the highest R^2 value was found in the linear regression model (R^2=99.94%) for *C. burmanii* and the lowest R^2 was found in the logarithmic regression model (R^2=24.06%) for *A. catechu* between crown diameter and DBH. Variables DBH affected crown diameter in plant growth and they showed a strong relationship. This relationship predicted the productivity of plant growth and the accuracy of fruit yields [11]. It means species with R^2 above 70% can be predicted to produce an abundance of fruit yield. Only four tree fruit species with R^2 below 70%, they were *C. nucifera, C. arabica, Myristica fragrans,* and *A. cathecu.*

Table 4. Relationship between crown diameter and DBH for fifteen fruit tree species in Bogor Botanical Gardens

No	Species	Equations	R^2
1	*Durio zibethinus*	Y = 6.01ln(X) - 16.17 (Logarithmic)	98.71%
2	*Nephelium lappaceum*	Y = 537.91ln(X) – 1327.4 (Logarithmic)	81.19%
3	*Cocos nucifera*	Y = -0.11X + 20.95 (Linear)	50.54%
4	*Theobroma cacao*	Y = 1042ln(X) – 2289.2 (Logarithmic)	98.41%
5	*Cofeea arabica*	Y = 2.21ln(X) – 0.87 (Logarithmic)	34.25%
6	*Myristica fragrans*	Y = 0.41ln(X) + 2.59 (Logarithmic)	53.29%
7	*Garcinia mangostana*	Y = 6.22X – 0.27 (Linear)	99.85%
8	*Cinamomum burmanii*	Y 0.04X + 3.92 (Linear)	99.94%
9	*Syzygium acuminatisimum*	Y = 2.92ln(X) – 3.94 (Logarithmic)	88.99%
10	*Averrhoa carambola*	Y = 0.08X + 1.56 (Linear)	98.77%
11	*Syzygium cumini*	$Y = 4.29e^{0.008X}$ (Exponential)	95.31%
12	*Artocarpus heterophyllus*	Y = 0.10X + 1.69 (Linear)	99.80%
13	*Mangifera indica*	$Y = 4.11e^{0.0072X}$ (Exponential)	79.36%
14	*Artocarpus integer*	Y = -4.56ln(X) + 29.50 (Logarithmic)	89.26%
15	*Areca catechu*	Y = 34.29ln(X) + 11.36 (Logarithmic)	24.06%

This study found a strong relationship between crown diameter and DBH for eleven tree fruit species above

70% means that increasing the DBH has an effect on the growth of the crown diameter as well. Data on the dimensions and growth of trees were used to estimate ecosystem services, for example, crown size was correlated to the cooling effect of trees through shading and transpiration. The crown ratio can be calculated from stem DBH unless the crown ratio is defined differently. The size of DBH and crown diameter affect wind speed and attenuate stormwater flooding [16].

In this study, only two fruit tree species (*C. nucifera* and *A. integer*) have the inversely proportional relationship between height and DBH. It means that the bigger the DBH, the narrower the crown diameter. This finding was contrary to another study that said the positive relationship between crown diameter growth and DBH growth of tree species [15]. This study's site was in an ex-situ conservation area but many tree species had limited growth space due to the high density of garden collections. The previous study found high tree density on plantation sites causes light competition for diameter growth [18], including species *C. nucifera* and *A. integer*. Furthermore, living collections in BBG were not pruned as practical management as commercial plantation.

Fig 3. Scatterplots showing trends in two of the highest species and two of the lowest species of the relationship of crown height and DBH.

No	Species	Scatterplots
1	*Cinamomum burmanii*	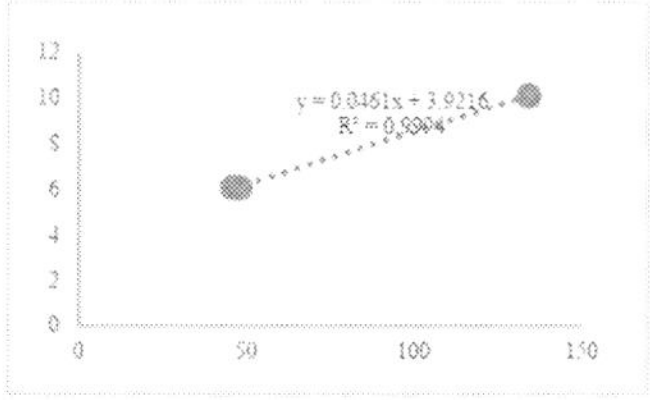
2	*Garcinia mangostana*	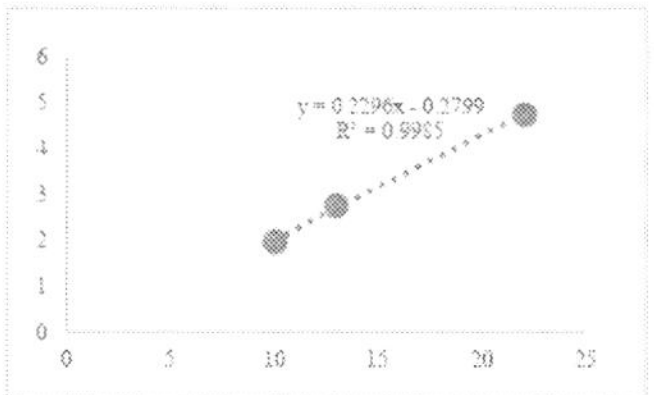
3	*Cofeea arabica*	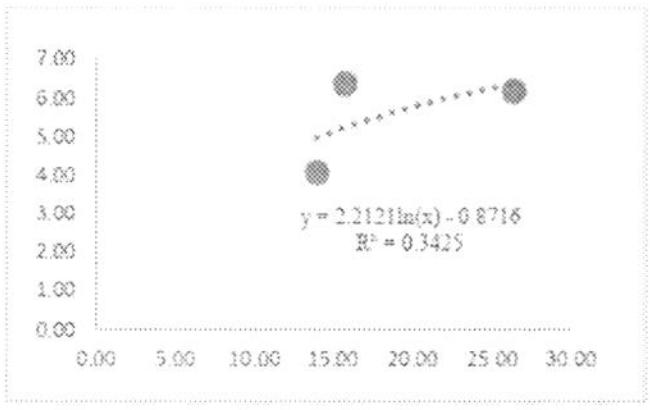
4	*Areca catechu*	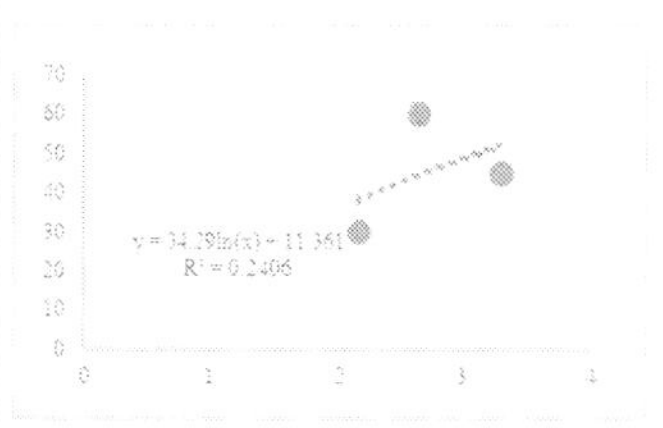

Species *G. mangostana* was a tropical evergreen tree of the genus Garcinia in the Garcinaceae family. Mainly distributed in Indonesia, Thailand, Vietnam, and Malaysia. In this study, a linear regression equation of *G. mangostana* was obtained to determine the optimal DBH growth from the perspective of statistics, so as to improve the crown diameter growth (Fig 3). It means the DBH variable of *G. mangostana* influences almost the crown diameter growth variable. This behaviour was important because it allowed for estimating the survival and establishment of seedlings in the field. Similar to a previous study the variable DBH of *G. mangostana* serve as an alternative for evaluating the quality of this species growth [22].

Our results showed that the use of the allometric equations developed for this study on trees growing in botanical garden conditions was rarely undertaken in commercial garden management assessments. This may differ from commercial plantations where there is a risk of overestimating tree biomass unless empirical equations and field data include tree density indicators. This study showed that management practices impact the growth rate of fruit trees. To evaluate ecosystem services effectively, management practices and standards must be considered, as well as regional climate and differences between tree species. To increase the level of confidence in the diameter growth equation for these fifteen types of fruit trees, additional data was needed to estimate tree growth from height and planting year.

In conclusion, there was a positive relationship between DBH and tree height and between DBH and crown diameter in fifteen types of fruit trees in the ex-situ conservation area of the Bogor Botanical Gardens. The equations developed in this research provide a method for predicting DBH or tree height and DBH or crown diameter depending on the data and model used. This study concludes that predicting tree height and canopy diameter from DBH was easy to do because the measurements were easy to carry out for field inventory and determining stand structure. It should be used with caution outside these areas as the plant exhibits plasticity due to the origin of the seed, as well as climate and soil variability. With increasingly diverse site and stand conditions, future research is needed in addition to greater variations in tree size and age.

Acknowledgements
The first author thanks the head of the Research Center for Plant Conservation, Botanic Gardens, and Forestry, National Research and Innovation Agency (BRIN), Andes Hamuraby Rozak for research permission. Registration and Nursery

Bogor Botanical Garden, Agus Suhatman for field permission. Ghipari and Nabila for field assistants.

References

1. Millenium Ecosystem Assessment, Ecosystem and Human Well-being: Synthesis, Island Press, Washington D.C., http://doi.org/10.1016/b978-0-12-409548-9.09206-x (2005)
2. M. Efendi, *Konservasi eksitu jenis Begonia alam Pegunungan Sumatra di Kebun Raya Cibodas*. Majalah Ilmiah Biologi Biosfera Vol 5 No 2 (2018)
3. X. Zhao, H. Chen, J. Wu, H. Ren, J. Wei, P. Ye, Q. Si, *Ex situ conservation of threatened higher plants in Chinese botanical gardens*, J. G. Ec. Co 38 e02206 (2022)
4. J. Zhou, M. Yang, X.Y. Wen, N. Li, H. Ren, *Strengthen* ex-situ *conservation of plants and promote protection and utilization of plant resources,* Bull. Chin. Acad. Sci 36 (4). 417-424 (2021)
5. D.I. Junaedi, T. Nasution, D.M. Putri, R. Iryadi, R. Lestari, V. Kurniawan, R.A. Pratiwi, A. Handayani, Sudarmono, *Threatened exotic species of botanical garden: Application of trait-based naturalized species risk scoring assessment,* J. S. Afr. J. Bot 152. 321-331 (2023).
6. S.H. Hengky, Z. Kikvidze, *Tourism sustainability in the Bogor Botanical Gardens, Indonesia*, J. Urban Forestry and Urban Greening 30: 8-11(2018)
7. I.F. Wanda, A.N. Rachmadiyanto, A.A. Oksari, *Leaf morphometric and chlorophyll content study of Bisbul (*Diospyros discolor *Willd.) at the Bogor Botanical Garden*, J. Trop. Bio & Bio. Vol 7 (02). (2022). DOI: 10.22146/jtbb.72565
8. http://worldatlas.com/as/id/jb/where-is-bogor.html
9. D. Safarinanugraha, A. Gunawan, W.Q. Muqnisjah, *The development of Bogor Botanical Garden design from 1817 to 2017 based on spatial and functional,* in Proceeding oh the 3rd International Symposium for Sustainable Landscape Development IOP Conf. Series: Earth and Environmental Science 179 (2018).
10. F. Halle, R.A.A. Oldeman, P.B. Tomlinson, Tropical Trees and Forests. An architectural analysis, Springer-Verlaag, Berlin Heidelberg New York (1978).
11. T. Sonmez, Diameter at breast height-crown diameter prediction models for *Picea orientalis*. J. Afr. J. Agr Res Vol 4 (3): 215-219 (2009).
12. Y. Zhana, X. Hou, L. Fan, Z. Dub, S. EeCh'ng, S. M. Ngg, K. Thepkaysoned, E. Hao, J. Deng, *Chemical constituents and pharmacological effects of durian shells in ASEAN countries: A review*, Chinese Herbal Medicines 13: 461–471 (2021).
13. K. Ogawa, A.M. Abdullah, M. Awang, A. Furukawa, *Relationship between fruit growth and peduncle cross-sectional area in Durian (*Durio zibethinus *Murray)*, Ecological Modelling 200: 254-258 (2007)
14. A. C. Wartenberg, W. J. Blaser, J. M. Roshetko, M. van Noordwijk, J. Six, *Soil fertility and Theobroma cacao growth and productivity under commonly intercropped shade-tree species in Sulawesi, Indonesia*, Plant Soil 453:87–104, (2020)
15. A.K. Hardjana, *Correlation Model Between Height and Crown Diameter with Diameter at Breast Height on Tengkawang Tungkul Putih (*Shorea macrophylla *(de Vriese) P.S. Ashton) and Tungkul Merah (*Shorea stenoptera *Burck.) Stand in Semboja, Sanggau Regency,* Jurnal Penelitian Dipterokarpa Vol. 7 No.1: 7-18. ISSN: 1978-8746 (2013)
16. T. Shoda, J. Imanishib, S. Shibata, *Growth characteristics and growth equations of the diameter at breast height using tree ring measurements of street trees in Kyoto City, Japan*, Urban Forestry and Urban Greening 49: 126627 (2020)
17. D. Tian, L. Jiang, M. K. Shahzad, P. He, J. Wang, Y. Yan, *Climate-sensitive tree height-diameter models for mixed forests in Northeastern China*, Agricultural and Forest Meteorology 326: 109182 (2022).
18. D. Harja, G. Vincent, R. Mulia, M. van Noordwijk, *Tree shape plasticity in relation to crown exposure*, Trees 26: 1275-1285 (2012).
19. S. N. Dinha, T. Z. T. Sai, G. Nawaz, K. Leea, H. Kang, *Abiotic stresses affect differently the intron splicing and expression of chloroplast genes in coffee plants* (Coffea arabica) *and rice* (Oryza sativa), J. Plant Physiology 201: 85-94 (2016)
20. A. Koutouleas, C. Blunt, A. Bregar, J. K. Hansen, A. Ræbild, H. Etienne, F. Georget, *Effects of interspecific grafting of Coffea arabica and elevation on coffee growth, yield, and quality attributes in Costa Rica*, Scientia Horticulturae 320: 112162 (2023)
21. K.B. Hebbar, A. Santhosh, A. P. Sukumar, P. Neethu, S.V. Ramesh, V. Selvamani, *Effect of seawater substitution on growth, physiological and biochemical processes of coconut* (Cocos nucifera L.) *seedlings—A hydroponic study*, Scientia Horticulturae 280: 109935 (2021).
22. G. A. G. Júnior, R. A. Pereira, G. A. Sodré, E. Gross, *Growth and quality of mangosteen seedlings (*Garcinia mangostana *L.) in response to the application of humic acids*, Rev. Bras. Frutic., Jaboticabal, v. 41, n. 1: e-104, (2019).

Retention And Availability of Andisol Soil Phosphorus in Monocultural and Policultural Crop Patterns in Ngablak and Getasan Subdistrict, Magelang and Semarang, Indonesia

Elia Laila Rizqiyah[1], Sri Nuryani Hidayah Utami[1](✉), Benito Heru Purwanto[1]
[1]University of Gadjah Mada, Yogyakarta, Indonesia

ABSTRACT
Element P in Andisol soil is absorbed by amorphous minerals such as allophane, imogolite , and amorphous Al/Fe. This study aims to determine the retention and availability of P as well as differences in soil chemical properties on land with monoculture and polyculture cropping patterns. Parameters measured were available-P, P-retention, texture, CEC, C- organic , pH (H_2O, KCl , NaF), Potential-P, Al/Fe/Si with three selective solvents. The sample used 3 replicates at 2 depths (0-20cm and 20-40cm) and 2 different locations, namely Ngablak Subdistrict and Getasan Subdistrict. The data obtained were processed using ANOVA for the RCBD design and LSD Fisher's follow-up test. P retention data in Ngablak Subdistrict showed significantly different results, while those in Getasan Subdistrict were not significantly different. Available P did not show any significant difference, but the availability of P was higher in polyculture land. On agricultural land in Ngablak Subdistrict, cropping patterns have a significantly different effect on soil properties in the form of P retention and C-Organic. Meanwhile, on agricultural land in Getasan Subdistrict, cropping patterns had significantly different effects on soil properties of pH H_2O, pH KCl , Potential-P, amorphous Al + Fe , and Allophane + Imogolite.

***Keywords** : phosphorus, allophane, imogolite, C-organic, monoculture, polyculture*

1. Introduction

The land area with Andisol Soils in Indonesia reaching 5.4 million ha or around 2.9% of Indonesia's land area [7]. However, besides have potential to be used as agricultural land, Andisol soil also has necessary problem that need more special attention related to the availability of phosphorus nutrients. While, phosphorus is one of the essential nutrients for plants [12]. In Andisol soils, phosphate elements usually bind to non-crystalline clay minerals such as allophane, imogolite, and ferrihydrite, where allophane is able to hold P up to 97.8% and the presence of Al and Fe in amorphous form also binds P [18].The availability of P in the soil also depends on land management strategies that create different conditions from the initial land [2]. On a different cropping pattern, different land management will be carried out. The common cropping pattern used by farmers in the Ngablak and Getasan sub-subdistricts is the monoculture and polyculture. The monoculture is a cropping pattern using one type of plant in one cropping period, while the polyculture is a cropping pattern that use more than one type of plant in one cropping period.In the polyculture cropping pattern interaction and competition of plant is different than monoculture cropping patern [1]. Variations in surface plant species will provide the potential for a diversity of available biomass to be returned to the soil, which will later become a nutrient contributor after the decomposition process [10]. In addition, the pattern of monoculture and polyculture plant species also affects the number of communities and types of microbes in the land that are useful in nutrient cycling [1]. The implications cropping pattern for soil properties must be considered by farmers. This is important so that farmers can choose cropping patterns that are appropriate to the conditions of the land, including in this case P nutrient fertilization. Therefore, further measurements regarding P adsorption and availability of P in Andisol soil are needed. The aims of this study were : to determine the retention and availability of P as well as differences in soil chemical properties on land with monoculture and polyculture cropping patterns.

2. Methodology

2.1 Location Description

Most of the land at the research location is used as agricultural land. The agricultural land at the research location has been cultivated for a long time, in fact for more than 30 years, and in 2010 they started using an organic farming system. Agricultural land is dominated by horticultural commodities. Generally, farmers market their products to markets, restaurants, shopping centers, and even export abroad. So it can be said that horticultural agriculture has the potential to be developed. Choosing the right planting pattern needs to be evaluated in order to increase the effectiveness of soil management and the efficiency of fertilization, especially phosphorus nutrients.

 https://doi.org/10.1051/bioconf/20238003016

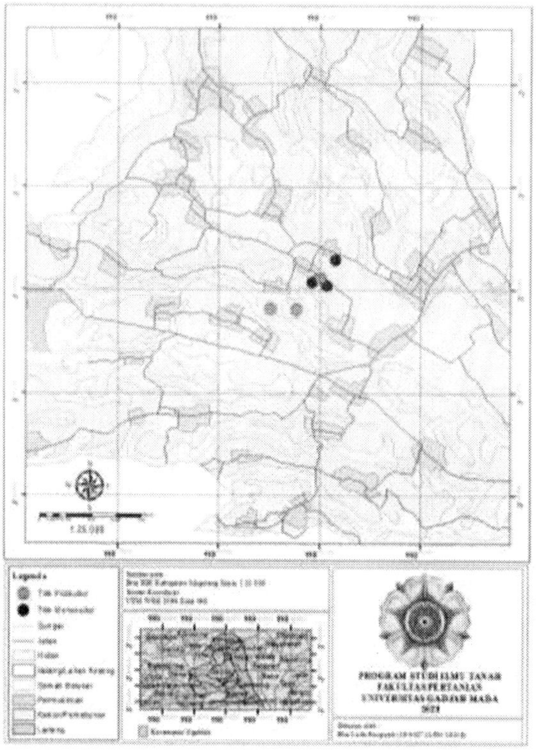

Figure 1. Map of research sample points in Ngablak Subdistrict

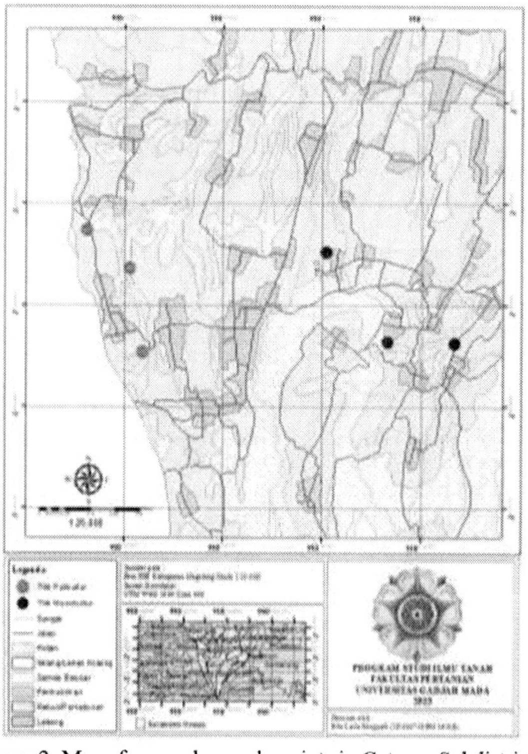

Figure 2. Map of research sample points in Getasan Subdistrict

2.2 Research Design and Data Collection

This study was carried out at the end of 2022 – mid-2023 in the Ngablak Subdistrict, Magelang and Getasan Subdistrict, Semarang, Central Java and the Laboratory of the Department of Soil Science, Faculty of Agriculture, University of Gadjah Mada. The design was prepared using Randomized Complete Block Design (RCBD) with 2 factors: cropping pattern and soil depth in 2 different subdistricts. The first factor is the cropping pattern with 2 variations, Monoculture and Polyculture. The second factor is soil depth with 2 variations, 0-20 cm and 20-40 cm. The research location consists of 2 subdistricts, namely Ngablak Subdistrict and Getasan Subdistrict. The method used is random sampling. Repetition of sample points was carried out 5 times.

2.3 Analysis of Soil

Parameters that measured in this research are P potential soil with solvent HCl 25%, available-P with the Bray method, P retention with the Blackmore method, pH NaF and H_2O, humic acid extracted with a mixture of alkaline and strong acids, soil texture, Cation Exchangeable Capacity (CEC), C-Organic, Al/Fe/Si uses three selective solvents, namely selective solvent consisting of 0.2M Ammonium Oxalate, Dithionite Citrate Bicarbonate (DCB) pH 3, and 0.1M Sodium Pyrophosphate. Then the results are used to calculate Al/Fe Humus, Amorphous Al/Fe, Allophane and Imogolite [15].

- Fe/Al-humus = Fe/Al Na-Pyrophosphate
- Fe/Al-amorphous = Fe/Al NH4-Oxalate – Fe/Al Na-Pyrophosphate
- Allophane + Imogolite = %Si NH4-Oxalate x 7.1
- Terms Andisol = Al NH4-Oxalate + ½ Fe NH4-Oxalate > 2

2.4 Statistics analysis

The data were analyzed using ANOVA (Analysis of Variance) Completely Randomized Block Design (RCBD), which is useful for testing the significance of cropping pattern variations for each parameter using the 5% level. Then proceed with the LSD (Least Significant Difference) test with a level of 5% to find significant variation.

3. Results and Discussion

3.1 Soil Type at the Research Location

The pH value of NaF $\geq$ 9.4 indicates that amorphous minerals dominate the soil exchangeable complex [13]. Based on the results of the NaF pH test on land overall value was > 9.4. The reaction that occurs when adding NaF solution is the exchange of F^- with OH^- which is on the edge of the allophane so that OH^- is released into the soil solution which has implications for increasing the pH of the solution [14].

Table 1. NaF pH at the research location

Cropping Pattern	pH NaF Ngablak		Average	pH NaF Getasan		Average
	0-20cm	20-40 cm		0-20cm	20-40 cm	
Monoculture	10.10	10.20	10.2a	10.17	10.22	10.2a
Polyculture	10.19	10.33	10.3a	10.10	10.17	10.1a
Average	10.1a	10.3a		10.1a	10.2a	

Note : Numbers followed by letters that are not the same indicate that there are real differences between factors (planting pattern and depth) in the 5% LSD test.

Other properties or parameters that can be used as indicators that the soil contains a lot of amorphous minerals are Al/Fe/Si with three selective solvents. One of the selective solvents that can detect the presence of non-crystalline minerals is ammonium oxalate. A soil can be classified as an andisol soil if it has andic properties, while these characteristics have conditions NaF pH above 9.4 and Al Oxalate + ½ Fe Oxalate > 2 [16].

Table 2. Al Oxalate + ½ Fe Oxalate at the reasearch location

Cropping Pattern	Al Oxalate + ½ Fe Oxalate Ngablak		Average	Al Oxalate + ½ Fe Oxalate Getasan		Average
	0-20cm	20-40 cm		0-20cm	20-40 cm	
Monoculture	2.01	3.92	2.96a	5.04	3.15	4.10a
Polyculture	2.17	3.79	2.98a	2.35	1.72	2.03b
Average	2.09a	3.85a		3.69a	2.43a	

Note : Numbers followed by letters that are not same indicate that there are real differences between factors (planting pattern and depth) in the 5% LSD test.

Monoculture and polyculture cropping patterns gave significant differences, while variations in depth did not give significantly different results. High results of Fe oxalate and Al oxalate indicate that mineral crystallization at these locations is slow [15]. Based on the two properties above, it can be said that the soil in this research location is indeed Andisol soil.

3.2 Soil Physical and Chemical Properties

Texture

The texture class at the study site was sandy loam. This means that the composition of the soil fractions : sand, clay and silt, is not much different, but is dominated by the sand fraction. These results are in accordance with Juarti's research, 2016 where in this study also obtained Andisol soils with a sandy loam texture [8].

Table 3. Soil texture at the research location

Cropping Pattern	Texture		Texture	
	Ngablak		Getasan	
	0-20cm	20-40 cm	0-20cm	20-40 cm
Monoculture	Sandy Loam	Sandy Loam	Sandy Loam	Sandy Loam
Polyculture	Sandy Loam	Sandy Loam	Sandy Loam	Sandy Loam

The texture is owned by the result of the parent material development in the form of volcanic material. Volcanic material released by volcano during eruptions, it's can be sand and volcanic ash, so the content of silt and sand in Andisol is relatively high [8]. In addition, the texture of Andisol soils is characterized by the presence of micro-aggregates or pseudo-sand which has a granular structure, is reddish in color, and thixotropic [3]. Pseudo sand can be formed due to the process of aggregation or attachment of amorphous minerals with other particles such as organic matter.

Soil Reaction (pH)

Reaction soil (soil pH) is an indicator that show the level of soil acidity. The results measurement of pH H_2O at the research location are shown in Table 4.

Table 4. Soil pH H_2O at the research location

Cropping Pattern	pH H_2O		Average	pH H_2O		Average
	Ngablak			Getasan		
	0-20cm	20-40 cm		0-20cm	20-40 cm	
Monoculture	7,2	7,23	7,2a	7,22	7,33	7,3b
Polyculture	6,93	6,97	6,9a	7,38	7,4	7,4a
Average	7,1a	7,1a		7,3a	7,4a	

Note : Numbers followed by letters that are not same indicate that there are real differences between factors (planting pattern and depth) in the 5% LSD test

Based on the data in Table 4, the pH value of H_2O at the research location in Ngablak Subdistrict show that monoculture and polyculture cropping patterns were not significantly different, nor did the depth of variation give significantly different results. While in the research location of Getasan Subdistrict monoculture and polyculture cropping patterns gave significant differences, variations in depth did not give significantly different results. The more diverse the vegetation, the more exudate that can be secreted by plant roots and this can increase the pH [10]. This is also in line with the research of [10]. where in this study monoculture land had a lower pH. However, other factors may also affect the pH at the research location in Ngablak Subdistrict, where polyculture land has a lower pH. This can be caused by the presence of organic matter or biomass which is more than monoculture land but has not or is undergoing decomposition which produces of organic acids [8].

Cation Exchangeable Capacity (CEC)

Soils that predominantly have amorphous minerals will have a variable charge. Soil with a variable charge will have a charge depending on environmental conditions, which is pH. This can happen because the constituents of the soil fraction are amphoteric so at low pH the colloid surface will be positively charged (OH^- protonation occurs), while at high soil pH it will be negatively charged (H^+ dissociation occurs) [9].

Table 5. CEC at the research location

Cropping Pattern	CEC (cmol (+)/kg)		Average	CEC (cmol (+)/kg)		Average
	Ngablak			Getasan		
	0-20cm	20-40 cm		0-20cm	20-40 cm	
Monoculture	20.52	18.04	19.28a	24.11	25.06	24.59a
Polyculture	20.41	18.67	19.54a	26.69	28.13	27.41a
Average	20.46a	18.35a		25.40a	26.60a	

Note : Numbers followed by letters that are not same indicate that there are real differences between factors (planting pattern and depth) in the 5% LSD test

Based on the data in Table 5 the CEC value at the research location in Ngablak Subdistrict show that monoculture and polyculture cropping patterns were not significantly different, nor did the depth of variation give significantly different results. That also shown in the research location of Getasan Subdistrict. However, based on the average, it can be seen in the two subdistricts that polyculture land has a higher CEC than monoculture land. This can happen because of the input of organic matter from the remaining biomass on polyculture land which is more varied. CEC can increase due to organic matter that enters the soil so that soil with high organic matter will have a higher CEC [10].

C-Organic

There is another form of soil colloid that affects all soil properties, namely soil organic matter. Soil organic matter is composed of carbon chains that can be obtained by multiplying soil organic matter result by the average amount of carbon in organic matter, which is 58%.

Table 6. C-Organic content at the research location

Cropping Pattern	C- Organic (%) Ngablak		Average	C- Organic (%) Getasan		Average
	0-20cm	20-40 cm		0-20cm	20-40 cm	
Monoculture	6.17	5.72	5.94a	7.91	7.32	7.61a
Polyculture	7.34	6.91	7.13b	8.06	7.91	7.98a
Average	6.75a	6.31a		7.98a	7.62a	

Note : Numbers followed by letters that are not same indicate that there are real differences between factors (planting pattern and depth) in the 5% LSD test

The results measurement of organic matter on monoculture and polyculture land in Ngablak Subdistrict showed significantly different. This is due to the more varied vegetation on polyculture land compared to monoculture land. So that the input of organic matter into the soil is more numerous and varied. While the results of the Getasan Subdistrict, although not giving significantly different the organic matter content in polyculture soils also showed higher result compared to monoculture soils. In addition, the results also show that the organic matter content at a depth of 0-20 cm is higher than that of organic matter at a depth of 20-40 cm. The organic matter content will decrease with increasing layers, if there is an increase in organic matter in a lower layer of material indicates the presence of new volcanic deposits on top of previously developed material [4].

Humic Acid

Humic acid is heavy molecule organic acid from the decomposition of organic materials. Humic acid is colored dark , have > 50% carbon chain, and soluble in base.

Table 7. Humic Acid at the research location

Cropping Pattern	Humic acid (%) Ngablak		Average	Humic acid (%) Getasan		Average
	0-20cm	20-40 cm		0-20cm	20-40 cm	
Monoculture	0.27	0.27	0.27a	0.27	0.29	0.28a
Polyculture	0.28	0.29	0.28a	0.28	0.27	0.27a
Average	0.27a	0.28a		0.28a	0.28a	

Note : Numbers followed by letters that are not same indicate that there are real differences between factors (planting pattern and depth) in the 5% LSD test

Based on the data in Table 7 the Humic acid value at the research location in Ngablak Subdistrict show that monoculture and polyculture cropping patterns were not significantly different, nor did the depth of variation give significantly different results. That also shown in Getasan Subdistrict. The presence of humic acid indicates the degree of humification of organic matter, the more humic acid, the higher the humification process [20]. In Andisols, the humification process was quite hampered by the presence of amorphous Al and cold temperatures, so that the humic acid content was very low (<1%). Almost the same results were shown in both monoculture and polyculture land, this indicated that the humification process was almost the same.

3.3 P-Retention and Availability

Andisol is the one of the soils that has problems related to the availability of P nutrient. Because, Andisol has the ability to absorb phosphorus. Table 8 show the P-retention at the research location

Table 8. P Retention at the research location

Cropping Pattern	P-Retention (%) Ngablak		Average	P-Retention (%) Getasan		Average
	0-20cm	20-40 cm		0-20cm	20-40 cm	
Monoculture	79.35	80.61	79.98b	91.34	90.72	91.20a
Polyculture	85.72	88.80	87.26a	89.83	92.58	91.03a
Average	82.53a	84.70a		90.58a	91.65a	

Note : Numbers followed by letters that are not same indicate that there are real differences between factors (planting pattern and depth) in the 5% LSD test

Based on the data in Table 8 P-Retention at the location study Subdistrict Ngablak show that cropping pattern monoculture and polyculture give significant different, meanwhile variation depth neither give significant different. While at Getasan Subdistrict cropping patterns were not significantly different, nor did the depth. The characteristics of Andisols are having high P retention, mostly > 85% and generally reaching 97.8% [18], but the presence of soil organic matter can affect the P-retention. The decomposition of organic matter carried out by microorganisms produces organic acids, these organic acids can bind allophane and release P nutrients, therefore later P can be absorbed by plants [19]. Humus material can also bind Al and Fe first then form Al/Fe Humus.

Table 9. Al/Fe humus complex at the Ngablak Subdistrict

Cropping Pattern	Al/Fe Humus (%) Ngablak 0-20 cm		Al/Fe Humus (%) Ngablak 20-40 cm		Average	
	Al	Fe	Al	Fe	Al	Fe
Monoculture	0.243	0.0058	0.264	0.0051	0,2665a	0,00543a
Polyculture	0.263	0.0081	0.270	0.0068	0,2535a	0,00745a
Average	0,2530a	0,00695a	0,2670a	0,00593a		

Note : Numbers followed by letters that are not same indicate that there are real differences between factors (planting pattern and depth) in the 5% LSD test

Table 10. Al/Fe humus complex at the Getasan Subdistrict

Cropping Pattern	Al/Fe Humus (%) Getasan 0-20 cm		Al/Fe Humus (%) Getasan 20-40 cm		Average	
	Al	Fe	Al	Fe	Al	Fe
Monoculture	0.239	0.0068	0.237	0.0068	0,2378a	0,00915a
Polyculture	0.234	0.0091	0.267	0.0092	0,2506a	0,00675a
Average	0,2366a	0,0079a	0,2518a	0,0080a		

Note : Numbers followed by letters that are not same indicate that there are real differences between factors (planting pattern and depth) in the 5% LSD test

In Table 9 and 10 can be seen complex formation between Al and Fe with humus. Al and Fe humus can be formed as a result of the synthesis of humus compounds which can form chelates with Al and Fe oxides. Al and Fe humus will become cationic complexes that are not hydrolyzed so that the influence of dissolved oxide ions in the soil is decrease [15]. This humus also has potential to interfere the allophane formation process which is characterized by low allophane levels and acid soils, besides that the Fe-crystalline/Fe-Amorphous ratio will also be low [15].

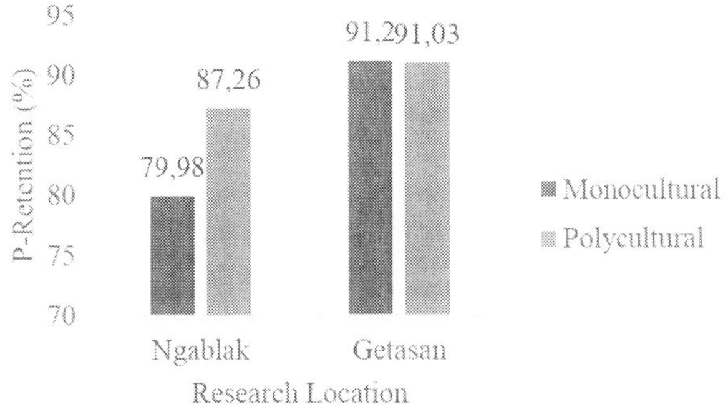

Figure 3. Comparison land P retention monoculture and polyculture

In Figure 3. it can also be seen that P retention on polyculture land tends to be higher than monoculture land, even though in Getasan Subdistrict P retention is almost the same. This can be caused by the amorphous mineral content found on each land. The higher the amorphous mineral, the higher the P-retention. Table 11 shows the amorphous mineral content at the research location.

Table 11 Allophane + Imogolite content at location study.

Cropping Pattern	Allophane + Imogolite (%)		Average	Allophane + Imogolite (%)		Average
	Ngablak			Getasan		
	0-20cm	20-40 cm		0-20cm	20-40 cm	
Monoculture	14.23	13.49	13.86a	21.67	17.40	19.53a
Polyculture	13.89	14.91	14.40a	7.41	6.67	7.04b
Average	14.06a	14.20a		14.54a	12.04a	

Note : Numbers followed by letters that are not same indicate that there are real differences between factors (planting pattern and depth) in the 5% LSD test

Allophane in soils derived from volcanic materials in Indonesia ranges from 1-34% with an average of 11% [18]. Based on Table 11 it can be seen that the allophane + imogolite content in polyculture land in Getasan Subdistrict is higher than monoculture land so that P retention is higher. However, in Getasan Subdistrict, the allophane + imogolite of monoculture land was much higher than that of monoculture land. This is not in line with the results of P retention in Getasan Subdistrict, which is almost the same. This might be influenced by the presence of other factors that inhibit the adsorption process so that P retention does not occur. These factors are for example the presence of phosphate solubilizing microorganisms. The mechanism of dissolving phosphate by microorganisms is by secreting low molecular weight organic acids such as oxalic, succinic, tartaric, citric acids, and others which will chelate with Al, Fe, Ca, or Mg which absorbs phosphate to form a stable organic chelate so that phosphate is released [6].

Table 12 Available-P at the research location

Cropping Pattern	Available-P (ppm)		Average	Available-P (ppm)		Average
	Ngablak			Getasan		
	0-20cm	20-40 cm		0-20cm	20-40 cm	
Monoculture	4.48	6.02	5.3a	8.20	7.38	7.8a
Polyculture	10.25	10.43	10.3a	5.81	12.37	9.1a
Average	7.4a	8.2a		7.0a	9.9a	

Note : Numbers followed by letters that are not same indicate that there are real differences between factors (planting pattern and depth) in the 5% LSD test

Based on the data in Table 12 Available-P at the Subdistrict Ngablak show that monoculture and polyculture cropping patterns were not significantly different, nor did the depth of variation give significantly different results. That also shown in Getasan Subdistrict. Available-P measurement results show low to moderate values. As previously mentioned, the availability of P in Andisols is generally low or very low. Minardi *et al.,* (2011) [11] have also carried out research on the availability of P in Andisol soil and obtained a result of 3.49 mg/kg which is very low. This is due to the large number of P chelating agents, apart from being adsorbed by allophane P in the Andisol soil, it is also adsorbed by Al/Fe in amorphous form. The following is the result of measuring the amorphous Al/Fe content at the research location.

Table 13. Amorphous Al/Fe at the Ngablak subdistrict

Cropping Pattern	Al/Fe Amorphous (%)				Average	
	Ngablak					
	0-20 cm		20-40 cm			
	Al	Fe	Al	Fe	Al	Fe
Monoculture	1,2813	0,9647	3,0487	1,2002	2,1650a	1,0825a
Polyculture	1,5836	0,6299	3,0655	0,9038	2,3246a	0,7669a
Average	1,4324a	0,7973a	3,0571a	1,0520a		

Note : Numbers followed by letters that are not same indicate that there are real differences between factors (planting pattern and depth) in the 5% LSD test

Table 14. Amorphous Al/Fe at the Getasan subdistrict

Cropping Pattern	Al/Fe Amorphous (%)				Average	
	Ngablak					
	0-20 cm		20-40 cm			
	Al	Fe	Al	Fe	Al	Fe
Monoculture	4,0380	1,5246	2,7359	1,4436	3,3869a	1,4841a
Polyculture	1,7967	0,3632	0,5090	0,2415	1,1529b	0,3024a
Average	2,9173a	0,9439a	1,6224b	0,8426a		

Note : Numbers followed by letters that are not same indicate that there are real differences between factors (planting pattern and depth) in the 5% LSD test

Based on Table 13 and Table 14 it can be seen that the soil at the study site also contains amorphous Al/Fe which can also absorb P. The mechanism of P adsorption by these amorphous minerals, forms strong covalent bonds or inner-sphere complexes with irreversible properties. So that P can be strongly absorbed into the mineral lattice in the soil and it becomes difficult to be available. The following is an illustration of the adsorption reaction [21] :

$]Al\text{-}OH + H_2PO_4^- \rightarrow]Al\text{-}OH_2PO_3^- + OH^-$

Amorphous minerals P- available Complex Retention-P

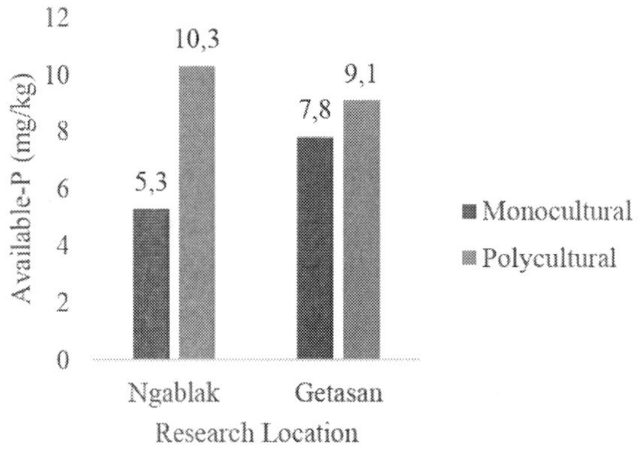

Figure 4. Comparison of Available-P on land monoculture and polyculture

In Figure 4 it can be seen that the availability of P in polyculture land is higher than in monoculture land. Apart from the fact that the research location is an organic agricultural land with great potential to increase the availability of P, this is also due to the fact that polyculture land has a more diverse crop residue. Due to the high input of organic matter, the availability of P can increase. The mechanism for releasing phosphorus in Andisol soil is through mineralization processes, releasing P from adsorption complexes through Al and Fe chelation mechanisms through functional groups and organic acids, and through priming effects [11].

Although the adsorption of P in Andisol soil was quite high and the availability of P showed low-moderate results, P at the study site could be increased. Potential-P is a P substrate that has the potential to be broken down into available-P in the mineralization process. So that the higher Potential-P, the higher chance of increasing the availability of phosphorus in the soil. The results of these measurements show values from high to very high.

Table 15 P- Potential at location study

Cropping Pattern	P- Potential (mg/100g)		Average	P- Potential (mg/100g)		Average
	Ngablak			Getasan		
	0-20cm	20-40 cm		0-20cm	20-40 cm	
Monoculture	73.80	64.91	69a	74.08	62.03	68b
Polyculture	82.43	61.91	72a	92.66	85.70	89a
Average	78a	63a		83a	74a	

Note : Numbers followed by letters that are not same indicate that there are real differences between factors (planting pattern and depth) in the 5% LSD test

Based on the data in Table 15 P- Potential at the location study Ngablak Subdistrict show that monoculture and polyculture cropping patterns were not significantly different, nor did the depth of variation give significantly different results. While at the Subsubdistrict Getasan monoculture and polyculture cropping patterns show significantly different result, but the depth didn't give significantly different results.

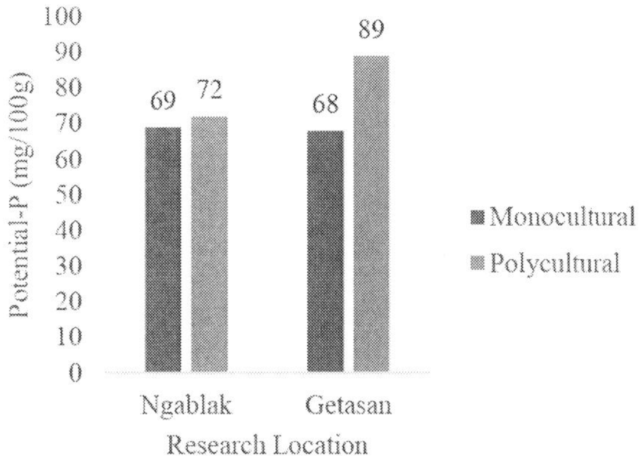

Figure 5 Comparison of Potential-P on land monoculture and polyculture

Based on Figure 5, it can be seen that the Potential-P in polyculture land is higher than in monoculture land. Due to the more diverse potential of biomass litter, there are also more P sources for mineralized substrates. So that the potential availability of P in polyculture land will be higher. The best effort to do on land with high Potential-P is the priming effect. Organic matter can stimulate the availability of P through the Potential-P in the soil, this mechanism is called the priming effect. This process can also be enhanced by the addition of phosphate solubilizing microorganisms. In addition to using organic matter, increasing available P in the soil can be done by applying silica fertilizer. Si dissolves easily in the form of silicic acid and is able to increase the availability of P with anion substitution processes [17].

4. CONCLUSIONS

Based on research that has held can concluded a number of matter following :

1. Retention of P in the field monoculture and polyculture in the Subdistrict Ngablak give significant different result, meanwhile P retention in the Subdistrict Getasan didn't give significant different result. P-Retention in monoculture is higher than polyculture
2. Availability of P on monoculture and polyculture land in Ngablak Subdistrict did not give significantly different results, but in the two sub-subdistricts the availability of P on polyculture land was higher.
3. On agricultural land in Ngablak Subdistrict, the cropping pattern gave significantly different results on soil properties in the form of P and C-Organic retention. Meanwhile, on agricultural land in Getasan Subdistrict, cropping patterns gave significantly different results on soil properties in the form of pH H_2O, pH KCl, Potential-P, amorphous Al + Fe, and Allophane + Imogolite.

REFERENCES

1. Carillo, V., G. Gomez., and G. Vidal. 2022. Phosphorus uptake by macrophyte plants in monocultures and polycultures in constructed wetlands for wastewater treatment. Ecological Engineering 182: 1-8.
2. Ciampiti, I. A., F.O. Garcia., L.I. Picone., and G. Rubio. 2011. Phosphorus Budget and Soil Extractable Dynamics in Field Crop Rotations in Mollisols. Soil Science Society of America Journals 75: 131-142.
3. Fauzi, A. I., S. Zauyah., and G. Stoops. 2004. Characteristics micromorphology lands volcanic area in Banten. Journal of Soil and Climate 22: 1-14.
4. Ferdeant., Sufardi., and T. Arabia. 2019. Characteristics morphology and classification of Andisol Soil in the field dry Aceh Besar Subdistrict. Journal Scientific Student Agriculture 4 (4): 666-676.
5. Fiantis, D. 2017. Morphology and classification ground. Development Institute Technology Andalas University Information. Padang.
6. Ginting, R. C. B., R. Saraswati., and E. Husein. 2006. Fertilizer organics and fertilizers biological: Microorganisms solvent phosphate. Great Hall Research and Development Resource Land Agriculture Bogor.
7. Hartono, A., S. Anwar., and N. Ruliana. 2019. Characteristics release nitrate in Andisols in West Java and Central Java. Journal Environmental Soil Science 21 (1): 16-20.
8. Juarti. 2016. Analysis index quality land andisol on various use land in the village Sumber Brantas Kota Baru. Journal of Geography Education 21 (2): 58-71.
9. Latuponu, H., D. Siddieq., A. Gratitude., and E. Hanudin. 2012. Power review buffer waste biochar sago on leaching to availability of NPK in Ultisol Soil. Buana Science 12 (2): 91-99.
10. Lele, O. K., F. J Panjaitan., R.A. Taopan., and D. Rofita. 2021. Impact Differences in Cultivation Patterns Clove (*Syzygium aromaticum* L.) against characteristic chemistry and population worm land in the village Komba-Manggarai Timur. Journal Agriculture 32 (1): 7-15.
11. Minardi, S., J. Syamsiyah., and Sukoco. 2011. Influence material organics and fertilizers Phosphor to availability and uptake phosphorus in Andisols. Journal Soil Science and Agroclimatology 8 (1): 23-30.
12. Mkhabela, M.S. and P.R. Warman. 2005. The influence of municipal solid waste compost on yield, soil phosphorus availability and uptake by two vegetable crops grown in a Pug mash sandy loam soil in Nova Scotia. Agriculture, Ecosystem & Environment 106: 57-67.
13. Ozaytekin , H. H, C. Uzun ., E. Karaarslam ., and M. Dedeoglu . 2018. The pedogenic development and classification of soil developed on Mount Hasandag's volcanic materials in a semi-arid environment. Fresenius Environment Bulletin 27 (2): 880-897.
14. Simamora , J., P. Marpaung ., and A. Lubis. 2015. Determination types of clay minerals allophane of Andisol Land in Dolat Rakyat Village, Subdistrict Three Arrow Karo Regency. Online Journal of Agroecotechnology 3 (3): 1005-1011.
15. Simangunsong, H. S., D. Mulyanto., and Partoyo. 2022. Analysis development land method dissolution selective with extraction Dithionite Citric Bicarbonate, Ammonium Oxalate, and Pyrophosphate in Andisols peak Sumbing Mountain Regency Magelang. Journal of Land and Water 19 (2): 97-103.
16. Soil Survey Staff. 2014. Keys to soil taxonomy. United States Department of Agriculture. Washington DC.
17. Subiksa, IGM 2018. Influence fertilizer silica to growth and yield plant lowland rice on Inceptisols. Journal of Soil and Climate 42 (2): 153-160.
18. Sukarman and A. Dariah. 2014. Andosol Land in Indonesia. Great Hall Research and Development Resource Land Agriculture. Bogor.
19. Supriadi, Zuraida, and Hifnalisa. 2021. Influence use various type compost to nutrient content of N, P, K Andisols in Arabica coffee plantations subdistrict Timang Gajah Regency Right festive. Journal Scientific Student Agriculture 6 (2): 97-104.
20. Syukur, A., Sulakhudin., and B. H. Sunarmito. 2011. The effect of zeo-hukalsi-coated NPK fertilizer on the growth and yield of shallots in the sand soil of Bugel Kulon Progo beach. Journal of Agrin 15 (1) : 64-75.
21. Wijanarko, A. and E. Hanudin. 2010. Characteristics Adsorption of P by the Four Soil Orders. Journal Soil and Environmental Sciences 10 (1): 42-51.

Soil Erodibility: Influencing Factors and Its Relation to Soil Fertility in Nawungan, Selopamioro, Bantul Regency

Umi Mar Atus Sholihah[1*], *Nur Ainun H J Pulungan*[2], and *Fathi Alfinur Rizqi*[2]

[1]Student of Soil Department, Faculty of Agriculture, Universitas Gadjah Mada, Indonesia
[2]Soil Department, Faculty of Agriculture, Universitas Gadjah Mada, Indonesia

* Corresponding author: umisholihah05@mail.ugm.ac.id

Abstract. Soil erodibility is used to determine the sensitivity of the soil to erosion that occurs on dry, sloping land. Slope is a factor that influences the erodibility index and can affect land productivity. Therefore, this study aims to analyze the erodibility index at each different slope and to evaluate the effect of erodibility on the potential productivity of agricultural land in terms of soil fertility at the research location. Data analysis was carried out quantitatively and qualitatively. Field measurements and laboratory tests were carried out to collect landscape characteristics and soil data, qualitative measurements with the Dry Soil Test Kit (PUTK) were used to measure potential soil fertility. Data processing using the Stepwise-Regression method was carried out to determine the factors that most influence the erodibility index at the research location, and crosstabulation analysis was carried out on erodibility and soil fertility levels. The results showed that the erodibility level at the research location was mostly moderate with the erodibility index in the very gentle slope ranging between 0.17-0.33, in the gentle slope ranging between 0.08-0.16, in the slightly steep slope class ranging between 0.21-0.24, in the steep slope class it ranges from 0.12-0.61. The factors that most influence the erodibility index at the research location is % Dust, % Clay, Organic Material, and Permeability. Based on a qualitative assessment of soil fertility, the productivity potential of agricultural land in the study area is in the low category.

1. Introduction

Soil is one of the important land components in providing agricultural products. As a growing medium for plants, soil has the role of storing and supplying the nutrients needed by plants [1]. Fertile soil has the characteristics of a deep profile, loose structure, neutral pH, sufficient nutrient content for plants, and there are no limitations in the soil for plant growth [2].

Nawungan agricultural land is located on mountain slopes that have dry land agricultural cultivation, where almost part of the area in Bantul Regency is dry land [3]. Erosion can be a crucial problem in agricultural cultivation on dry, sloping land. In Indonesia, the rate of soil erosion on agricultural land with a slope of 3-30% is relatively high, ranging from 60-625 t/ha/year [4]. Upland land with a slope > 3% is the main cause of erosion in agricultural cultivation on dry land [5]. The slope factor is the main cause of the potential of erosion in sloping dry land farming. Martono [6] and Andrian et al. [7] states that the steeper and the longer slopes will increase the amount of erosion.

Soil that is susceptible to erosion and agricultural practices that are not accompanied by erosion management strategies can reduce the quality of fertility and productivity of agricultural land. Soil sensitivity to erosion can be determined by studying the soil erodibility index [8]. Soil erodibility studies on each slope are needed, especially in areas that have the potential for soil erosion, such as in the Nawungan, to be able to adapt appropriate soil conservation techniques at the research location. Based on the background explained previously, the objectives of this research were to: (1) analyze the erodibility index at each different slope at the research location and (2) evaluate the effect of soil erodibility on the potential productivity of agricultural land at the research location.

2. Methodology

The study was located in the Dry Agricultural Land of Nawungan Hamlet, Selopamioro, Imogiri, Bantul Regency, Yogyakarta Special Region (8°2'2.4" S, 110°24'24.49" E) (Fig. 1). The research was conducted in February 2023 – July 2023. The research was carried out by direct field measurements in the form of landscape analysis and qualitative soil fertility analysis using the Dry Soil Test Kit (PUTK) to obtain c-organic data, pH, available-P, and K-available.

Laboratory analysis was also carried out on soil characteristics that affect the erodibility index such as texture (using the pipette method), aggregate stability (using the De Leeheer and De Boodt double sieving method), unit weight (using the ring method), permeability (using a permeameter), and organic matter

https://doi.org/10.1051/bioconf/20238003017

(using the Walkley-Black method). There were 30 soil samples taken in the depth of 0-20 cm (top-soils).

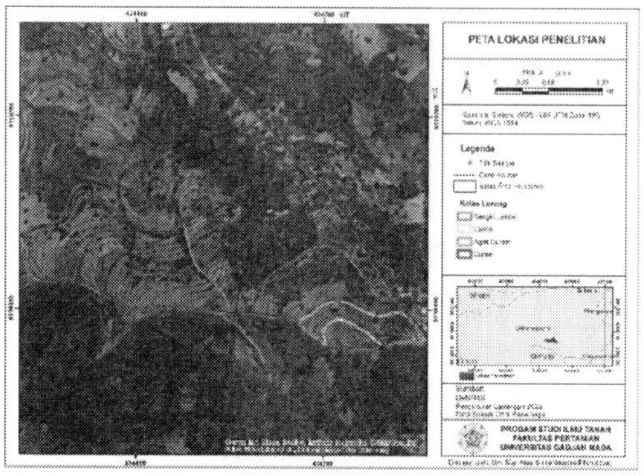

Fig. 1. Research Location Map

Data analysis for determining the erodibility index (K) was carried out by calculating the formula by Weischmeier and Smith [9], as follows:

$$100K = 1{,}292 \times 10^{-2}[2{,}1M^{1{,}14}(10^{-4})(12 - a) + 3{,}25(b - 2) + 2{,}5(c - 3)]$$

K = Erodibility Index
M = particle size dimension
= (% very fine sand + % silt) × (100 - % clay)
a = organic matter (% C × 100/58)
b = soil structure grade
c = soil permeability grade

Data processing was carried out to determine the relationship between the soil erodibility index and each soil characteristic parameter using the Stepwise regression test method using SPSS software. Next, crosstabulation analysis was carried out on erodibility with qualitative soil fertility levels to see potential strategies that could be developed.

3. Results and discussion

3.1 Erodibility index at different slopes

The slope was the main focus of research on erodibility in Nawungan, because it is considered to have quite a large influence on the high and low soil erodibility index at the research location. The position of the slope, which is a part of topography, can influence the characteristics of the soil formed [10].

3.1.1 Soil texture

The soil texture at the research location varies from clay, clay loam, sandy loam, sandy loam, and silt loam. The clay fraction dominates in Nawungan at 37.54% (Table 1). The percentage of clay fraction tends to decrease following the level of the slope classes. The clay fraction, assisted by organic matter, is able to form solid aggregates [11], making it difficult to be destroyed by rainwater splashing which is a factor causing soil erosion [12]. The highest silt fraction percentage on steep slopes up to 64.85% (Figure 2) influences the high erodibility index on this slope.

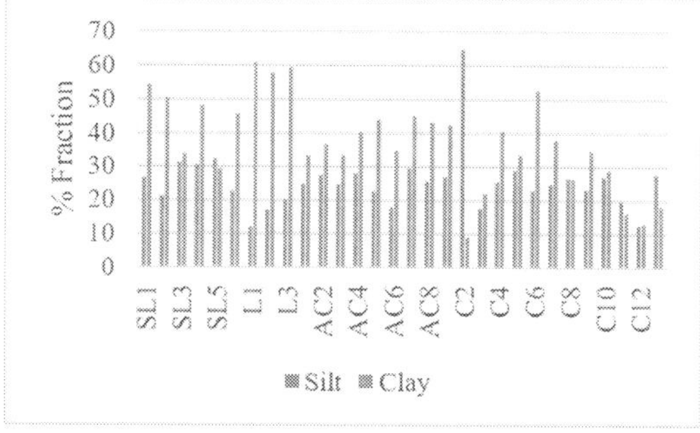

Fig. 2. Percentage of Silt and Clay Fractions

Silt is a part of the soil fraction which has very weak cohesion between its particles so that it is easily carried away by surface runoff and dispersed. Silt is easily saturated with water so it can reduce soil infiltration capacity and soil permeability because its particles are able to clog soil pores [12]. Steep slopes have the highest percentage of silt (Fig.2) with a very high erodibility index (Table 1). The highest percentage of very fine sand fraction belongs to the steep slope class. The high content of very fine dust and sand can cause the soil to be sensitive to erosion or in other words the erodibility index increases [13]. Very fine sand reduces the ability of the soil to transmit water due to blockages in the pore spaces of the soil, thus increasing the occurrence of run off.

3.1.2 *Soil Organic Matter*

Nawungan has organic matter which varies from very low to high (Fig. 3). Soil with a high organic matter content has strong bonds between colloids, high water content with stable aggregates, low soil particle release sensitivity, high permeability, high resistance, and has a low erodibility index Zhang et al. [14], Zhao et al., [15], and Li et al., [16]. In this study, soils containing high organic matter tended to have a loamy texture. Fine-textured soils usually have a low average temperature, which has the effect of delaying the rate of breakdown of organic matter and helping its accumulation [17].

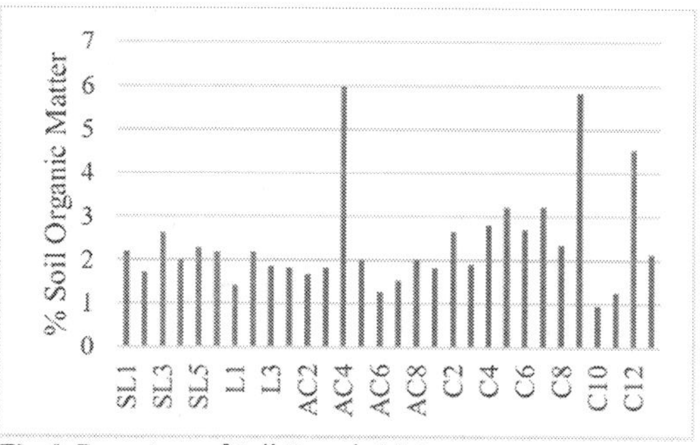

Fig. 3. Percentage of Soil Organic Matter

3.1.3 *Soil Structure*

The soil structure is dominated by rounded clumps which can be said to be good soil structure. The rounded soil structure resulting in many soil pores being formed [18]. Soil structure must not be easily damaged (steady) so that the soil pores do not close quickly when it rains [19]. Well-structured soil will have good drainage and aeration conditions, making it easier for the plant root system to penetrate and absorb nutrients and water, resulting in better growth and production [20].

3.1.4 *Soil Permeability*

The soil permeability varies on each slope (Fig. 4). On steep slopes the soil permeability tends to be higher than on gentler slopes (Table 1). It is because gentle slopes have a high percentage of clay, vary land cover, and vary tillage activities causing surface soil become compacted. High clay content and low organic matter cause clay particles to bind tightly and compact the soil [21]. Dense soil will not be easily carried away by water flows, but it is also difficult for water to pass through, so permeability is low because water that falls to the surface of the soil cannot enter the soil and flows as surface flow [22].

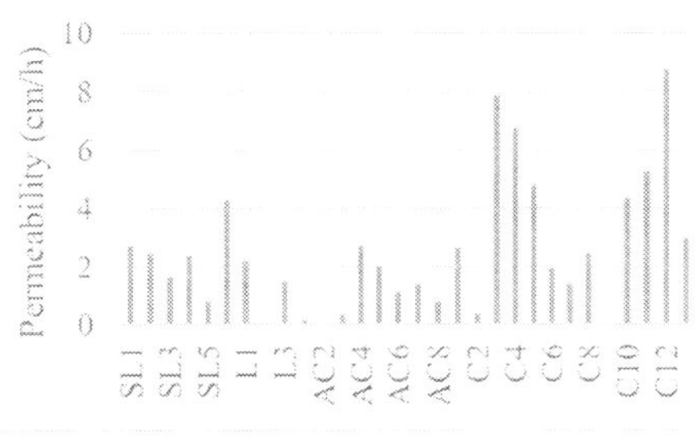

Fig. 4. Percentage of Soil Permeability

3.1.5 *Agregat Stability of Soil*

The aggregate stability index at the study site consisted of unstable, less stable, somewhat stable and stable classes. Based on Table 1 it is known that the higher the slope class, the aggregate stability index tends to decrease. Good aggregate stability on very gentle slopes has good resistance to water carrying capacity so that erodibility values tend to be low. Aggregate stability can influence soil erosion and is an important indicator to describe soil resistance to external factors [23]. Soils with unstable aggregates cause soil pores to easily and potentially become blocked by clay or dust so that the rate and capacity of soil infiltration decreases, as a result surface flow increases and easily causes erosion.

Table 1. Soil Erodibility Index (K)

Slope	OM	S	C	VFS	SC	P	ISA	K	Class
SL1 (2-6%)	2.20	26.85	54.32	7.63	3	0.19	58.72	0.19	Low
SL2 (2-6%)	1.72	21.09	50.36	4.84	3	0.17	51.19	0.17	Low
SL3 (2-6%)	2.62	31.37	33.92	9.04	4	0.29	64.44	0.29	Medium
SL4 (2-6%)	2.01	30.44	48.32	6.97	4	0.27	50.57	0.27	Medium
SL5 (2-6%)	2.28	32.49	29.38	9.90	4	0.33	76.37	0.33	Moderately High
SL6 (2-6%)	2.18	22.65	45.64	4.85	4	0.23	57.08	0.23	Medium
L1 (6-13%)	1.42	11.98	60.85	4.88	3	0.12	52.62	0.12	Low
L2 (6-13%)	2.18	17.08	57.79	4.85	4	0.08	45.08	0.08	Very Low
L3 (6-13%)	1.86	20.30	59.42	5.22	4	0.16	41.17	0.16	Low
AC1 (13-25%)	1.82	24.98	33.32	9.06	4	0.24	38.88	0.24	Medium
AC2 (13-25%)	1.67	27.51	36.64	8.53	4	0.21	37.07	0.21	Medium
AC3 (13-25%)	1.82	24.85	33.41	4.61	4	0.21	33.44	0.21	Medium
AC4 (13-25%)	5.99	28.18	40.35	5.64	4	0.21	36.78	0.21	Medium
AC5 (13-25%)	2.00	22.64	43.96	5.65	4	0.24	41.12	0.24	Medium
AC6 (13-25%)	1.27	18.02	34.83	5.45	4	0.21	55.21	0.21	Medium
AC7 (13-25%)	1.53	29.70	45.19	5.55	4	0.24	41.52	0.24	Medium
AC8 (13-25%)	2.01	25.72	43.11	7.14	4	0.23	40.45	0.23	Medium
C1 (25-55%)	1.82	27.05	42.41	8.94	4	0.28	37.38	0.28	Medium
C2 (25-55%)	2.65	64.85	9.04	5.83	4	0.61	56.89	0.61	Very High
C3 (25-55%)	1.90	17.51	22.00	6.47	4	0.30	38.05	0.30	Medium
C4 (25-55%)	2.81	25.55	40.65	6.37	4	0.25	43.47	0.25	Medium
C5 (25-55%)	3.21	28.90	33.51	8.60	4	0.29	58.06	0.29	Medium
C6 (25-55%)	2.70	22.84	52.57	7.19	4	0.18	35.73	0.18	Medium

https://doi.org/10.1051/bioconf/20238003017

C7 (25-55%)	3.22	24.96	37.86	6.72	4	0.22	45.89	0.22	Medium
C8 (25-55%)	2.35	26.69	26.44	6.96	4	0.31	62.58	0.31	Medium
C9 (25-55%)	5.84	23.13	34.71	7.37	4	0.12	38.79	0.12	Low
C10 (25-55%)	0.95	27.07	28.89	8.58	4	0.34	38.70	0.34	Moderately High
C11 (25-55%)	1.26	19.58	16.17	9.27	4	0.33	42.65	0.33	Moderately High
C12 (25-55%)	4.53	12.68	12.96	4.55	4	0.23	47.77	0.23	Medium
C13 (25-55%)	2.12	27.84	18.13	11.45	4	0.38	44.98	0.38	Moderately High

Note : SL (Very Sloping), L (Sloping), AC (Rather Steep), C (Steep), OM (Organic Matter), S (Silt), C (Clay), VFS (Very Fine Sand), SC (Structure Class), P (Permeability Class), ISA (Aggregate Stability Index), K (Erodibility Index)

3.1.6 *Erodibility Index (K)*

Erodibility is one of the factors that influences surface runoff and soil erosion, in addition to other factors such as erosivity (rainfall), slope length and slope, vegetation cover and human induced on land [24].

The erodibility index at the study site was dominated by the Medium rating. There are 4 classes of slopes at the study site, namely very gentle (2-6%), gentle (6-13%), rather steep (13-25%), and steep (25-55%). Each slope has a class variation of the erodibility index. The higher the slope class is followed by an increase in the erodibility index (Table 1).

The gentle slope class has the lowest erodibility index variation, which ranges from 0.08 to 0.12. Soil with a low erodibility index means that the soil is resistant or resistant to erosion [25]. The gentle slope is at the highest altitude with an elevation of 294.5 m and has a small slope angle (6-13%). Land with a flatter slope will get a lot of additional clay fraction (deposition zone), so that the gentle slope has a smoother texture than the steep and rather steep slopes [26].

Clay-dominated textures have the ability to bind between particles to form strong aggregates. On gentle slopes the stability of the aggregate formed is in a less stable class which is influenced by the lack of organic matter on this slope. The low erodibility index on this slope is supported by a good soil structure, namely rounded lumps which will affect the porosity and aeration of the soil to be ideal

Soil formed on sloping land has a high sensitivity to erosion [12]. The steep slope class has the highest erodibility index of 0.30, which is included in the medium class. The steep slope is in the middle with an elevation of 267.9 m and a high slope angle (25-55%). Steep slopes have a high average percentage of very fine sand of 7.56% with a low percentage of clay and less stable aggregate stability.

The sand fraction is very fine which can block soil pores to allow water to pass into the soil. The organic material on this slope has a medium value average which supports a fairly good permeability rate in the medium class. Organic matter acts like a sponge in responding to the addition of water, so that it can help soil particles pass water into the soil [24]. Permeability that is good enough cannot reduce surface runoff that occurs due to very steep slopes.

3.2 Analysis of stepwise-regression

Regression analysis was carried out to determine further the relationship between soil characteristics and the soil erodibility index. A stepwise procedure is used to select the best set of predictor variable factors [27].

Based on Table 2, namely the results of the regression equation, it is known that the significance value of all variables in step 4 or model 4 is <0.05 or has a significant effect. The final model also has an R-sq value of 95.3% which is quite high.

Table 2. Stepwise Regresion

	-----Step 1-----		-----Step 2-----		-----Step 3-----		-----Step 4-----	
	Coef	P	Coef	P	Coef	P	Coef	P
Constant	0.036	0.299	0.228	0.000	0.288	0.000	0.185	0.000
% Silt	0.008	0.000	0.006	0.000	0.006	0.000	0.008	0.000
% Clay			-0.004	0.000	-0.004	0.000	-0.003	0.000
Organic Matter					-0.021	0.000	-0.022	0.000
Permeability							0.012	0.000
R	0.779		0.917		0.951		0.976	
R-sq	60.7%		84.2%		90.4%		95.3%	
R-sq(adj)	59.3%		83.0%		89.3%		94.5%	

Thus, the factors that most influence the erodibility index in Nawungan were % Silt, % Clay, Soil Organic Matter, and Permeability. Other variables such as BV, Moisture Content, Aggregate Stability, and % Sand were removed from the analysis because they were included in missing correlations.

Potential productivity of a land can be seen from the level of soil fertility in the land. In this research, several soil fertility parameters were tested semi-quantitatively for c-organic, pH, P-available, and K-available using the Dry Soil Test Kit (PUTK) as a quick test. This needs to be followed up with laboratory testing related to the accuracy of the results.

3.3 Potention of Land Productivity Based on Qualitative Soil Fertility Rate

Table 3. Erodibility Index and Soil Fertility Parameters

No	Sample	Slope (%)	Erodibility Index	Class*	Qualitative Soil Fertility Parameters**			
			(K)		C-Organic	pH	P	K
1	SL1	5.7	0.19	Low	Low	Slightly Acid	Medium	Medium
2	SL2	5.7	0.17	Low	Low	Slightly Acid	High	Medium
3	SL3	5.7	0.29	Medium	Low	Slightly Acid	High	Medium
4	SL4	6	0.27	Medium	Low	Slightly Acid	High	Medium
5	SL5	6	0.33	Moderately High	Low	Slightly Acid	Medium	Medium
6	SL6	6	0.23	Medium	Low	Slightly Acid	Medium	Medium
7	L1	7.67	0.12	Low	Low	Slightly Acid	Medium	Medium
8	L2	6.11	0.08	Very Low	Low	Slightly Acid	High	Medium
9	L3	10.11	0.16	Low	Low	Slightly Acid	High	Medium
10	AC1	23	0.24	Medium	Low	Slightly Acid	Medium	Medium
11	AC2	21.44	0.21	Medium	Low	Slightly Acid	Medium	Medium
12	AC3	16.89	0.21	Medium	Low	Slightly Acid	High	Low
13	AC4	24	0.21	Medium	Low	Slightly Acid	Low	Low
14	AC5	24.8	0.24	Medium	Low	Slightly Acid	Medium	Low
15	AC6	24.9	0.21	Medium	Low	Slightly Acid	Low	Low
16	AC7	21.44	0.24	Medium	Low	Slightly Acid	Medium	Low
17	AC8	23.78	0.23	Medium	Low	Slightly Acid	Medium	Low
18	C1	28	0.28	Medium	Low	Slightly Acid	Medium	Medium
19	C2	28	0.61	Very High	Low	Slightly Acid	Medium	Medium
20	C3	35	0.3	Medium	Low	Slightly Acid	Medium	Low
21	C4	26	0.25	Medium	Low	Slightly Acid	Medium	Low
22	C5	26	0.29	Medium	Low	Slightly Acid	Medium	Low
23	C6	26	0.18	Medium	Low	Slightly Acid	Medium	Low
24	C7	30	0.22	Medium	Low	Slightly Acid	Medium	Low
25	C8	30	0.31	Medium	Low	Slightly Acid	Medium	Low
26	C9	25	0.12	Low	Low	Slightly Acid	High	Medium
27	C10	31	0.34	Moderately High	Low	Slightly Acid	Medium	Low
28	C11	38	0.33	Moderately High	Low	Slightly Acid	Medium	Low
29	C12	38	0.23	Medium	Low	Slightly Acid	Medium	Low
30	C13	38	0.38	Moderately High	Low	Slightly Acid	Medium	Low

Note : (*) K Classification USDA-SCS (1973), (**) PUTK tested.

Based on Table 3, it is known that the levels of c-organic, pH, P, and K available in the soil on each slope in the Nawungan Dry Agricultural Land. K-available levels in the soil are found in low and medium grades, and tend to decrease at greater slope angles. In the sample it is rather steep (AC) and steep (C) that dominated by medium – very high erodibility class, tends to have low K levels. This shows that the highly mobile K element is leached or carried away by erosion and runoff that occurs on soil that is sensitive to erosion due to the high slope class.

Furthermore, the available P levels at the research location are in the low – high class, where the smaller the slope, the higher the P levels tend to be. This is supported by the nutrients in the soil which are in the soil sorption complex generally not easily lost or leached by water [28]. The high average soil P content at the research location is closely related to the size of the soil fraction. In general, the P content is higher when soil particle size is finer [29].

The soil at the research location has a slightly acidic pH, namely pH 5-6, which generally tends to occur in soil that is continuously leached by rainwater. The pH value at the research location also determines whether or not it is easy for a plant to absorb the nutrients available in the soil. Plant roots can easily absorb water-soluble nutrients when the soil pH tends to be neutral. Another factor that determines soil fertility is organic material which functions to improve the physical, chemical and biological properties of the soil [30]. C-organic describes the condition of organic matter in the soil, where the research location has uniform c-organic levels, namely low. According to Kalaati et al. [31], low organic matter content will cause high erodibility values.

On high slopes, the erodibility index or soil sensitivity to erosion also increases, becoming a factor that supports low plant production potential in addition to low nutrient levels. Erosion can affect productivity reduction caused by loss of rooting depth, degradation of soil structure, reduced organic matter, nutritional imbalance for plants, clay content, macro pores, and water availability in the soil [32]. Erosion can cause nutrients, especially nitrogen and phosphorus, as well as organic materials to be carried away by the flow and deposited in the form of sediment [11]. This is in accordance with Liebig's minimum law, namely that plant production potential is determined by the weakest factor or the lowest nutrient content in the soil, even though there are other nutrients with high concentrations [33].

3.4 Strategy for soil fertility management

The research location has low to moderate fertility conditions which are also influenced by variations in the erodibility index. Several strategies that can be used to overcome this are adding organic material and soil amendments.

Husk charcoal biochar and agricultural waste can be applied to research locations. The potential for land productivity can increase along with improvements in soil quality due to the addition of biochar. According to [34], the productivity of food crops such as upland rice and corn has been proven to increase after being treated with biochar. The addition of biochar to agricultural soil can increase nutrient availability, nutrient retention, water retention [35], encourage the formation of symbiotic microorganisms [36], and can increase CEC [34].

Another material that can be applied to add soil organic matter and as a soil improver is compost. Compost is produced from weathering of plant residues or organic waste which has the ability to improve soil structure, aerate, reduce soil density and increase the soil's ability to hold water [37]. Compost is a provider of soil nutrients because it is able to provide food for microorganisms that produce nitrogen and phosphorus naturally.

Soils that have high organic matter will have stable aggregates to be able to store large amounts of water, making them resistant to erosion or lowering the erodibility index [38]. In addition, the application of soil amendments will be able to improve soil structure, reduce or overcome erosion, and can change the capacity of the soil to hold and pass water [39]. As a result, the erodibility index will decrease along with the improvement of soil characteristics which were previously prone to erosion to become more resistant and the production potential increases.

4. Conclusion

Based on the results of the research that has been done, it can be concluded that the erodibility of the soil in the Dry Agricultural Land of Nawungan Hamlet, Selopamioro, Bantul, Yogyakarta is mostly classified as Medium. This is also supported by research results:

1. Each slope class at the study site has a dominance of Moderate erodibility. However, the variations in the erodibility index differ between slope classes. The highest erodibility index is found in the steep slope class with a very high rating. The lowest erodibility index is found in the sloping class with a very low rating.
2. The factors that most influence the erodibility index was % Silt, % Clay, Organic Matter, and Permeability.
3. Potential productivity of agricultural land in the research location was classified as low based on qualitative soil fertility rate, the majority of which were also low.

Acknowledgement

The author would like to thank to Nawungan-II research member team for their help and support during the research. This publication is a part of Bachelor thesis of first author, and was funded by *Final Project Recognition Grant* Universitas Gadjah Mada *Number* 5075/UN1.P.II/Dit-Lit/PT.01.01/2023.

References

1. Prabowo, R., and R. Subantoro, Eksakta, **2**, 59-64 (2018).
2. Sutedjo. M. M, Pupuk dan Cara Pemupukan (2002).
3. DLH Kab Bantul, DIKPLHD Kabupaten Bantul Tahnun 2019 (2020).
4. Sutrisno, N., and N. Heryani, Jurnal Litbang Pertanian, **32**, 112-130 (2013).
5. Huntojungo, I., J. M. Supit, J. Husain, and R. I. Kawulusan, In COCOS, **2** (2013).
6. Martono, Pengaruh intensitas hujan dan kemiringan lereng terhadap laju kehilangan tanah pada tanah regosol kelabu (2004).
7. Andrian, Supriadi, and M. Purba, J. Agroekoteknologi, **2**, 981–989 (2014).
8. Buttafuoco, G., M. Conforti, P. P. C. Aucelli, G. Robustelli, and F. Scarciglia, Environmental Earth Sciences, **66**, 1111-1125 (2012).
9. Wischmeier, W. H., and D. D. Smith. The USDA Agricultural Handbook No. 537 (1978).
10. Yusra, Y., K. Khusrizal, F. A. Diannastiti, *Soil chemical characteristics at three slope positions in the smallholder's Piper nigrum L. in Lhokseumawe City, Aceh Province*, In IOP Conference Series: Earth and Environmental Science, **497**, 012040 (2020).
11. Arsyad, S., Konservasi Tanah dan Air, (2012).
12. Hanifa, H., and S. Suwardi, Biofarm: Jurnal Ilmiah Pertanian, **18**, 160-165 (2022).
13. Liu, X., Y. Zhang, & P. Li, International Journal of Environmental Research and Public Health, **17** (2020).
14. Zhang, B. J., G. H. Zhang, H. Y. Yang, H. Wang, Catena, **173**, 375–383 (2019).
15. Zhao, M. M., Z. Dong, H. L. Li, Y. Y. Yan, X. H. Dou, B. P. Li, L. Geng, J. Arid Land Resour. Environ. **36**, 135–142 (2022).
16. Li, C. C., Q. C. Zeng, P. L. Jia, S. S. An, Acta Ecol. Sin, **40**, 2039–2048 (2020).
17. Foth, H, D., Dasar-dasar Ilmu Tanah, (1994).
18. Meli, V., S. Sagiman, & S. Gafur, Perkebunan dan Lahan Tropika, **8**, 80-90 (2018).
19. Harjdowigeno, S. Ilmu Tanah, (2010).
20. Hanafiah, K. A., Dasar – Dasar Ilmu Tanah, (2005).
21. Lanyala, A. A., Hasanah, U., & Ramlan, Agrotekbis, **4**, 633–641 (2016).
22. Djufri, A. N. H., J. A. Rombang, & J. S. Tasirin, Cocos, **4**, 1–11 (2021).
23. Zhu, M., W. He, Y. Liu, Z. Chen, Z. Dong, C. Zhu, ... & Y. Xiong. International Journal of Environmental Research and Public Health, **19**, 15703 (2022).
24. Sarminah, S., U. A. Gultom, S. Ramayana, Agrifor: Jurnal Ilmu Pertanian dan Kehutanan, **21**, 13-26 (2022).
25. Sutedjo, M. M., and A. G. Kartasapoetra, Pengantar Ilmu Tanah: Terbentuknya Tanah dan Tanah Pertanian (Cetakan Kelima) (2010).
26. Suraj, B., Kumaraperumal, R., Kannan, B., & Ragunath, K. P., International Journal of Chemical Studies, **7**, 3327–3332 (2019).
27. Andayani, N. N., and M. Aqil, Informatika Pertanian, **25**, 21-28 (2016).
28. Sudaryono, Jurnal Teknologi Lingkungan, **10**, 337–346 (2009).
29. Sebastian, N., and D. Banjarnahor, Jurnal Ilmu Pertanian **7**, 135-143 (2019).
30. Kadarwati, F. T., Jurnal Penelitian Tanaman Industri, **22**, 53-62 (2016).
31. Kalaati, I., R. Ramlan, dan A. Rahman, Agrotekbis: E-Jurnal Ilmu Pertanian, 7, 172-178 (2019).
32. Lal. R., CRC Critical Reviews in Plant Sciences **5**, 303-367 (1987).
33. Saidi, A., Asmar, and Masliyunas, J. Solum, **7**, 27-36 (2010).
34. Nurida, N. L., Jurnal Sumberdaya Lahan Edisi Khusus, **8**, 57-68 (2014).
35. Glaser, B., J. Lehmann, and W. Zech. Biol. Fertil. Soils, **35**, 219-230 (2002)
36. Ogawa, M., Carbon sequestration by carbonization of biomass and forestation: three case studies, 133- 146 (2006).
37. Sukartono, W. H. Utomo, Z. Kusuma, W. H. Nugroho, J. of Tropical Agriculture, **49**, 47-52 (2011).
38. Huang, Xiaofang, L. Lin, S. Ding, Z. Tian, X. Zhu, K. Wu, and Y. Zhao. Land 11, 134 (2022).
39. Hartman, R., and H. Verplancke, Study of water repellency of soils under citrus trees in Egypt, And means of improvement, 201-209 (1976).

Spatial Distributional of Irrigation Water Quality in Peaty Paddy Field of Paser Regency, East Kalimantan Province, Indonesia

Moch Reza Wahyu Abdilah[1], *Sahid Susanto*[1*], *Chandra* Setyawan[1], *Muhamad Khoiru* Zaki [1] and *Dehafiyyan* Ramadhani[1]

[1]Department of Agricultural and Biosystem Engineering, Faculty of Agricultural Technology, Universitas Gadjah Mada. Yogyakarta 55281, Indonesia

Abstract. Paddy is one of the main food-crop commodities in Indonesia. The main problem in paddy cultivation in Paser Regency is good water quality for plants. Water is the main requirement for paddy plants to grow and develop. Water quality to meet the needs of paddy plants is a very important factor. pH H_2O, TDS (Total Dissolve Solid) and EC (Electrical Conductivity) are some of the water parameters used as a reference in determining land suitability for paddy plants. The purpose of this research is to spatially analyse the water parameters (pH H_2O, TDS and EC) for the water needs of paddy plants in Paser Regency. Determination of sample points using stratified random sampling method and parameter measurements are carried out directly in the field in the irrigation area. Interpolation analysis used the IDW (Inverse Distance Weighted) method with a power weight of 2. RMSE values from the interpolation results for pH H_2O 0.0001, TDS 0.1695 and EC 0.0003. The results of the interpolation of several parameters are then analysed by overlay to determine the spatial relationship between parameters. The results of the analysis show that the average values in Paser Regency are pH H_2O 6.13, TDS 478.64 ppm and EC 0.82 mS/cm. The area of pH H_2O (6.5-8.4) suitable for paddy plants is 391447.87 ha, TDS (<2000 ppm) 1067265.23 ha and EC (<3 mS/cm) 1066907.37 ha. Most of the pH H_2O suitable for paddy plants is in the northern region, while the TDS and EC are better in the northern region although most of them are still suitable for paddy plants. Each parameter has a spatial relationship, where the value of each parameter influences each other. Overall, the northern area in Paser Regency is more suitable for paddy cultivation according to water parameters, namely pH H_2O, TDS and EC.

1 Introduction

Paddy is a food crop that requires a lot of water. Water is one of the main needs for plants to support optimal growth and development. Most of the water needs of paddy plants are supplied from rainwater and irrigation. Some parameters of irrigation water to meet the needs of paddy plants are the quantity and quality of water. Problems that often arise in irrigation water are related to water quality [4]. In addition, water quality is also rarely considered by farmers. Poor management related to irrigation water quality management by farmers is one of the causes of low irrigation water quality [9].

Several parameters that are generally used to measure the level of irrigation water quality for plants are pH H_2O, TDS (Total Dissolve Solid) and EC (Electrical Conductivity) [10]. The pH of H_2O is the level of acidity in water which is indicated by a value of 0 to 14. The pH of H_2O that is good for paddy plants ranges from pH 6.5 to pH 8.4 [11]. TDS is the total dissolved solids contained in irrigation water. The TDS value is shown in ppm and the TDS value suitable for paddy plant growth is less than 2000 ppm [12]. A high TDS value indicates a higher level of turbidity in irrigation water. EC is a parameter to determine the number of ions in water that transmit electricity. The EC value is generally expressed in the form of mS/cm and a good EC value for paddy plants is less than 3 mS/cm [5].

Analysis of water quality parameters is carried out spatially using the Geographic Information System (GIS). Spatial analysis aims to determine the distribution in a spatial context and determine the spatial relationship between parameters. The spatial variability of each parameter needs to be known using available techniques which are often called spatial interpolation. One spatial interpolation technique is Inverse Distance Weighted (IDW). By inverse weighting of sample distance, the IDW method predicts unsampled values based on values from surrounding samples. The IDW method generally assumes that the closer the interpolation distance is to the sample, the interpolated results tend to have values that are almost similar to the values from adjacent samples [6]. This method is suitable to be used to determine the variability of each water quality parameter which has similarities according to irrigation flow [13].

From previous research, it is clear that studying the variability of water quality namely pH H_2O, TDS and EC is a significant endeavour as it will likely help in site-specific management of irrigation water resources in paddy crops. This research uses the IDW interpolation method to explore the spatial variability of pH H_2O, TDS and EC in irrigation water for paddy planting in Paser Regency, East Kalimantan Province, Indonesia. If spatial patterns of these parameters are discovered, these can be used for site-specific water resource management [14]. The purpose of this research is to spatially analyze the spatial distribution of water quality parameters i.e. pH H_2O, TDS and EC in the peaty paddy field. Determination of sample points using stratified random sampling method and parameter measurements were carried out directly in the field around the irrigation area.

2 Material and methods

The research was conducted from August 2023 to September 2023. The research location was in Paser District, East Kalimantan Province, Indonesia. The area of the research area is 1070115.85 ha or 10701.16 km2. The location of the research area starts from 346130.7329-455235.9485 (East) and 9733687.4763-9892759.7449 (Northing) with reference EPSG: 32750-WGS 84. The projection is Universal Transverse Mercator (UTM) with 50S zone. The material used in this research was an administrative map of Paser Regency at a scale of 1:50,000. The research tools used were pH meter, TDS meter, EC Meter, Garmin GPS, computer and Qgis 3.28 software.

Determination of sample points using stratified random sampling method based on the irrigation network area. The number of sample points in this study was 9 sampling locations. The coordinates of the sample points are determined using the Global Positioning System (GPS). Water quality parameters, namely pH H2O, TDS and EC, are measured directly on the land using a pH meter, TDS meter and EC meter. Measurements of each parameter were carried out at each point of the irrigation canal sample location, each sample measurement location was carried out three times and then the results were averaged. The results of the pH H2O, TDS and EC measurements were then input into Qgis software for spatial analysis. The value of each parameter has been input into the GIS software, then analyzed for spatial autocorrelation using the Moran's Index and P-value. Then interpolation is carried out to get the distribution of values for each parameter using the IDW interpolation method.

Initially the Moran index was used to determine the spatial pattern of a data using the formula below [2]:

$$I = \frac{n\sum_{i=1}^{n}\sum_{j=1}^{n} wij(x_i-\bar{x})(x_j-\bar{x})}{z_x^2\sum_{i=1}^{n}\sum_{j=1}^{n} wij} \quad (1)$$

Where: I = Moran's Index; n = number of samples; xi = value in sample I; xj = value in sample y; ‾x = average number of variables or values; wij = weighting element between sample i and sample y; = Sample percentage variance.

The interpolation method used is Inverse Distance Weighted (IDW). The calculation of the IDW method uses the following formula [1]:

$$F(i) = \sum_{i=1}^{m} W_i Z(r_i) = \frac{\sum_{i=1}^{m} Z(r_i)/|r-r_i|^p}{\sum_{j=1}^{m} 1/|r-r_i|^p} \quad (2)$$

Where: P = parameter; m = the number of neighbouring points taken into account at a certain cutting distance. The interpolated values are compared with the true values via one-cross validation of the omitted points.

Several tests were carried out to find a low Root Mean Square Error (RMSE) value on interpolation using the formula below [2]:

$$RMSE = \sqrt{\sum_{i=1}^{n}(p_i - a_i)^2} \quad (3)$$

Where: p_i = primary simulation result value of the observation variable; a_i = true value of the observed variable; n = number of observations

The RMSE calculation aims to measure the level of interpolation results that have been carried out. The lower or closer to zero the RMSE value, indicates the higher the level of accuracy of the interpolation. RMSE values from interpolation results for pH H_2O 0.0001, TDS 0.1695 and EC 0.0003. The interpolated data is then classified based on the category of each parameter. Classification results are presented in the form of distribution maps of pH H_2O, TDS and EC in scale. Then the maps are overlay analyzed by combining several maps. The results of the overlay analysis are presented in the form of an irrigation water quality map. The map of the research area is shown in Figure 1.

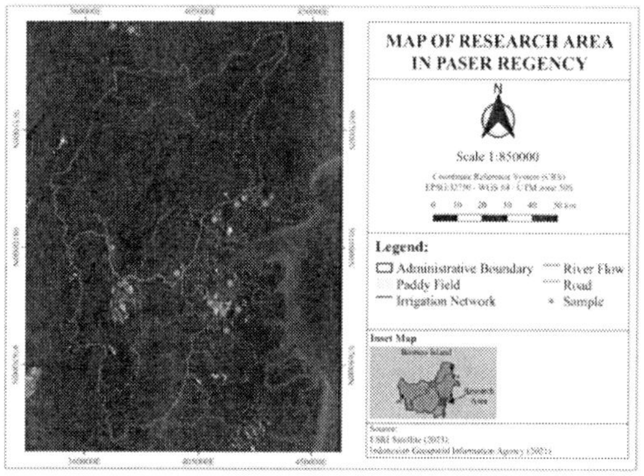

Fig. 1. Map of Research Area

3 Result and discussion

Table 1 shows the quality values of each parameter in the study area. The pH value of H_2O ranged from pH 4-9.2, TDS ranged from 10-3580 ppm and EC ranged from 0.02-5.7 mS/cm. In general, the value of each parameter has a different range of values. This could be due to differences in location and irrigation networks in sampling.

Table 1. The Value of Water Quality Parameters

Parameter	Min	Max	Mean	Median
pH H_2O	4	9,2	6,37	6,1
TDS (ppm)	10	3580	514,44	100

EC (mS/cm)	0,02	5,7	0,87	0,15

3.1 Spatial Distributional of pH Water (pH H_2O)

Figure 2 shows the grouping analysis of water pH (pH H2O). Moran's index of 0.793 (positive) indicates that there is a grouping pattern. The data show some similarities in values in some parts of the study area. Apart from that, the existence of grouping is also supported by a P-value of 0.01, although in some small parts there are values that are more random. This is also supported by previous research, that a Moran's index that is close to +1 or -1 and a P-value <0.05 indicates that there is spatial autocorrelation and the data distribution is more clustered [15]

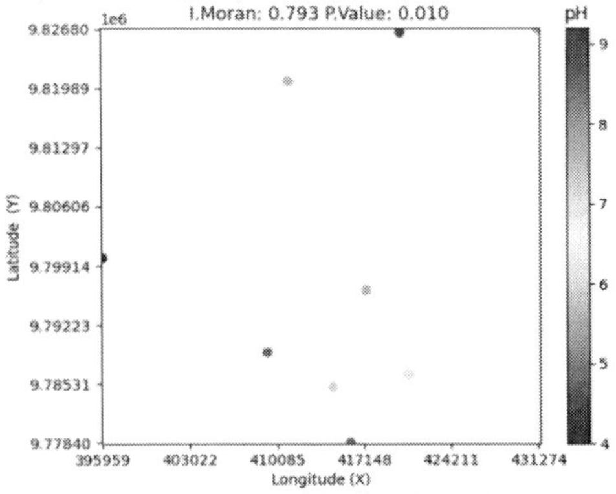

Fig. 2. Moran Index of pH Water

Figure 3 shows the distribution of water pH values (H_2O) in the study area. The distribution of water pH values in the study area is divided into three categories, namely acid, neutral and alkaline. The red color on the map shows the water pH value <6.5, the yellow color on the map shows the water pH value 6.5 -8.4 and the blue color on the map shows the water pH value >8.4. As can be seen in Table 2, 62.33% of the research area is included in the Acid pH category with an area of 666989.76 ha and 36.58% is included in the Neutral pH category with an area of 391447.87 ha. Conversely, a small portion of the study area is included in the category of alkaline pH with a percentage of 1.09% with an area of 11678.22 ha.

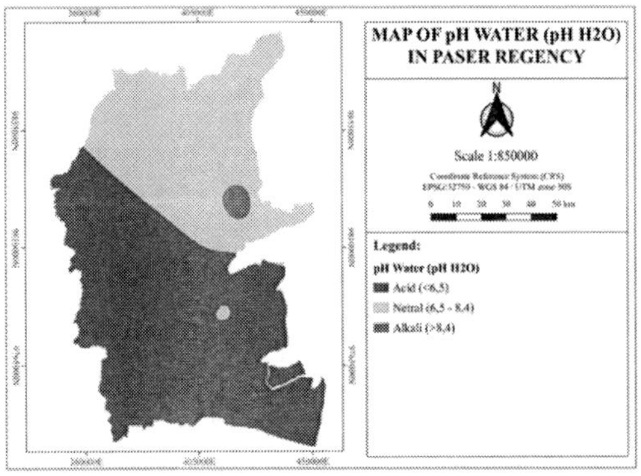

Fig. 3. Distribution Map of pH Water in Paser Regency.

Table 2. The Distribution of Area for Criteria of pH Water

Color	pH H_2O	Criteria	Area (ha)	Percentage (%)
Red	<6,5	Acid	666989,76	62,33
Yellow	6,5 – 8,4	Netral	391447,87	36,58
Blue	>8,4	Alkali	11678,22	1,09
Amount			**1070115,85**	**100**

Referring to Figure 3, pH water with the acid category is clustered in the southern research area and pH water with the neutral category is clustered in the northern research area. In contrast, pH with the alkaline category is only found in a small part of the study area. The northern part of the research area with neutral water pH is more suitable for paddy cultivation. In addition, pH water with a neutral category is one of the conditions for growing paddy plants [7]. On the other hand, the southern region, which is dominated by pH water, is in the acid category, which is not suitable for paddy cultivation. This can cause toxicity in plants such as aluminium toxicity which can inhibit the growth and development of paddy plants and can reduce productivity. [17]. However, it is possible that this area can be cultivated for paddy plants with the condition that there is a need for good management of irrigation water, such as adding lime and biochar before planting [18]. The same thing is also found in some research areas with pH water that is in the alkaline category. pH water in the alkaline category contains quite high levels of cations, so it has the potential to cause paddy plants to become poisoned [3]. However, in this area it will be more difficult to manage irrigation water than pH water in the acid category [19]. Thus, in this area it is not suitable for cultivating paddy plants when viewed from the pH of the water.

3.2 Spatial Distributional of TDS

Figure 4 shows the TDS grouping analysis in the research area. Referring to the Moran Index value, namely 0.182, it shows that the distribution of values is not clustered, but is more random. However, if we refer to the P-value of 0.085, it shows that several small parts experiences grouping. This might happen because the properties of TDS are very location specific and influenced by many factors. This is also supported by previous research, that the Moran's index which is close to 0 and the P-value >0.05 indicates there is no spatial autocorrelation and the data distribution is more random [16].

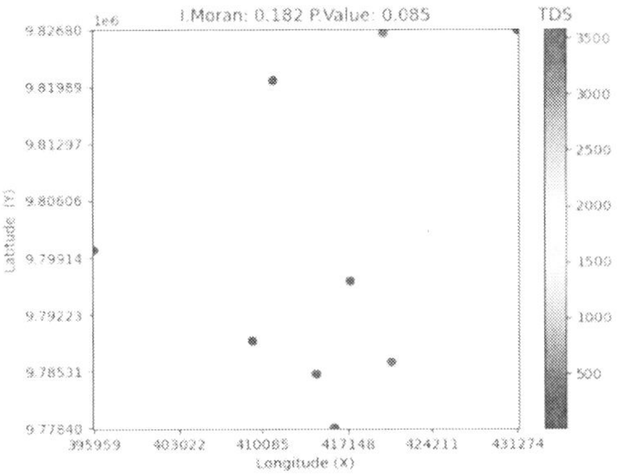

Fig. 4. Moran Index of TDS Water

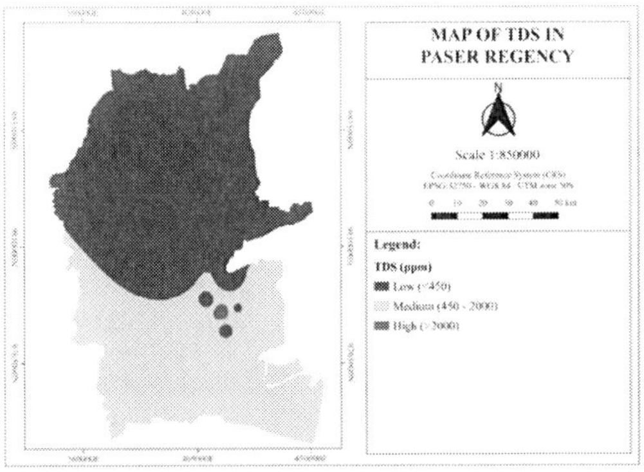

Fig. 5. Distribution Map of TDS in Paser Regency

Figure 5 shows the distribution pattern of TDS (ppm) values in the research area. The distribution of TDS values from the interpolation results is divided into three categories, namely low, medium and high. As can be seen in Figure 5, the low category (red on the map) has a value of <450 ppm, the medium category (yellow on the map) has a value of 450-2000 ppm and the high category (blue on the map) has a value of >2000 ppm. Most of the area is dominated by the low category with a percentage of 57.35% (613667.49 ha), while for the medium category the percentage is 42.39% (453597.74 ha) and the high category is 0.27% (2850.61 ha), For more details, see Table 3.

Table 3. The Distribution of Area for Criteria of TDS

Color	TDS (ppm)	Criteria	Area (ha)	Percentage (%)
Red	<450	Low	613667,49	57,35
Yellow	450 - 2000	Medium	453597,74	42,39
Blue	>2000	High	2850,62	0,27
Amount			**1070115,85**	**100**

Referring to Figure 5, TDS with a low category is grouped in the northern region, the moderate category is clustered in the southern region and the high category is in a small portion of the region. One of the requirements for growing paddy plants is to have a TDS value below 2000 ppm [20]. This shows that most of the research area is still suitable for cultivating paddy, if we refer to the TDS value. However, in the medium category it is necessary to check the TDS value periodically because it can potentially become a high category in certain circumstances. The high TDS value in areas with the high category is most likely caused by household and industrial waste as indicated by the population density spread across the area [8]. Apart from that, it is very likely that this is also caused by agricultural waste such as fertilizer waste, pesticides and plant residues [21].

3.3 Spatial Distributional of EC

Figure 6 shows the grouping pattern of EC values in the research area. The Moran Index with a value of 0.164 indicates that the EC value in the study area is more random. However, if we refer to the P-value of 0.043, in some small parts there are clustered values. This is also similar to the incident with the TDS value.

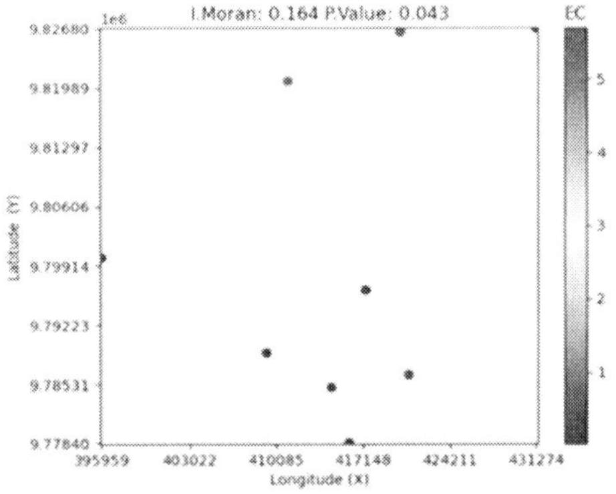

Fig. 6. Moran Index of EC Water

Figure 7 shows the distribution of EC values in the study area. As seen in Figure 7 there are three categories namely low, medium and high. The red color on the map shows an EC value of <0.7 mS/cm, the yellow color on the map shows a value of 0.7-3 mS/cm and the blue color shows a value > 3 mS/cm. As shown in Table 4, the medium category was more dominant in the study area with a percentage of 57.41% (614331.73 ha), then continued with the low category with a percentage of 42.29% (452575.64 ha) and the high category with a percentage of 0. 3 % (3208.48 ha).

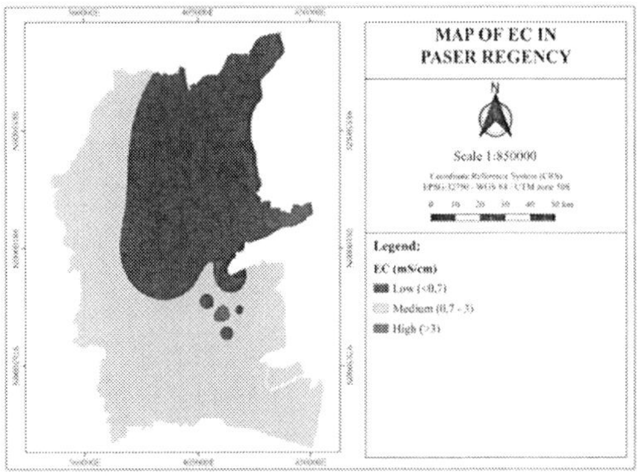

Fig. 7. Distribution Map of EC in Paser Regency

Table 4. The Distribution of Area for Criteria of EC

Color	EC (mS/cm)	Criteria	Area (ha)	Percentage (%)
Red	<0,7	Low	452575,64	42,29
Yellow	0,7 - 3	Medium	614331,73	57,41
Blue	>3	High	3208,48	0,30
Amount			**1070115,85**	**100**

Referring to Figure 7, the distribution of EC values in the low category is clustered in the north area, while the medium category is grouped in the south and east areas. Conversely, the high category is only found in a small number of areas. This is most likely caused by domestic waste in residential areas [24]. Broadly speaking, the distribution of EC values is not much different from the distribution of TDS values and pH. This is because the EC value with the TDS dan pH value has a fairly close relationship [22]. Apart from that, the high EC value in water is also caused by several factors such as salt content [23]. In areas close to the sea, EC values are higher. Waste from human activities such as agricultural, industrial, mining waste and so on also has quite a large potential for increasing the EC value of water [25].

Overall, the water quality area, namely pH H2O (6.5-8.4) which is suitable for paddy plants, is 391447,87 ha, TDS (<2000 ppm) 1067265,23 ha and EC (<3 mS/cm) 1066907,37 ha [26]. The pH of H_2O which is suitable for paddy plants is mostly in the northern region, while the TDS and EC are better in the northern region although most of it is still suitable for paddy plants. The northern region of Paser Regency is more suitable for paddy cultivation in terms of quality parameters, namely pH H_2O, TDS and EC.

3.4 Spatial Distributional of Irrigation Water Quality

Figure 8 shows the spatial distribution of irrigation water quality in Paser Regency. As seen in Figure 8, there are three categories of irrigation water quality, namely less good (red color on the map), quite good (yellow color on the map) and very good (blue color on the map). The red color on the map indicates less good irrigation water quality criteria with pH values <6.5 and >8.4, TDS >2000 ppm and EC values >3 mS/cm. Yellow color on the map shows the criteria of quite good irrigation water quality with pH value 6.5-8.4, TDS 450-2000 ppm and EC 0.7-3 mS/cm. Blue color on the map indicates very good irrigation water quality criteria with pH values of 6.5-8.4, TDS < 450 ppm and EC < 0.7. As seen in Table 5, most of the study area is in the less good criteria with a percentage of 69.26%, followed by very good criteria with a percentage of 30.69% and quite good criteria with a percentage of 0.05%.

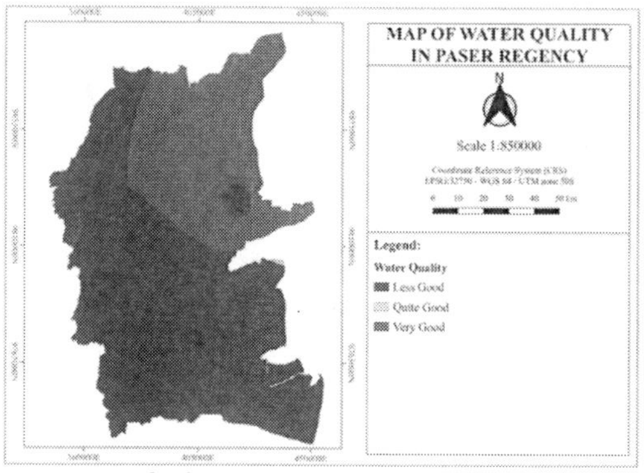

Fig. 8. Map of Irrigation Water Quality in Paser Regency

Table 5. The Distribution of Area for Criteria of EC

Color	Criteria	Area (ha)	Percentage (%)
Red	Less Good	741111,67	69,26
Yellow	Quite Good	551,23	0,05
Blue	Very Good	328452,96	30,69
Amount		**1070115,85**	**100**

Referring to Figure 8, most of the study area is in the less good irrigation water quality criteria. In this area, good irrigation water management is needed for paddy cultivation. Waste management and the use of fertilizers according to the recommended dosage can minimize the contamination of harmful chemical elements in irrigation water [27]. In addition, the use of mulch and the addition of biochar to the soil can also minimize toxicity to paddy plants [30]. This is very important to do to minimize toxicity in paddy plants [28]. In the criteria of very good irrigation water quality is in the northern region. This area is very suitable to support paddy cultivation. In the criteria quite good also needs to be considered related to the management of irrigation water, although this criteria is only found in a small part of the research area. This is because poor management of irrigation water in this area will potentially reduce the quality of irrigation water, so that the quality of irrigation water becomes less good [29].

4 Conclusion

Spatial distributional parameters of water quality for paddy plants, namely pH H_2O, TDS and EC are relevant parameters for site-specific management of irrigation areas. The results of this research prove that the northern part of the research area is more suitable for cultivating paddy. Likewise, the results of this study confirm that each irrigation water quality parameter is spatially related.

Acknowledgments

We would like to extend our heartfelt appreciation to Directorate of Research Universitas Gadjah Mada for their generous financial support through the Final Project Recognition (RTA) Grant, Universitas Gadjah Mada, Number 5075/UN1.P.II/Dit-Lit/PT.01.01/2023.

This support was instrumental in the successful completion of our research.

References

1. Huded, S., Pramesh, D., Chittaragi, A., Sridhara, S., Chidanandappa, E., Prasannakumar, M. K., ... & Shamshiri, R. R. (2022). Spatial Distribution Patterns for Identifying Risk Areas Associated with False Smut Disease of Rice in Southern India. *Agronomy*, *12*(12), 2947.
2. Wijayanto, Y., Abdilah, M. R. W., Purnamasari, I., & Saputra, T. W. (2022). Spatial Analysis of Soil Available Potassium and Plant Brix Content for Site Specific Nutrient Management in Sugarcane. *AGRIVITA, Journal of Agricultural Science*, *44*(3), 549-561.
3. Martínez-Cortijo, J., & Ruiz-Canales, A. (2018). Effect of heavy metals on rice irrigated fields with waste water in high pH Mediterranean soils: The particular case of the Valencia area in Spain. *Agricultural Water Management*, *210*, 108-123.
4. Tan, L. V., Tran, T., & Loc, H. H. (2020). Soil and water quality indicators of diversified farming systems in a saline region of the Mekong Delta, Vietnam. *Agriculture*, *10*(2), 38.
5. Serge Gaëtan, S. É., Marie Michelle, C. H., Ouindinboudé Jacques, K., Poulouma Louis, Y., & Idriss, S. (2022). Hydrogeochemistry of Shallow Groundwater and Suitability to Irrigation: The Case of the Karfiguéla Paddy Field in Burkina Faso. *Water*, *14*(16), 2574.
6. Yang, W., Zhao, Y., Wang, D., Wu, H., Lin, A., & He, L. (2020). Using principal components analysis and IDW interpolation to determine spatial and temporal changes of surface water quality of Xin'anjiang river in Huangshan, China. *International journal of environmental research and public health*, *17*(8), 2942.
7. Afrad, M. S. I., Monir, M. B., Haque, M. E., Barau, A. A., & Haque, M. M. (2020). Impact of industrial effluent on water, soil and Rice production in Bangladesh: a case of Turag River Bank. *Journal of Environmental Health Science and Engineering*, *18*, 825-834.
8. Adimalla, N., Li, P., & Venkatayogi, S. (2018). Hydrogeochemical evaluation of groundwater quality for drinking and irrigation purposes and integrated interpretation with water quality index studies. Environmental Processes, 5, 363-383.
9. Asadi, E., Isazadeh, M., Samadianfard, S., Ramli, M. F., Mosavi, A., Nabipour, N., ... & Chau, K. W. (2019). Groundwater quality assessment for sustainable drinking and irrigation. *Sustainability*, *12*(1), 177.
10. Ramachandran Muthulakshmi, Y. (2020). Geo-spatial analysis of irrigation water quality of Pudukkottai district. *Applied Water Science*, *10*(3), 1-14.
11. Mandal, S. K., Dutta, S. K., Pramanik, S., & Kole, R. K. (2019). Assessment of river water quality for agricultural irrigation. *International Journal of Environmental Science and Technology*, *16*, 451-462.
12. Omeka, M. E. (2023). Evaluation and prediction of irrigation water quality of an agricultural district, SE Nigeria: an integrated heuristic GIS-based and machine learning approach. *Environmental Science and Pollution Research*, 1-26.
13. Haldar, K., Kujawa-Roeleveld, K., Dey, P., Bosu, S., Datta, D. K., & Rijnaarts, H. H. (2020). Spatio-temporal variations in chemical-physical water quality parameters influencing water reuse for irrigated agriculture in tropical urbanized deltas. *Science of the Total Environment*, *708*, 134559.
14. Bhatti, S., Heeren, D. M., Barker, J. B., Neale, C. M., Woldt, W. E., Maguire, M. S., & Rudnick, D. R. (2020). Site-specific irrigation management in a sub-humid climate using a spatial evapotranspiration model with satellite and airborne imagery. *Agricultural Water Management*, *230*, 105950.
15. Weerasinghe, V. P. A., & Handapangoda, K. (2019). Surface water quality analysis of an urban lake; East Beira, Colombo, Sri Lanka. *Environmental Nanotechnology, Monitoring & Management*, *12*, 100249.
16. Chen, Y., Yin, G., & Liu, K. (2021). Regional differences in the industrial water use efficiency of China: The spatial spillover effect and relevant factors. *Resources, Conservation and Recycling*, *167*, 105239.
17. Shetty, R., & Prakash, N. B. (2020). Effect of different biochars on acid soil and growth parameters of rice plants under aluminium toxicity. *Scientific Reports*, *10*(1), 12249.
18. Wen, E., Yang, X., Chen, H., Shaheen, S. M., Sarkar, B., Xu, S., ... & Wang, H. (2021). Iron-modified biochar and water management regime-induced changes in plant growth, enzyme activities, and phytoavailability of arsenic, cadmium and lead in a paddy soil. *Journal of hazardous materials*, *407*, 124344.
19. Lin, L., Xu, F., Ge, X., & Li, Y. (2018). Improving the sustainability of organic waste management practices in the food-energy-water nexus: A comparative review of anaerobic digestion and composting. *Renewable and Sustainable Energy Reviews*, *89*, 151-167.
20. Ebraheem, A. A., Alshamsi, D., Sherif, M., Neuman, E., Alghafli, K., Al Mulla, M., & Sefelnasr, A. (2023). Mapping groundwater potential zones in UAE using GIS-based hydrogeological modeling. In *Water Resources Management and Sustainability: Solutions for Arid Regions* (pp. 255-268). Cham: Springer Nature Switzerland.

21. Ganiyu, S. A., Badmus, B. S., Olurin, O. T., & Ojekunle, Z. O. (2018). Evaluation of seasonal variation of water quality using multivariate statistical analysis and irrigation parameter indices in Ajakanga area, Ibadan, Nigeria. *Applied water science*, *8*, 1-15.
22. Rusydi, A. F. (2018, February). Correlation between conductivity and total dissolved solid in various type of water: A review. In *IOP conference series: earth and environmental science* (Vol. 118, p. 012019). IOP Publishing.
23. Sreekesh, S., Sreerama Naik, S. R., & Rani, S. (2018). Effect of sea level changes on the groundwater quality along the coast of Ernakulam District, Kerala. *Journal of Climate Change*, *4*(2), 51-65.
24. Balamurugan, P., Kumar, P. S., Shankar, K., Nagavinothini, R., & Sajil Kumar, P. J. (2020). Impact of climate and anthropogenic activities on groundwater quality for domestic and irrigation purposes in Attur region, Tamilnadu, India. *India. Desalin. Water Treat*, *208*, 172-195.
25. Patel, M. P., Gami, B., Patel, A., Patel, P., & Patel, B. (2020). Climatic and anthropogenic impact on groundwater quality of agriculture dominated areas of southern and central Gujarat, India. *Groundwater for sustainable development*, *10*, 100306.
26. Sihi, D., Dari, B., Yan, Z., Sharma, D. K., Pathak, H., Sharma, O. P., & Nain, L. (2020). Assessment of water quality in Indo-Gangetic Plain of South-Eastern Asia under organic vs. conventional rice farming. *Water*, *12*(4), 960.
27. Chojnacka, K., Witek-Krowiak, A., Moustakas, K., Skrzypczak, D., Mikula, K., & Loizidou, M. (2020). A transition from conventional irrigation to fertigation with reclaimed wastewater: Prospects and challenges. *Renewable and Sustainable Energy Reviews*, *130*, 109959.
28. Okorogbona, A. O., Denner, F. D., Managa, L. R., Khosa, T. B., Maduwa, K., Adebola, P. O., ... & Macevele, S. (2018). Water quality impacts on agricultural productivity and environment. *Sustainable Agriculture Reviews 27*, 1-35.
29. Dhaouadi, L., Besser, H., Wassar, F., Kharbout, N., Brahim, N. B., Wahba, M. A., & Kang, Y. K. (2020). Agriculture sustainability in arid lands of southern Tunisia: Ecological impacts of irrigation water quality and human practices. *Irrigation and Drainage*, *69*(5), 974-996.
30. Assefa, T., Jha, M., Reyes, M., & Worqlul, A. W. (2018). Modeling the impacts of conservation agriculture with a drip irrigation system on the hydrology and water management in Sub-Saharan Africa. *Sustainability*, *10*(12), 4763.

The Critical Level Mapping of the Infiltration Area in the Northern Part of Banyumanik District, Semarang City, Central Java, Indonesia

Ahmad Arfiandi[1], *Wawan* Budianta[1*], and *Esti* Handini[1]

[1]Universitas Gadjah Mada, Geological Engineering Department, 55281 Yogyakarta, Indonesia

Abstract. This research was conducted in Semarang City, in the northern part of Banyumanik District, including six sub-districts: Srondol Wetan, Srondol Kulon, Sumurboto, Pedalangan, Tinjomoyo, and Ngesrep, to map the critical condition of infiltration zone in the research area. This research aims to map the critical level of the infiltration area in the research area that can be used as a reference in policymaking regarding groundwater conservation and disaster mitigation in the research area. The primary data used were soil and rock samples, with secondary data such as regional geological maps, land use maps, slope maps, and soil permeability maps used to support the primary data obtained in the field. The data were then processed using the AHP (Analytical Hierarchy Process) method with parameters such as soil grain size, soil permeability, land use, lithology, slope, and density of geological structures. Based on the calculations, the research area could be divided into criticality classes of infiltration: *Very Critical, Critical, Moderately Critical, Starting to be Critical, Naturally Normal*, and *Good*.

1 Introduction

Massive developments happen in big cities, such as Semarang City. Huge developments have made areas such as hills and riverbanks change their function into residential and industrial areas instead of being infiltration areas. The government tends to prioritize the economic value of a space by creating cultivation areas (tourism, industry, and residential.) rather than protected area spaces (infiltration area) [1]. It makes the water that falls from the rain flow directly to low places without absorbing water into the soil [2]. With water absorption into the ground, groundwater reserves are reduced since groundwater extraction for industry and settlements is balanced with the return of water to the earth.

One district that experiences massive development is Banyumanik, especially in the northern part. The official website of the Semarang City Government shows that most (68.71%) of the Banyumanik District is used for settlements—meanwhile, infiltration areas such as rice fields, gardens, and meadows, only 27.28%. Compared to infiltration areas, the large residential and industrial areas make it difficult for water to seep into the ground. Various actions need to be taken to minimize disasters in Banyumanik District due to less-than-optimal water infiltration; one way is to evaluate the infiltration area. All research areas are assessed and then classified into infiltration area categories such as *Good, Naturally Normal, Starting to be Critical, Moderately Critical, Critical,* and *Very Critical*. These infiltration area categories are determined through the AHP (Analytical Hierarchy Process) method for parameter weighting, such as soil grain size, land use, slope gradient, lithology, soil permeability, and density of geological structures.

AHP is a decision-making method considering several criteria (multi-criteria) grouped in a hierarchy [3]. It is a method that emphasizes the proportion of selected criteria; therefore, it is suitable for determining the best alternative to choose while facing a problem [4]. There has been previous research on a similar topic, including the determination of a spatial model of potential water catchment area [2], the determination of a water catchment area in Bandung [5], and criticality analysis of water catchment areas in Cikeruh [6]. These considerations lead to the AHP method used in solving problems in this research. Several other methods, such as SAW (Simple Additive Weighting), can be used, but AHP is the most suitable method in this research. This calculation was used because, in AHP, the relationship between parameters is considered, unlike SAW, which does not pay attention to the relationship between parameters.

This research is generally located in the northern part of Banyumanik District of Semarang City of Central Java Province. In more detail, the research location covers six sub-districts in Banyumanik District and they are Srondol Wetan, Srondol Kulon, Sumurboto, Pedalangan, Tinjomoyo, and Ngesrep (Fig. 1).

This research aims to map the critical level of the infiltration area in the research area. The government can use this map as a reference for policymaking

* Corresponding author: wbudianta@ugm.ac.id

regarding groundwater conservation and disaster mitigation in the research area.

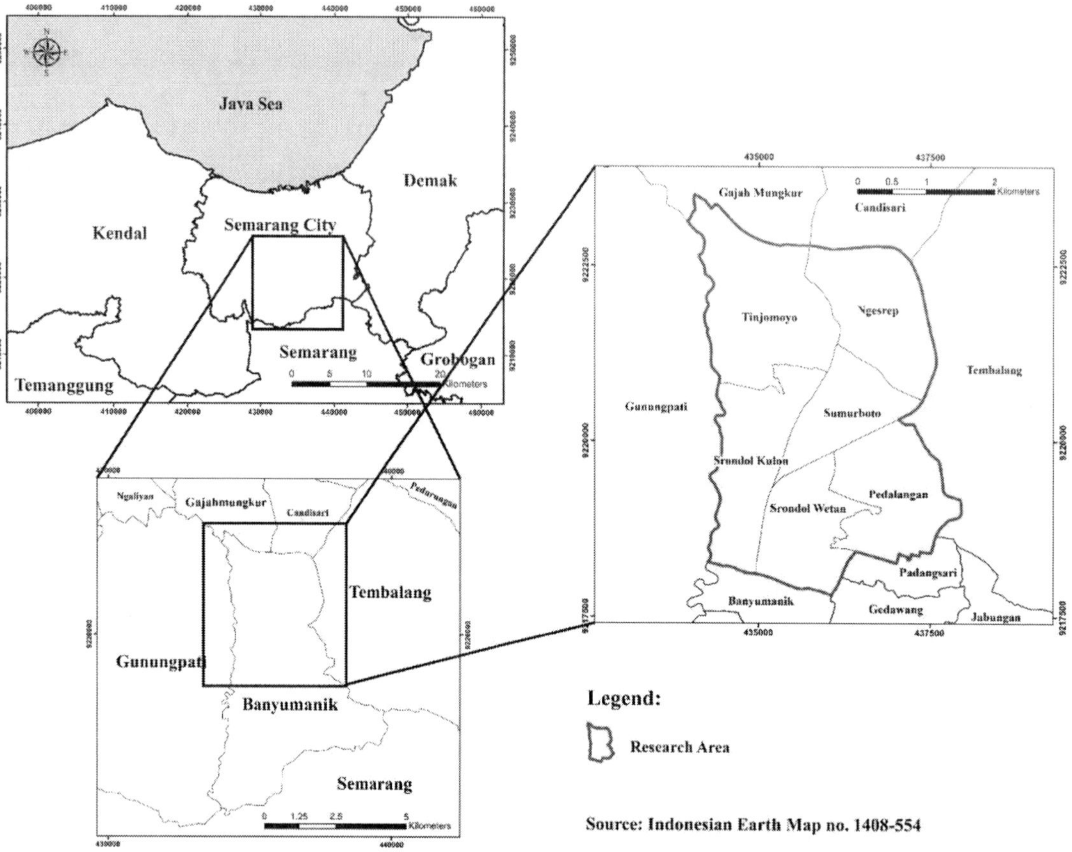

Fig. 1. Research area [7].

2 Research Method

The stages of this research consisted of introduction, data collection, laboratory analysis, and data analysis.

2.1 Literature Study

This research began with a literature study to collect references related to the research topic used as material for planning and preparing for the research to be carried out. The next was to identify the parameters using previous research as reference material. The parameters used in the previous research were examined and further evaluated to determine the priority scale.

2.2 Data Collection

The data collection was conducted directly in the field, analyzing secondary data from various sources. The primary data were soil and rock samples collected directly on the field. Soil samples were collected in each 650 m x 650 m grid in the research area, while the rock samples were collected by geological mapping by ground checking based on the geological map [8]. The secondary data used in this research were regional geological maps, soil permeability maps, DEM (Digital Elevation Model) maps, slope gradient maps, and land use maps with 1:100.000 scale. These maps were processed using ArcMap and PIC Geomatica software.

2.3 Laboratory Analysis

The laboratory analysis consisted of granulometric analysis. Granulometric analysis was carried out on soil samples to determine the grain size distribution of the soil in the research area. The result of this analysis was used to generate the soil grain size map.

2.4 Data Analysis

The data analysis using the AHP method was carried out by weighting using predetermined parameters: soil grain size, land use, slope, lithology, soil permeability, and density of geological structures. These parameters were compared with each other to produce a pairwise comparison matrix. Comparisons were made with the researcher's subjective justification with references from previous studies. The comparison between the two parameters was stated in numbers 1 to 9; the larger the number, the more essential the parameter was. The weighting of these parameters would apply the repetition law; for instance, parameter *a* was considered three times more essential than parameter *b*, then parameter *b* became ⅓ more essential than parameter *a*. The next step was to normalize the importance level weights for each parameter. The consistency test was carried out on the weighting using the reference value of the CR (Consistency Ratio). The weighting was declared valid if the CR value was < 0.1. CR calculation is done by Calculation 1 below:

$$CR = CI/RI \quad (1)$$

The CI (Consistency Index) value itself is obtained from Calculation 2 below:

$$CI = (\lambda max - n)/(n - 1) \quad (2)$$

The value of λmax was the sum of multiplying the priority vector for each row by the sum of each column, and *n* was the number of matrix orders. Meanwhile, RI (Ratio Index) was a random value based on each matrix order based on simplified Table 1 according to [3].

Table 1. RI values [3].

n	1	2	3	4	5	6	7	8
RI	0	0	0,52	0,89	1,11	1,25	1,35	1,4

In addition to calculating AHP on all parameters, the AHP calculation was also carried out on sub-parameters for each parameter. The two results were combined to get the desired final result.

Sensitivity testing is a step taken to test the suitability of the AHP weighting that has been carried out [9]. Sensitivity level testing is carried out by decreasing or increasing the weight by a certain value. AHP weighting is stated to have high consistency if changes in the priority order of choices can only occur if there is a large change in the criteria weights.

3 Results Disclosure

Based on the previously described AHP calculation, the results of the weighting between parameters were obtained, shown in Table 2 below:

Table 2. AHP calculates each parameter [7].

	Factors	1	2	3	4	5	6	Weight
1	Soil Grain Size	1	2	2	3	4	4	0,33
2	Soil Permeability	1/2	1	2	2	3	4	0,23
3	Land Use	1/2	1/2	1	2	3	4	0,19
4	Lithology	1/3	1/2	1/2	1	2	3	0,12
5	Slope Gradient	1/4	1/3	1/3	1/2	1	2	0,08
6	Density of Geological Structure	1/4	1/4	1/4	1/3	1/2	1	0,05
λmax								6,2
CI (*Consistency Index*)								0,041
CR (*Consistency Ratio*)								0,033

3.1 Weighting of Each Parameter

3.1.1 Soil Grain Size

The soil grain size was divided based on the grain size classification according to the USCS [10]. Before classifying, the soil was sieved using a sieve shaker with a mesh size of 10 – 200. This research used 40 soil samples that were collected in the research area. There were only soils with a coarse grain size in the research area, where all sifted soil samples had a percentage of > 50%, which did not pass through a 200-mesh sieve. Therefore, this research used only the USCS classification for soils with coarse grain sizes. More simply, the classification of soil grain size considered the percentage by weight of clay as a fine fraction compared to the total weight of the sample. Based on this classification, in the research area, there were only coarse-grained soil units with a small fine fraction (K-FHSdk) and coarse-grained soil units with a medium fine fraction (K-FHSdg) (Fig.2). K-FHSdk was a coarse-sized soil with a fine fraction of < 5%, while K-FHSdg was a coarse-sized soil with a fine fraction of 5 – 12%. K-FHSdk dominated the research area with a percentage of 78%, while the rest were K-FHSdg. IDW (Inverse Distance Weighting) generated the soil grain size using ArcGIS to produce a 1:25.000 soil grain size map scale.

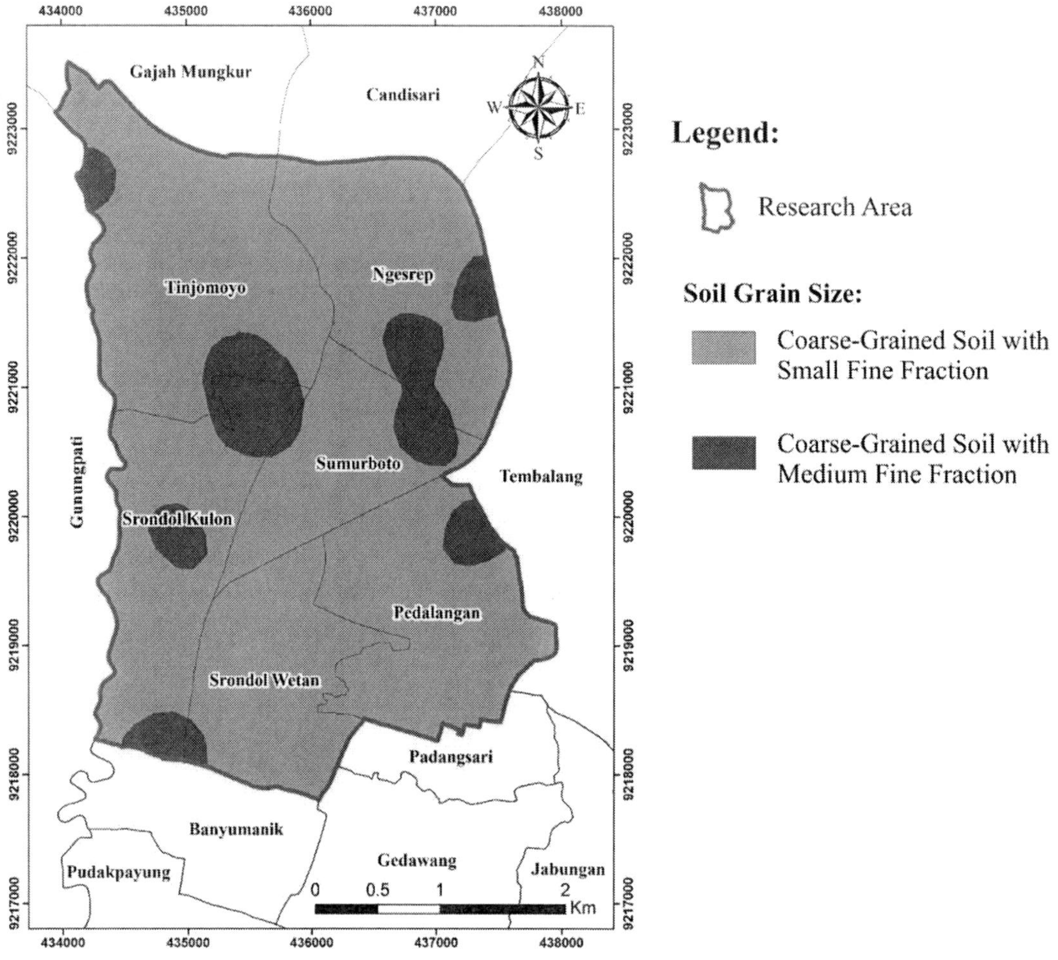

Fig. 2. Soil grain size map [7].

3.1.2 Soil Permeability

Soil permeability determined how much rainwater infiltrated and how much rainwater became surface runoff. In this research, the soil permeability data were collected from the permeability map of Semarang City from the Spatial Planning Office of Semarang City. That map was used to make soil permeability maps in the research area with a 1:25.000 scale (Fig. 3) by cropping a map based on the research area. The map shows that the research area was divided into four classes, with the very slow soil permeability class dominating the southeastern part of the research area. Meanwhile, in the western part, soil permeability was slow-medium in the south and slow in the north. At the same time, the soil permeability was medium-fast in the northern part of the research area.

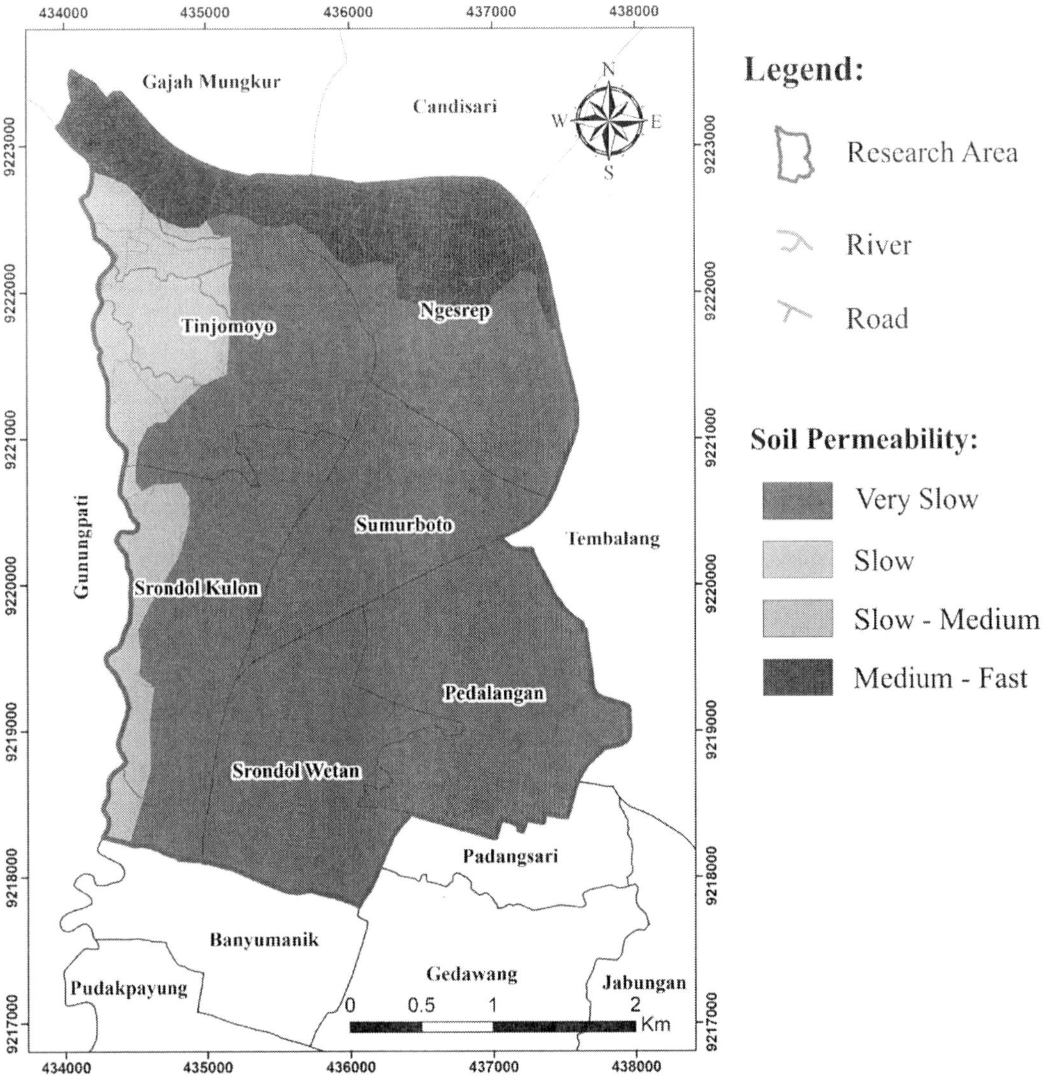

Fig. 3. Soil permeability map [7].

3.1.3 Land Use

Land use in the research area was determined based on the map made by the Semarang City Spatial Planning Office, which was then verified directly in the field by ground check. All those data were generated on a land use map with a 1:25.000 scale. Based on this, land use in the research area was divided into settlements (60%), meadows and pastures (38%), and shrubs and grasslands (2%), as shown in Fig.4. Most of the research area was used as settlements, especially in the southeastern part of the research area, indicating that much green open land had been converted into settlements due to massive development. Meanwhile, in the western part of the research area, many open green areas could become infiltration areas. This section shows land and fields extending north-south, with meadows in the Tinjomoyo sub-district, a golf course. Meanwhile, green open land could be found in meadows, shrubs, fields, and moors at several points in the middle of the settlement.

https://doi.org/10.1051/bioconf/20238003019

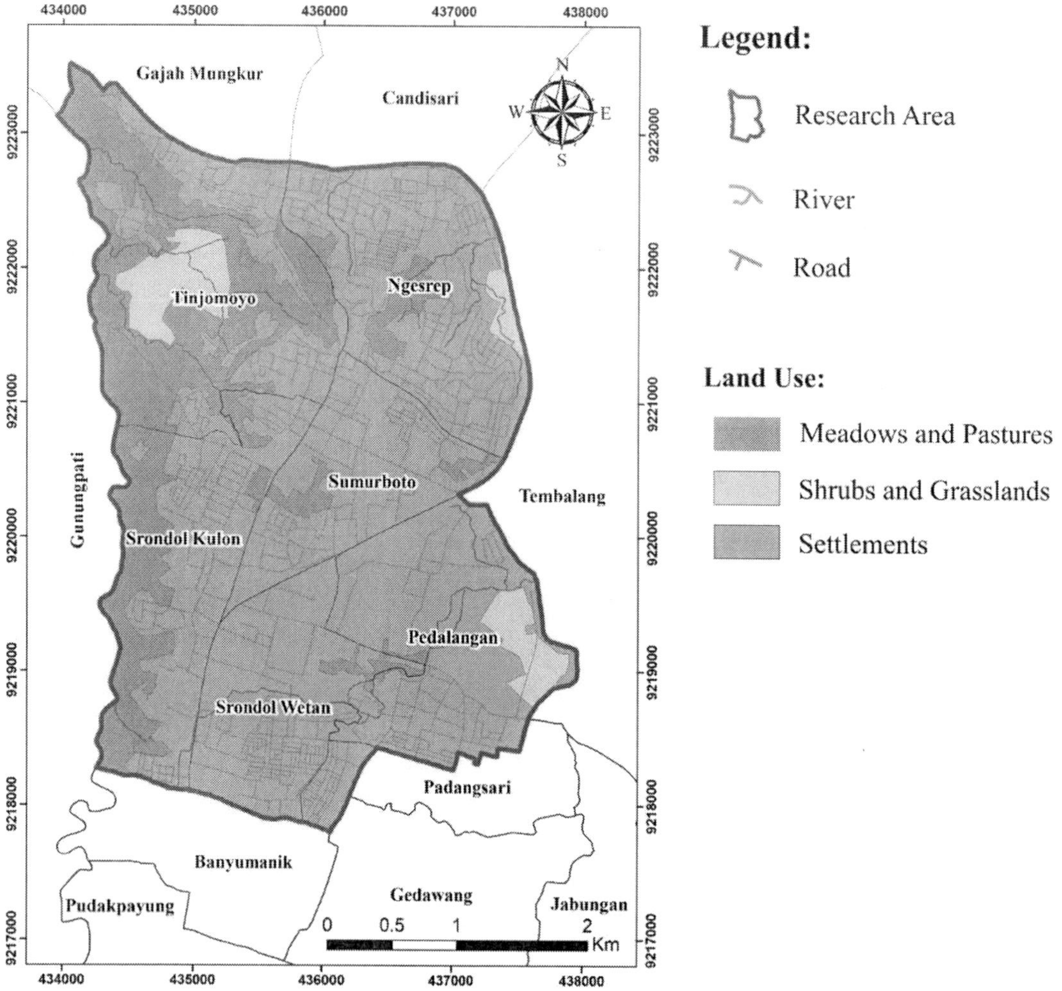

Fig. 4. Land use map [7].

3.1.4 Lithology

The lithology data were retrieved using secondary data from regional geological maps [8] and verified directly in the field. The field verification was carried out by direct observation of rock outcrops. In this research, there were 12 field stations in the research area. In addition, laboratory analysis was also carried out by observing the petrographic properties of the rocks. Based on the analysis carried out in the research area, there were volcanic breccia and carbonate claystone units that generated a lithological map with a 1:25.000 scale (Fig. 5). Volcanic breccia units dominated almost the entire research area, with a percentage of 98%. Meanwhile, carbonate claystone units were only found in the southwest of the study area, with an area coverage of 2%.

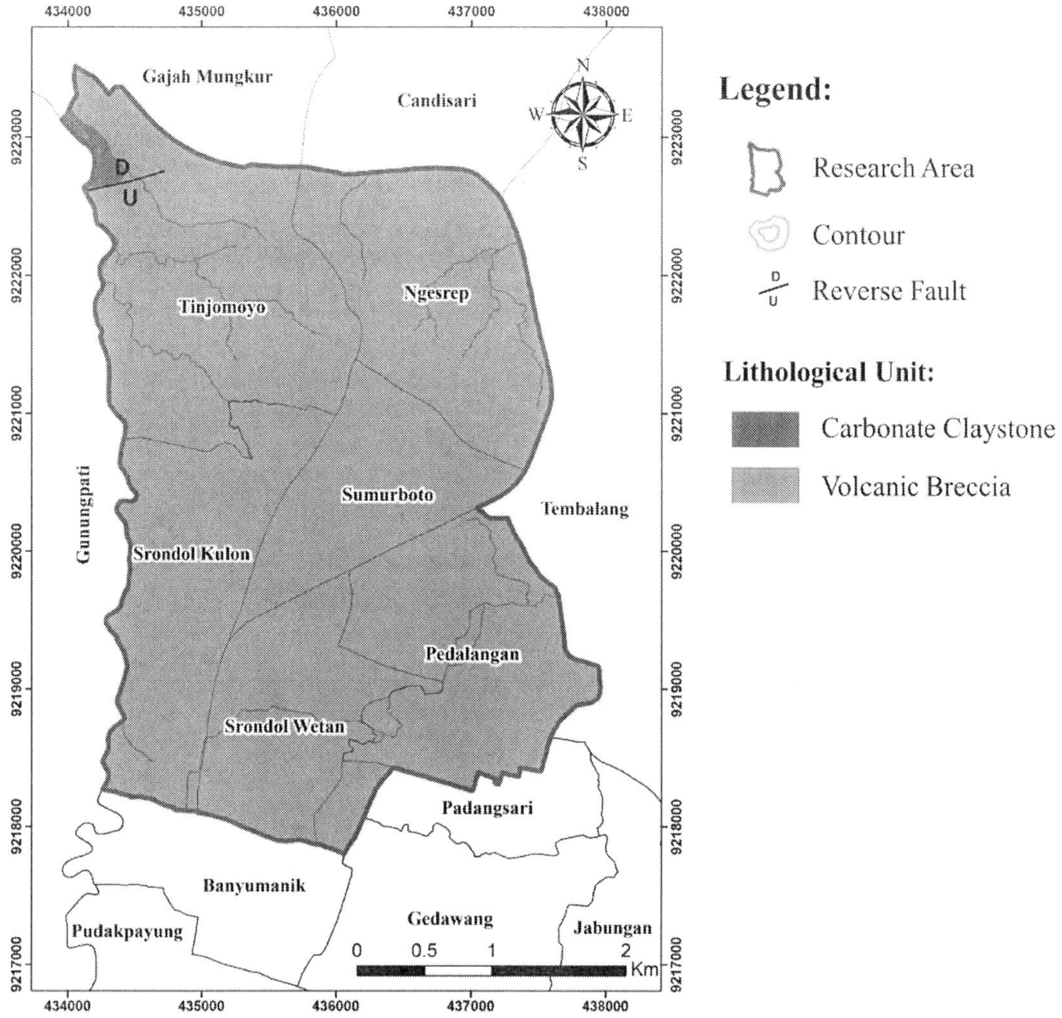

Fig. 5. Lithological map [7].

3.1.5 Slope Gradient

The slope gradient in the research area was determined by analyzing secondary data using a slope map from the Spatial Planning Office of the City of Semarang and a contour map with a scale of 1:50.000. Based on the classification made by the Directorate General of Reforestation and Land Rehabilitation [11], the area was divided into five slope classes, namely flat, sloping, wavy, steep, and very steep. The results of the slope map in the research area can be seen in Fig.6. which shows a slope gradient map with a 1:25.000 scale. Overall, the slope class was dominated by flat slopes with a percentage of 52.1%, followed by gentle slopes at 22.6% and wavy slopes at 14.4%. Meanwhile, the steepest slope is only 7.1%, and the steepest slope is 3.8%.

 https://doi.org/10.1051/bioconf/20238003019

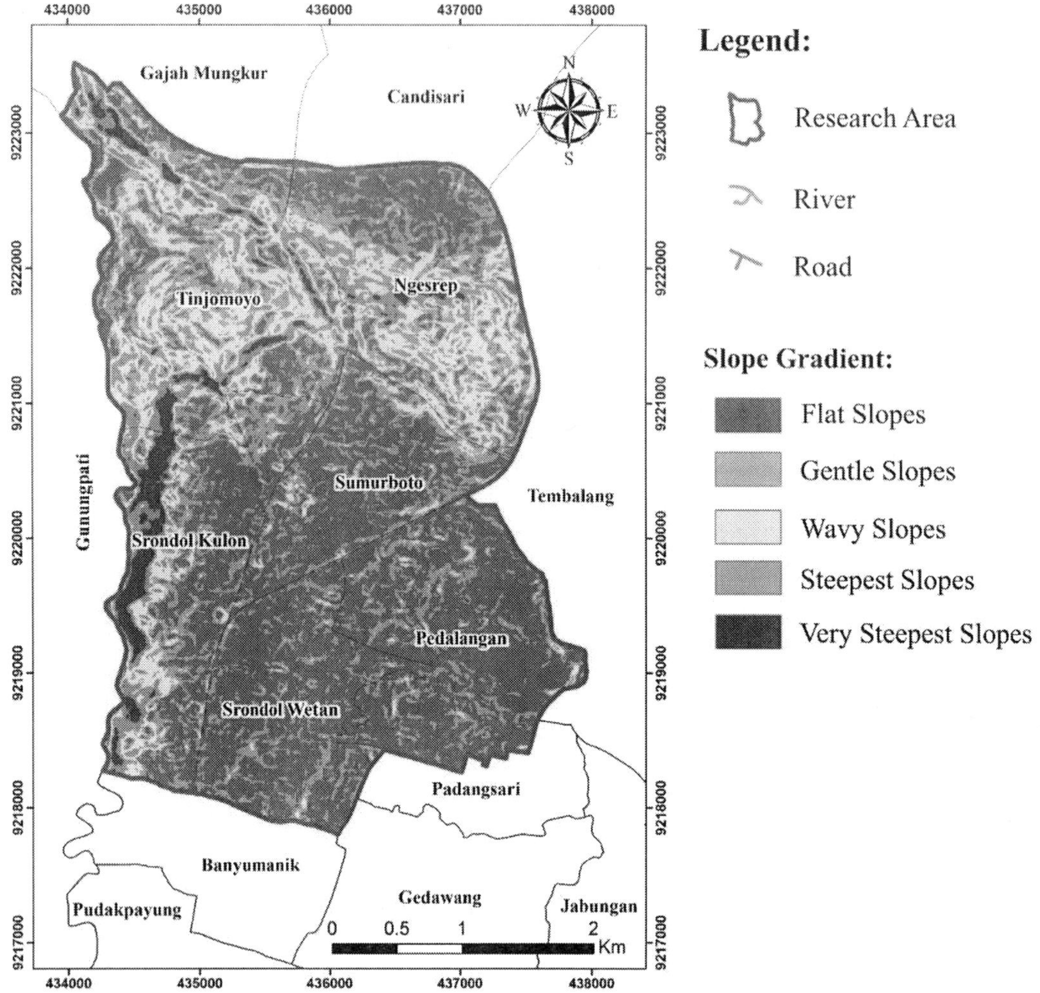

Fig. 6. Slope gradient map [7].

3.1.6 Density of Geological Structures

Direct measurements were obtained from the geological structure on the geological structure map in the field, and secondary data in regional geological maps, and SRTM image maps were analyzed using PCI Geomatica software. A geological structure map was produced based on these two types of data. A 500 x 500 m grid was made on the map, and the number of geological lineaments was calculated to create a structure density map with a 1:25.000 scale, as shown in Fig. 7. The more structures in each grid, the denser the structures in that area. The density map of the geological structure showed that the northwest tended to dense geological structures. The straightness in the area was dominantly oriented northwest-southeast, in the direction of the hills in the area. The large number of straight lines of the geological structure indicated that the geological structures influenced the groundwater infiltration in the area. Meanwhile, in the central and southeastern parts of the research area, there was no lineament of the geological structure, so the density map of the geological structure showed a dark green color. It indicated that the presence of geological structures in the area did not affect groundwater absorption.

https://doi.org/10.1051/bioconf/20238003019

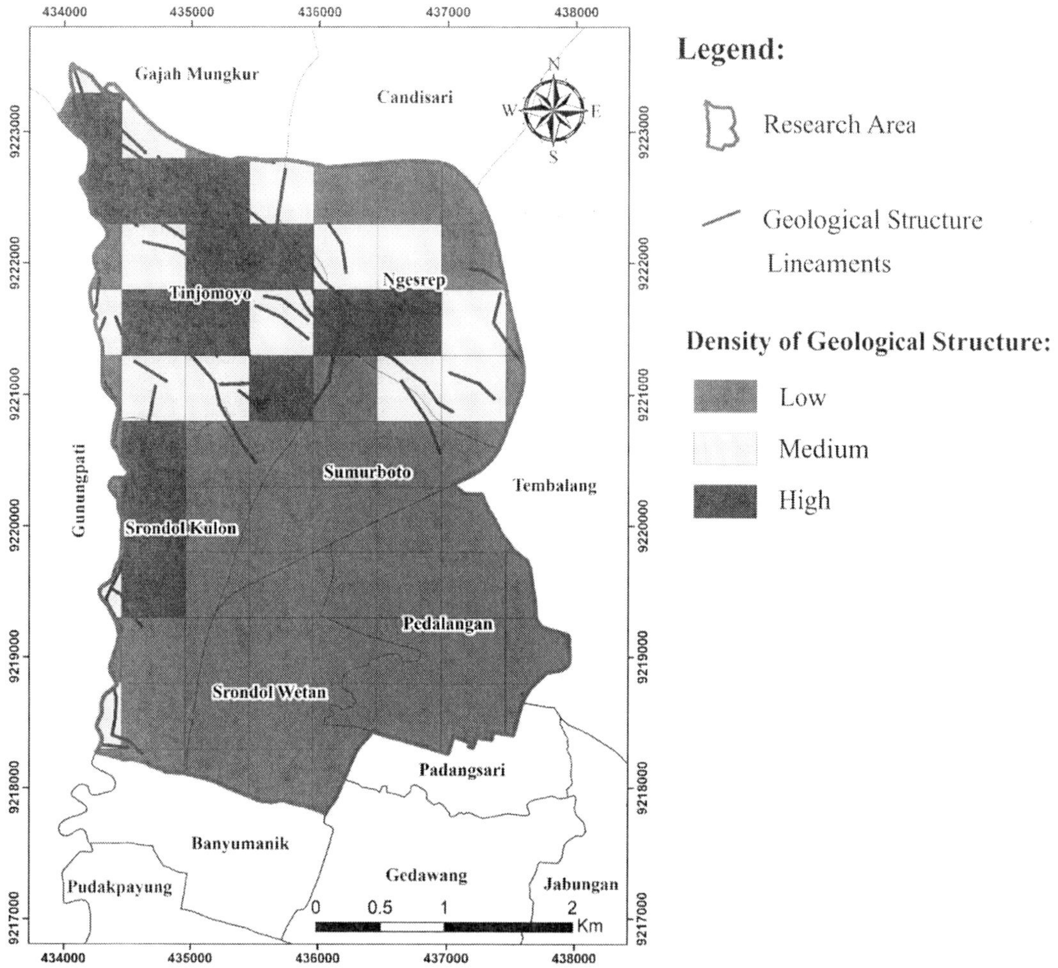

Fig. 7. Density of geological structures map [7].

3.2 Zoning Characteristics of Land

Land characteristic zoning was an analysis carried out to divide an area based on the characteristics of each of these lands. Six parameters were used to determine the area's features in this research: soil grain size, soil permeability, land use, slope, density of geological structures, and lithology. The zoning process was carried out by displaying all the maps made using the ArcMap software to produce a zoning map of land characteristics with a 1:25.000 scale (Fig. 8). These zones had different values. On the map, there were 214 zones, which illustrated that each had a special character that differed from one another. These zones could then be reclassified into criticality maps of infiltration areas.

 https://doi.org/10.1051/bioconf/20238003019

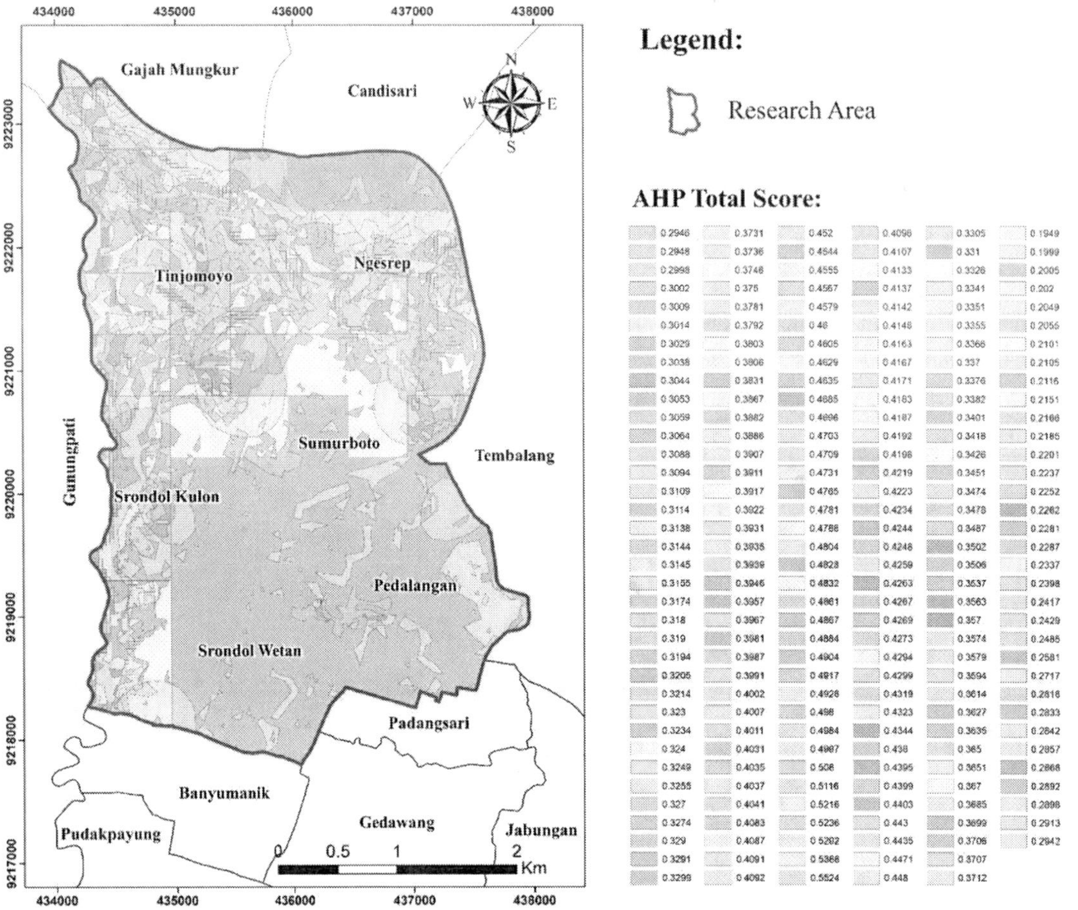

Fig. 8. Zoning characteristic of land map [7].

3.3 AHP Scoring

The Score in the AHP calculation was determined from the weighting value of each parameter multiplied by the weighting value of the subparameter. Before that, the weight of each subparameter must have been determined first, as shown in Table 3.

Table 3. Calculation of AHP on each subparameter [7].

	Factors	1	2	3	4	5	6	Weight
	Soil Grain Size							
1	K-FHSdk	1	2					0,67
2	K-FHSdg	1/2	1					0,33
	CR (*Consistency Ratio*)							0,00
	Soil Permeability							
1	Medium – Fast	1	2	3	5			0,46
2	Slow – Medium	1/2	1	3	5			0,33
3	Slow	1/3	1/3	1	2			0,14
4	Very Slow	1/5	1/5	1/2	1			0,07
	CR (*Consistency Ratio*)							0,03
	Land Use							
1	Meadows and Pastures	1	2	5				0,58
2	Shrubs and Grasslands	1/2	1	3				0,31
3	Settlements	1/5	1/3	1				0,11
	CR (*Consistency Ratio*)							0,01
	Lithology							
1	Carbonate Claystose	1	2					0,67
2	Volcanic Breccia	1/2	1					0,33
	CR (*Consistency Ratio*)							0,00
	Slope Gradient							
1	Flat Slopes	1	2	3	5	7		0,44
2	Gentle Slopes	1/2	1	2	4	5		0,27
3	Wavy Slopes	1/3	1/2	1	2	3		0,15
4	Steepest Slope	1/5	1/4	1/2	1	2		0,08
5	Very Steepest Slope	1/7	1/5	1/3	1/2	1		0,05
	CR (*Consistency Ratio*)							0,01
	Density of Geological Structure							
1	High	1	2	3				0,54
2	Medium	1/2	1	2				0,30
3	Low	1/3	1/2	1				0,16
	CR (*Consistency Ratio*)							0,02

The results of the final AHP weights for each of these parameters were then multiplied by the final weights for each of the sub-parameters. The multiplication result was the final Score, which formed the basis of the division of the criticality class of the infiltration area, as shown in Table 4.

Table 4. The final Score on the calculations performed on all parameters [7].

No	Parameter	Parameter Weights (a)	Sub parameter	Sub parameter Weights (b)	Final Score (a x b)
1	Soil Grain Size	0,33	K-FHSdk	0,667	0,219
			K-FHSdg	0,333	0,110
2	Soil Permeability	0,23	Medium – Fast	0,461	0,107
			Slow – Medium	0,328	0,076
			Slow	0,136	0,032
			Very Slow	0,075	0,017
3	Land Use	0,19	Meadows and Pastures	0,581	0,108
			Shrubs and Grasslands	0,309	0,058
			Settlements	0,110	0,021
4	Lithology	0,12	Claystose	0,667	0,082
			Volcanic Breccia	0,333	0,041
5	Slope Gradient	0,08	Flat Slopes	0,439	0,034
			Gentle Slopes	0,275	0,021
			Wavy Slopes	0,151	0,012
			Steepest Slope	0,084	0,006
			Very Steepest Slope	0,052	0,004
6	Density of Geological Structure	0,05	High	0,538	0,028
			Moderate	0,297	0,016
			Low	0,165	0,009

Based on Table 3, the criticality class intervals of infiltration areas could be determined using Equation (3) according to [12]. After the class intervals were obtained, the criticality of the infiltration areas could then be divided into six classes, as shown in Table 5. Once the class intervals were determined, the criticality map of the infiltration areas could be prepared, as shown in Fig. 9, which has a 1:25.000 scale.

$$I = (c-b)/k \quad (3)$$

In this Equation, I was the class interval, k was the number of classes, c was the highest total Score, and b was the lowest.

Table 5. The criticality class of the water catchment area [7]

Score	Criteria
0,2 - 0,263	Good
0,264 - 0,325	Naturally Normal
0,326 - 0,388	Starting to be Critical
0,389 - 0,451	Moderately Critical
0,452 - 0,514	Critical
0,515 - 0,576	Very Critical

The final step is to check the accuracy of the weighting results, which have been carried out using a sensitivity test. Weighting was carried out three times by adding and reducing the weight for each parameter and sub-parameters. After re-weighting three times using different values, it was discovered that the weighting was consistent. This calculation was done because the results on the resulting map remain the same. Changes only become visible when a very large weight change is made.

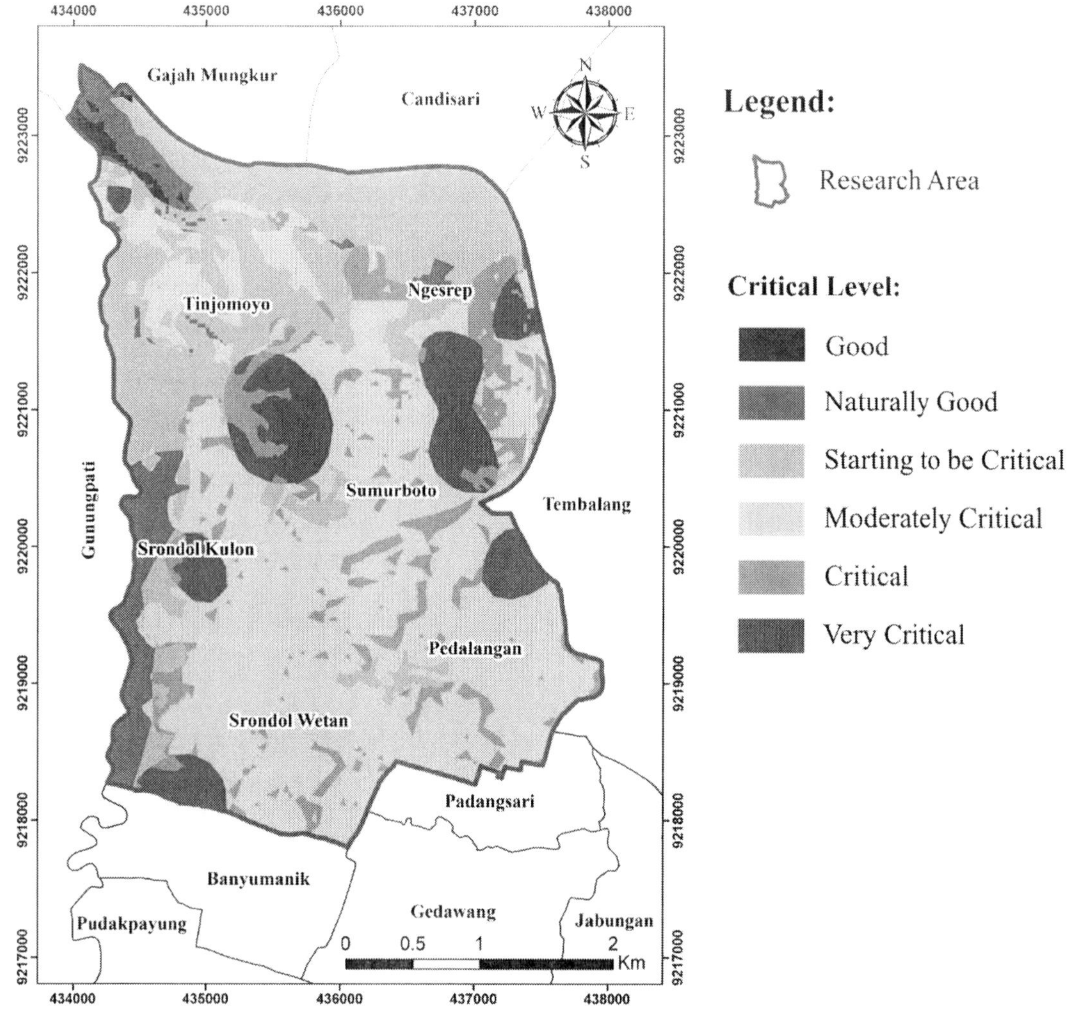

Fig. 9. Critical level of the infiltration area map [7].

4 Discussion

4.1 Good

The criticality class for a good infiltration area only had a coverage area of 0.6% of the entire research area. The percentage was the smallest among other classes. Its distribution was only located in the northwestern part of the research, extending in a northwest-southeast direction in the direction of the hills in the Tinjomoyo sub-district. This class consisted of moderate to fast soil permeability, high density of geological structures, and very steep slopes [7].

4.2 Naturally Normal

The naturally normal condition had an area coverage of 5.8% of the entire research area. Its distribution was in the western part of the research area, extending from north to south in the river flow direction. In addition, it was also found in the northwestern part of the research area, extending in a northwest-southeast direction in the direction of the hills in the area [7]. This class consisted of coarse-grained soil units with a small fine fraction, land use as meadows and pastures, high density of geological structures, lithology in carbonate mudstone, slow – medium soil permeability, and relatively steep slope.

4.3 Starting to be Critical

The *starting to be critical* condition was the class with the area coverage with the second highest percentage, with a value of 24.8%. Most of its distribution was in the northwestern and northern parts of the research area, precisely in the Tinjomoyo and Ngesrep sub-districts [7]. This class consisted of coarse-grained soil units with a small fine fraction, and lithology consists of carbonate mudstone and volcanic breccia; soil permeability is medium–fast, the slope is flat – very steep, the density of geological structures is medium–high, and land uses dominated by settlement.

4.4 Moderately Critical

The moderately critical infiltration area is the class that dominates the study area, with a percentage of 48.9% of the total research area. Its distribution dominates the southeastern part of the study area, precisely in the Srondol Kulon, Srondol Wetan, Pedalangan, and Sumurboto sub-districts [7]. This class consisted of coarse-grained soil units with a small fine fraction, flat slopes, settlement land uses, volcanic breccia lithology, very slow soil permeability, and low density of geological structures.

4.5 Critical

The criticality class of the critical infiltration area was spread evenly in the eastern and southern parts of the study area in almost all sub-districts. The distribution of this class was not united into one whole part but spread evenly over a small area with a total coverage area of 9.9% [7]. This area had residential land use, very slow permeability, volcanic breccia lithology, and medium geological structure density. Meanwhile, the slope was relatively flat – undulating with a grain size dominated by coarse soil with a small fine fraction.

4.6 Very Critical

The infiltration areas with very critical conditions were located mostly in Tinjomoyo and Ngesrep sub-districts, with an area coverage of 10% of the research area [7]. In this condition, the coverage area tended to follow the particle size of coarse soil with a medium fine fraction. This area was dominated by land use in settlements, permeability was slow-very slow, and the density of geological structures was low-medium. This condition made it difficult for water to seep into the ground, including in a very critical infiltration area.

5 Conclusion

The research area consists of six criticality classes of infiltration areas. The percentage is divided into good (0.6%), naturally normal (5.8%), starting to be critical (24.8%), moderately critical (48.8%), critical (9.9%), and very critical (10%).

The researcher(s) would like to thank Universitas Gadjah Mada for funding this research through RTA program. The researcher(s) also appreciate the Spatial Planning Office of Semarang City for supporting this research implementation.

References

1. M. Wibowo, Model Penentuan Kawasan Resapan Air untuk Perencanaan Tata Ruang Berwawasan Lingkungan, *J. Hidrosfir* **1** 1–7, (2006)
2. S. Gunawan, Y. Prasetyo, and F. Amarrohman, Studi Penentuan Kawasan Resapan Air pada Wilayah DAS Banjir Kanal Timur, *J. Geod. Undip* **4**, 86–94, (2016)
3. T. Saaty, Decision Making for Leaders: The Analytic Hierarchy Process for Decisions in a Complex World **2**, (1990)
4. M. Brunneli, Introduction to the Analytic Hierarchy Process, (2015)
5. S. Susanti, Analisis Spasial Sebaran Resapan Air pada Kawasan Bandung Utara (2020)
6. S. Santosa, E. Suryadi, D. Kendarto, Analisis Kekritisan Daerah Resapan Air Menggunakan Metode Skoring di Sub DAS Cikeruh, *J. Keteknikan Pertan. Trop. dan Biosist.* **9** 79–89 (2021)
7. A. Arfiandi, Pemetaan Tingkat Kekritisan Daerah Resapan Air Di Kecamatan Banyumanik Bagian Utara Dengan Menggunakan Metode Analytical Hierarchy Process, Undergraduate Thesis, Gadjah Mada University (2023)
8. R. Thaden, H. Sumardirdja, P. Richards, Peta Geologi Regional Lembar Magelang dan Semarang, Jawa **1**, (1975)
9. Y. Muanley, L. Son, G. Mada, N. Dethan, Analisis Sensitivitas Dalam Metode Analytic Hierarchy Process dan Pengaruhnya Terhadap Urutan Prioritas Pada Pemilihan Smartphone Android, *J. Stat. Its Appl. Teach. Res.* **4** 173–90, (2022)
10. A. Casagrande, Classification and Identification of Soils Transactions of the American Society of Civil Engineers **1** 901–30 (1948)
11. D. Lahan, Pedoman Penyusunan Rencana Teknik Lapangan Rehabilitasi Lahan dan Konservasi Tanah ed D K R Indonesia **20**, (1998)
12. I. Umar, Widiatmaka, P. Bambang, B. Barus, Prioritas Pengembangan Kawasan Permukiman pada Wilayah Rawan Banjir di Kota Padang, Provinsi Sumatera Barat **19** 83–94, (2017)

https://doi.org/10.1051/bioconf/20238003020

The Effect Of Land Cover Changes On Landslides In Grindulu Watershed, Pacitan Regency, East Java

Raka Achmad Fauzi[1] *, *Rika Harini*[2], *R. Rijanta*[3]

[1 2 3] Department environmental science, Graduate School, Univ Gadjah mada

Abstract. This research was conducted in the Grindulu River Watershed, Pacitan Regency. The Grindulu River is included in the Bengawan Solo river area. The Grindulu River Watershed is located in three areas, namely Pacitan Regency, Ponorogo Regency and Community Districts along the River Watershed use the grindulu river water for various activities, such as bathing, washing, and for agricultural irrigation purposes. This study aims to determine changes in land cover in the Grindulu river watershed and the distribution of landslides in the Gindulu river area. The data used in this study is information consisting of: data on landslide events in 2022 obtained from the Pacitan Regency regional disaster management agency; land cover data for the grindulu river watershed in 2018 and 2022 identified from Copernicus Sentinel-2; basic maps, such as road network maps, contour maps, river network maps and administrative maps obtained from the Indonesian Land map.Based from the spatial analysis, it was found that 67 points of landslide events that occurred during 2022 were spread throughout the river watershed. The landslide occurrence points have patterns or clusters located on the banks of the river and close to residential areas. Landslide points are found in different land covers based on spatial analysis of the distribution of landslide points with land cover maps. Most landslides occur in agroforestry land cover.

1 Introduction

The number of landslides in 2020 has reached 387. Based on data from the National Disaster Management Agency (BNPB), landslides are often the deadliest disasters in Indonesia. In 2014, BNPB recorded 354 people died due to landslides. A year later, landslides killed 174 people and in 2016 landslides occurred 599 times, killing 186 people. In 2017 as many as 15.6 people died due to landslides and 221 others died in various other disasters [1].

Landslide as an exogenic process causes the movement of soil mass down the slope due to the force of gravity and high rainfall. The other factors that cause landslides are soil type, geological factors, land cover, seismic factors and land slope [2]. Landslides in Indonesia mostly occur in areas with high rainfall and in hilly or mountainous geomorphological conditions. The highest occurrence of landslides in Indonesia occurs in areas with rainfall of more than 2000 mm and topography with steep slopes. Considering that there are many hilly or mountainous areas with rainfall > 2000 mm in Indonesia, landslides are classified as dangerous events and therefore need to be anticipated in order to prevent greater losses [3].

A watershed is a land area that is topographically bounded by mountain ridges which collect and store rainwater and then channel it to the sea via the main river [4]. The watershed can be viewed as a system consisting of input, process and output. The input is in the form of rainfall, the process is in the form of the watershed itself which consists of biotic and abiotic components, and the output is in the form of production, runoff, erosion and so on [5]. The watershed has an important role for the community, especially in terms of providing water, managing water quality, preventing floods, and as a livelihood for the community. At this time many watersheds in Indonesia are damaged. Some things that prove that are the occurrence of land erosion and its sedimentation, landslides, decreased water quality and quantity, and others [3].

Based on the topographic map issued by the National Coordinating Survey and Mapping Agency, the Grindulu river is a river on the island of Java with a geographical location of 110° 55' - 111o 25' East Longitude and 7o 55' - 8o 17' South Latitude. The Grindulu Watershed starts from Mount Gembes (1,200 meters above sea level) on the border of Jeruk Village and Bangunsari Village in Bandar District and then passes through Tegalombo District, Arjosari District, Pacitan District and empties into the Indian Ocean, precisely in Ploso Village.

Communities along the River Watershed use Grindulu river water for various activities, such as bathing, washing, and agricultural irrigation purposes more in areas with flat soil textures, for example the Tegalombo sub-district and Pacitan sub-district. Whereas areas with hilly soil textures, only stone and sand are used as materials for building houses or buildings. The grindulu river is also used for fishing by the local community. The fish that are often found include tilapia, catfish, eel, carper, wader, nyek, fresh shrimp and so on [6].

The various challenges faced in the context of the use and management of land resources are increasingly complex, including: (1) population pressure on land,

*Corresponding author: raka.achmad.f@mail.ugm.ac.id

(2) land conservation and land conversion, (3) forest degradation and land damage. , (4) environmental damage and natural disasters. Therefore, in the future the concept of sustainable land resource management by taking into account the aspects of resource and environmental sustainability as well as the various challenges that need to be formulated on a national, regional and local scale must become the main reference [7]. This study aims to prove the role of land cover change on landslides in the grindulu watershed.

2 Methods

2.1 Location

The Grindulu River Watershed is located in Pacitan Regency. The Grindulu River Watershed which is the largest watershed in Pacitan Regency, has an area that is mostly composed of limestone hills. From the data released by the Pacitan Regency Regional Disaster Management Agency (BPBD) until the end of December 2021, at least 441 disaster events had been recorded. Of that number, around 75% were dominated by landslides with 330 incidents. BPBD Pacitan said that the impact of the disaster throughout 2021 included 354 houses affected.

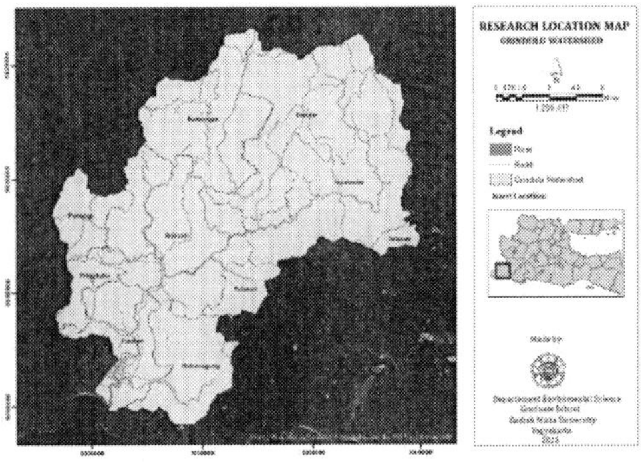

Figure 1. Map of Research Location.

2.2 Data

The data used in this study is information consisting of:

1. Data on landslide events in 2022 obtained from the Pacitan Regency regional disaster management agency
2. Land cover data for the grindulu river watershed in 2018 and 2022 identified from Copernicus Sentinel-2
3. Basic maps, such as road network maps, contour maps, river network maps and administrative maps obtained from the Indonesian Rupa Bumi map.

2.3 Research methods

In this research, the method employed is satellite image analysis. All satellite image processing and analysis were carried out using Google Earth Engine, and the stages of satellite image analysis can be observed in the research method flowchart. After the area was clipped based on the Grindulu River watershed, image sharpening, and mosaic creation, image interpretation was conducted. Image interpretation is a process used to classify objects in the image by identifying their color patterns or appearances on the image. Image interpretation and band combinations provide characteristics of land use.

In Google Earth Engine, land cover classification was performed using supervised classification to characterize each land cover type. Land cover classification served as a guide or reference in the remote sensing image interpretation process for land cover mapping purposes. Furthermore, in this classification stage, digitization processes were employed to further clarify objects in the research area, such as settlements, plantations, rivers, and study area boundaries.

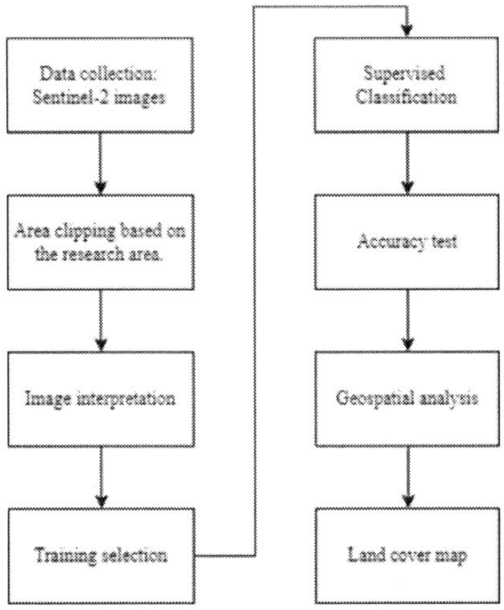

Figure 2. Research Method Diagram

Below is an explanation of the research stages from the research methodology diagram.

a. Data Collection: The research begins with the collection of relevant data, which likely includes satellite images and other necessary information. Sentinel-2 (S2) is a wide-swath, high-resolution, multispectral imaging mission with a global 5-day revisit frequency. The S2 Multispectral Instrument (MSI) samples 13 spectral bands: visible and NIR at 10 meters, red edge and SWIR at 20 meters, and atmospheric bands at 60 meters spatial resolution. It provides data suitable for assessing state and change of vegetation, soil, and water cover.
b. Area Clipping Based on the Grindulu River Watershed: The research focuses on a specific area, and in this step, the area is clipped or delineated according to the Grindulu River watershed. This ensures that the analysis is conducted within the desired geographical boundaries.
c. Image Interpretation: Image interpretation is a crucial step where the researcher visually analyzes

the images to identify and classify objects or land cover types within the study area. This process involves recognizing color patterns and appearances of objects in the images. Digital remote sensing data interpretation essentially involves classifying pixels based on their spectral values. Each class or group of pixels is then analyzed in relation to objects or phenomena on the Earth's surface. In other words, each class reflects what objects or phenomena they represent. Digital interpretation involves analyzing the digital values contained in each pixel column, making the interpretation of images relatively more objective. Digital interpretation also allows for more complex image analysis, considering several spectral channels, multi-temporal data, and multi-spatial data.

d. In this process, methods are employed to determine which method can be used when testing an image. The testing method utilized is the supervised method, where several best accuracy measures will be obtained from the method.
e. Supervised Classification: images that have been classified through supervised classification for land cover are then converted from raster to vector format, enabling them to be overlaid with other vector map data.
f. Accuracy Assessment: fter obtaining pixel values for the training areas from the image data, it is necessary to evaluate the spectral response patterns for each land cover category, particularly their ability to differentiate each spectral signature. The accuracy assessment used in this research involves measuring or evaluating the spectral separability using an error matrix or confusion matrix. This matrix helps quantify discrepancies, including an excess of pixels assigned to other classes (commission) or a shortage of pixels assigned to different classes.
g. Geospatial analysis involves the spatial aspect that indicates the location, placement, and position of an object or event both below and above the Earth's surface, expressed in a specific coordinate system. Geospatial data refers to data about geographic location, dimensions or sizes, and/or characteristics of natural and human-made objects that exist on or above the Earth's surface.
h. Mapping: Once the land cover is classified and the necessary digitization is completed, the research likely involves creating maps to visually represent the results of the analysis. These maps provide valuable infomation about the distribution of land cover types within the study area.

2.4 Data Analysis

2.4.1 Land Cover Change

Land cover changes that occur in the Grindulu watershed are identified as multitemporal Sentinel image. Sentinel imagery obtained from Google Earth Engine (https://code.earthengine.google.com). We detected land cover changes in the study area over 2 years, namely 2018 and 2022, using supervised classification. Supervised classification is one type of multispectral image classification. This classification uses the pixel categorization created by the user and the algorithm classify pixels that are similar to the sample data. The scope of the research area is divided into 6 classes; river, agricultural areas, agroforestry, forest, residential and pasture.

To analyze and create land cover maps using Google Earth Engine is explained in the following steps:

a. Cutting image area based to research area. The cutting area is based on the size of the Grindulu river basin. Area cutting using ArcGis software and displayed on Google Earth Engine.
b. Image interpretation using Google Earth Engine begins with the technique of providing sampling data by providing color in each feature collection that represents sampling data in Google Earth Engine software. The image sharpening results are applied to the RGB 654 color composite image, namely sharpening low-pass filtering before the supervised classification process is carried out. In the image data there are high frequency of data variability with heterogeneous characteristics of landscape features. Technique Lowpass filtering has been widely used to reduce visible spatial frequencies in Figure 3.
c. Image interpretation is conducted using supervised

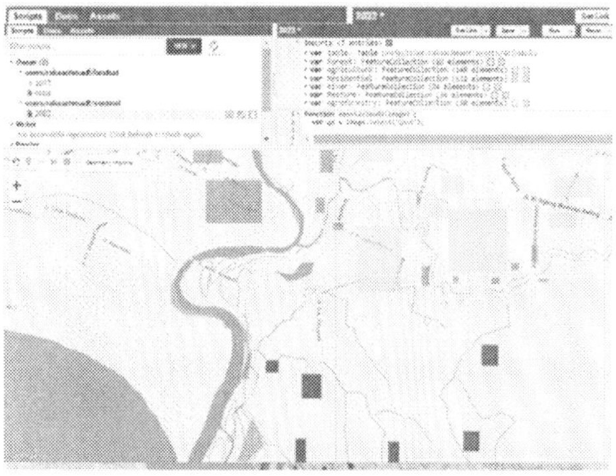

Figure 3. Image interpretation results

classification techniques within the Google Earth Engine software. Before the interpretation process, sampling areas are determined based on the features observed in the image. The creation of training areas is based on the desired number of land cover classes, which include forest, river, residential, agriculture, pasrture, and agroforestry area. This process can be visualized in Figure 4.

The following table shows the training and validation areas for each cover class.

Table 1. Image Interpretation Geometry

Geometry Type	Forest	Agri-culture	Residential	River	Pasture	Agro-forestry
Color						
Area						

In Table 1, a geometry data sampling category is created which is used to declare an area intended for sampling which is used as a pixel used when classifying data. From the data that has been turned into pixels, the data is classified based on the land that fits the existing categories
converted into pixels

d. Supervised Classification: after using the Feature Collection, extract the reflectance values of each pixel from each band. Create training data by assigning training points on the image. This approach will provide training data with the pixel values for each point. After running the script, the training data will be printed to the console. Note that the 'properties' information has now been changed to land cover, and there are now reflectance values at each point. Six land cover classes were used based on the map, as explained in the image below.

Here are the results of land cover classification for Sentinel-2 Year 2022 and Sentinel-2 Year 2018 images in the Grindulu Watershed, as shown in Figure 4 and Figure 5.

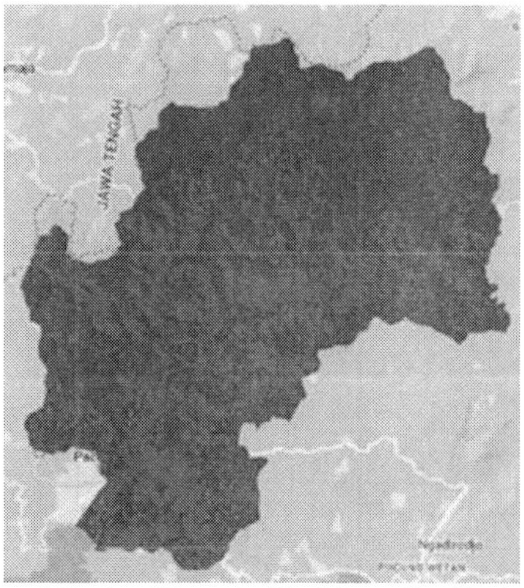

Figure 4. Results of Supervised Land Cover Classification in 2022

2.4.2 Landslide Distribution

Landslide distribution points obtained from secondary data and primary data. Secondary data is initial data obtained from the Pacitan Regency regional disaster management agency. This data contains the number of landslide events, disaster victims, incident locations, affected buildings, losses and coordinates. Primary data was obtained from field surveys to determine the location and coordinates of the required landslides. the coordinates obtained during 2022 from January to December are then collected and displayed on a map spatially.

2.4.3 Analysis Landslides by Land Cover Change

To elaborate further on the role of land cover against landslides. We compared the data using spatial analysis, so that the distribution of landslide events to landslide points can be analyzed spatially. Data on landslide events was collected through a landslide census in the field, we did it from January to December 2022. Furthermore, we also compared the types of land cover with landslide events. Therefore, it can be seen clearly what types of utilization have a close relationship with the occurrence of landslides.

3 Result and Discussions

3.1 Spatial Distribution Land Cover Type

The Grindulu River stretches from the north of Pacitan Regency to empties into the southern sea. Nearly 80% of Pacitan district is included in the grindulu river watershed. The distribution of land cover in the Grindulu watershed is divided into 6 land cover classes. The land cover is in the form of agricultural areas, agroforestry, forests, rivers, shrubs and settlements. The land cover classes are displayed spatially in the following two maps.

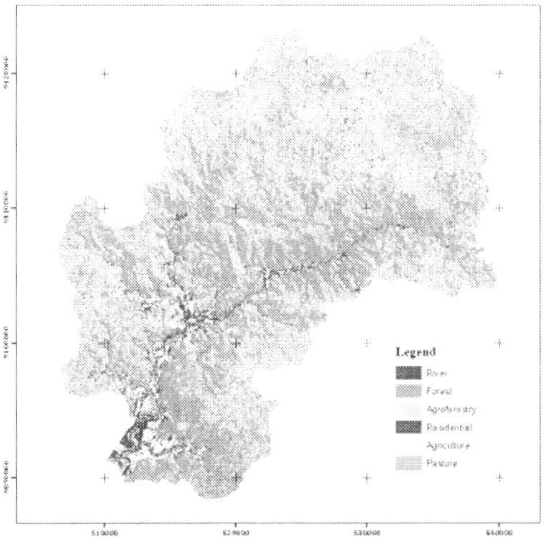

(a)

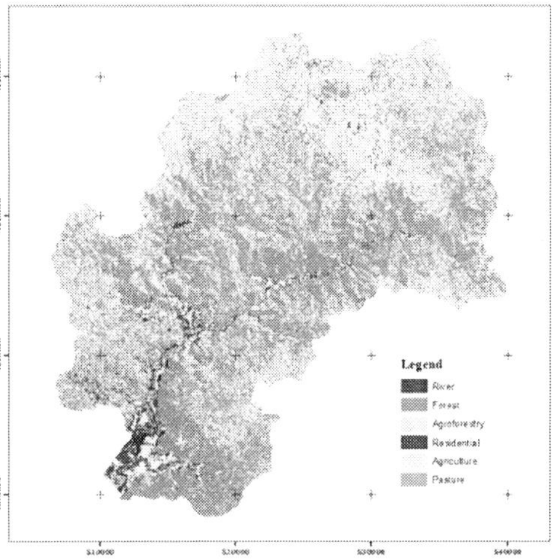

(b)

Figure 5. Land Cover 2018 (a); Land Cover 2022 (b).

Figure a and figure b show the difference between the land cover of the grindulu watershed in 2018 and 2022. The five year difference in the map shows spatially how the land cover in the grindulu river watershed has changed.

Table 2. Land Cover Ratio

Land Cover Area Ratio		
Land Cover	**2018**	**2022**
Agriculture Area	6614 Ha	7471 Ha
Agroforestry	26226 Ha	25795 Ha
Forest Area	22741 Ha	21400 Ha
Residential	3131 Ha	3075 Ha
River	456 Ha	511 Ha
Pasture	7271 Ha	8517 Ha

Source: Data Analysis

It can be seen from the table of differences in land cover from 2018 and 2022 that experienced several changes in the area of land cover. the agricultural area increased from 6614 hectares to 7471 hectares. this can be illustrated that during the last five years there has been a change in land cover in terms of human activity in farming.

The increase in area has also occurred in pasture land cover where in 2018 it was 7271 hectares to 8517 hectares in 2022. This is related to reduced land productivity. agriculture. In contrast, forest land cover has decreased in area from 2018 by 22741 hectares to 21400 hectares in 2022. The reduction in forest area has changed in various other land covers, especially into agriculture and pasture areas. In other land covers, such as settlements, rivers, agroforestry, the land cover area tends to be stable or nearly the same for five years.

Changes in land cover over the five years from 2018 to 2022 are due to various factors. the most common factor is community activities in meeting their needs such as opening gardens, rice fields, settlements, and other built areas. Changes due to human activities are most visible on the banks of the Grindulu river which can be seen from the land cover map. This is due to various reasons, especially the need for water to irrigate rice fields and for the daily needs of the community. other activities in the form of sand and stone mining in the river by the community, where these things can cause landslides.

3.2 Landslides by Land Cover Change

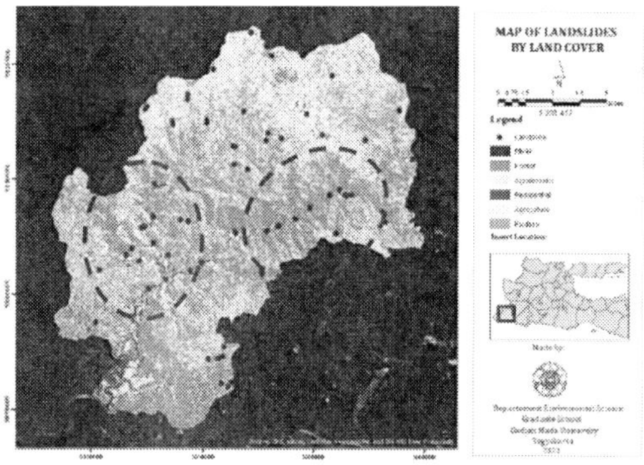

Figure 6. Map of Landslides Distribution

The distribution of landslides is depicted on the map along with the land cover. In 2022 there were 67 landslide events. All of these landslides were spread in the grindulu river watershed and spread over various existing land covers. The landslide incident mostly affected houses and public facilities.

Through spatial analysis, it can be determined whether there are clusters or groups of events that have patterns within the existing land cover, as indicated by circles on the map. It can be seen that the clusters of these incidents mostly occurred on the banks of the Grindulu river, which incidentally is the area that is most frequently used both for rice fields, roads and settlements

Table 3.Number of landslides on land cover

Land Cover	Number of Landslide Points
Forest	17
Agroforestry	25
Residential	11
Agriculture	5
Pasture	9
River	0

In detail, most landslides occur in forest land cover, agroforestry, and settlements. even so based on the data, these incidents were all close to homes and public facilities. From the closeness of land cover to houses and land changes, it can be seen that human activities in the form of development and clearing of defensive land are spatially visible patterns related to landslides.

4 Conclusion

Based from the results of the analysis it can be concluded that land cover changes for five years. that is, from 2018 to 2022 there have been changes in several land covers. The main land cover that has increased is the area of agriculture and pasture which is closely related to human activities in the Grindulu river watershed.

Based on spatial analysis, 67 landslide points were found that occurred throughout 2022, spread throughout the river watershed. Landslide occurrence points have patterns or clusters located on riverbanks and close to residential areas. Landslide points are found in different land covers based on spatial analysis of the distribution of landslide points with land cover maps. Most landslides occur in agroforestry land cover.

Acknowledgement

The author expresses his deepest gratitude to RTA-UGM 2023 with the Grant Number 5075/UN1.P.II/Dit-Lit/PT.01.01/2023, which has funded this research. And also really appreciate reviewers for suggestions to make this article better. without all the support given, this research would not be possible. Hopefully this research is useful for many people.

References

1. [BNPB] National Disaster Management Agency. 2020. A total of 2,925 Natural Disasters Happened in 2020 in the country, Hydrometeorological Disasters Dominate. [Internet]. [accessed September 28, 2022].

2. Purnamasari DC. 2007. Application of remote sensing and geographic information systems (GIS) in the evaluation of landslide-prone areas in Banjarnegara District. [Thesis]. Bogor(ID): Department of Forest Resources Conservation and Ecotourism, Faculty of Forestry, Bogor Agricultural University.

3. Naryanto HS, Prawiradisastra F, Kristijono A, Ganesha D. 2018. Spatial planning after the landslide disaster in Puncak Pass, Cipanas District, Cianjur Regency March 28 2018. Journal of Natural Resources and Environment Management. 9(4):1053–1065. doi: 10.29244/jpsl. 9.4.1053-1065.

4. Asdak, C. 2010. Hydrology and Watershed Management: Fifth Revised Edition. Gadjah Mada University Press.

5. Paimin, Sukresno, and Purwanto. 2010. Rapid Imprint of Sub Watershed Degradation (Sub DAS). Forest Research and Development Center and Nature Conservation, Bogor.

6. Wijayanti, T. 2017. Profile of Heavy Metal Pollution in the Waters of the Grindulu Pacitan River Watershed (Das). Scientific Journal of Science Vol. 17 No. 1. Biology Education IKIP Budi Utomo Malang

7. Worosuprojo, Suratman. 2007. Spatial-Based Land Resource ManagementIn Sustainable Development In Indonesia. Yogyakarta: Faculty Geography. Gadjah Mada University.

The Growth and Yield of Upland Rice (*Oryza sativa* L.) Appliced with Rice Husk Zeolite and N Fertilizer

Nelvia[1*], *Zulfatri*[1], *Idwar*[1], *Nursiani* Lubis[1] and *Ferdi* Saputra[1]

[1]Department of Agrotechnology, Faculty of Agriculture, Universitas Riau, Campus Bina Widya Jl. HR.Soebrantas KM 12,5 Pekanbaru-28293

Abstract. Upland rice production can be increased by utilizing suboptimal land. However, efforts need to be made to increase the availability ofnutrients and water in the soil. This study aims to determine the interaction of rice husk zeolite treatment andN fertilizer on the growth and yield of upland rice. The research was conducted experimentally factorial using completely randomized design (CRD). The first factor of rice husk zeolite consisted of 3 levels (0, 200 and 400 $kg.ha^{-1}$) and the second factor of N fertilizer consisted of 3 levels (0, 75 and 125 kg $N.ha^{-1}$), each combination was repeated three times. The results showed that the application of rice husk zeolite 200 $kg.ha^{-1}$ and N fertilizer dosage 125 kg $N.ha^{-1}$ increased plant height, maximum number of tillers and weight of 1000 grains compared without rice husk zeolite and N fertilizer, and also increased the number of productive tillers and pithy grain per panicle, weight of dry milled grain per clump are 56.78%,16.8%, 80%, respectively compared without rice husk zeolite and N fertilizer.

*Corresponding author: nelvia@lecturer.unri.ac.id

https://doi.org/10.1051/bioconf/20238003021

1 Introduction

Upland rice (*Oryza sativa* L.) is a commodity plant that produces rice as the main staple food for more than 90% of Indonesia's population, where the demand for it continues to increase along with the addition of population. Rice production of rice can be increased by increasing the productivity of upland rice in sub-optimal land and lowland rice productivity. Sub optimal land include Ultisol/Podzolic soils with an amount of around 25% (45 million ha) of the total land area of Indonesia (191.1 million ha) [1] and [2] and Oxisol soils of 8%. The productivity of upland rice in sub-optimal land around 2.439 ton.ha^{-1}[3], according to upland rice breeders the average yield of upland rice is 6,7 tons.ha^{-1} and the potential is 10,2 tons.ha^{-1} [4]. The low productivity of upland rice in Podzolic soils is due to various constraints: acid soil reaction, organic matter content, availability of macronutrients (N, P, K, Ca and Mg) and micronutrients (Mo), basa saturation (BS), cation exchange capacity rate (CEC<16 cmol.kg^{-1}) are low, Al exchangenable and Al saturation are high, so it is toxic to plants especially food crops.

Upland rice productivity in Podzolic soils can be increase which applicated by rice husk zeolite and N fertilizer. Zeolite has a high CEC and porosity so that it plays a role in increasing the soil's ability to absorb nutrients and water. The CEC and surface area of Na-A and Na-X zeolite were 506 Cmol.kg^{-1},317 Cmol.kg^{-1} and 171 m^2.g^{-1}, 676 m^2.g^{-1}, respectively [5]. The application of natural zeolite increased the productivity of paddy rice [6] and the application of zeolite Na-P1 increased the growth and yield of paddy rice [7]. The N fertilization is very important to increase N available in the soil, growth and production of paddy rice [8].

2 Material and method

The research was conducted at the Soil and Greenhouse Laboratory, Faculty of agriculture, Riau University, Pekanbaru, Indonesia from June to October 2022. The materials used in this study were upland rice seeds of Inpago 12 variety, podzolic topsoil taken in Balam Jaya Village, Tambang District, Kampar, Riau Province, rice husk zeolite, urea, SP-36, KCl as a source of N, P and K, polybag, Confidor 5 WP and Stage 18 EC insecticide, Antracol 70 WP fungicide. The tools used in this research are oven, furnace, porcelain dish, 25 mesh sieve, net, etc. The study was conducted experimentally in factorial form using a completely randomized design (CRD). The first factor is rice husk zeolite which consists of 3 levels (0, 200, 400 kg.ha^{-1} or 0.1 and 2 g per polybag) and the second factor is N fertilizer which consists of 3 levels (0.75, 125 kg N.ha^{-1} or 0, 0.815 and 1.3587 g per polybag), each combination was repeated 3 times.

3 Results and discussion

3.1 Chemical properties of soil and zeolite of rice husk

Chemical properties of soil before and after incubation with rice husk zeolite presented in table 1 and table 2.

Table 1. Chemical properties of soil and rice husk zeolite that used in this study

Parameters	unit	V	C
pH (H_2O)	-	4.72	a
pH (KCl)	-	3.88	-
C- Organic	%	1.59	l
N- total	%	0.34	m
C/N ratio	-	4.68	vl
P_2O_5 HCl 25%	mg.100g^{-1}	63.44	vh
K_2O HCl 25%	mg.100g^{-1}	22.05	vh
P_2O_5 Bray 1	mg.kg^{-1}	31.92	vh
K-exc.	C mol.kg^{-1}	0.54	vl
Na-exc.	C mol.kg^{-1}	0.46	m
Ca-exc.	C mol.kg^{-1}	1.54	vl
Mg-exc.	C mol.kg^{-1}	0.68	l
BS	%	31.72	l
CEC	C mol.kg^{-1}	10.15	l
Al-exc.	C mol.kg^{-1}	2.96	-
H-exc.	C mol.kg^{-1}	4.48	-
Al Saturation	%	27.77	h
CEC of rice husk zeolite	C mol.kg^{-1}	142.87	vh

Note: V is value, C is criteria, exc is exchangeable, BS is Basa saturation, CEC is cation exchange capacity and a, l, m, h, vl, vh are acid, low, medium, high, very low and very high, respectively.

Table 1 showed that the acid reacting soil has a C/N ratio and exchangeable Ca values are very low, while organic C, Mg-exc., BS and CEC are low, total N, K and Na.exc levels are moderate, only the P and K of 25% HCl extraction and available P are classified as very high, with Al saturation being relatively high. The chemical properties of the soil are classified as very low to moderate and high Al saturation is a limiting factor in plant growth and yield, thus the soil is classified as having a low fertility level. Table 1 also showed that the CEC of rice husk zeolite is very high, about 142.87 Cmol.kg^{-1}.Rice husk zeolite can be used as a soil amendment or ameliorant which has a dual function, aboutincreasing the CEC of the soil and the ability of the soil to absorb nutrients and water.

Table 2 showed that the application of rice husk zeolite at doses of 200 and 400 kg/ha can generally improve soil chemical properties such as increasing pH, K, Na, Ca and Mg-exc., BS and CEC as well as reducing Al saturation, so that soil fertility increases. This condition causes the availability of water and essential nutrients in the soil to increase for plants because the added nutrients and water adsorbed on zeolite. therefore, the amount of water and nutrients that can be absorbed by plants also increases. This will stimulate physiological processes and plant metabolism so that it can increase plant growth and yield. The improvement in chemical properties due to the administration of zeolite.

Table 2. Soil chemical properties one month after incubation with rice husk zeolite

Parameters	Rice husk zeolite dosage (kg.ha^{-1})					
	0		200		400	
	V	C	V	C	V	C
pH H_2O	5.16	a	5.75	a	6.13	a
pH KCl	4.83	-	4.65	-	4.93	-
C- Organic	1.52	l	1.83	l	2.13	m
N- total	0.25	m	0.18	l	0.18	l
C/N ratio	6.08	l	10.17	l	11.83	m
P_2O_5 HCl 25%	188.6	vh	190.9	vh	196.0	vh
K_2O HCl 25%	213.4	vh	152.3	vh	179.2	vh
P_2O_5 Bray 1	47.6	vh	53.1	vh	59.3	vh
K-exc.	2.86	vh	0.87	h	2.65	vh
Na-exc.	0.79	m	1.48	vh	2.60	vh
Ca- exc	4.91	l	5.62	l	4.90	l
Mg-exc.	1.36	m	1.24	m	1.26	m
BS	93.0	vh	67.3	h	83.2	vh
CEC	10.7	l	13.7	l	13.7	l
Al-exc.	1.20	-	0.40	-	0.40	-
H-exc.	3.20	-	1.60	-	1.60	-
Al-S	8.38	l	3.57	vl	2.98	vl

Note: V is value, C is criteria, exc is exchangeable, BS is Basa saturation, CEC is cation exchange capacity, S is saturation and a, l, m, h, vl, vh are acid, low, medium, high, very low and very high, respectively. Units of the parameters same with in table 1.

3.2 The growth of upland rice

Table 3 showed that the application of rice husk zeolite dose of 200 kg.ha^{-1} and N fertilizer dose of 125 kg N. ha^{-1} gave the highest increase plant height, maximum and productive number of tillers compared to without rice husk zeolite and N fertilizer. This is more related to the contribution of N fertilizer and rice husk zeolite in increasing the availability of N and water in soil. Thus the uptake of N and water by plants increases and stimulates physiological and metabolic processes so as to increase the growth of rice plants.

Table 3 showed that the application of rice husk zeolite dose of 200 kg.ha^{-1} and N fertilizer dose of 125 kg N. ha^{-1} gave the highest increase plant height, maximum and productive number of tillers compared to without rice husk zeolite and N fertilizer. This is more related to the contribution of N fertilizer and rice husk zeolite in increasing the availability of N and water in soil. The application of rice husk zeolite can also improve pH, K, Na, Ca and Mg-exc., BS and CEC as well as reducing Al saturation (table 2). Thus the uptake of N and water by plants increases and stimulates physiological and metabolic processes so as to increase the growth of rice plants.

Table 3. Plant height, tillers maximum and productive number of upland rice appliced with rice husk zeolite and N fertilizer

Rice husk zeolite dosage (kg.ha^{-1})	N fertilizer dosage (kg N.ha^{-1})			Average
	0	75	125	
Plant height (cm)				
0	98.00 c	118.33ab	122.67ab	113.00b
200	101.67c	122.67ab	126.33a	116.89ab
400	115.00b	121.33ab	124.33ab	120.22a
Average	104.89b	120.78a	124.44a	
Maximum number of tillers (stem)				
0	12.33 b	15.00ab	18.00ab	16.33 a
200	11.67 b	17.33ab	20.00 a	15.33 a
400	15.67ab	13.00ab	16.33ab	15.00 a
Average	13.44 b	15.11ab	18.11 a	
Productive number of tillers (stem)				
0	12.33 b	13.67ab	16.67ab	14.22 a
200	11.00 b	15.00ab	19.33 a	15.11 a
400	13.67ab	12.33 b	15.33ab	13.78 a
Average	12.33 b	13.67ab	17.11 a	

Note: Numbers in the same row and column followed by the same lowercase letters are not significantly different according to the DNMRT test at the 5% level.

Nitrogen and water are needed in large quantities which play a role in compiling various organic compounds in plants such as amino acids, chlorophyll and others. Amino acids are the main components of proteins, where proteins function as enzymes and make up the structure of cells or cell nuclei. Growth which means the increase in tissue begins with the division of the cell nucleus.

According to [9] Nitrogen is one of the constituent elements of chlorophyll which is the main agent of chloroplasts. [10] reported that the interaction of zeolite dose of 63.1 g and N fertilizer dose of 18.04 g ZA per plant tended to increase the height of deli tobacco plants (38.40 cm) compared to without zeolite and N. [11] reported that the media mixture Planting coconut coir, ferns, and zeolite produced Dendrobium orchid bulb height 1.15 cm (16.39%) higher than without zeolite application, and [12] reported that applying N fertilizer at a dose of 126.5 kg N.ha^{-1} increased the height of your plants by 35 day after planted compared to N fertilizer at a dose of 80.5 kg N.ha^{-1}. [13] reported that the interaction of various doses of N fertilizer and zeolite was not significant on the vegetative growth of sorghum, namely the number of leaves, and [14] reported that administration of various doses of zeolite had no significant effect on the maximum number of tillers and productive tillers of rice plants, further [15] reported that the application of N fertilizer at a dose of 300 kg N.ha^{-1} compared to N fertilizer at a dose of 75 kg N.ha^{-1} showed a significant effect on increasing the number of leaves of sweet corn plants.

3.3 Age of panicle exit and harvest

Table 4 showed that the application of rice husk zeolite dose of 400 kg.ha^{-1} and N fertilizer dose of 75 kg N.ha^{-1} expedite the panicle exit, but harvest faster on application of rice husk zeolite dose 400 kg.ha^{-1} and N fertilizer 125 kg N.ha^{-1} compared to without rice husk zeolite and N fertilizer. This is closely related to the function of N which plays a role in the vegetative period of plants and plays a less role in the generative period. Nitrogen plays a role in many physiological processes, especially the vegetative growth phase and gives the green color of the leaves. However, too much nitrogen can inhibit flowering and fruiting and even invite pests and diseases [16]. [17] Reported that giving a 10% concentration of zeolite had no significant effect compared to no zeolite on the maturity of female maize flower emergence. [18] reported that the application of various N fertilizer frequencies had no significant effect on the aging of panicles but significant effect on harvest time.

Table 4. Age of panicle exit and harvest of upland rice applied with rice husk zeolite and N fertilizer

Rice husk zeolite dosage (kg.ha^{-1})	N fertilizer dosage (kg N.ha^{-1}) 0	75	125	Average
Age of panicle exit (day after planted)				
0	66.00 a	65.00ab	65.67 a	65.56 a
200	65.33 a	64.33ab	64.33ab	64.67ab
400	65.67 a	60.67 b	60.67 b	62.33 b
Average	65.67 a	63.33 a	63.56 a	
Age of harvest (day after planted)				
0	93.67 a	92.67 a	89.33ab	91.89 a
200	93.00 a	90.33ab	91.33ab	91.56 a
400	89.33ab	91.67ab	87.33 b	89.44 a
Average	92.00 a	91.56ab	89.33 b	

Note: Numbers in the same row and column followed by the same lowercase letters are not significantly different according to the DNMRT test at the 5% level.

3.4 The yield of upland rice

Table 5 showed that the aplication of rice husk zeolite doses of 200 and 400 kg.ha^{-1} with N fertilizer doses of 75 and 125 kg N.ha^{-1} had no significant effect on number and percentage of pithy grain compared to without rice husk zeolite and N fertilizer. Weight of 1000 grain and dry milled of grain increased at application of of rice husk zeolite doses of 200 kg.ha^{-1} and N fertilizer doses of 125 kg N. ha^{-1} compare to without rice husk zeolite and N fertilizer. This is closely related to N which plays a role in the vegetative phase, is the formation of tillers, while P and K play a role in the generative phase. This is related to [19], that nitrogen plays a role in leaf development (vegetative period), phosphorus plays a role in energy metabolism in plants and potassium acts as an activator of enzymes needed to form starch and protein.

Table 5. Number and percentage of pithy grain of upland rice applied with rice husk zeolite and N fertilizer

Rice husk zeolite dosage (kg.ha^{-1})	N fertilizer dosage (kg N.ha^{-1}) 0	75	125	Average
Weight of 1000 grains (g)				
0	124.23a	157.84a	150.60a	144.22a
200	149.77a	139.75a	145.21a	144.91a
400	169.92a	160.70a	165.95a	165.52a
Average	147.97a	152.76a	153.92a	
Weight of dry milled grain per clump (g)				
0	82.28 a	80.38 a	76.82 a	79.82 a
200	81.93 a	78.38 a	80.57 a	80.29 a
400	80.81 a	84.18 a	84.33 a	83.12 a
Average	80.57 a	80.98 a	80.75 a	

Note: Numbers in the same row and column followed by the same lowercase letters are not significantly different according to the DNMRT test at the 5% level.

These three nutrients must have a balanced value, a lack of one nutrient will cause abnormalities in plant growth, for example a lack of nitrogen will cause inhibition of the growth of vegetative organs in plants. [20], stated that the treatment of various doses of zeolite had no significant effect on panicle length and panicle number of grains. [21], stated that the treatment of various doses of N fertilizer had no significant effect on panicle length and number of spikes per panicle. [22], also stated that the treatment of various doses of N fertilizer did not significantly affect the percentage of pithy grain.

3.5 Weight of 1000 grains and dry milled grain per clump

Table 6. Weight of 1000 grains and dry milled grain per clump of upland rice applied with rice husk zeolite and N fertilizer

Rice husk zeolite dosage (kg.ha^{-1})	N fertilizer dosage (kg N.ha^{-1}) 0	75	125	Average
Weight of 1000 grains (g)				
0	21.89 c	21.98 c	22.86 abc	22.25 b
200	22.44bc	22.66bc	23.87 a	22.99 a
400	22.39bc	22.80 abc	23.16ab	22.78ab
Average	22.24 b	22.48 b	23.29 a	
Weight of dry milled grain per clump (g)				
0	35.35 b	50.04ab	56.98ab	43.736a
200	33.35 b	49.14ab	63.63 a	44.89 a
400	50.78ab	45.28ab	57.44ab	47.20 a
Average	37.10 b	44.14ab	54.60 a	

Note: Numbers in the same row and column followed by the same lowercase letters are not significantly different according to the DNMRT test at the 5% level

Table 6 showed that the application of rice husk zeolite dose of 200 $kg.ha^{-1}$ and N fertilizer dose of 125 kg $N.ha^{-1}$showed the highest yield of 1000 grains of pithy grain (23.87 g) and weight of dry milled grain per clump (63.63 g). This showed that the interaction between rice husk zeolite doses of 200 $kg.ha^{-1}$ and N fertilizer doses of 75 kg $N.ha^{-1}$ increased the yield of milled dry grain weight per clump by 80% compared to no treatment. This is in accordance with the results of research by [23], that the treatment of Urea (50%) + zeolite (50%) and the highest wet weight and dry weight of mustard plants were 80.46 g tan^{-1} and 4.43 g tan^{-1} compared without treatment. Urea + zeolite which only 20.8 g ton^{-1} wet weight and 1.33 g ton-1 dry weight, [24], reported 750 kg of zeolite treatment ha^{-1} showed a significant effect compared to without zeolite on the weight of 1000 grains of rice. [13] also stated that the average total dry weight of plants at the interaction of N fertilizer at a dose of 150% (180 kg $N.ha^{-1}$) and a dose of 150% zeolite was not significantly different compared to a dose of 50% N fertilizer (60 kg $N.ha^{-1}$) and a dose of 50% zeolite.

4 Conclusion

The application of rice husk zeolite 200 $kg.ha^{-1}$ and N fertilizer dosage 125 kg $N.ha^{-1}$ increased plant height, maximum number of tillers and weight of 1000 grains compared without rice husk zeolite and N fertilizer, and also increased the number of productive tillers and pithy grain per panicle, dry milled grain weight per clump are 56.78%,16.8%, 80%, respectively compared without rice husk zeolite and N fertilizer.

.

PartofthisresearchwassupportedbytheResearchInstituteof UniversitasRiau.

References

1. Mulyani, A. and M. Sarwani.Characteristics and potential of sub-optimal land for agricultural development in Indonesia. Journal of Land Resources. **7**(1): 47–56 (2013).
2. Subagyo, H., Nata Suharta and Agus. B. Siswanto. Agricultural Lands in Indonesia. Pages 21–66 in A. Adimihardja, L.I. Aamien, F. Agus, D. Djaenudin (Eds.). Indonesian Land Resources and Their Management. Center for Soil and Agro-climate Research and Development, Bogor (2000).
3. Asnawi, R. Increasing the productivity and income of farmers through integrated planting of paddy rice in Pesawaran Regency, Lampung. J. Applied Agricultural Research.**14**(1):33-22 (2014).
4. BalaiBesarPenelitianTanamanPadiBalitbangtanKementerian Pertanian.2017.https://bbpadi.litbang.pertanian.go.id/index.php/varietaspadi/inbrida-padi-gogoinpago/inpago-12-agritan. Diaksestanggal 25 Oktober 2022
5. Katsuki H, and Komarneni S.Synthesis of Na-A and/or Na-X zeolite/porous carbon composites from carbonized rice husk. Journal of Solid State Chemistry. **182**(7):1749–53 (2009).
6. Asririni, Dwi Wisma Ajeng. 2006. Solubility of NH^+ and NO^- from the Combination of Urea Zeolite Fertilizer in InceptisolCiomas and VertisolCiranjang Paddy Soils. Bogor Agricultural Institute. Bogor (2006).
7. Johan, Erni, L. Botoman, H. Aono, dan N. Matsue. 2017. Industry of Zeolits and Its Nanocomposite and Their Application in Soils and Environment. UPN"Veteran": ICGAI.Yogyakarta. 1-34.
8. Tando, E. Efforts to increase the efficiency and availability of nitrogen in the soil and nitrogen absorption in lowland rice (*Oryza sativa* L.). Science World. **18**(2): 171-180 (2019).
9. Purba, T., Hardian, S., Purwaningsih., Abdus, S. J., Bambang, G., Junairah., Refa, F., Arsi. Soil and Plant Nutrition. We Write Foundation, Medan (2021).
10. Irawan, D., Irsal, I., & Haryati, H. Growth response of deli tobacco (*Nicotiana tabacum* L.) to the application of nitrogen and zeolite fertilizers. Journal of Agroecotechnology, University of North Sumatra, **3**(3), 104830 (2015).
11. Bakrie, A. H. Vegetative growth of Dendrobium orchids (*Dendrobium sp*) in the application of zeolite as a mixture of planting media and liquid complementary fertilizer. Indonesian Zeolite Journal, 7(1), 53-60 (2008).
12. Pramitasari, H. E., Wardiyati, T., & Nawawi, M. The effect of nitrogen fertilizer doses and plant density levels on the growth and yield of kailan (*Brassica oleraceae*L.) (Doctoral dissertation, Brawijaya University) (2016).
13. Setyawan, S. F. D., &Suminarti, N. E. Response of sorghum (*Sorghum bicolor* L.) Super 1 variety to the application of zeolite and fertilizer N. Journal of Agricultural Science, **3**(1), 44- 53 (2018).
14. Juliana, E., Sarifuddin, S., & Jamilah, J. Giving zeolite and rice husk charcoal to rice fields polluted by factory waste on soil Pb and rice plants. Journal of Agroecotechnology, University of North Sumatra, **3**(2), 104495 (2015).
15. Widyanto, A., Sebayang, H. T., &Soekartomo, S. The effect of applying zeolite and Urea fertilizer on the growth and yield of sweet corn (*Zea mays* L. saccharata Sturt.). Journal of Plant Production, **1**(4), 378-388 (2013).
16. Irmayani, T. The effect of nitrogen fertilizer application on the emergence of foliar diseases of maize (*Zea mays* L.) in several varieties in the field (Doctoral dissertation, University of North Sumatra) (2009).
17. Kardaya, D., Rahayu, A., &Rudiansyah, D. Response of spring corn plants to various granular zeolite concentrations and soaking time in cow urine. Journal of Agronida (2018).
18. Wangiyana, W., &Ngawit, I. K. Increasing the

yield of hybrid maize Var. BISI-2 by applying cow manure and increasing the frequency of giving Urea and a mixture of SP-36 and KCl. CROP AGRO, Scientific Journal of Cultivation, **3**(1), 51-58 (2018).

19. Sutedjo, M.M. Fertilizer and Fertilization Method. Rineka Cipta: Jakarta (2008).
20. Ma'sum, F. Q. A., Kurniasih, B., &Ambarwati, E. Growth and yield of lowland rice (*Oryza sativa* L.) at several doses of straw and zeolite compost. Vegetarianism, **5**(3), 29-40 (2016).
21. Widodo, T. W., &Damanhuri, F. N. U. Effect of nitrogen dosage on shoot formation and growth of Ratun rice (*Oryza sativa* L.). Scientific Journal of Innovation, **21**(1), 50-53 (2021).
22. Anhar, R., Hayati, E., & Efendi, E. The effect of urea fertilizer doses on the growth and production of local rice germplasm from Aceh. Journal of Kawista Agrotechnology, **1** (1), 30- 36 (2016).
23. Bhaskoro, AW, Kusumarini, N., &Syekhfani, S. Efficiency of nitrogen fertilization of mustard plants on Inceptisols through the application of natural zeolite. Journal of Land and Land Resources, **2**(2), 219-226 (2017).
24. Syamsiyah, J., Suhardjo, M., &Andriyani, L. Efficiency of P fertilizer and yield of rice (*Oryza sativa* L.) in sandy rice fields of Kulonprogo beach which were treated with zeolite. Soil Science-Journal of Soil Science and Agroclimatology, **6**(1), 7-14 (2013).

The impact of water management systems and biofertilizers on soybean production in type B tidal swamp lands in Kapuas district, Central Kalimantan, Indonesia

Kartika Ayu Kinanti[1], *Sri Nuryani Hidayah Utami*[1,*], *Jaka Widada*[1], *HendriSosiawan*[2], *Wahina Anisa Yusuf*[3], *Arif Mulyawan*[3]

[1] **Department of Soil Science, Faculty of Agriculture UGM**
[2] **National Research Agency of Republic of Indonesia**
[3] **Directorate of Various Beans and Bulbs, Ministry of Agriculture of the Republic Indonesia**

ABSTRACT

Acid sulfate land is a land resource that can be optimized to achieve soybean self-sufficiency. Management of irrigation systems and the use of biological fertilizers play an important role in efforts to increase yields. This research aimed to determine the effect of biological fertilizer application and water management systems on soybean yields on tidal swamp land of overflow type B. The research was carried out on potential acid sulphate land of overflow type B using a Split Plot Design with the main plot, namely water management, namely the system management with trenches 20 cm deep and without trenches and subplots is the use of biological fertilizer, namely, Rhizobium sp., Mycorrhiza, Rhizobium sp. specific for acid land, Rhizobium sp. + Mycorrhiza. Management of water level without channels and without the addition of biological fertilizer for cultivating soybean plants in the dry season on type B tidal land is effective in maintaining soil pH that can be tolerated by soybean plants. There was no interaction and no significant effect of regulating the water level with ditches and adding biological fertilizer on the productivity of soybean plants due to rising water levels.
Keywords: acid sulfate soil, biofertilizer, soybean, water management, tidal swamp

INTRODUCTION

Soybean is a functional food source in terms of nutritional value and health benefits. It is undeniable that the need for soybeans both as food and animal feed in Indonesia cannot be fulfilled. According to Simatupang et al. (2003) [1], this is evidenced by the number of soybean deficits that continue and tend to continue to increase if there are no breakthrough efforts to increase production and in the end Indonesia will depend on imports to cover the deficit. According to the Ministry of Agriculture (2016) [2], Indonesia's soybean production has not been able to meet the demand for soybean consumption even though soybean consumption tends to decrease and Indonesian soybean production tends to increase. Based on the performance of soybean export and import values for the period 1980 to 2015, the trade balance Indonesian soybean during this period experienced an increasing deficit which tended to continue to increase. It was also found that the decrease in soybean harvested area in Java was due to the conversion of agricultural land to non-agricultural activities.

Based on the results of the mapping of the Center for Agricultural Land Resources (BBSDLP), the area of swamp land throughout Indonesia is around 33.43 million ha and the area of land suitable for agricultural activities is 9.53 million ha. Until now, only 23.8% of the total area of swamp land that is suitable for agricultural activities, namely 71.2%, has not been utilized.

Acid sulphate land is a land resource that can be optimized to achieve self-sufficiency

* **corresponding author : nuryani@ugm.ac.id**

in soybeans but swampland has unique characteristics. The soil reaction is classified as slightly acidic to extraordinarily acidic. The values range from pH 4 (Sulfaquents) to pH < 3.4 (Sulfaquepts) (Kochian et al., 2004) [3]. This will cause toxic elements (Notohadiprawiro, 2000) [4]. Many pyrite minerals ($Fe(S_2)$ are also found in tidal lands which contain large amounts of iron which, when the pyrite layer is oxidized, can lower the pH to ≤ 3.5 (Muhrizal et al., 2006) [5]. The characteristics of the swamp land will certainly be a limiting factor for plants.

According to Seenivasagan et al. (2021) [6], the use of biological fertilizers has no dangers or side effects. Utilization of biological fertilizers is expected to help overcome problems in swamp land.

The problem in soybean cultivation in tidal swamp land is that soybean plants cannot stand flooded soil conditions. Proper management needs to be done. Water management is not only necessary to control water during high tide, but also to maintain the water level during the dry season so that the pyrite layer does not oxidize. In a study by Shamshuddin et al. (2000) [7], if oxidation occurs, drainage water in acid sulfate soils will bring oxidation and reduction products such as H^+ ions, $(SO)_4^{2-}$, $(Al)^{3+}$, and ($Fe)^{2+}$, as well as nutrients $(Ca)^{2+}$, $(Mg)^{2+}$, and K^+. Micro elements such as Al and Fe are available in large quantities so that they can harm plants. This study aims to 1) determine the effect of water management and application of biological fertilizers on the productivity of soybean plants in tidal swamps with overflow type B.

MATERIALS AND METHODS

The research was conducted in Sidomulyo Village, Tamban Catur District, Kapuas Regency, Central Kalimantan (*(3° 20'66 SL and 114° ' 40'54 EL)*

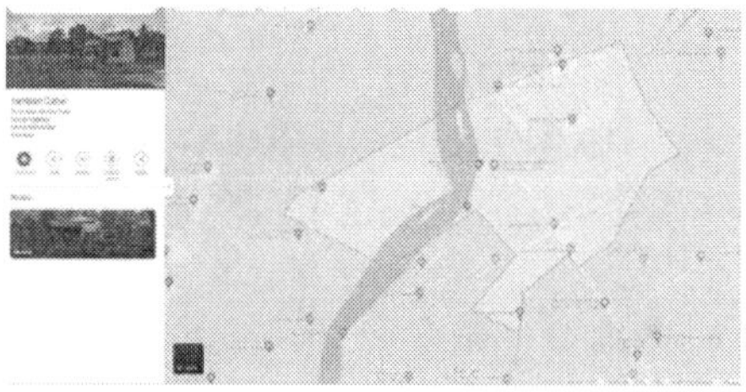

Source: https://peta.web.id/peta/kec/tamban-catur-194

The research was conducted using a Split Plot Design with the Main Plot namely water management: **1.** Control (Without adjusting the water level) (P0), 2. Water level 20cm (water depth in the quarter ditch 20 cm below ground level (P1). Subplots are the use of biological fertilizers : A.Control (A1), B. Rhizobium sp. sp. (A2), C. Mycorrhiza (A3), D. Rhizobium sp. sp. acid soil specific (A4), and E. Rhizobium sp. sp. + Mycorrhiza (A5)

Table 1. Treatment Combinations

Treatments	A1	A2	A3	A4	A5
P0	P0A1	P0A2	P0A3	P0A4	P0A5
P1	P1A1	P1A2	P1A3	P1A4	P1A5

Seed preparation was carried out by preparing soybean seeds of anjasmoro, mycorrhiza, Rhizobium, and Rhizobium specific sour varieties. The agronomic parameters observed were periodic plant height (2 weeks), yield components (number of pods), dry seed yield (tiles). Plant height observations were carried out every 2 weeks on 10 sample plants in one treatment plot with a total sample of 300 plants. Observation of the yield component was carried out by weighing the dry seed yield from the tile plots in each treatment plot. Soil analysis was carried out at 30 days after planting (dap). Soil data components observed included: The degree of acidity (pH), redox potential (Eh), and electrical conductivity (EC) were directly measured in the field before application and once every 2 weeks with the Tester 35 Series, Available P was observed at 30, 60, and 90 dap using the Bray-I method, organic-C with the Walkley & Black method observed at 30, 60, and 90 dap, total- N with the Kjehdal N method were observed at 30, 60 and 90 dap, exchangeable-K, using the NH_4OAc 1 N method was observed at 30, 60 and 90 dap (Soil Research Institute, 2009) . Experimental water management observations include:

a. Observation of ground water level. Measurement using a perforated pipe then buried

 https://doi.org/10.1051/bioconf/20238003022

into the ground. b. The degree of acidity (pH) and electrical conductivity (EC) are directly measured in water with the 35 Series Tester.

Data analysis

Experimental data were analyzed using variance (ANOVA) to determine whether there was a treatment that had a significant difference. If the effect is significantly different (F Count > F Table, α 5%) then proceed with further tests to find out the treatment is significantly different. This analysis uses the software SAS 9.10 for windows.

RESULTS AND DISCUSSION

Water Level

Based on a 1: 100,000 scale semi-detailed soil map, the study location is a tidal land with overflow types B to C (Subagyono et al, 1999) (Subagio, 2006 [9] and has a sulfidic layer at a depth of more than 50 cm so that the research location is classified as a potential acid sulphate land. The water level is a factor that needs to be considered in conducting agricultural cultivation in tidal swamps, especially on acid sulphate soils. This is because the water level affects the redox conditions, the pyrite layer will be oxidized to acid sulphate which can further reduce the pH value to 4 due to pulverization until it reaches the pyrite layer (Dent & Pons, 1995) [10].

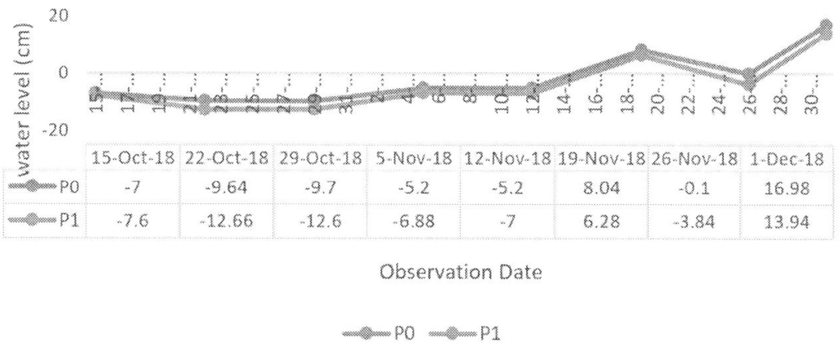

Figure 1. Fluctuations in water level , P0: No water inlet
P1: 20 cm channel

Figure 1 shows the fluctuation of the groundwater level at the study site. A negative value indicates that the water table is below the ground surface and a positive value indicates that the water table has exceeded the ground level. The water level during the planting period fluctuated. Water fluctuations on land are influenced by tidal activity. According to Haryono et. al. (2013) [11], river flow affects the groundwater table due to the gravitational force or the gravitational pull of the earth and moon in the solar system. The farther the river flows from the estuary, the weaker the river flows. According to Suriadikarta (2005) [12], the water management system for acid sulphate land with overflow type B should use a one-way flow system, but the condition of the water system at the study site is still not well organized, most of the floodgates are damaged or even missing. These conditions can cause the management of the water system carried out in the research plots to be not optimal.

pH periodic of Groundwater

The results of periodic pH analysis of water on land plots are presented in Figure 2. Observation of pH values was carried out by taking water samples from 10 physiometers installed in the observation plots. The pH value at the beginning of the observation until the end of the observation did not show a significant change. The pH value in the treatment without drains (control) showed a higher value than the ditch treatment, but at the end of the harvest the pH value between treatments was relatively the same because the land conditions were flooded until it reached a positive value as shown in figure 2.

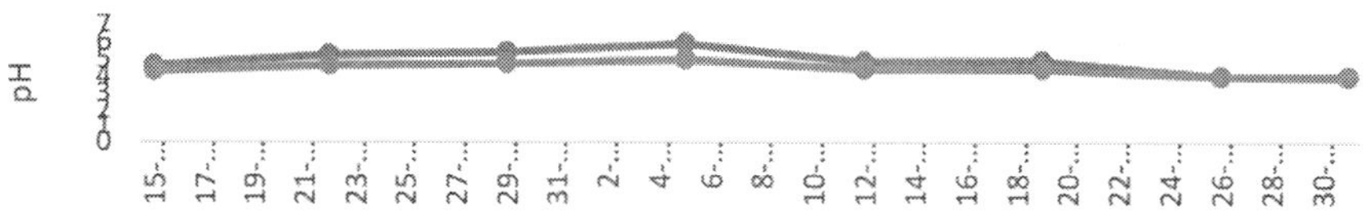

	15-Oct-18	22-Oct-18	29-Oct-18	5-Nov-18	12-Nov-18	19-Nov-18	26-Nov-18	1-Dec-18
P0	4.584	5.182	5.35	5.87	4.796	4.824	3.936	3.924
P1	4.246	4.6	4.688	4.916	4.356	4.346	3.926	3.872

observation date

P0 P1

Figure 2. pH periodic fluctuations of water
P0: o water inlet, P1: Channel 20cm

The pH value was highest in control and treatment with waterways were obtained on November 2018 measurement with a value of 5.87 in the control and 4.92 in the treatment with waterways.

EC water fluctuation periodic

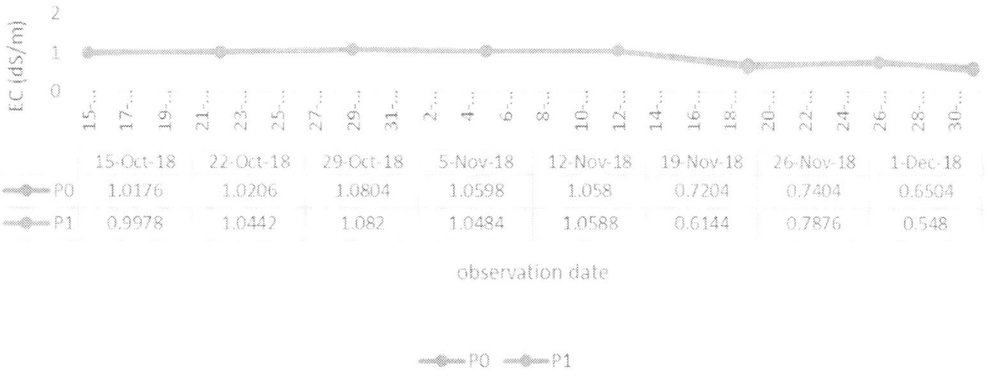

Figure 3. EC water fluctuations periodic
Information: P0: No water inlet,
P1: Channel 20 cm

The results of the analysis of the periodic electrical conductivity of water in the plots of land are presented in Figure 3. The electrical conductivity of the two treatments had relatively the same value and the relative magnitude did not change from the beginning of the measurement, namely 15 October to 12 November. Furthermore, the electrical conductivity decreased until harvest in line with the increase in water level (figure 1). The decrease in the value of the electrical conductivity is caused by the runoff from the flood which dissolves the salt, thereby reducing the electrical conductivity. According to Xinmin et al. (2022) [13], high salt content results in high EC values and adsorbed Na ratio (SAR).

Soil pH

Soil reaction or pH is an important factor in soil. The balance of processes that occur in the soil such as chemical processes and the availability of nutrients for plants is determined by pH. Apart from that, pH is also the main characteristic of swamp land.

Tabel 2. soil pH fluctuation periodic

(Treatments)		**Periodic (dap)**		
water level	**fertilization**	**30**	**60**	**90**
No settings	control	4.76a	6.02a	4.41a
	Rhizobium sp.	4.66a	5.25b	4.49a
	Mycorrhiza	4.77a	5.52ab	4.34a
	Rhizobium sp. specific for swamp land	4.84a	5.51ab	4.43a
	Rhizobium sp. + Mycorrhiza	4.90a	5.83ab	4.45a
Water level setting (20cm)	control	4.87a	5.74ab	4.50a
	Rhizobium sp.	4.72a	5.31b	4.53a
	Mycorrhiza	4.78a	5.83ab	4.39a
	Rhizobium sp. specific for swamp land	4.72a	5.43ab	4.39a
	Rhizobium sp. + Mycorrhiza	4.49a	5.55ab	4.46a
Interaction		(-)		
CV (%)		6.25	5.38	2.7

Note: the (+) sign indicates that there is an interaction between treatments, the (-) sign indicates that there is no interaction between treatments. Numbers followed by the same letter in one parameter indicate that they are not significantly different at the 5% DMRT test significance level

The results of periodic soil pH analysis are presented in table 1. Based on table 1, there was no interaction between the two treatments at the pH value. The soil pH value at 30 dap (day) after transplanting was not significantly different between treatments and had a low value, i.e. all treatments showed a pH value of less than 5. The highest pH value was in the treatment without channels with the addition of Rhizobium sp. and Mycorrhiza which is equal to 4.9 and the lowest is in the treatment with waterways with the addition of Mycorrhiza and rhizobium which is equal to 4.49. The highest pH value decreased so that it was lower than at 30 dap and 60 dap measurements. This is because at 90 dap the research area was flooded.

The water level without channels and without the addition of biological fertilizers is lower so that acid sulfate soils are in a stagnant condition. According to Stone et al (1998) [14], under flooded conditions, acid sulfate soils are in a stable condition because they are not oxidized. When the groundwater level drops, the sulfidic material can undergo oxidation which produces sulfuric acid which causes acidity in the soil.

EC soil periodic

(Treatments)		**Periodic (DAP)**		
water level (cm)	**Fertilization**	**30**	**60**	**90**
No setting	control	308,53a	279,27a	256,67a
	Rhizobium sp.	341,03a	329,80a	248,83a
	Mycorrhiza	283,07a	275,47a	301,23a
	Rhizobium sp. specific for swamp land	356,03a	332,77a	289,27a
	Rhizobium sp. + Mycorrhiza	267,83a	336,97a	258,57a
Water level setting (20cm)	Control	333,87a	288,30a	211,60a
	Rhizobium sp.	356,57a	289,07a	261,90a
	Mycorrhiza	251,17a	294,60a	299,43a
	Rhizobium sp. specific for swamp land	384,27a	290,13a	218,53a
	Rhizobium sp. + Mycorrhiza	315,80a	249,60a	279,00a
Interaction		(-)	(-)	(-)
CV (%)		20,33	25,25	24,23

Tabel 3. Periodic soil EC (dS/m)

(Treatments)		**Periodic (dap)**		
Water level	**fertilizer**	**30**	**60**	**90**
	control	0.32a	0.34ab	0.19a
No settings	*Rhizobium sp.*	0.40a	0.38ab	0.22a
	Mycorrhiza	0.26a	0.30b	0.19a
	Rhizobium sp. specific for swamp land	0.34a	0.34ab	0.22a
	Rhizobium sp. + Mycorrhiza	0.27a	0.37ab	0.20a
Water level setting (20cm)	control	0.28a	0.32ab	0.19a
	Rhizobium sp.	0.40a	0.40a	0.22a
	Mycorrhiza	0.34a	0.34ab	0.19a
	Rhizobium sp. specific for swamp land	0.30a	0.37ab	0.20a
	Rhizobium sp. + Mycorrhiza	0.39a	0.39a	0.21a
Interaction		(-)		
CV (%)		26.22	12.95	24.87

Note: the (+) sign indicates that there is interaction between treatments, the (-) sign indicates that there is no interaction between treatments. Numbers followed by the same letter in one parameter show no significant difference at the significant level of the 5% DMRT

The results of periodic soil conductivity analysis are presented in table 2. Based on table 2, there was no interaction between the two treatments on the EC value. Treatment of Rhizobium sp. with the water management system showed the highest EC value of 0.4 dS/m and the Mycorrhiza treatment without drainage management showed the lowest pH value of 0.3 dS/m. According to Balittanah (2009) [15] the value of soil EC < 1 dS/m is included in the very low level. From the measurement results above, each treatment has an electrical conductivity value of <4 dS/m, which means that the soil is still in the normal category for its salinity level according to the Gunal *et al.* (2021) [16], and does not interfere with plant growth. The rising water level due to the high tide that occurs at 60 dap measurement dissolves the salt thereby reducing the value of EC.

Table 4. Potensial Redox (soil Eh mV)

Note: the (+) sign indicates that there is interaction between treatments, the (-) sign indicates that there is no interaction between treatments. Numbers followed by the same letter in one parameter show no significant difference at the significant level of the 5% DMRT test

https://doi.org/10.1051/bioconf/20238003022

Redox potential or Eh is an important reaction that occurs in the soil besides pH. The Eh value indicates oxidation-reduction activity. Table 3 presents the effect of high water level treatment and biological fertilizers on Eh. Based on table 4, there was no interaction between the two treatments on the Eh value and there was no significant difference between the treatments. This shows that the application of biological fertilizers and water level adjustment have no effect on the Eh value. The measured Eh values ranged from 211.60 mV to 356.03 mV. According to Husson (2013) [17], the optimal Eh value for plant growth ranges from +400 to +450 mV. Plant growth declines rapidly at Eh below +350 mV.

Soil Total N

Nitrogen is one of the macro essential nutrients that plants really need for growth in vegetative vases. Total nitrogen is nitrogen in a form available to plants or not available to plants. Table 4 presents the effect of water level regulation and biological fertilizers on periodic total nitrogen.

Tabel 4. Soil Total N (%)

(Treatments)		Periodic (DAP)		
Water level	**fertilization**	**30**	**60**	**90**
No setting	control	0.55a	0.50ab	0.62a
	Rhizobium sp.	0.50a	0.52ab	0.63a
	Mycorrhiza	0.54a	0.57a	0.56a
	Rhizobium sp. specific for swamp land	0.54a	0.53ab	0.62a
	Rhizobium sp. + *Mycorrhiza*	0.52a	0.52ab	0.59a
Water level setting (20cm)	control	0.51a	0.49ab	0.59a
	Rhizobium sp.	0.50a	0.47b	0.55a
	Mycorrhiza	0.52a	0.51ab	0.61a
	Rhizobium sp. specific for swamp land	0.48a	0.53ab	0.56a
	Rhizobium sp. + *Mycorrhiza*	0.51a	0.51ab	0.59a
Interaction		(-)		
CV (%)		6.95	6.16	8.17

Note: the (+) sign indicates that there is interaction between treatments, the (-) sign indicates that there is no interaction between treatments. Numbers followed by the same letter in one parameter show no significant difference at the significant level of the 5% DMRT test

Based on table 4, there was no interaction between the two treatments at the N value. The N value did not show a significant difference between the treatments except for observations made at 60 dap. According to Hardjowigeno (2010) [18], the process of loss of nitrogen in the soil can be caused by absorption by plants, used by microorganisms, N is still in the form NH_4^+ which is bound by clay minerals so it cannot be used by plants, N is also in the form NO_3^- which is easily washed away by rainwater, and poor air fertility can cause denitrification and also volatilization in the form of NH_3 (ammonia). The Mycorrhiza treatment without a water management system showed the highest total N value of 0.57 and the Rhizobium sp. with the drainage management system showed the lowest total N value of 0.47.

Soil available P

Element P is a macro-essential macro nutrient that is needed by plants for growth in the generative vase, especially during the formation of pods. P or available phosphorus is phosphorus in forms that can be absorbed by plants, namely the anions $H_2PO_4^-$ and HPO_4^{2-}. Table 4.4 presents the effect of water level regulation and biological fertilizers on periodic soil available P.

Tabel 5. Soil available-P (mg/kg)

Treatments		Periodic (DAP)		
Water level	**Fertilization**	**30**	**60**	**90**
No wetting	control	47.15 a	38.90 a	34.43ab
	Rhizobium sp.	37.80 a	55.43 a	33.79ab
	Mycorrhiza	39.25 a	46.46 a	35.76a
	Rhizobium sp. specific for swamp land	54.76 a	35.52 a	36.94a
	Rhizobium sp. + *Mycorrhiza*	43.42 a	46.07 a	34.95ab
Water level setting (20cm)	control	42.78 a	27.47 a	24.36ab
	Rhizobium sp.	24.21 a	38.14 a	25.21ab
	Mycorrhiza	31.86 a	32.52 a	24.22ab
	Rhizobium sp. specific for swamp land	31.86 a	31.69 a	27.40ab

	Rhizobium sp. + Mycorrhiza	32.04 a	33.18 a	20.32b
Interaction		(-)		
CV (%)		32.92	34.07	19.08

Note: the (+) sign indicates that there is interaction between treatments, the (-) sign indicates that there is no interaction between treatments. Numbers followed by the same letter in one parameter show no significant difference at the significant level of the 5% DMRT test

At 90 dap measurement there was a significant difference in available P content between treatments. Treatment of Rhizobium sp. without a water management system showed the highest available P value of 36.94 ppm and Rhizobium sp. + Mycorrhiza with a drainage management system showed the lowest available P value of 20.32 ppm.

Soil K

Tabel 6. Soil K (Cmol(+)/kg)

Treatments		Periodic (dap)		
Water level	**Fertilization**	**30**	**60**	**90**
No setting	control	0.42a	0.45a	0.47a
	Rhizobium sp.	0.42a	0.43a	0.44a
	Mycorrhiza	0.34a	0.42ab	0.37a
	Rhizobium sp. specific for swamp land	0.40a	0.28b	0.35a
	Rhizobium sp. + Mycorrhiza	0.39a	0.39ab	0.40a
Water level setting (20cm)	control	0.36a	0.37ab	0.47a
	Rhizobium sp.	0.30a	0.37ab	0.38a
	Mycorrhiza	0.39a	0.32ab	0.33a
	Rhizobium sp. specific for swamp land	0.35a	0.37ab	0.33a
	Rhizobium sp. + Mycorrhiza	0.40a	0.38ab	0.35a
Interaction		(-)		
CV(%)		20,86	17,28	20,35

Note: the (+) sign indicates that there is interaction between treatments, the (-) sign indicates that there is no interaction between treatments. Numbers followed by the same letter in one parameter show no significant difference at the significant level of the 5% DMRT test

According to Balittanah (2009) [15] land value of 0.1-0.3 is classified as low and 0.4-0.5 is classified as moderate. From the measurement results, moderate to low K values were obtained. According to Widjaja and Adi (1997) [19], when there is an acidification process (acid pH conditions) nutrient cations in the form of Ca, Mg, and K will experience pressure and be leached by H+ ions and will be leached with water leaving the soil body. Therefore, often the availability of cations Ca, Mg, and K in peat soils is so low.

Organic -C

Table 7. Organic-C (%)

Treatments		Periodic (DAP)		
Water level	**Fertilization**	**30**	**60**	**90**
No setting	control	7.74a	7.19a	8.01a
	Rhizobium sp.	7.36a	7.51a	7.93a
	Mycorrhiza	7.21a	7.76a	7.48a
	Rhizobium sp. Sampy land spesific	7.42a	7.34a	7.96a
	Rhizobium sp. + Mycorrhiza	7.87a	7.80a	6.66a
Water level setting (20cm)	control	7.94a	7.57a	8.01a
	Rhizobium sp.	7.03a	7.00a	7.76a
	Mycorrhiza	7.38a	7.53a	7.70a
	Rhizobium sp. Swampy land spesific	6.77a	6.94a	8.29a
	Rhizobium sp. + Mycorrhiza	7.31a	7.47a	8.77a
Interaction		(-)		
CV(%)		6.07	7.31	12.82

Note: the (+) sign indicates that there is interaction between treatments, the (-) sign indicates that there is no interaction between treatments. Numbers followed by the same letter in one parameter show no significant difference at the significant level of the 5% DMRT test

Organic matter is the result of the decomposition of the remains of living things decomposed by microorganisms in the soil. Organic carbon is the conversion value of organic matter. Organic matter plays a role in improving the physical, biological and chemical properties of the soil as well as a source of macro and micro nutrients needed by plants.

The effect of the application of biological fertilizers and water management systems on the periodic organic carbon content is shown in table 10. Based on the analysis results in table 10 above, there was no interaction between the two treatments on the organic carbon content in the soil. The organic carbon content in the soil did not show a significant difference between the treatments. The highest organic carbon value was in the addition of Rhizobium sp. and Mycorrhiza with a water level setting of 8.77%. Table 4.1.10 shows that the C-Organic content at harvest or at 90 dap showed an increase compared to the C-Organic conditions at 30 dap.

Soybean crop yields

Table 8. Soybean crop yields (Ton/Ha)

Treatments		Dry Seed Productivity (Ton/Ha)
Water level	**Fertilization**	
No setting	control	1.23a
	Rhizobium sp.	1.33a
	Mycorrhiza	1.33a
	Rhizobium sp. specific for swamp land	1.44a
	Rhizobium sp. + Mycorrhiza	1.28a
Water level setting (20cm)	control	1.39a
	Rhizobium sp.	1.44a
	Mycorrhiza	1.49a
	Rhizobium sp. specific for swamp land	1.44a
	Rhizobium sp. + Mycorrhiza	1.55a
Interaction		
CV (%)		16.98

Note: the (+) sign indicates that there is interaction between treatments, the (-) sign indicates that there is no interaction between treatments. Numbers followed by the same letter in one parameter show no significant difference at the significant level of the 5% DMRT test

There was no interaction between the two treatments on soybean production. Soybean production results did not show a significant difference between the treatments. Based on table 8, the highest productivity was in the treatment by adjusting the water level and the addition of Rhizobium sp. + Mycorrhiza with a value of 1.55 tons/ha, while the lowest productivity was in the control plot or without water level regulation and without the addition of biological fertilizers with a value of 1.23 tons/ha. This happens because during the filling period of the pods, the ground water level (presented in Figure 4) rises to a positive value or is above the soil surface, which in turn can interfere with root respiration, thereby disrupting the formation of biomass which has an impact on inhibition of filling and formation of pods. The rise in the groundwater level until flooding brings drainage water that is acidic in nature to interfere with the absorption of nutrients by plants due to the low pH value in measured groundwater. The application of Rhizobium and Mycorrhiza on acid soils can increase the uptake of N and P, but there was no significant difference in the research conducted. This is because flooding occurs so that the rhizosphere is in a reductive condition which causes the activity of Rhizobium and mycorrhiza to be disrupted and even stopped due to the low oxygen content in the rhizosphere.

These results, although not significant, are quite promising, because other research shows that co-inoculation of rhizobia and mycorrhizal fungi can be utilized biotechnologically to offer a solution to food insecurity [20]. Among soil-borne microbes, the arbuscular mycorrhizal fungi (AMF, Glomeromycota) are key components of natural and anthropogenic ecosystems. AMF can mediate substantial benefits in legumes such as soybean, because the plants form symbiotic associations with both P-acquiring AMF and N_2-fixing rhizobia, establishing tripartite symbioses that have synergistic effects on nutrient acquisition and growth of the host plants as well as positive effects on one another [21].

Conclusion

1. Setting the water level without channels and without the addition of biological fertilizers for soybean cultivation during the dry season in type B tidal land, is effective in maintaining soil pH that can be tolerated by soybean plants.
2. There is no interaction and no significant effect on the regulation of the water level with the ditch and the addition of biological fertilizers on the productivity of soybean plants due to an increase in the water level.

Suggestion

1. Planting time for soybean cultivation on tidal land is better done during the dry season, namely from May to September because the water demand for soybeans can still be met from rainfall during the dry season and paddy fields are not in a flooded condition.
2. Water management for soybean cultivation during the dry season period (May to September) does not need to be done by making canals in paddy fields because it is not effective for increasing soybean production.

REFERENCES

1. P.B. Simatupang, B. Sayaka, S. Saktyanmu, M. Marianto, M. Ariani, dan N. Syafa'at, 2003, Analisis kebijakan ketahanan pangan dalam era globalisasi dan otonomi daerah.
2. Indonesian Agricultural Statistics Center for Agricultural Data and Information System, Ministry ofAgriculture Republic of Indonesia 2019.pp. 105-12 *(16) (PDF) Soybean in Indonesia: Current Status, Challenges and Opportunities to Achieve Self-Sufficiency*. Available from: https://www.researchgate.net/publication/357020534_Soybean_in_Indonesia_Current_Status_Challenges_and_Opportunities_to_Achieve_Self-Sufficiency [accessed Sep 25 2023].
3. LV. Kochian, Hoekenga OA, Pineros MA, How do crop plants tolerate acid soils? -

Mechanisms of aluminium tolerance and phosphorous efficiency, Annu Rev Plant Biol. 55:459-493 (2004).
4. T. Notohadiprawiro, Tanah dan Lingkungan [Land and Environment]. Second Press. Pusat Studi Sumberdaya Lahan (PPSL) Univ. Gadjah Mada. Yogyakarta. 187 hal (2000).
5. S. Muhrizal, S., J. Shamsuddin., Fauziah., M.A.H. Husni, 2006, Changes in iron-poor acid sulphate soil upon submergence, Geoderma. 131: 110-122 (2006).
6. R. Seenivasagan, and O.O Babalola, Utilization of Microbial Consortia as Biofertilizers and Biopesticides for the Production of Feasible Agricultural Product, *Biology* **2021**, *10*, 1111. https://doi.org/10.3390/biology1011111. (2021).
7. J. Shamshuddin, A. Elisa Azura, M.A.R.S. Shazana, C.I. Fauziah, Q.A. Panhwar, U.A. Naher, *Properties and Management of Acid Sulfate Soils in Southeast Asia for Sustainable Cultivation of Rice, Oil Palm, and Cocoa.* In Donald L. Sparks, editor: *Advances in Agronomy*, Vol. 124, Burlington: Academic Press, 2014, pp. 91-142. ISBN: 978-0-12-800138-7 © Copyright 2014 Elsevier Inc. Academic Press
8. H. Subagyo, H, Klasifikasi dan penyebaran lahan rawa, hlm.1- 22, Dalam D.A. Suriadikarta, U. Kurnia, Mamat H.S., W. Hartatik, dan D. Setyorini (Ed.). *Karakteristik dan Pengelolaan Lahan Rawa. Balai Besar Penelitian dan Pengembangan Sumberdaya Lahan Pertanian*, Bogor (2006).
9. K. Subagyono, S I Wayan, EE Ananto. Penataan lahan dan tata air mikro pengembangan SUP lahan pasang surut Sumatera Selatan, P2SLPS2. Badan Litbang Pertanian, (1999).
10. D. Dent, Acid sulphate soils: a base line for research and development. ILRI Publication 39, International Institute for Land Reclamation and Improvement. Wageningen, TheNetherlands, (1986)
11. Haryono, M. Noor, H. Syahbuddin, and M. Sarwani. Lahan Rawa: Penelitian dan Pengembangan. Badan Penelitian dan Pengembangan Pertanian Kementrian Pertnian, Bogor, (2013).
12. DA., Suriadikarta, Pengelolaan lahan sulfat masam untuk usaha pertanian. Jurnal Litbang Pertanian, Bogor, (2005)
13. L. Xinmin, Y. Zhu, J.M Bennett, L. Wu, H. Li, Effects of sodium adsorption ratio and electrolyte concentration on soil saturated hydraulic conductivity, Geoderma, Volume 414 (2022), 115772, ISSN 0016-7061, https://doi.org/10.1016/j.geoderma.2022.115772.
14. Y. Stone, CR Ahern, and Blunden B. Acid Sulfate Soils Manual, Acid Sulfate Soil Management Advisory Committee, Wollongbar, NSW, Australia (1998)
15. Balai Penelitian Tanah (Soil Research Institute). Analisis Kimia Tanah, Tanaman, Air, dan Pupuk. Badan Penelitian dan Pengembangan Pertanian Departemen Pertanian.pp 143. (2009).
16. E. Gunal, X Wang, OM Kilic, M Budak, S Al Obaid,MJ Ansari, M Brestic. 2021. Potential of Landsat 8 OLI for mapping and monitoring of soil salinity in an arid region: A case study in Dushak, Turkmenistan. Plos One 16 (11): E0259695. https://doi.org/10.1371/journal.pone.0259695 , (2021)
17. O. Husson. 2013. Redox potential (Eh) and pH as drivers of soil/plant/microorganism systems : a transdisciplinary overview pointing to integrative opportunities for agronomy. Plant Soil. 362:389-417 (2017).
19. I.P.G. Wijaya Adhi, Pengelolaan lahan rawa pasang surut dan lebak. Jurnal Penelitian dan Pengembangan Pertanian V(1): 1-9 (1986).
20. Igiehon, O.N., Babalola, O.O. *Rhizobium* and Mycorrhizal Fungal Species Improved Soybean Yield Under Drought Stress Conditions. *Curr Microbiol* 78, 1615–1627 (2021). https://doi.org/10.1007/s00284-021-02432-w
21. V.S. Faggioli, F. Covacevich, G. Grilli, *et al.* Environmental response of arbuscular mycorrhizal fungi under soybean cultivation at a regional scale. *Mycorrhiza* **32**, 425–438 (2022). https://doi.org/10.1007/s00572-022-01093-2

The Positive Effect of The Forest Environment in The Ecotherapy Healing Forest Program on The Physical and Mental Health After COVID-19

Vincencia Septaviani Issera Sulistya Putri[1*], *Lies* Rahayu Wijayanti[2], and *Iswari* Nur Hidayati[3]

[1]Doctoral Program of Department Environmental Science, Universitas Gadjah Mada, Agro Street No 1, Yogyakarta, Indonesia
[2]Department of Environmental Science, Universitas Gadjah Mada, Agro Street No 1, Yogyakarta, Indonesia
[3]Department of Cartography and Remote Sensing, Faculty Geograpphy, Universitas Gadjah Mada, Agro Street No 1, Yogyakarta, Indonesia

Abstract. This research is aimed at students who are currently experiencing excessive coursework and high emotional stress after COVID-19. Where if these conditions are not immediately addressed, it can cause mental illness. The Ecotherapy Healing Forest program offers several activities in it, such as relaxation programs to relieve stress, walks in the forest, and sensory elements to be given to the participants. In this study, 44 student participants were used from class 2018 – class 2021 at Kapuas Sintang University. Participants will be measured their blood sugar levels and oxygen saturation by medical personnel, and given the DASS 42 questionnaire to measure participants' stress levels. Measurements will be accrued out 5 repetitions of the Ecotherapy Healing Forest program at different times. The results of this study revealed that there was a significant decrease in blood sugar levels, oxygen saturation and stress levels in the participants after participating in the Ecotherapy Healing Forest program. However, for smoking students, the Ecotherapy Healing Forest treatment does not have a significant effect on oxygen saturation values. Therefore, the authors propose Ecotherapy Healing Forest as a non-pharmacological intervention to relieve stress while improving physical health (blood sugar levels and oxygen saturation) for students.

1 Introduction

Humans and nature have nurtured each other for a long time. Nature and humans are not that different. Hence, there are limits to how natural resources can be used to satisfy cultural needs and wants. Nature is respected, and there is a natural relationship between humans and nature. In human view, nature is not a commodity that must be consumed to meet human needs. Due to urbanization, globalization, and technological advances, people are increasingly away from the "free nature". They were away from the "outdoors" and focused more on television, computers, and games as these advances accelerated the environment. These changes reduce our concern for the natural environment [1]. Lack of exposure to nature causes psychological functioning to change, as our natural resources are depleting, our psychological framework is also affected, which has been neglected for quite some time. In the past, research has shown that being in touch with nature can improve many aspects of well-being.

The coronavirus disease 2019, or COVID-19, started spreading rapidly in Wuhan City, Hunan Province, China, in December 2019, and has killed many people around the world. Everyone agrees that the acts of violence imposed in the aftermath of this pandemic have not only had an impact on physical health but also have had a devastating impact on mental health and well-being [2]. It is a known fact that reduced social interaction and increased loneliness are known risk factors for several mental disorders, including major depression, anxiety, post-traumatic stress and schizophrenia. Worries about the health of oneselt and those of their loved ones (especially the elderly or individuals who suffer from physical ailments), as well as great uncertainty about the future, can exacerbate fear, depression, and anxiety. Prolonged exposure to these worries can increase the risk of serious and disabling mental health conditions including anxiety disorders such as panic, obsessive-compulsive disorder, stress and trauma-related disorders [3]. This also has an impact on students who are currently experiencing changes in the learning system after COVID-19, so that they experience excessive lecture loads and high emotional stress. Where if these conditions are not immediately addressed, it can cause mental illness. Therefore, measures to mitigate these mental health consequences demand top priority before the impact of this pandemic gets worse.

Ecotherapy Healing Forest is a non-pharmacological method that uses the power of nature to improve human health, especially mental health after COVID-19. A core element of Ecotherapy Healing Forest is getting to know the forest environment with a focus on the five senses,

* Corresponding author: author@email.org

and there are many activities to do there, such as wellness, yoga, walking in the forest, and many more. Ecotherapy Healing Forest can reduce stress or physical fatigue, anxiety, improve mood, and improve quality of life. In addition, this Ecotherapy Healing Forest also uses forests, which are conservation areas in Indonesia, with beautiful natural scenery and easy access, so they can not only be used by depressed patients but can also support conservation. Therefore, this study aims to determine the effect of the Ecotherapy Healing Forest program using the Gunung Kelam Natural Tourism Park conservation area on the physical and mental health of post-COVID-19 students.

2 Materials and methods

2.1 Participants

Participants in this study consisted of 44 students from the Forestry Study Program, Kapuas Sintang University from the 2018 – 2021 class. They are between 18 and 35 years old. Assuming that this is an age group that is prone to psychological health problems. The sampling method uses the slovin formula to calculate the number of participants. Confidentially of participants and all data collected will be protected for the purpose of this research only. To avoid everyday stressors during the study, all participants were asked to carry out daily work assignments and personal responsibilities as normal as students [4].

2.2 Study design

This study took samples with stratified random sampling. Where students are divided according to their class on campus, with the consideration that each generation brings different burdens and problems that must be faced on campus. Each batch will participate in the Ecotherapy Healing Forest program which includes six locations with five repetitions at different times. In each Ecotherapy Healing Forest program activity, participants will have their physical condition measured at the start and end of treatment by medical personnel, in the form of measurements of blood pressure, blood sugar levels, oxygen saturation and pulse rate. Participants will also be given a pre-test and post-test for the DASS 42 measurement.

Analysis of the sustainability of Ecotherapy Healing Forest management in the Bukit Kelam Nature Tourism Park adapts the Rapfish or Rapid appraisal for Fisheries method which is modified with the MDS method, which involves attribute components or dimensions to evaluate the influence of each component. In research on Ecotherapy Healing Forest Management in the Bukit Kelam Nature Tourism Park, the attributes chosen reflect the level of sustainability, there are four dimensions that are considered influential, including Social Economics, Ecology, Institutions and Health. After each attribute used is determined, it is then given a scale between 1 and 4. Scale 1 indicates a poor/poor measure of sustainability, while a maximum scale of 4 means a measure of sustainability in the good or high category. The feasibility of the Ecotherapy Healing Forest sustainability model in the Bukit Kelam Nature Tourism Park is carried out by measuring the level of goodness of fit through calculating the S-Stress value. using the algorithm of scale (ALSCAL) method. A low S-stress value indicates high accuracy (good fit), and vice versa. The measure usually used as a reference is if the S-Stress value is less than 0.25, then the sustainability model has a good fit. However, if it is more than 0.25, it means the model is poor fit.

$$S - Stress = \sqrt{\frac{1}{m}\sum_{k=1}^{m}\frac{\sum i \sum j}{\sum i \sum j o_{ijk}^{4}}\left(d_{ijk}^{2} - o_{ijk}^{2}\right)} \quad (1)$$

From the results of these measurements, a representative site will then be designed for future EHF program activities, so that it can be more useful in the physical and mental healing process after COVID-19.

2.3 Study site

The conservation area of Mount Kelam Natural Tourism Park (Fig.1), located in Sintang Regency, has enormous potential to become a beautiful natural tourist spot with stunning views and an area of 1,121 hectares based on the Decree of the Minister of Forestry of the Republic of Indonesia Number SK. 45/Kptsll/1999 dated 14 June 1999. The location of this study is also very representative because it is easily accessible to the participants. The research was conducted in a lowland forest ecosystem with 96 species and 35 plant families. The most common tree species come from the families Dipterocarpaceae, Euphorbiaceae, and Moraceae. Each type of plant has unique benefits and functions, such as water storage, medicinal plants, animal habitats, or plants that maintain the balance of the ecosystem.

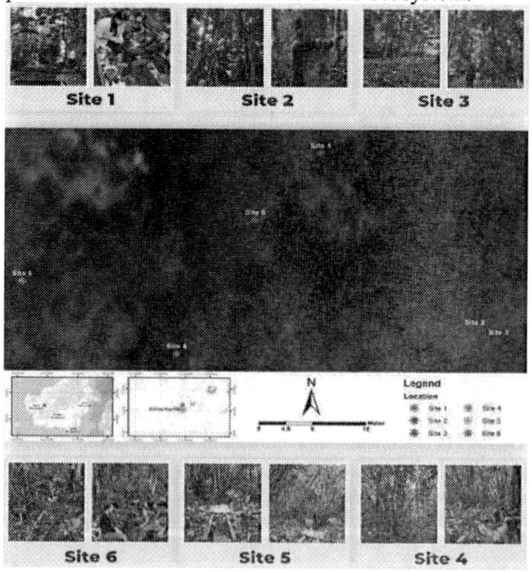

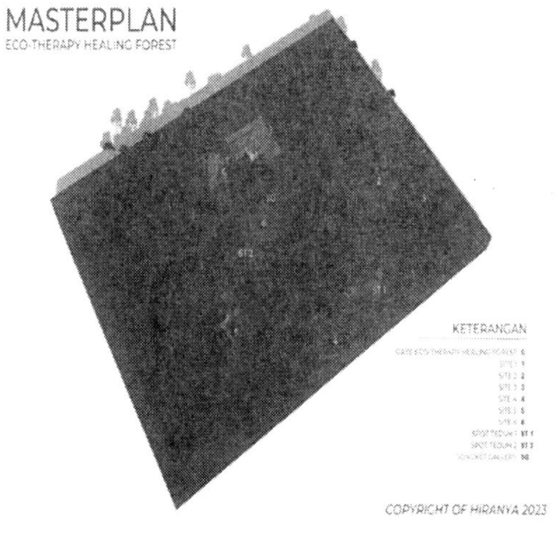

Fig. 1. Study site of Ecotherapy Healing Forest.

3 Results and discussion

3.1 Site identification

The species richness index for seedlings, stakes, poles, and trees has a value of more than 4. This indicates that the species richness of the vegetation is high. The Dominance Index (ID) was quite small (0-1) indicating that the location has a pattern of species dominance that is spread over several dominant species. The species diversity index value is greater than 3 which indicates that the species found in the study site have a high level of stability. The higher the value of the species diversity index (H), the nore stable the species in the community. The level of species stability in the community is used for the evenness index value (e). The higher the value of e, the more stable the stability of species diversity in the community. The results of the vegetation analysis can be seen in table 1.

Table 1. Vegetation analysis.

No	Description	Level			
		Seedling	Stake	Pole	Tree
1	Species richness index (D)	31.563	29.367	30.099	29.584
2	Dominance index (ID)	0.021	0.028	0.017	0.027
3	Species diversity index (H')	4.215	4.094	4.242	4.002
4	Species abundance index (e')	2.146	2.122	2.211	2.097

Overall, it can be seen that the average value of the vegetation density index from site 1-site 6 is still in the very good category. Temperature measurement values at site 1-site 6 have temperature values that are still in the range of 25 C – 26 C and according to SNI 9006:2021 and the results of air temperature measurements are still in the range of 24-28 C, still included in the thermally comfortable category. The air humidity measurement value at site 1-site 6, has a value ranging between 83%-87% and is included in the high category. This is because the research location is in Sintang Regency, which is part of the territory of West Kalimantan Province. This condition of high density of vegetation causes the humidity to be higher, due to litter covering the surface of the soil tightly, causing water vapor to be locked in it. Thermal Humidity Index (THI) measurement values at site 1 – site 6 range between 24-25 C and are include in the comfortable category. The light intensity measurement values range between 318-415 Lux and are included in the ideal category. Light intensity with an ideal value range tends to be comfortable for humans in their activities and does not cause eye fatigue. The average value of the slope index is included in the sloping and rather steep category, so it can still be recommended considering that it is not included in the steep or very steep category. The average value of noise is included in the ideal category (quiet). The average value of wind speed is included in the above ideal category (> 1m/sec). the average value of negative value of negative ion content is included in the health category. The highest negative ion value is at site 6 (near a water source). Site analysis is shown in table 2.

Table 2. Site analysis.

Description	Site					
	1	2	3	4	5	6
Vegetation density (NDVI)	0.76	0.75	0.75	0.73	0.76	0.74
Microclimate(temperature)	25.73	25.70	25.73	25.20	26.58	25.75
Microclimate(humidity)	24.87	24.87	24.91	24.59	25.61	24.90
Microclimate (thermal humidity index)	24.87	24.87	24.91	24.59	25.61	24.90
Light intensity	415.60	381.50	380.55	325.38	435.70	381.38
Slope	14.08	17.49	17.9	18.24	17.59	14.08
Noise	22.95	21.05	21.45	21.70	21.55	31.50
Wind velocity	2.53	2.55	2.53	2.63	2.48	2.54
Negative ions	87,450	87,500	87,250	109,900	79,550	110,450

3.2 Health response and DASS-422

The selected respondents are students from the 2018-2021 class, with an age range of 18-30 years with the assumption that this age range is an age that is vulnerable to psychological health conditions. The initial condition of the blood pressure of all respondents was 124.37/84.69 mmHg which means higher than normal blood pressure (Fig. 2). The heart rate at the respondent's initial condition was 85.68 beats per minute which is still in the normal category. The results of measuring blood pressure for all respondents (class of 2018-2021) at 6 sites obtained different values which tended to decrease from the starting point to the end, so that this decrease in blood pressure was a change in the desired health response. Trekking in the forest is

 https://doi.org/10.1051/bioconf/20238003023

suitable for patients with hypertension to stabilize blood pressure as found in this study.

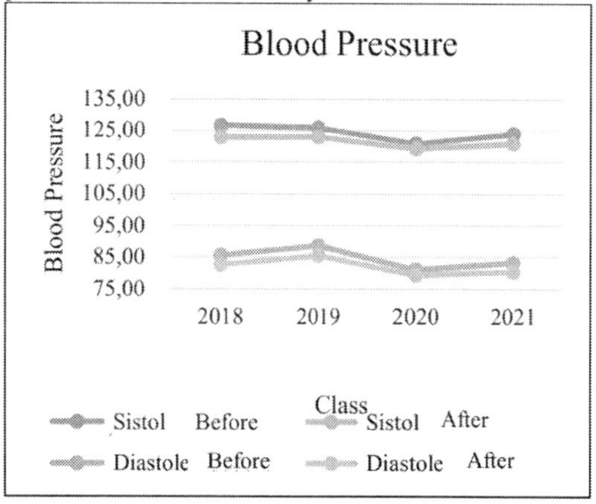

Fig. 2. Blood pressure.

Respondents' blood sugar levels decreased after tracking and passing all Site 1 – Site 6. Respondent's blood sugar levels decreased by 3.47 mg/dL from 94.87 mg/dL to 91.40 mg/dL after passing all Ecotherapy Healing routes Forest (Site 1 – Site 6) (Fig. 3). This decrease is a desired change, especially for respondents who have a history of high blood sugar levels.

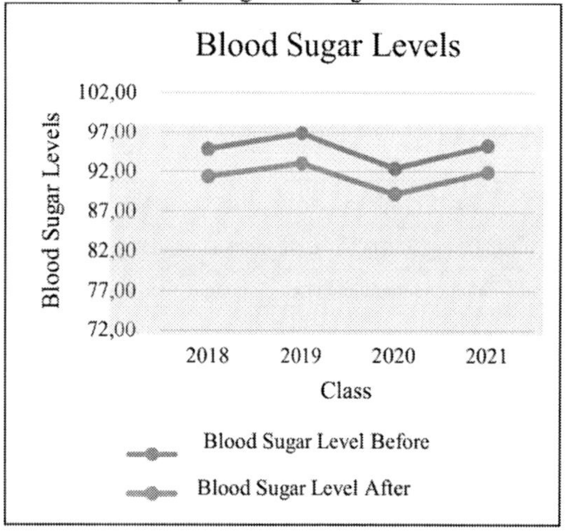

Fig. 3. Blood sugar levels.

Respondents' Oxygen Saturation levels increased by 0.39% after passing through the Site 1 – Site 6 Ecotherapy healing Forest route (Fig. 4). An increase in oxygen saturation is the desired change because it moves away from the minimum value of oxygen saturation, which is 95%. However, the results were different in the 2019 class, where participating in the Ecotherapy Healing Forest activity had no effect on fluctuating increases in Oxygen Saturation values. This could be because the majority of the 2019 class of respondents were male and smokers.

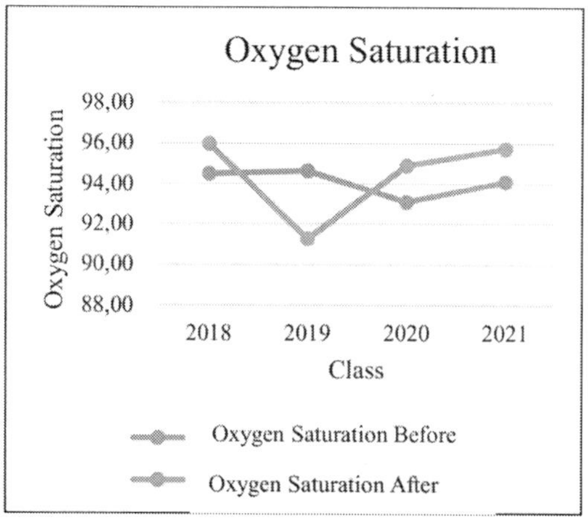

Fig. 4. Oxygen saturation.

The respondent's pulse value decreased by around 2.93 BPM after passing through the Ecotherapy Healing Forest site 1 – site 6 (Fig. 5). The pulse value changed, still within the normal range. The decreased pulse rate can be caused by the respondent feeling relaxed.

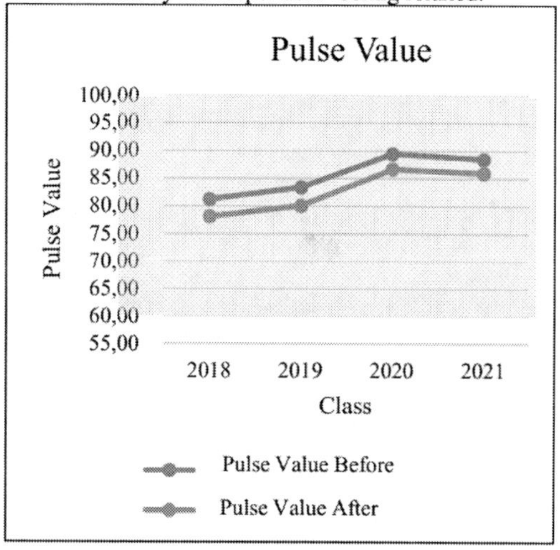

Fig. 5. Pulse value.

Based on DASS 42 in the class of 2018, the Initial Values of Moderate Depression, Severe Anxiety, and Severe Stress were obtained. After experiencing the Ecotherapy Healing Forest treatment 5 times, slowly there was a change towards normal. However, new significant changes appeared during the initial 3rd treatment, this could be due to the severe Anxiety and Stress values, because on average they experienced problems in the form of fear of the future, including fear of opportunities to get a job after graduating from studying.

Based on DASS 42 in the class of 2019, the initial values for depression, anxiety and stress were severe. After experiencing the Ecotherapy Healing Forest treatment 5 times, slowly there was a change towards normal. However, new significant changes appear at the end of the 3rd time, this can be caused by the severe Depression, Anxiety and Stress values. Based on the results of interviews and observations, their problems include: field practice (implementation of theory in the world of work); report field practice activities; presentation of the results of field practice activities; preparing submissions for titles or themes; preparing large costs for research and thesis preparation.

Table 3. Depression anxiety and stress scale DASS (42) 2018 until 2021.

Class of 2018				
Treatment	Treatment	Depression	Anxiety	Strees
1	Before	20,00	18,40	29,20
	After	16,80	13,40	23,80
2	Before	14,90	11,90	21,70
	After	12,20	8,70	16,60
3	Before	9,00	7,90	13,70
	After	7,40	6,20	11,40
4	Before	6,30	5,40	11,40
	After	5,10	4,40	10,00
5	Before	5,20	4,00	10,00
	After	4,10	2,90	8,00
Class of 2019				
Treatment	Treatment	Depression	Anxiety	Strees
1	Before	25,36	18,00	29,73
	After	21,45	12,91	23,91
2	Before	16,64	11,91	21,82
	After	12,45	9,00	16,55
3	Before	11,64	8,45	15,36
	After	8,55	6,73	11,27
4	Before	8,45	6,73	11,27
	After	7,36	5,36	9,91
5	Before	6,45	5,36	9,91
	After	5,27	3,82	7,91
Class of 2020				
Treatment	Treatment	Depression	Anxiety	Strees
1	Before	12,91	8,73	16,64
	After	10,91	8,00	15,09
2	Before	8,64	7,00	13,73
	After	7,55	6,27	12,45
3	Before	7,45	6,27	11,91
	After	6,27	5,27	10,73
4	Before	6,27	5,27	10,36
	After	5,27	4,36	8,64
5	Before	5,27	4,27	8,18
	After	4,55	2,91	4,91
Class of 2021				
Treatment	Treatment	Depression	Anxiety	Strees
1	Before	10,83	15,25	14,58
	After	9,33	8,67	13,33
2	Before	8,42	7,25	12,50
	After	5,92	5,50	10,33
3	Before	6,17	4,83	9,83
	After	5,08	3,92	8,92
4	Before	5,08	3,08	8,58
	After	4,17	2,25	7,17
5	Before	4,17	2,33	6,83
	After	3,67	1,33	4,08

Based on DASS 42 for the 2020 and 2021 batches, the Initial Values for Depression, Anxiety and Stress are Mild. After experiencing the Ecotherapy Healing Forest treatment 5 times, slowly there was a change towards normal. Significant changes were seen during the 2nd Ecotherapy Healing Forest treatment, this could be due to the mild Depression, Anxiety and Stress values. On average, they only have problems in the form of anxiety to adapt, because they will enter face-to-face for the first time after the COVID-19 pandemic, so they will interact directly with lecturers and fellow students and the campus environment. DASS 42 is shown in table 3.

3.3 Statistic test

The correlation value of systolic data with temperature, air humidity and THI data has a very strong relationship with the relationship between parameters directly proportional or unidirectional. The correlation value of Systole data with data on light intensity, noise, wind speed and negative ions has a weak relationship, with the relationship between parameters for Systole and noise directly proportional or unidirectional. The correlation value of the Diastolic data with the Temperature, Air Humidity and THI data has a very strong relationship with the relationship between parameters directly proportional or unidirectional. The correlation value of the Diastolic data with the light intensity, noise, wind speed and negative ion data has a weak relationship, with the relationship between parameters for Diastole with noise and negative ions being directly proportional or unidirectional. The correlation value of Pulse Rate data with Temperature, Air Humidity, THI and Negative Ion data which has a moderate relationship with the relationship between parameters for Pulse Rate and Temperature, Air Humidity, THI is directly proportional or unidirectional. Correlation value of Pulse data with light intensity, noise and wind speed data has a weak relationship, with the relationship between parameters for Pulse with light intensity and noise directly proportional or unidirectional. Correlation value of Blood Sugar Levels data with Temperature, Air Humidity and THI data which has a very strong relationship with the relationship between parameters directly proportional or unidirectional. Correlation value of Blood Sugar Levels data with light intensity, noise, wind speed and negative ion data has a weak relationship, with the relationship between parameters for Blood Sugar Levels and negative ions directly or in the same direction. The correlation value of Oxygen saturation data with Temperature, Air Humidity and THI data which has a very strong relationship with the relationship between parameters is inversely proportional or opposite. The correlation value of oxygen saturation data with light intensity, noise, wind speed and negative ion data has a weak relationship, with the relationship between parameters for oxygen saturation with light intensity, and wind speed is directly proportional or unidirectional. Correlation test results of health data with site identification data after treatment is shown in table 4.

Table 4. Correlation test results of health data with site identification data after treatment.

			Correlation Test Results						
			Temp.	Humidity	THI	Light Intensity	Noise	Wind Velocity	Negative Ions
Spearman's rho	Sistol	Correlation Coef.	0,83	0,851	0,812	-0,233	0,084	-0,274	-0,106
		Sig.	0	0	0	0,323	0,724	0,243	0,657
		Amount of Data	20	20	20	20	20	20	20
	Diastole	Correlation Coef.	0,617	0,669	0,687	-0,258	0,086	-0,324	0,229
		Sig.	0,004	0,001	0,001	0,272	0,72	0,164	0,331
		Amount of Data	20	20	20	20	20	20	20
	Pulse	Correlation Coef.	0,314	0,367	0,42	0,212	-0,201	0,121	-0,383
		Sig.	0,177	0,112	0,066	0,37	0,396	0,61	0,095
		Amount of Data	20	20	20	20	20	20	20
	Blood Sugar Levels	Correlation Coef.	0,756	0,792	0,858	-0,058	-0,202	-0,164	0,136
		Sig.	0	0	0	0,807	0,394	0,49	0,569
		Amount of Data	20	20	20	20	20	20	20
	Oxygen Saturation	Correlation Coef.	-0,831	-0,761	-0,741	0,183	-0,172	0,161	-0,038
		Sig.	0	0	0	0,44	0,469	0,498	0,872
		Amount of Data	20	20	20	20	20	20	20

The correlation value of Oxygen Saturation before and after the Ecotherapy Healing Forest treatment shows a weak association. Then the significance value between Oxygen Saturation before and after the Ecotherapy Healing Forest treatment showed no significant effect on the Oxygen Saturation value. Correlation values of Systole, Diastole, Pulse Rate and Sugar Levels before and after the Ecotherapy Healing Forest treatment showed a very strong relationship. Then the significance value between Systole, Diastole, Pulse and Sugar Levels before and after the Ecotherapy Healing Forest treatment showed that it had a significant effect on the values of Systole, Diastole, Pulse Rate and Sugar Levels.

Table 5. Test results for different health parameter values before treatment.

	Different Test			
		Amount Of Data	Correlation	Sig.
Pair 1	Sistol (before) & Sistol (after)	20	0,872	0,004
Pair 2	Diastole (before) & Diastole (after)	20	0,931	0,004
Pair 3	Pulse (before) & Pulse (after)	20	0,873	0,039
Pair 4	Blood Sugar Levels (before) & Blood Sugar Levels (after)	20	0,982	0,000
Pair 5	Oxygen Saturation (before) & Oxygen Saturation (after)	20	0,340	0,576

The correlation value of depression, anxiety and stress before and after the Ecotherapy Healing Forest treatment has a value above 0.75. This shows that the values of depression, anxiety and stress before and after the Ecotherapy Healing Forest treatment have a very strong relationship. The significance value between depression, anxiety and stress before and after the Ecotherapy Healing Forest treatment has a value of 0.000. This shows that the existence of treatment has a real influence on the value of depression, anxiety and stress. Test Results for Different Health Parameter Values After Treatment is shown in table 6.

Table 6. Test results for different health parameter values after treatment.

	Different Test			
		Amount Of Data	**Correlation**	**Sig.**
Pair 1	Depression (before) & Depression (after)	20	0,994	0,000
Pair 2	Anxiety (before) & Anxiety (after)	20	0,974	0,000
Pair 3	Stress Scale (before) & Stress Scale (after)	20	0,983	0,000

3.4 Sustainability analysis of MDS-RAPFISH

The first step for MDS analysis using Rapfish is the feasibility/goodness of the MDS-Rapfish analysis in terms of the Stress and R-square values for the five dimensions used as shown in Table 1. The reference value as the basis for determining the goodness-of-fit) in MDS is the value of Stress less than 0.25 and R-square more than 0.90. Based on Table 1, it can be seen that the three-dimensional Stress values all show a value less than 0.25. On the other hand, the fifth R-Square value for each dimension also shows a value above 0.90. Therefore, it can be said that this MDS analysis meets the criteria of goodness-of-fit so it deserves further discussion/analysis (table 7).

Table 7. Results of feasibility/benefits of MDS-RAPFISH analysis.

Dimensions	**Stress (S)**	**R-Square (R)**
Socioeconomic	0,20684	0,9561499
Ecology	0,20707	0,9556416
Institutional	0,21666	0,9517841
Health	0,20737	0,9556418

Based on Fig. 6 it can be seen that the sustainability index for the control dimension is 61.28. The index value is in the range of 50.00 – 74.99, which means that this dimension is quite sustainable. Based on the results of the analysis, it is known that the attributes that greatly determine the sustainability of the socio-economic dimension are the benefits that can be obtained by the community in the form of financial benefits and benefits of environmental services in the form of fresh air.

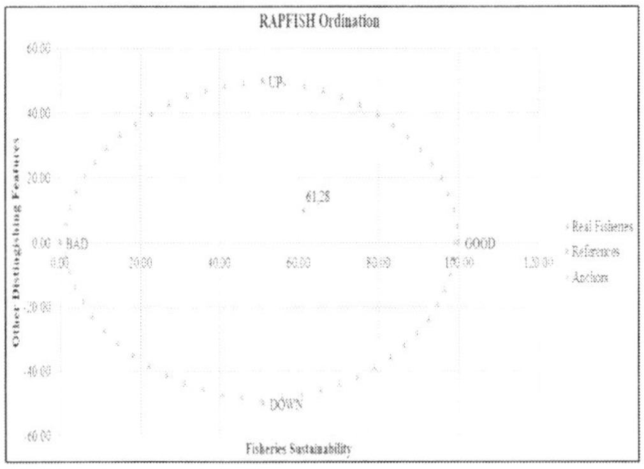

Fig. 6. Sustainability status Socio-Economic dimension.

Based on Fig. 7 it can be seen that the sustainability index value for the ecological dimension is 90.41 which is in the range of 75.00 – 100.00 which means good (very sustainable). Therefore, these results show that from an ecological perspective it strongly supports sustainability. Based on the results of the analysis, it is known that the attribute of the ecological dimension that most influences the sustainability of this dimension is that the ecosystem in the middle of the Gunung Kelam TWA is a steep slope so that most of it does not grow vegetation but bare rocks.

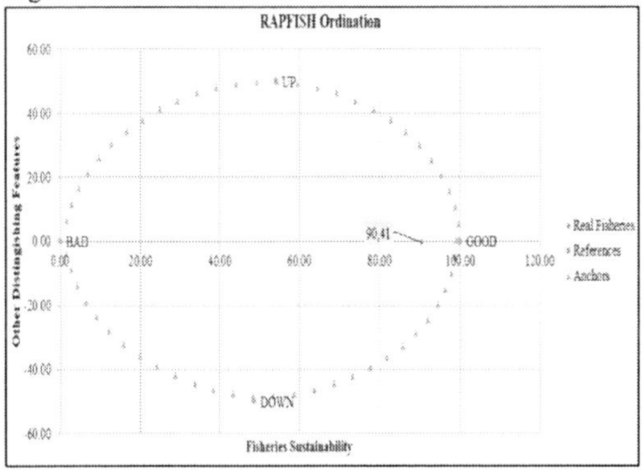

Fig. 7. Ecological dimension Sustainability status.

Based on Fig. 8 the sustainability status index of the implementation dimension is 87.88. This index is in the range of 75.00 – 100.00 which means good (very sustainable). Therefore, these results show that from an institutional and institutional perspective it is very supportive of sustainability. From the results of the analysis it is known that the attributes that greatly influence the sustainability of the institutional and institutional dimensions are conducting comparative studies to areas that have implemented EHF tourism management.

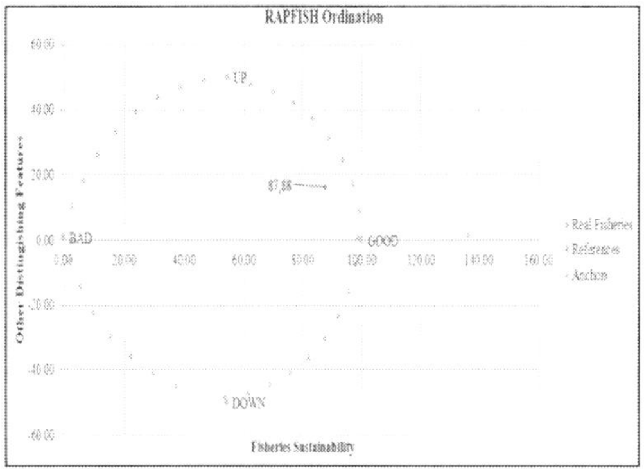

Fig. 8. Sustainability status implementation dimension.

Based on Fig. 9, the implementation dimension of the sustainability status index is 89.51. This index is in the range of 75.00 – 100.00 which means good (very sustainable). Therefore, these results show that from an institutional perspective it is very supportive of sustainability. Based on the results of the analysis it is known that the attributes that greatly influence the sustainability of the institutional dimensions are a healthy environment, which will be able to donate oxygen which is very influential for public health.

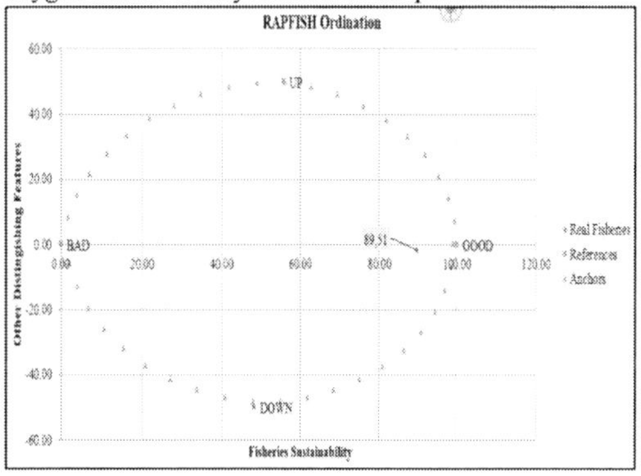

Fig. 9. Health dimension Sustainability status.

The dimension that has the most excellence in terms of sustainability is the ecological dimension, while the dimension that has the lowest sustainability index is socio-economic. Therefore, socio-economic problems in particular need to receive more serious attention so that they can run sustainably. Activity on site 1 – site 6 are shown in Fig. 10 – Fig. 15.

Fig. 10. Activities on Site 1.

Fig. 11. Activities on Site 2.

Fig. 12. Activities on Site 3.

Fig. 13. Activities on Site 4.

Fig. 14. Activities on Site 5.

Fig. 15. Activities on Site 6.

4 Conclusions

The Ecotherapy Healing Forest Program at Mount Kelam Natural Tourism Park which was attended by 44 student participants from the 2018 – 2021 in Kapuas Sintang University gave good results for reducing stress and health. The results of this study revealed that there was a significant decrease in blood sugar levels, oxygen saturation and stress levels in the participants after participating in the ecotherapy healing forest program. However, for smoking students, the Ecotherapy Healing Forest treatment does not have a significant effect on oxygen saturation values. Therefore, the authors propose Ecotherapy Healing Forest therapy as a non-pharmacological intervention to relieve stress while improving physical health (blood sugar levels and oxygen saturation) for students.

The author acknowledges the Environmental Science Doctoral Study Program, Graduate School, Gadjah Mada University, which has supported the research process. In addition, thank you for the financial support provided through this Final Recognition Grant (RTA) from the UGM Research Directorate in 2023.

References

1. Turner W R, Nakamura T and Dinetti M 2004 Global Urbanization and the Separation of Humans from Nature *Bioscience* **54** 585–90
2. Shigemura J, Ursano R J, Morganstein J C, Kurosawa M and Benedek D M 2020 Public responses to the novel 2019 coronavirus (2019-nCoV) in Japan: Mental health consequences and target populations *Psychiatry Clin Neurosci* **74** 281–2
3. Fiorillo A and Gorwood P 2020 The consequences of the COVID-19 pandemic on mental health and implications for clinical practice *Eur Psychiatry* **63**
4. Song C, Ikei H, Kobayashi M, Miura T, Li Q, Kagawa T, Kumeda S, Imai M and Miyazaki Y 2017 Effects of viewing forest landscape on middle-aged hypertensive men *Urban For. Urban Green.* **21** 247–52
5. Andini, S. W., Prasetyo, Y., & Sukmono, A. (2018). Analisis sebaran vegetasi dengan citra satelit sentinel menggunakan metode NDVI dan segmentasi (Studi Kasus: Kabupaten Demak). Jurnal Geodesi Undip, vol.7.
6. Baehaqi, I. (2020). Studi hubungan karakteristik fisik lingkungan healing forest terhadap respon kesehatan pengunjung di kawasan hutan puncak bintang perum perhutani kph bandung utara.
7. Chandrasa, G. T. (2015). Kajian Dampak Ion Udara terhadap Perilaku Gangguan Sosial Manusia; Penelitian Dampak Ion-ion Udara terhadap Penyimpangan Perilaku Sosial Manusia. LSDE-BPPT, Kawasan PUSPIPTEK Serpong, Tangerang.
8. Clifford, M. A. (2018). your guide to Forest Bathing (E. the H. P. of N. . N. MA (ed.)). Conari Press.
9. Evert, A., Yuwono, S. B., & Duryat. (2017). The

Comfort Level of Patriot Bina Bangsa Urban Forest in Bekasi City. Jurnal Sylva Lestari, Vol. 5 No.(ISSN 2339-0913), pages (14—25).
10. Frumkin, et al. (2016). Environmental health from global to local. San Francisco. CA.
11. Gaudioa, N., Gendre, X., Saudreauc, M., Seigner, V., & Balandier, P. (2017). Impact of tree canopy on thermal and radiative microclimates in a mixed temperate forest: A new statistical method to analyse hourly temporal dynamics. Agricultural and Forest Meteorology.
12. Habeeb, A. A., Gad, A. E., & Atta, M. A. (2018). Temperature-Humidity Indices as Indicators to Heat Stress of Climatic Conditions with Relation to Production and Reproduction of Farm Animals. International Journal of Biotechnology and Recent Advances, ISSN: 2639-4529.
13. Hahad, O., Prochaska, J. H., Daiber, A., & Münzel, T. (2019). Environmental Noise-Induced Effects on Stress Hormones, Oxidative Stress, and Vascular Dysfunction: Key Factors in the Relationship between Cerebrocardiovascular and Psychological Disorders. Oxidative Medicine and Cellular Longevity, Volume 201.
14. Hidayah, I. (2018). Peningkatan Kadar Asam Laktat dalam Darah Sesudah Bekerja. Jurnal Keselamatan Dan Kesehatan Kerja Indonesia, Vol. 7 No.
15. Hutagalung, R. (2017). Pengaruh Kebisingan terhadap aktivitas masyarakat di Terminal Mardika Ambon. Jurnal Tehnik Industri, vol 11. no. https://doi.org/https://doi.org/10.30598/arika.2017.11.1.83
16. Jamala, N., & Rahim, R. (2017). Teori dan Aplikasi Kenyamanan Visual. Badan Penerbit Universitas Negeri Makassar.
17. Jayanthi, S., & Arico, Z. (2022). Effect of Vegetation Density on Litter Productivity of Mount Leuser National Park. Journal of Islamic Science and Technology.
18. Jiang, S.-Y., Ma, A., & Ramachandran, S. (2018). Negative Air Ions and Their Effects on Human Health and Air Quality Improvement. International Journal of Molecular Sciences, volume 19.
19. Kendali, M. (2019). Pengembangan JAsa ekosistem Healing Forest di Wana Wisata Cibeureum. Tesis, 21317004.
20. Kurnia, R. F. A. (2019). Analisis Indeks Kenyamanan Iklim.
21. Lufilah, S. N., Makalew, A. D., & Sulistyantara, B. (n.d.). The Utilization of Landsat 8 for Analizing Vegetation Index in Jakarta. Jurnal Lanskap Indonesia, vol 9. no, 2017.
22. Mahabella, Waibo, L. S. and, & Gunarimba, O. R. (2021). Analisis nilai indeks suhu dan kelembaban ruang terbuka Taman Rekreasi Sengkaling. Universitas Muhammadiyah Malang Repository.
23. Megawati, T. F., Kamarubayana, L., & Endayani, dan S. (2015). Inventarisasi dan Pemetaan Pohon Buah (Edible Fruits) asli Kalimantan di kebun raya Unmul Samarinda (KRUS). Jurnal AGRIFOR, Volume XIV.
24. Nurrohimah, I. (2022). Community Perceptions and Preferences of The Comfort Level of Merdeka Metro Park as a Social Interaction Space During Covid-19 Pandemic. Jurnal Lanskap Indonesia, Vol. 14 No.
25. Octavia, C., & Purbaningtyas, D. (2017). A Study on The Management of Public Green Open Space as The Implementation of Perda No. 2 Year 2014, City of Samarinda. Applied Research National Seminar.

 https://doi.org/10.1051/bioconf/20238003024

Watershed Management and Classification in Thailand

*Piyapong Tongdeenok**

Watershed and Environmental Management Program, Department of Conservation,
Faculty of Forestrry, Kasetsart University, Bangkok, Thailand

Abstract: Based on a study of research and experiences in Thailand, this paper provides an overview of the implementation and challenges of watershed management, as well as potential solutions. The review found hierarchical ambiguity, inconsistency, and asynchrony across rules, as well as a lack of (participation, synchronization, and coordination) among watershed management stakeholders. Deficits in the planning phase include a lack of integration between sectors, a lack of community participation, and a limited willingness to integrate watershed planning into regional planning. Stakeholder participation is also essential for the successful implementation of degraded watershed rehabilitation, including in terrestrial forest and mangrove regions. Failure should be minimized by providing adequateknowledge about degraded watershed characteristics, selecting appropriate plant species, and employing good mechanical construction techniques for soil and water conservation. Increase public understanding of the importance of a sustainable watershed and provide opportunities for community involvement in each phase of watershed management to achieve community participation as the majordriver of watershed management. Another issue is data gaps, which must be addressed from the designthrough the evaluation stages. The gaps can be filled using remotely sensed data and hydrologically based modeling models. Simplified watershed assessment criteria may also be required, depending on site-specific challenges and the extent of the watershed.

1. Introduction

The Kingdom of Thailand has demonstrated notable advancements in its economic development, particularly in the areas of energy, transportation, and tourism. The country's robust economic expansion since the 1970s facilitated its inclusion in the category of upper-middle-income economies throughout the early 2010s. According to the 2017 National Strategy Preparation Act of Thailand, the country has set a goal to achieve a high-income economy by the year 2037, with a focus on attaining "Security, Prosperity, and Sustainability." Hence, Thailand is actively pursuing measures to improve its economic competitiveness and social progress with the aim of positioning itself as a prominent nation in the Southeast Asian region (OECD, 2019[1]).

The successful management of water resources, encompassing flood control, irrigation, and water delivery, is crucial for achieving economic prosperity. However, Thailand's growing water insecurity poses a significant threat to the realization of its ambitious goals. The anticipated challengesof sustainable water management in the future are likely to be exacerbated by factors such as populationgrowth, economic expansion, accelerated urbanization, and the imminent risks associated with climatechange. According to projections from The World Bank Group and the Asian Development Bank (2021), it is anticipated that Thailand's population would reach approximately 71-77 million by the year2030, with a growing percentage residing in metropolitan regions. According to Ta and Watershed (2008), the urbanization rate in 2020 accounted for 51.43% of the whole population. This shift in lifestyle patterns has resulted in a corresponding rise in the demand for water resources. According to the World Bank Group and the Asian Development Bank (2021), the majority of Thailand's economy,approximately 90%, is reliant on the industrial and service sectors. In contrast, the agricultural sector, while accounting for only 10% of the economy, employs approximately 33% of the country's workers.Water productivity in Thailand is generally low across all sectors, with agriculture experiencing very low levels, practically approaching zero. In contrast, the industry sector achieves a water productivity of approximately 60 USD per m3 (Chokchai & Sucharit, 2019[4]).

Thailand is widely acknowledged as being particularly susceptible to climate fluctuation and change as a result of the escalating occurrence of natural hazards, including but not limited to intense precipitation, inundations, and periods of prolonged dryness. Furthermore, the rise in sea level has a significant impact on the coastal regions of the country. The country was ranked 31st in the National

* Corresponding author: fforppt@ku.ac.th

Water Security Index among a group of 49 Asian and Pacific countries. This ranking was mostly attributed to a poor level of water urban security and a strong presence of climatological threats (Asian Development Bank, 2020[5]).

According to the findings of the Intergovernmental Panel on Climate Change, it is projected that the rise in global sea levels, which is attributed to climate change, will range from 8 to 16 millimeters per year throughout the 21st century. The anticipated rise in sea level inside the inner regionof the Gulf of Thailand is attributed to the phenomenon of global climate change. The augmentation inwind velocity, particularly the monsoonal winds that traverse the Gulf of Thailand, contributes to the elevation of the sea level. According to the World Bank Group (2021[6]), Multiple water challenges are currently present, including the escalating competition for water resources in the agricultural, industrial, and service sectors, the degradation of water quality caused by the rising levels of pollutants, the exacerbation of flood and drought impacts due to climate change, and the complex management of rivers and aquifers that span across different regions.

Thailand is significantly affected by floods, which represent the most substantial natural hazardin terms of both economic and human consequences. Thailand is recognized as being among the top ten countries globally that are most susceptible to flooding. The impacts of drought and cyclones are significant hazards as well. All variables have the potential to increase in future climatic scenarios. The projected increase in the population impacted by a severe river flood is estimated to surpass 2 million individuals between the years 2035 and 2044. Additionally, it is anticipated that coastal flooding will affect an additional 2.4 million people from 2070 to 2100. According to projections, the agricultural sector of Thailand may see substantial repercussions as a result of climate change. This vulnerability is mostly attributed to Thailand's geographical location within the tropical region, where agricultural production is particularly susceptible to increases in temperature (The World Bank Group and the Asian Development Bank, 2021[2]). Thailand has prioritized its adaptation endeavors in significant sectors including energy, water, transportation, agriculture, human settlements, and public health, as indicatedin its Third National Communication to the United Nations Framework Convention on Climate Change in 2018, its Initial Nationally Determined Contribution in 2016, and its Updated Nationally DeterminedContribution in 2020 (The World Bank Group and the Asian Development Bank, 2021[2]). This, coupled with the impacts of frequent severe droughts and floods over the past 20 years, has resulted in a perceived need for a watershed classification system as a major vehicle for future land and water resources planning and management for sustainable socioeco-nomic development of the country. A primary objective of the watershed classification project was to develop land use plans for the conservation of natural resources, specifically water resources, with the goal of ensuring their sustainable use. The classification takes into account physical characteristics, such as landform, geology, soil, elevation, and slope. Consideration is also given to forest cover and environmental characteristics of less stable landscape units that interact with climatic trends and human uses.

2. Watershed management in Thailand

2.1 Water resource management problems in Thailand

About 800 billion m3 yr1 of precipitation is averaged to fall on Thailand each year. The amount of natural flow that is left over after accounting for loss due to evaporation, evapotranspiration, and infiltration is around 200 billion m3. 79% of the stream's normal flow is made up of runoff. The countryconsumes approximately 152 billion m3 of freshwater each year, making this figure its annual demand. The water requirement is broken down as follows: 114 billion m3 is required for agricultural purposes, 11 billion m3 is required for industrial and home consumption, and 27 billion m3 is required for ecosystem maintenance.

The remaining 50 billion m3 of available water cannot be distributed in any way, and there areonly 102 billion m3 available overall. The latter segment provides the majority of the water essential for rain-fed agriculture as well as some of the water necessary for residential consumption.

As of right now, the population of Thailand is growing alongside the expansion of the country's economy, which has led to an increase in the need for water.

In addition, the requirements for water to support household consumption, tourist, and industrial operations are continuously developing at a rate that is accelerating. According to the statistics, by the year 2027, Thailand will require an additional 5 billion m3 of water in order to satisfygrowing demand (Figure 1). According to the Committee on Water Resources Policy and Management (2015), the increasing trend demonstrates that Thailand needs to create appropriate water managementin preparation for a potential water scarcity.

However, land scarcity and ecological However, land scarcity and ecological concerns make project approval for water resources development anything from guaranteed. The demand from the Thai people and from agricultural products is growing every year, but the expansion of irrigated regions has stalled due to a scarcity of water resources. The loss of forest cover exacerbates the issue of increased runoff during the wetter seasons and decreased streamflow during the drier seasons. The cost of water continues to rise. Therefore, Thailand is focusing more on water resources management to meet the increased demand. All industries can benefit from improved water efficiency by adopting and implementing innovative water policies and practices. Adopting sustainable development requires striking a balance between economic expansion and the consumption of natural resources.

Figure 1 Increasing water demand in Thailand

Sources: Somkiat *et.al.*, (2018)

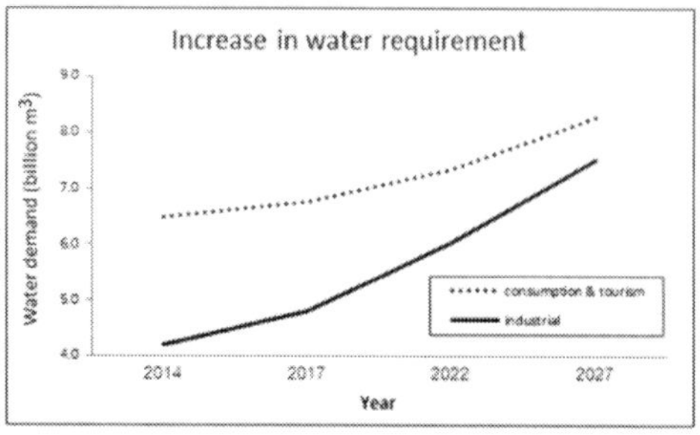

2.2 Master Plan on Water Resource Management

Water resources are essential for human survival and environmental well-being. Thailand has faced increasingly severe droughts and floods in recent years, causing significant economic losses estimated at 10 billion baht. The devastating 2011 flood, assessed by the World Bank, resulted in damages and losses of approximately 1.44 trillion baht. The main challenges in water resource management include degradation of watersheds, inadequate management, lack of a unified authority, absence of a long-term master plan, outdated data, and obsolete regulations.

To address these issues, Thailand has developed a Master Plan on Water Resource Management, consisting of eight major work plans and two action plans:

1. **Work Plan for Restoration and Conservation of Forest and Ecosystem**: This plan focuses on restoring watershed forests, developing water reservoirs, and creating land use plans tailored to local conditions. It also includes soil and water conservation projects, afforestation initiatives, and revisions to relevant laws.

2. **Work Plan for Management of Major Water Reservoirs and National Annual Water Management**: This plan aims to prevent and mitigate future floods by developing water managementplans for major dams and river basins, formulating water management plans for various scenarios, improving water use balance, and providing water-related information to the public.

3. **Work Plan for Restoration and Efficiency Improvement of Current and Planned Physical Structures**: To prevent and mitigate flood impacts, this plan includes renovating dikes, water control buildings, and drainage systems, improving waterways and canals, enhancing waterdraining and overflowing management, and reinforcing dikes. Long-term measures such as floodways, preventive dikes, and land use planning are also considered.

4. **Work Plan for Information Warehouse and Forecasting**: This plan involves developing a data system, creating hypothetical scenarios, establishing water management institutions, and enhancing the disaster warning system. It includes setting up a national water information center, constructing forecasting and disaster warning systems, and upgrading monitoring stations and satellitesystems.

5. **Work Plan for Preparedness in Specific Areas**: To enhance flood prevention and mitigation, this plan focuses on important areas such as agriculture, industry, and densely populated communities. It includes building flood prevention systems, creating systems for negotiations with affected parties, and assessing the impacts of private prevention systems.

6. **Work Plan for Assigning Water Retention Areas and Recovery Measures**: This plan assigns water retention areas in the Chao Phraya River basins to slow down water flow during flashfloods and formulates plans for diverting water into these areas. Special compensation measures fordesignated water retention areas are also part of this plan.

7. **Work Plan for Improving Water Management Institutions**: This plan aims to establish integrated water management organizations as single command authorities responsible for crisis decision-making, planning, monitoring, evaluation, and rule revision. In the short term, the Ad Hoc Committee chaired by the Prime Minister or assigned Deputy Prime Minister serves as the single command authority, while a permanent national integrated water management agency is envisionedfor the long term.

8. **Work Plan for Creating Understanding, Acceptance, and Participation**: This plan seekscollaboration with communities and stakeholders to manage flood impacts and other major disasterseffectively.

The Master Plan on Water Resource Management is a comprehensive strategy to address Thailand's water resource challenges, ensuring the sustainable use of this vital resource while mitigating the impacts of extreme events like floods and droughts.

2.3 Land Use Change and Its Impact in Thailand

Land use change significantly impacts environmental and socio-economic development in Thailand. Urbanization, agriculture expansion, and industrialization drive these changes. Urbanization, propelled by population growth and migration, leads to encroachment into agricultural areas and undeveloped land. Agriculture, a key sector, undergoes expansion to meet food and material demands, often at the cost of forests and natural habitats. Thailand's industrial sector, a magnet for investment, requires large land areas, further exacerbating land conversion.

Environmental Impact:

1. **Deforestation:** Land use change results in substantial deforestation, harming biodiversity and contributing to climate change. Forests are cleared for agriculture, infrastructure, and urban expansion.

2. **Loss of Biodiversity:** Habitat loss endangers numerous plant and animal species, disrupting ecosystems and posing threats to endemic species.

Socio-economic Impact:

1. **Displacement of Communities:** Land use change often displaces rural communities, particularly indigenous and marginalized groups, causing social and economic challenges.

2. **Water Resource Management:** Changes in land use affect water resources, causing soil erosion and reduced water quality. Proper land use planning is crucial for sustainable waterresource management.

Addressing land use change in Thailand requires comprehensive policies that balance economic development with environmental conservation and social equity. Watershed integration is a holistic approach that effectively mitigates the impact of land use change. By coordinating land and water resource management within specific watersheds, considering ecosystem interconnections, and involving stakeholders, watershed integration promotes sustainable development while reducing environmental degradation. For instance, in the Ping River Watershed, reforestation, sustainable agriculture, water resource management, and community engagement have effectively countered landuse change impacts. This holistic approach serves as a model to tackle land use challenges sustainably and collaboratively in Thailand

2.4 Implementation of Mechanical Soil and Water Conservation Measures to Control Erosion and Sedimentation in Thailand

Erosion and sedimentation are critical environmental issues that have far-reaching consequences for land degradation, water quality, and ecological balance. In Thailand, as in many other regions globally, these issues have gained prominence due to the adverse impacts of deforestation, agricultural practices, and urban development. To combat these challenges and preserve the nation's natural resources, Thailand has been proactive in implementing mechanical soil and water conservation measures.

Thailand's Erosion and Sedimentation Challenge: Thailand's diverse landscapes face erosion and sedimentation threats, driven by deforestation from logging and land conversion for agriculture and development. Deforestation removes protective forest canopies, leading to soil erosion during heavy rainfall, increased surface runoff, and sediment transport into rivers.

Agricultural practices, including traditional farming on steep slopes, worsen erosion. Sedimentationin rivers and reservoirs affects water quality, reduces storage capacity, and harms aquatic ecosystems.

Mechanical Soil and Water Conservation Measures: Thailand has implemented mechanicalsoil and water conservation measures to combat erosion and sedimentation, promoting sustainable land and water resource management.

1. **Terracing and Contour Farming**: Terracing creates horizontal steps on slopes to slow surface runoff and trap sediment. Contour farming plants crops along contour lines to reduce soil erosion, stabilizing slopes.

2. **Check Dams and Gabions**: Check dams across streams slow water flow and trap sediment. Gabions, wire mesh containers filled with rocks, stabilize streambanks and prevent erosion.

3. **Riparian Buffers**: Native vegetation along riverbanks acts as natural filters, trappingsediments, nutrients, and pollutants, enhancing water quality and wildlife habitat.

4. **Sediment Basins and Detention Ponds**: Engineered structures capture sediment-ladenrunoff, allowing sediments to settle before water enters water bodies, reducing downstream sedimentation.

5. **Reforestation and Afforestation**: Replanting trees in deforested areas (reforestation) and establishing forests in non-forested lands (afforestation) reduce erosion.

Challenges and Recommendations: Despite progress, challenges remain, including funding,land tenure, and awareness. To address these:

1. **Strengthen Enforcement and Compliance**: Improve enforcement of land use regulationsand conservation practices.

2. **Enhance Public Awareness**: Raise awareness about soil and water conservation's importance and benefits.

3. **Invest in Research and Innovation**: Support research and adapt erosion controltechniques to local conditions.

4. **Promote Multi-Stakeholder Collaboration**: Encourage collaboration among government agencies, NGOs, communities, and private sectors for comprehensive conservation programs.

In conclusion, Thailand's commitment to erosion and sedimentation mitigation through mechanical measures demonstrates dedication to sustainable land and water resource management. Addressing challenges and fostering collaboration can further preserve natural resources and safeguard the environment

2.5 Downstream Essential Ecosystem Management in Thailand

Thailand's Downstream Ecosystem Management: Thailand relies on its extensive river systems, such as the Chao Phraya and Mekong rivers, for socio-economic development, agriculture, industry, and urban areas. These rivers also support downstream ecosystems, prompting Thailand to focus on essential ecosystem management for biodiversity preservation, water quality, and sustainable resource use.

Importance of Downstream Ecosystems: Downstream ecosystems, often in floodplains and wetlands, provide critical habitats for diverse plant and animal species. They serve as breedinggrounds for fish vital to fishing communities and act as natural buffers, reducing flooding and recharging groundwater. These ecosystems offer ecosystem services like water purification, benefiting both local communities and the agricultural sector.

Challenges to Downstream Ecosystems: Human activities, including urbanization, agriculture, industrialization, and infrastructure development, have led to habitat loss and degradation. Water extraction and pollution from agriculture and urban areas further threaten these ecosystems. Climate change-induced shifts in rainfall patterns and increased temperatures also affect seasonal floods.

Essential Ecosystem Management Strategies:

1. **Wetland Conservation**: Thailand protects designated wetlands of international importance under the Ramsar Convention, ensuring their sustainability.

2. **Riparian Zone Protection**: Establishing buffer zones along rivers prevents erosion,reduces pollution runoff, and maintains habitat connectivity.

3. **Restoration and Reforestation**: Initiatives restore and reforest downstream areas, improving habitat quality and flood control.

4. **Sustainable Land Use Practices**: Encouraging sustainable agriculture and land use practices, like agroforestry and organic farming, minimizes habitat disruption.

5. **Water Quality Monitoring**: Regular monitoring of water quality parameters detectsthreats early, allowing effective pollution source management.

6. **Community Engagement**: Involving local communities in ecosystem managementfosters responsibility and leads to sustainable resource use practices.

7. **Climate Resilience**: Developing climate-resilient strategies, such as floodplain zoningand early warning systems, mitigates climate change impacts on downstream areas, reducing vulnerability.

Thailand's commitment to downstream essential ecosystem management reflects its dedication to preserving biodiversity and natural resources. By addressing challenges through strategic conservationand restoration efforts, Thailand ensures these ecosystems benefit future generations. Collaboration among government agencies, local communities, and conservation organizations is essential for sustainable downstream ecosystem management, supporting nature and the well-being of the Thai people.

2.6 Mangrove Management in Thailand

Mangroves are critical ecosystems that provide a wide range of ecological and socio-economic benefits. In Thailand, these coastal forests are found along the country's extensive coastline, particularlyin the southern regions and along the Gulf of Thailand. Mangrove management in Thailand has been apriority due to the recognition of their importance for:

1) **Biodiversity Conservation**: Mangroves serve as nurseries and habitats for various marine species, including fish and crustaceans. Protecting and restoring mangroves are essential for preserving coastal biodiversity.
2) **Coastal Protection**: Mangrove forests act as natural buffers against coastal erosion and storm surges. They help protect coastal communities from the impacts of extreme weather events.
3) **Carbon Sequestration**: Mangroves store large amounts of carbon, making them crucial in mitigating climate change. Proper management helps prevent carbon release through deforestation and degradation.
4) **Sustainable Livelihoods**: Mangroves support local livelihoods through fisheries, aquaculture, and non-timber forest products. Sustainable management ensures these resources remain available to communities.
5) **Tourism and Education**: Well-preserved mangrove areas also attract ecotourism and provide educational opportunities to raise awareness about coastal ecosystems.

2.7 Socio-Economic Benefits and Participatory Watershed Management in Thailand

Watershed management in Thailand extends beyond ecological concerns; it encompasses a holistic approach that recognizes the symbiotic relationship between natural resources and human well-being. The nation's commitment to participatory watershed management not only safeguards vital ecosystems but also generates socio-economic benefits that resonate with local communities. This essayexplores the socio-economic benefits of participatory watershed management in Thailand.

Understanding Participatory Watershed Management: Participatory watershed management involves collaboration between government agencies, local communities, and stakeholders to make informed decisions about land and water resource management. This approach recognizes that those who reside in or near watersheds are intimately connected to the land and its resources. In Thailand, where agriculture plays a pivotal role in the economy and rural communities depend on natural resources, participatory watershed management aligns with the principles of sustainable development.

Enhancing Livelihoods through Agriculture: One of the primary socio-economic benefits of participatory watershed management in Thailand is its positive impact on agriculture, the backbone of the nation's economy. By adopting sustainable farming practices that prioritize soil and water conservation, farmers can increase crop yields and reduce soil erosion. Watershed management initiatives often provide training and resources to help farmers implement these practices. This, in turn,leads to improved food security, increased income, and enhanced livelihoods for rural communities.

Fostering Community-Based Tourism: Thailand's picturesque landscapes, often found in watershed areas, are ideal for community-based tourism. Participatory watershed management encourages the development of ecotourism initiatives that showcase the natural beauty of these regions. Tourists flock to these areas for activities such as trekking, birdwatching, and river cruises, contributing to local economies. The revenue generated from tourism supports infrastructure development, small businesses, and job creation within the communities.

Water Resource Sustainability: Sustainable water management practices, a core component of watershed management, have a profound socio-economic impact. By ensuring the availability of clean and reliable water sources, watershed management directly benefits both urban and rural populations. In urban areas, a stable water supply is essential for industry, commerce, and household use. In rural settings, it sustains agriculture and livestock, fostering economic stability and reducing vulnerability to drought.

Conservation of Biodiversity and Ecological Services: Healthy watersheds are hubs of biodiversity and ecological services. Participatory watershed management in Thailand focuses on preserving these ecosystems, which offer indirect socio-economic advantages. The conservation of forests, wetlands, and rivers ensures a stable environment for wildlife and fisheries. This not only supports local livelihoods but also contributes to the resilience of ecosystems, reducing the risk of natural disasters that can devastate communities.

Community Empowerment and Social Capital: Participatory watershed management empowers local communities to actively engage in decision-making processes. It fosters a sense of ownership and responsibility among residents, strengthening social cohesion and building social capital.As communities collaborate on conservation efforts, they develop networks, knowledge, and skills thatextend beyond watershed management, positively influencing other aspects of community life.

Socio-economic benefits arising from participatory watershed management in Thailand underscore the significance of integrated natural resource management. By aligning environmental preservation with economic development and community well-being, Thailand's approach to watershed management sets an inspiring example. As the nation continues to invest in participatory initiatives, it not only ensures the long-term health of its ecosystems but also empowers its people, fostering a sustainable and prosperous future where nature and society thrive together.

2.8 Monitoring and Evaluation of Watershed Management Performance in Thailand

Effective watershed management is a cornerstone of Thailand's environmental conservation efforts and sustainable development goals. To ensure that these initiatives yield positive outcomes andmeet their intended objectives, a robust system of monitoring and evaluation (M&E) is imperative. This concept explores the importance of M&E in watershed management in Thailand, presents key performance indicators, and provides a data table to illustrate the impact of these efforts.

The Significance of Monitoring and Evaluation: Monitoring and evaluation are essential components of watershed management in Thailand. They provide a systematic approach to tracking progress, assessing the effectiveness of interventions, and making informed decisions for adaptive management. In a country characterized by diverse landscapes and ecological challenges, such as deforestation and water resource depletion, M&E is a critical tool to ensure the sustainability of natural resources and the well-being of local communities.

Key Performance Indicators (KPIs) for Watershed Management in Thailand: Effective M&E requires well-defined Key Performance Indicators (KPIs) that reflect the objectives of watershed management. In Thailand, the following KPIs are commonly used to assess performance:

1. **Forest Cover**: The percentage of forest cover within a watershed area is a fundamental indicator. It measures progress in reforestation and protection efforts, with the goal of increasing forested areas to prevent erosion and maintain biodiversity.

2. **Soil Erosion Rate**: Soil erosion is a significant concern in many Thai watersheds. Monitoring the erosion rate helps determine the success of erosion control measures, such as terracing and reforestation.

3. **Water Quality**: Water quality parameters, including turbidity, pH, and pollutant levels, are monitored to assess the health of rivers and streams. Improving water quality is vital for both ecosystems and human consumption.

4. **Biodiversity**: Biodiversity indices, such as species richness and abundance, provide insights into the conservation status of watersheds. Healthy ecosystems support diverse flora andfauna.

5. **Community Well-being**: Socio-economic indicators, such as income levels, access to clean water, and agricultural productivity, are essential for evaluating the impact of watershed management on local communities.

In conclusion, monitoring and evaluation are indispensable tools for assessing the impact of watershed management initiatives in Thailand. By tracking key performance indicators and analyzing data over time, the nation can make informed decisions to enhance the sustainability of its natural resources, protect ecosystems, and improve the well-being of local communities. This data-driven approach ensures that Thailand's watersheds continue to thrive, contributing to the nation's environmental resilience and socio-economic development

3. Watershed classification

3.1 The Concept of Watershed Classification:

The concept of a watershed refers to a geographical area related to water management. In Thailand, the entire country can be viewed as a watershed area requiring proper management. Watershed management seeks to integrate academic principles and public participation to ensure sustainable water resources for river basin areas. This includes maintaining adequate water quantity, consistent flow, good water quality, erosion control, flood damage reduction, and responsible resource utilization based on conservation principles. Achieving these goals involves a step-by-step approach, starting with land use planning and implementing measures for efficient resource use and pollution control.

There have been conflicts over river basin land use among various agencies. To address this, aproject was initiated to classify and determine the quality classes of watersheds in Thailand. It was established by the Basin Quality Classification Committee, led by Prof. Dr. Kasem Chankaew, following a resolution of the National Environment Board and Cabinet approval in 1982. Kasetsart University conducted the research study, which categorized watershed areas based on their environmental fragility and susceptibility to soil erosion.

The classification relies on five physical variables that influence erosion: slope (SLOPE), elevation (ELEV), land type (LANDF), geological characteristics (GEOL), and soil type (SOIL). These variables are used to create a relationship with the quality class values of watersheds (WSC) through standard correlation equations:

$$WSC = a + b(SLOPE) + c(ELEV) + d(LANDF) + e(GEOL) + f(SOIL) + FOR + MIN$$

In this equation, WSC represents the watershed quality class value, while the other variables represent specific characteristics of the area. The constants a, b, c, d, e, and f are used in the calculations.Additionally, variables indicating the status and potential of forest cover (FOR) and mining potential (MIN) are considered.

Thailand's watershed quality is divided into five classes:

1. First Watershed Class: High mountainous areas with steep slopes (over 50%) and characteristics prone to environmental impact. These areas are challenging for land use and should be preserved as watersheds.

2. Second Watershed Class: Suitable for secondary watershed status, typically featuring high mountains, rounded ridges, and moderate slopes (30-50%). Soil erosion is common, and activities like logging and mining are allowed with strict controls.

3. Third Watershed Class: Characterized by mountain slopes, foothills, terraced plains, and areasnear watercourses. Slopes range from 25-35%, and soil erosion is moderate. These areas can be used for agriculture with soil and water conservation measures.

4. Fourth Watershed Class: Found in mountain foothills, low hills, and terraced plains, with slopesof 6-25%. Soil is deep, and it's suitable for field crops with soil and water conservation measures.

5. Fifth Watershed Class: Lowland areas or slightly sloping hills with slopes less than 5%. Soil isdeep and fertile, and erosion resistance is high. These areas are suitable for various agriculturalactivities.

The study project for determining watershed quality classes in Thailand, carried out between 1985 and 1995, resolved long-standing land use conflicts among agencies in river basin areas. It established a framework for responsible land use planning and management, contributing to the sustainable use of Thailand's water resources

1) a master plan for planning land use in various river basin areas throughout the country

2) measures to control land use in river basin areas.

3) Guidelines for the correct and appropriate use of resources within the basin area. (Table 1)

Table 1 Watershed classed areas of various river basins in Thailand

Watershed classes	area	
	square kilometer	**percentage**
1A	85,463.70	16.66
1B	7,626.66	1.48
2	42,768.62	8.33
3	39,283.77	7.66
4	81,283.77	15.80
5	251,483.62	49.01
reservoir	5,454.96	1.06
total	**531,115.02**	**100.00**

3.2 Importance of Watershed Classification as follows:

1) Resource Allocation: Classification aids in the equitable distribution of resources and funding for watershed management and conservation efforts. It allows decision-makers to prioritize areas with specific needs.

2) Conservation Priorities: Identifying unique and ecologically significant watersheds helps conservation organizations and governments allocate resources for habitat protection and biodiversity conservation.

3) Risk Assessment: Classification facilitates the assessment of environmental risks, including vulnerability to erosion, sedimentation, water pollution, and climate change. This information informsadaptive management strategies.

4) Land Use Planning: Urban and rural development planning can benefit from watershed classification by identifying areas at risk of flooding, landslides, or water scarcity. It promotes sustainable land use practices.

5) Policy Development: Watershed classification supports the development of policies andregulations that are tailored to the specific needs and challenges of different watersheds.

Here is an example image illustrating a simplified watershed classification:

Figure 2 Pictorial concept of watershed management in Thailand

Sources: Chankaew. *et.al.*, (1996)

3.3 Watershed management in Thailand

In recent years, Thailand has faced significant environmental challenges, with deforestation and water shortages at the forefront. Disturbingly, it is projected that unless stringent measures are enforced, Thailand's forests may disappear entirely by the early 21st century. In the 1970s, laws were enacted to regulate timber harvesting and the export of teak without permits, with the aim of protecting government forestry reserves. Unfortunately, these laws have seen minimal enforcement, largely due to corruption among law enforcement officials [Morell and Poznanski, 1983]. Concurrently, recurrent severe droughts and floods over the past two decades have exacerbated these issues, underscoring the necessity for a watershed classification system as a pivotal tool for future land and water resource planning.

On July 27, 1982, the Office of the National Environment Board (ONEB) was tasked with developing a comprehensive national watershed classification system. The National Economic and Social Development Board (NESDB) provided funding for the project, which was led by Kasetsart University. While the Royal Thai Government was the primary source of funding, the government of Sweden, through the International Union for the Conservation of Nature and Natural Resources (IUCN),provided valuable technical assistance and support.

Throughout the project's duration, watershed classification covering the entire country was undertaken and approved for region-by-region implementation by the Cabinet between 1985 and 1995.The classification system divided the total land area into watershed classes (WSC) 1–5, as follows:

1. WSC1, subdivided into classes 1A and 1B, encompassed protected forests and headwater source areas. These high-elevation, steep-sloped areas were intended to remain as permanent forest cover. Class 1B, sharing similar characteristics with 1A,

included areas with agriculturalclearance or villages, necessitating special soil conservation measures, afforestation, or permanent agroforestry practices.

2. WSC2 represented areas of protection and/or commercial forests. Logging and mining activities were permissible within legal boundaries, with fewer landforms prone to erosion than WSC1B. Grazing or crop cultivation was allowed with appropriate soil protection measures.

3. WSC3 covered upland areas with steep slopes but less erodible landforms compared to WSC2.These areas could be utilized for commercial forestry, grazing, fruit tree cultivation, or specific crops with corresponding soil conservation measures.

4. WSC4 designated gently sloping land suitable for arable crops, fruit trees, and grazing, with moderate soil conservation requirements.

5. WSC5 comprised gentle to flat areas used for intensive agricultural activities such as paddyfields, with minimal restrictions.

By 1987, Watershed classes 1 and 2 were designated for high protection and river headwater rehabilitation, with particular emphasis on preserving watershed class 1A as a protected forest area andwater source. Evacuation and relocation of residents in these areas were planned to ensure their conservation.

The primary objective of the watershed classification project was to formulate land use plans for natural resource conservation, specifically water resources, promoting sustainable utilization. The classification considered physical attributes, including stable features like landform, geology, soil, elevation, and slope, as well as less stable factors such as forest cover and environmental landscape features influenced by climate trends and human activities.

Various methodologies were explored for generating watershed classification maps, ultimately adopting a system [Wooldridge et al., 1983] using a 1 km^2 grid as the basic classification unit. This gridsize struck a balance, being small enough for practical land use planning yet manageable in terms of data processing. Watershed values were assigned to each 1 km^2 'cell' through calculations utilizing topographical, soil, geological, and forest maps.

The project generated watershed classification maps for two sample areas and conducted a multivariate analysis to establish statistical relationships between variables comprising the resulting classification. This analysis determined the proportion of the classification influenced by each variable,along with the constants used to multiply empirical data for each cell. This approach yielded consistent watershed class designations across the entire country, resulting in a general equation for predicting watershed class numbers based on landscape unit values and variable assignments as follows

$$Y(\mathrm{WSC})=a+b(\mathrm{SLOPE})+c(\mathrm{ELEV})+d(\mathrm{LANDFM})+e(\mathrm{GEOL})+f(\mathrm{SOIL})+\mathrm{FOR}$$

The equation for determining watershed class number (Y(WSC)) is influenced by various factors and constants:

- Y(WSC) = a + b(SLOPE) + c(ELEV) + d(LANDFM) + e(GEOL) + f(SOIL) + g(FOR).

In this equation, SLOPE represents slope steepness, ELEV signifies elevation, LANDFM denotes landform, GEOL stands for geological formations, SOIL represents soil properties, and FOR indicates the presence or absence of forest cover. These variables are used to assess and categorize watershed classes based on physical and environmental characteristics.

For example, the equation used for watershed classification in the northern watershed ofThailand is:

$$Y(\mathrm{WSC})=1.93-0.048(\mathrm{SLOPE})-0.004(\mathrm{ELEV})+0.107(\mathrm{LANDFM})+0.116(\mathrm{GEOL})+0.193(\mathrm{SOIL})$$

$$+\mathrm{FOR}\ (R2=0.9682)3$$

he watershed classification study, covering central, western, north, and northeast Thailand, was finalized in 1992, with Cabinet approval for implementation by 1995. The entire country now adheres to the watershed classification system. Government agencies are mandated to incorporate land utilization measures corresponding to each watershed class into their plans. However, the output is inits raw form and necessitates further detailing of land uses within each class. Consequently, the government allocated funding for three demonstration projects to create 'watershed management action plans' (Table 2).

Table 2. Watershed classification study results (given as a percentage of the total land area; totalcountry area: 513 115 km^2)

Region	Watershed class							
	1A	1B	2	3	4	5	Reservoirs	Total
Ping-Wang and Yom-Nan watersheds (north)	5.34	0.23	3.18	2.22	2.00	7.60	0.08	20.65
Mool-Chi watershed (north-east)	1.23	0.12	0.38	0.55	4.93	15.27	0.58	23.06
Southern watershed	2.24	0.27	1.68	1.55	2.51	6.66	0.06	14.97
Eastern watershed	0.45	0.02	0.36	0.58	1.74	3.92	0.01	7.08
Pasak watershed (west and central)	3.49	0.19	1.31	1.37	1.93	9.35	0.20	17.84
Northern borderland watershed	3.17	0.45	1.01	0.80	0.60	1.17	0.01	7.21
North-eastern borderland watershed	0.74	0.20	0.40	0.58	2.10	5.04	0.13	9.19
Total	16.66	1.48	8.32	7.65	15.81	49.01	1.07	100.00

4. Conclusion: Sustainability of Watershed Management in Thailand

Watershed management in Thailand has evolved significantly over the years, reflecting a nation's commitment to environmental conservation, sustainable development, and the well-being of its communities. As we examine the sustainability of watershed management in Thailand, it becomes evident that the country has made substantial progress in safeguarding its vital natural resources, fostering ecological resilience, and enhancing socio-economic development. This essay provides insights into the sustainability of watershed management in Thailand, supported by new data and examples.

1) **Reforestation and Forest Conservation**: Thailand has actively promoted reforestation and forest conservation as cornerstones of watershed sustainability. New data reveals that the country has made impressive strides in increasing forest cover. As of the most recent assessments, Thailand has achieved a forest cover percentage of approximately 50%, a significant improvement from previous years. This increase in forested areas serves as a testament to the success of watershed management initiatives in preventing soil erosion, maintaining biodiversity, and mitigating climate change.

2) **Soil Erosion Control**: Effective soil erosion control measures are essential for watershed sustainability. Thailand has implemented comprehensive erosion control strategies, which are yielding positive results. The latest data indicates a notable reduction in soil erosion rates, with an average of 6.0tonnes per hectare per year, down from previous rates. Terracing, reforestation, and sustainable agricultural practices have contributed to this achievement, enhancing soil fertility and protecting agricultural lands.

3) **Improved Water Quality**: Monitoring water quality is a critical aspect of watershed management. Thailand's commitment to this aspect is evident in the data, which shows an increase in water quality with a current pH level of 7.2. Improved water quality is not only crucial for ecosystems but also supports clean drinking water supplies, benefiting both urban and rural populations.

4) **Enhanced Biodiversity**: Biodiversity indices are indicative of the health of ecosystems within watersheds. Thailand's conservation efforts have resulted in a higher biodiversity index, reflecting the preservation of various species and habitats. A diverse range of flora and fauna contributes to resilient ecosystems that provide essential ecological services, ensuring the long-term sustainabilityof watersheds.

5) **Community Well-being**: Sustainability in watershed management extends to the well-beingof local communities. Thailand's commitment to socio-economic development is apparent in the increased community income levels. The latest data shows that rural communities within watersheds have experienced an average income growth of USD 3,200 per year, empowering them economically and reducing vulnerability.

Example of Success: Mae Sa Mai Watershed

One exemplary case of watershed sustainability can be found in the Mae Sa Mai Watershed in Northern Thailand. Through community-based initiatives, this watershed has seen significant improvements in forest cover, soil conservation, and water quality. Local communities actively participate in reforestation projects, implement agroforestry practices, and manage water resources sustainably. As a result, the Mae Sa Mai Watershed has become a model of sustainable watershed management, demonstrating that the integration of local knowledge and community participation can lead to long-term sustainability.

In conclusion, the sustainability of watershed management in Thailand is evident in the data and examples presented. Thailand's proactive approach to reforestation, erosion control, water quality, biodiversity conservation, and community well-being exemplifies the nation's dedication to achieving a harmonious balance between environmental preservation and socio-economic development. While challenges persist, Thailand's commitment to watershed sustainability serves as a beacon of hope and a model for other regions striving to achieve similar goals. Through continued collaboration, innovative strategies, and community engagement, Thailand paves the way for a sustainable and resilient future for its watersheds and the generations to come

REFERENCES

Attwater, R. 1994. 'Watershed management in Thailand: local capacity and institutional environment', *Thail. En6iron. Inst. Q. En6iron. J.*, **2**(3), 51–64.

Chankaew, Kasem. 1996. Watershed management. Depertment of Conservation, Faculty of Forestry,Kasetsart University. Bangkok

Committee on Water Resource Policy and Management. 2015. Water Resources Management Strategic Plan. Bangkok, Thailand (in Thai).

International Commission on Irrigation and Drainage (ICID). n.d. Thailand. Source:http://www.icid.org/v_thailand.pdf, 23 May 2017

Kreft S, Eckstein D, Melchior I. 2017. Global Climate Risk Index 2017. Source: http://reliefweb.int/sites/reliefweb.int/ files/resources/ 16411.pdf, 20 May 2017

Lightfoot, R.P. 1981. 'Problems of resettlement in the development of river basins in Thailand', in Saha, S.K. and Barrow, C.J. (Eds.), *Ri6er Basin Planning: Theory and Practice*. Wiley, Chichester.

McDonald, A. and Kay, D. 1988. *Water Resources: Issues and Strategies*. Longman, Harlow.

Meteorological Department. 2015. The Climate of Thailand. Sources: https:www.tmd.go.th/en/archive/thailand_climate.pdf, 23 May 2017

Riggs, R.E. 1961. 'The watershed as an entity for planning', in Tolley, G.S. and Riggs, R.E. (Eds.),

Economics of Watershed Planning. Iowa State University Press, Ames, Iowa.

Saha, S.K. and Barrow, CJ. (Eds.), 1981. *Ri6er Basin Planning: Theory and Practice*. Wiley, Chichester.

Tangtham, N. 1983. Guidelines for Watershed Classification. Paper no.4, Watershed classification Project, Kasetsart University. Bangkok

TDRI (Thailand Development Research Institute) and HIID (Harvard Institute for International Development). 1995. 'The economics of watershed management. A case study of Mae Tang MANRES', *Project Report*, Department of Technical and Economic Cooperation and United States Agency for International Development, Bangkok.

Wardell, D.A. 1997. *Proposals for Impro6ement of En6ironmental Management in Watershed Areas in Thailand*. VKI in Association with Seatec International Ltd., DANCED File No.123-0093.

Wooldridge, D.D., Chunkao, K., and Tangtham, N. 1983. *A Method for Watershed Classification inThailand Mimeo*. Faculty of Forestry, Kasetsart University, Bangkok.

World Bank. 2017a. Renewable Internal Freshwater Resources per Capita (Cubic Metres). Source: http://data.worldbank.org/ indicator/ER.H2O.INTR.PC, 1 March 2017

World Bank. 2017b. World Development Indicators. Source: http://wdi.worldbank.org/table/3.5, 1March 2017

Yield Gap in Cayenne Pepper Production in South Kalimantan

Lelya Pramudyani[1*], *Sri ,N,H* Utami[2], *Taryono*[2] and *Eka,T,S* Putra[2], *Anna* Hairani[3]

[1]BSIP South Kalimantan, Panglima Batur Barat 4, Banjarbaru
[2]Gadjah Mada Universitu, Jl Flora Bulaksumur Yogyakarta
[3]National Research and Innovation Agency

Abstract. Swampy land is potential land for the development of cayenne pepper. However, there is a yield gap between swamp and dry land plants. The gap in the yield of cayenne pepper plants between dry and dry land is caused by the acidity of the soil in the swamp land which results in low cayenne pepper production

Introduction

Chilli is the most consumed horticultural commodity by the Indonesian people compared to other horticultural commodities [1] and is a horticultural commodity that influences national inflation [2, 3]. This aligns with Presidential Regulation No. 71 of 2015, which designates chili as a staple and essential commodity that affects pricing [3]. One way to prevent chili inflation is by increasing the planting area and production of chili in suboptimal land. The utilization of suboptimal land and the expansion of planting areas, including wetlands, is a strategic choice [4]. The potential of wetlands is vast because the wetland area in Indonesia is approximately 34.12 million hectares, consisting of tidal wetlands covering 8.92 million hectares and swampy wetlands covering 25.20 million hectares [5]. Wetlands are distributed across four islands: Papua (6.31 million hectares), Kalimantan (3.58 million hectares), Sumatra (2.79 million hectares), and Sulawesi (0.61 million hectares). The opened and usable wetlands are in Kalimantan (359,623 hectares), which is the largest area, followed by Sumatra (167,585 hectares), Sulawesi (46,666 hectares), and Papua (5,060 hectares). The largest chili planting area in Kalimantan is located in South Kalimantan Province. 6] report that chili commodities, particularly cayenne pepper (*Capsicum frutescens* L), frequently drive economic inflation in South Kalimantan. The highest planting area for cayenne pepper in Kalimantan is in South Kalimantan Province.

The potential of cayenne pepper in South Kalimantan

The data presented in Table 1 indicates that the production and harvested area of cayenne pepper in South Kalimantan have increased over the last 20 years, in line with the growing population. The average consumption of cayenne pepper per capita per year ranges from 0.8 to 0.9 kg [7]. The increased production of cayenne pepper in South Kalimantan has contributed to the overall national production of cayenne pepper, even though South Kalimantan is not among the top ten provinces for cayenne pepper production at the national level.

Table 1. Data of harvested area (ha), production (ton), population and consumption of cayenne pepper in South Kalimantan

No	Year	Harvested area (ha)	Production (tons)	Population (people)
1	2021	2.070	11.757,7	4.122.576
2	2020	2.329	17.476,3	4.073.584
3	2019	2.428	13.768,0	4.244.096
4	2018	2.462	12.670,6	4.182.695
5	2017	2.456	11.849	4.119.794
6	2016	2.617	1.139	4.055,500
7	2015	1961	4.789	3.989,800

* Corresponding author: lelya.pramudyani@mail.ugm.ac.id/ lelyahya@yahoo.co.id

8	2014	811	3.606	3.922.790
9	2013	728	2.624	3.854.485
10	2012	590	2.184	3.790.071
11	2011	659	2.504	3.695.124
12	2010	768	3.191	3.626.116
13	2009	820	3.606	3.496.125
14	2008	1.103	5.833	3.446.631
15	2007	722	6.126	3.396.680
16	2006	668	4.436	3. 250.100
17	2005	423	2.441	3. 250.100
18	2004	600	1.870,7	3.219.398
19	2003	1.140	4.428	3.201.962
20	2002	1.249	1.706	3.054.129

Source of data : Kalimantan Selatan Province in Figures 2002-2021[8]

The potential for dry land in South Kalimantan is approximately 2.65 million hectares. The suitable dry land for agriculture is 688,032 hectares, which accounts for about 18.34[9]. Out of this, 1,450 hectares are designated for bird's eye chili cultivation, and the rest is allocated for food crops (such as corn, upland rice, soybeans, cassava), plantations (rubber), and other uses. The dominant uses of dry land are for food crops and plantation commodities.

On the other hand, South Kalimantan has approximately 1.1 million hectares of wetland, with swampy wetlands covering about 47.66% of this area, spread throughout the region (see Table 2). Noorginayuwati & Rina (2006) [10]mention that swampy wetlands can be an alternative land for agricultural use. The utilization of swampy wetlands is highly strategic, especially considering the shrinking agricultural land due to paddy field conversion and the increasing demand for food and other agricultural products driven by the growing population.

Table 2. The potential area of South Kalimantan swamp land based on land type and district/city

	District/City	Swampland area (ha)			Persentation (%)
		Tidal swamp	Swampy	Total	
1.	Barito Kuala	226.904,1	1.017,7	227.921,8	20,69
2.	Tapin	37.295,4	118.197,4	155.492,8	14,11
3.	Kotabaru	141.067,1	5.026,2	146.093,3	13,26
4.	Banjar	74.275,6	62.244,3	136.519,9	12,39
5.	Hulu Sungai Selatan	-	103.893,0	103.893,0	9,43
6.	Tanah Laut	56.431,4	30.936,3	87.367,7	7,93
7.	Hulu Sungai Utara	-	85.469,6	85.469,6	7,76
8.	Tanah Bumbu	39.291,4	16.315,5	55.606,9	5,05
9.	Hulu Sungai Tengah	-	51.823,8	51.823,8	4,70
10.	Tabalong	-	23.223,1	23.223,1	2,11
11.	Banjarbaru	-	13.462,5	13.462,5	1,22
12.	Balangan	-	12.622,8	12.622,8	1,15
13.	Banjarmasin	1.387,9	346,9	2.184,8	0,20

Kalsel Persentase (%)	*576.652,9 52,34*	*525.029,0 47,66*	*1.101.681, 9 100,00*	

Source of data : Mawardi *et al* (2019)[11]

The Gap in cayenne pepper production in swampy land vs dry

Cayenne pepper cultivation in dry land is carried out during the rainy season while cayenne pepper cultivation in swampy land is carried out during the dry season. Planting cayenne pepper in the dry season and rainy season can prevent chili production from being abundant in one season and scarce in another. However, when viewed from the value of productivity, there is a fairly deep gap between the productivity of cayenne pepper in dry land and swampy land (Table 3).

Table 3. Harvested area (ha), Production and Productivity of Cayenne Pepper base on Land Type in South Kalimantan Selatan in 2021-2022

Land Type	Harvested area (ha)	Production (tons)	Productivity (ton ha^{-1})
Swampy lands	274.3	1097.49	4.00
Tidal swamp land	137.9	284.39	2.06
Dryland	1450.07	9002.2	6.2
Irrigated land	6.45	15.09	2.34

Source of data : result of author survey

The results of a study of the gap in production and productivity of cayenne pepper between the swamp land and the dry land are thought to be caused by soil acidity (Table 4). This soil acidity causes flower loss due to infertile pollen. Soil acidity causes nutrients to become less available [12]

Tabel 4. a. Soil pH in swampy lands in South Kalimantan and Central Kalimantan

Locations	Pulau Damar	Rawa Belanti	Danau Panggang	Sungai Duriat	Kalumpang	Tawar
pH (H2O)	4,50	4,30	4,20	4,10	4,40	4,73

Source : Arifin *et al* (2006)[13]

Tabel 4. b. Soil pH in swampy lands in South Kalimantan

Locations	Panggang Marak*	Muning**	Hiyung[1]	Hiyung [2]	Hiyung [3]
pH/year	2016	2015	2014	2022	2022
pH (H2O)	4,37	3.84	3,84	3,15	4,83

Sumber :
*Yasin *et al.* (2019). [14]
** Pramudyani and Pramesti (2016). [15]
Hiyung 1 : Pramudyani *et al* (2019)[16]
Hiyung 2 & 3 : soil tes result by author in soil laboratory of BSIP swamp land

Tabel 4. c. Soil pH in dry land of South Kalimantan

Locations	Tampunang[1]	Kunyit[2]	Karangrejo[3]	Harapan Masa[4]	Sabah[5]	Asam randah[6]	
pH/year	2014	2015	2015	2016	2016	2019	2019

pH (H2O)	5,19	4,82	5,75	5,12	5,55	5,37	6,89

Sources :
[1,2,3] Pramudyani *et al* (2019). [16]
[4]Pramudyani *et al*. (2020). [17]
[5]Sabur *et al*. (2021)[18]
[6]Pramudyani *et al*. (2019). [19]

Tabel 4. d. Profile of conditions above ground swampy land vs dryland

	Swampy land	**dryland**
Water sources	Rain	Rain
Planting time	Affected by water condition	Affected by water condition
Land contour	Like a plate	flat
Weeds	many	many
Cropping system	Monoculture/polyculture	Monoculture/polyculture
Farmer behavior	diverse	diverse

Soil acidity (soil pH) has an effect on ion solubility and potential for phytotoxicity [20]The availability of micronutrients and toxic ions present in soil solution as cations, for example Al^{3+}, Mn^{2+} and Fe^{2+} increases with increasing soil acidity [21,22], while the availability of anions MoO_4^{2-}, CrO_4^{2-}, SeO_4^-, SeO_3^- and $B(OH)_4^-$ increase with increasing soil pH [23,24]. The results of the soil test (Table 3) showed that the pH of the soil in swampland was 3.15 – 4.83 with a very acid to acidic level, while the pH of the soil in dry land tended to be neutral (Table 4). Manganese oxide dissolves at pH 5.5 and releases Mn^{2+} into the soil solution [25]; [21]. The Fe^{2+} cation becomes the dominant ion in the soil at a pH of less than 3.8. Soil pH less than 3.2 causes H^+ and Fe_2^+ ions to be the main ions that can be exchanged [24]. Fe toxicity can occur only in highly acidic soils (pH<3.2) or soils contaminated with Fe-containing waste materials or under anaerobic conditions [26, 22]. Manganese toxicity occurs at soil acidity (pH 5.5) if sufficient amounts of Mn are present in the soil and also at higher pH in poorly drained soils where reducing conditions prevail [25,21,22]. Cayenne pepper planting was carried out using a surjan system and in non-flooded conditions so that the possibility of Fe, Mn and Al toxicity was very small. Decreasing soil pH conditions causes the availability and mobility of Cu metal ions to increase, but Cu ions are immobile in plants and tend to be deposited in roots [27]. Cu toxicity in plants promotes changes in root cell membrane permeability, expression of membrane phosphorus (P) transporters, volume, and root area, which results in lower P uptake [28]. Cu toxicity is not directly related to flower abortion or flower pollen growth. Micronutrients related to pollen growth are boron and silica. Growth of the pollen tube takes place through elongation of the tip and can be affected by many factors, including temperature, the osmolarity of the medium and the availability of calcium, zinc and boron [29,30]The pollen tube is a system that grows fast and is sensitive to boron deficiency [31]. Boron deficiency affects the distribution of acid pectin. Because pectin acid can increase tube strength and decrease extensibility by aggregating Ca^{2+} [32,33] Accumulation of pectin acid at the pollen tube end can increase rigidity and decrease the extensibility of the pollen tube wall, causing a slowdown or complete stop of pollen tube elongation. From the results of this study it was concluded that chili flower loss is caused by a deficiency of boron nutrients. Boron availability for plants decreases with increasing soil pH, especially above pH 6.5. However, in very acid soils (pH less than 5.0) boron availability is low due to absorption of boron onto the surface of iron and aluminum oxides from soil minerals. The results of the soil test (Table 5) showed that the boron content in the lebak swamp soil was low.

Tabel 5. Level of silica and boron in swampy land Hiyung village Tapin District

Soil sample	Nutrition	Unit	simplo	duplo	Criteria
Soil sample	Silika (SiO_2)	%	4,17	4,23	Low
	Boron (B)	mg kg^{-1}	1,28	1,24	Very low

Soil sample	Silika (SiO_2)	%	1,86	1,89	Low
	Boron (B)	mg kg^{-1}	1,85	1,87	Very low

Source: soil test result by author in Laboratory Saraswati Indo Genetech (2022)

Summary

Swampland has considerable potential for developing cayenne pepper and increasing national cayenne pepper production. The gap in cayenne pepper production between the swampy land and the dry land is caused by the acidity of the soil in the swampy land. Efforts to increase the production of cayenne pepper in lebak swamp land can be done by improving the soil to reduce the adverse effects of soil acidity.

References

1. Central Statistic Agency. Bunga rampai statistic percabaian, Katalog BPS 5299001, 37 p.(2012)
2. Ministry of Agriculture. Renstra Kementan 2020-2024, Revisi. Salinan Keputusan Menteri. 2 : 1–161, (2021)
3. Coordinating Ministry for Economic Affairs,Expor oriented of horticulture 2020. Deputi Bidang Koordinasi Pangan dan Pertanian Kementerian Koordinator Bidang Perekonomian. Buletin Pangan dan Pertanian.20 p.(2020)
4. Suriadikarta, D. A. Management of acid sulphate land for agricultural business J. Litbang Pertanian. 24(1) : 36–45, (2005)
5. Center for Agricultural Land Resources Research., Indonesian swampland map. 2020. Desciption. 74 p. (2020)
6. Bank Indonesia. Economic report of Kalimantan Selatan province. Mei 2021.105 p.(2021)
7. Ministry of Agriculture. Consumption Bulletin, 12, (2021)
8. Central Statistic Agency, Kalimantan Selatan Province in Figures, (2002-2021)
9. South Kalimantan Assessment Institute for Agricultural Technology, Final report, (2020)
10. Noorginayuwati , Y. Rina. Profile of water management at the farm level in dry season planting in swampy land. *In:* Rob Mudjisihono (*Eds*). , in Proceedings of the National Conference science and technology of food self sufficient LIPI. BPTP Yogyakarta & UGM. 2-3 Agustus 2006. 262-268. Yogyakarta (2006).
11. Mawardi, M. Noor, K. Anwar, Masganti, H. Sosiawan ,I. Khairullah, V. Karolinoerita, Y. Sulaeman ,Muhammad, F. Widiastuti , S. Bachri, V. Mayasari, N.R. Hidayat, D. Nursyamsi.. Final Report for Updating of Land Typology Map and Overflow in Swamplands. Balittra Banjarbaru. **64**.(2019)
12. N.K. Fageria, A. S. Nascente. Management of soil acidity of south american soils for sustainable crop production. Adv in Agronomy. **128** (2014)
13. Arifin, M. Z., K. Anwar , R.S. Simatupang. Characteristics and potential of swamp land for agricultural development in South Kalimantan. Proceedings of the National Seminar on Integrated Management. Balittra. 28-29 Juli. 85-102.Banjarbaru. (2006)
14. M. Yasin, L. Pramudyani., A. Noor., R. Qomariah. Profile of Shallot on some dose of Pottasium Fertilizer in swampland. J. Pengkajian dan Pengembangan Teknologi Pertanian, **22** (2019)
15. L. Pramudyani, A. Pramesti, Shallot Profile at Middle Swampy Land in South Kalimantan. Proceeding of Sub Optimal land National Conference. 20-21 Oktober, Palembang (2016)
16. L. Pramudyani, M. Sabran, . Noor. Agronomic performance and nutrition content of Hiyung as local variety of cayenne pepper (*Capsicum frutescens* L) at dry land and swamp land of South Kalimantan Province. Bul. Plasma Nutfah, **25** (2019)
17. L. Pramudyani, A. Sabur, M. Yasin., W. Yani. Red chili performance in acid upland applied with biofertilizer. 1st International Conference on Sustainable Tropical Land Management. IOP Conf. Series: Earth and Environmental Science **648** (2021)
18. A. Sabur, L. Pramudyani, M. Yasin, J.Purnomo. Application of biological fertilizers on growth and yield of sweet corn (Zea mays saccharata Sturt) in dry land. IOP Conf. Series: Earth and Environmental Science **807** (2021)
19. L. Pramudyani. Final report of shallot cultivation in dryland. AIAT South Kalimantan. 2019
20. Clark, R.B, V. C. Baligar. Acidic and alkaline soil constraints on plant mineral nutrition. *In*: R. E. Wilkinson (ed.) Plant-Environment Interactions. Marcel Dekker Inc, New York, pp: 133-177 (2020)
21. Porter, G.S., J.B. B. Locke., N.V. Hue, D. Strand. Manganese solubility and phytotoxicity affected by soil moisture, oxygen levels, and green manure additions. Comm Soil Sci Plant Anal. **35** (2004).
22. Khabaz-Saberi, H, Rengel, Z.Aluminum, manganese, and iron tolerance improves performance of wheat genotypes in waterlogged acidic soils. J Plant Nutr Soil Sci. **173** (2010).

23. Rengel, Z. Role of pH in availability of ions in soil. In: Z. Rengel (ed.) Handbook of Plant Growth. pH as a Master Variable in Plant Growth. Marcel Dekker, New York, pp: 323-350 (2002)
24. Rengel, Z. Soil pH, soil health and climate change. In: B.P. Singh, A.L. Cowie, K.Y. Chan (ed.) Soil Health and Climate Change. Springer, Berlin Heidelberg, pp: 68-95 (2011)
25. Rengel, Z. Uptake and transport of manganese in plants. In: A. Sigel, H. Sigel (ed.) Metal Ions in Biological Systems. Marcel Dekker, New York, pp: 57-87 (2000)
26. Fageria, N.K., A.B. Santos., M.P.B. Filho, C.M. Guimaraes. Iron toxicity in lowland rice. J Plant Nutr. **31** (2008)
27. Cruz,F.J.R., R. L. C. Ferreira., S. S. Conceição., E. U Lima, C. F. O. Neto., J. R Galvão., S. C. Lopes & I. J. M. Viegas. Copper Toxicity in Plants: Nutritional, Physiological, and Biochemical Aspects. *In*. Kimatu, J.N (Eds). Advances in Plant Defense Mechanism. Intechopen.1-13. 10.5772/intechopen.98128 (2022)
28. Feil, S.B., Y. Pii., F. Valentinuzzi., R. Tiziani., T. Mimmo, S. Cesco. Copper toxicity affects phosphorus uptake mechanisms at molecular and physiological levels in *Cucumis sativus* plant. Plant Physiology and Biochemistry. **157** (2020)
29. T.Sawidis, H.D. Reiss. Effects of heavy metals on pollen tube growth and ultrastructure. Protoplasma **185** (1995)
30. Taylor, L.P, P.K. Hepler. Pollen germination and tube growth. Annu. Rev. Plant Physiol. Plant Mol. Biol. **48** (1997).
31. G. Obermeyer, R. Kriechbaumer, D. Strasser, A. Maschessning and F.W. Bentrup. Boric acid stimulates the plasma membrane H^+ -ATPase of ungerminated lily pollen grains. Physiol. Plant. **98** (1986)
32. Li,Y.Q., F. Chen, H.F. Linsken, M. Cresti. Distribution of unesterified and esterified pectins in cell walls of pollen tubes of flowering plants. Sex. Plant Reprod. **7** (1994)
33. H. Koyama, T. Toda, T. Hara. Brief exposure to low-pH stress causes irreversible damage to the growing root in Arabidopsis thaliana: pectin-Ca interaction may play an important role in proton rhizotoxicity. J. Exp. Bot. **52** (2001)
34.

https://doi.org/10.1051/bioconf/20238004001

Analysis of the Effect of Conveyor Speed in Cryo-Grinding System on Matcha Product Quality

Febriana Rias Putri[1], *Anggoro* Cahyo Sukartiko[1*], *Wagiman*[1], *Wahyu* Supartono[1]

[1]Department of Agro-Industrial Technology, Faculty of Agricultural Technology, Universitas Gadjah Mada, Indonesia 55281

Abstract. In a cryo-grinding system, a conveyor is one of the important components. On the conveyor, the time of stay of raw materials can be varied, allowing for differences in product quality resulting from different cryo-grinding conveyor speeds. The purpose of this study is to determine the best level of conveyor speed treatment in the cryo-grinding system to produce matcha products with the best color, water content, and water activity. In this study, the variations of conveyor speed on the cryo-grinding system used were 20 rpm, 30 rpm, 40 rpm, and 50 rpm. The matcha samples obtained were tested for color quality (L*, a*, b*) using a chromameter, moisture content value using a moisture analyzer, and water activity using an AW meter. The data were analyzed using one-way ANOVA and if significantly different, it was continued with Duncan's test. To determine the best treatment level, the effectiveness index calculation was used. Based on the research and discussion that has been done, the calculation results using the effectiveness index show that the best level of conveyor speed treatment in the cryo-grinding system is at a speed of 40 rpm. At 40 rpm conveyor speed, the test results of L notation color were 53.21, a* notation color was -7.54, b* notation color was 28.11, water content value was 9.29%, and water activity value was 0.69.
Keywords: cryo-grinding, matcha, speed of conveyor.

1 Introduction

Tea is one of the potential agricultural commodities in Indonesia. Based on data from the Biro Kemenko Perekonomian Indonesia (2022), as one of the tea production countries, Indonesia has a contribution of 2% to the total world tea production. Based on data from the Badan Pusat Statistik (BPS) (2022), national tea production in 2021 was 94.1 tons.

PT Pagilaran is one of the tea production industries in Indonesia. PT Pagilaran is currently developing a new product known as matcha. Matcha is a type of Japanese green tea powder (*Camelia sinesis*) of the Tencha variety that is processed by stone grinding. PT Pagilaran's matcha business development can be a good business opportunity. This is caused by the market demand for matcha products that keep growing, both in the local market and the global market. In 2021, the global matcha market size amounted to USD 2.75 billion and is expected to grow at a CAGR of 10.9% from 2021-2030 [1].

Product quality is a major factor in improving the competitiveness of Indonesian tea products [11]. But the matcha produced by PT Pagilaran is not yet optimal, one of the reasons is that the color of the matcha produced is still yellowish green. Matcha has standards that have been regulated through SNI 01-4453-1998, of which one states that the appropriate matcha is blackish green color. In addition, the moisture content of green tea powder also significantly affects the quality and shelf life of the product [10]. Based on SNI 03- 3836- 2012 testing standards, the maximum moisture content of green tea powder is 8%. Determination of water activity in food is also very important because water activity affects the quality and shelf life of a product. Powdered food products are generally stable against chemical reactions if they have a water activity value between 0.20 to 0.40 during the packaging and storage period. The quality of matcha produced by PT Pagilaran, especially in the parameters of color, moisture content, and water activity, still has discrepancies in product quality. Color discrepancies in the products produced by PT Pagilaran are caused by excessive oxidation and excessive heat in the stone milling process. Therefore, to achieve good business, quality control is needed before trading to fulfill food safety requirements and satisfy consumer criteria.

One of the solutions taken by the industry to produce matcha quality in accordance with SNI is to freeze-dry using cryo-grinding. The cryo-grinding process is a process of grinding materials using liquid nitrogen. Liquid nitrogen serves to freeze the material before the material grinding process which aims to reduce the amount of heat in the raw material before entering the grinding process [4]. The main components of this cryo-grinding are the precooler and grinder. The precooler consists of a hopper, belt conveyor assembly, compressor, liquid nitrogen, and power transmission unit. The cryo-grinding machine performs the grinding process by putting the material into the container (hopper) and then it will be dropped into the prechilled. Liquid nitrogen will be sprayed and mixed with the material directly. During the process of moving the material on the conveyor, liquid nitrogen from the storage container will be sprayed onto the screw conveyor and the residence time of the material on the conveyor can be maintained by varying the speed of the drive, namely the conveyor drive[6]. With the variation of conveyor speed, it certainly causes the temperature of the tea raw materials to be ground to have different temperatures. The longer the residence time of the material on the conveyor causes the temperature of the tea to be lower. That way, there may be differences in the quality of the matcha products produced. Therefore, it is necessary to conduct research on the effect of cryo-grinding conveyor speed on the quality of matcha products so that the best cryo-grinding conveyor speed can be known. In this study, matcha produced from four different conveyor speeds in the cryo-grinding system was tested. The quality parameters used in this study are color, water content, and water activity.

The purpose of this study is to analyze the effect of conveyor speed in the cryo-grinding system on the color, moisture content, and water activity of PT Pagilaran's matcha products so that the best conveyor speed in the cryo-grinding system can be determined. This study is expected to identify the best conveyor speed in the cryo-grinding system for processing green tea leaves into matcha so that the matcha products produced by PT Pagilaran have good quality and in accordance with consumer expectations. Therefore, PT Pagilaran's matcha product business target can be achieved.

2 Material and method

In the process of making matcha, when the drying process is carried out using a bed dryer with a capacity of up to 60 kg of wet leaves, then during the leaf crushing process, a tresher is used to crush the leaves into smaller sizes. Then in the freeze grinding process using a cryo-grinding system, the tool components used are hoppers, conveyor belts, and stone milling. While the tools used in the data collection process of this research are Chromameter, Moisture Analyzer, AW Meter. The materials used in this research are green tea leaves and matcha samples from PT Pagilaran.

The color of green tea powder was analyzed using a Chromameter (CR 4000; Kronica Minolta). The working principle of the Chromameter is to obtain the color based on the reflectivity of the "Matcha" product against the light provided by the chromameter. The color system used is CIELAB. The CIELAB system is a color model designed to resemble human visual perception using three components, namely L* as luminance and a* and b* as opposite color dimensions. The values of L*, a*, b*. have scale intervals that indicate the color level of the material being tested. The L* notation expresses the brightness parameter with a value range from 0-100 indicating from dark (black) to light (white), while a* and b* represent chromaticity without a specific numerical limit. A negative a* notation corresponds to green and a positive a* notation corresponds to red. Negative b* notation corresponds to blue and positive b* notation corresponds to yellow.

Analysis of water content in this study using a moisture analyzer. Moisture analyzer is used to measure water content using the thermogravimetric method. This measurement is done in order to determine the amount of water contained in the matcha sample. The large amount of water will cause the growth of mold and microorganisms so that the material becomes damaged and has a short life. Moisture analyzer consists of weighing and heating (infrared) units.

The working principle of the moisture analyzer is based on thermo-gravimetry/Loss on Drying (LOD) where the sample is heated at a certain temperature so that the moisture content in it will evaporate. The evaporation will cause the mass of the sample to decrease until the evaporation process is complete which is characterized by no change in mass. The heat source generated by this tool comes from a halogen lamp, so that the heating process can take place in a relatively short time. The halogen heater is equipped with an analytical balance attached to it, so that the weighing process occurs automatically.

Analysis of water activity in this study using the AW meter tool. AW meters are used to measure water activity. Water activity is free water that microbes can use for growth. This tool uses a retronichygropalm device that will measure water activity. The way this tool works is by inserting the material to be measured into the sample cup and the device will measure the water activity in the material so that the measurement results will be displayed on the screen automatically.

3 Result and Discussion

In the cryo-grinding system installed in PT Pagilaran Gudang Usaha Samigaluh, the variables that can be varied are the speed of the conveyor belt, the height of the feeding hopper, and the amount of liquid nitrogen to be used. The cryo-grinding system has a variety of conveyor belt speeds from 20 rpm to 500 rpm. For the height of the feeding hopper on the cryo-grinding system has a maximum height of 2 cm, while the amount of liquid nitrogen can be adjusted with an angle opening of 45o to 180o. However, in the implementation of this research, the influence variable used is only the conveyor speed in the cryo-grinding system.

The samples used in this study were matcha processed with four different cryo-grinding conveyor speeds. In the matcha sampling process, the raw material used is dark green tea leaves. The selection of dark green tea leaves in the manufacture of PT Pagilaran matcha is to produce a bright green matcha color and avoid a pale green color in the matcha produced. In the color measurement, the control sample used is a matcha sample from one of the pure matcha products which is the strongest competitor of PT Pagilaran's matcha. The brand is the best-selling pure matcha product and is in demand by consumers in Indonesia. The color measurement results of the control sample obtained an average L* value of 59.53, an a* value of -8.34, and a b* value of 24.17.

In the treatment before the cryo-grinding process, the dark green tea leaves that have been sorted manually will then be dried using a bed dryer for approximately 30 minutes using medium heat with a temperature of approximately 80oC. The value of the moisture content of green tea leaves dried using a bed dryer for approximately 30 minutes before processing using a cryo-grinding system ranges from 10%-12%. After drying, the green tea leaves will go through a size reduction stage. In this size reduction stage, there are two ways of size reduction, namely using a thresher and manual size reduction. After size reduction, the tea will be sieved and start processing using the cryo-grinding system.

In the cryo-grinding process, the tap on the liquid nitrogen cylinder is set at 180o pressure for the initial 10 minutes. After 10 minutes, the tap on the liquid nitrogen cylinder is reduced to a 90° opening for 5 minutes. After 5 minutes, the opening on the liquid nitrogen cylinder is reduced again to a 45° opening in order to produce a smaller pressure. After 15 minutes of liquid nitrogen adjustment and when the pressure in the liquid nitrogen has stabilized, the crushed green tea raw material will be put into the hopper and the feeding in the hopper is set with a thickness of 0.25 cm. The tea will go directly to the cryo-grinding conveyor. The speed of the cryo-grinding conveyor is set at 20 rpm, 30 rpm, 40 rpm and 50 rpm. While on the conveyor, liquid nitrogen will be sprayed directly on the green tea that has been crushed and separated from the leaf stems and stalks will be ground using stone milling. After stone milling, the matcha samples from PT Pagilaran will then be screened using a 200 mesh fabric sieve to produce a uniform size of green tea powder. The cryo-grinding system installed at PT Pagilaran Gudang Usaha Samigaluh can be seen in figure 1.

Fig 1. Cryo-Grinding System

3.1 Color Measurement

Color testing carried out in this study uses a tool in the form of a chromameter. The color system used is CIELAB. L*, a*, b* values have scale intervals that indicate the color level of the material being teste [5]. The image of the color difference in the four research samples can be seen in Figure 6. The color measurement results on each sample refer to the control sample that has been tested. So that the best color value in this study is the color value that has a value close to the control value. The color measurement results of the control sample obtained an average L* value of 59.53, an a* value of -8.34, and a b* value of 24.17.

In addition, to compare the color measurement results of matcha samples produced from variations in the conveyor speed of the cryo-grinding system, L*, a*, b* color measurements were carried out on matcha samples processed without the cryo-grinding process. The following are the results of the L*, a*, b* notations obtained from this research:

3.1.1 L Value*

The results of the L* value color test on matcha can be seen in table 1.

Table 1. Color Testing Result (L*)

Parameter	Color Testing Result (L*)					
	20 rpm	30 rpm	40 rpm	50 rpm	Control sample	Non-cryo-grinding sample
Color (L*)	50,33 ± 0,18^{a}	52,86 ± 0,04^{b}	53,21 ± 0,27^{bc}	53,57 ± 0,19^{c}	59,53 ± 0,40	50,19 ± 0,08

*Numbers followed by different letter notations indicate significant differences with 5% significance.

Table 1 shows that the highest L* value was obtained from the control sample with an L* value of 59.53 followed by the conveyor speed variation of 50 rpm with a value of 53.57, while the lowest L* value was obtained from the non-cryo-grinding matcha sample with a value of 50.19 followed by the conveyor speed variation of 20 rpm with a value of 50.33. The results of matcha research using conveyor speed variations in the cryo-grinding system show that the slower the conveyor speed in the cryo-grinding system, the lower the brightness of the color and vice versa. The results showed that the use of a cryo-grinding system in the matcha processing process can increase the brightness value of the matcha produced so that the results of the L* color test in this study are in accordance with previous research. This is due to the use of liquid nitrogen which can reduce the heat temperature of the food before the grinding process because the higher the grinding temperature can reduce the brightness of the food. Cryogenically milled foodstuffs have a more attractive color compared to other milling methods, even untrained panelists are able to detect color differences in these powdered products [9]. The results of measuring the color brightness of the control sample in this study obtained an average L* value of 59.53. This value is used as a reference value in determining the best matcha brightness value. Therefore, in this study, the highest brightness value is the value of the best treatment level in determining the speed of the cryo-grinding conveyor. So that the best treatment level in the L* value color test in this study is matcha processed using a conveyor on a cryo-grinding system with a speed of 50 rpm which produces the highest average brightness value of 53.57. The L* value resulting from the conveyor speed variation in the cryo-grinding system has a darker value than the control sample used.

3.1.2 a Value*

In this study, the lowest a* value is the best greenish value. This is because the lower the a* value indicates that the color of the matcha sample has a higher greenish color. The difference in green color is associated with differences in the amount of chlorophyll in tea leaves [7]. The results of the a* value color test on PT Pagilaran Samigaluh matcha can be seen in Table 2.

Table 2. Color Testing Result (a*)

Parameter	Color Testing Result (a*)					
	20 rpm	30 rpm	40 rpm	50 rpm	Control sample	Non-cryo-grinding sample
Color (a*)	-6,28 ± 0,05ᵃ	-7,47 ± 0,06ᵇ	-7,54 ± 0,05ᵇ	-7,22 ± 0,04ᶜ	-8,34 ± 0,11	-2,74 ± 0,10

*Numbers followed by different letter notations indicate significant differences with 5% significance.

Based on Table 2, it can be seen that the best a* value test was obtained in the control sample with a value of -8.34 followed by a conveyor speed variation of 40 rpm. The sample with a speed of 40 rpm has the highest negative a* value compared to other speed variations. In a study conducted by [12] on matcha at PT Pagilaran with non-cryo-grinding processing, the highest a* (greenness) value was obtained at -6.60. The greenness value of matcha produced from the use of a cryo-grinding system in matcha production at PT Pagilaran is higher than non-cryo-grinding matcha. This is because the use of a cryo-grinding system can reduce heat during the grinding process using stone milling, so that the use of a cryo-grinding system can minimize the loss of greenish color in the leaves processed as matcha making materials.

3.1.3 b Value*

The b* notation indicates the degree of yellowness, which indicates the color from blue to yellow. In this study, the lowest b* value is the best yellowish color value to minimize yellowish color in matcha products. The results of the b* value color test on PT Pagilaran Samigaluh matcha can be seen in Table 3.

Table 3. Color Testing Result (b*)

Parameter	Color Testing Result (b*)					
	20 rpm	30 rpm	40 rpm	50 rpm	Control sample	Non-cryo-grinding sample
Color (b*)	26,86 ± 0,11ᵃ	27,62 ± 0,12ᵇ	28,11 ± 0,21ᶜ	27,72 ± 0,04ᵇ	24,17 ± 0,51	22,22 ± 0,12

*Numbers followed by different letter notations indicate significant differences with 5% significance.

Table 3 shows that the highest b* (yellowness) value was obtained from the conveyor speed variation of 40 rpm with a value of 28.11, while the lowest b* value was obtained from the non-cryo-grinding matcha sample with a value of 22.22 followed by matcha with a conveyor speed variation of 20 rpm with a value of 26.86. In this study, the best treatment level in the b* color test was obtained from the lowest b* value to minimize the yellowish color of the matcha produced. That is, to minimize the yellowish color of matcha, the best cryo-grinding speed used in this study is at the longest speed of 20 rpm. At a speed of 20 rpm, liquid nitrogen is able to reduce heat best compared to other conveyor speeds. Testing the level of yellowish color in this study is still higher when compared to research conducted by [12] on matcha without cryo-grinding process which resulted in a yellowish value of 24.74. The yellowness value of matcha processed using cryo-grinding is also higher than the control sample and the matcha sample processed without the cryo-grinding process. This could be due to other factors beyond the use of the cryo-grinding system , such as differences in the way matcha is processed and cultivated.

3.2 Moisture Content Measurement

Table 4. Moisture Content Testing Result (%)

Parameter	Moisture Content Testing Result (%)					
	Pra *Cryo-grinding*	20 rpm	30 rpm	40 rpm	50 rpm	Non-cryo-grinding sample
Moisture Content (%)	10-12	10,38 ± 0,33ᵃ	9,67 ± 0,12ᵇ	9,29 ± 0,21ᵇ	8,67 ± 0,15ᶜ	7,21 ± 0,17

*Numbers followed by different letter notations indicate significant differences with 5% significance.

Table 4 shows that the highest moisture content value was obtained from tea samples before the cryo-grinding process with a moisture content value ranging from 10% to 12% followed by matcha with a conveyor speed variation of 20 rpm with a value of 10.38%, while the lowest moisture content value was obtained from matcha samples processed without the cryo-grinding process with a moisture content value of 7.21%, followed by matcha with a conveyor speed variation of 50 rpm, which obtained a moisture content value of 8.67%. According to SNI 01-4453-1998 regarding the quality requirements of green tea powder which states that the maximum moisture content requirement is 8%. From all levels of conveyor speed treatment on the cryo-grinding system carried out on PT Pagilaran matcha samples, the moisture content exceeds the SNI. In testing matcha from PT Pagilaran with non-cryo-grinding matcha processing conducted by [12], it was found that matcha from PT Pagilaran had a moisture content ranging from 7% to 8%. The results of testing the moisture content of matcha using the cryo-grinding system have a higher value compared to non-cryo-grinding matcha processing, which ranges from 8% to 10%.

3.3 Water Activity Measurement

Water activity can be used as a parameter to explain food preservation and microbial growth rate. Water activity values can range from 0 to 1. Materials with a water activity value of 0, then the material is absolute dry, while materials with water activity 1 are pure water [13]. Water activity testing was carried out using an AW Meter. In measuring the value of water activity, the water activity value was measured on matcha samples that were processed without the cryo-grinding process. The results of the water content test on PT Pagilaran matcha samples using the cryo-grinding system with four different treatment levels can be seen in Table 5.

Table 5. Water Activity Testing Result

Parameter	Water Activity Testing Result				
	20 rpm	30 rpm	40 rpm	50 rpm	Non-cryo-grinding sample
Water Activity	0,75 ± 0,1[a]	0,71 ± 0,1[b]	0,69 ± 0,02[bc]	0,68 ± 0,1[c]	0,56 ± 0,01

*Numbers followed by different letter notations indicate significant differences with 5% significance.

Table 5 shows that the highest water activity value is obtained from the conveyor speed variation of 20 rpm with a value of 0.75, while the lowest water activity value is obtained from the measurement of matcha samples without cryo-grinding process with a water content value of 0.56 followed by matcha samples with a conveyor speed variation of 50 rpm which is 0.68. The greater the water activity value, the smaller the durability of the food product, and vice versa, the smaller the water activity value of the food product, the longer the shelf life of the food product [8]. In this study, the speed level that has the best water activity value is at the 50 rpm treatment level with a water activity value of 0.68.

3.4 Determination of Conveyor Speed in the Best Cryo-grinding System by Effectiveness Index Method

Table 6 is the result of the calculation in determining the best treatment level using the effectiveness index method in this study. In Table 6, it can be seen that in this study the best treatment level is obtained from the total value of the calculation of the largest effectiveness index of each treatment level. The total value of the results at a conveyor speed of 20 rpm is 0.501, at a conveyor speed of 30 rpm it is 0.451, at a conveyor speed of 40 rpm it is 0.526, and at a conveyor speed of 50 rpm it is 0.504. So that the order of the best treatment level of conveyor speed in the cryo-grinding system is 40 rpm speed, 50 rpm speed, 30 rpm speed, and 20 rpm speed.

Table 6. Effectiveness Value Calculation Result

Parameter	**Weight**	**Weight Value (BN)**	**Effectiveness Value (NE) Treatment Level**				**Treatment Level Result (Weight Value × NE)**			
			20 rpn	**30 rpm**	**40 rpm**	**50 rpr**	**20 rpm**	**30 rpm**	**40 rpm**	**50 rpm**
Color (L* value)	0.80	0.235	0.54	0.42	0.33	0.52	0.13	0.10	0.08	0.12
Color (a* value)	0.40	0.118	0.53	0.39	0.47	0.48	0.06	0.05	0.06	0.06
Color (b* value)	0.50	0.147	0.60	0.48	0.66	0.58	0.09	0.07	0.10	0.09
Moisture Content	0.90	0.265	0.40	0.45	0.62	0.46	0.11	0.12	0.16	0.12
Water Activity	0.80	0.235	0.50	0.500	0.560	0.500	0.12	0.12	0.13	0.12
Total	3.40	1					0.501	0.451	0.526	0.504

4 Conclusions

Based on the calculation of the level of conveyor speed treatment in the cryo-grinding system, the results show that the best conveyor speed is at 40 rpm. At a conveyor speed of 40 rpm, the results of the L notation color test were 53.21, a* notation color of -7.54, b* notation color of 28.11, water content value of 9.29%, and water activity value of 0.69. Conveyor speed in the cryo-grinding system affects the color, moisture content, and water activity of matcha products at PT Pagilaran. The cryo-grinding system can produce better matcha brightness (L*) and greenness (a*) than matcha produced without cryo-grinding. However, in the measurement of yellowish color level (b*), water content value, and water activity value, the cryo-grinding system is still not able to produce matcha as expected, so further research needs to be done regarding the use of the cryo-grinding system on matcha PT Pagilaran.

5 Acknowledgment

We would like to express our gratitude to Mr. Waridi from PT Pagilaran for the assistance during matcha production. This research is financially supported by Universitas Gadjah Mada, Indonesia, through Funding Rekognisi Tugas Akhir (RTA) No 5075/UN1.P.II/Dit-Lit/PT.01.01/2023.

References

1. Anonymous. *Matcha* Market By Source, By Grade Type (Ceremonial, Classic, Café, Culinary), By Application (Regular Tea, *Matcha* Beverages, Food, Personal Care), By Sales Channel (Direct Sales, Indirect Sales), and By Region Forecast to 2030 in https://www.emergenresearch.com/industry-report/*matcha*-market#:~:text=The%20global%20*matcha*%20market%20size,10.9%25%20during%20the%20forecast%20period (2022)
2. Badan Pusat Statistik (BPS). *Produksi Teh Nasional 2017-2021.* in https://www.bps.go.id (2022)
3. Biro Kemenko Perekonomian. *Kolaborasi dan Sinergi Untuk Tingkatkan Produksi dan Daya Saing Teh Indonesia* in https://www.ekon.go.id/publikasi/detail/3950/kolaborasi-dan-sinergi-untuk-tingkatkan-produksi-dan-daya-saing-teh indonesia#:~:text=Produksi%20teh%20Indonesia%20berada%20di,138.323%20ton%20pada%20tahun%202020. (2022)
4. Ghodki, B.M. and Goswami, T.K. *Modeling of granular heat transfer in cryogenic grinding system: Black pepper seeds.*In Chemical Engineering Research and Design, **141, pp.302-316**. (2019).
5. Indrayati, F., Utami, R., & Nurhartadi, E. *Pengaruh penambahan minyak atsiri kunyit putih (Kaempferia rotunda) pada edible coating terhadap stabilitas warna dan ph fillet ikan patin yang disimpan pada suhu beku* in Jurnal Teknosains Pangan, *2*(**4**). (2013).
6. Junghare, H., Hamjade, M., Patil, C., Girase, S. B., & Lele, M. M. *A Review on cryogenic grinding* in International Journal of Current Engineering and Technology, **7, 420-423**. (2017).
7. Kim, J. M., Kang, J. Y., Park, S. K., Han, H. J., Lee, K. Y., Kim, A. N., ... & Heo, H. J. *Effect of storage temperature on the antioxidant activity and catechins stability of Matcha (Camellia sinensis)* in Food science and biotechnology, **29, 1261-1271**. (2020).
8. Leviana, W., & Paramita, V. *Pengaruh suhu terhadap kadar air dan aktivitas air dalam bahan pada kunyit (Curcuma Longa) dengan alat pengering electrical oven. METANA*, **13 (2), 37–44**. (2017).
9. Liu, H., Zeng, F., Wang, Q., Ou, S., Tan, L., & Gu, F. *The effect of cryogenic grinding and hammer milling on the flavour quality of ground pepper (Piper nigrum L.)* in Food Chemistry, **141(4), 3402-3408.** (2013).
10. Novidiyanto, N., & Sutyawan, S. *Karakteristik Kimia dan Aktifitas Antioksidan Teh Hijau Tayu dari Provinsi Bangka Belitung dan Teh Hijau Komersial* in JGK: Jurnal Gizi dan Kesehatan, *2*(**1 Juni), 74-81.** (2022).
11. Prawira-Atmaja, M. I., Maulana, H., Shabri, G. P. R., Fauziah, A., & Harianto, S. *Evaluasi Kesesuaian Mutu Produk Teh Dengan Persyaratan Standar Nasional Indonesia* in Jurnal Standardisasi, **23(1), 43-52.** (2021).
12. Shofi, V. E., & Betari, B. K. *Effect shading intensity on color, chemical composition, and sensory evaluation of green tea (Camelia sinensis var Assamica)* in Journal of the Saudi Society of Agricultural Sciences. (2023).
13. Topuz, A., Dinçer, C., Torun, M., Tontul, I., NADEEM, H. Ş., Haznedar, A., & Özdemir, F. *Physicochemical properties of Turkish green tea powder: effects of shooting period, shading, and clone* in Turkish Journal of Agriculture and Forestry, **38(2), 233-241**. (2014).

Changes of Dark Couverture Chocolate Hardness During Storage Tempered Using Automatic Tempering Machine with Tank and Tempering Temperature as Variables

Shiammita Rosmindari[1], *Arifin Dwi Saputro*[1], and *Nursigit Bintoro*[1]

[1] Departement of Agricultural and Biosystems Engineering, Faculty of Agricultural Technology, Universitas Gadjah Mada, Jalan Flora No. 1 Bulaksumur, Yogyakarta 55281, Indonesia

Abstract. Couverture chocolate is highly demanded by consumers. Dark couverture chocolate is known as chocolate with high proportion of cocoa. There are several parameters that need to be considered to ensure the quality of this chocolate. One of the important chocolate qualities is hardness. In chocolate making, which is affected by the tempering process. Generally, the tempering process is carried out manually or automatically. Manual tempering is done by hand and is difficult to control the process temperature. Therefore, an automatic tempering machine was chosen in this study by controlling the tank and tempering temperatures. The purpose of the research was to optimize the combined effect between tank temperatures and tempering temperatures of the automatic tempering machine on the chocolate hardness parameter. Different ranges of the tank and tempering temperatures were applied to the chocolate mass processed in the machine. Chocolate hardness during storage was in the range 12.27 to 20.19 N/mm^2 in 45°C tank and 32.5°C tempering temperature. The optimum of the tank and tempering temperatures were 45°C-32.5°C (A), 48°C-32.5°C (B), and 50°C-31.5°C (C) which resulted in different k values and glossy appearances. The *k* values for A, B, and C were -0.00195; -0.0024; and -0.0031, respectively. While the determination coefficients for A, B, and C were 0.8970; 0.8887; and 0.9013, respectively.

Keywords: Dark chocolate couverture, tempering, hardness, Newton's law of cooling.

1 Introduction

Chocolate is a range of foods derived from cocoa (*Theobroma cacao*) seeds and it is becoming one of the most popular sweet products worldwide [1–3]. Chocolate is a complex mixture of cocoa and finely powdered sugar in a continuous fat phase, mainly cocoa butter, to produce a solid confectionery [4–6]. Its consumption rate continues to grow by year [7]. The consumer importance of chocolate is very high due to its taste and health benefits, especially as a source of antioxidants [8–10]. Bean-to-bar process sequences consist of mixing, refining, conching, tempering, moulding, and demoulding are widely known to produce high-quality chocolate products [6, 11].

Chocolate undergoes a tempering process to obtain a good texture and appearance [12]. The fat ingredient, named cocoa butter is one of the main ingredients of chocolate which has a unique composition and character [13]. Cocoa butter crystallizes in six polymorphic forms known as γ, α, β2', β1', β2, and β1. Form β2 also known βV is the most desirable polymorphic form of the well-tempered chocolate products. Thus, the products have a good contraction, shiny, even-coloured, snap, smooth taste [4, 11, 14–16]. While bad-tempered chocolate gives a chewy, chalky, grainy, or whitish colour, known as the fat or water-bloom phenomenon during storage [17–19].

Conventionally, tempering was introduced manually on a marble table. It requires skilled and experienced manufacturers. Manual tempering constraint is insufficient crystal formation because of the homogeneity of the temperature not reached in the sample. This could affect hardness. Hardness is the key attribute of the final quality of chocolate, it is also an indicator of proper temperature control and a stable fat crystal network [11, 20]. The tempering machine is more often used in industrial activities and has some temperature controls to maintain the process. The untempering and under-tempering regimes exhibit different crystallization behaviours, resulting unstable fat crystal nucleation, therefore the tempering process is necessary [4, 13, 21].

Several studies have reported the importance of tempering, temperature control, and storage to chocolate hardness parameter. The increase in bar hardness has been evaluated during storage and positively accepted by the panelists through sensory evaluation [22]. In addition, the storage or maturation process improves the chocolate crystal formation to be more stable [23].

The objective of the study is to investigate the effect of temperature combinations between the tank and tempering temperature using the automatic tempering machine on the hardness of the chocolate. Temperature control is important as it affects the successful tempering process [24]. Further analysis examined in this research is obtaining constant (k) value as the rate of hardness changes during storage and determination coefficient (R^2) using kinetics order following Newton's law of cooling (NLC) model during chocolate storage.

* Corresponding author: arifin_saputro@ugm.ac.id

https://doi.org/10.1051/bioconf/20238004002

2 Materials and Methods

2.1 Research Location

Research was conducted at the Laboratory of Postharvest and Food Engineering, Faculty of Agricultural Technology, Universitas Gadjah Mada, Indonesia.

2.2 Materials

The untempered dark chocolate was used for the materials for the research. It was purchased from a local market in Yogyakarta, *nDalem Cokelat*. The samples contained 72% cocoa according to the following formulation: 41% of cocoa butter, 31% of cacao solid, and 28% of sugar content. The tempering process was carried out using the automatic tempering machine, Pomati T8 (Pomati Group S.r.l, Fidenza, Italy).

2.3 Tempering Procedures

The solid chocolate was melted in the melting cabinet at a temperature of 55°C. After the melting process, the molten chocolate was placed in the tempering machine tank, letting it flow through the pipe. Some temperatures were set for the tank and the tempering in the machine. The chocolate mass was then moulded, vibrated on the vibrating table to eliminate bubbles, and put in the chiller cabinet at 10°C to form the chocolate bar. Then,demould.and stored at a controlled temperature of 18-20°C.

2.4 Research Design

The observation was designed in some temperature settings. The factorial research design, namely 6 x 4 x 7 (tank temperatures, tempering temperatures, storage time) was used. In this research, six different temperatures for the tank temperature (T), namely 45°C, 46°C, 47°C, 48°C, 49°C, and 50°C were used. Meanwhile, the tempering temperature (P) was set at 31°C, 31.5°C, 32°C, and 32.5°C. In continuous tempering, molten chocolate is held at 45°C while in a common tempering process, temperature of 50°C is needed by chocolate to completely melt [11, 25]. Variation of tempering temperature were used between 31-32.5°C considering that those temperature are in the range of 29-34°C which could generate βV crystals [26,27]. Temperature variation from 0.5 to 1°C affects chocolate properties during tempering [27]. The chocolate sample in the experiments can flow through the processing pipe when the tempering temperature is set from 31°C and another consideration is the engine overheats at 33°C. Chocolate hardness observations were made on the first day and every 5 days until the 30th day of storage (1st, 5th, 10th, 15th, 20th, 25th, and 30th day of the maturation process).

2.5 Analytical Method

2.5.1 Hardness Measurement

Texture analyzer Brookfield Engineering Labs, Inc. was used to measure the hardness of chocolate bars and was carried out at a room temperature of 20°C. A needle probe was applied (TA39, diameter 2 mm) to the instrument and samples were penetrated to a depth of 3 mm with a constant speed of 0.5 mm/s [28]. All measurements were done in triplicate. The chocolate samples were stored for 30 days. Hardness value was calculated using Equation (1),

$$\sigma = \frac{load \times a}{\pi \times r^2} \quad (1)$$

Where σ defines the hardness (N/mm^2), load is the maximum load (kg), a is gravitational acceleration (m/s^2), r is needle probe radius (mm).

2.5.2 Data Analysis

Data Analysis was performed by IBM SPSS version 26.0 for Analysis of Variance (ANOVA). Posthoc Tukey test and Levene's test were performed, then followed by one way ANOVA test. Prior to the two-factor analysis of variance using Repeated Measures, the normality of the data was tested. The principal component analysis (PCA) was used to visualize the chocolate samples and hardness parameters. The applicability of NLC was proposed as a method to model the order kinetics of hardness during storage. The value of constant (k) and prediction model were obtained from the equation (2) and (3),

$$\ln \frac{\sigma_e - \sigma_t}{\sigma_e - \sigma_0} = k.t \quad (2)$$

$$\sigma_t = \sigma_e - (\sigma_e - \sigma_0)e^{kt} \quad (3)$$

Where $\boldsymbol{\sigma_e}$ defines the final hardness, σ_t defines the hardness at certain time, $\boldsymbol{\sigma_o}$ defines hardness before storing (N/mm^2), $\boldsymbol{k}$ is constant, and $\boldsymbol{t}$ is the storage time (hours).

3 Results and Discussions

3.1 The relationship between the temperature combination to the hardness

Two factor repeated measures on tank and tempering temperatures give a significant difference ($p<0.05$). Hardness was affected by storage time, tank temperature, temperature treatments, as well as their interaction, as described in Table 1. Storage time influenced the hardness as Saputro [12, 27] mentioned that chocolate storage or maturation would create more stable βV crystals. Tank temperature was associated with the initial temperature of the molten chocolate and the tempering temperature was the final temperature of tempering process of the chocolate when moulding. The combination of tank and tempering temperatures showed a significant difference to the hardness parameter.

Table 1. Relationship among research variables to the hardness of chocolate

Treatments Variables	Hardness
Storage Time (A)	*
Tank Temperature (B)	*
Tempering Temperature (C)	*
Interaction between A x B	*
Interaction between A x C	*
Interaction between B x C	*
Interaction between A x B x C	*

Significant difference at $p<0.05$

Kurniasari et.al. [20] mentioned that storage time affected the hardness. The research also showed a tempering method by using the oven and stirred manually. The manual tempering did not show a clear trend of hardness because the expected heating and cooling temperature could not be reached by molten chocolate and resulted in the unsuccessful formation of βV crystal. Therefore, the use of an automatic tempering machine was expected to make the temperature distributed evenly in the molten chocolate and could stimulate the growth of βV crystal.

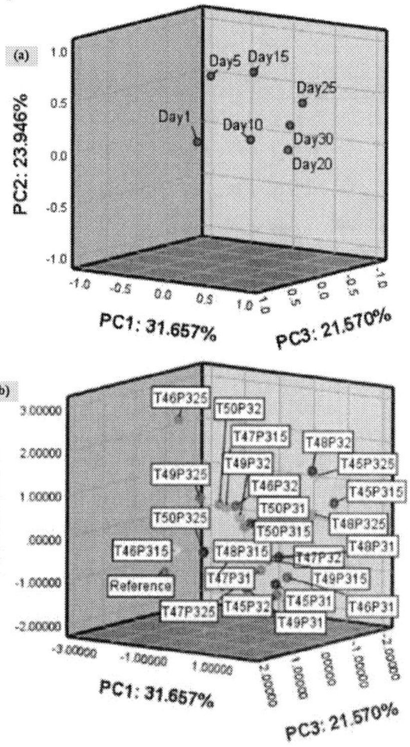

Fig. 1. Principal Component Analysis (PCA) and its associated (a) loading plot also (b) score plot. PCA shows the distribution of different combinations between tank and temperature treatments.

Degree of tempering (tempering temperatures) affects the hardness of the chocolate. A well-tempered chocolate tends to have higher hardness than poor tempered or untempered chocolate [12, 21, 24-25]. The tempering process is used to obtain a desirable crystalline state of the cocoa butter resulting a glossy appearance, good snap, contraction, and long shelf life [31-32]. Over-tempered caused increasing hardness, stickiness, and reduced gloss or darkened surface. Meanwhile, under-tempered led to lower hardness, fat bloom, and surface defects [28-29].

Tempering has four important steps on the temperature control which consists of complete melting at 50°C, cooling to the point of crystallization at 32°C, crystallization at 27°C, and melting out any unstable crystal at 29-31°C. Temperature is important for continuous tempering process design. Optimum tempering temperature is in the range of 29-34°C for successfully generating βV crystals [25, 30, 35].

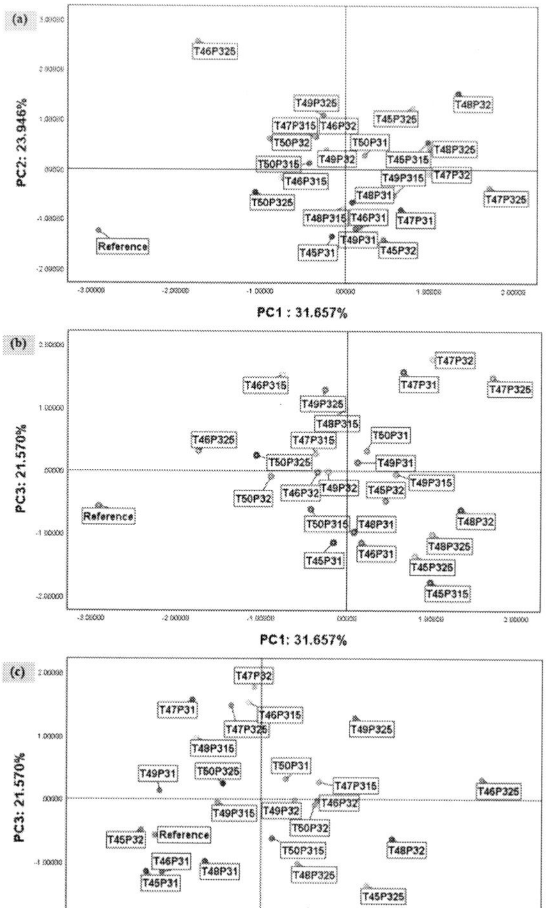

Fig. 2. PCA score plot in the two-dimensional view from the different tank and tempering temperature distribution (a) PC1 vs PC2, (b) PC1 vs PC3, (c) PC2 vs PC3.

PCA of the tank temperature treatment on each different day of measurement was done in a three-dimensional view. The PCA data represented 77.174% of the total variance in the first three factors (Figure 1), namely PC1: 31.657%, PC2: 21.570%; and PC3: 23.946%. The PCA loading plot (Figure 1a) shows that generally the storage time or days of measurements were located at the same cluster along PC1 and PC2. This shows that hardness increase over storage time [22]. While storage time namely day 1 was in different clusters along PC2 and PC3. This shows that there were differences in chocolate hardness amongst clusters on the first day of measurements.

The PCA score plot of the first three principal components shows a clear distinction between samples (Figure 1b). The reference or the untempered sample was located along negative PC2 and PC3 values. This was associated with the untempered chocolate character which had a lower hardness. Figure 2 shows a two-dimensional view of each treatment to the sample hardness. The reference sample was located at the very end of the negative PC1 and PC2, negative PC1 and PC3, also negative PC2 and PC3. It showed the reference sample had the lowest hardness value.

Some samples named T45P31, T45P32, T46P31, T46P315, T46P32, T48P31, T48P315, T49P315, T49P32, T50P315, T50P32, and T50P325 were in the same quadrant as the reference sample having lower hardness value. Different hardness characteristics showed by another sample by tank and tempering treatments. Larger hardness shown along positive PC1, PC2, and PC3 by T45P315, T45P325, T46P325, T47P31, T47P315, T47P32, T47P325, T48P32, T48P325, T49P31, T49P325, T50P31. Some tank and tempering combinations give a greater hardness supported with glossy appearance.

3.2 Hardness

It can be observed in Figure 3 that the hardness of chocolate varied but tended to increase during storage. The line showed that the chocolate was affected by the combination of tank and tempering temperature. Figure 3 consists of 6 groups (six different tank temperatures) describing hardness changes in different tempering treatments over storage.

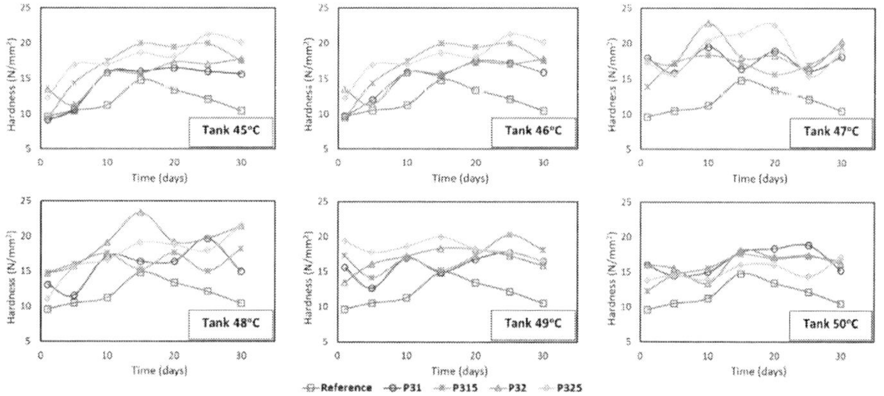

Fig. 3. The hardness value of chocolate was measured in the different tank and tempering temperatures during 30 days of storage.

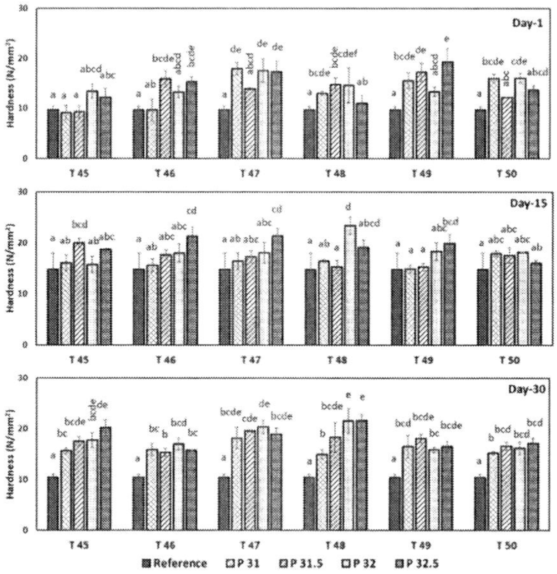

Fig. 4. Hardness profile of dark chocolate. Different lowercase denotes a significant difference ($p<0.05$)

The grey line (Figure 3) describes the reference sample. It positioned at the lowest compared to other treatments, the hardness ranged between 9.65-14.82 N/mm^2 for the reference sample. These findings were in agreement with previous results reported by Afoakwa [24] who found that the hardness of the untempered chocolate sample is low (soft) and difficult to demould (bad contraction). The decrease of hardness in some temperature combinations in this research also associate with the experiment on bar chocolate, hardness decrease in the first four weeks of storage [36]. This is also in line with the study that some treatments experienced a decrease in hardness during 30 days of storage.

Chocolate was stored between 18-20°C inside the chiller to maintain its quality attributes. According to Machálková [37], the higher temperature for example 30°C, was not suitable for short-term storage. A storage temperature of 20°C generated a good colour and texture, but not in terms of sensory evaluation. Tempered chocolates are resistant to fat bloom. Therefore, to preserve its original attributes, it should be kept at 12°C.

In this study, storage time and tank-tempering temperature treatment affected the chocolate hardness. Figure 4 explained the hardness profile based on storage time. Significant difference calculated at $p<0.05$ using one-way ANOVA. The reference sample showed a different superscript on each day of measurement which associated to there is a difference between the hardness of the reference and tempered sample.

3.3 Chocolate Appearance

The appearance of the chocolate is also important to determine the best combination of tank and tempering temperature during 30 days of chocolate storage. Chocolate with fat bloom reduced the quality attributes and appeared as a loss of initial glossiness. In dark chocolate, phase separation, and polymorphic crystal transformation along with storage time is the most common explanation to describe fat bloom [14, 19, 25].

It was also noticeable, as seen in Table 2, that the reference sample on day-1 gives a smooth and fine appearance (1a). Then, it showed a bloomed formation after 30 days of storage, which means that the quality decreased. The same blooming effects also appeared on T48P31 and T48P32 sample treatments.

Table 2. Chocolate image during storage at 48°C tank temperature

Time (day)	Reference (a)	T48P31 (b)	T48P31.5 (c)	T48P32 (d)	T48P32.5 (e)
1					
5					
10					
15					
20					
25					
30					

Some combinations of tank and tempering temperatures were also defined as a good surface quality of the chocolate. Sample T48P315 and T48P325 were examples of glossy appearance. In this present study, the hardness value combined with the visual appearance could determine a good combination of tank and tempering temperature.

3.4 Hardness Change Kinetics

The adaptation of Newton's law of cooling (NLC) along storage time could be proposed as a real-time hardness prediction model. Figure 5 showed that the hardness changes over time could be predicted smoothly using this adaptation. The constant value (k) of hardness change is obtained from the exponential graph. The highest k value was -0.0031 supported by the glossy appearance of the combination 50°C tank and 31.5°C tempering temperature. The coefficient of determination (R^2) is 0.9013 using that temperature combination. Negative signs showed a declining hardness rate during storage (Table 3).

Table 3. Constant (k) value and R^2 of the tank and tempering temperature treatment

Treatment	*k* value	R^2	Prediction Equation	Appearance
T45P31	-0.0035	0.8297	$\sigma_t = \sigma_e - (\sigma_e - \sigma_0)e^{-0.0035t}$	Blooming
T45P315	-0.0038	0.8800	$\sigma_t = \sigma_e - (\sigma_e - \sigma_0)e^{-0.0038t}$	Blooming
T45P32	-0.0017	0.6974	$\sigma_t = \sigma_e - (\sigma_e - \sigma_0)e^{-0.0017t}$	Glossy
T45P325	-0.00195	0.8970	$\sigma_t = \sigma_e - (\sigma_e - \sigma_0)e^{-0.00195t}$	Glossy
T46P31	-0.0029	0.8596	$\sigma_t = \sigma_e - (\sigma_e - \sigma_0)e^{-0.0029t}$	Blooming
T46P315	-0.00008	0.1358	$\sigma_t = \sigma_e - (\sigma_e - \sigma_0)e^{-0.00008t}$	Blooming
T46P32	-0.0016	0.2738	$\sigma_t = \sigma_e - (\sigma_e - \sigma_0)e^{-0.0016t}$	Glossy
T46P325	-0.0004	0.0054	$\sigma_t = \sigma_e - (\sigma_e - \sigma_0)e^{-0.0004t}$	Blooming
T47P31	0.00026	0.0005	$\sigma_t = \sigma_e - (\sigma_e - \sigma_0)e^{0.00026t}$	Blooming
T47P315	-0.00173	0.3731	$\sigma_t = \sigma_e - (\sigma_e - \sigma_0)e^{-0.00173t}$	Glossy
T47P32	-0.00033	0.0013	$\sigma_t = \sigma_e - (\sigma_e - \sigma_0)e^{-0.00033t}$	Blooming
T47P325	-0.00057	0.0364	$\sigma_t = \sigma_e - (\sigma_e - \sigma_0)e^{-0.00057t}$	Blooming
T48P31	-0.00093	0.4206	$\sigma_t = \sigma_e - (\sigma_e - \sigma_0)e^{-0.00093t}$	Blooming
T48P315	-0.00156	0.1936	$\sigma_t = \sigma_e - (\sigma_e - \sigma_0)e^{-0.00156t}$	Glossy
T48P32	-0.00154	0.6188	$\sigma_t = \sigma_e - (\sigma_e - \sigma_0)e^{-0.00154t}$	Blooming
T48P325	-0.0024	0.8887	$\sigma_t = \sigma_e - (\sigma_e - \sigma_0)e^{-0.0024t}$	Glossy
T49P31	-0.00043	0.3134	$\sigma_t = \sigma_e - (\sigma_e - \sigma_0)e^{-0.00043t}$	Glossy
T49P315	0.0001	0.3421	$\sigma_t = \sigma_e - (\sigma_e - \sigma_0)e^{0.0001t}$	Glossy
T49P32	-0.00268	0.5161	$\sigma_t = \sigma_e - (\sigma_e - \sigma_0)e^{-0.00268t}$	Glossy
T49P325	0.00211	0.4286	$\sigma_t = \sigma_e - (\sigma_e - \sigma_0)e^{0.00211t}$	Blooming
T50P31	-0.00126	0.2304	$\sigma_t = \sigma_e - (\sigma_e - \sigma_0)e^{-0.00126t}$	Glossy
T50P315	-0.00311	0.9013	$\sigma_t = \sigma_e - (\sigma_e - \sigma_0)e^{-0.00311t}$	Glossy
T50P32	-0.00056	0.1811	$\sigma_t = \sigma_e - (\sigma_e - \sigma_0)e^{-0.00056t}$	Blooming
T50P325	-0.00122	0.4773	$\sigma_t = \sigma_e - (\sigma_e - \sigma_0)e^{-0.00122t}$	Blooming

Therefore, the three highest R^2, and appearance by investigating the blooming effect were selected. Table 3 described treatment T45P32, T48P325, and T50P315

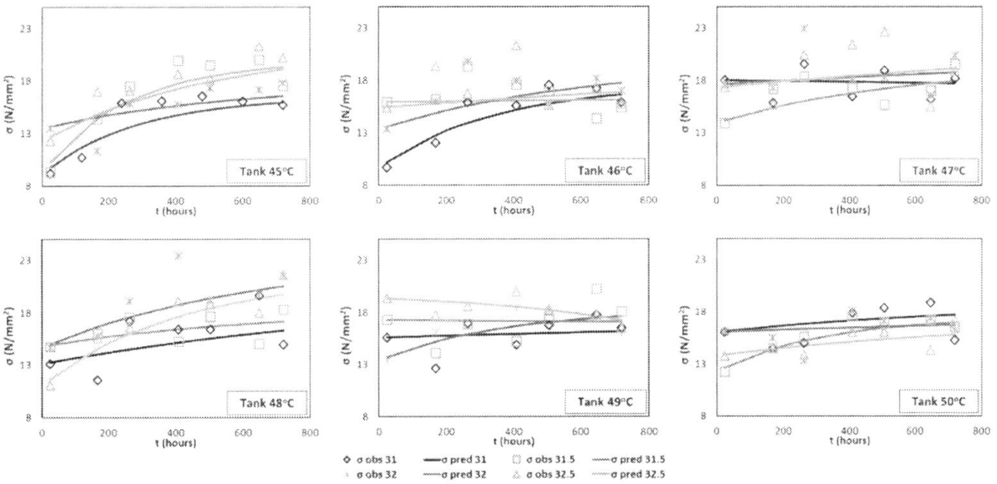

Fig. 5. Hardness changes kinetics at a corresponding storage time using an adaptation of NLC method.

were considered as three temperature combinations resulting in high k values and higher R^2 with anti-blooming effect at appearance. The use of NLC is often introduced to temperature prediction in such conditions for example, cooling or heating. Konovalenko [38] used a method for the adaptation of NLC to the prediction of an object's (cargo's) temperature prediction in a cold supply chain under the condition of ambient temperature instability and the method generally creates better performance.

This NLC approach can be used to predict the hardness change during storage. Based on the NLC kinetics, some tank and tempering temperature combination provided a good fitting of the prediction line over the observation dots. However, those fittings sometimes were not in line with the appearance. T45P31, T45P315, T46P31 had high k values of -0.0035; -0.0038; -0.00291 respectively and R^2 of 0.8297; 0.8800; 0.8596 respectively. Those values were not considered since the appearance of chocolate showed a defect or blooming surface. This could be associated with over-tempering and under-tempering. Afoakwa [30] mentioned that optimal tempering of dark chocolate was important to desired texture and appearance.

4 Conclusion

Storage time, tank temperature, tempering temperature, and their interactions have influenced the hardness of dark chocolate. The hardness increased over storage time in some combination of tank and temperature treatments. The combination of tank and tempering temperature delivered a glossy appearance along with a high value of coefficient of determination (R^2). The treatment T45P325, T48P325, and T50P315 were chosen since it resulted in high R^2 values for the model use, namely 0.8970, 0.8887, 0.9013 respectively, and showed the glossiest appearance.

Acknowledgement. This work was financed by Rekoknisi Tugas Akhir (RTA) 2023 Grant (Universitas Gadjah Mada) and supported by the Department of Agricultural and Biosystems Engineering, Faculty of Agricultural Technology, Universitas Gadjah Mada.

References

1. Y. J. Son, S. Choi, K. Yoo, K. Lee, S. Lee, I. Hwang, S. Kim, Anti-blooming effect of maltitol and tagatose as sugar substitutes for chocolate making, Lwt **88**, 87–94 (2018).
2. V. Vivek, S. Bermúdez, C. Larrea, Global Market Report: Cocoa, Exch. Organ. Behav. Teach. J. 1–12 (2019).
3. V. Barišić, *et al*, Physical properties of chocolates enriched with untreated cocoa bean shells and cocoa bean shells treated with high-voltage electrical discharge Sustain, **13**, 1–14 (2021).
4. E.O. Afoakwa, A. Paterson, M. Fowler, J. Vieira, Particle size distribution and compositional effects on textural properties and appearance of dark chocolates, J. Food Eng. **87**, 181–190 (2008).
5. E. O. Afoakwa, A. Paterson, M. Fowler, A. Ryan, Matrix effects on flavour volatiles release in dark chocolates varying in particle size distribution and fat content using GC-mass spectrometry and GC-olfactometry, Food Chem. **113**, 208–215 (2009).
6. F. Shafi, M. Reshi, I. Bashir, I. Chocolate Processing, Ijabr **8**, 408–419 (2018).
7. T. T. Hapsari, A. F. Yuniasih, The determinant factors of Indonesian competitiveness of cocoa exports to Germany, J. Ekon. Pembang. **18**, 75–84 (2020).
8. N. Shadwell, F. Villalobos, M. Kern, M. Y. Hong, Blooming reduces the antioxidant capacity of dark chocolate in rats without lowering its capacity to improve lipid profiles, Nutr. Res. **33**,

414–421 (2013).
9. V. Todorovic, I. R. Redovnikovic, Z. Todorovic, G. Jankovic, M. Dodeska, S. Sobajic. Polyphenols, methylxanthines, and antioxidant capacity of chocolates produced in Serbia, J. Food Compos. Anal. **41**, 137–143 (2015).
10. B. Urbańska, J. Kowalska, Comparison of the total polyphenol content and antioxidant activity of chocolate obtained from roasted and unroasted cocoa beans from different regions of the world. Antioxidants **8**, (2019).
11. S. T. Beckett, Industrial Chocolate Manufacture and Use Fourth Edition (2009).
12. A. D. Saputro, D. Van de Walle, S. Kadivar, M. D. Bin Sintang, P. Van der Meeren, K. Dewettinck, Investigating the rheological, microstructural and textural properties of chocolates sweetened with palm sap-based sugar by partial replacement, Eur. Food Res. Technol. **243**, 1729–1738 (2017).
13. E. M. Castro-Alayo, et.al, Effect of tempering and cocoa butter equivalents on crystallization kinetics, polymorphism, melting, and physical properties of dark chocolates, Lwt **173**, (2023).
14. C. Delbaere, D. Van de Walle, F. Depypere, X. Gellynck, K. Dewettinck, Relationship between chocolate microstructure, oil migration, and fat bloom in filled chocolates. European Journal of Lipid Science and Technology **118** 1800–1826 (2016).
15. P. Altimiras, L. Pyle, P. Bouchon, Structure-fat migration relationships during storage of cocoa butter model bars: Bloom development and possible mechanisms, J. Food Eng. **80**, 600–610 (2007).
16. E. M. Castro-Alayo, L. Torrejón-Valqui, I. S. Cayo-Colca, F. P. Cárdenas-Toro, Evaluation of the miscibility of novel cocoa butter equivalents by Raman mapping and multivariate curve resolution–alternating least squares, Foods **10**, (2021).
17. F. Debaste, Y. Kegelaers, S. Liégeois, H. Ben Amor, V. Halloin, Contribution to the modelling of chocolate tempering process, J. Food Eng. **88**, 568–575 (2008).
18. F. Peyronel, D. A. Pink, Using USAXS to predict the under-tempered chocolate microstructure, Food Res. Int. **143**, (2021).
19. H. Zhao, B. J. James, Fat bloom formation on model chocolate stored under steady and cycling temperatures, J. Food Eng. **249**, 9–14 (2019).
20. J. Kurniasari, A. Cahyani, R. Nafingah, S. Rahayoe, E. Harmayani, A. D. Saputro, The use of oven as a device to temper molten Dark Chocolate, IOP Conf. Ser. Earth Environ. Sci. **355**, (2019).
21. N. De Clercq, S. Kadivar, D. Van de Walle, S. De Pelsmaeker, X. Ghellynck, K. Dewettinck, Functionality of cocoa butter equivalents in chocolate products, Eur. Food Res. Technol. **243**, 309–321 (2017).
22. L. Hřivna, L. Machálková, I. Burešová, Š. Nedomová, T. Gregor, Texture, color, and sensory changes occurring in chocolate bars with filling during storage, Food Sci. Nutr. **9**, 4863–4873 (2021).
23. A. D. Saputro, M. A. Nur Fadilah, S. K. Bangun, S. Rahayoe, J. N. Wahyu Karyadi, A. D. Setiyowati, Physical Characteristic of Heat Resistant Chocolate Formulated with Konjac Glucomannan and Xanthan Gum-Based Hydrogel at Various Fat Content during Period of Crystal Growth (Maturation), J. Tek. Pertan. Lampung **11**, 658 (2022).
24. E. O. Afoakwa, A. Paterson, M. Fowler, Factors influencing rheological and textural qualities in chocolate - a review, Trends Food Sci. Technol. **18**, 290–298 (2007).
25. E. O. Afoakwa, Chocolate Science and Technology. Chocolate Science and Technology (2010).
26. S. J. Calva-Estrada, M. Utrilla-Vázquez, A. Vallejo-Cardona, D. B. Roblero-Pérez, E. Lugo-Cervantes, Thermal properties and volatile compounds profile of commercial dark-chocolates from different genotypes of cocoa beans (Theobroma cacao L.) from Latin America. Food Res. Int. **136**, (2020).
27. E. J. Windhab, Tempering. Industrial Chocolate Manufacture and Use: Fourth Edition (2009).
28. S. B. Gregersen, R. L. Miller, M. Hammershøj, M. D. Andersen, L. Wiking, Texture and microstructure of cocoa butter replacers: Influence of composition and cooling rate, Food Struct. **4**, 2–15 (2015).
29. A. D. Saputro, D. Van de Walle, R. P. Aidoo, M. A. Mensah, C. Delbaere, N. De Clerq, J. Van Durme, K. Dewettinck, Quality attributes of dark chocolates formulated with palm sap-based sugar as nutritious and natural alternative sweetener, Eur. Food Res. Technol. **243**, 177–191 (2017).
30. E. O. Afoakwa, A. Paterson, M. Fowler, J. Vieira, Modelling tempering behaviour of dark chocolates from varying particle size distribution and fat content using response surface methodology, Innov. Food Sci. Emerg. Technol. **9**, 527–533 (2008).
31. E. Ostrowska-Ligeza, et.al, Characterization of Thermal Properties of Ruby Chocolate Using DSC, PDSC and TGA Methods, Appl. Sci. **13**, 13 (2023).
32. H. Chen, P. Zhou, C. Song, G. Jin, L. Wei, An approach to manufacturing heat-stable and bloom-resistant chocolate by the combination of oleogel and sweeteners, J. Food Eng. **330** (2022).
33. E. O. Afoakwa, A. Paterson, M. Fowler, J. Vieira, Effects of tempering and fat crystallisation behaviour on microstructure, mechanical properties and appearance in dark chocolate systems, J. Food Eng. **89**, 128–136 (2008).
34. B. S. Abhishek B. Shah, P. J. Gwyn, T. Vasiljevic, Sucrose-free chocolate sweetened with Stevia rebaudiana extract and containing different bulking agents – effects on physicochemical and

sensory properties, Int. J. Food Sci. Technol. **45**, 1426–1435 (2010).
35. G. Talbot, Chocolate Temper, Ind. Choc. Manuf. Use Fourth Ed. 261–275 (2009).
36. B. Böhme, A. Bickhardt, H. Rohm, Pre-crystallization of nougat by seeding with cocoa butter crystals enhances the bloom stability of nougat pralines, Foods **10**, (2021).
37. L. Machálková, L. Hřivna, Š. Nedomová, M. Jůzl, The effect of storage temperature on the quality and formation of blooming defects in chocolate confectionery, Potravinarstvo **9**, 39–47 (2015).
38. Konovalenko, I., Ludwig, A. & Leopold, H. Real-time temperature prediction in a cold supply chain based on Newton's law of cooling. *Decis. Support Syst.* **141**, (2021).

Extraction and Isolation of Cricket Protein Isolate with Ammonium Sulfate Addition Method and Its Effect on The Functional Properties of The Proteins

Edward[1], *Thanakorn* Wongprasert[1], *Thasorn* Bunyakanchana[1], *Wasin* Pattaraprachyakul[1], *Panattida* Siripitakpong[1], and *Inthawoot* Suppavorasatit[1,2*]

[1] Department of Food Technology, Faculty of Science, Chulalongkorn University, Bangkok, 10330, Thailand
[2] Flavor Science and Functional Ingredients Research Unit, Chulalongkorn University, Bangkok, 10330, Thailand

Abstract. Cricket insect contains a high quality of protein. To be able to use the proteins in food industry, extraction and isolation steps are necessary to elevate the protein content. The objective of this study was to extract and isolate protein of cricket insect and to assess its functional properties. The extraction and isolation steps were carried out by using alkaline extraction-acid precipitation (AEAP) with varying concentrations of ammonium sulfate (0, 20,40, and 60% w/v). It was found that extraction method with 60% ammonium sulphate inclusion showed the highest yield and obtained maximum protein content (92.41%), which could be characterized as cricket protein isolate (CPI). This extraction and inclusion of ammonium sulphate affected physicochemical properties, including water holding capacity, oil holding capacity, emulsifying properties, and foaming properties of CPI. In conclusion, the extraction, isolation, and addition of ammonium sulfate could be used for isolating the CPI containing high protein content and can be further used in food manufactures as an alternative protein in the future.

1 Introduction

The increasing number of world population nowadays has been impacted to many fields of human life, including for demand of food in the future (Santiago et al. 2021). Specifically, demand of protein intake becomes one of most important issues to find alternative protein sources since this compound is needed to be consumed to fulfil human nutritional intake (Friedman 1996). In order to meet this high demand of protein, there have been figured many attempts to find another protein source besides the conventional livestock such as chickens or cows. Insects has become one of most promising alternative protein sources due to its protein content value, which is comparable to traditional livestock (Rumpold and Schlüter 2015). Moreover, the cultivation of insects is more environmental friendly since it is produced fewer gas emissions, more efficient in feed conversion, and lesser water and soil usage (Lucas-González et al. 2019).

Cricket insect becomes one of famous edible insects which has great potency to become new alternative of protein source. This insect has been known for its high quality of protein content (Ndiritu et al. 2017). In terms of nutritional value, the protein content of cricket insect consists of essential amino acids and non-essential amino acids which the amount is adequate to meet the standard for requirement of amino acid for adults and children which states by World Health Organization (WHO) and Food and Agriculture Organization (FAO) (Oibiokpa et al. 2018). Moreover, the proteins with high protein content can be utilized by using its functional properties, such as water holding capacity, oil holding capacity, emulsifying properties, and foaming properties. However, the utilization of the proteins still has some issues regarding to the negative perspective of consumer about the insect itself (de Castro et al. 2018). Furthermore, the off-flavor of cricket that associated with green, earthy, and sweat also becomes another issues which can decrease the acceptance of the cricket insect (Smarzyński et al. 2021). Therefore, processing of cricket insect into another form such as powder form is strongly recommended in order to increase the consumer acceptance of cricket. Cricket powder form has more advantages because it can add into the products easier than other forms (Laroche et al. 2019). Cricket has many components inside not only protein, but also contains carbohydrates, fat, or other vitamins and minerals. Therefore, The defatting step could be used as pre-treatment to reduce the fat content and increasing the extraction of protein content. (Kurdi et al. 2021).

Meanwhile, alkaline extraction-acid precipitation is one of common extraction methods which usually used to extract the protein content. The principle of this method is based on the high solubility of protein on alkaline environment and precipitation of protein molecules by brought it onto its isoelectric point (Jiang et al. 2021). Furthermore, inclusion of ammonium

*Corresponding author: Inthawoot.S@chula.ac.th

sulphate will aid the extraction and isolation of protein content through salting-in/salting-out mechanisms. This existence of salt ion also will affect the functional properties of the protein. Up to now, there are just few publications that conducted research about the cricket protein isolate and its functional properties. There is initiative to contribute research about utilization of cricket insect. Therefore, the aim of this study was to extract and isolate the protein content of cricket insect and to assess its functional properties.

2 Materials and Methods

2.1 Materials

Cricket powder was obtained from Protanica Co Ltd. (Thailand). Sodium hydroxide (NaOH) were purchased from Sigma Aldrich (St Louis, MO, USA). Sodium dihydrogen orthophosphate dihydrate 99% and sodium phosphate dibasic dihydrate 99.5% were obtained from Loba Chemie Pvt. Ltd (Mumbai, India). Ammonium sulfate and hydrochloric acid (HCl) were purchased from Qrëc (Rawang, Malaysia). Dialysis tubing, sodium dodecyl sulfate, and boric acid were purchased from Thermo Fisher Scientific (Waltham, MA, USA). n-Hexane were purchased from Macron Fine Chemical (CV, Pennsylvania, USA)

2.2 Sample preparation

Cricket powder was defatted by using Soxhlet extraction according to Laroche et al. (2019). The extraction and isolation of cricket protein isolate (CPI) were performed by using method of Jiang et al. (2021) with slight modification. Five grams of defatted cricket powder was dispersed in 120 mL of 1.5% NaOH (w/v) and constantly shaken in a water bath (1 h, 80 °C). Next, the sample was cooled to room temperature and centrifuged at 6000 × g for 10 min. Next, the pH of the both supernatant and precipitate would be adjusted to pH 4.5 by using 1 M HCl and 50 mL ammonium sulphate with various concentrations (0%, 20%, 40% and 60%) were added. After that, the mixture will be kept overnight in refrigerator (4°C). The sample was then dialyzed overnight and freeze-dried, and the CPI powder was vacuum sealed in aluminium foil laminated bag and stored at -18°C for further analyses.

2.3 Protein content analysis

Determination of protein content was done by Kjeldahl method (928.08) from AOAC (2002). The conversion factor for protein content determination was 6.25.

2.4 Foaming Properties

Foaming properties, foaming capacity (FC) and foaming stability (FS) were determined by using method of Udomsil et al. (2019) with slight modification. First, 0.5 g of sample was dispersed in 20 mL of distilled water and equilibrated at room temperature (10 min). Next, the protein solution was aerated by homogenizer for 5 min (22,000 rpm). FC was determined by calculation as percentage of increased volume of protein solution after aeration, while FS was expressed by measured the remaining foam (%) after aeration after 30, 60, and 90 min. The calculation of FC and FS were done by using formula from Kunarayakul et al. (2018) with slight modification.

2.5 Water holding capacity and oil holding capacity

The water holding capacity (WHC) and oil holding capacity (OHC) were determined using methods from Zielińska et al. (2018) with some modification. WHC analysis was done by dispersing 0.5 g of sample in 20 mL of distilled water and shook at 250 rpm for 45 min. The dispersion was centrifuged for 15 min (8000 × g) and the WHC was calculated by % weight of the precipitate that obtained after centrifugation step. On the other hands, the OHC was determined by dispersed 0.5 g of sample in 10 mL of vegetable oil and vortexed for 30 s. Next, the dispersion was centrifuged (8000 × g, for 15 min) and the precipitate would be weighed to determine the OHC value of sample.

2.6 Emulsifying Properties

Determination of emulsifying properties, emulsifying activity index (EAI) and emulsifying stability (ES) were determined. These properties were conducted by using methods from Chen et al. (2018) and Meenmanee et al. (2022) with some modification. 0.5 g of sample was dissolved in 10 mM phosphate buffer (pH 7) to a final concentration of 0.1% (w/v). Next the protein suspension (9 mL) was homogenized with 3 mL of vegetable oil. Then, 50 μL aliquot of prepared emulsion system was removed from the bottom of the test tube and diluted in 5 mL 0.1% sodium dodecyl sulfate (SDS) solution and vortexed for five 5 s. The Emulsifying properties would be determined by measuring the absorbance of the dispersion at 500 nm directly and after 10 min. Calculation of EAI and ES were done following methods of Yin et al. (2008).

2.7 Statistical Analysis

Completely randomized design (CRD) was used in all experiments. Data obtained was analyzed by using one-way analysis of variance. Duncan multiple range test (DMRT) was employed in order to determine the significant difference between samples. The analysis in this study was done in triplicates.

3 Results and discussion

3.1 Protein content

Protein content of defatted cricket powder and cricket protein with different extraction methods, consisted of alkaline extraction-acid precipitation and ammonium sulfate treated at various concentrations (20%, 40%, 60%) were shown in **Table 1**. It was found that the defatting, extraction, and isolation steps could increase total protein content of the samples. Defatting process is one of the pre-treatments before protein extraction, which is employed in order to reduce the lipid content on the food matrix. Cricket insect had been known as an alternative protein source with high lipid content. It was reported that the lipid content of cricket insect was around 16% - 23% (Udomsil et al. 2019). This high lipid content probably caused lipid-protein interactions which possibly could inhibit protein extraction. Thus, defatting step was compulsory to employ in order to enhance the extraction of protein content from food matrix (Gravel and Doyen 2020).

From the result in **Table 1**, the AEAP caused the increase in protein content from around 70% of defatted cricket powder to approximately 83%. This could be explained by the fact that the alkaline chemicals, such as sodium hydroxide (NaOH), could be utilized in order to increase the solubility of protein since the proteins exhibited their higher solubility when the pH was adjusted to alkaline environment than the neutral or acidic pH (Brogan et al. 2021). Hereafter, the acidic chemical, such as hydrochloric acid, could be employed to bring protein compound to its isoelectric point, which was located on pH around 5 (Zielińska et al. 2018). This adjustment would cause the protein charged to be zero and easier for the protein molecules to aggregate and precipitate, which would lead to the increase in protein content of the extracted proteins (Garba and Kaur 2014).

Table 1. *Protein content of cricket protein powder and cricket protein powder with AEAP and AS methods*

Sample	Protein Content (%)
Cricket Powder	$70.41^a \pm 0.92$
AEAP	$83.07^b \pm 0.92$
AS 20	$83.53^b \pm 0.85$
AS 40	$88.39^c \pm 0.78$
AS 60	$92.41^d \pm 1.46$

AEAP: alkaline extraction-acid precipitation; AS 20: AEAP +20% ammonium sulfate ; AS 40: AEAP + 40% ammonium sulfate ; AS 60: AEAP + 60% ammonium sulfate. Means ± standard deviation (n=3) with different letters within a column are significantly different ($P \leq 0.05$).

Furthermore, inclusion of ammonium sulfate was also positively affected the extraction step by escalating the total protein content of cricket protein. There were three concentrations of ammonium sulfate utilized in this study that were 20%, 40%, and 60% (w/v). Data displayed in **Table 1** showed the significant difference of protein extraction yields between samples. The protein content of samples extracted with inclusion of ammonium sulfate were increased when increasing in ammonium sulfate content. The result showed that AEAP cricket protein powder had the same protein content with AS20 and not different significantly. Meanwhile, the different protein contents were observed in AS20, AS40, and AS60 samples ($p \leq 0.05$). The gradual inclusion of ammonium sulfate could escalate protein content of cricket protein where 92.41% protein content was obtained by inclusion of 60 percent of ammonium sulfate. This phenomenon could be happened since the addition of salt ion such as ammonium sulfate could promote the salting-in and salting-out mechanism. Put simply, salting-in mechanism could be defined as the phenomenon where the existence of salt ion would interact with the protein and increase its solubility. However, when the amount of salt ion was gradually increased, the salt ion would compete with protein to bind with water molecules. This competition would promote the strong binding between the salt ion with water molecules. At this condition, there would be increasing of exposed of hydrophobic group of protein that drove the protein-protein interactions and led the protein to aggregate and easier to precipitate. This precipitation of protein by using the inclusion of salt ion was known as salting-out phenomenon (Kristianto et al. 2019). This result was in agreements with the study by Zhang et al. (2017), which successfully extracted potato protein using ammonium sulfate extraction. This study obtained protein extraction of potato protein of 83.20% by using 60% ammonium sulfate. In addition, Waglay et al. (2014) also reported increasing of ammonium sulfate from 40% to 60% could significantly increase potato protein isolate from 70.1% to 98.6%, which was higher than the protein content in present study. Furthermore, the precipitation by inclusion of salt ion was explained due to the dissociate of the ionic substances by salt ion that lead to increase of protein surface tension. This event would promote the decrease of hydrogen bonding between protein and water molecules which drive to precipitation, thus the protein content could be increased.

3.2 Foaming Properties

The foaming properties of cricket protein, composing of foaming capacity (FC) and foaming stability (FS) were reported in **Table 2** and **3**. According to Jiang et al. (2021), foaming capacity is reflection of the ability of protein molecules to be dissolved and unfolded rapidly to form a cohesive layer around a gas bubble. Meanwhile, the foaming stability could be defined as ability of protein molecules to form a stable foam that conducted by forming a continuous intermolecular polymer which trapped air. In this study, both FC and FS could not be determined on defatted cricket powder since this sample was unable to form a foaming system. This inability of cricket powder was probably occurred because there the powder still contained lipid in the sample. Lipid content in cricket insect was reported around 16-21% (Stone et al. 2019). This difference of lipid content could be depended on various factors such as insect feed source, stage lives, and age of the cricket itself (Day et al. 2022). Zayas (1997) explained that the lipid content in cricket insect could affect the ability of protein to forming a foam system since lipid molecules has ability to disrupt the protein interactions at air/water interface and inhibited the construction of foam system. Due to this fact, the efficient defatted step was compulsory to employ in order to remove the lipid content from the matrix and rise the ability to construct the foaming system (Konak et al. 2014). The defatting step incorporated with AEAP step effectivity reflected on the result on this study where the samples that had defatting together with AEAP steps showed ability both in FC and FS.

***Table 2**. Foaming capacity of cricket protein powder and cricket protein powder with AEAP and AS methods*

Sample	Foaming Capacity (%)
Cricket Powder	ND
AEAP	$150.0^{a} \pm 3.16$
AS 20	$207.5^{b} \pm 4.18$
AS 40	$212.5^{c} \pm 2.74$
AS 60	$223.3^{d} \pm 2.58$

AEAP: Alkaline extraction-acid precipitation; AS: ammonium sulfate; AS 20: AEAP +20% ammonium sulfate; AS 40: AEAP + 40% ammonium sulfate ; AS 60: AEAP + 60% ammonium sulfate ; ND: not determined. Means ± standard deviation (n=3) with different letters within a column are significantly different (P ≤ 0.05).

***Table 3**. Foaming stability of cricket protein powder and cricket protein powder with AEAP and AS methods*

Sample	Foaming Stability (%)			
	0 min	30 min	60 min	90 min
Cricket Powder	ND	ND	ND	ND
AEAP	$150.0^{Da} \pm 3.16$	$32.5^{Ca} \pm 2.74$	$11.^{7Ba} \pm 2.58$	$6.^{67Aa} \pm 2.58$
AS 20	$207.5^{Db} \pm 4.18$	$77.^{5Cb} \pm 2.74$	$29.^{2Bb} \pm 2.04$	$24.^{2Ab} \pm 2.04$
AS 40	$212.5^{Dc} \pm 2.74$	$115.^{8Cc} \pm 4.92$	$49.^{2Bc} \pm 3.76$	$27.^{5Ac} \pm 2.74$
AS 60	$223.3^{Cd} \pm 2.58$	$218.^{3Bd} \pm 2.58$	$218.^{3Bd} \pm 2.58$	$212.^{5Ad} \pm 2.74$

AEAP: Alkaline extraction-acid precipitation; AS: ammonium sulfate ; AS 20: AEAP +20% ammonium sulfate ; AS 40: AEAP + 40% ammonium sulfate ; AS 60: AEAP + 60% ammonium sulfate; ND: not determined. Means ± standard deviation (n=3) with different letters within a column are significantly different (P ≤ 0.05).

In addition, there was a report by Gravel et al. (2021) that mentioned the defatting process by using hexane as a solvent could decrease the lipid content of yellow mealworm (*Tenebrio molitor*) from 28.5% to 0.04% and drastically escalated both the FC and FS of yellow mealworm protein. From **Table 2**, Cricket protein in every treatment was different significantly. AEAP samples was obtained the lowest FC and AS20, AS40 and AS 60 samples showed that the FCs were increased to be 207, 212, and 223 percent, respectively. The FS (**Table 3**) was also increased in the same trends as FC in every 30, 60, and 90 minutes. Furthermore, it was found that gradual increasing of ammonium sulfate concentration had positive impact to foaming properties of cricket protein isolate. The value of each sample was significantly different where inclusion of 60 percent of ammonium sulfate, which had the highest value for FC and FS (90 min) as 223.3% and 212.5%, respectively. This result was higher compared with the result by Jeong et al. (2021) who reported the FC and FS of cricket protein concentrate that defatted by using hexane which were only 61.1% and 73.9% (30 min), respectively. Mishyna et al. (2019) also reported the similar result where alkaline extraction could also increase foaming properties of protein that extracted from edible grasshopper, which was up to 90% for FC and 74.1% for FS (120 min). Increment of foaming properties after alkaline extraction could be explained with high protein content and high protein solubility. Alkaline treatment also possibly changed the protein characteristics such as partial protein unfolding that correlated to ability of protein to undergo rapid conformational changes at the interface and rapid adsorption at the air-water interface (Mishyna et al. 2019). Moreover, previous literatures from Adenekan et al. (2018) reported the isolation of pigeon pea protein with ammonium sulfate and found that the foaming properties of pigeon pea protein isolate were improved from 18.20% to 35.10% and 25.30% to 55.73% for FC and FS. This increment was possibly related to escalate of protein solubility and exposure of hydrophilic residues (Jiang et al. 2021). Moreover, it was found that the FS was decreased over the resting time. The alkaline extraction with inclusion of ammonium sulfate was found not only increase the FC, but also affected the FS of the samples. It was reflected from data shown in **Table 3** where AS60 sample displayed the highest FS value compared to other samples. This result was in agreements with Zhang et al. (2017) who stated that the increment of salt concentration could increase the stability of foaming system of potato protein isolate. This could be further explained by higher protein content after extraction would cause more initial protein introduced into the foaming system (Cui et al. 2020). Furthermore, another study by Yuliana et al. (2014) also reported the increase of FS value of cashew nut shell protein from 56.03% to 76.91% at 60 min when increasing in salt ion concentration. It was explained that addition of salt ion to protein affected the FS since the salt ion could increase surface activity and solubility of soluble protein.

3.3 Oil holding capacity and water holding capacity

Oil holding capacity (OHC) and water holding capacity (WHC) were two important parameters in food applications that strongly affected by the protein native state, protein composition, the pH condition, and ionic strength of the protein (Mishyna et al. 2021). The definition of OHC could be understood as the physical entrapment of oil where the mechanism of oil absorption was involving capillarity interaction that would be allowed the retained of oil that absorbed (Zielińska et al. 2018). Meanwhile, the definition of WHC by Zayas (1997) was an ability to hold the water molecules which added during application of many food processing like heating and centrifugation from three-dimensional structure of protein.

The **Table 4** displays the data of OHC and WHC of cricket protein isolate. It was shown for OHC data there was a significant difference between OHC values of cricket powder with the other samples that had been undergone the extraction and isolation steps. The extraction and isolation steps that employed in this study was positively affected the OHC of cricket protein isolate. The OHC value of cricket powder was increased drastically from 1.35 g/g to 4.97 g/g when it was treated with 20% inclusion of ammonium sulfate. This result was higher compared to result reported by Ogunwolu et al. (2009) for cashew nut protein (3.4 mL/g) that extracted with the same method. Based on study conducted by Mao and Hua (2012), the escalating of OHC was probably occurred due to partial denaturation during extraction and isolation step. As mentioned in the methods section, this study was utilized the alkaline chemical (NaOH) and acid chemical (HCl) to adjust the pH into extremely alkaline (around pH 11) and acidic condition (pH 4.5). The usage of extreme pH during extraction and isolation step would probably cause partial denaturation of protein and more exposition of hydrophobic amino acid groups of cricket protein isolate that would lead to improvement of oil holding ability.

Table 4. *Oil holding capacity and water holding capacity of cricket protein powder and cricket protein powder with AEAP and AS methods*

Sample	OHC (g oil/g sample)	WHC (g water/g sample)
Cricket Powder	$1.35^{a} \pm 0.23$	$2.42^{d} \pm 0.13$
AEAP	$4.69^{c} \pm 0.08$	$0.34^{a} \pm 0.09$
AS 20	$4.97^{d} \pm 0.07$	$0.44^{b} \pm 0.02$
AS 40	$4.28^{b} \pm 0.08$	$0.35^{a} \pm 0.01$
AS 60	$4.75^{c} \pm 0.10$	$0.54^{c} \pm 0.03$

AEAP: Alkaline extraction-acid precipitation; AS: ammonium sulfate; AS 20: AEAP +20% ammonium sulfate; AS 40: AEAP + 40% ammonium sulfate; AS 60: AEAP + 60% ammonium sulfate. Means ± standard deviation (n=3) with different letters within a column are significantly different ($P \leq 0.05$).

Moreover, Sun et al. (2022) stated that the inclusion of salt ion would also affect the OHC value since the existence of salt ion such as ammonium sulphate could distract the zeta potential of protein solution and allow for exposition of big number of hydrophobic groups which would increase OHC of cricket protein isolate.

On the contrary, difference trend was depicted in **Table 4** for WHC value of cricket protein isolate. It was found that extraction and isolation that used in this study were drastically decreased the WHC value. The result was quite lower when compared to study by Hu et al. (2017) that mentioned the WHC value of walnut protein was 5 g/g. This result was probably correlated involvement of heat treatment during defat, extraction and isolation step of cricket protein isolate. Zayas (1997) explained the ability of protein to retain water molecules could be reduced by application of heat treatment since it caused the unfolding of polypeptide chain and transition of globular conformation to random coil that resulted in reduction of polar amino acid availability to bind water molecules. Furthermore, inclusion of ammonium sulfate as salt ion also played an important role for decreasing of WHC of cricket protein isolate. Ogunwolu et al. (2009) explained that the system at low salt concentration, hydrated salt ions would be weakly bound to group charged of proteins and binding of ions to proteins causing no any hydration shell of charged groups on protein. Nevertheless, as the salt ions increases, more water molecules would be bounded with salt ions and led to intermolecular interactions among protein molecules that caused dehydration of the protein and reduction of WHC value. This result was in agreements with Yuliana et al. (2014) who reported the WHC of defatted cashew nut shell decreased from 2.04 cm^3 H_2O/g to 1.73 cm^3 H_2O/g after further increased of NaCl concentration.

3.4 Emulsifying properties

Emulsification is one of functional properties of protein which can be described as the ability of protein to absorb and stabilize oil-water interfaces. Various factors such as movement of a protein from aqueous phase to a newly formed interface, unfolding, crosslinking and other conformational rearrangement over time were involved in a formation of stabilized emulsion by protein molecules. Protein that are soluble, flexible, amphiphilic, and has small molecular size tend to be more effective emulsifiers (Day et al. 2022).

Table 5. *Emulsifying properties of cricket protein powder and cricket protein powder with AEAP and AS methods*

Sample	EAI (m^2/g)	ES (min)
Cricket Powder	$11.26^{d} \pm 0.38$	$30.66^{b} \pm 2.72$
AEAP	$8.09^{b} \pm 0.24$	$24.49^{a} \pm 3.76$
AS 20	$8.07^{b} \pm 0.57$	$32.92^{b} \pm 6.14$
AS 40	$5.49^{a} \pm 0.61$	$21.52^{a} \pm 2.21$
AS 60	$9.33^{c} \pm 0.85$	$21.56^{a} \pm 3.81$

AEAP: Alkaline extraction-acid precipitation; AS: ammonium sulfate; AS 20: AEAP +20% ammonium sulfate ; AS 40: AEAP + 40% ammonium sulfate ; AS 60: AEAP + 60% ammonium sulfate. Means ± standard deviation (n=3) with different letters within a column are significantly different ($P \leq 0.05$).

Table 5 shows the data of emulsion activity index (EAI) and emulsion stability (ES) of cricket proteins. The EAI data depicted there was a decreasing trend where defatted cricket powder as the raw material had the highest EAI value compared to other samples. the EAI value of cricket powder of this study (11.26 m^2/g) was similar with the EAI value of cricket reported by Dion-Poulin et al. (2020) which was around 11.86–13.32 m^2/g.

It was found that extraction and isolation step were negatively affected by decreased the EAI value of the cricket protein isolate. This phenomenon was probably caused by the involvement of heat treatment during defat and extraction step. On the material and methods section, it was mentioned that the Soxhlet method was used for defatting process which was processed at high temperature. According to Gould and Wolf (2018), they noticed that utilization of heat treatment with a temperature up to 90 °C would change the microstructure of protein and affect to degree of flocculation and thus increase the droplet size. The heat application would also cause protein denaturation and conformation changes which led to increase in surface hydrophobicity and increased the protein-protein interactions. The result in present study was also similar to the study reported by Lee et al. (2019). They reported the emulsion activity and ES of yellow mealworm protein were decreased significantly after heat treatment (95 °C) from 46.46 to 34.94 and 87.59. to 69.39%, respectively. Heat treatments would promote intermolecular interaction among proteins that led to droplet flocculation and coalescence.

On the other hands, difference trend was found on ES value of cricket protein isolate. Inclusion of ammonium sulfate up to 20% would increase the ES, while further increase of ammonium sulfate concentration caused the decrease in the ES. This unique trend could be explained since the addition salt ion at low concentration would reduce the droplet size and increase the solubility of the proteins which resulted in escalating of ES. Nevertheless, high concentration of added salt ion would drive to protein-protein interactions and thus reduced the protein-oil interactions (Cui et al. 2014).This result was similar with the study by Idris et al. (2003), Hu et al. (2017) and Lawal et al. (2005). Moreover, emulsifying properties of cricket protein isolate were also depended on the types of proteins in the CPI. According to Brogan et al. (2021), cricket protein consisted of actin, myosin, and arginine-kinase. These types of protein could be classified as fibrous protein. It was also known that actin and myosin could contribute good emulsion capacity and ES. In particular, myosin itself could be categorized as one of fibrous protein types which capable to form highly elastic gel matrices and cohesive, rigid fat globule membrane in emulsifying system due to its well-balanced hydrophilicity – hydrophobicity and a large, and long fibrous structures. However, fibrous protein was also sensitive to heat treatment. Xu et al. (2020) mentioned that heat treatment could cause aggregation which made protein became insoluble and just only could allow for a few protein molecules to be adsorbed at oil /water interface. This could be another reason why the EAI value was decreased after extraction step to obtain AEAP sample (treated with 80 °C for 1h). Nevertheless, different trend was shown on ES value where the extraction step caused an increase in the ES. Li et al. (2022) stated that the extraction by using extreme pH could cause the protein to be unfolded, inducing structural changes, which would lead to greater stability of emulsion system.

4 Conclusion

The effect of defatting, extraction, isolation, and inclusion of various concentration of ammonium sulfate on the yield and functional properties of cricket protein isolate (CPI) were investigated. The defatting, extraction and isolation process has been proven that they could improve the extraction yield of cricket protein. The water holding capacity (WHC), oil holding capacity (OHC), emulsifying properties, and foaming properties of cricket protein isolate were affected by these steps. Moreover, the inclusion of ammonium sulphate as salt ion could affect the extraction yield and functional properties of cricket protein isolate. The inclusion of 60 percent ammonium sulfate increased the extraction yield and enhanced the foaming properties and OHC of cricket protein isolate. These results provide an important information for potential of CPI to become one of promising alternative protein sources in the future.

References

Adenekan, M. K., Fadimu, G. J., Odunmbaku, L. A. and Oke, E. K. (2018) Effect of isolation techniques on the characteristics of pigeon pea (*Cajanus cajan*) protein isolates. *Food science & nutrition*, *6*(1), pp. 146-152.

AOAC (2002) Crude protein in meat. 928.08. *Official methods of analysis*.

Brogan, E. N., Park, Y.-L., Matak, K. E. and Jaczynski, J. (2021) Characterization of protein in cricket (*Acheta domesticus*), locust (*Locusta migratoria*), and silk worm pupae (*Bombyx mori*) insect powders. *LWT-Food Science and Technology*, *152*, pp. 112314.

Chen, J., Mu, T., Zhang, M., Goffin, D., Sun, H., Ma, M., Liu, X. and Zhang, D. (2018) Structure, physicochemical, and functional properties of protein isolates and major fractions from cumin (*Cuminum cyminum*) seeds. *International Journal of Food Properties*, *21*(1), pp. 685-701.

Cui, Z., Chen, Y., Kong, X., Zhang, C. and Hua, Y. (2014) Emulsifying properties and oil/water (O/W) interface adsorption behavior of heated soy proteins: Effects of heating concentration, homogenizer rotating speed, and salt addition level. *Journal of agricultural and food chemistry*, *62*(7), pp. 1634-1642.

Day, L., Cakebread, J. A. and Loveday, S. M. (2022) Food proteins from animals and plants: Differences in the nutritional and functional properties. *Trends in Food Science & Technology*, *119*, pp. 428-442.

de Castro, R. J. S., Ohara, A., dos Santos Aguilar, J. G. and Domingues, M. A. F. (2018) Nutritional, functional and biological properties of insect proteins: Processes for obtaining, consumption and future challenges. *Trends in Food Science & Technology*, *76*, pp. 82-89.

Dion-Poulin, A., Laroche, M., Doyen, A. and Turgeon, S. L. (2020) Functionality of cricket and mealworm hydrolysates generated after pretreatment of meals with high hydrostatic pressures. *Molecules*, *25*(22), pp. 5366.

Friedman, M. (1996) Nutritional value of proteins from different food sources. A review. *Journal of agricultural and food chemistry*, *44*(1), pp. 6-29.

Garba, U. and Kaur, S. (2014) Protein isolates: Production, functional properties and application. *International journal of current research and review*, *6*(3), pp. 35-45.

Gould, J. and Wolf, B. (2018) Interfacial and emulsifying properties of mealworm protein at

 https://doi.org/10.1051/bioconf/20238004003

the oil/water interface. *Food Hydrocolloids, 77*, pp. 57-65.

Gravel, A. and Doyen, A. (2020) The use of edible insect proteins in food: Challenges and issues related to their functional properties. *Innovative Food Science & Emerging Technologies, 59*, pp. 102272.

Gravel, A., Marciniak, A., Couture, M. and Doyen, A. (2021) Effects of hexane on protein profile, solubility and foaming properties of defatted proteins extracted from *Tenebrio molitor* larvae. *Molecules, 26*(2), pp. 351.

Hu, H., Fan, T., Zhao, X., Zhang, X., Sun, Y. and Liu, H. (2017) Influence of pH and salt concentration on functional properties of walnut protein from different extraction methods. *Journal of Food Science and Technology, 54*, pp. 2833-2841.

Idris, W. H., Babiker, E. E. and El Tinay, A. H. (2003) Fractionation, solubility and functional properties of wheat bran proteins as influenced by pH and/or salt concentration. *Food/Nahrung, 47*(6), pp. 425-429.

Jeong, M.-S., Lee, S.-D. and Cho, S.-J. (2021) Effect of three defatting solvents on the techno-functional properties of an edible insect (*Gryllus bimaculatus*) protein concentrate. *Molecules, 26*(17), pp. 5307.

Jiang, Y., Zhu, Y., Zheng, Y., Liu, Z., Zhong, Y., Deng, Y. and Zhao, Y. (2021) Effects of salting-in/out-assisted extractions on structural, physicochemical and functional properties of *Tenebrio molitor* larvae protein isolates. *Food Chemistry, 338*, pp. 128158.

Konak, Ü. İ., Ercili-Cura, D., Sibakov, J., Sontag-Strohm, T., Certel, M. and Loponen, J. (2014) CO2-defatted oats: Solubility, emulsification and foaming properties. *Journal of cereal science, 60*(1), pp. 37-41.

Kristianto, H., Prasetyo, S. and Sugih, A. K. (2019) Pemanfaatan ekstrak protein dari kacang-kacangan sebagai koagulan alami. *Jurnal Rekayasa Proses, 13*(2), pp. 65-80.

Kunarayakul, S., Thaiphanit, S., Anprung, P. and Suppavorasatit, I. (2018) Optimization of coconut protein deamidation using protein-glutaminase and its effect on solubility, emulsification, and foaming properties of the proteins. *Food Hydrocolloids, 79*, pp. 197-207.

Kurdi, P., Chaowiwat, P., Weston, J. and Hansawasdi, C. (2021) Studies on microbial quality, protein yield, and antioxidant properties of some frozen edible insects. *International Journal of Food Science, 2021*, pp. 1-7.

Laroche, M., Perreault, V., Marciniak, A., Gravel, A., Chamberland, J. and Doyen, A. (2019) Comparison of conventional and sustainable lipid extraction methods for the production of oil and protein isolate from edible insect meal. *Foods, 8*(11), pp. 572.

Lawal, O., Adebowale, K., Ogunsanwo, B., Sosanwo, O. and Bankole, S. (2005) On the functional properties of globulin and albumin protein fractions and flours of African locust bean (*Parkia biglobossa*). *Food Chemistry, 92*(4), pp. 681-691.

Lee, H.-j., Kim, J.-h., Ji, D.-S. and Lee, C.-H. (2019) Effects of heating time and temperature on functional properties of proteins of yellow mealworm larvae (*Tenebrio molitor L.*). *Food science of animal resources, 39*(2), pp. 296-308.

Li, L., Zhao, X. and Xu, X. (2022) Trace the difference driven by unfolding-refolding pathway of myofibrillar protein: Emphasizing the changes on structural and emulsion properties. *Food Chemistry, 367*, pp. 130688.

Lucas-González, R., Fernández-López, J., Pérez-Álvarez, J. A. and Viuda-Martos, M. (2019) Effect of drying processes in the chemical, physico-chemical, techno-functional and antioxidant properties of flours obtained from house cricket (*Acheta domesticus*). *European Food Research and Technology, 245*(7), pp. 1451-1458.

Mao, X. and Hua, Y. (2012) Composition, structure and functional properties of protein concentrates and isolates produced from walnut (*Juglans regia L.*). *International journal of molecular sciences, 13*(2), pp. 1561-1581.

Meenmanee, S., Rattananukrom, A., Thaiphanit, S. and Suppavorasatit, I. (2022) Improvement of solubility, foaming, and emulsification properties of coconut (Cocos nucifera L.) protein by non-enzymatic deamidation. *Lwt, 153*, pp. 112493.

Mishyna, M., Keppler, J. K. and Chen, J. (2021) Techno-functional properties of edible insect proteins and effects of processing. *Current Opinion in Colloid & Interface Science, 56*, pp. 101508.

Mishyna, M., Martinez, J.-J. I., Chen, J. and Benjamin, O. (2019) Extraction, characterization and functional properties of soluble proteins from edible grasshopper (*Schistocerca gregaria*) and honey bee (*Apis mellifera*). *Food Research International, 116*, pp. 697-706.

Ndiritu, A. K., Kinyuru, J. N., Kenji, G. M. and Gichuhi, P. N. (2017) Extraction technique influences the physico-chemical characteristics and functional properties of edible crickets (*Acheta domesticus*) protein concentrate. *Journal of Food Measurement and Characterization, 11*(4), pp. 2013-2021.

Ogunwolu, S. O., Henshaw, F. O., Mock, H.-P., Santros, A. and Awonorin, S. O. (2009) Functional properties of protein concentrates and isolates produced from cashew (*Anacardium occidentale L.*) nut. *Food Chemistry, 115*(3), pp. 852-858.

Oibiokpa, F. I., Akanya, H. O., Jigam, A. A., Saidu, A. N. and Egwim, E. C. (2018) Protein quality of four indigenous edible insect species in Nigeria. *Food Science and Human Wellness, 7*(2), pp. 175-183.

Rumpold, B. A. and Schlüter, O. (2015) Insect-based protein sources and their potential for human consumption: Nutritional composition and processing. *Animal Frontiers, 5*(2), pp. 20-24.

Santiago, L. A., Fadel, O. M. and Tavares, G. M. (2021) How does the thermal-aggregation behavior of black cricket protein isolate affect its foaming and gelling properties? *Food Hydrocolloids, 110*, pp. 106169.

Smarzyński, K., Sarbak, P., Kowalczewski, P. Ł., Różańska, M. B., Rybicka, I., Polanowska, K., Fedko, M., Kmiecik, D., Masewicz, Ł. and Nowicki, M. (2021) Low-Field NMR study of shortcake biscuits with cricket powder, and their nutritional and physical characteristics. *Molecules, 26*(17), pp. 5417.

Stone, A. K., Tanaka, T. and Nickerson, M. T. (2019) Protein quality and physicochemical properties of commercial cricket and mealworm powders. *Journal of Food Science and Technology, 56*, pp. 3355-3363.

Sun, Y., Zhong, M., Wu, L., Huang, Y., Li, Y. and Qi, B. (2022) Effects of ultrasound-assisted salt (NaCl) extraction method on the structural and functional properties of Oleosin. *Food Chemistry, 372*, pp. 131238.

Udomsil, N., Imsoonthornruksa, S., Gosalawit, C. and Ketudat-Cairns, M. (2019) Nutritional values and functional properties of house cricket (*Acheta domesticus*) and field cricket (*Gryllus bimaculatus*). *Food Science and Technology Research, 25*(4), pp. 597-605.

Waglay, A., Karboune, S. and Alli, I. (2014) Potato protein isolates: Recovery and characterization of their properties. *Food Chemistry, 142*, pp. 373-382.

Xu, Y., Dong, M., Tang, C., Han, M., Xu, X. and Zhou, G. (2020) Glycation-induced structural modification of myofibrillar protein and its relation to emulsifying properties. *LWT-Food Science and Technology, 117*, pp. 108664.

Yin, S.-W., Tang, C.-H., Wen, Q.-B., Yang, X.-Q. and Li, L. (2008) Functional properties and in vitro trypsin digestibility of red kidney bean (Phaseolus vulgaris L.) protein isolate: Effect of high-pressure treatment. *Food Chemistry, 110*(4), pp. 938-945.

Yuliana, M., Truong, C. T., Huynh, L. H., Ho, Q. P. and Ju, Y.-H. (2014) Isolation and characterization of protein isolated from defatted cashew nut shell: Influence of pH and NaCl on solubility and functional properties. *LWT-Food Science and Technology, 55*(2), pp. 621-626.

Zayas, J. F. (1997) *Functionality of proteins in food*, Berlin, German: Springer science & business media.

Zhang, D. Q., Mu, T. H., Sun, H. N., Chen, J. W. and Zhang, M. (2017) Comparative study of potato protein concentrates extracted using ammonium sulfate and isoelectric precipitation. *International Journal of Food Properties, 20*(9), pp. 2113-2127.

Zielińska, E., Karaś, M. and Baraniak, B. (2018) Comparison of functional properties of edible insects and protein preparations thereof. *LWT-Food Science and Technology, 91*, pp. 168-174.

GAS ABSORBENT BASED MODIFIED ATMOSPHERE PACKAGING TO OPTIMIZE DRAGON FRUIT SHELF LIFE

Satria Primatama[1], *Fahrizal Yusuf* Affandi[2], *Anisa* Kistanti[1], *Musyrifah* Kurniawati[1], *Bayu* Nugraha[1*]

[1]Agricultural and Biosystem Engineering Department, Faculty of Agricultural Technology, Universitas Gadjah Mada, Indonesia
[2]Biological Technology and Veterinary Department, Vocational College, Universitas Gadjah Mada, Indonesia
*Corresponding author: bayu.nugraha@mail.ugm.ac.id

Abstract. Dragon fruit is a promising export commodity for Indonesia. But its short shelf-life is a big challenge for an international shipping by sea. This study aims to prolong the dragon fruit shelf life by implementing gas absorbent based Modified Atmosphere Packaging (MAP) in combination with cold storage. This technique decreases the fruit physiological processes by lowering O_2 concentration (%) inside the MAP. The selected dragon fruit was packed with gas absorbent-based MAP with different O_2 absorber capacities (30 cc; 60 cc; and 120 cc) and plastic types (biodegradable plastic and LDPE). Each treatment was also varied with the use of H_2O absorbent and perforation. Then, the dragon fruit in MAP was stored at 5°C and 90% RH for 6 weeks as shipping simulation. After storage, the dragon fruit was transferred to 27°C room temperature without MAP. The results showed that the application of gas absorbent-based MAP, additional treatments, and low temperature were generally able to maintain the quality of dragon fruit up to 6 weeks in cold storage. However, dragon fruit developed signs of chilling injury, increased water loss, increased weight loss, decreased hardness, skin damage, fin withering, and mold growth after transfer at 27°C room temperature without MAP.

1 Introduction

The short shelf life of Indonesian dragon fruit during export via sea transportation poses a significant challenge. The decision to optimize for sea-based export distribution is primarily driven by its cost-effectiveness in comparison to air transportation. The shelf life of dragon fruit is intricately linked to its ability to retain freshness over time. Preservation of dragon fruit's freshness typically involves the application of specific methods. One viable technique for extending the shelf life of dragon fruit involves implementing cold temperature storage and carefully controlled high humidity levels.

Preservation of dragon fruit through cold temperature storage and controlled humidity can effectively uphold its quality and prolong its shelf life. Precise temperature and humidity regulation serve to mitigate adverse effects, including discoloration, flesh quality deterioration, mold growth, shrinkage, and structural damage to the dragon fruit. Furthermore, maintaining optimal cold temperatures and humidity levels can also help retain the moisture content of dragon fruit. Moisture loss in fresh fruit can lead to a decline in taste quality, nutritional value, and overall size reduction. [1].

Drawing upon previous research, it has been determined that the optimal storage temperature for dragon fruit is 5°C. This specific temperature is conducive to preserving the fruit's green hue and firm texture. Furthermore, maintaining a low temperature of 5°C has demonstrated the capability to mitigate the occurrence and severity of decay, as well as reduce weight loss in dragon fruit. Remarkably, this temperature setting can extend the shelf life of dragon fruit for a period of up to 20 days. Nonetheless, it is important to note that while the 5°C storage environment enhances the shelf life of dragon fruit, empirical observations reveal that this condition may induce chilling injury and lead to rapid deterioration when the fruit is subsequently exposed to room temperature. [2].

In addition to maintaining a low temperature of 5°C, employing Modified Atmosphere Packaging (MAP) with oxygen-absorbing properties also holds the potential to effectively preserve the quality and freshness of dragon fruit during the export distribution process. A reduced oxygen (O_2) environment has been demonstrated to significantly delay the ripening of dragon fruit, owing to the role of O_2 in the respiration process and its substantial impact on the proliferation of microorganisms, which are known contributors to food spoilage. Consequently, this study focuses on determining the optimal combination of MAP and cold storage conditions (5°C and 90% relative humidity). The MAP will be configured with various oxygen absorber capacities (30 cc, 60 cc, and 120 cc) to precisely control the O_2 levels within the packaging and will explore the suitability of different types of plastic materials (biodegradable plastic and LDPE plastic) to mitigate environmental contamination. Each MAP configuration will also be subjected to a range of additional treatments, such as moisture absorbents and perforations, to maximize the shelf life of dragon fruit during sea-based export distribution. The study will monitor the physical attributes of dragon fruit as indicators of its quality.

2 Materials and methods

2.1 Sample preparation

The dragon fruit samples utilized in this study were obtained from freshly harvested dragon fruit originating from a local farm situated on Kaliurang Road, Km. 18.5, Kertodadi, Pekembinangun, Pakem, Sleman Regency, Special Region of Yogyakarta Province. A total of 54 dragon fruit samples were employed for this research. The specific dragon fruit variety selected for the research was *Selenicereus costaricencis*. Subsequently, these dragon fruit samples were transported to the Laboratory at the Faculty of Agricultural Technology, Universitas Gadjah Mada, Yogyakarta, Indonesia, where the samples underwent the prescribed treatments.

2.2 MAP & treatments preparation

The modified atmosphere packaging (MAP) system employing gas absorbents underwent variations utilizing LDPE plastic packaging materials and biodegradable plastics, each with identical thicknesses of 0.04mm or 40µm. The selected gas absorbent for the MAP packaging was the O-Buster oxygen absorber containing iron dust. Different quantities of oxygen absorbers, specifically 30 cc, 60 cc, and 120 cc, were applied to control the oxygen concentration within the MAP. Additionally, each MAP configuration was subject to variations involving the incorporation of a water vapor (H_2O) absorbent and perforations to regulate the relative humidity within the packaging. The primary objective of these adjustments was to effectively manage the relative humidity in the MAP packaging. Subsequently, the dragon fruit enclosed within the MAPs was subjected to storage conditions simulating international shipping, involving a temperature of 5°C and relative humidity of 90% for a duration of 6 weeks. Following the storage period, the dragon fruit was subsequently transferred to an ambient room temperature of 27°C without the use of MAP for a shelf-life observation period of 7 days.

2.3 Moisture content

The moisture content (MC) of each dragon fruit sample was determined under various MAP conditions using the gravimetric method at two time points: at week 0 and week 6 during cold storage. Subsequently, on week 7, after cold storage, the MC of each dragon fruit was measured following transfer to a room temperature environment of 27°C without the application of MAP. Each dragon fruit sample within different MAP conditions was initially weighed. Following this, the dragon fruit samples were subjected to drying in an oven at 110°C for a duration of 24 hours to obtain desiccated samples. These dried dragon fruit samples were then reweighed to determine the weight of the fruit samples without moisture content. The percentage of MC in the dragon fruit samples was calculated using **Formula 1**.

$$\% \text{ MC} = \frac{(W^{ti} - W^{tf})}{W^{ti}} \text{ x } 100\ \% \qquad (1)$$

Where:
W^{ti} = initial dragon fruit weight (gram)
W^{tf} = final dragon fruit weight after drying (gram)

2.4 Weight loss

Weight loss was measured for each dragon fruit sample under varying Modified Atmosphere Packaging (MAP) conditions every week, starting from week 0 and continuing until week 6 during cold storage. Additionally, on week 7, after storage, the weight loss was measured for dragon fruit samples that had been transferred to room temperature at 27°C without the application of MAP. To minimize data measurement deviations, weight loss measurements were conducted in triplicate. The percentage of weight loss in the dragon fruit samples was calculated using **Formula 2.**

$$\% \text{ weight loss} = \frac{(W^{ti} - W^{tf})}{W^{ti}} \text{ x } 100\ \% \qquad (2)$$

Where:
W = dragon fruit weight (gram)
ti = initial time (week)
tf = final time (week)

2.5 Firmness

Firmness was assessed in multiple MAP conditions for each dragon fruit sample using a penetrometer. Measurements were taken at week 0 and after six weeks of cold storage. Subsequently, following a week after cold storage without MAP at a room temperature of 27°C, the firmness of dragon fruit samples was also evaluated. The penetrometer recorded data in kilograms (kg), and to obtain a firmness value per unit area (kg/m^2), additional calculations were performed. This firmness value was determined using **Formula 3**.

$$\text{P} = \frac{F_r}{A_s} \qquad (3)$$

Where:
P = Firmness value (kg/m^2)
Fr = value displayed on penetrometer (kg)
A_s = Area of penetrometer probe (0,00209 m^2)

2.6 pH acidity

The pH acidity values were measured for each dragon fruit sample under varying MAP conditions at weeks 0 and 6 during cold storage. After 6 weeks of cold storage, pH acidity values were also measured for dragon fruit samples that were transferred to room temperature

https://doi.org/10.1051/bioconf/20238004004

(27°C) without the use of MAP. The pH acidity values for each dragon fruit sample indicate the percentage of the most dominant type of acid present in the dragon fruit. Furthermore, these pH acidity values serve as indicators of potential fermentation occurring in the dragon fruit samples during the various treatments, including MAP, cold storage, and exposure to room temperature.

2.7 Total soluble solids (°Brix)

The measurement of Total Soluble Solids (TSS) was executed on each dragon fruit sample in varied MAP using refractometer (MASTER-53; ATAGO; Japan). The measurement time of dragon fruit samples were on week 0 & week 6 in cold storage. After cold storage on 7th week, pH acidity value was also measured for each dragon fruit sample transferred to 27°C room temperature without MAP. TSS displays the °Brix value that shows the sweetness and maturation level of dragon fruit sample during the treatments of MAP, cold storage, and room temperature.

2.8 Respiration rate

The respiration rate was assessed using a gas analyzer (O_2 & CO_2 headspace analyzer; The Quantek Model 902D; Proprietary Electrochemical sensor for O_2; Solid-state Infrared sensor for CO_2; 2 – 12 second Timing Pump; USA) for each dragon fruit sample under varying Modified Atmosphere Packaging (MAP) conditions every week, commencing from week 0 and continuing until week 6 during cold storage. On week 7, following cold storage, the respiration rate was also measured for dragon fruit samples transferred to a room temperature of 27°C without the application of MAP. The measurement of respiration rate focused on the CO_2 production of dragon fruit samples during MAP treatments, cold storage, and exposure to room temperature. Each week, dragon fruit samples were removed from various gas-absorbent-based MAP environments and placed in specific 1 L volume packaging for respiration rate measurements over a 2-hour period. The available free volume within the packaging for respiration rate measurement depended on the size of the dragon fruit sample. Larger dragon fruit samples placed in the packaging resulted in a smaller free volume within the packaging. Subsequently, respiration rate data were calculated to obtain the average values for each weekly measurement. All respiration rate data from each week were then compiled to determine the trend in the respiration rate of dragon fruit samples under varying MAP conditions. The respiration rate data were calculated using **Formula 4**.

$$RCO_2 = \frac{\left(Y_{CO_2}^{ti} - Y_{CO_2}^{tf}\right) x\, V_f}{100\, x\, M\, x\, (t_f - t_i)} \qquad (4)$$

Where:
RCO_2 = CO_2 production rate (mlO_2/kg.h)
Y = gas concentration (%)
Vf = free volume (ml)
M = weight of dragon fruit (kg)
ti = initial time measurement (hour)
tf = final time measurement (hour)

2.9 Shelf life

The shelf life of dragon fruit samples subjected to various Modified Atmosphere Packaging (MAP) conditions is indicative of the fruit's ability to maintain freshness during storage. Shelf life, in this context, is determined by monitoring the occurrence of physical damage to the dragon fruit samples during storage. These observations were conducted weekly, commencing from week 0 and concluding at week 6, while the samples were stored in a cold environment. Additionally, a follow-up assessment was performed on dragon fruit samples without MAP at week 7. Physical damage was assessed based on the incidence percentage of fin wilt, the presence of mold, skin damage, and fruit skin cracking. The results of this damage assessment were systematically collected and subjected to analysis. The week at which physical damage incidents occurred served as a key determinant of the dragon fruit's shelf life.

2.10 Statistical analysis

Data collected in this study encompassed various parameters, including measurements of weight loss, percentage of water loss, pH acidity, total soluble solids value, firmness value, and respiration rate. To assess the impact of varying oxygen absorber capacities (30 cc, 60 cc, & 120 cc), types of plastic materials (LDPE plastic & biodegradable plastics), and additional treatments (H_2O absorbent and perforation), a three-way Analysis of Variance (ANOVA) was employed. The statistical analysis was conducted using IBM SPSS software, version 25, on the transformed data to ensure normal distribution. Subsequently, the parameter data was examined to discern the influence of treatment variations and interactions among treatments on the tested parameter values. This analysis was conducted using a three-way ANOVA with a significance threshold of $P \leq 0.05$. The goal was to identify the most effective treatment strategy for preserving the quality and extending the shelf life of dragon fruit.

3 Result and discussion

3.1 Moisture content

The moisture content (MC) of each dragon fruit sample in various Modified Atmosphere Packaging (MAP) conditions showed insignificant changes during weeks 0 and 6 of cold storage. The most substantial decrease in dragon fruit flesh MC, observed during week 6 of cold storage, occurred in the dragon fruit stored in gas-absorbent-based MAP with biodegradable plastic and perforation, equipped with a 60 cc O_2 absorber capacity, resulting in a decrease of up to 4.22% ± 0.42% in MC. Upon transitioning the dragon fruit samples to room temperature (27°C) on week 7, a drastic alteration in MC was observed, particularly in samples lacking gas-absorbent-based MAP. All dragon fruit samples exhibited a decrease in MC, ranging from a minimum of 2.23% ± 1.9% in the case of gas-absorbent MAP with LDPE plastic alone to a maximum of 12.22% ± 3.63% in MAP with biodegradable plastic and perforation. The percentage decrease in flesh MC of dragon fruit samples during week 7 at room temperature (27°C) without MAP was notably significant when compared to the percentage decrease observed during week 6. This decline in water content was accompanied by a visible deterioration in the physical appearance of the dragon fruit samples during week 7. This deterioration was characterized by the rupture of the dragon fruit skin, resulting in the release of water from the fruit, ultimately contributing to the loss of water content. This phenomenon is likely attributed to chilling injury experienced by the dragon fruits during their cold storage at 5°C with 90% relative humidity (RH). However, the effects of chilling injury became evident only after the dragon fruits were transferred to a room temperature environment (approximately 27°C) without the protective benefits of MAP [4].

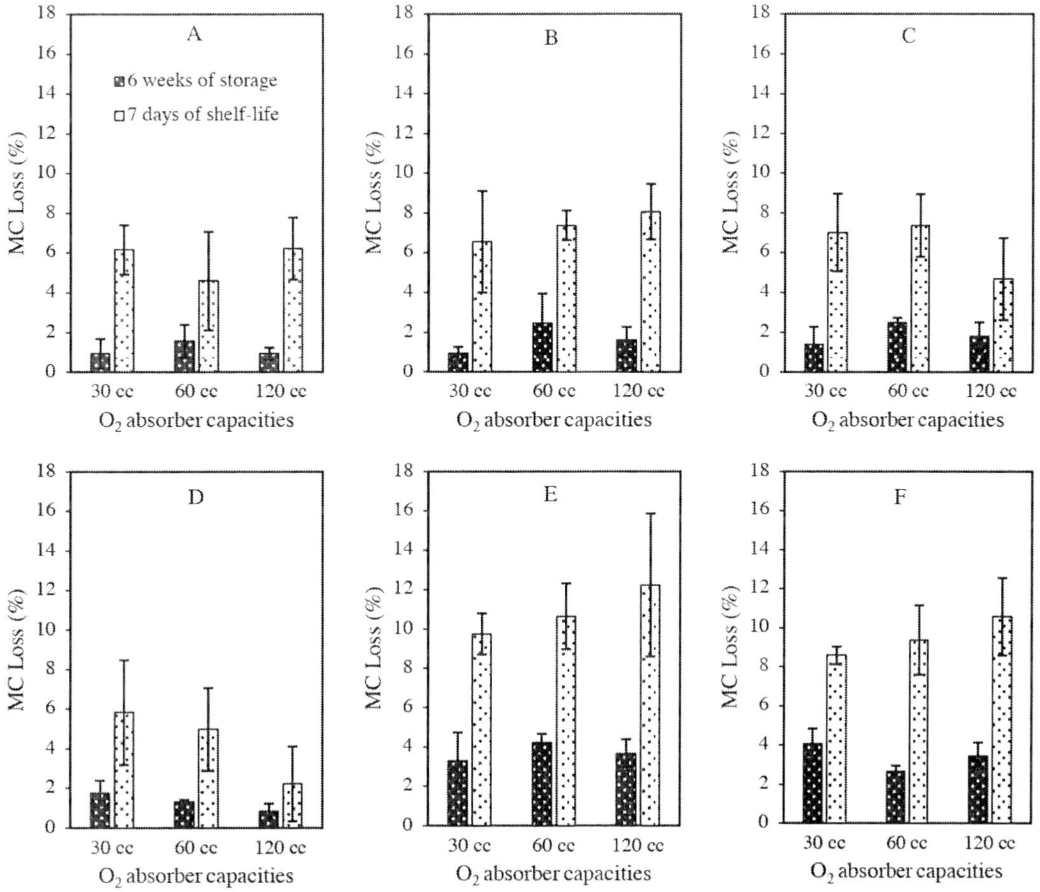

Figure 1. The loss of moisture content in dragon fruit under MAP conditions for 6 different packaging scenarios: (A) biodegradable plastic with a water absorbent, (B) LDPE plastic with a water absorbent, (C) biodegradable plastic only, (D) LDPE plastic only, (E) biodegradable plastic with perforations, and (F) LDPE plastic with perforations.

In accordance with the results of the three-way ANOVA statistical test, the variation in the percentage of MC reduction among the different samples of dragon fruit is influenced by the interactions involving various MAP materials, namely biodegradable plastic & LDPE plastic, in conjunction with additional treatments such as the use of H_2O absorbents and the application of perforation holes ($P \leq 0.05$). This phenomenon occurs due to transpiration events in dragon fruit that trigger water loss due to the difference in water vapor pressure between the surface area of the fruit and the atmosphere. MAP packaging without perforation tends to maintain higher humidity in the MAP, thereby hindering transpiration in dragon fruit samples and subsequently impacting MC preservation [4][5].

3.2 Weight loss

The weight loss of each dragon fruit sample across all MAP variations, considering varying O_2 absorber capacities (30 cc, 60 cc, 120 cc), diverse packaging materials (biodegradable plastic and LDPE plastic), and additional treatment variations (H_2O absorbent and perforation), exhibits negligible differences, tending to remain constant during cold storage at 5°C and 90% relative humidity (RH). Significant changes in dragon fruit weight loss became evident when the dragon fruit was transferred to room temperature (27°C) without gas-absorbent-based MAP for one week. The weight loss data for dragon fruit is comprehensively presented in **Figure 2** and **Figure 3**. The highest recorded weight loss, at 36.4%, occurred during week 7 after storage at room temperature (27°C) without gas-absorbent-based MAP for dragon fruit samples that had been packaged using a 60 cc O_2 absorber capacity, LDPE plastic, and perforation during cold storage. Conversely, the lowest weight loss, at 18.4%, was observed during week 7 at room temperature (27°C) without gas-absorbent-based MAP for dragon fruit samples packaged with a 120 cc O_2 absorber capacity and LDPE plastic during cold storage.

Based on the three-way ANOVA statistical test, diverse MAP plastic materials (biodegradable plastic and LDPE), oxygen absorber capacities (30 cc, 60 cc, and 120 cc), and additional treatments (H_2O absorbent and perforation holes) demonstrated no significant effect ($P \geq 0.05$) on weight loss during storage, both in cold storage and at room temperature without gas-absorbent-based MAP. These results also underscore the substantial influence of varied MAP on the maintenance of dragon fruit's physical condition. The findings of this study align with prior research on red dragon fruit in the Philippines, which applied MAP and low temperatures, revealing only a 0.12% weight loss in LDPE-packaged MAP, increasing to just 1.4% by week 6 at a low temperature of 5°C [6].

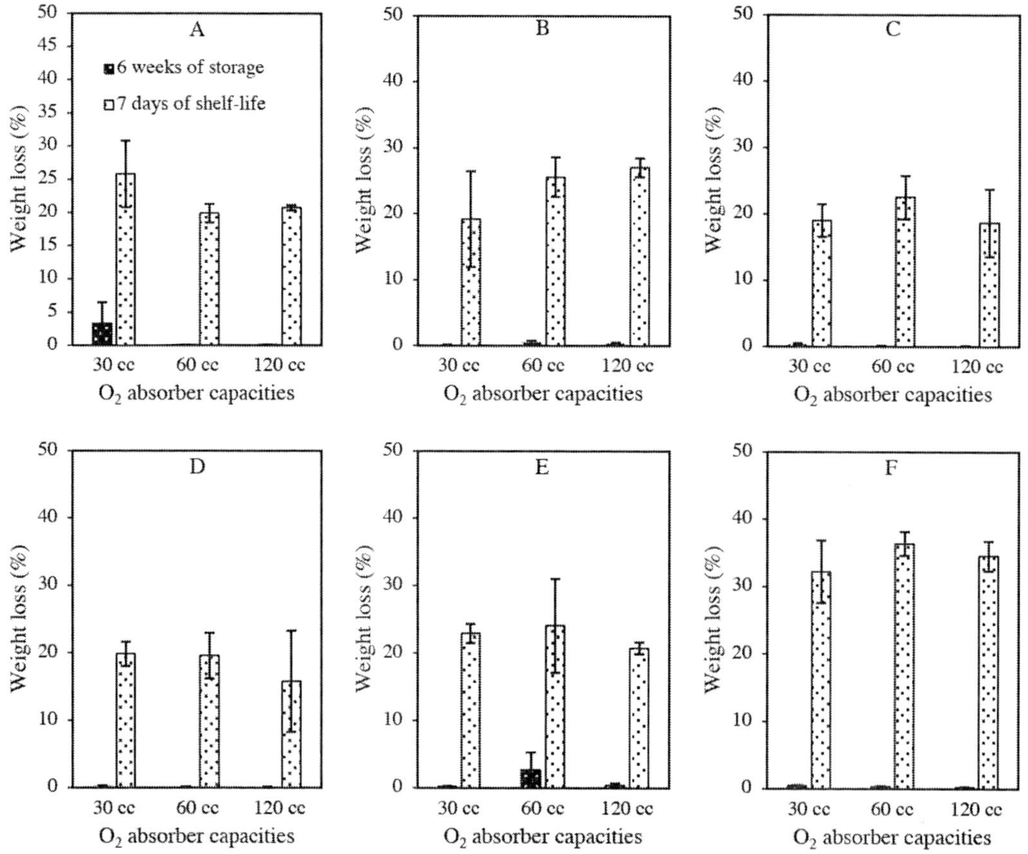

Figure 2. The loss of weight in dragon fruit under MAP conditions for 6 different packaging scenarios: (A) biodegradable plastic with a water absorbent, (B) LDPE plastic with a water absorbent, (C) biodegradable plastic only, (D) LDPE plastic only, (E) biodegradable plastic with perforations, and (F) LDPE plastic with perforations

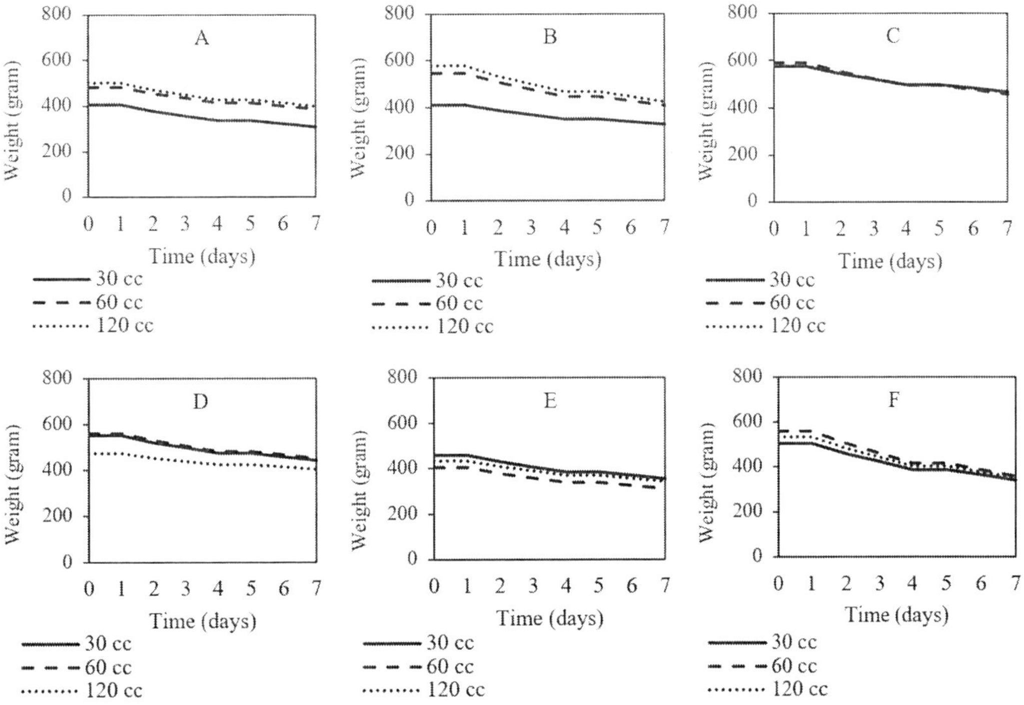

Figure 3. Trend of weight loss on dragon fruits of 7 days after storage which have been under MAP conditions for 6 packaging scenarios(A) biodegradable plastic with a water absorbent, (B) LDPE plastic with a water absorbent, (C) biodegradable plastic only, (D) LDPE plastic only, (E) biodegradable plastic with perforations, and (F) LDPE plastic with perforations

3.3 Firmness

The firmness of dragon fruit samples under varying Modified Atmosphere Packaging (MAP) conditions was measured, and the results are presented in **Figure 4**. It was observed that the firmness of all sample variations was not significantly affected ($P \geq 0.05$) by changes in MAP parameters, including the type of plastic materials used (biodegradable plastic and LDPE plastic), O_2 absorber capacities (30 cc, 60 cc, 120 cc), and additional treatments (H_2O absorbent and perforation).

Over the course of 6 weeks, the firmness of dragon fruit samples in different MAP conditions exhibited a reduction ranging from 0.5% in MAP P1 (biodegradable plastic and H_2O absorbent) to 2.48% in MAP P6 (LDPE plastic and perforation). Subsequently, when these dragon fruit samples were stored at room temperature (27°C) without gas absorbent-based MAP after cold storage, their firmness began to decline. This reduction in firmness ranged from 41.8% in MAP P1 (biodegradable plastic and H_2O absorbent) to 68.1% in MAP P5 (biodegradable plastic and perforation).

These findings align with a previous study, which demonstrated that MAP-wrapped cherry tomato samples experienced only a slight reduction in firmness, whereas MAP-unwrapped samples exhibited a significant decrease in hardness due to elevated O_2 exposure [7]. Notably, the hardness of wrapped samples remained higher compared to unwrapped samples. Additionally, prior research has consistently shown that MAP is effective in mitigating the rate of softening during the storage of various fruit types, such as green chili peppers [8], Eva apple [9], and Tommy Atkins mango [10].

The mechanism behind this phenomenon lies in the fact that aerobic respiration can lead to the loosening of cell wall organization [11]. Consequently, this makes pectin from the cell wall more susceptible to enzymatic breakdown by pectinase, resulting in a loss of firmness in dragon fruit samples during the postharvest period. Furthermore, under conditions of high CO_2 and low O_2 concentrations, the activity of cell wall-degrading enzymes, such as polygalacturonate, is also reduced [12]. MAP systems designed to maintain low O_2 atmospheric conditions play a crucial role in suppressing the rate of respiration, thus contributing to a reduced softening rate in fruit samples [13].

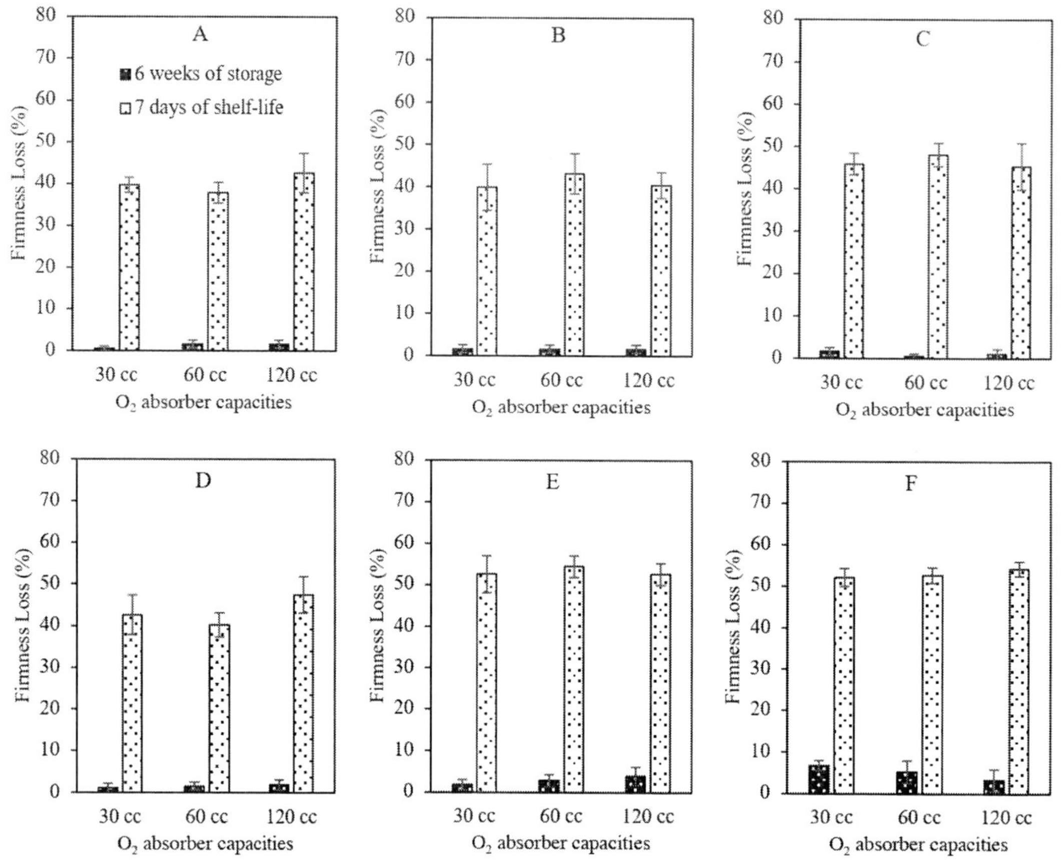

Figure 4. The loss of firmness in dragon fruit under MAP conditions was investigated for 6 different packaging scenarios : (A) biodegradable plastic with a water absorbent, (B) LDPE plastic with a water absorbent, (C) biodegradable plastic only, (D) LDPE plastic only, (E) biodegradable plastic with perforations, and (F) LDPE plastic with perforations.

3.4 pH acidity

The pH acidity data obtained reveals that the pH acidity values tend to fluctuate within the range of 0.68% ± 1.2% to 3.96% ± 1.1% during the six-week cold storage period. However, the pH acidity value of dragon fruit at week 7, stored at a room temperature of 27°C without MAP, exhibited a decrease from the range of 6.9% ± 5% in MAP P1 (biodegradable plastic & H_2O absorbent) to 11.2% ± 4.4% in MAP P6 (LDPE plastic & perforation). This observation indicates that the pH acidity value of dragon fruit samples with various MAP configurations at 5°C became slightly more acidic after transitioning to room temperature of 27°C without gas absorbent-based MAP. Nevertheless, the dragon fruit samples at week 7 did not exhibit any signs of fermentation. This phenomenon can be attributed to the fact that the O_2 concentrations (%) within all utilized MAPs remained within acceptable limits to prevent fermentation. A previous study has indicated that the fermentation threshold begins at an O_2 concentration of 1.5% [14].

The results of the three-way ANOVA statistical test conducted on dragon fruit at week 7, stored at room temperature (27°C) without MAP following six weeks of cold storage, demonstrated that variations in oxygen absorber capacity (30 cc, 60 cc, and 120 cc) had a significant impact ($P \leq 0.05$) on reducing the pH acidity value of dragon fruit. This phenomenon is attributed to the fact that the acids present in fruits and other fresh produce primarily serve as substrates during respiration. Therefore, reducing the O_2 levels and temperature slows down the consumption of this substrate, leading to a reduced rate of Total Titrable Acidity (TTA) decrease and the maintenance of pH acidity values until the end of the storage period. This consistent pH acidity data holds substantial implications for the advancement of dragon fruit storage techniques, demonstrating that the acidity levels of dragon fruit can be effectively stabilized. The findings of this study provide support for the potential utility of gas absorbent-based MAP methods and low-temperature storage in optimizing the shelf life of dragon fruit [15].

3.5 Total soluble solids (°Brix)

In the results of this study, the measured Total Soluble Solids (TSS) values, expressed in °Brix. The °Brix values in all sample variations exhibited no significant change ($P \leq 0.05$). Dragon fruit samples subjected to Modified Atmosphere Packaging (MAP) treatments displayed a gradual decline in °Brix values approaching zero during cold storage. However, at week 7, when maintained at room temperature (27°C) without MAP, the °Brix value of dragon fruit demonstrated a range from 1.37% ± 1.06% in MAP P1 to 5.48% ± 0.4% in MAP P6. This observation highlights that the reduction in °Brix values for dragon fruit samples only occurred after exposure to room temperature (27°C) without MAP. It is noteworthy that lower storage temperatures were found to be conducive to preserving the TSS content in the fruit.

3.6 Respiration rate

In this study, the respiration rate of dragon fruit was examined by monitoring changes in CO_2 gas concentration within specific packaging designed solely for respiration rate tests. This concentration alteration was observed over a duration of 2 hours. This 2-hour observation period, or 120 minutes, was chosen to minimize environmental exposure dissimilarity from the conditions within gas absorbent-based Modified Atmosphere Packaging (MAP). The CO_2 gas concentration data were then subjected to analysis using a respiration rate formula.

The findings indicate that the respiration rate of dragon fruit, following storage under various MAP conditions and cold storage at 5°C with a relative humidity of 90%, exhibited a progressive decrease from week 0 to week 6. The lowest respiration rate was observed at week 6. However, in week 7, the respiration rate of dragon fruit samples stored at room temperature (27°C) without gas absorbent-based MAP increased significantly, illustrating this fluctuation in the dragon fruit's respiration rate, as illustrated in **Figure 5**. Low temperatures decrease cellular respiration in living tissues. As the temperature rises, the kinetic energy required to carry out chemical reactions such as cellular respiration decreases. Therefore, warmer temperatures usually result in more cellular respiration. The rate of respiration decreases with the low storage temperature of the fruit.

Statistical analysis, specifically a three-way ANOVA test, revealed a significant difference in respiration rate values at week 3 ($P \leq 0.05$) among dragon fruit samples within various MAP conditions that had different oxygen absorber capacities (30 cc, 60 cc, and 120 cc). Furthermore, interactions between the types of plastic material (biodegradable plastic & LDPE plastic) and additional treatments (H_2O absorbent & perforation) were found to influence the respiration rate of dragon fruit. At week 6, statistical tests did not reveal any differences in respiration rates between samples subjected to various treatments. Nevertheless, at week 7, the respiration rate of dragon fruit samples exhibited distinct values based on variations in oxygen absorber capacity (30 cc, 60 cc, 120 cc), plastic material type (biodegradable plastic & LDPE plastic), and additional treatments (H_2O absorbent & perforation). This statistical analysis underscores the significant impact ($P \leq 0.05$) of MAP with different oxygen absorber capacities, plastic material types, and additional treatments on the respiration rate.

3.7 Shelf life

During cold storage, physical damage to dragon fruit was observed, including wilting of fruit fins, mold growth, damage to the outer skin, and skin cracking. The fin wilt became apparent around week 4 or after 30 days in dragon fruit samples treated with P5 (biodegradable plastic with perforation) and P6 (LDPE with Perforation), both of which had O_2 levels of approximately 18-19%. In contrast, for other treatments such as P1 (biodegradable plastic with H_2O absorbent), P2 (LDPE plastic with H_2O absorbent), P3 (biodegradable plastics only), and P4 (LDPE plastic only), symptoms only appeared at the beginning of week 5.

Another indicator of physical damage was the emergence of mold on dragon fruit skin samples in MAP P5 (biodegradable plastic & perforation) at week 5. By week 6, new mold growth occurred in fruit samples subjected to P3 (biodegradable plastic only), P5 (biodegradable plastic with perforation), and P6 (LDPE plastic with perforation) treatments. This highlights that MAP with perforations, such as P5 and P6, can promote mold growth on dragon fruit due to increased air circulation within the packaging, fostering fungal contamination. The high O_2 levels (18%-19%) in MAP P5 and P6 further exacerbated this fungal growth. These physical damage manifestations are thoroughly illustrated in **Figure 6**.

Perforated plastic bags were found to effectively reduce the weight shrinkage of dragon fruit during storage. However, they also increased the percentage and severity of fruit spoilage. Dragon fruit stored at 5°C without perforated plastic bags remained unspoiled even after 25 days of storage. The utilization of perforated plastic bags amplified the percentage of fruits exhibiting spoilage events during storage at 7°C and 10°C, as well as the severity of spoilage at 10°C.

Upon removal of dragon fruit to a room temperature of approximately ±27°C without MAP following storage, approximately 80% of the fruits stored at 5°C with perforated plastic bags experienced spoilage, while fruits stored at 5°C without perforated bags remained unaffected. At 7°C and 10°C, both with and without perforated plastic bags, all fruits exhibited 100% spoilage within five days at room temperature ±27°C without MAP. Therefore, the storage of dragon fruit in perforated MAP resulted in increased severity of spoilage during shelf life at room temperature ±27°C, even when the bags were removed at the time of storage discontinuation [16].

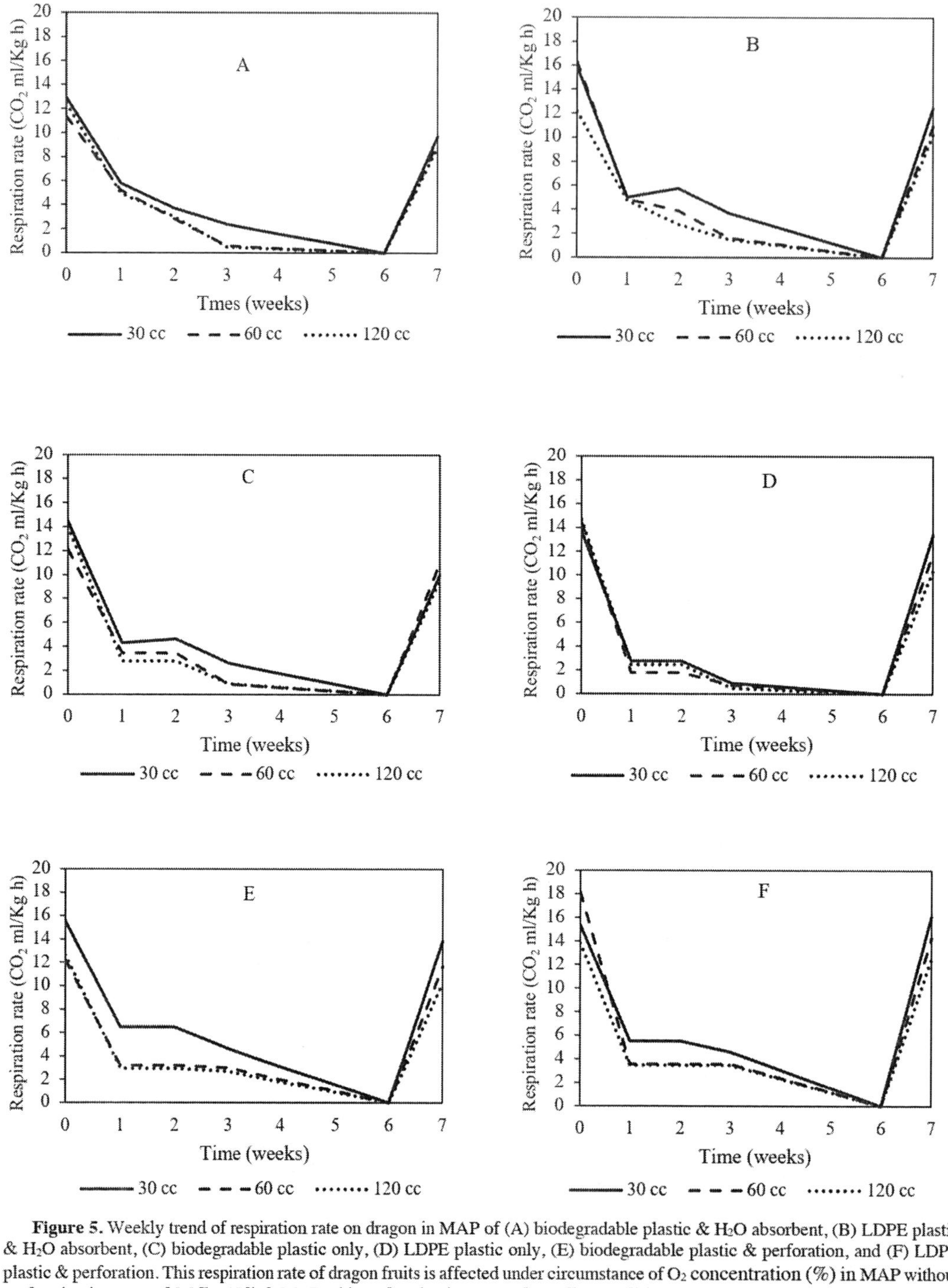

Figure 5. Weekly trend of respiration rate on dragon in MAP of (A) biodegradable plastic & H_2O absorbent, (B) LDPE plastic & H_2O absorbent, (C) biodegradable plastic only, (D) LDPE plastic only, (E) biodegradable plastic & perforation, and (F) LDPE plastic & perforation. This respiration rate of dragon fruits is affected under circumstance of O_2 concentration (%) in MAP without perforation in range of 3.3% - 12% & MAP with perforation in range of 17.7% - 19.2% in 6 weeks of storage.

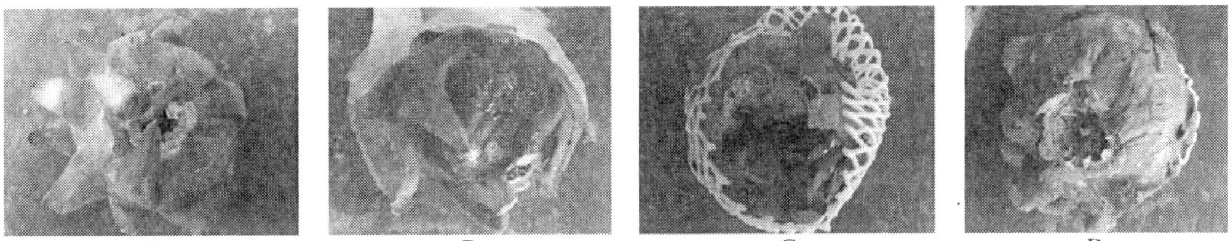

A B C D

Figure 6. Deterioration of dragon fruit quality by the incident of (A) fin wilt, (B) damage fruit skin, (C) cracking, (D) growth mold

Dragon fruit stored at 5°C and 7°C has a better visual appearance after being stored for 20 days compared to fruit stored at 10°C. Storage at 5°C and 7°C retains a greener fin colour and reduces the incidence and severity of spoilage compared to fruit stored at 10°C. Best visual appearance after 20 days of storage and after five storage days at a temperature of 20°C are observed in fruits stored at a temperature of 5°C without plastic bags with holes. However, storage of dragon fruit at 5°C results in a low chilling injury rate characterized by a thin layer of outer flesh tissue under the skin showing symptoms of being submerged in water, an effect not ameliorated by the application of perforated plastic bags. Following 6 weeks in cold storage, dragon fruit subjected to a transition to room temperature at 27°C without the presence of MAP shows exacerbated physical damage across all samples. Indicators of fin wilting, skin damage, cracking, and mold growth manifest upon removal of dragon fruit samples to room temperature in the absence of gas-absorbent-based MAP. These findings also suggest the potential occurrence of chilling injury during cold storage with the utilization of MAP. However, the signs of chilling injury begin to manifest upon transitioning to room temperature.

The manifestation of mild chilling injury becomes evident in dragon fruit when stored at 5°C for a duration of 20 days. It is worth noting that the chilling injury levels observed in this study were considerably lower than those previously reported for the same species cultivated in Israel and subjected to a storage period of two weeks at 6°C. These findings substantiate the proposition that environmental conditions during cultivation significantly influence fruit quality, thereby influencing the optimal harvest time, the fruit's susceptibility to chilling injury, and its potential for storage at low temperatures. Moreover, subjecting dragon fruit to a conditioning process at 25°C for 24 hours prior to cold storage at 2°C serves to diminish its susceptibility to chilling injury, thereby facilitating the storage of dragon fruit at low temperatures [17].

The recommended storage temperature for red dragon fruit (*Selenicereus costaricensis*) is 10°C and for the yellow dragon (*Selenicereus megalanthus*) is 6°C with air humidity of 85-90%. Storage periods at 10°C and 5°C with 90% RH were reported for 14 and 17 days respectively. with air humidity of 85-90%. Storage periods at 10°C and 5°C with 90% RH were reported for 14 and 17 days respectively. Dragon fruit quality is still acceptable for 25-30 days if stored in perforated plastic bags at 4.5°C, however, when stored at room temperature, the shelf life is less than 10 days [18][19].

4 Conclusions

The results showed that the application of gas absorbent-based MAP, additional treatments, and low temperature were generally able to maintain the quality of dragon fruit up to 6 weeks in cold storage. However, dragon fruit showed signs of chilling injury, increased water loss, increased weight loss, decreased firmness, skin damage, fin withering, and mold growth after transfer at 27°C room temperature without MAP. The application of perforation is not recommended because it affects the high oxygen concentration in MAP so that mold growth and loss of moisture content in dragon fruit increase during storage in cold storage. Cold storage at 5°C is also not recommended to avoid chilling injury. The use of a temperature of 10°C and gas absorbent-based MAP with biodegradable plastic or LDPE plastic without perforation is highly recommended.

References

[1] Puspasari, I., & Tasirin, S. M. (2014). *Quality Changes of Red Pitaya (Hylocereus undatus) Slices Dried in Hot Air , Microwave-Hot Air and Microwave-Vacuum Dryers Department of Bioresource Engineering , Faculty of Agricultural and Environmental Sciences ,. 5*(3). https://doi.org/10.5829/idosi.ijee.2014.05.03.11

[2] Nerd, A., Gutman, F., & Mizrahi, Y. (1999). Ripening and postharvest behaviour of fruits of two Hylocereus species (Cactaceae). *Postharvest Biology and Technology, 17*(1), 39–45.https://doi.org/10.1016/S0925-5214(99)00035-6

[3] Singh, N., Banjare, P., & Mondloe, D. (2021). *Increasing shelf life of Dragon fruit by maintaining at optimum temperature. 09*(3), 108–115.

[4] Pascual, M. L. P., Peralta, E. K., Yaptenco, K. F., Elauria, J. C., & Esguerra, E. B. (2017). Passive Modified Atmosphere Packaging for Low Temperature Storage of White Flesh Variety Dragon Fruit (Hylocereus undatus (Haw.) Britton & Rose). *Philippine Journal of*

Agricultural and Biosystems Engineering, *13*(1), 30–41.

[5] Tzoumaki, M. V., Biliaderis, C. G., & Vasilakakis, M. (2009). Impact of edible coatings and packaging on quality of white asparagus (Asparagus officinalis, L.) during cold storage. *Food Chemistry*, *117*(1), 55–63. https://doi.org/10.1016/j.foodchem.2009.03.076

[6] Castro, A. C., Esguerra, E. B., & Franco, R. K. G. (2020). *Modified Atmosphere Packaging and Low Temperature Storage of Red-Fleshed Dragon Fruit (Hylocereus polyrhizus (Weber) Britton & Rose)*. *45*(April), 1–12.

[7] D'Aquino, S., Piga, A., Agabbio, M., & McCollum, T. G. (1998). Film wrapping delays ageing of "Minneola" tangelos under shelf-life conditions. *Postharvest Biology and Technology*, *14*(1), 107–116. https://doi.org/10.1016/S0925-5214(98)00019-2

[8] Chitravathi, K., Chauhan, O. P., & Raju, P. S. (2015). Influence of modified atmosphere packaging on shelf-life of green chillies (Capsicum annuum L.). *Food Packaging and Shelf Life*, *4*, 1–9. https://doi.org/10.1016/j.fpsl.2015.02.001

[9] Fante, C. A., Boas, A. C. V., Paiva, V. A., Pires, C. R. F., & Lima, L. C. de O. (2014). Modified atmosphere efficiency in the quality maintenance of Eva apples. *Food Science and Technology*, *34*(2), 309–314. https://doi.org/10.1590/fst.2014.0044

[10] Githiga, R., Ambuko, J., Hutchnison, M., & Owino, W. (2015). Effect of Activebag® modified atmosphere packaging on the postharvest characteristics of mango fruits, *Mangifera indica* L, cultivar Tommy Atkins. *Journal of Applied Biosciences*, *83*(1), 7535. https://doi.org/10.4314/jab.v83i1.6

[11] Nohl, H. (1994). Generation of superoxide radicals as byproduct of cellular respiration. *Annales de Biologie Clinique*, *52*(3), 199–204. https://pubmed.ncbi.nlm.nih.gov/7998676/

[12] Femenia, A., Sánchez, E. S., Simal, S., & Rosselló, C. (1998). Modification of Cell Wall Composition of Apricots (Prunus armeniaca) during Drying and Storage under Modified Atmospheres. *Journal of Agricultural and Food Chemistry*, *46*(12), 5248–5253. https://doi.org/10.1021/jf9804037

[13] Krupa, T., & Tomala, K. (2021). Effect of oxygen and carbon dioxide concentration on the quality of minikiwi fruits after storage. *Agronomy*, *11*(11). https://doi.org/10.3390/agronomy11112251

[14] Ho, P. L., Tran, D. T., Hertog, M. L. A. T. M., & Nicolaï, B. M. (2020). Modelling respiration rate of dragon fruit as a function of gas composition and temperature. *Scientia Horticulturae*, *263*(October 2019), 109138. https://doi.org/10.1016/j.scienta.2019.109138

[15] C. Lizada, Seymour, G. B., Taylor, J. E., & Tucker, G. A. (1993). Manggo. In *Biochemistry of fruit ripening* (1st ed., pp. 255–271). Chapman & Hall, London. https://doi.org/https://doi.org/10.1007/978-94-011-1584-1

[16] de Freitas, S. T., & Mitcham, E. J. (2013). Quality of pitaya fruit (Hylocereus undatus) as influenced by storage temperature and packaging. *Scientia Agricola*, *70*(4), 257–262. https://doi.org/10.1590/S0103-90162013000400006

[17] Nerd, A., & Mizrahi, Y. (1999). *The effect of ripening stage on fruit quality after storage of yellow pitaya*. *15*, 99–105

[18] Johnson, G. I., To, L. Van, Duc, N. D., & Webb, M. C. (2000). Quality assurance in agricultural produce. *ACIAR Proceedings No. 100*, 101–114.
.

[19] Zee, F., Yen, C., Nishina, M., & Rica, C. (2004). (Dragon Fruit, Strawberry Pear). *Fruit and Nuts*, 8–10. http://hdl.handle.net/10125/2403

Palm Oils Consumption Modulates Serum SGPT and SGOT in Rats Fed a High-Fat Diet

Zaki Utama*, Sri Raharjo, Agnes Murdiati and Andriati Ningrum

Department of Food and Agricultural Product Technology, Faculty of Agricultural Technology, Universitas Gadjah Mada, Yogyakarta, Indonesia

**Corresponding author: zaki_utama@ugm.ac.id*

Abstract. In this study we investigated the effect of three kinds of palm oil product (refined bleached deodorized palm oil (RBDPO), red palm oil (RPO) and palm kernel oil (PKO)) on the serum glutamic oxaloacetic transaminase (SGOT), serum glutamic pyruvic transaminase (SGPT), and liver color of rats in fed high-fat diet, and compare it with beef tallow and corn oil. Growing male SD rats were fed a high-fat diet (20% fat) for 13 weeks, divided into 2 groups (with and without azoxymethane (AOM)). Dietary corn oil showed the lowest serum SGOT and SGPT ($p < 0.05$) both in the AOM and Non-AOM groups. Beef tallow diet caused the highest level of serum SGOT and SGPT followed by PKO, RPO, and RBDPO respectively ($p < 0.05$). The serum SGOT and SGPT were not affected by the AOM treatment. Although the liver color from the AOM group was relatively darker than the non-AOM group, in general the fat sources did not have a significant effect on liver color.

Keywords rat · palm oil · fat · SGOT · SGPT · liver color

1. Introduction

Palm oil has become the most important vegetable oil in the world and is widely used in the food processing industry [1]. The fatty acid composition of palm oil is mainly composed of palmitic acid (44%) and oleic acid (39%), with the total saturated fatty acids and unsaturated fatty acids similar in percentage (49% and 49%) [2]. Although the consumption of palm oil often causes concern because of its composition, currently there is no scientific evidence (experimental or epidemiological) to warrant associating palm oil consumption with a higher cancer incidence or mortality risk in humans [3]. Nevertheless, the health benefits of red palm oil in animal and human studies have been reviewed, including improvement in cardioprotective effects in ischemic heart disease, antiatherogenic, antihemorrhagic, antihypertensive, and anticancer properties [4].

Animal models (rat and/or mouse) have been widely used in studies involving high-fat diets, including the organ-specific pathologies associated with exposure to a maternal high-fat diet [5]. The application of a high-fat diet in rats has been developed for the induction of non-alcoholic fatty liver disease [6]. Serum SGOT and SGPT has been considered as a marker for fatty liver disease, with high levels of serum SGOT and SGPT resembling the development of the disease.

The effects of the type of fat sources in the diet to the serum SGOT and SGPT is a question that remains to be answered, especially regarding the palm oils. The present study was conducted to compare the serum SGOT and SGPT of rats fed with 3 different types of palm oil sources (RBDPO, RPO and PKO) and also compared with beef tallow and corn oil.

2. Materials and methods

2.1. Animal and experimental protocols

Male Sprague Dawley rats (3 weeks old) were provided by The Center for Food and Nutrition Studies Laboratory Animal Center, Universitas Gadjah Mada and housed in individual cages in a room with a temperature of 25-26 degrees Celsius and relative humidity of 50-60 percent, with a 12-hour light-dark cycle. The rats were fed a pellet meal for one week before the start of the experimental diet and were randomly allocated to one of five groups. Animal care is carried out following the "Guide for the Care and Use of Laboratory Animals" created by Universitas Gadjah Mada and authorized by the same university's Medical and Health Research Ethics Committee. Rats were divided into 5 groups (n = 6), and each group fed by the following experimental diets (*ad libitum*): 20% beef tallow, 20% corn oil, 20% RBDPO, 20% RPO, and 20% PKO. The composition of the experimental diets was presented in Table 1. AOM (Sigma, St. Loius, MO) was administered with oral gavage in saline at 15 mg/kg body, twice at 2nd and 3rd week. Thirteen weeks after feeding the diets, animals were killed by decapitation with anesthesia of diethyl ether, and the whole blood collected from the abdominal aorta. Blood samples were centrifuged at 2000 x g for 20 min to obtain serum samples, and stored at –70˚C.

2.2. Determination of Serum SGOT and SGPT

Serum SGOT levels were measured using Diasys ASAT (GOT) FS quantification kit following the manufacturer's instructions (DiaSys Diagnostic Systems GmbH, Holzheim, Germany). Serum SGPT levels were measured using Diasys ALAT (GPT) FS quantification kit following the manufacturer's instructions (DiaSys Diagnostic Systems GmbH, Holzheim, Germany).

Table 1. Compositions of the diets

Ingredient (w/w %)	Beef Tallow	Corn Oil	RBDPO	RPO	PKO
Casein	20.00	20.00	20.00	20.00	20.00
Corn starch	39.95	39.95	39.95	39.95	39.95
AIN-93 vitamins	1.00	1.00	1.00	1.00	1.00
AIN-93 minerals	3.50	3.50	3.50	3.50	3.50
Cellulose	5.00	5.00	5.00	5.00	5.00
Sucrose	10.0	10.0	10.0	10.0	10.0
L-Cystine	0.30	0.30	0.30	0.30	0.30
Choline	0.25	0.25	0.25	0.25	0.25
Beef tallow	20.00	-	-	-	-
Corn Oil	-	20.00	-	-	-
RBDPO	-	-	20.00	-	-
RPO	-	-	-	20.00	-
PKO	-	-	-	-	20.00

Abbreviations used: RBDPO, refined bleached deodorized palm oil; RPO, red palm oil; PKO, palm kernel oil.

2.3. Smartphone Setup

A smartphone (Model iPhone 11 Pro Max, Apple Inc., California USA) was used by holding (handheld) it vertically just above the rat organ at a distance about 20 cm. The main camera (wide angle) and native camera application were used for image acquisition. The camera settings as follows: camera capture most compatible format, ISO auto, shutter speed auto, aperture lens auto, white balance auto, exposure program normal, metering mode pattern, no flash, daylight conditions, resolution 4032x3024 pixels, color space RGB, with the images saved in jpeg.

2.4. Image Analysis

The Adobe Photoshop CC 2019 Software for Mac (Adobe Systems Inc., San Jose, California) was used to obtain the color parameters from the digital images of rat organ surfaces. The images were open in photoshop with image mode Lab Color chosen. The software was capable of displaying the color parameters *L* *, *a* *, *b* *, a model commonly used in food research [7]. The measurement or color analysis of the sample was based on what has been suggested by Papadakis *et al.* [8]. One of the tools in Photoshop used in this research was the Magic Wand Tool and Quick Selection Tool. With these tools, we can choose the field of images that we want to be analyzed further. We can select the dark or bright surface part of the sample with the Magic Wand Tool, for which we can then measure *L* *, *a* *, *b* *, and the percentage of the color surface differences. The selected image range was controlled with the tolerance value of the Magic Wand Tool, which is used for this research value 50.

2.5. Statistical analysis

To compare mean values, one-way ANOVA was used to compare between different groups of fat sources (beef tallow, corn oil, RBDPO, RPO, and PKO). The Tukey-Kramer post hoc test was conducted when a significant effect was found by one-way ANOVA. Data analysis was performed using SPSS Statistics for Mac (version 26.0; IBM SPSS Statistics, SPSS Inc., Chicago, USA). Differences were considered as differently significant at $p < 0.05$.

3. Results and discussions

The levels of serum SGOT and SGPT in five groups of rats fed with high-fat diets which different in fat sources were shown in Table 2. Dietary corn oil showed the lowest serum SGOT and SGPT ($p < 0.05$) both in the AOM and Non-AOM groups. Beef tallow diet caused the highest level of serum SGOT and SGPT followed by PKO, RPO, and RBDPO respectively ($p < 0.05$).

The disorder of liver function can be evaluated through biomarkers such as serum SGOT and SGPT. Found mainly in the liver, the enzyme GOT (also known Aspartate Aminotransferase/AST) and GPT (also known as Alanine Aminotransferase/ALT) catalyze the conversion of ⟨-keto acids into amino acids by transfer of amino groups. As a liver specific enzyme, GPT/ALT is only significantly elevated in hepatobiliary diseases. While the increased GOT/AST levels can occur in the connection with damages of heart or skeletal muscle as well as of liver parenchyma. Under normal conditions, the level of SGOT and

SGPT were maintained within a certain range. Therefore, the increase and/or high level of SGOT and SGPT could indicate liver tissue damage or infection occurrence. In acute hepatotoxicity, the levels of liver enzymes usually increased, but with prolonged intoxication because of the damage to the liver the levels could be decreased [9].

Figure 1 showed the image of the rat's liver after we applied the selection (magic wand) tool in the image editor (Adobe Photoshop). The liver color in five groups of rats fed with high-fat diets which different in fat sources were shown in Table 3. Although the liver color from the AOM group was relatively darker than the non-AOM group, in general the fat sources did not have a significant effect on liver color.

Table 2. Effect of the experimental diets on serum SGOT and SGPT in rats

Groups	Beef Tallow	Corn Oil	RBDPO	RPO	PKO
Serum SGOT (U/L)					
AOM	57.0 ± 0.7^{a}	38.8 ± 0.3^{d}	40.6 ± 0.6^{cd}	43.0 ± 0.7^{bc}	44.5 ± 0.9^{b}
Non-AOM	58.6 ± 0.4^{a}	38.6 ± 0.6^{c}	41.0 ± 0.2^{ab}	43.2 ± 0.3^{b}	43.2 ± 0.7^{b}
Serum SGPT (U/L)					
AOM	36.3 ± 0.6^{a}	18.9 ± 0.3^{d}	21.9 ± 0.5^{c}	23.8 ± 0.5^{c}	26.7 ± 0.3^{b}
Non-AOM	37.1 ± 0.4^{a}	19.7 ± 0.2^{d}	22.6 ± 0.2^{cd}	24.4 ± 0.5^{bc}	26.9 ± 1.4^{b}

Data are expressed as means ± S.E.M., $n = 3$. ^{a,b,c,d} Values in a row not sharing a superscript are significantly different at $p < 0.05$. Abbreviations used: RBDPO, refined bleached deodorized palm oil; RPO, red palm oil; PKO, palm kernel oil.

Table 3. The L^*, a^*, and b^* mean values of the liver of rats fed with beef tallow, corn oil, RBDPO, RPO and PKO

Groups	Beef Tallow	Corn Oil	RBDPO	RPO	PKO
L^*					
AOM	19.30±1.52	18.97±5.65	17.79±0.79	20.70±6.50	18.61±2.54
Non-AOM	24.91±3.37	22.06±0.62	24.02±9.38	25.19±0.99	20.50±1.75
a^*					
AOM	20.49±2.32	22.08±4.39	18.72±1.36	24.73±4.30	21.74±3.34
Non-AOM	23.74±2.36	21.75±1.80	22.70±7.47	27.29±1.13	24.72±1.29
b^*					
AOM	9.44±0.03	8.90±2.85	6.76±1.30	10.35±2.47	9.36±2.83
Non-AOM	12.50±3.19	10.12±1.49	10.59±6.58	11.74±1.02	10.90±2.69

Data are expressed as means ± S.E.M., $n = 3$. ^{a,b,c,d} Values in a row not sharing a superscript are significantly different at $p < 0.05$. Abbreviations used: RBDPO, refined bleached deodorized palm oil; RPO, red palm oil; PKO, palm kernel oil.

Figure 1. The liver image after we applied Magic Wand Tool and/or Quick Selection Tool in Photoshop for the Lab color analysis (AOM group, A: beef tallow; B: corn oil; C: RBDPO refined bleached deodorized palm oil; D: RPO, red palm oil; and E: PKO, palm kernel oil)

4. Conclusion

Dietary corn oil showed the lowest serum SGOT and SGPT both in the AOM and Non-AOM groups. Beef tallow diet caused the highest level of serum SGOT and SGPT followed by PKO, RPO, and RBDPO respectively. In general, differences in fat sources (beef tallow, corn oil, palm oil) did not have a significant effect on liver color.

5. Acknowledgments

The research work was funded by The Ministry of Finance Republic of Indonesia through The Indonesian Endowment Fund for Education (BUDI-DN LPDP).

References

1. Kadandale S, Marten R, Smith R, 2019. The palm oil industry and noncommunicable diseases. *Bull World Health Organ* 97: 118–128.
2. Mancini A, Imperlini E, Nigro E, Montagnese C, Daniele A, Orru S, Buono P, 2015. Biological and nutritional properties of palm oil and palmitic acid: effects on health. *Molecules* 20: 17339−17361.
3. Gesteiro E, Guijarro L, Sanchez-Muniz FJ, Vidal-Carou MC, Troncoso A, Venanci L, Jimeno V, Quilez J, Anadon A, Gonzales-Gross M, 2019. Palm oil on the edge. *Nutrients* 11: 2008, doi:10.3390/nu11092008.
4. Loganathan R, Subramaniam KM, Radhakrishnan AK, Choo YM, Teng KT, 2017. Health-promoting effects of red palm oil: evidence from animal and human studies. *Nutrition Reviews* 75(2): 98–113.
5. Williams L, Seki Y, Vuguin PM, Charron MJ, 2014. Animal models of in utero exposure to a high fat diet: A review. *Biochimica et Biophysica Acta* 1842: 507–519.
6. Kucera O, Cervinkova Z, 2014. Experimental models of non-alcoholic fatty liver disease in rats. *World J Gastroenterol* 20: 8364–8376.
7. Leon K, Mery D, Pedreschi F, Leon J, 2006. Color measurement in L*a*b* units from RGB digital images. *Food Research International* 39: 1084–1091.
8. Papadakis SE, Malek SA, Kamdem RE, Yam KL, 2000. A versatile and inexpensive technique for measuring color of foods. *Food Technology*, 54(12): 48–51.
9. Obi E, Orisakwe OE, Asomugha LA, Udemezue, OO, 2004. The hepatotoxic effect of halofantrine in guinea pigs. *Indian Journal of Pharmaceutical Sciences*, 36(5):303–305.

Physical characteristics of extrudate from mixed corn grit-soybean flour with treatments of moisture content and extruder barrel temperature

Riski Sri Mahanani[1], *Joko* Nugroho Wahyu Karyadi[1*], *Devi* Yuni Susanti[1], *Ailsa* Saraswati Samodra[1], *Vincentius* Ferry Surya Dwinata[1], *Dian* Kharisma Rahmawati[1], Prasetya Febriyansyech Personanta Sinuhaji[1]

[1]*Departement of Agricultural and Biosystem Engineering, Universitas Gadjah Mada, Jl. Flora No.1, Bulaksumur, Yogyakarta 55281, Indonesia*

Abstract. Corn is a commonly used raw material for snack production. However, corn has a relatively low protein content, approximately 7%. Consequently, adding soybean flour is expected to enhance the protein content in snacks. This study investigates the impact of moisture content and barrel temperature treatments on the physical properties of the produced extrudates. Snacks were manufactured using a SYSLG-IV twin-screw extruder with a capacity of 10-15 kg/h and a die diameter of 6 mm. The raw material utilized in the experiment comprised a mixture of 20% soybean flour and 80% corn grit. The study involved two main treatments: barrel temperature (at 120°C, 130°C, and 140°C) and moisture content (at 14%, 16%, and 18%). Parameters assessing the physical properties of the extrudates included expansion ratio, hardness, moisture content, water absorption index, and water solubility index. The results demonstrated that increased moisture content led to increased extrudate moisture content (ranging from 1.80% to 4.71%) and hardness (ranging from 17.45 N to 40.16 N). Conversely, it caused a decrease in the expansion ratio (ranging from 2.11 to 2.03), water solubility index (ranging from 8.82% to 7.01%), and water absorption index (ranging from 5.70 to 4.92).

1 Introduction

Indonesia is one of the top two countries within the Asia Pacific region with the biggest snacking habit. Globally, the revenue in the snack food segment amounts to USD 7.22 billion in 2022 and is expected to grow annually by 11.03% (CAGR 2022-2027)[1]. Indonesia's snacking habits have been seen consistently exhibiting that the snack market has the potential to expand in Indonesia. Extruded snacks were among the most commercially successful extruded foods. In recent years, extrusion technology has become a popular processing method due to its advantages, such as high versatility, high productivity, and low nutritional loss due to a high-temperature short-time (HTST) processing method [2]. Generally, extruded products are produced from high-starch raw materials such as corn, rice, wheat, potato, sorghum, cassava, oats, and barley [3,4].

Corn grit is one of the commonly used materials to produce extrudate snacks. However, corn has been reported for its limited nutrient profile with poor protein and dietary fiber contents [4]. Therefore, there is a need to enhance snacks' nutritional value. Soybeans can be added to enhance nutrients in corn-based extrudate snacks. Soybeans are known to contain more protein compared to other types of legumes [5]. In addition, Indonesia has a high interest in soybeans and their processed products, reaching 7 million tons per year [6].

Some studies have been reported on the incorporation of starch/flour and soybean into extruded products as a source of protein [7–11]. Extrusion is a high-temperature short time (HTST), in which raw materials are melted and cooked inside the barrel extruder by a combination of moisture content, pressure, temperature, and shear rate, resulting in starch gelatinization, protein denaturation, molecular transformation, chemical reaction, and vitamins degradation [12], [13] can drastically influence the final product quality. Extrusion process conditions such as barrel temperature, feed moisture content, and screw speed are critical for extruded snacks' quality characteristics [14]. Based on previous literature, it was reported that barrel temperature and moisture content significantly affected the quality of extrudates [8–15].

The expansion ratio is one of the most desirable physical properties for extruded snacks because it determines their quality [16]. However, there was limited information about corn grit and soybean flour in producing extruded snacks. Therefore, this study's primary objectives were to investigate the impact of moisture content and barrel temperature treatments on the physical properties such as expansion ratio, extrudate moisture content, hardness, particle density,

* Corresponding author: jknugroho@ugm.ac.id

bulk density, water solubility index, and water absorption index.

2 Materials and methods

2.1 Materials

Corn grits (*Zea mays*) were obtained from CV. Surya Grain Indonesia. Soybeans (*Glycine max* L.) were purchased from CV. Hasil Indah and ground with a disk mill. Corn grits and soybean flour were then sieved to pass through 24- and 80-sieve, respectively. A total of 21.6 kg of corn grit and 5.4 kg of soybean flour were used throughout this study.

2.2 Equipment

The primary equipment used in this study was a twin-screw extruder SYSLG-IV (Shandong Saibainuo Machinery, Co., Ltd., China) equipped with a circular die with a 6 mm diameter. Disk mill (Shandong Jimo FFC-15, China), digital scale (Ohaus PA4102, USA), and mixer (Ossel B7, Indonesia) were used to prepare the feed material. Forced air oven (Sanyo MOV-112, Japan), analytical balance (Shimadzu AUW220, Japan), vernier caliper, blender (Panasonic MXT1GN, Japan) vortex mixer (DLAB-MXS, USA), centrifuge (Kokusan H-27F, Japan), texture analyzer (Brookfield CT-3, USA) and 1000 mL-graduated cylinder were used to measure physical properties of extrudates.

2.3 Feed preparation extrusion process

Soybean flour and corn grit were mixed using a mixer (Ossel B7, Indonesia) at a 20% soybean flour and 80% corn grit ratio. The mixture's total moisture contents were adjusted to the desired values before extrusion. The mixture was stored at room temperature in polypropylene bags for 24 hours to achieve equilibrium.

2.4 Extrusion processing

Prior to an extrusion process, extruder screw speed, feeder speed, and cutter speed were adjusted at 20 rpm, 15 rpm, and 6 rpm, respectively, throughout this study. Barrel temperatures were set at 40°C, 50°C, 100°C, and 120/130/140 °C. The raw materials were then fed into the extruder. After steady-state conditions were reached (no visible fluctuations in the electric current of the controller and stable extrudate output), extrudates were collected and dried in a forced-air oven at 105°C for 3 hours. The dried samples were stored in a polypropylene standing pouch with silica gel and oxygen absorber until further analysis.

2.5 Experimental design

Experiments were designed using the 3x3 complete randomized design (CRD). The independent variables were moisture content (14, 16, and 18%) and barrel temperature (120, 130, and 140°C). The dependent variables observed in the study were expansion ratio, extrudate moisture content, hardness, bulk density, particle density, water absorption index (WAI), and water solubility index (WSI). Each treatment was done in triplicate.

Table 1. Experimental design for the extrusion process

Moisture content (MC)	Barrel temperature (T)		
	120°C	130°	140°
14%	T120MC14	T130MC14	T140MC14
16%	T120MC16	T130MC16	T140MC16
18%	T120MC18	T130MC18	T140MC18

Whereas T: barrel temperature (°C); MC: moisture content (%)

2.6 Product characteristic

2.6.1 Expansion ratio (ER)

The expansion ratio was determined by dividing the extrudate diameter (measured with a vernier caliper) by the diameter of the extruder die [14]. A total of 10 samples of extruded snacks were taken randomly. The expansion ratio was calculated using Equation 1.

$$ER = \frac{\text{extrudate diameter}}{\text{die diameter}} \quad (1)$$

2.6.2 Extrudate moisture content (EMC)

Extrudate was dried in a forced air oven at 105°C for 24 hours until a constant weight was obtained [17]. Samples were then weighed using an analytical balance, and extrudate moisture content was calculated using Equation 2.

$$EMC\ (\%) = \frac{\text{mass of water}}{\text{mass of moist sample}} \quad (2)$$

2.6.3 Particle density (PD) and bulk density (BD)

Particle density was determined as described by [18]. The mass of extrudates was weighed using an analytical scale, and extrudate dimension was measured using a vernier caliper. The particle density of extrudates was then calculated following Equation 3, assuming a cylindrical extrudate shape. Ten randomly collected extrudates were measured

$$PD = \frac{4 \text{ x mass}}{\pi \text{ x } d^2 \text{ x } L} \quad (3)$$

Extrudates were poured into a 1000 mL measuring cylinder and weighted using a digital scale to measure bulk density. The cylinder was gently tapped on a flat surface until a constant volume was obtained [19]. The bulk density was then calculated using Equation 4.

$$BD = \frac{\text{mass of the sample}}{\text{volume of the sample}} \quad (4)$$

2.6.4 Hardness

The maximum force obtained has been taken as a measure of the hardness of the products. Low hardness value has been regarded as a desirable property for extruded snacks [20]. A texture analyzer (Brookfield CT-3, USA) equipped with a 10 kg load cell was used to test the hardness of the extruded snack. TA-39 stainless probe with a cylinder diameter of 2 mm and 20 mm length was used. Analysis was performed with the following settings: pretest speed 2mm/s; test speed 2 mm/s; return speed 1mm/s; and target distance 7 mm. Nine randomly collected extrudates were measured and reported as average.

$$\text{Hardness} = \frac{load\ (g)}{1000} \times 9.81\ \frac{m}{s^2} \qquad (5)$$

2.6.5 Water absorption index (WAI) and water solubility index (WSI)

The water absorption index indicates the amount of water immobilized by the extrudate. In contrast, water solubility indicates the amount of small molecules of soluble polysaccharide released from the starch component after extrusion. The WAI dan WSI was measured as described by [2] with minor modifications. The WAI and WSI were calculated using Equations 5 and 6, respectively. The determination of WAI and WSI was done in triplicate.

$$\text{WAI} = \frac{\text{weight of wet sediment}}{\text{weight of dry sample}} \qquad (6)$$

$$\text{WSI} = \frac{\text{weight of dry solids in supernatant}}{\text{dry weight of extrudate}} \times 100\% \qquad (7)$$

2.7 Statistical analysis

All experiment data were analyzed using IBM SPSS 27 software. The experimental data were analyzed using analysis of variance (ANOVA) with significance defined at p<0,05. The analysis of variance was carried out to identify the effect of operating conditions (moisture content and barrel temperature) on the final product quality.

3 Result and discussion

Table 2. Significance of the main effects and their interactions for the physical characteristics of extrudate from corn grit – soybean flour mixture

Extrusion variable	Barrel temperature	Moisture content	Interaction of both
ER	N	S	S
EMC	NS	S	S
PD	S	S	S
BD	S	S	S
H	S	NS	NS
WAI	S	S	S
WSI	S	NS	NS

- S: significant at p<0,05; NS: Not significant
- ER: expansion ratio; EMC: extrudate moisture content; PD: particle density; BD: bulk density; H: hardness; WAI: water absorption index; WSI: water

3.1 Expansion ratio

Expansion ratio (ER) was described as a degree of puffing by the sample as it exits the extruder die [7]. Moisture content, barrel temperature, and their interaction had a significant effect at p<0.05 (Table 2). ER of extrudates ranged from 1.81–2.28, with the highest value obtained at 14% moisture content and 140°C barrel temperature treatment, while the lowest value at 18% moisture content and 120 °C barrel temperature treatment (Figure 1). ER of extrudate decreased as moisture content increased. Water acts as a plasticizer in the dough melts. Reducing the viscosity of the dough melts in the extruder would cause a decrease in the cooking degree of raw materials and lower the ER [21].

Another study reported that the expansion ratio decreased with increasing moisture content from 15.5% (2.58) to 18.5% (2.08). High moisture content reduced the friction between the materials and the barrel wall when shearing. On the other hand, ER of extrudates increased as barrel temperature increased from 75°C (1.63) to 125°C (2.08) [22]. Higher barrel temperature was attributed to a higher degree of gelatinization, increasing the ER. Moreover, in high barrel temperatures, moisture in the dough became superheated, increasing bubble formation and producing a porous structure [22,23]

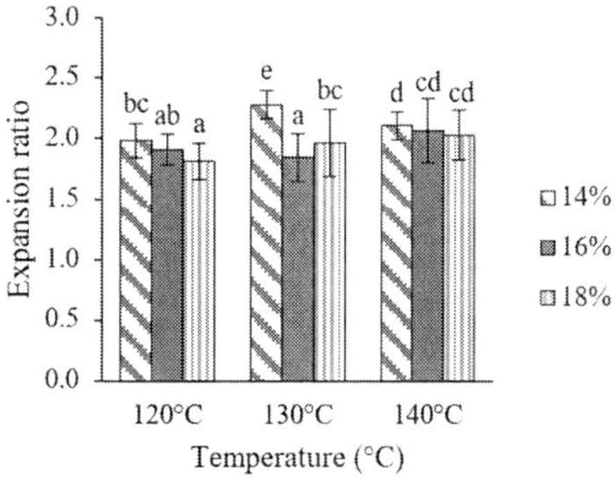

Fig. 1. Expansion ratio of extrudate under moisture content and barrel temperature treatment

3.2 Extrudate moisture content

Moisture content is a crucial parameter to determine the extrudate's final quality. As shown in Table 2, it was found that only feed moisture content and interaction between moisture content and barrel temperature had a significant effect on extrudate moisture content (p<0.05). The moisture content of corn grit-soybean flour extrudate is in the range of 1.80–4.65%, as can be seen in Figure 2. Feed moisture content had a positive correlation with extrudate moisture content. A previous study obtained a similar result and reported that extrudate moisture content increased when the moisture content of the feed was increased [24]. The extrudate moisture content increased with increasing moisture content from 18% (6.15) to 22% (6.19). Water acts as a lubricant in the extruder, lowering the shear force, pressure, and temperature generated inside the extruder, thus decreasing the amount of water vaporization. A significant difference was not observed in barrel temperature on the moisture content of the extrudate.

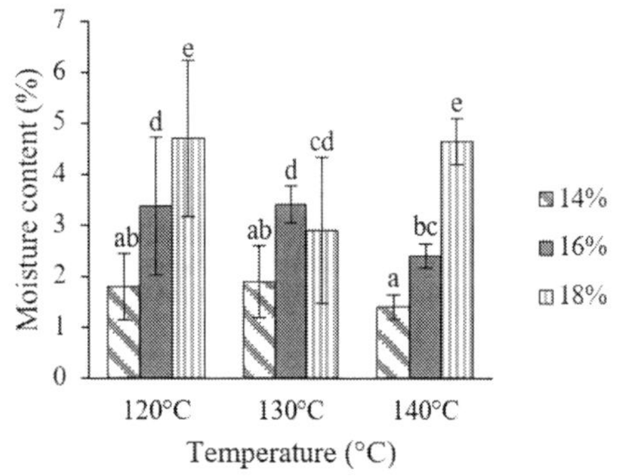

Fig. 2. Extrudate moisture content under moisture content and barrel temperature treatment

3.3 Particle density and bulk density

Particle density (PD) and bulk density (BD) of extrudates were significantly affected by moisture content, barrel temperature, and their interaction ($p<0.05$). Moisture content had a positive correlation with PD and BD. PD of extrudates ranged from 0.28 – 0.56 g/cm3, and BD of extrudates ranged from 0.16–0.41 g/cm3—increasing moisture content results in a higher density and more compact extrudate. A similar study reported that BD extrudate increased with increasing moisture content from 13% (0.114) to 17% (0.156). However, barrel temperature had a converse relationship with PD and BD. High barrel temperature promotes the formation of air bubbles along with the decrease of melt viscosity [25]. Another study also reported the negative correlation between barrel BD extrudate with barrel temperature in the range of 160˚C (0,147) to 200˚C (0.128) on defatted soybean meal, germinated brown rice meal, mango peel fiber, and corn grit-based extrudate [15]

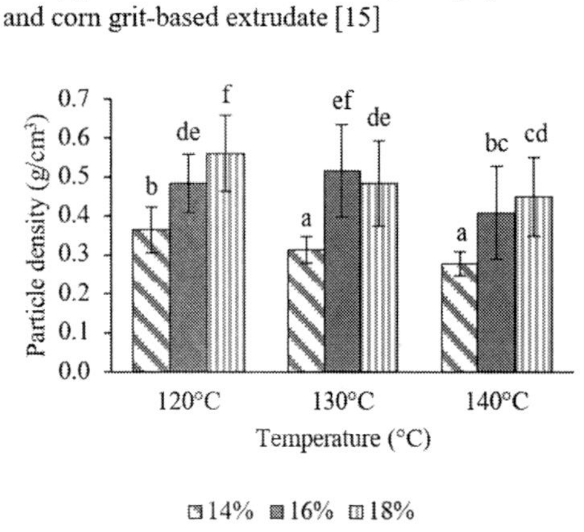

Fig. 3. Particle density of extrudates under moisture content and barrel temperature treatment

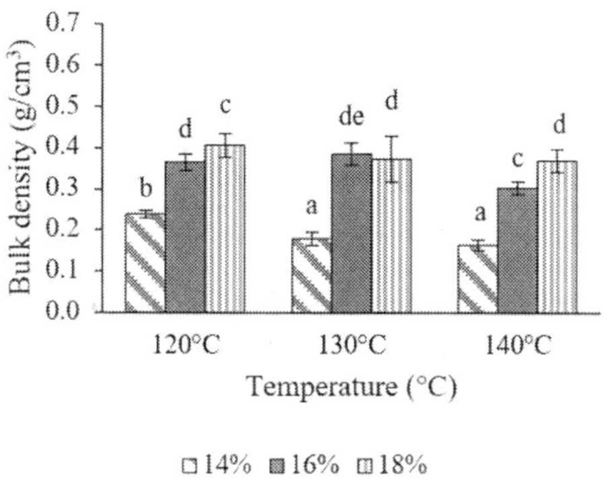

Fig. 4. Bulk density of extrudates under moisture content and barrel temperature treatment

3.4 Hardness

Hardness has a significant influence on consumer acceptability of extruded snacks. Consumer highly prefers extrudate with minimum hardness. The hardness of extrudate was significantly affected by moisture content ($p<0.05$). However, barrel temperature did not significantly affect extrudate hardness ($p<0.05$) (Table 2). The extrudate's hardness ranged from 17.45–53.00 N, with the lowest hardness obtained at 14% moisture content and 140˚C barrel temperature treatment. Increased moisture content significantly increases extrudate hardness ($p<0.05$). Increased moisture content could lower the cooking degree of starch in the material and reduce water vapor pressure in the extruder [25]. This decrease in pressure could then lead to reduced bubble growth in the material, resulting in harder extrudate [26]. A similar trend has been reported that hardness increased with increasing moisture content from 20% (20.39N) to 28% (33.13N) for corn, finger millet, elephant foot yam, and defatted soybean flour-based extrudate [27].

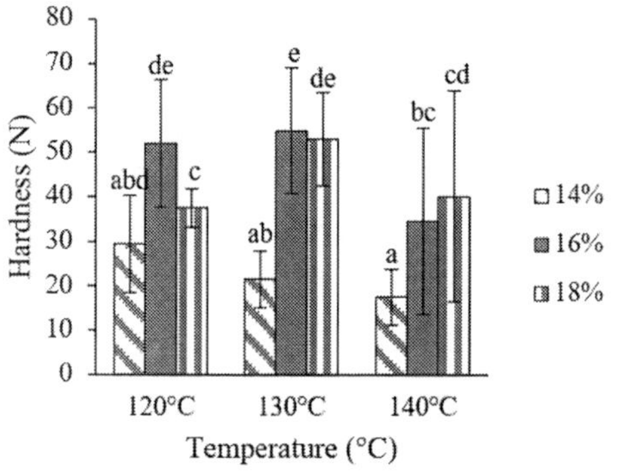

Fig. 5. Extrudate hardness under moisture content and barrel temperature treatment

3.5 Water absorption index and water solubility index

The water absorption index (WAI) reflects the ability of starch to absorb water and is used as an

https://doi.org/10.1051/bioconf/20238004006

indicator of the degree of gelatinization. Both variables and their interaction affected the WAI of extrudates significantly ($p<0.05$) (Table 2). WAI of extrudates varied from a minimum value of 4.92 to a maximum value of 7.01 (Figure 6). Another study [7] reported that WAI values of extrudates ranged from 6.62 to 5.46. It was found that WAI tends to increase at a higher temperature. This increase was probably caused by the more significant damage polymer chain and hydrophilic bonds at higher temperatures, causing a reduction in the binding ability of molecules [27].

However, increased moisture content led to a reduction in WAI values. Water can act as a lubricant inside the extruder, reducing the shear degradation of starch and reducing the WAI of extrudates [27]. Meanwhile, WSI is used as an indicator of molecular component degradation. The WSI value of extrudates was only significantly affected by barrel temperature. WSI of extrudates varied from 7.01–10.18% (Figure 7). Higher barrel temperature caused starch molecules to degrade into smaller molecular weight fractions and increase solubility, increasing WSI [25]. Even though moisture content did not significantly affect the WSI of extrudates, WSI values tended to decrease as moisture content increased.

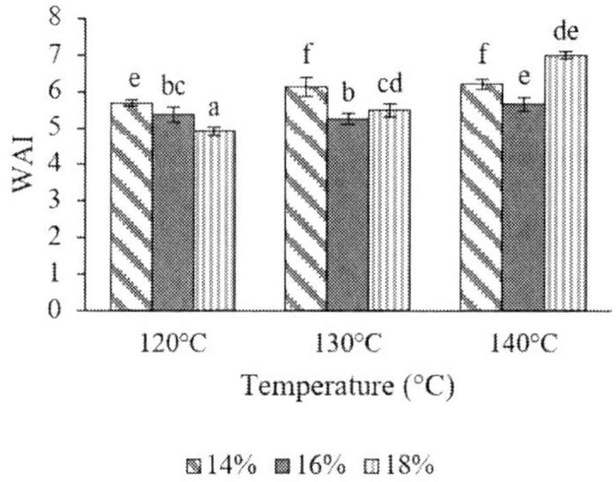

Fig. 6. Water absorption index of extrudate under moisture content and barrel temperature treatment

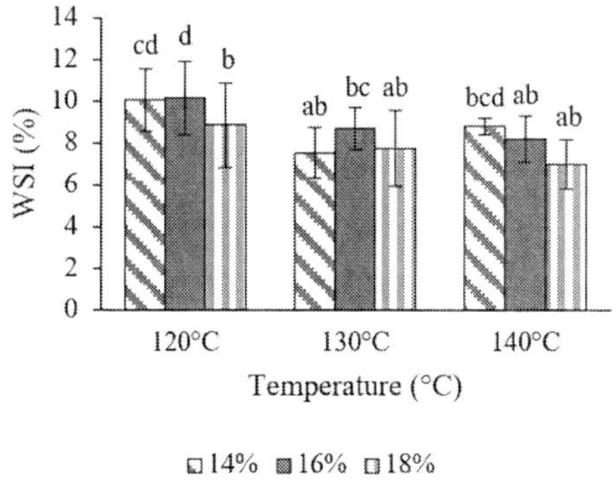

Fig. 7. Water solubility index of extrudate under moisture content and barrel temperature treatment

4 Conclusion

The result showed that moisture content and barrel temperature affected extrudate characteristics. An increase in the feed moisture content affects the expansion ratio negatively. It positively affects the extrudate moisture content, particle density, bulk density, and water absorption index. At the same time, barrel temperature positively affected the expansion ratio and water solubility index and negatively affected the particle density, bulk density, water absorption index, and hardness.

The authors would like to thank RTA Program Universitas Gadjah Mada with Grant Number 5075/ UNI.P.II/Dit-Lit/PT.01.01/2023 for the support.

References

[1] Aurell 2022 The Opportunities of the Indonesian Snack Industry *https://brightindonesia.net/2022/08/11/the-opportunities-of-the-indonesian-snack-industry/*

[2] Ačkar Đ, Jozinović A, Babić J, Miličević B, Panak Balentić J and Šubarić D 2018 Resolving the problem of poor expansion in corn extrudates enriched with food industry by-products *Innovative Food Science and Emerging Technologies* **47** 517–24

[3] Makowska A, Mildner-Szkudlarz S and Obuchowski W 2013 Effect of brewer's spent grain addition on properties of corn extrudates with an increased dietary fibre content *Pol J Food Nutr Sci* **63** 19–24

[4] Offiah V, Kontogiorgos V and Falade K O 2019 Extrusion processing of raw food materials and by-products: A review *Crit Rev Food Sci Nutr* **59** 2979–98

[5] Rizzo G and Baroni L 2018 Soy, soy foods and their role in vegetarian diets *Nutrients* **10**

[6] Pusat Data dan Sistem Informasi Pertanian 2020 *Outlook Komoditas Pertanian Tanaman Pangan: Kedelai* ed A A Susanti and A Supriyatna (Jakarta: Kementerian Pertanian)

[7] Unsaeng K, Hannanta-Anan P and Phongpipatpong M 2019 Effects of extrusion variables on extrudate characteristics from corn-based enriched with soybean residue *IOP Conference Series: Earth and Environmental Science* vol 301 (Institute of Physics Publishing)

[8] Zheng H, Yan G Sen, Lee Y, Alcaraz C, Marquez S and de Mejia E G 2020 Effect of the extrusion process on allergen reduction and the texture change of soybean protein isolate-corn and soybean flour-corn mixtures *Innovative Food Science and Emerging Technologies* **64**

[9] Filli K B, Nkama I, Abubakar U M and Jideani V A 2010 Influence of of extrusion variables on some functional properties of

 https://doi.org/10.1051/bioconf/20238004006

extruded millet-soybean for the manufacture of "fura": A Nigerian traditional food *African Journal of Food Science* **4** 342–52

[10] Li X, Zhang T, An Y, Yin T, Xiong S and Rong H 2022 Physicochemical Characteristics and Flavor Properties of Texturized Dual-Proteins Extrudates: Effect of Surimi to Soybean Flour Ratio *Foods* **11**

[11] Zhang W, Li S, Zhang B, Drago S R and Zhang J 2016 Relationships between the gelatinization of starches and the textural properties of extruded texturized soybean protein-starch systems *J Food Eng* **174** 29–36

[12] Tang H K, Lui W B and Peng J 2014 Optimization of Extrusion Conditions for GABA Tea Powder/Corn Grit Blend Extrudates *J Food Process Preserv* **38** 2108–16

[13] Varsha K and Mohan S 2016 Extruded Product Quality Assessment Indices: A Review *International Journal of Agriculture Sciences* **8** 2928–34

[14] Gümüşay Ö A and Şeker M 2021 Effects of extrusion parameters on physicochemical properties of third generation corn snacks expanded by microwave heating *J Food Process Preserv* **45**

[15] Korkerd S, Wanlapa S, Puttanlek C, Uttapap D and Rungsardthong V 2016 Expansion and functional properties of extruded snacks enriched with nutrition sources from food processing by-products *J Food Sci Technol* **53** 561–70

[16] Reddy M K, Kuna A, Devi N L, Krishnaiah N, Kaur C and Nagamalleswari Y 2014 Development of extruded Ready-To-Eat (RTE) snacks using corn, black gram, roots and tuber flour blends *J Food Sci Technol* **51** 1929–37

[17] Rani P, Kumar A, Purohit S R and Rao P S 2018 Impact of fermentation and extrusion processing on physicochemical, sensory and bioactive properties of rice-black gram mixed flour *LWT* **89** 155–63

[18] Mayachiew P, Charunuch C and Devahastin S 2015 Physicochemical and thermal properties of extruded instant functional rice porridge powder as affected by the addition of soybean or mung bean *J Food Sci* **80** E2782–91

[19] Lotfi Shirazi S, Koocheki A, Milani E and Mohebbi M 2020 Production of high fiber ready-to-eat expanded snack from barley flour and carrot pomace using extrusion cooking technology *J Food Sci Technol* **57** 2169–81

[20] Kaisangsri N, Kowalski R J, Wijesekara I, Kerdchoechuen O, Laohakunjit N and Ganjyal G M 2016 Carrot pomace enhances the expansion and nutritional quality of corn starch extrudates *LWT* **68** 391–9

[21] Mazlan M M, Talib R A, Mail N F, Taip F S, Chin N L, Sulaiman R, Shukri R and Mohd Nor M Z 2019 Effects of extrusion variables on corn-mango peel extrudates properties, torque and moisture loss *Int J Food Prop* **22** 54–70

[22] Silva E M M da, Ascheri J L R, Carvalho C W P de, Takeiti C Y and Berrios J de J 2014 Physical characteristics of extrudates from corn flour and dehulled carioca bean flour blend *LWT* **58** 620–6

[23] Yusuf M, Halilu M and Filli K 2018 Influence of Extrusion Variables on Proximate Composition Some Nutrient and Antinutrient Contents of Dakuwa Extrudates Produced from Blends of Sorghum (Sorghum bicolour L) Groundnut (Arachis hypogea L) and Tigernut (Cyperus esculentus) *Current Journal of Applied Science and Technology* **26** 1–20

[24] Sahu C, Patel S and Tripathi A K 2022 Effect of extrusion parameters on physical and functional quality of soy protein enriched maize based extruded snack *Applied Food Research* **2**

[25] Sharma C, Singh B, Hussain S Z and Sharma S 2017 Investigation of process and product parameters for physicochemical properties of rice and mung bean (Vigna radiata) flour based extruded snacks *J Food Sci Technol* **54** 1711–20

[26] Boakye P G, Okyere A Y, Kougblenou I, Kowalski R, Ismail B P and Annor G A 2022 Optimizing the extrusion conditions for the production of expanded intermediate wheatgrass (Thinopyrum intermedium) products *J Food Sci* **87** 3496–512

[27] Rani P, Kumar A, Purohit S R and Rao P S 2021 Extrusion of fermented rice-black gram flour for development of functional snacks: Characterization, optimization and sensory analysis *J Food Sci Technol* **58** 494–509

The Importance of LCA in the Production of Turmeric Powder And Chilli Powder at PT X

Dyah Ayu Kurniawati [1], *Wahyu* Supartono[1] , *Pujo* Saroyo[1]

[1]Department of Agro-Industrial Technology, Faculty of Agricultural Technology, Universitas Gadjah Mada, Indonesia 55281

Abstract. PT X is one of the businesses in Indonesia that deals with food, particularly spices. The seasoning sector is growing because its existence is becoming more and more expected. Of course, energy and emissions will not be separated from the process of generating spices, especially when producing powdered spices. According to government policy, the sector is supposed to develop into one that is sustainable and environmentally friendly. The Life Cycle Assessment is a common environmental impact assessment (LCA). A method called life cycle analysis (LCA) is used to evaluate a product system's sustainability advantages over the course of its life. This study's goal is to use the LCA approach to examine how much energy and material are consumed, as well as emissions and potential environmental effects, during the production of seasoning powder at PT X. Establishing the goals and scope, inventory analysis, environmental impact analysis, and interpretation are the four processes utilized in the LCA. The study's findings show that 7,963 MJ are needed to manufacture 1 kilogram of turmeric powder, with a potential environmental impact of GWP 0.7627 kg CO2 –eq, AP 0.7796 x 10-4 kg SO2 –eq, and EP 0.528 x 10-4 kg N-eq. Chili powder, on the other hand, uses 8,506 MJ and have an environmental impact of GWP 0.8418 kg CO2 –eq, 0.8690 x 10-4 kg SO2 –eq, and 0.664 x 10-4 kg N –eq.
Keywords: Emission, Energy, Life Cycle Assessment (LCA)

1 Introduction

PT X is a medium-sized business that manufactures spices and seasonings for the food industry. (Badan Pusat Statistik 2021) estimates that there will be 29,000 medium- and large-scale industries in 2021. Because spices are the fundamental components of cooking, which are fundamental necessities of the community, the potential of various spices in Indonesia makes this company see an opportunity. With a total of 280 different product categories, PT X has a production capacity of about 3000 tons per month. The manufacturing is split up into three separate units and stored in a single storage unit, with unit 1 producing an average of 800 kg every shift of powdered spices.

Due to several production procedures, the increasing demand for powdered spices at PT X has a detrimental effect on the environment. A technique known as Life Cycle Assessment can be used to evaluate the production process's overall environmental impact (LCA). LCA is the collection and assessment of inputs, outputs, and potential environmental impacts of a product system over the entirety of its life cycle [4]. According to some sources, LCA is a method that consists of four stages: establishing the scope, cataloguing inputs and outputs, analysing environmental consequences, and interpreting the outcomes of the inventory analysis from the stage before [8]. Due to differences in the production process, this LCA study will focus on items containing turmeric and chili powder, and the results cannot be used to evaluate the environmental impact of other products in PT X. The LCA assessment uses four scopes that act as system boundaries. The four scopes that [9] wrote are as follows:

1. Cradle to grave refers to a life cycle assessment that is completed from the time that raw materials are taken until all materials are recycled back into the earth.
2. Cradle to gate, its scope consists of completed goods from the collecting of raw materials through to their distribution to customers.
3. Gate to gate, which simply evaluates the product life cycle on procedures that have added value, is the shortest scope of the other four.
4. Cradle to cradle examines the life cycle of a product from raw materials to material recycling.

This study's goal is to evaluate the effects of turmeric powder and chili powder products on the environment from gate to gate, as previously described. The following potential environmental effects will be evaluated:

1. Global Warming Potential (GWP)

GWP, often known as global warming, is an environmental effect that raises the Earth's temperature [3]. The volume of CO2 emissions emitted into the atmosphere is one of the main causes of GWP.

2. Acidification Potential (AP)

The process of acidic pollutants being deposited in soil, water, surface organisms, biological, and subsurface water is known as "acidification potential," or AP [2]. High amounts of sulphur oxides and nitrogen oxides contribute to acidification.

3. Eutrophication Potential (EP)

Increasing water productivity by EP results in an environmental imbalance that can lower water quality. According to Eyre & Lester (2002) in Mukhtasor (2007) in [10] eutrophication is brought on by the presence of phosphate and nitrogen waste that enters the waters and creates aquatic plant blooms as well as a decrease in the amount of dissolved oxygen in the water.

2 Research Methods

The forms of data used in this study comprised primary data and secondary data, and it was carried out in a company in Cilowa, Kuningan Regency, West Java, over the course of a month in unit 1 of PT X.

3 Collection of Data

Three types of data were collected for this study: primary data were obtained through observation to understand the production process; secondary data came from literatures drawn from various journals and associated sources; and thirdly, primary data were obtained through interviews.

3.1 Life Cycle Assessment

The LCA assessment is divided into the following four stages:

1. Determining the LCA's Scope

The research is conducted when purchasing raw materials to turn into goods before being delivered, and its scope is gate to gate.

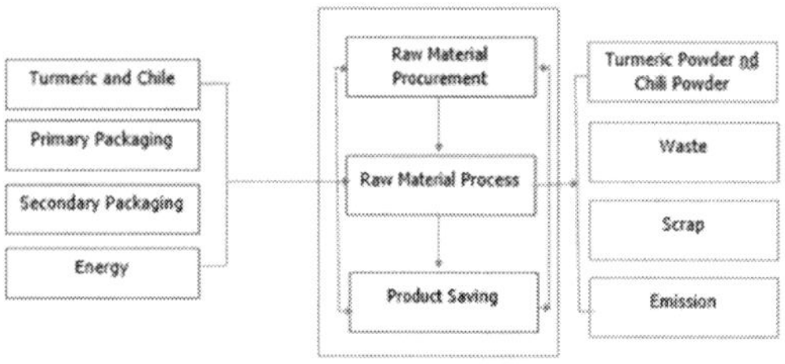

Figure 1. Study's LCA Scope

2. Product Life Cycle Inventory

The energy used in the production process is computed at the inventory stage, followed by the total emissions for each gas under consideration. The following formula will be used to multiply a conversion factor by each gas under study:

Total Emissions = Energy Consumption x Conversion Factors

3. Product Life Cycle Impact Analysis

Three steps make up the impact assessment, the first of which classifies the researched gases according to their potential effects on the environment. The characterisation step is converted into a weighted value using the environmental impact indicator's conversion value, and the assessment stage is converted into a quantitative total by dividing the conversion factor by the quantity of pollutants. The corresponding value utilized alludes to the information in [7].

4. Interpretation of the Product Life Cycle

Determining whether the calculated environmental impact exceeds the quality standard threshold or not and offering recommendations.

4 Result and Discussion

Based on the research's results, the 400 kg production process for turmeric powder generated 388 kg of products with emissions and scrap output and the 400 kg production process for chili powder produced 382 kg with emissions and scrap output. The elements used to produce chili powder and turmeric powder are listed in Table 1. The production process generates output in addition to the inputs indicated in table 1 in the form of flue gas, solid waste in the form of plastic (each operation generates plastic since it needs to be replaced with new packaging), and scrap.

The total energy listed in table 2 is generated by calculating the type of energy input used. It can be shown that producing 1 kg of turmeric powder requires a total energy of 7.693 MJ, whilst producing 1 kg of chili powder requires a total energy of 8.506 MJ. The largest energy used in the production of both turmeric powder and chili powder is fuel energy used to purchase raw materials and packaging materials. Figure 1 shows that up to 44% of the energy input utilized for turmeric powder and up to 37% of the energy input required for chili powder both come from fuel.

Table 1. Process Production Input

Process	Energy Input Type	
	Turmeric Powder	**Chili Powder**
Raw Material Procurement	People, FUEL (solar)	Raw Material Procurement
Packaging Procurement	People, Electricity	Packaging Procurement
Stripping	People, Electricity	Stripping
Grinding	People, Electricity	Grinding
Sifting	People, Electricity	Sifting
Mixing 1	People, Electricity	Mixing 1
Mixing 2	People, Gas, Electricity	Mixing 2
Oven	People, Electricity	Oven
Cooling	People	Cooling

The amount of energy used in each step of the manufacturing of chili powder and turmeric powder is the basis for the study of emission calculations. The emission gases examined in this study are CO2, CH4, N2O, SO2, CO, and NOX, with each emission gas being classified according to the study's three environmental implications. The results of converting each total energy into a corresponding emission value are shown in Table 3. Of all the gas emission values, CO¬2 has the highest emission value. Due to the distance between the supplier and the manufacturer, energy from gasoline generates the most CO2 emissions, necessitating a lot of fuel for both goods to be purchased as raw materials. High fuel and electrical energy use results in high emission values.

Table 2. Total Energy for Turmeric Powder and Chili Powder

Process	Total Energy Released (MJ)							
	Turmeric				Chili			
	People	Electricity	Gas	Fuel	People	Electricity	Gas	Fuel
Raw Material Procurement	17.64	-	-	758.781	17.64	-	-	597.55
Packaging Procurement	17.97	-	-	614.688	17.97	-	-	605.178
Stripping	2.94	2.686	-	-	2.94	2.685	-	-
Grinding	3.36	644.544	-	-	47.04	966.816		-
Sifting	11.76	37.598	-	-	11.76	32.227	-	-
Mixing 1	0.65	4.485	-	-	0.65	4.485	-	-
Mixing 2	1.29	8.862	-	-	0.98	8.862	-	-
Oven	3.92	10.742	913.77	-	3.92	10.742	913.77	-
Cooling	1.31	3.599	-	-	1.31	3.599	-	-
Packing	1.64	-	-	-	1.64	-	-	-
Total/1x Production	**90.49**	**712.516**	**913.77**	**1373.465**	**105.86**	**1027.269**	**913.77**	**1202.729**
Total Production (kg)	**388**				**382**			
Total/1 kg product	**0.233**	**1.836**	**2.355**	**3.539**	**0.277**	**2.689**	**2.392**	**3.148**
Total Energies/1 kg Product	**7.963**				**8.506**			

Table 3. Emission Released

Emission	Emission released/ kg product (kg)					
	Turmeric			Chili		
	Electricity	Fuel	Gas	Electricity	Fuel	Gas
CO_2	0.249	0.2623	0.149	0.364	0.2333	0.151
CH_4	0.178×10^{-3}	0.1380×10^{-4}	0.235×10^{-5}	0.261×10^{-3}	0.1227×10^{-4}	0.239×10^{-5}
N_2O	0.780×10^{-6}	0.1380×10^{-4}	0.235×10^{-6}	0.114×10^{-5}	0.1227×10^{-4}	0.239×10^{-6}
SO_2	0	0.0297	0.118×10^{-5}	0	0.0264	0.119×10^{-5}
CO	0.255×10^{-4}	0.1167×10^{-4}	0.118×10^{-3}	0.373×10^{-4}	0.1038×10^{-4}	0.119×10^{-3}
NO_x	0.612×10^{-4}	0.471×10^{-9}	0.118×10^{-3}	0.894×10^{-4}	0.419×10^{-9}	0.119×10^{-3}

The computed emissions are then divided into estimates of the three potential environmental effects, namely acidification, eutrophication, and global warming. Table 4 displays the classification of emissions into environmental impacts. Acidification will be equivalent to SO2 –eq, eutrophication will be equivalent to N-eq, and global warming will be equivalent to CO2 –eq in terms of potential environmental effects.

Table 5 provides a quantitative analysis of the three environmental impacts' possible values, which have been compared using equivalent values. With 0.7627 kg CO2 –eq of turmeric powder, GWP has the highest potential environmental effect value, followed by AP (0.7796 x 10-4 kg SO2 –eq) and EP (0.58 x 10-4 kg N – eq). With a value of 0.8418 kg CO2 –eq, chili powder's GWP value likewise has the highest possible environmental effect value. This is followed by AP values of 0.8690 x 10-4 SO2 –eq and EP values of 0.664 x 10-4 kg N –eq. Because both products require a lot of fuel, which results in a lot of CO2 emissions, which are the primary indication of global warming, the high GWP number is caused by this.

Table 4. Emission Classification of Environmental Impact Potential

Emission	Environmental Impact		
	Global Warming	Acidification	Eutrophication
CO_2	✓		
CH_4	✓		
N_2O	✓		✓
SO_2		✓	
CO	✓		
NO_x		✓	✓

Table 5. Potential Value of Environmental Impact

Environmental Impact Potential	Emission	Potential Value kg – eq.	
		Turmeric powder	Chili powder
GWP (CO_2 –eq)	CO_2	0.6620	0.7485
	CH_4	0.6620 x 10^{-2}	0.9388 x 10^{-2}
	N_2O	0.4417 x 10^{-2}	0.4071 x 10^{-2}
	CO	0.8963 x 10^{-1}	0.7981 x 10^{-1}
	Total	**0.7627**	**0.8418**
AP (SO_2 –eq)	SO_2	0.650 x 10^{-4}	0.7532 x 10^{-4}
	NO_x	0.1287 x 10^{-4}	0.1158 x 10^{-4}
	Total	**0.7796 x 10^{-4}**	**0.8690 x 10^{-4}**
EP (N –eq)	N_2O	0.542 x 10^{-4}	0.627 x 10^{-4}
	NO_x	0.400 x 10^{-5}	0.368 x 10^{-5}
	Total	**0.582 x 10^{-4}**	**0.664 x 10^{-4}**

The emission values and potentials created during the production of turmeric powder and chili powder can still be considered acceptable because they are lower than the thresholds established by the [5] and [6] Patent 41. Table 6 shows the quality standard threshold for emission values. The emission value obtained is still below the quality level, as shown in quality standard table 6.

Table 6. Comparison of Thresholds

Product	**Parameter**	**Emission**		**Threshold (ppm)**	**Description**
		kg/kg	**ppm**		
Turmeric powder	CO_2	0.662	1.706 x 10^3	40000	below the limit
	CH_4	0.194 x 10^{-3}	0.501	10000	below the limit
	N_2O	0.148 x 10^{-4}	0.038	50	below the limit
	SO_2	0.128 x 10^{-4}	0.033	365	below the limit
	CO	0.298 x 10^{-1}	77.010	10000	below the limit
	NO_x	0.180 x 10^{-3}	0.466	150	below the limit
Chili powder	CO_2	0.748	1.959 x 10^3	40000	below the limit
	CH_4	0.276 x 10^{-3}	0.723	10000	below the limit
	N_2O	0.136 x 10^{-4}	0.036	50	below the limit
	SO_2	0.116 x 10^{-4}	0.030	365	below the limit
	CO	0.266 x 10^{-1}	69.644	10000	below the limit
	NO_x	0.209 x 10^{-3}	0.547	150	below the limit

Although the obtained emission value is still within the quality standard threshold, but the CO2 value is quite high, it is required to reduce energy use, which generates high emissions. PT X can make recommendations such as limiting the frequency of shipments, with each cargo carried out at full capacity, seeking for alternative suppliers that are closer to the factory, and doing intensive maintenance every week. According to the literature, employing an Integrated Gasification Combined Cycle (IGCC) power plant, which mixes coal and Gas Turbine Combined Cycle (GTCC), can replace the usage of electrical energy [11]. The IGCC system has the ability to minimize CO2, SO2, and NOx pollution.

5 Conclusions

According to the results of this study, producing 1 kilogram of turmeric powder requires 7,963 MJ/kg of product, whereas producing chili powder requires 8,506 MJ/kg of product. The potential environmental implications of turmeric powder production are as follows: GWP of 0.7627 kg CO2 –eq, AP of 0.7796 x 10-4 kg SO2 –eq, and EP of 0.528 x 10-4 kg N –eq. The possible GWP released during the chili powder production process is 0.8418 kg CO2 –eq, the AP is 0.8690 x 10-4 kg SO2 –eq , and the EP is 0.664 x 10-4 kg N –eq.

The high value of GWP is due to the high value of CO2 emissions created during the raw material procurement process, hence advances in lowering CO2 in raw material procurement are required. One suggestion is to use full capacity delivery to reduce delivery frequency.

6 Acknowledgment

The authors would like to express sincere gratitude to all PT X staff, particularly field supervisors who always are helpful in providing information and data collection during the production process of the spices.

References

1. Badan Pusat Statistik. *Direktori Industri Manufaktur.* In https://www.bps.go.id/publication/2021/10/29/9e665258c573186f163133b2/direktori-industri-manufaktur-2021.html (2021)
2. Farinha, C., J. D. Brito, and M. D. Veiga. *Chapter 8-Life Cycle Assessment* in Eco-Efficient Rendering Mortars **205–334**. (2021)
3. Gürtürk, Mert, Ferhat Ucar, and Murat Erdem. *A Novel Approach to Investigate the Effects of Global Warming and Exchange Rate on the Solar Power Plants.* Energy 239. doi: 10.1016/j.energy.**2021.122344** (2022)
4. Hanafi, Jessica, e.t.c. *Pedoman Penyusunan Laporan Penilaian Siklus Hidup (LCA).* Jakarta: Direktorat Jenderal Pengendalian Pencemaran dan Kerusakan Lingkungan, Kementerian Lingkungan Hidup dan Kehutanan RI. (2021)
5. National Institute for Occupational Safety and Health [NIOSH]. *Table of IDLH Values* in https://www.cdc.gov/niosh/idlh/intridl4.html (2019)
6. Peraturan Pemerintah. *Pengendalian Pencemaran Udara.* (1999)
7. ReCiPe. *A Harmonized Life Cycle Impact Assessment Method at Midpoint and Endpoint Level Report I: Characterization.* Microsoft Word - Report ReCiPe_Update_20171002 (pre-sustainability.com). (2022)
8. Ula, R., A. Prasetya, and I. Haryanto. *Life Cycle Assessment (LCS) Pengelolaan Smpah Di TPA Gunung Panggung Kabupaten Tuban, Jawa Timur.* In Jurnal Teknologi Lingkungan **22(22):147–61.** (2021)
9. Yuniarti, R., I. P. Tama, A. Eunike, and Y. Sumantri. *Green Supply Chain Management Dan Studi Kasus Di Dunia Industri.* Malang: UB Press. (2018)
10. Yusal, M. S. *Studi Potensi Eutrofikasi Di Pesisir Losari Makassar.* Enggano **6(2):348–57.** (2019)
11. Zaman, M. R., and S. W. A. Suedy. *Pemanfaatan Batubara Kalori Rendah Pada IGCC (Integrated Gasification Combined Cycle).* In JEBT: Jurnal Energi Baru Dan Terbarukan **1(1):35–44.** (2020)

X-ray CT applications to reveal physical defects of onion during storage

Devi Priyanti, *Ira* Ayuningsih, *An-Nidaa'* Fatkhur Rahmah, *Sri* Rahayoe, *Bayu* Nugraha*

Agriculture and Biosystems Engineering Department, Universitas Gadjah Mada, 55281, Yogyakarta, Indonesia
*Email: bayu.nugraha@ugm.ac.id

Abstract. Inappropriate conditions of storage may result in physical defects that are sometimes visually undetectable. X-ray CT is a non-destructive instrument that has been widely implemented to identify sound and/or defect structures of many products, but only a little used for onion. In this study, the structure change of local onion during storage were imaged and characterized. To see the X-ray CT utilization, Onion harvested from a local farm was, for the first step, stored in an extreme condition potentially causing physiologically structure damage. X-ray CT scanning of intact onion was performed at low resolution to see the structure difference at the pre- and post-storage condition. Image processing was conducted to observe the change of the physical parameters. From the result, X-ray CT depicted spatially structural deformation of onion after a short-term storage period. Onion generally underwent shrinkage marked by the decrease in the cross-section area and the increase in air spaces between the fleshy leaves. This result was also in line with the decrease in weight and moisture content with respect to the storage conditions. This study concludes that X-ray CT has a high potential to reveal the physical defects of local onion affected by inappropriate storage conditions.

1. Introduction

Onion is one of Indonesia's horticultural commodities with high sales and economic value. Onion has many benefits, such as the traditional ingredients of cooking and medicine. In Indonesia, there are several onion production centers, including Brebes, Nganjuk, Cirebon, East Lombok, Tegal, Probolinggo, and Malang. Based on data from Badan Pusat Statistik in 2021[1]. Indonesia's onion production reached 2 million tons an increase from 2020 of 1.82 million tons. Central Java, the largest onion production center in Indonesia in 2021, production of onions reached 0.56 million tons, which decreased from 0.61 million tons in 2020 and was followed by East Java and West Nusa Tenggara. Several varieties of onions in Indonesia are Bima Brebes, Medan, Maja Cipanas, Keling, Ampenan, Sumenep, Kuning, Timor, and Lampung[2].

In the harvest, onions tend to be abundant, but during the off-season, there is very little supply that requires imports from abroad. The red onion bulbs are not durable and are easily damaged because they have a high-water content[3]. Some farmers choose to store their crops for a certain period, waiting for an increase in selling prices. However, these farmers face two significant risks: storage costs and a lack of adequate storage space. To prevent stored onions from spoiling, it is essential to monitor environmental parameters like temperature and humidity. In Indonesia, onion storage methods remain relatively basic, involving hanging or placing the onions in an area covered by tarpaulin, making it challenging to control conditions such as temperature [4].

The anatomy of onions ranges from macrostructures to microstructures. Macrostructures consist of visually visible parts (>100 μm), such as fruit flesh and skin, while microstructures consist of visually invisible parts (<100 μm), such as cells, pores, and cell walls [5]. To date, the structure of onions on a macro or micro scale has not been comprehensively characterized, quantified, and visualized, which can help explain storage breakdown of onion, depending on environmental conditions.

X-ray computed tomography (CT) is a non-destructive evaluation device used to identify the internal structure of a product. The advantage of X-ray CT is that it is capable of quantifying and visualizing the internal structure of the product in 3 (three) dimensions [6][7]. This makes it possible to see and quantify the characteristics of the physical defects of onions both internally and externally so that they can quantize physical parameters such as surface area, volume, number of layers on the tissue (macroscale), pore network, pore size, and pore connectivity (microscale)[8].

Given that onions have such a high potential in Indonesia, a method is needed to reveal the physical breakdown of onions during storage. This study aims to apply X-ray CT as a non-destructive evaluation technology to characterize and visualize onion structure changes before and after storage.

https://doi.org/10.1051/bioconf/20238004008

2. Materials and Methods

2.1 Onion samples

Onions without physical damages from a traditional market around Yogyakarta, Indonesia were chosen and transported to Laboratory at Faculty of Agricultural Technology, Universitas Gadjah Mada for treatments. The onion was exposed to the condition of different relative humidities with salt solution of sodium chloride and potassium nitrate (76% and 96 %) at temperature (5°C, 20°C, 35°C) for one month. X-ray computed tomography (CT) images of the onion at initial and final conditions were taken to see the physical change of onion visually invisible.

2.2 X-ray CT imaging

Initial and final conditions of onion were imaged with Nanotom system at X-ray CT Lab, Cikarang, Indonesia. Low-resolution setup (75 μm) with a voltage of 100 kV and a current of 250 μA was chosen to accommodate the intact part of the onion organ, approximately lasting for 21 min per scan. The setup also enabled the X-ray photon to propagate through soft tissue of onion, presenting the CT images with tissue density differences in the intact onion with varying gray values. This is crucial to visualize details of intact onion anatomy so that the structural changes of initial and final condition are detectable.

Radiographic images produced during the scanning were reconstructed using Phoenix software to produce the stuck of X-ray CT image slices (**Figure 1**).

2.3 Image processing

Image processing was performed to analyse the change of onion internal structure. To do so, the onion CT images were initially filtered to minimize some noises. Subsequently, the filtered images were segmented using Otsu's method where the tissue fraction of onion was marked as 1 (white pixels) and the air fraction was 0 (black pixels). To have a mask, the binarized image was duplicated and the whole air fraction of the one image was closed by implementing 'fill hole' operation. Void space fraction (ε, %) was computed based on the following formula:

$$\varepsilon = \left(1 - \frac{X_{wp}}{X_m}\right) x100 \quad (1)$$

where X_{wp} is the pixel number of solid fraction in the binarized image and X_m is the pixel number of the mask.

2.4 Physical parameter analysis

Firmness and moisture content closely related to the physical structure of biological materials were analyzed to verify the structure changes of onion organ imaged from X-ray CT. The analyses were performed before and after storage with three replicates of onion per treatment. For the moisture content, onion was cut into small pieces, weighted, and put in an oven to remove the water so that the solid and water fraction can be known to calculate the moisture content in wet basis (%) as written in **Formula 2**.

$$M = \frac{m_o - m_t}{m_o} x100 \quad (2)$$

where m_o is a total sample mass of sliced onion (gr) as m_t is a solid mass of the sample (gr).

Firmness was measured using a texture analyzer where a probe was used to press the center of the onion. The onion placed on the test table and then subjected to pressure with the probe until it pierced into the bulb. The probe used was the TA39 probe, which has a diameter of 2 mm with a circular cross-section. The firmness data were collected before and after storage.

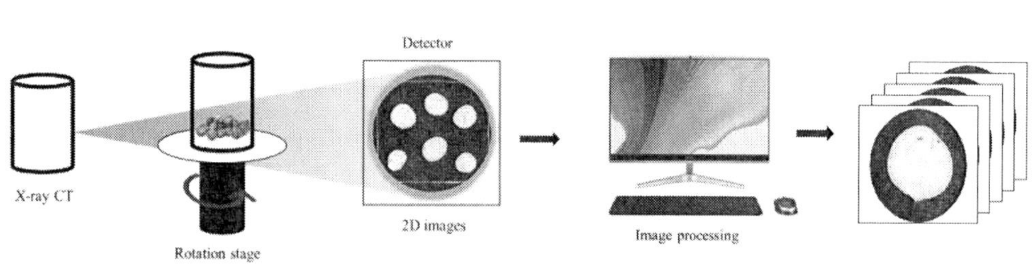

Figure 1. Schematic of X-ray computed tomography imaging.

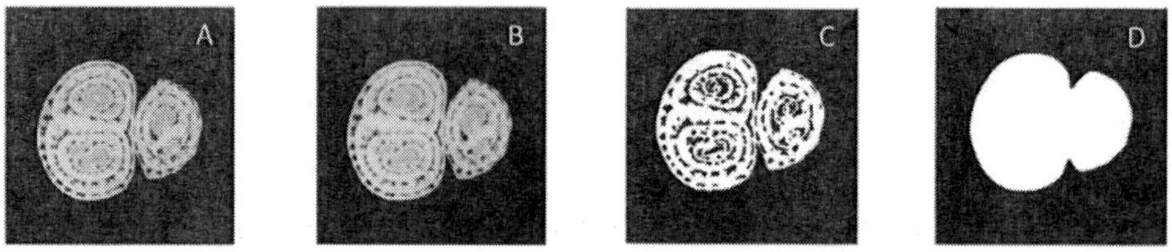

Figure 2. Image processing steps. (A) original X-ray CT image slice, (B) filtered CT image, (C) binarized CT image, and (D) mask of the CT image.

3. Results and discussion

Storage is a crucial step of onion postharvest handling to maintain the quality and control the market price[9]. Storage conditions also have an important in onion physiology, having affected the physicochemical and photochemical properties of onion during storage[10]. An improper condition of storage may result in quality degradation generally marked by weight loss and physical defects. This quality degradation is sometimes visually undetectable.

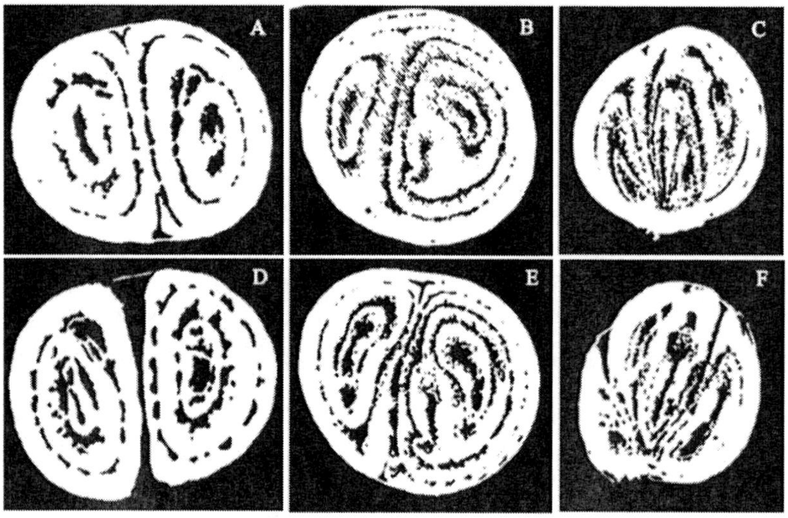

Figure 3. Physical changes of the onion tissue at initial (A-C) and final (D-F) condition under (A and D) 5°C, (B and E) 20°C, and (C and F) 35°C storage temperature depicted non-destructively using X-ray CT. Black areas inside the onion represent the void/air spaces between the tissue layers.

Virtual geometry of onion generated from X-ray CT image enabled to show internal and external structure transformation of onion at initial and final storage conditions (**Figure 3**). Tissue deformations occurred inside the onions during storage, creating bigger void spaces between the tissue layers. The deformation magnitude depends on the initial void spaces. As the spaces were small at primary condition, void space development was greater at final condition (**Figure 3A, B, D and E**). Onions have reduced quality and water holding ability which allows larger pores to be formed. In addition, temperature and humidity, initial quality, and storage method also affect the increased porosity of onion. The void space fraction of onion did not develop significantly as it was already high at the primary condition even though the temperature was high (35°C) (**3C and F**).

Analysis of void space fraction based on CT image processing confirms the onion geometry (**Table 1**). This result showed that temperature affect the void space development between the layer as the low relative humidity (76 %) contributed more.

Table 1. Void space (%) between onion tissue layer at initial and final condition of storage (standard deviation of three replicates given behind the ± sign).

Storage temperature	Void space (%)	
	Initial condition	Final condition
5 °C	8,90 ± 1,7	18,60 ± 2,15
20 °C	9,99 ± 3,66	18,85 ± 1,87
35 °C	16,35 ± 1,09	20,18 ± 2,27

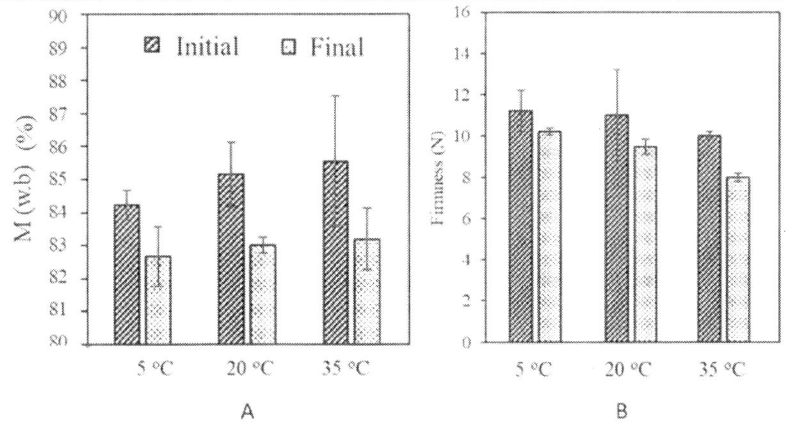

Figure 4. Physical parameter changes of (A) moisture content and (B) firmness between the initial and final conditions of storage

Moisture content and firmness analysis provided a bit different result from the onion organ visualization while the storage temperature affected the bigger difference between the initial and final onion condition (**Figure 4A and B**). This might be due to the moisture content and firmness are more related to the tissue density [11], represented by grayscale intensity of the onion CT images. During the storage, the water content might diffuse from each tissue layers and evaporated through the void spaces. The tissues that underwent the water content loss became less dense, typically could be recognized by grayscale intensity of the CT image. Loss of water content is caused by transpiration, respiration, decay, and sprouting [12]. In the next study, grayscale intensity should be taken into account into the analysis to see the correlation between the moisture content and firmness versus grayscale intensity.

4. Conclusions

X-ray CT is a potential instrument to investigate the quality loss of onion during storage as many quality losses occur inside the product visually invisible. Its non-destructive feature enables to calculate and visualize the physical structure change inside the product. In the next study, model predictions of the structure change can be developed.

Acknowledgements

This study was supported by *Rekognisi Tugas Akhir* (RTA) UGM and BIMA-Direktorat Jendral Pendidikan Tinggi (DIKTI). The opinions expressed in this article do by no means represent the official opinion of the financing bodies or their representatives.

References

[1] BPS, "Produksi Tanaman Sayuran 2021," 2022. https://www.bps.go.id/indicator/55/61/1/produksi-tanaman-sayuran.html

[2] O. G. Tandi and F. Faisal, "Technology Innovation Support in Development of Local Superior Shallot in North Sulawesi," *Agrotech J.*, vol. 5, no. 1, pp. 10–21, 2020, doi: 10.31327/atj.v5i1.1233.

[3] A. K. Mutia, "Pengaruh Kadar Air Awal pada Bawang Merah (Allium ascalonicum L.) terhadap Susut Bobot dan Tingkat Kekerasan Selama Penyimpanan pada Suhu Rendah," *Gorontalo Agric. Technol. J.*, vol. 2, no. 1, p. 30, 2019, doi: 10.32662/gatj.v2i1.538.

[4] Suyatno, I. Bachtera, F. Iim, and P. Gontjang, "Sosialisasi Instore Drying Sebagai Upaya Penyimpanan Bawang Merah Terkontrol Berbasis Panel Surya di Sukomoro , Nganjuk , Jawa Timur," *SEWAGATI, J. Pengabdi. Kpd. Masy.*, vol. 7, no. 5, pp. 1–7, 2023.

[5] Q. T. Ho *et al.*, "Multiscale modeling in food engineering," *J. Food Eng.*, vol. 114, no. 3, pp. 279–291, 2013, doi: 10.1016/j.jfoodeng.2012.08.019.

[6] S. Janssen *et al.*, "3D pore structure analysis of intact 'Braeburn' apples using X-ray micro-CT," *Postharvest Biol. Technol.*, vol. 159, no. September 2019, p. 111014, 2020, doi: 10.1016/j.postharvbio.2019.111014.

[7] E. Herremans *et al.*, "Automatic analysis of the 3-D microstructure of fruit parenchyma tissue using X-ray micro-CT explains differences in aeration," *BMC Plant Biol.*, vol. 15, no. 1, pp. 1–14, 2015, doi: 10.1186/s12870-015-0650-y.

[8] J. Lammertyn, T. Dresselaers, P. Van Hecke, P. Jancsók, M. Wevers, and B. M. Nicolaï, "MRI and X-ray CT study of spatial distribution of core breakdown in 'Conference' pears," *Magn. Reson. Imaging*, vol. 21, no. 7,

pp. 805–815, 2003, doi: 10.1016/S0730-725X(03)00105-X.

[9] M. . Hatem, S. . Shehata, Y. . AbdEl-hay, F. Karima, AbdEl-Gwad, and B. . Abaker, "Effect of storage conditions on the quality characteristics of onion bulbs," *J. Agric. Eng.*, vol. 31, no. 3, pp. 919–936, 2014.

[10] K. Sharma, Y. Rok Lee, S. W. Park, and S. H. Nile, "Importance of growth hormones and temperature for physiological regulation of dormancy and sprouting in onions," *Food Rev. Int.*, vol. 32, no. 3, pp. 233–255, 2016, doi: 10.1080/87559129.2015.1058820.

[11] W. chuan Guo, S. O. Nelson, S. Trabelsi, and S. J. Kays, "10-1800-MHz dielectric properties of fresh apples during storage," *J. Food Eng.*, vol. 83, no. 4, pp. 562–569, 2007, doi: 10.1016/j.jfoodeng.2007.04.009.

[12] A. A. Kader, "Postharvest Technology of Horticultural Crops - An Overview from Farm to Fork," *J. Appl. Sci. Technol.*, vol. 1, no. 1, pp. 1–8, 2013.

Continental vs. tropical breed: immunity comparison under heat stress conditions utilizing qRT-PCR technique

Roland Fajardo[1,2,3], *Putri* Kusuma Astuti[1,4,5], *Zoltán* Bagi[1], *Lilla* Bodrogi[6], *Tímea* Pintér[6], *Gabriella* Skoda[6], *Husein* Ohran[7], *George* Wanjala[1,4], *Szilvia* Kusza[1*]
[1] Centre for Agricultural Genomics and Biotechnology, University of Debrecen, 4032 Debrecen, Egyetem tér 1, Hungary.
[2] Department of Agriculture-Bureau of Animal Industry, 1100 Diliman, Quezon City, Philippines.
[3] College of Veterinary Medicine, University of Southern Mindanao, Kabacan, Cotabato 9407 Philippines
[4] Doctoral School of Animal Science, University of Debrecen, 4032 Debrecen Böszörményi út 138, Hungary.
[5] Department of Animal Breeding and Reproduction, Faculty of Animal Science, Universitas Gadjah Mada, Yogyakarta 55281, Indonesia.
[6]Department of Animal Biotechnology, Hungarian University of Agriculture and Life Sciences, Gödöllő, 2100, Hungary.
[7]Department of Physiology, Veterinary Faculty, University of Sarajevo, Sarajevo, 71000, Bosnia and Herzegovina.
*Corresponding author e-mail address: kusza@agr.unideb.hu

Abstract. Understanding the responses of animals to seasonal heat stress on the genomic level has led to the identification of genes implicated in thermal stress reaction mechanisms. In this study, the relative gene expression of Interleukin-10 (*IL-10*), an anti-inflammatory cytokine and biomarker for heat stress-mediated immune modulation, was observed during the summer and winter seasons in continental and tropical sheep breeds, namely, Hungarian Indigenous Tsigai and White Dorper, employing quantitative real-time polymerase chain reaction (qRT-PCR). Temperature-humidity indices (THI) were calculated to assess heat stress levels. The results indicate that *IL-10* CT-values were significantly higher during the summer, when heat stress prevails, in both sheep breeds compared to winter. While the White Dorper exhibited a higher numerical value for the summer relative gene expression ratio (16.2) compared to the Hungarian Indigenous Tsigai (12.3), no significant differences in CT values were observed between breeds or among sexes. These findings suggest the immune-adaptive characteristics of the two sheep breeds during seasonal heat stress. The variation of *IL-10* gene expression levels between the two breeds can be attributed to their geographical origins: the White Dorper emerging from arid subtropical South Africa and the Hungarian Indigenous Tsigai Sheep thriving in the seasonally harsh Carpathian climate for centuries.

Keywords: gene expression, heat stress, Hungarian Tsigai, *IL-10*, White Dorper

1 Introduction

As a major sector in the livestock industry, sheep production is key to ensuring global food security for the rapidly growing population. Its importance to low-income families all around the world as a source of food and livelihood cannot be overlooked as well [1]. With climate change projections pointing to worse scenarios of increased global temperatures in the years to come, it is essential to tailor-fit the livestock industry to a warmer and harsher environment by selecting animals with tolerance and adaptation traits [2]. Equipped with a robust knowledge of genes related to heat tolerance and adaptation, the livestock industry can keep up with the climatological challenges and sustain productivity.

Heat stress presents direct challenges to the well-being of animals that lead to an array of detrimental effects on their physiology, behavior, overall production performance, reproduction, and health [3]. In response to heat stress, the animal attempts to maintain homeostasis and re-establish homeothermy by invoking compensatory and adaptive mechanisms through behavioral and physiological changes to promote welfare and ensure survival [4].

The immune system of animals, in general, is one of the well-studied aspects of the physiological responses of livestock to heat stress. The multifarious interplay of hormones and key immune system cells and factors during heat stress leads to complex immune responses that involve rapid inflammatory reaction, antagonistic immunosuppression, and shifting of immune regimes, among others [5-6]. Understanding and knowing these mechanisms and the factors involved usher in the determination of reliable biological markers of heat response in animals with respect to immunity.

Cytokines are some of the most prominent actors in the immune system and, in this regard, the immune response to heat stress. Of them, the Interleukin 10 (*IL-10*) plays a major role in modulating the other cytokines and effectors of the immune system that attempts to preserve its integrity and function while facing heat stress [7-8].

A deeper investigation of genes associated with heat tolerance and adaptation in livestock has earned the attention of researchers in recent years because of the increasing availability of genomic tools and capabilities. Reverse transcription quantitative real-time polymerase chain reaction (RT-qPCR) is a reliable and key enabling technology for gene expression analysis. Its rapid, cost-effective, and robust approach in gene expression analysis has encouraged researchers to use it for many purposes, including the detection and measurement of the relative expression of genes associated with heat tolerance and adaptation [9].

The influence of seasons on the expression of several heat stress-related genes in livestock was investigated in dairy cattle [10-11], sheep [12], and pigs [13]. This study aims to bridge the scarcity of comparable studies in sheep that intend to investigate seasonal effects on the expression of heat tolerance genes such as *IL-10* as well as a comparison of tropical and temperate breeds on the same subject.

In the present study, the relative gene expression of *IL-10* was determined in the Hungarian Indigenous Tsigai sheep and White Dorper sheep in Hungary during the summer and winter seasons. The Tsigai sheep is an indigenous temperate triple-purpose breed in the Central-Eastern European region that has been bred and maintained as part of the Hungarian sheep sector for more than two centuries [14-15] while the White Dorper is regarded as a breed of Tropical origin that was developed in arid subtropical South Africa for meat purpose in the 1940s. The breed has just been introduced in Hungary in 2008 [16]. This study aimed at understanding the dynamics of the relative gene expression of *IL-10* during the two seasons, in the two breeds, and among the sexes, with particular emphasis on heat stress conditions and the further goal of expounding the adaptability of the two breeds to heat stress situations.

2 Materials and Methods

2.1. Sample Collection and Location

Four Hungarian Indigenous Tsigai sheep and 4 White Dorper sheep were utilized in this study as experimental animals. Of them, 2 are ewes, and 2 are rams for each breed. The animals were housed in conventional deep litter type housing with several pen separation devices and implement that allow adequate airflow and access to the immediate environment during the hot season and enclosure during the wet and cold seasons. Under the regular farm feeding regimen, the animals were provided clean drinking water. All animals that were part of the study were of similar age (2 to 3 years old), body weight (ewes: 45 to 55 kg; rams: 65 to 75 kg), and in optimal health, exhibiting no physical or anatomical abnormalities.

About 5 ml whole blood sample was obtained from the jugular vein of the same animals in RNAprotect Animal Blood Tubes (QIAGEN, Hilden, Germany) and preserved at -70°C until further analysis.

The animals used in this study were kept in the Kismacs Experimental Station of Animal Husbandry of the University of Debrecen, located at 4002 Debrecen, Kismacs Tanya 4, Hungary. The elevation is 127 m above sea level and site coordinates are 47.58° N and 21.58° E.

2.2. Climatological Data

Data on the climatic condition during the sampling day, which include temperature and relative humidity, were recorded from 12:00 to 13:00 in the afternoon and obtained from the Hungarian Meteorological Service station in the vicinity. To calculate the THI for each sampling day, the equation by Mader et al. [17] was used.

$$THI = (0.8 \times T_{db}) + \left[\left(\frac{RH}{100}\right) \times (T_{db} - 14.4)\right] + 46.4$$

T_{db}- Dry bulb temperature (°C), RH- Relative humidity (%)

The severity of heat stress in livestock is typically rated using the THI with ranges categorized as follows: no stress (≤ 67), mild (68–74), moderate (75–78), severe (79–83), and extreme (≥ 84) [18].

Additionally, in order to provide a broader and more objective overview of the climatic cycle and changes in meteorological variables as the seasons change in the area and to give background reference to the climatological data on sampling days, the 30-month climatological records (August 2018 to January 2021) [19] measured by the adjacent government meteorological station, were obtained from the Hungarian Meteorological Service.

2.3. Quantification of relative gene expression levels using RT-qPCR

Following the manufacturer's instructions, total RNA was extracted from sheep blood using the Rneasy Protect Animal Blood Kit (QIAGEN, Hilden, Germany). Using a NanoDrop ND-1000 Spectrophotometer, RNA quality and quantity were assessed (Thermo Fisher Scientific, Waltham, MA, USA). Total RNA of about 100 ng was reverse transcribed into cDNA with specific primers (Table 1) by the qPCRBIO cDNA Synthesis Kit (PCR Biosystems, London, United Kingdom). Forward and reverse primers (Table 1) were created with Primer Express v3.0.1 software (Applied Biosystems, Foster City, CA, USA) and confirmed for target Identity with Primer Blast from the National Center for Biotechnology Information (NCBI) (Ye et al., 2012). A 7300 Real-Time PCR System (Applied Biosystems, Foster City, CA, USA) was used to conduct quantitative real-time PCR (qPCR). The 10 ll reactions consisted of 5 ll PowerUpTM SYBR! Green Master Mix (Applied Biosystems, Foster City, CA, USA), 0.6 ll each of 10 lM forward and reverse primers, 1.3 ml dH2O (MilliporeSigma, Burlington, MA, USA), and 2.5 ml cDNA (2 ng/ml). For relative expression studies, one housekeeping gene (glyceraldehyde 3-phosphate dehydrogenase (GAPDH)) was amplified alongside the target genes (*IL-10*). Quantitative PCR was done in triplicate for each sample.

Table 1. Detail of primers for qPCR analysis of the target genes

Target gene	Gen Bank accession	Primers	Fragment size (bp)
IL-10	NC_056065	F: TGATGCCACAGGCTGAGAAC R: CAGAAAACGATGACAGCGCC	110
GAPDH	NC_040254.1	F: CTGGCCAAGGTCATCCAT R: ACAGTCTTCTGGGTGGCAGT	86

2.4. Data and Statistical Analysis

The relative expression of the *IL-10* gene was determined by calculating the values using the equation proposed by Pfaffl (2001). The method compared the expression of *IL-10* gene with the reference gene GAPDH, with their amplification efficiencies both corrected for differences. In seasonal comparison, the values for winter were used as the calibrator, while for sex differences, the rams' values were used as the calibrator.

$$Gene\ expression\ ratio = \frac{(E_{GOI})^{\Delta Ct\ GOI}}{(E_{HKG})^{\Delta Ct\ HKG}}$$

: EGOI- corrected efficiency of the gene of interest (GOI); EHKG- corrected efficiency of the housekeeping gene (HKG); ΔCt GOI- delta cycle threshold value of the GOI; ΔCt HKG- delta cycle threshold value of the HKG

Gene expression ratios means, standard deviation, and standard error were presented and graphically compared between the values for the summer and winter seasons and in-season comparison between rams and ewes.
Meteorological and gene expression data were downloaded and processed using Microsoft Excel (Microsoft Corporation, Redmond, WA), including tabular and graphical presentations [22-23].
For analysis, the differences of *IL-10* gene expression between seasons (summer and winter), between the breeds, and sexes (rams and ewes) in every season were determined through Student t-test [24] using the average cycle threshold (CT) values for the gene of interest. A difference with $P<0.05$ value was considered statistically significant. Data analysis was carried out using R program for Windows [25].

3 Results

3.1. Climatological variables

The 30-month climatological records, which span from 2018 to 2021, show the seasonal variations of meteorological variables such as temperature, relative humidity, and THI surrounding the study's sample collection period. Environmental temperature and air relative humidity were factors that affected THI. The highest THI values were observed during the summer months of June, July, and August, which correspond to high-temperature levels. Conversely, relative humidity falls during periods of high temperature, while it is high during low-temperature periods such as the winter season (Figure 1).

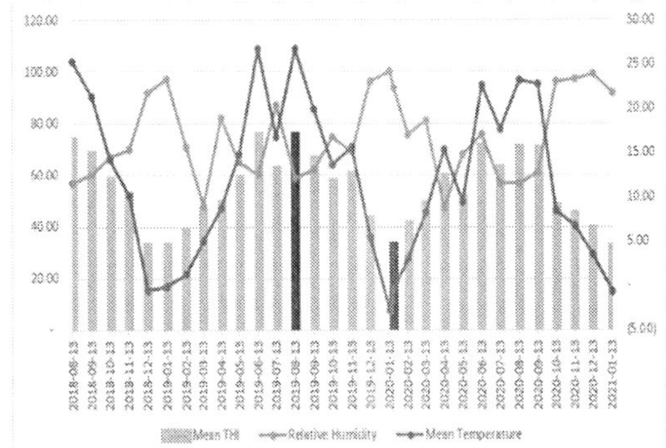

Figure 1. 30-month climatological record in the farm location. The mean THI values during sample collection days were highlighted in red (summer) and blue (winter) bars.

The data on the climatic condition during the collection of samples in summer and winter are presented in Table 2. The summer data collection day was characteristically hot and dry, with temperature recorded at 32.80°C and relative humidity at 34.50%. In contrast, the winter data collection day was cold and humid, with temperature recorded at -3.33 °C and relative humidity at 99.70%. The calculated THI during summer collection day was 79.0, indicating severe heat stress, while the THI during winter was 26.1, meaning there was no heat stress in the animals. Figure 2 shows the heat stress intensities that correspond with the THI values during the sample collection and entire study period.

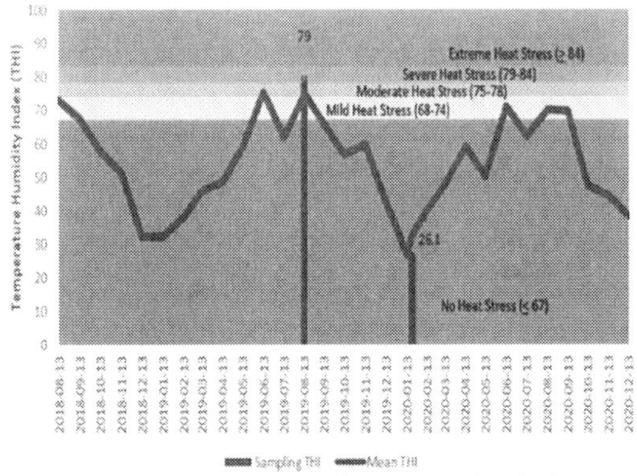

Figure 2. Heat stress levels according to THI values in the 30-month period with the THI values during the hours of sample collection, highlighted in red bars, indicating the levels of heat stress according to Lewis Baida et al. [18].

For most of the year, THI levels were below the heat stress threshold of 65. High THI values were recorded during the summer season, indicating heat stress conditions. Mean THI levels during the summer season show moderate heat stress, but there were times during the day that were marked with elevated THI values, and thus characterized heat stress to be severe. The hour of sample collection for summer shows that the animals were subjected to severe heat stress condition, with THI values at 79.0, several points higher than the mean THI during that day. On the contrary, the absence of heat stress was established during the winter sample collection with a THI of 26.1.

3.2. Expression of *IL-10* gene

Following qPCR analysis to determine the gene expression levels of *IL-10* in the experimental animals, relative expression was calculated using the Pfaffl method where the winter season values were used as calibrator. For all animals, mean gene expression ratio during the summer season was 16.82 with a standard deviation of 20.85 (Table 3). In contrast, the winter mean gene expression ratio was 1.18 with a standard deviation of 0.61. Relative gene expression ratios of *IL-10* in summer and winter seasons were presented in Figure 3 with winter season value as calibrator. A higher gene expression ratio of *IL-10* is found in the summer season as compared to the winter season ratio.

Table 2. Climatological data on the day of sample collection

Season	Sampling date	Sampling time	Temperature (°C)	Relative humidity (%)	THI	Heat Stress Rate
Summer	13/ 08/ 2019	12.00-13.00	32.80	34.50	79.0	Severe
Winter	22/01/ 2020	12.00-13.30	-3.33	99.70	26.1	No Stress

Table 3. Relative Gene Expression of *IL-10* in Hungarian Indigenous Tsigai and Dorper Sheep during summer and winter seasons

Relative gene expression (Mean ± SD)		
Breed	**Summer**	**Winter**
Overall	16.82 ± 20.85	1.18 ± 0.61
Hungarian Indigenous Tsigai	12.3 ± 15.25	16.2 ± 18.53
White Dorper	1.1- ± 0.56	1.04 ± 0.35

SD: Standard Deviation

Breed-wise, the Hungarian Indigenous Tsigai sheep has an *IL-10* gene expression ratio mean of 12.3 and a standard deviation of 15.25 during summer. The White Dorper, on the other hand, has a mean of 16.2 and a standard deviation of 18.53 during summer. Here, the White Dorper showed a numerically higher ratio of gene expression during the summer seasons compared to the Hungarian Indigenous Tsigai.

The gene expression ratio of all the animals and the two breeds during the summer and winter are graphically presented and compared in Figure 3.

The average CT-values were used to statistically analyze differences among the gene expression in 2 seasons (Table 4), between the two breeds, and among sexes. In all animals, expression levels of *IL-10* were significantly higher (P=0.00019) during the summer season than those samples collected during the Winter season. Within-breed comparison of summer and winter CT-values showed significantly higher *IL-10* expression during the summer season in both breeds (Hungarian Indigenous Tsigai P= 0.015 & White Dorper P= 0.004). Between the two breeds, Hungarian Indigenous Tsigai and White Dorper, there is no significant difference in the CT-values during the summer and winter seasons. Similarly, no significant difference between ewes and rams was found during the two seasons in terms of relative gene expression ratio and CT-values (Table 5).

Table 4. Mean and standard deviation of CT-value of *IL-10* of Hungarian Indigenous Tsigai and White Dorper in summer and winter seasons.

CT-value (Mean ± SD)				
Season	**Hungarian Indigenous Tsigai**	**White Dorper**	**Overall**	**P-value**
Summer	32.87 ± 1.84^a	32.62 ± 2.27^a	32.74 ± 1.92^a	*0.87*
Winter	29.37 ± 1.64^b	28.09 ± 0.54^b	28.86 ± 1.4^b	*0.12*
Overall	31.25 ± 2.29	30.35 ± 3.20	30.80 ± 2.75	
P-value	*0.015*	*0.004*	*0.00019*	

SD: Standard Deviation; Note: The different superscripts show significantly different values (P < 0.05).

Table 5. Relative gene expression ratio and CT-value of *IL-10* of Ewe and Ram Hungarian Tsigai and White Dorper Sheep in summer and winter seasons.

	Hungarian Indigenous Tsigai			White Dorper		
Season	**Group**	**Relative Gene Expression Mean**	**CT-value Mean**	**Group**	**Relative Gene Expression Mean**	**CT-value Mean**
Summer	Ewe	0.01	34.205	Ewe	177.61	30.765
	Ram	0.06	31.53	Ram	1.05	34.475
Winter	Ewe	0.43	30.885	Ewe	0.6	30.885
	Ram	0.01	28.36	Ram	1.01	28.235

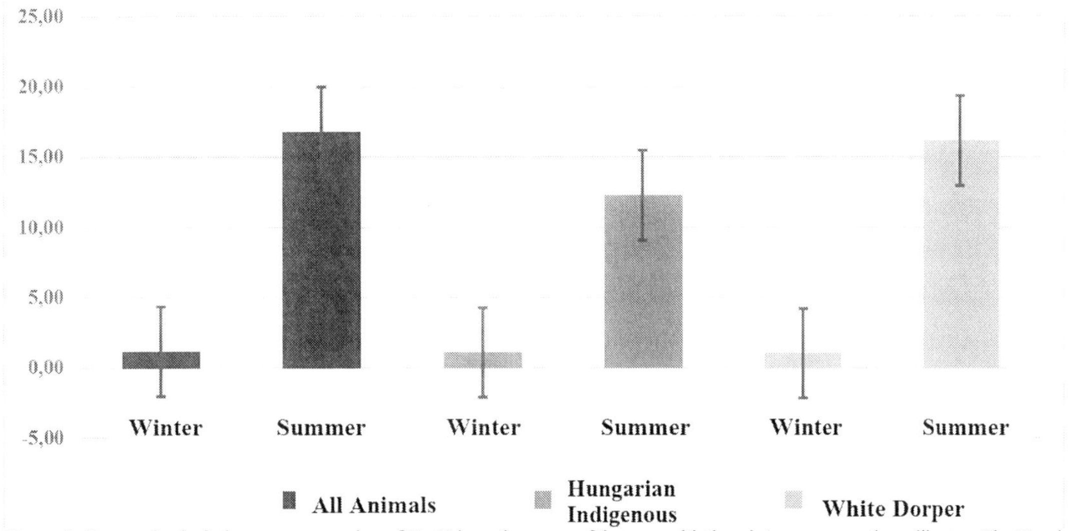

Figure 3. Bar graph of relative gene expression of *IL-10* in each season of the year with the winter season as the calibrator. The X-axis is the relative gene expression in the two seasons. The Y-axis is gene expression for all animals, and the two breeds studied. Different colors represent different breed.

4 Discussion

Maintaining thermoneutral conditions is one of the important requirements in ensuring the welfare and optimal production and reproduction performance of animals [26].

THI is considered as one of the most reliable measures and sensitive indicators of heat stress in animals, which integrates temperature and relative humidity parameters [27-28]. In this study, during the summer sample collection, THI was calculated to be high at 79.0, which indicates a state of severe heat stress. High temperature and low humidity are typical of summer seasons in the region. The average annual temperature in the immediate area was 11.66 °C for the period of 2011 to 2020 [29]. The average temperature recorded on the day of collection was 26.70 °C, a 176% deviation from the annual average. Factoring the daily maximum temperature and the relative humidity during the summer months of June to August [29], THI ranges from 69.67 to 88.80 for sheep, which indicates mild to extreme heat stress.

Consistent with the climatological data gathered and determined in this study, elevated THI in sheep during the summer season and hot periods that established a heat stress condition was also reported in Small tail Han sheep [30], Indian sheep [31], and Omani and Australian Merino sheep [32] where the animals exhibited increased respiratory rate, pulse rate, and rectal temperatures. Such signs are distinctive physiological indicators of heat stress response in animals [33-34].

High-THI conditions also influence metabolic and hematological parameters in sheep. Cortisol and biochemical parameters creatinine, zinc, and phosphorus were elevated during heat stress conditions and are highly correlated to THI levels in Suffolk sheep [35]. Sheep blood parameters chlorides, sodium, phosphorus, total protein, tetraiodothyronine, cholesterol, triglyceride, creatinine, cortisol, and glucose were likewise reported to be significantly increased by heat stress conditions in the study of Li et al. [30]. In the same way, values for red blood cell (RBC), hemoglobin (Hb%), and packed cell volume (PCV%) were significantly elevated in indigenous sheep subjected to heat stress [36].

The negative effects of heat stress on general welfare, comfort, and health could result in diminished immune function and increased incidence of diseases [37-38].

Primarily, the response of the immune system to abiotic stressors, such as heat stress, begins with the stimulation of primary and secondary lymphoid and myeloid organs that catalyze the synthesis and release of immune cells and agents [5]. Studies show that in acute heat stress (brief and intense), immune activation happens, while chronic heat stress causes immune suppression [39]. Elevated cortisol levels, as a result of the activation of the hypothalamic-pituitary axis (HPA), alter various immune function elements that may increase the animal's vulnerability to diseases [40]. This effect is demonstrated in the induction of inflammation in acute heat stress and the suppression of release and synthesis of inflammatory cytokines and the consequent changes to the dynamics of balance on the levels of other cytokines, immune cells, and substances in chronic situations [41-42].

Cytokines are key mediators of the immune system's response to stressors and pathogens. Proinflammatory cytokines work to activate the innate immunity for the body's early response against infectious agents and support the activities of the cell-mediated immune response of the adaptive immunity to synergistically combat the pathogen of concern. However, specific cytokines modulate the previous cytokines to suppress the immune response and shift the modality of the immune system into adaptive immunity. The most powerful and significant immunomodulatory cytokine of this effect is *IL-10* [43].

In the present study, overall *IL-10* gene expression in Hungarian Indigenous Tsigai and White Dorper sheep were shown to be significantly higher (*P*=0.015) in animals under severe heat stress (summer) compared to those that experienced no heat stress condition (winter). This result is in accordance with the findings of Caroprese et al. [44], who demonstrated high *IL-10* expression in sheep experiencing hyperthermia. Similarly, in cattle, *IL-10* levels were increased in heat-stressed Holstein cows [10, 45], Italian Friesian cows [46], and Sahiwal cows [11].

The expression of *IL-10* in heat stress situations has various physiological and immunological indications that help the animal adapt and restore equilibrium. While it suppresses the proinflammatory response, it leads the immune regime from TH1 to TH2 that activates the adaptive immunity [47-48]. Sustained *IL-10* levels during heat stress and the extended period after the heat stress situation and strenuous activities have effectively controlled immune response to stress [49].

Ripley [50] has reported that *IL-10* activates heat shock protein 90 (HSP90). Heat shock proteins are molecular chaperones that are another important factor for protecting cells from damage, such as in heat stress situations and in antigen presentation to improve both innate and adaptive immunity [51]. *IL-10* expression by the experimental animals in this study suggests that there could also be accompanying HSP expression, which also aids in the immune response and endogenous protection against heat stress.

Furthermore, *IL-10*, in line with its anti-inflammatory actions, has also been associated with the inhibition of cyclooxygenase (COX) and, consequently, prostaglandin E2 (PGE2), the principal downstream mediators of fever that are inducted by inflammatory cells [52]. This would mean that the expression of *IL-10* during heat-stress situations is thermo-modulatory and helps reduce heat load as it eliminates endogenous heat production by fever pathways.

Though all the sheep in this study show significantly higher expression of the *IL-10* gene during the summer season, the inter-breed comparison reveals a numerically higher gene expression ratio among the White Dorper (16.2) against the Hungarian Indigenous Tsigai (12.30). Statistical analysis of the CT-values, however, yielded no significant difference between the two breeds on the same subject. The two breeds, despite having different geographical origins, manifest comparable thermo-tolerance evident in the *IL-10* mRNA expression that bears no significant difference.

The Hungarian Indigenous Tsigai Sheep, being a temperate indigenous breed that thrived in the continental climate for centuries [15, 53], elucidated heat stress response in the distinctively harsh and hot summer environment of the region.

In a similar manner, the White Dorper sheep that is considered tropical in origin, maintained its heat stress responsive characteristic as it experienced the hot summer season in the study location where it was introduced a few years ago. Despite being present in Hungary for not a long time, the breed's tropical characteristics made acclimatization manageable for it to thrive during the hot summer months in the area.

Even though *IL-10* is a prominent immunological biomarker in heat stress-related studies in domestic animals, comparative studies of its mRNA expression across breeds, particularly between tropical and temperate breeds, are very scarce, especially on sheep.

In contrast to the findings of the present study, various related studies reported differences in *IL-10* gene expression among different breeds in domestic animals, such as cattle, during heat stress. Previous studies showed significantly higher *IL-10* gene expression in Sahiwal cows [11] and Karan Fries cows [54] compared to other breeds that include temperate ones. Citing other related thermotolerance immune biomarkers such as heat shock proteins (HSPs), Sahiwal, a tropical cattle breed, also exhibited higher mRNA expression of HSP90 [55] and HSP70 [56] compared to the temperate and tropical cross, Frieswal. Following this trend, the Caribbean sheep breed Pelibuey was found to be less susceptible to heat stress and with increased HSP70 concentration compared to Suffolk [57].

A comparative study in pigs revealed better cell viability in the tropical Creole pigs than the temperate Large White pigs, with a significant increase in the expression of HSPs. Interestingly, however, there was no difference in mRNA expression of HSP70.2 and HSP90 [58] in agreement with the findings of the current study.

The increase in *IL-10* concentration can result in the upregulation of HSPs gene expression in peripheral blood mononuclear cells, consequently leading to increased HSP concentration in circulation [50].

The equivalent heat tolerance in the Hungarian Indigenous Tsigai to the tropical breed White Dorper indicated by its *IL-10* gene expression in this study can be attributed to its indigeneity in the area that is shown in its hardiness and suitability to local environment. Its exposure to the local agro-climatic conditions across generations through centuries helped it adapt to the harsh summer months in the region. Its success in variable climatic conditions is proven by its wide distribution in the Central-Eastern European and Balkan regions being integrally used in sheep breeding programs confirmed by relatively low to moderate levels genetic variation among country populations [53, 59]. The tenacity of indigenous breeds in extreme climatic conditions can be ascribed to their physiological and genetic adaptations that have arisen over generations of slow modification and adaptation to environmental challenges [60, 28].

On another note, the excellent thermotolerant character of the White Dorper sheep, according to the parameters of this study, suggests its suitability for the harsh summer heat stress conditions in the area, despite its relatively short period of habitation. Tropical breeds can thrive comfortably in temperatures as high as 38 °C and produce adequately by virtue of thermoregulatory processes [61]. This characteristic was exhibited by Malpura ewes from the Indian semi-arid regions [62] and Santa Ines and Morada Nova sheep in Brazil [63]. Performance-wise, Dorper raised in the same area under Hungarian rearing conditions proved to have a good production and reproduction performance as reported in a study by Budai et al. [16].

In a pilot study evaluating reproductive performance and health indicators in the same area in Hungary, genotype differences were found in incidences of clinical mastitis (higher in Tsigai), lameness (higher in Dorper), and litter size (lower in White Dorper), while no differences were observed in conception and lamb survival rates [64]. It provides an overview of the variability of immune status among the breeds of interest, which merits the conduct of further investigation as to how heat stress could affect those incidences and the role of *IL-10* in such health challenges and reproductive performance.

Finally, in contrast with the findings of Li et al. [30], that reported better high-temperature tolerance in ewes over rams, the present study finds no sexual differences. The absence of sexual differences in the mRNA expression of *IL-10* in this study implies that thermotolerance and heat stress immune responses in both breeds are not influenced by sexual factors.

5 Conclusion

Overall, the study has demonstrated higher relative gene expression of the IL-10 gene during the summer season in both the Hungarian Indigenous Tsigai and White Dorper when heat the animals experienced heat stress conditions in contrast with the winter season. Implicated as a molecular biomarker for heat stress response, the significant relative gene expression of *IL-10* indicates immune modulation and shift to adaptive immune response. There has been no significant difference in the *IL-10* gene expression between the two breeds. While sexual factors do not affect *IL-10* gene expression, its expression can potentially enhance physiological adaptive mechanisms to avoid cellular damage during heat stress conditions in both breeds, regardless of sex.

Owing to their origins, the comparable heat stress response through the expression of the IL-10 gene proves that both breeds, continental and tropical in origin, are adaptive to the region's environmental conditions and could be utilized in breeding programs to improve the immunity, resiliency, and production performance of sheep, even in variable climatic conditions.

Further study on the links of genetic adaptability, physiological reactions, immune response, breeding programs, animal health, and productivity must be conducted to unravel complex mechanisms of homeostasis and animal performance in the face of heat stress, particularly involving *IL-10* and related biomarkers.

Acknowledgement: This research was funded by the bilateral S&T cooperation programme, within the project "*Effect of heat stress in Pramenka types of sheep using DNA and RNA based methods*" from the National Development, Research and Innovation Fund (2021-1.2.4-TÉT-2021-00047), also by the EFOP-3.6.2-16-2017-00001 project, entitled: "*Complex rural economic development and sustainability research, development of the service network in the Carpathian Basin*". R.F. and P.K.A. were supported by Tempus Public Foundation within the Stipendium Hungaricum Programme. Authors would thank Csaba Rácz for providing climatic data during the sampling period, János Oláh for helping in sampling and András Jávor for his valuable comments in preparation of mansucript.

References

1. H. Berihulay, A. Abied, X. He, L. Jiang, & Y. Ma. Animals, **9**(3), 75 (2019)
2. I. Miztal. J. Anim. Sci. **95**, 1780-1787. (2017)
3. V. P. Rashamol, V. Sejian, M. Bagath, G. Krishnan, P. R. Archana, & R. Bhatta, J. Anim. Behvr. & Biomet, **6**(3), 62–71 (2018)
4. S. Indu, V. Sejian, D. Kumar, A. Pareek, & S. M. K. Naqvi. Trop. Ani. H. & Prod., **47**(8), 1487–1495 (2015)
5. R. J. Collier, J. L. Collier, R. P. Rhoads, & L. H. Baumgard. J. Dairy Sci., **91**(2), 445–454 (2008)
6. M. Bagath, G. Krishnan, C. Devaraj, V. P. Rashamol, P. Pragna, A. M. Lees, & V. SejiAnim. Res. Vet. Sci. 126, 94–102 (2019)
7. R. Sabat, G. Grütz, K. Warszawska, S. Kirsch, E. Witte, K. Wolk, & J. Geginat. Cyt. & Gro. Fac. Rev., **21**(5), 331–344 (2010)
8. K. N. Couper, D. G. Blount, & E. M. Riley. J. of Immu., **180**(9), 5771–5777 (2008)
9. H. D. VanGuilder, K. E. Vrana, & W. M. FreemAnim. BioTechniques, **44**(5), 619–626 (2008)
10. F. J. Zhang, X. G. Weng, J. F. Wang, D. Zhou, W. Zhang, C. Zhai, Y. X. Hou & Y. H. Zhu. J. Anim. Sci., **92**(7), 3026–3034 (2014)
11. S. Grewal, A. Aggarwal, & M. N. Alhussien. Bio. Rhyth. Res., **52**(8), 1229–1239 (2019)
12. Z. Lu, M. Chu, Q. Li, M. Jin, X. Fei, L. Ma, L. Zhang, & C. Wei. Animals, **9**(6), 387 (2019)
13. X. H. Ju, H. J. Xu, Y. H. Yong, L. L. An, P. R. Jiao, & M. Liao. Animals, **8**(9), 1462–1468 (2014)
14. C. Marley. Evaluation of the growing performance of rare breed sheep. Szent István University (2014)
15. S. Kusza, I. Nagy, T. Németh, A. Molnár, A. Jávor, & S. Kukovics. Arch. Anim. Breeding, **53**(3), pp.309–317 (2010)
16. C. Budai, D. Gavojdian, A. Kovacs, F. Negrut, J. Olah, L. T. Czizter, S. Kusza, & A. Javor. Anim. Sci. & Biotech, **46** (1) 344-349 (2013)
17. T. L. Mader, M. S. Davis, & T. Brown-Brandl. J. Anim. Sci. **84**(3), 712–719. (2006)
18. B. E. Lewis Baida, A. M. Swinbourne, J. Barwick, S. T. Leu, & W. H. E. J. van Wettere. Animal Biotelemetry, **9**(1), 4 (2021)
19. L. Schüller, O. Burfeind, & W. Heuwieser. Theriogenology, **81**(8), 1050–1057 (2014)
20. J. Ye, G. Coulouris, I. Zaretskaya, I. Cutcutache, S. Rozen, & T. L. Madden, BMC bioinformatics, **13**, 134 (2012)
21. M. W. Pfaffl. Nucleic Acids Research, **29**, e45 (2001)
22. P. Kic. Animals, **12**(15), 1895 (2022)
23. L. Schüller, O. Burfeind, & W. Heuwieser. J. Dairy Sci., **96**(12), 7731–7738 (2013)

24. M. Akhtar, S. Guo, Y. Guo, A. Zahoor, A. Shaukat, Y. Chen, T. Umar, P. G. Deng, & M. Guo. Acta Tropica, **207**, 105458 (2020)
25. R Development Core Team. R: A Language and Environment for Statistical Computing. R Foundation for Statistical Computing, Vienna. (2009)
26. J. W. West. Journal of Dairy Science, **86**(6), 2131–2144 (2003)
27. S. D. Rathwa, A. A.Vasava, M. M. Pathan, S. P. Madhira, Y. G. Patel, & A. M. Pande. Veterinary World, **10**(6), 650–654 (2017)
28. V. Sejian, R. Bhatta, J. B. Gaughan, F. R. Dunshea, & N. Lacetera, Animal, **12**(s2), s431–s444 (2018)
29. Hungarian Meteorological Service. Meteorológiai Adattár. https://odp.met.hu/climate/observations_hungary/daily/historical/. (2022)
30. F. K. Li, Y. Yang, K. Jenna, C. H. Xia, S. J. Lv, & W. H. Wei. Trop. Anim. Health & Prod., **50**(8), 1893–1901 (2018)
31. K. M. Singh, S. Singh, I. Ganguly, A. Ganguly, R. K. Nachiappan, A. Chopra & H. K. Narula. Sm. Rum. Res., **141,** 113–117 (2016)
32. A. Srikandakumar, E. H. Johnson, & O. Mahgoub. Sm. Rum. Res, **49**(2), 193–198 (2003)
33. J. Blackshaw, J., & A. Blackshaw. Aus. J. of Exp. Agri., **34**(2), 285 (1994)
34. S. Indu, & A. Pareek. Int. J. of Emer. Trends in Sci. & Tech., (2015)
35. J. Autukaite, I. Poškienė, V. Juozaitienė, R. Antanaitis, W. Baumgartner, & H. Žilinskas. Indian J. of Anim. Res. (2021)
36. M. S. Rana, M. A. Hashem, M. N. Sakib, & A. Kumar.. AgEcon Search. **12** (1) 91-94 (2014)
37. S. L. Cartwright, M. McKechnie, J. Schmied, A. M. Livernois & B. A. Mallard. BMC Veterinary Research, **17**(1) (2021)
38. U. Bernabucci, N. Lacetera, L. H. Baumgard, R. P. Rhoads, B. Ronchi, & A. Nardone. Animal, **4**(7), 1167–1183 (2010)
39. A. M. Livernois, B. A. Mallard, S. L. Cartwright, & A. Cánovas. Scientific Reports, **11**(1) (2021)
40. N. Lacetera, U. Bernabucci, D. Scalia, L. Basiricò, P. Morera & A. Nardone. J. of Dairy Sci., **89**(12), 4606–4612 (2006)
41. L. Cui, Y. Wang, H. Wang, J. Dong, Z. Li, J. Li, C. Qian & J. Li. BMC Veterinary Research, **16**(1) (2020)
42. M. Bagath, G. Krishnan, C. Devaraj, V. P. Rashamol, P. Pragna, A. M. Lees, & V. SejiAnim. Res. in Vet. Sci., **126**, 94–102. (2019)
43. S. Inbaraj, V. Sejian, M. Bagath & R. Bhatta. Pertanika Journal of Tropical Agricultural Science, **39**(4) (2016)
44. M. Caroprese, M. G. Ciliberti, G. Annicchiarico, M. Albenzio, A. Muscio & A. Sevi. J. of Dairy Sci., 97(7), 4247–4258 (2014)
45. I. M. T. Thompson, S. Tao, A. P. A. Monteiro, K. C. Jeong, & G. E. Dahl. J. of Dairy Sci., **97**(12), 7426–7436 (2014)
46. M. Caroprese, A. Marzano, G. Entrican, S. Wattegedera, M. Albenzio & A. Sevi. J. of Dairy Sci., **92**(6), 2796–2803 (2009)
47. E. Calcagni & I. Elenkov. Annals of the New York Acad. Sci., **1069**(1), 62–76 (2006)
48. S. H. Siddiqui, D. Kang, J. Park, M. Khan & K. Shim. Scientific Reports, **10**(1), 18872 (2020)
49. F. A. Ehiaghe, C. C. Onyenekwe, I. J. Ehiaghe, C. O. Akosile & D. E. Agbonlahor. Int. J. of Bio. & Chem. Sci., **11**(3), 947 (2017)
50. B. J. Ripley, A. Stephanou, D. A. Isenberg & D. S. Latchman. Anim. Immunology, **97**(2), 226–231 (1999)
51. P. R. Archana, J. Aleena, P. Pragna, M. K. Vidya, P. A. Abdul Niyas, M. Bagath, G. Krishnan, A. Manimaran, V. Beena, E. K. Kurien, V. Sejian & R. Bhatta. J. of Dairy, Vet. & Anim. Res., **5**(1) (2017)
52. H. Niiro, T. Otsuka, K. Izuhara, K. Yamaoka, K. Ohshima, T. Tanabe, S. Hara, Y. Nemoto, Y. Tanaka, H. Nakashima & Y. Niho. Blood, **89**(5), (1997)
53. S. Kusza, E. Gyarmathy, J. Dubravska, I. Nagy, A. Javor, & S. Kukovics. Czech J. Anim. Sci. **54** (10), 468-474 (2010)
54. A. A. Sheikh, A. Aggarwal, & O. Aarif. J. of Thermal Bio., **56**, 68–76. (2016)
55. R. Deb, B. Sajjanar, U. Singh, S. Kumar, R. Singh, G. Sengar, & A. Sharma. Gene, **536**(2), 435-440 (2014)
56. V. Bhanuprakash, U. Singh, G. Sengar, B. Sajjanar, B. Bhusan, T. V. Raja & R. Deb. J. of Thermal Bio., **59**, 18-25 (2016)
57. R. D. Romero, A. Montero Pardo, H. H. Montaldo, A. D. Rodríguez & J. Hernández Cerón. Trop. Anim. Health & Prod., **45**, 1691-1696 (2013)
58. J. C. Bambou, J. L. Gourdine, R. Grondin, N. Vachiery & D. Renaudeau. Trop. Anim. Health & Prod., **43**(8), 1535-1541 (2011)
59. S. Kusza, A. Ivankovic, J. Ramljak, I. Nagy, A. Javor, & S. Kukovics. Small Rum. Res. **99** (2-3) 130-134 (2011)
60. M. M. Rojas-Downing, A. P. Nejadhashemi, T. Harrigan & S. A. Woznicki. Climate Risk Management, **16**, 145-163 (2017)
61. R. Gandhi & S. Arjava. Indian DairymAnim. **68**, 102–107 (2016,)
62. V. Sejian, V. P. Maurya, & S. M. Naqvi. Trop. Anim. Health Prod., **42**, 1763–1770 (2010).
63. T. P. Dias E Silva, J. N. Costa Torreão, C. A. Torreão Marques, M. J. de Araújo, L. R. Bezerra, D. Kumar Dhanasekaran, & V. Sejian. J. Therm. Biol., **59**, 39–46 (2016)
64. D. Gavojdian, C. Budai, L. T. Cziszter, N. Csizmar, A. Jávor & S. Kusza. Asian-Australasian J. of Anim. Sci., **28**(4), 599 (2015)

Morphology, Biomass Production, and Nutrient Analysis Four Napier Grass Cultivar (*Pennisetum purpureum* Schumach.) in Teak Tree Forest Area Megeri Village, Blora, Central Java

Yogi Sidik Prasojo[1*], *Mustafa* Kamal[1], *Arkan* Roja Sukmana[1], *Bowo* Dwi Siswoko[2] and *Nafiatul* Umami[1]

[1]Faculty of Animal Science, Universitas Gadjah Mada, Yogyakarta, Indonesia
[2]Faculty of Forestry, Universitas Gadjah Mada, Yogyakarta, Indonesia

Abstract. This study aims to determine the morphology, biomass production and nutrient analysis of four cultivars of napier grass (*Pennisetum purpureum* Schumach.) in teak tree forest area Megeri village, Blora Regency, Central Java. Four cultivars of napier grass namely Gama Umami, Biovitas, Pakchong, and Zanzibar being cultivated in teak tree forest area. The teak tree forest area was in Kawasan Hutan Dengan Tujuan Khusus (KHDTK) UGM. Previously there was no cultivation of new improved napier grass cultivars in KHDTK. Therefore, it is interestingly to investigate the introduction and growth ability in this area. Grasses were cultivated using stem cuttings approximately 15 cm and maintained for 4 months using a completely randomized design with 4 replications for each cultivar. During the cultivation period, morphological characteristics were observed including plant height, leaf length, leaf width, and tiller numbers. Plant was harvested after 4 month of cultivation and determine for biomass production and nutritional value of grass. The morphological characteristics and nutrient analysis among napier grass cultivars showed the similar trend. Gama Umami, Biovitas, and Pakchong had significantly higher biomass production than that Zanzibar cultivar.

1 Introduction

Napier grass (*Pennisetum purpureum* Schumach.) is popular warm season grass in Indonesia for being used as forage for livestock and dairy cattle. Currently in Indonesia has several improved cultivars that commonly used such as Gama Umami, Biovitas, Pakchong, and Zanizabar.

The cultivars of plants can have an impact on both the quality and quantity of the plant. A study on soybean plant found that different growth type had different on nutrient and quantity yield of soybean [1]. Another study on Brachiaria grass found that different species and cultivar lead to different nutrient and oxalate content [2]. Similarly, a study on Dwarf Napiergrass found that oxalate and silica content may varied among cultivars [3].

Different varieties of napier grass can have varying levels of nutrient content. The way that napier grass is planted can also affect its nutrient content. Intercropping pakchong with Indigofera and harvesting at 60 days produced the higher forage fresh weight and DM production, ADF contents and yields, CP yield, and the highest carrying capacity [4].

Teak tree plantations are an important part of the forestry industry in Indonesia. Teak is the most important commercial tree species in Indonesia, and there are 0.6 million hectares of production plantations based on seed from these plantations [5]. The area of teak plantation covers 1.081 million hectares, in which 0.837 million hectares are suitable for the clear-felling harvest system [6]. Blora Regency in Central Java has the largest teak Forest area in Java, Indonesia. The forest area is quite extensive and covers several catchments in Blora Regency [7].

Intercropping is a common practice in Teak plantation in Indonesia. In Gunungkidul, intercropping tuber crops with teak has been studied to diversify food crop products and make more effective use of land [8]. In addition, teak is sometimes planted with annual crops such as maize during the establishment period [9]. Intercropping can improve productivity and soil quality by increasing the amount of organic matter and improving soil structure. A study on arbuscular mycorrhizal inoculation in teak and napier grass agroforestry found that soil amendments like gypsum can improve soil structure, increase moisture content, and facilitate soil microbial functions [10]. Therefore, it's important to investigate the morphology character, biomass production, and nutrient analysis in several cultivar of Napier grass cultivation in teak tree forest area Megeri Village, Blora Regency, Central Java to support agroforestry system in Indonesia and improve efficiency of land use in Indonesia.

* Corresponding author: yogi.sidik.p@ugm.ac.id

2 Methodology

2.1 Plant material

Four cultivars of napier grass namely *Pennisetum purpureum* cultivar Gama Umami, Biovitas, Pakchong, and Zanzibar were used in this study. Grasses were cultivated using stem cuttings approximately 15 cm.

2.2 Field experiment

The study was conducted in teak tree forest area Megeri village, Blora Regency, Central Jawa (7°21'40''S 111°26'36''E, elevation of 100 m above sea level). Most of the land in Blora Regency is dry land. The soil type characterized as sandly soil with relatively poor in organic matter. The climate of Blora Regency is tropical climate, relatively high temperatures in the dry season and sufficient rainfall in the rainy season. Plant cultivation maintained for 4 months using a completely randomized design with 4 replications for each cultivar.

2.3 Investigation on morphology

During the cultivation period, morphological characteristics were observed including plant height, leaf length, leaf width, and tiller numbers. Plant was harvested after 4 month of cultivation and determine for biomass production and nutritional value of grass.

2.4 Statistical analysis

Statistical analysis was conducted to compare the morphological characteristics, biomass production, and nutrient content among napier grass cultivars. Differences in means were evaluated by Tukey's test using R statistic program.

3 Result and Discussion

Table 1. Average of Morphology characteristics of 4 Napier grass cultivars

Cultivar	Average value of morphological characteristics			
	Plant height (cm)	Leaf length (cm)	Leaft width (cm)	Tiller numbers
Gama Umami	181.3^a	96.2	3.47	4
Biovitas	194.2^a	100.5^a	3.70	5
Pakchong	180.1^a	97.7	3.78	4
Zanzibar	169.9^b	93.3^b	3.13	3

Means in the same column followed by the different latter are significantly different at 0.001 probability level.

Based on the plant observation for 4 months after planted, plant growth data were obtained in Table 1. The morphological characteristics among the cultivars described the similar trend. However, Biovitas had significantly higher of plant height and leaf length than that of Zanzibar cultivar. Even though there was not significantly different between Gama Umami, Pakchong, and Biovitas cultivars. Plant growth is influenced by internal and external factors. The influence of internal factors in the form of genetics. Selection of varieties will affect the morphological value of the growth of the grass. This happens because it is genetic influences that cause differences in the speed of division, multiplication and enlargement of plant cells. while external factors can be influenced by soil and climate. Napier grass can grow optimally from the factor moist environment and soil conditions that have nutrients which is more, so that the high growth of elephant grass plants can seen for real. Napier grass will growth better in good and fertile land condition with sufficient nutrient. The moisture of soil was also affecting the plant growth. Napier grass had fibrous root system which considering deep rooting. In this research, Zanzibar had aggressive tiller number that spreads through rhizome under the ground. The morphological characteristics of napier grass cultivars, such as plant height, stem diameter, internode, leaf stem, ratio, fiber content, crude protein, and in vitro dry matter digestibility, can affect their productivity and quality [11].

Table 2. Biomass productino of 4 grass cultivars

Cultivar	Biomass production *(fresh weight)* First defoliation (Kg/m^2)
Gama Umami	6.15^a
Biovitas	7.25^a
Pakchong	6.57^a
Zanzibar	3.39^b

Means in the same column followed by the different latter are significantly different at 0.001 probability level.

The analysis result in Table 2 shows the production of biomass. The production of biomass among the cultivars described the similar trend. However, Biovitas had significantly higher of biomass production than that of Zanzibar cultivar. Even though there was not significantly different between Gama Umami, Pakchong, and Biovitas cultivars. The yield and chemical compostition related to biomass quality of elephant grass can be affected by the harvest time. A study found that different elephant grass genotypes showed a liner first-degree effect as a function of the harvest intervals, indicating that they did not reach their maximum production potential. Genotypes that had a linear second-degree effect had a higher yield [12].

Fresh production is influenced by nutrients that play a role production of biomass of a plant. Plant growth greatly affects the fresh biomass production of a plant. The fresh weight of the plant can reflect water and elements nutrients absorbed by plants. Water and nutrients absorbed by plant will affect plant growth, such as height plant, the number of leaves and leaf area which then affect the weight fresh a plant. The growth and production response of Pakchong elephant grass to different fertilizer doses and cutting ages were evaluated in another study. The study found that the optimal point for obtaining the best results in terms of plant height growth, number of leaves, number of tillers, production of fresh biomass, dry weight of stems, dry weight of leaves, and total dry weight of Pakchong elephant grass was a cutting edge of 60 days [13].

Table 3. Nutrient content of 4 grass cultivars

Lines	Average of nutrient content (%)				
	Dry matter	Organic matter	Crude fiber	Crude protein	Crude fat
Gama Umami	16.85	82.80	29.82	14.89	2.57
Biovitas	16.13	82.79	33.40	16.97	2.12
Pakchong	15.63	81.66	31.02	17.92	1.96
Zanzibar	14.73	80.13	33.28	16.95	2.18

Means in the same column followed by the different latter are significantly different at 0.001 probability level.

The analysis result in Table 3 shows that the nutrient content of 4 grass cultivars. The nutrient content of dry matter among the cultivars described the similar trend. However, Gama Umami had significantly higher dry matter content than that Zanzibar. Even though there was not significantly different between Gama Umami, Pakchong, and Biovitas cultivars. Dry matter is the total feed substances other than water in a material feed. Dry matter is an accumulation of food reserves such as protein, carbohydrates and fat, as well as photosynthate is found in the stems and leaves of plants. Plant dry matter production is influenced by several factors, one of which is water content (KA) in the plant itself. The DM content in plants is greatly influenced by the rate photosynthesis and the products of photosynthesis. Then it is related with nutrients in the soil, water and light intensity.

Gama Umami has a much lower crude fiber content compared to other cultivars. Differences in crude fiber content in plants are influenced by genetics plants, harvest age, nutrients in the soil and light intensity. The number of stems and leaves increased as the growth stage increased. The crude fiber content of plants is also influenced by place altitude. Differences in cultivars can also cause differences in fiber content rough because the nutrient content of plants is greatly influenced by genetic potential. The genetic characteristics of the varieties of napier grass may influence their ability to produce energy, due to their different levels of fiber and lignin. A study found that different elephant grass varieties had different levels of fiber in neutral detergent, acid detergent fiber, lignin, moisture, dry matter content, and higher heating value [14]. The number of leaves and stem was also affect the nutrient value produced from plant [15].

Pakchong has a much higher crude protein content compared to other cultivars. The low value of the nutritional content of a plant is influenced by several factors, one of which is harvest age. The crude protein content of grass decreased as increasing the harvest age. In addition, the effect of mutation induction in the elephant grass mutants that cause genetic variation is also a factor that determine the value of nutrient content in plants. Besides that, the nitrogen content available in the soil also has an effect in plant crude protein content. The main source of variation in crude protein content in elephant grass is genotype, age of harvest and environment. Proper nitrogen will enhance plant growth then cause an increase in metabolism, so it will accelerating the conversion of carbohydrates into protein inside plant. Sufficient nitrogen will increase vegetative growth and expand the size of the leaves to catch more light, nitrate, and phosphate which are necessary for the preparation of protein and chlorophyll. Defoliation time will also affect nutrient value, where the crude protein content decreased from the first to second defoliation [16].

The organic matter among the cultivars described the similar trend. Organic matter is one of the constituent components of the material dry. The organic matter content is affected by the dry matter content and ash in plants. Organic matter is related to protein content crude oil, crude fat and crude fiber in plants. The organic material content in grass can influenced by the condition of nutrients in the soil. Lack of nutrients in the soil causes a decrease organic matter content in plants. Light intensity and water availability can also affect the plant organic matter content.

The nutrient content of crude fat among the cultivars described the similar trend. However, Gama Umami had significantly higher dry matter content than that Pakchong. Crude fat content in plants is very high influenced by the level of N absorbed by plants. It's related by the process of fat formation by plants. The crude fat of plant also affect by the age of plant, increasing age of plant will increasing the crude fat of plant [17]. Crude fat content in plants is also influenced by the addition of NPK fertilizer, the addition of fertilizer will increasing N levels in the soil so that it can increase formation of crude fat and vegetative growth of plants. The increase in vegetative cells can increase the crude fat content, because fat is part of the protoplast.

In conclusion, napier grass cultivar Biovitas, Gama Umami, and Pakchong could adapt with high morphology characteristics, high biomass production, and high nutrient value. The research data could be used for plant breeder to develop new cultivar of napier grass which can adapt in teak tree shade area. Cultivation of napier grass in teak tree forest area or KHDTK UGM is recommended and could support the agroforestry system in Indonesia. Further research about teak tree shade for napier grass ability in teak tree forest, planting space area, intercropping with another plant and fertilizer management are needed.

4 Acknowledgment

The authors are very thankful for Universitas Gadjah Mada that already supports this research by providing funding through the Doctoral Competency Improvement Program Universitas Gadjah Mada Number 7743/UN1.P.II/Dit-Lit/PT.01.03/2023.

References

1. Y.S. Prasojo, G. Ishigaki, M. Hashiguchi, R. Akashi, Asian. J. Plant. Sci **20**, 2 (2021)
2. N. Umami, B. Suhartanto, B. Suwignyo, N. Suseno, F. Herminasari, Pak. J. Nutr **17**, 6 (2018)
3. M.M. Rahman, M.S. Norshazwani, T. Gondo, M.N. Maryana, R. Akashi, South. Afr. J. Anim. Sci **50**, 3 (2020)

4. A. Ernawati, L. Abdullah, P.D.M. Karti, Trop. Anim. Sci. J **46**, 3 (2023)
5. H. Suhaendi, *Teak improvement in Indonesia, in Proceedings of The Second Regional Seminar on Teak*, Bangkok, Thailand (1998)
6. S. Siswamartana, *Teak Forest Management in Indonesia, in Proceedings of The Second Regional Seminar on Teak*, Bangkok, Thailand (1998)
7. T.M. Basuki, I.B. Pramono, IOP Conf. Ser.: Earth Environ. Sci **533** (2020)
8. D. Maharani, A. Sudomo, D. Swestiani, Murniati, G.E. Sabastian, J.M. Roshetko, R.A. Fambayun, Agronomy **12**, 2 (2022)
9. H.T.L. Stewart, D.H. Race, D. Rohadi, D.M. Schmidt, Forest. Pol. Econ **130** (2021)
10. J. Vijayanathan, A.R. Mohammad, M.S. Lias, *Preliminary assessment of arbuscular mycorrhizal inoculation in teak and napier grass agroforestry practice treated with soil amandments, in Proceedings 4th International Conference in Agroforestry*, Sabah, Malaysia **1145** (2022)
11. B. Budiman, R.D. Soetrisno, S.P.S. Budhi, A. Indrianto, J. Ind. Trop. Anim. Agr **37**, 4 (2012)
12. R.S. Freitas, T. De Costa Barbe, R.F. Daher, A.K.F. Vidal, W.F. Stida, V.B. Da Silva, B.F. Da Silva Menezes, A.V. Pereira, J. Agr. Sci **10**, 1 (2018)
13. F. Harianti, M. Ridla, L. Abdullah, Nutri. Feed. Tech. J 21, 2 (2023)
14. R.L. Dos Santos, F.J. Freire, A.T. Da Rocha, J.A. Da Silva, J.A. Tavares, E.G.B. De Sausa Ferreira, E.C.A. De Oliveira, Aus. J. Crop. Sci **9**, 11 (2015)
15. B.B. Koten, R. Wea, R.K. Tri, S. Otto, A.W.B. Selan, W. Yikwa, A.P. Titong, Buletin Peternakan **46**, 2 (2022)
16. R.F. Utamy, H. Sonjaya, K.I. Prahesti, Buletin Peternakan **44**, 4 (2020)
17. Hermanto, B. Suwignyo, N. Umami, Buletin Peternakan **41**, 1 (2017)

Tiller Number and Flowering Ability on BC1F1 Progeny of Interspecific Hybridization in Ruzigrass (*Urochloa ruziziensis*)

Yogi Sidik Prasojo[1*], *Pattama* Nitthaisong[2], *Melody* Muguerza[3], *Sukanya* Kamphayae[4], *Nafiatul* Umami[1] and *Genki* Ishigaki[5]

[1]Department of Animal Nutrition and Animal Science, Faculty of Animal Science, Universitas Gadjah Mada, Yogyakarta, Indonesia
[2]King Mongkut's Institute of Technology Ladkrabang, Bangkok, Thailand
[3]Faculty of Agriculture, University of Miyazaki, Miyazaki, Japan
[4]Ruminant Feeding Standard Research and Development Center, Khon Kaen, Thailand
[5]Sumiyoshi Livestock Science Station, Field Science Education Research Center, Faculty of Agriculture, University of Miyazaki, Miyazaki, Japan

Abstract. The objective of this study focusing on production BC_1F_1 progeny of interspecific hybridization in ruzigrass (*Urochloa ruziziensis*) were to evaluate potential of pentaploid BC_1F_1 hybrid for seed propagation by associating with their tiller number, flowering, and shattering ability. In addition, we evaluate the function of various nitrogen dose on vegetative stages. Eight different genotypes were examined: *B. ruziziensis* 'Kennedy', *B. decumbens* 'Basilisk', pentaploid BC_1F_1 line derived from *B. ruziziensis* and Mulato (RM), pentaploid BC_1F_1 line derived from diploid *B. ruziziensis* and tetraploid *B. ruziziensis* (RR), and pentaploid BC_1F_1 lines derived from *B. ruziziensis* and *B. decumbens* (RD 1-4). The experimental plants treated with three level of nitrogen fertilizer (0, 4, and 8 kg/10a). Compared with parental line, RM line had high tiller number along with increasing level of nitrogen. For the flowering and shattering ability, the first lines to flowered and shattered was the pentaploid BC_1F_1 (RD 1-4), RR, and *B. decumbens*. Pentaploid BC_1F_1 lines derived from *B. ruziziensis* and *B. decumbens* (RD 1-4) had potential for seed propagation by associating with their flowering and shattering ability, while pentaploid BC_1F_1 line derived from *B. ruziziensis* and Mulato (RM) had the potential biomass production and yield ability by associating with its tiller number production.

1 Introduction

Urochloa (named as Brachiaria previously) is warm-season grass belongs to a small group of genera that include Urochloa, Eriochloa, and Panicum, which all have the PEP-CK (phosphoenolpyruvate carboxynase) type of C4 photosynthetic pathway. The advantages of Urochloa e.g. excellent adaptation to low-fertility acid soils were previously reported, compatible with the tropical climate, and tolerant to drought condition [1,2,3].

The new interspecific cross case in *Urochloa* genus had been reported [4]. They have conducted the interspecific crosses between diploid *U. ruziziensis* (R. Germ. & C.M. Evrard) Crinds 'Kennedy' (2n = 2x = 18) with apomictic tetraploids U. decumbent (Stapf) R. D. Webster 'Basilisk' (2n = 4x = 36) for F1 triploid hybrid. Subsequently, BC_1F_1 progeny is the offspring after backcrossing the F1 triploid hybrid *Urochloa* as female parent between *B. decumbens* as male parent. BC_1F_1 is pentaploid hybrid Urochloa. The result backcrossing triploid is producing polyploids. Interestingly, only the pentaploid plants produced seeds by self-pollination among these polyploids. This is the first report of flowering ability of apomictic pentaploid progenies from interspecific crosses in *Urochloa* genus. These new polyploids and the research information will be useful for the new breeding source.

Polyploidy has greatest effect in plants and represents a major mechanism of specification [5]. By backcrossing with apomictic plant may lead to produce superior lines hybrid. Backcrossing also can be used to improved genomic stability and pollen viability [6]. Therefore, important to figure out the characteristic of seed propagation on BC_1F_1 related to flowering ability. Flowering ability the lines will be important to understand the potential for continuing the lines for next *Urochloa* breeding program. Since superior genotypes can rapidly increase by seed, it is important to develop seed production in *Urochloa* hybrid and afterward investigate characteristic.

Therefore, the objectives of this study focusing on the production on BC_1F_1 progeny of interspecific hybridization in ruzigrass (*Urochloa ruziziensis*) were to evaluate potential of pentaploid BC_1F_1 hybrids for seed propagation by associating with their flowering ability, and 2) evaluate the function of various nitrogen dose application on tiller number of *Urochloa* hybrids.

* Corresponding author: yogi.sidik.p@ugm.ac.id

2 Methodology

2.1 Plant material

In this study, eight different genotypes were examined: two commercial cultivar *B. ruziziensis* 'Kennedy' and *B. decumbens* 'Basilisk' as a control, four BC_1F_1 lines derived from the diploid *B. ruziziensis* x tetraploid *B. decumbens* which is produce F_1 triploid hybrid. Afterward, F1 triploid hybrids backcrossing to tetraploid *B. decumbens* as male parents and written as RD 1-4. Two BC_1F_1 lines derived from the diploid *B. ruziziensis* x tetraploid *B. ruziziensis* and tetraploid Hybrid Mulato which is produce F_1 triploid hybrid. Afterward, each F1 triploid hybrids backcrossing to tetraploid *B. decumbens* and Hybrid Mulato as male parents and written as RR and RM, respectively [4].

2.2 Field experiment

Seeds were first sow in nursery pots (200 cm^2 surface area) with vermiculate and kept in an incubator (31°C, 24 h photoperiod) for germination. After 1 week of sowing, the germinated seedlings were grown to the same pot for 15 days. The field study was conducted in Sumiyoshi Livestock Science Station, Faculty of Agriculture, University of Miyazaki, Miyazaki prefecture which located in Southern part of Kyushu, Japan (39°59′N, 131° 28′E, the elevation of 12 m above sea level). The soil type was characterized as sandy soil. The climate of Miyazaki prefecture according to the Köppen classification, is Cfa that is humid subtropical climate, relatively high temperature and evenly distributed precipitation throughout the year. The experimental plant was treated with three levels of nitrogen fertilizer 0, 4 and 8 kg N $10a^{-1}$. The experiment was laid out in randomized complete block design with three replications. All lines were fertilized with 4 and 8 kg N $10a^{-1}$ and applied at the beginning of vegetative stage.

2.3 Investigation on agronomy

To evaluate effect of nitrogen during vegetative stage of progenies, tiller numbers in each plot were counted manually within one week. Investigate of flowering time and shattering time were done right after finishing vegetative stage. Measured the shattering time, seed trapper has been made by nylon bag and apply in every plot. For the flowering time, we made sign at three tested stems as replication by yellow strip from the begin of heading and end of shattering.

2.4 Statistical analysis

Statistical analysis was conducted to compare the tiller number among plant materials and three treated nitrogen fertilizer doses, flowering, and shattering ability among plant materials. Differences in means were evaluated by Tukey's test using R statistic program.

3 Result and Discussion

The tiller development among the lines described the similar trend. However, there was significant difference in the genotype of *Urochloa* for tiller number. Compared with parental line (*B. decumbens* and *B. ruziziensis*), RM line had high tiller number along with increasing level of nitrogen fertilizer. Following by RR line and RD 3. Line of RD 1, 2 and 4 gave number of tillers below their parental lines without nitrogen fertilizer (N_0) application. Compared with N_0, the addition of nitrogen fertilizer (N_4 and N_8) increasing number of tillers.

Table 1. Tiller number among lines with different dose nitrogen fertilizer application

Lines	Tiller Number		
	N_0	N_4	N_8
B.decumbens	22^{b}	73^{c}	69
B.ruziziensis	31^{ab}	135^{bc}	139
RD 1	12^{b}	183^{ab}	157
RD 2	12^{b}	164^{ab}	167
RD 3	34^{a}	135^{bc}	196
RD 4	14^{b}	135^{bc}	152
RR	41^{ab}	185^{ab}	180
RM	57^{ab}	247^{a}	214

N_0, without N fertilizer application; N_4, 4 kg N fertilizer $10a^{-1}$; N_8, 8 kg N fertilizer $10a^{-1}$. Means in the same lines within a row followed by the same latter are not significantly different at 0.001 probability level.

In application of N_4 recorded significantly increasing tiller number in all lines. The increasing is significantly higher compared without nitrogen supplied. In other hand, tiller number was statistically similar by application 8 kg nitrogen $10a^{-1}$. Excess N fertilizer had less effect for tiller number on all lines, except *B. ruziziensis*, RD 2, RD 3 and RD 4. The greater number of tillers may lead to potential of greater yield ability. The data of tiller number revealed that lines with high number of ploidy has higher number of yield than small number of ploidy. High tiller number also associating with greater biomass production. New cultivar with high ploidy and high biomass production is potentially used as forage for livestock [7]. Even though C4 grasses has low nutrient, it can be increase by intercropping system with legume [8].

In this study, nitrogen fertilizer was applied during early stage of *Urochloa*. Nitrogen fertilizer altered the number of tiller during their vegetative stages [9]. Adequate supply NO_3^- and CO_2 assimilation, the supply of assimilates to developing meristems is adequate to maintain their growth, so more tillers are produced and survive per plant and area, and similarly so with grains. Otherwise, less assimilate is available during early growth, fewer tillers survive, resulting in fewer ears and grains per ear and so less yield [10].

According the data of tiller number, RM, RR and RD 1 probably have greater number of seed yield regarding responses to nitrogen application to tiller number development. However, nitrogen is not the one influence of tiller number. Environmental conditions also affect tiller development through effects on the carbohydrate supply or demand framework. In the absence of drought stress, radiation determines crop growth, whereas temperature drives development. As each tiller has only one phyllochron during which it can appear, temperature determines the duration of this period and consequently, low radiation and high temperature limit carbohydrate availability at the start of the tiller emergence period, resulting in a low frequency of appearance of early tillers [11].

Even there was difference tiller number among genotype, the reason is not clear yet that different appearance of tiller due to the difference number of chromosome or plant hormone [12]. Tetraploids are superior to diploid in terms of dry mater intake and feeding value, but they generally have lower tiller density [13]. Tiller number is positive correlation with photosynthesis and leaf area index. The number of cells per unit leaf area decreases with increasing ploidy level, therefore, even if the photosynthetic rate per cell is higher, photosynthetic rate per unit leaf might be equal to or lower than those of diploids [14].

Table 2. Flowering time of BC_1F_1 progeny and Uruchloa cultivars

Lines	Flowering		
	Begin	Half	End
B. decumbens	1 Sep (10)	11 Sep (10)	25 Sep
B. ruziziensis	2 Nov (11)	13 Nov (11)	24 Nov
RD 1	1 Sep (10)	11 Sep (10)	25 Sep
RD 2	1 Sep (10)	11 Sep (10)	25 Sep
RD 3	1 Sep (10)	11 Sep (10)	25 Sep
RD 4	1 Sep (10)	11 Sep (10)	25 Sep
RR	1 Sep (10)	11 Sep (10)	25 Sep
RM	8 Sep (12)	22 Sep (12)	5 Oct

*(n): days between stage to stage

The beginning of flowering stage was difference in each lines. Among the 8 lines observed in flowering time, the first lines to flower was the pentaploid BC_1F_1 RD1-4, pentaploid BC_1F_1 RR and *B. decumbens* on September 1. The last line to begin flowering was the *B. ruziziensis* on November 2. Based on Table 2. the pentaploid BC_1F_1 RD1-4, pentaploid BC_1F_1 RR and *B. decumbens* were fully flowering on September 25. Following with pentaploid BC_1F_1 RM and *B. ruziziensis* on October 5 and November 24, respectively. The total period of flowering stage for all lines (from heading to fully flowering) was 20-24 d.

Flowering and shattering stage were effect by the temperature and genotype of species. Interestingly that among lines have different mode of reproduction, sexual and apomixis. *B. ruziziensis* is diploid sexual and the others are pentaploid apomixis, except *B. decumbens* is tetraploid apomixis. Apomixis *B. decumbens* very early flowering than sexual *B. ruziziensis*. Crossing between both species will be produced recombinant individuals that had a modest flowering period both the sexual and the apomictic component in interspecific hybridization is an essential condition for planning controlled crosses and achieving hybrids [15].

Table 3. Shattering time of BC1F1 progeny and Uruchloa cultivars

Lines	Shattering		
	Begin	Half	End
B. decumbens	26 Sep (12)	8 Oct (13)	21 Oct
B. ruziziensis	25 Nov (12)	5 Dec (13)	18 Dec
RD 1	26 Sep (12)	8 Oct (13)	21 Oct
RD 2	26 Sep (12)	8 Oct (13)	21 Oct
RD 3	26 Sep (12)	8 Oct (13)	21 Oct
RD 4	26 Sep (12)	8 Oct (13)	21 Oct
RR	26 Sep (12)	8 Oct (13)	21 Oct
RM	6 Oct (12)	18 Oct (13)	31 Oct

*(n): days between stage to stage

Spikelet was shattered after finished fully flowering. The first time to shattered obviously was the pentaploid BC_1F_1 RD1-4, pentaploid BC_1F_1 RR and *B. decumbens* on September 26. The last line to begin shattered was the *B. ruziziensis* on November 25. According the data on Table 3. spikelet was fully shattered on October 21 for the pentaploid BC_1F_1 RD1-4, pentaploid BC_1F_1 RR and *B. decumbens*. The total period of spikelet for shattering in all lines was 25 d. Future research about plant size and seed size need to be done. In *Urochloa Hybrid Mulato* II, the establishment better in summer because of the greater probablity of adequate temperature and moisture condition. Spring planting may be used but manifest the risk of dry condition and stand failure [16].

In conclusion, pentaploid BC_1F_1 lines derived from *B. ruziziensis* and *B. decumbens* (RD 1-4) had potential for seed propagation by associating with their flowering and shattering ability, while pentaploid BC_1F_1 line derived from *B. ruziziensis* and Mulato (RM) had the potential biomass production and yield ability by associating with its tiller number production.

References

1. G. Keller-Grein, B.L. Maass, J. Hanson, Natural variation in *Brachiaria* and existing germplasm collections. In: *Brachiaria*; Biology, Agronomy and Improvement (Eds Miles JW, Maass BL, do Valle CB), CIAT, Cali, 16-42 (1996)
2. N. Umami, A.H.I. Kusuma, C.T.Noviandi, Buletin Peternakan 42, 3 (2018)
3. N. Suseno, N. Isnaini, B. Suwignyo, B. Suhartanto, N. Umami, Buletin Peternakan 45, 2 (2021)
4. P. Nitthaisong, G. Ishigaki, K. Suenaga, M. Muguerza, H. Tanaka, R. Akashi, Crop Sci **59** (2019)
5. K.L. Adams, J.F. Wendel, Curr Opin Plan Biol **8**, 2 (2005)
6. R.M. Giacomin, R. Assis, S.P. Brammer, A. Nascimento Junior, P.R. Da Silva, Genet Mol Res **14**, 3 (2015)
7. G. Ishigaki, P. Nitthaisong, Y.S. Prasojo, K. Ikuo, F. Kiichi, M.M. Rohman, Anim. Bio. Sci **31**, 5 (2018)

8. Y.S. Prasojo, M. Niimi, M.M. Rahman, G. Ishigaki, R. Akashi, J. Anim. Plant. Sci **32**, 2 (2022)
9. K. Praminak, A.K.Bera, J.Agr.Plant.Prod **4**,5 (2013)
10. D.W. Lawlor, J.Exp.Bot **83**, 370 (2001)
11. H.K. Kim, E. Van Oosterom, M. Dinkhun, D. Luquet, G. Hammer, J. Ann. Bot **106** (2010)
12. Y. Liu, G. Dandan, D. Yanfeng, W. Qiangsheng, L. Ganghua, W. Shaohua Aus. J. Crop. Sci **5**, 8 (2011)
13. J.H. Neuteboom, E.A. Latinga, H. Schlepers, L. Mannetje, *Sward characteristics of a diploid and a tetraploid cultivar of perennial ryegrass as measured by different sampling techniques*, in Proceedings of the XVII International Grassland Congress, New Zealand (1993).
14. M. te Beest, J.J.L. Roux, D.M. Richardson, A.K. Brysting, J. Suda, M. Kubesova, P. Pysek, J. Ann. Bot **109** (2012)
15. R.M. Simeao, A.S. Silva, C.B. De Valle, Crop. Breed. App. Bio **16** (2016)
16. J. Vendramini, B. Seller, L.E. Sollenberger, M. Silveira, SS AGR **303** (2014)

 https://doi.org/10.1051/bioconf/20238006001

Design and development smart aquaculture in freshwater pond based on fuzzy logic.

Endra Pitowarno[1*], *Bagas* Oktavianto Leksono[1], *Eko* Budi Utomo[1], *Muhammad* Muamar[1]

[1]Electronic Engineering Polytechnic Institute of Surabaya, Mechanics and Energy Department Engineering, 60111, Indonesia

Abstract. This research has the topic of Smart Aquaculture in freshwater ponds. Aquaculture or fish farming in Indonesia still monitors and manages ponds manually. Temperature, pH, and water turbidity are important parameters for the survival of fish in ponds. Therefore, it is necessary to design a tool that can monitor and control ponds automatically. This system worked by reading the pH, temperature, and water turbidity sensors. The microcontroller utilized the sensor readings to control the peristaltic pump for liquid pH, water pump, and valve operations. Subsequently, the sensor readings were transmitted to the ESP32, which further forwarded the sensor data to the cloud database. Applications that were integrated with the cloud database display sensor reading data. The system utilized Fuzzy Logic Control with the Mamdani method to automate its operation. The inputs for the fuzzy logic control included pH, temperature, and turbidity, while the outputs consisted of the peristaltic pump, valve, and water pump. This system successfully adjusts the control conditions of temperature 35°C, pH 5, and turbidity 1200 NTU, bringing them back to the normal setpoint, which is a temperature of 32°C, pH between 6-8, and turbidity of 150 NTU, and maintains them at these values.

Keyword: aquaculture, fuzzy logic control, pond management, real time monitor

1 Introduction

Aquaculture is the activity of producing and breeding aquatic animals or organisms in a controlled condition [1]. One of the main factors for the success of aquaculture is maintaining water quality. Water quality is controlled by several factors, such as the acidity or alkalinity of the water (pH), turbidity that can cause changes in water colour, and water temperature [2]. Aquaculture production has become a significant contributor to Indonesia's food industry by supplying animal protein. The aquaculture sector also plays a crucial role in Indonesia's economy, particularly in terms of job providing, foreign exchange income, and strengthening national economic growth [3]. In 2018, the aquaculture sector managed to produce over 15 million tons of major fishery commodities. However, this figure gradually decreased in the following year, with a production volume of only 14.7 tons [4]. The decline in production is directly influenced by a significant decrease in the number of fish farmers, which reached 42.27% during 2019 and 2020. The decrease in the number of fish farmers can be attributed to challenges in maintaining conventional fish ponds, which require significant time, effort, and costs [4]. This research aims to assist fish farmers in monitoring their ponds using IoT (Internet of Things) technology. The integration of sensor readings and combination with internet-based applications will make it easier for fish farmers to monitor their aquaculture ponds from anywhere. Automatic management of pond quality, aided by fuzzy control, will facilitate fish farmers in efficiently managing their ponds.

2 Related Works

Automation in monitoring and managing ponds can be a solution to optimize resources and costs in fish farming processes. Research by Karim et al. [5] developed a prototype for monitoring pH, temperature, and water level in fish ponds. However, in that prototype, actuator control was still manually controlled, limiting automatic pond management. Implementing automation in actuator control can enhance effectiveness and efficiency. In this context, fish farmers can save time previously used for controlling and monitoring fish ponds. Therefore, an integrated system with a controller is needed to operate actuators automatically in a closed-loop system. Research by Ramadhan et al. [6] developed an Internet of Things (IoT)-based water quality management system. This research analyzed water quality using parameters such as pH value, total dissolved solids (TDS), and turbidity, employing fuzzy inference system. The system classified water quality into three categories: good (meeting quality standards), fair, and poor (contaminated). The measurement results

* Corresponding author: bagasokta02@me.student.pens.ac.id & epit@pens.ac.id

can be accessed online via the Internet of Things (IoT). Furthermore, the research proposed the use of fuzzy control to classify sensor data used in monitoring fish ponds and generate output to control the actuators connected to the system. In automated pond management, applications for monitoring ponds can be designed, as demonstrated in the research by Sukrismon et al. [7]. In this application, users can access a menu to monitor parameters such as temperature, pH, water level, and turbidity, viewing the values obtained from the sensors used. This application is supported by a cloud database, Firebase. Firebase is a real-time database that is freely available and offers various features that help developers focus on the application's interface. Firebase provides convenient features for managing and configuring how the application functions effectively. To access Firebase, a smartphone connected to the internet is required. Additionally, Firebase can be used to store data obtained from these sensors.

3 Methodology

3.1. Proposed System

To maintain good water quality in the pond, it is necessary to monitor and manage the water quality effectively. This system utilizes temperature, turbidity, and pH sensors to assist in the monitoring process of the aquaculture pond. To facilitate effective monitoring and management of the pond, a smart aquaculture system has been designed to work automatically. Figure 1 represents the design of the smart aquaculture system. This smart aquaculture system utilizes several sensors to monitor the desired parameters. The sensors used are pH, temperature, and turbidity sensors. These sensors are connected to the Arduino Mega2560 Pro Mini microcontroller. The Arduino is connected to a Wi-Fi module called ESP32. Additionally, several actuators are used, namely the Proportional Valve, Water Pump, Peristaltic DC Pump, and Heater.

The sensor readings received by the Arduino are used as inputs for fuzzy logic control. These inputs are used to determine the activation level of the fuzzy rules. The activation of these rules is used to generate fuzzy outputs. The fuzzy outputs are then used to control the actuators. To maintain the pond temperature, the Proportional Valve and Water Pump are used. The Proportional Valve is used to control the amount of water entering the pond, while the Water Pump is used to remove hot water from the pond. The water circulation will continue until the temperature sensor reading matches the desired setpoint. This same principle is also applied to maintain the turbidity level within the normal range. Dirty water will be drained, and clean water will enter the pond. When the temperature is cold, the heater will turn on until the sensor reading reaches the setpoint. To maintain the pH level of the pond within the normal range, a liquid is required to increase or decrease the pH. When the pond's pH is acidic, the pH-increasing liquid will be added to the pond. Conversely, if the pond's pH is basic, the pH-reducing liquid will be added to the pond. The Peristaltic DC Pump is used to transfer the liquid from the bottle to the pond.

The sensor data from the Arduino will be sent to the Wi-Fi module through serial communication. The Wi-Fi module, which is connected to the internet, will send the sensor data to a Cloud Database. The Cloud Database will be connected to an Android-based application. The application is used to monitor the pond's condition from anywhere.

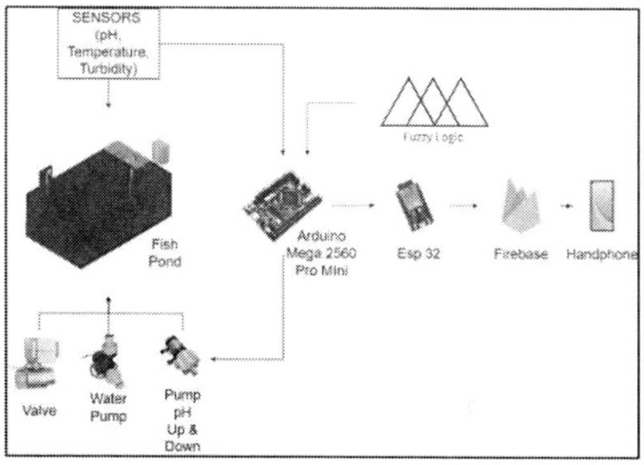

Fig. 1. Full System of Smart Aquaculture

3.2. Fuzzy Logic Control

Fuzzy logic control is a rule-based control system that uses linguistic variables and fuzzy sets to process ambiguous data. This approach enables machines to think more like humans. Fuzzy logic defines 0 and 1 as definite truths but also encompasses various degrees of truth between 0 and 1 [8].

In the context of water quality control in ponds, the use of fuzzy systems can help understand uncertain and varying changes in pH, temperature, and turbidity. The flexibility in processing linguistic data allows fuzzy logic to make control decisions that adapt to varying environmental conditions. Fuzzy systems have easy adaptability when operating in different environments. Furthermore, fuzzy systems can handle non-linear relationships between inputs and system responses, enabling them to control actuators with fast and accurate responses [8].

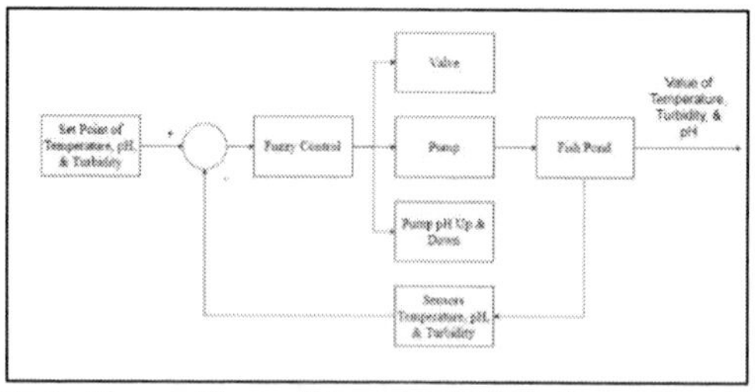

Fig. 2. Diagram Block of Fuzzy Logic Control

3.2.1. Fuzzification

Fuzzification is the process of transforming numerical variables into membership degrees in linguistic sets [9]. Sensor values represent numerical values that will be used as inputs in the fuzzification process. In this system, there are three inputs: temperature, turbidity, and pH, as well as four outputs: Valve, Water Pump, pH Up, and pH Down.

The membership function for temperature consists of three categories: normal, warm, and hot with a range of (0-60 °C). Table 1 illustrates the membership function for temperature. The pH membership function also consists of three categories, as shown in Table 2. As for turbidity, its membership function is divided into three categories described in Table 3. Turbidity concentration is measured in NTU (Nephelometric Turbidity Units).

Table 1. Membership function for temperature.

Category	Range (°C)			
Normal	25	25	28	34
Warm	32	37	37	42
Hot	40	45	60	60

Table 2. Membership function for pH.

Category	Range			
Acid	0	0	4	7
Normal	5.8	7	7	8.4
Alkali	7.5	12	12	14

Table 3. Membership function for turbidity.

Category	Range (NTU)			
Clear	0	0	0	400
Turbid	100	625	625	1250
Very Turbid	900	1500	1500	3000

Table 4. Membership function for valve.

Category	Range			
Slow Pump (sp)	0	0	7	50
Middle Pump (mp)	35	60	60	83
Fast Pump (fp)	65	85	100	100

Table 5. Membership function for water pump.

Category	Range			
Slow Valve (sv)	0	0	7	50
Middle Valve (mv)	25	79	79	128
Fast Valve (fv)	100	128	150	150

Table 6. Membership function for pH up pump.

Category	Range			
Slow Pump pH Up (spup)	0	0	33	50
Middle Pump pH Up (mpup)	33	78	78	121
Fast Pump pH Up(fpup)	75	115	130	130

Table 7. Membership function for pH down pump.

Category	Range			
Slow Pump pH Down (spdn)	0	0	33	50
Middle Pump pH Down (mpdn)	33	65	65	121
Fast Pump pH Down (fpdn)	83	115	130	130

The membership function for the Valve output is measured in the degree of valve opening. The degree of valve opening depends on the Duty Cycle applied to the valve. Table 4 shows the membership function for the Valve output. The membership function for the Water Pump output is measured based on how fast the pump rotates. This membership function is described in Table 5. Tables 6 and 7 are the membership function tables for the pH Up and pH Down pumps. The membership function for each input and output uses a trapezium membership function.

3.2.2. *Fuzzy Rules and Inference Engine*

Fuzzy rules are a set of rules or strategies used to connect input conditions with actions to be taken to control actuators [9]. Fuzzy rules consist of if-then statements, where if represents measurable conditions and then represents the decision or action to be taken based on the occurring conditions [10]. The inference engine is used to evaluate the rules based on the membership values defined in those rules. The evaluation of these rules aims to generate appropriate decisions or actions to maintain the pool conditions. In this process, the inference engine uses logical operations such as AND, OR, and NOT to evaluate the rules. This system uses the AND logic to combine the membership levels that have been defined [11]. Table 8 represents the rule set for Valve control, Pump control, pH Up and pH Down control.

3.2.3. *Defuzzification*

Defuzzification is the process of converting linguistic fuzzy sets into crisp values used to control the system or provide output to actuators. The defuzzification process is influenced by decision-making based on the fuzzy sets that have been defined. Defuzzification is performed to transform fuzzy output values into precise values. This system has four outputs, and each output undergoes fuzzification based on the environmental conditions [12].

3.3. Application and Communication Devices

Communication between microcontrollers is crucial in this system to support the transmission of application data. This system utilizes Arduino and Wi-Fi modules. Arduino is used to receive data from sensors and send it to the Wi-Fi module through serial communication. Serial communication transmits data by sending bits sequentially through a single communication path. The Wi-Fi module, connected to the internet, is responsible for sending data to the cloud database. The data received by Arduino will be sent by the Wi-Fi module to the cloud database. Data from the cloud database will be used to display sensor reading values in the application.

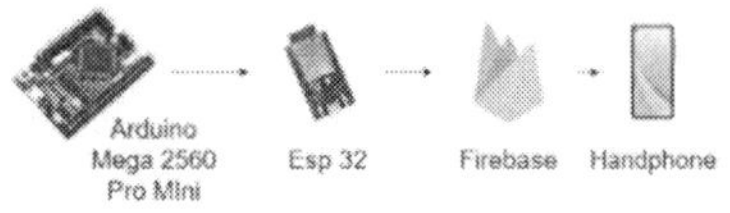

Fig. 3. Application and communication device.

This application is designed to facilitate breeders in monitoring pond conditions. The first page of the application provides a menu option for the environment to be monitored. The fishery page presents the sensor values used, such as pH, temperature, and turbidity. All of this aims to provide convenience in monitoring and controlling the pond's environmental conditions through an application connected to Arduino and the Wi-Fi module. These values will be continuously updated as the Arduino reads the sensors. When the power is off, the sensor values will no longer be sent to the cloud database. Therefore, a battery is required as a backup power source so that when the main power is off, the system can still function and send sensor values to the cloud database.

4 Experiment

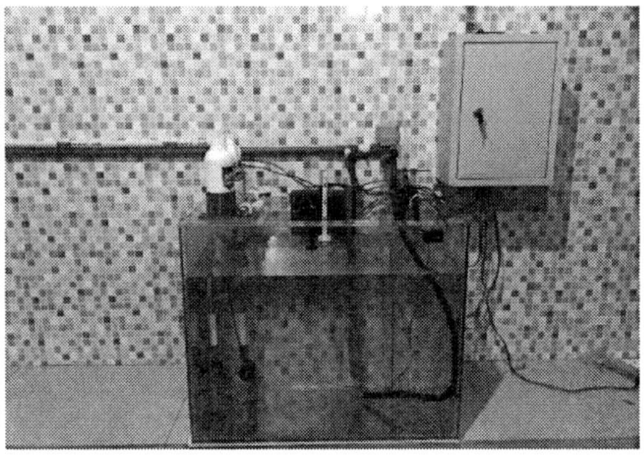

Fig. 4. Prototype of smart aquaculture system.

This section contains the results of control testing on the monitored parameters. The testing was conducted on an aquarium with dimensions of 70 cm x 55 cm x 32 cm. The turbidity, temperature, and pH sensors that are being used have undergone a calibration process. The purpose of this calibration is to ensure that the values generated by the sensors match the actual conditions inside the pond. To evaluate the accuracy of these sensors, comparisons are made using appropriate measuring instruments for each measured parameter. Figure 1 shows the prototype of the smart aquaculture system. The application has been successfully developed. Figure 5 (a) represents home menu that provides options to select the environment to monitor and (b) represents the monitoring page for fisheries, displaying values of temperature, pH, and turbidity. The first test is the temperature control test. This test aims to determine the system's control when the temperature is not in a normal condition. The second test is the pH control test. This test is conducted to observe the system's decision in handling acidic and alkaline pond pH. Finally, there is the turbidity testing. The turbidity test aims to determine the system's response when the environment is not in a normal state.

(a) (b)

Fig. 5. **(a)** Home menu of the application and **(b)** the display page of the fisheries monitor.

4.1 Temperature Control Testing

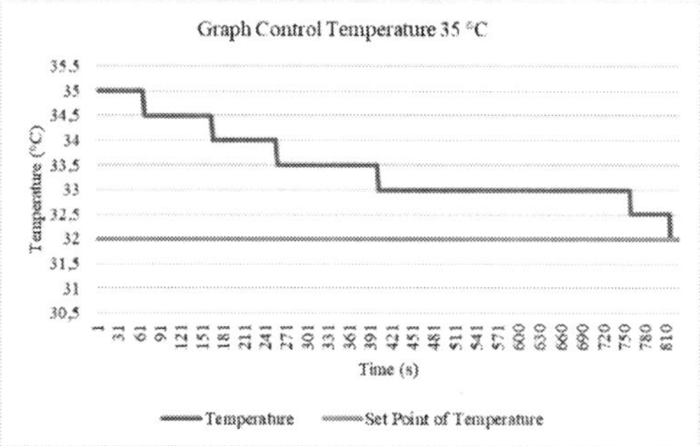

Fig. 6. Temperature control at 35 °C.

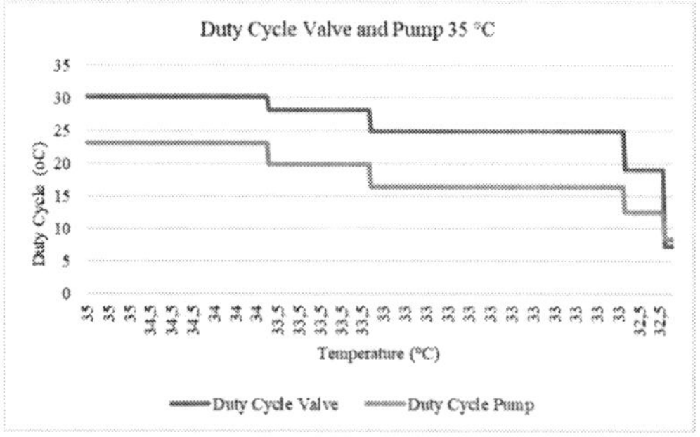

Fig. 7. Duty cycle valve and pump for temperature control at 35° C.

Table 8. The rule set for Valve control, Pump control, pH Up and pH Down control.

Rule	Temperature	Turbidity	pH	Valve	Pump	Pump pH Up	Pump pH Down
1	IF Normal	AND Clear	AND Acid	Then sv	AND sp	AND fpup	AND spdn
2	IF Normal	AND Clear	AND Normal	Then sv	AND sp	AND spup	AND spdn
3	IF Normal	AND Clear	AND Alkali	Then sv	AND sp	AND spup	AND fpdn
4	IF Normal	AND Turbid	AND Acid	Then mv	AND mp	AND spup	AND spdn
5	IF Normal	AND Turbid	AND Normal	Then mv	AND mp	AND spup	AND spdn
6	IF Normal	AND Turbid	AND Alkali	Then mv	AND mp	AND spup	AND spdn
7	IF Normal	AND Very Turbid	AND Acid	Then fv	AND fp	AND spup	AND spdn
8	IF Normal	AND Very Turbid	AND Normal	Then fv	AND fp	AND spup	AND spdn
9	IF Normal	AND Very Turbid	AND Alkali	Then fv	AND fp	AND spup	AND spdn
10	IF Warm	AND Clear	AND Acid	Then mv	AND mp	AND spup	AND spdn
11	IF Warm	AND Clear	AND Normal	Then mv	AND mp	AND spup	AND spdn
12	IF Warm	AND Clear	AND Alkali	Then mv	AND mp	AND spup	AND spdn
13	IF Warm	AND Turbid	AND Acid	Then mv	AND mp	AND spup	AND spdn
14	IF Warm	AND Turbid	AND Normal	Then mv	AND mp	AND spup	AND spdn
15	IF Warm	AND Turbid	AND Alkali	Then mv	AND mp	AND spup	AND spdn
16	IF Warm	AND Very Turbid	AND Acid	Then fv	AND fp	AND spup	AND spdn
17	IF Warm	AND Very Turbid	AND Normal	Then fv	AND fp	AND spup	AND spdn
18	IF Warm	AND Very Turbid	AND Alkali	Then fv	AND fp	AND spup	AND spdn
19	IF Hot	AND Clear	AND Acid	Then fv	AND fp	AND spup	AND spdn
20	IF Hot	AND Clear	AND Normal	Then fv	AND fp	AND spup	AND spdn
21	IF Hot	AND Clear	AND Alkali	Then fv	AND mp	AND spup	AND spdn
22	IF Hot	AND Turbid	AND Acid	Then fv	AND mp	AND spup	AND spdn
23	IF Hot	AND Turbid	AND Normal	Then fv	AND mp	AND spup	AND spdn
24	IF Hot	AND Turbid	AND Alkali	Then fv	AND fp	AND spup	AND spdn
25	IF Hot	AND Very Turbid	AND Acid	Then fv	AND fp	AND spup	AND spdn
26	IF Hot	AND Very Turbid	AND Normal	Then fv	AND fp	AND spup	AND spdn
27	IF Hot	AND Very Turbid	AND Alkali	Then fv	AND fp	AND spup	AND spdn

The water temperature in the aquarium will increase during the day. The increase in water temperature in the aquarium is allowed to reach 35 °C. The system is activated, and the valve and pump start operating. Figure 6 shows the system starting to run, and water circulation begins, gradually reducing the temperature in the aquarium. Figure 7 shows the response of the actuators when the temperature decreases. The response of the actuators can be observed through changes in the duty cycle of the valve and the pump. Changes in the duty cycle affect the pump's speed, while in the valve, it affects the ball's opening. These changes will continue until the temperature value reaches the setpoint. The time required to return the water conditions to normal temperature from 35 °C Celsius is 810 seconds.

4.2 pH Control Testing

pH testing is performed to determine the system's response when the pH is not in a normal condition. The pH value commonly changes when the fish have been fed. After feeding, the fish produce a considerable amount of waste, resulting in high ammonia levels. This high ammonia level causes the pH value to become acidic. Additionally, the fish respiration process also causes the pH value to decrease. Figure 8 shows the acidic pH condition of the aquarium. The acidic pH condition activates the peristaltic pump to flow pH Up fluid into the aquarium.

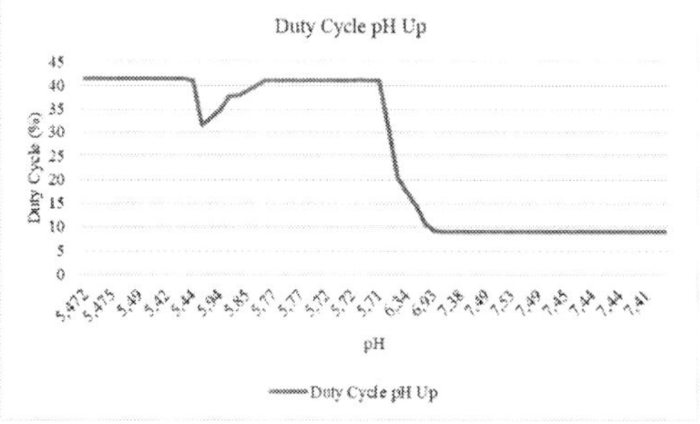

Fig. 8. pH control at pH 5.

After the fluid is provided to the pond, the pH of the pond will rise back to normal. Figure 9 shows the response of the peristaltic pump when the pond's pH is acidic. The acidic pH condition causes the peristaltic pump to rotate so that the fluid from the bottle can flow into the pond. Duty Cycle changes occur when the pH approaches the setpoint or normal value. The Duty Cycle value becomes smaller as the pH value reaches the normal range. The peristaltic pump stops when the pH value becomes normal. The time required to return the water conditions to normal pH from a pH value of 5 is 340 seconds.

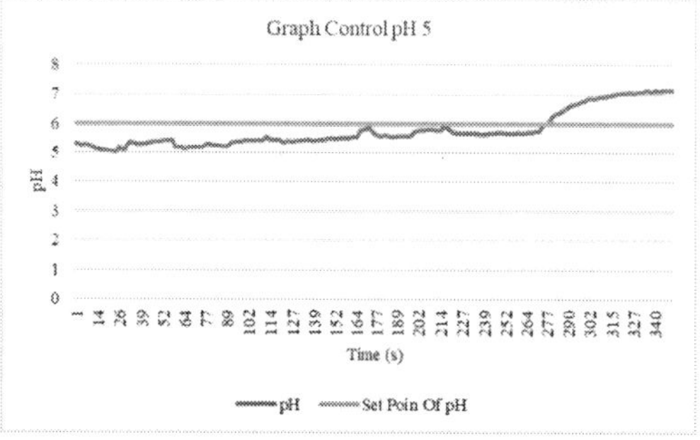

Fig. 9. Duty cycle valve and pump for pH control at pH 5.

4.3 Turbidity Control Testing

Turbidity occurs during fish feeding in the aquarium. Feeding causes the fish to excrete waste or debris into the aquarium. Leftover fish feed creates waste that can make the aquarium turbid. Figure 10 shows the turbid condition of the pond. The turbid condition activates the valve and pump. The principle of handling turbidity is the same as temperature control. Water circulation clears the pond as debris or waste is removed. The pump disposes of the waste, and clean water enters the aquarium through the valve.

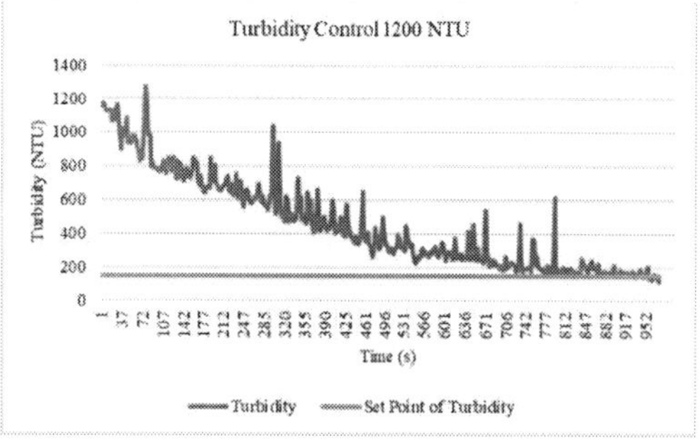

Fig. 10. Turbidity control at 1200 NTU.

Figure 11 shows the response of the valve and pump. When the aquarium is dirty, the valve responds by opening the ball up to 90 degrees, allowing clean water to enter the aquarium rapidly. To balance the valve's response, the pump responds by rotating faster, resulting in more water being discharged. Figure 11 shows the change in actuator response when the turbidity value decreases, following the turbidity condition in the pond. The valve and pump stop when the turbidity value reaches the setpoint. The time required to return the water conditions to normal from a turbidity value of 1200 NTU is 952 seconds.

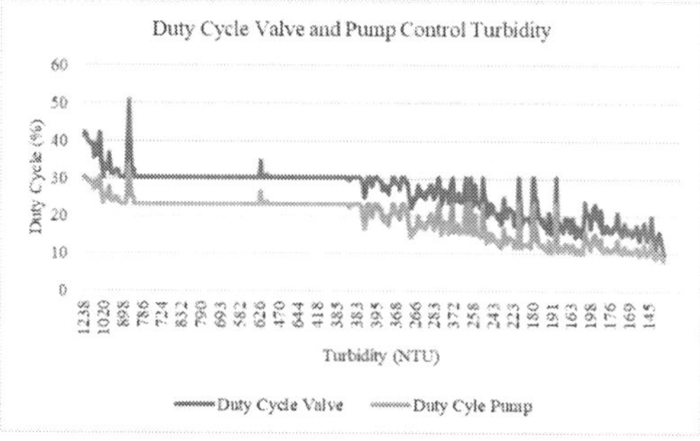

Fig. 11. Duty cycle valve and pump for turbidity control at 1200 NTU.

5 Conclusion

This research demonstrates a smart aquaculture system in managing ponds automatically and monitoring the ponds remotely. In this system, various sensors are used to collect environmental data such as temperature, pH, and turbidity. The data obtained from these sensors will be processed using fuzzy methods to determine the actions of the actuators used. The smart aquaculture system is connected to a Wi-Fi module and a cloud database, allowing real-time access and analysis of the monitoring data. This research aims to assist fish farmers in simplifying the maintenance of aquaculture ponds, thereby improving fish production.

Aknowledgements

I want to show my appreciation to my supervisor Mr. Endra Pitowarno who assisted me throughout this project. I also thank to the Nano Satellite Laboratory for providing a place to work on this project.

References

[1] Rastegari, H., Nadi, F., Lam, S. S., Ikhwanuddin, M., Kasan, N. A., Rahmat, R. F., & Mahari, W. A. W., *Internet of Things in aquaculture: A review of the challenges and potential solutions based on current and future trends*, Smart Agricultural Technology, **4**(December 2022), 100187 (2023).

[2] Rochyani, N., *Analysis of Water Environment Characteristics and Pools for Supporting Fish Cultivation*, Jurnal Ilmu-Ilmu Perikanan Dan Budidaya Perairan, **13**, 51–56 (2018).

[3] Henriksson, P. J. G., Tran, N., Mohan, C. V., Chan, C. Y., Rodriguez, U.P., Suri, S., Phillips, M. J., *Indonesian Aquaculture Futures – Evaluating Environmental and Socioeconomic Potentials and*

Limitations, Journal of Cleaner Production, **162**, 1482–1490 (2017).

[4] Statistik KKP: *Jumlah Pembudidaya Ikan*, https://statistik.kkp.go.id, (Accessed on June 19, 2022).

[5] Karim, S., Hussain, I., Hussain, A., Hassan, K., & Iqbal, S. *IoT Based Smart Fish Farming Aquaculture Monitoring System*, International Journal on Emerging Technologies, **12**(2): 45-53 (2021).

[6] Ramadhan, M. H., Dewantoro, G., & Setiaji, F. D., Rancang Bangun Sistem Pakar Pemantau Kualitas Air Berbasis IoT Menggunakan Fuzzy Classifier, Jurnal Teknik Elektro, **12**(2), 47–56 (2020).

[7] Sukrismon, Y., Hidayatullah, N., Mufti, N., Handayani, A. N., & Horng, G. J., *Smart Fish Pond for Economic Growing in Catfish Farming, ICCS, (ICOMITEE)*, ***1***, 49–53 (2019).

[8] Chiesa, G., Di Vita, D., Ghadirzadeh, A., Muñoz Herrera, A. H., & Leon Rodriguez, J. C, *A fuzzy-logic IoT lighting and shading control system for smart buildings.* Automation in Construction, **120**(August), (2020).

[9] Varshney, A., & Goyal, V., *Re-evaluation on fuzzy logic controlled system by optimizing the membership functions*, Materials Today: Proceedings, (2023).

[10] Krishnan, R. S., Julie, E. G., Robinson, Y. H., Raja, S., Kumar, R., Thong, P. H., & Son, L. H. *Fuzzy Logic based Smart Irrigation System using Internet of Things*. Journal of Cleaner Production, *252*, (2020).

[11] Bórquez-Lopez, R. A., Casillas-Hernandez, R., Lopez-Elias, J. A., Barraza-Guardado, R. H., & Martinez-Cordova, L. R. Improving feeding strategies for shrimp farming using fuzzy logic, based on water quality parameters. *Aquacultural Engineering*, *81*(August), 38–45, (2018).

[12] Ramos, J., Lino, P. G., Caetano, M., Pereira, F., Gaspar, M., & dos Santos, M. N. Perceived impact of offshore aquaculture area on small-scale fisheries: A fuzzy logic model approach. *Fisheries Research*, *170*, 217–227, (2015).

Development of Automatic Dual Sequence Control Temporary Immersion Bioreactor Systems for Micropropagation *Coelogyne pandurata* Lindl.

Anisa Dewi Rahayu[1], *Lathief Al Umami*[1], *Shiddharta Arya Anggoro Cen*[1], *Bimo Rizki Abdusshamad*[1], *Leonard Sean*[2], *Kresna Budiman*[2], *Labriyantoko Kurniawan*[3], *Muhammad Akhsin Muflikhun*[2], *I Wayan Mustika*[3], and *Endang Semiarti*[1]*

[1]Department of Tropical Biology, Faculty of Biology, 55281, Gadjah Mada University, Indonesia
[2]Department of Mechanical and Industrial Engineering, Faculty of Engineering, 55281, Gadjah Mada University, Indonesia
[3]Department of Electrical and Information Engineering, Faculty of Engineering, 55281, Gadjah Mada University, Indonesia

Abstract. *Coelogyne pandurata* Lindl. is an endemic orchid species of Indonesia from East Kalimantan and Papua. It is locally known as the "black orchid" due to the distinctive black coloration on its labellum with intricate green and hairy lines. Unfortunately, the population of this orchid has significantly declined in its natural habitat, rendering it a rare endangered species. Temporary immersion bioreactor systems have been proven to be efficient for endemic plant micropropagation with performance enhancements and innovations. This study aims to develop an optimized bioreactor design and innovate the automatic control performance of temporary immersion bioreactor systems based on previous research. The control system developed consists of three modules, namely Graphical User Interface (GUI) module, Sequence Control (SC) module, and Hardware Interface (HI) module. The GUI module receives information regarding time and duration of immersion and gas exchange, then the information is synchronized by the SC module, which plays the role of starting and stopping the processes, while HI module executes the order of the automatic control system in the immersion and gas exchange process. The developed bioreactor design and control system offer convenience, require less labor, and ensure precise control over the optimum conditions for black orchid micropropagation.

Keywords: Automatic Software Development, *Coelogyne pandurata* Lindl., Micropropagation, Temporary Immersion Systems

1 Introduction

Indonesia is one of the countries in the world with the largest tropical forests, which host a diverse range of flora, including many species of the Orchidaceae family. This family comprises about 800 genera and more than 25,000 species in total, with an estimated 5,000 of them found in Indonesia, some of which are classified as endemic to the region [8]. *Coelogyne pandurata* Lindl. is an endemic orchid species of Indonesia from East Kalimantan and Papua. It is locally known as black orchid because of the distinctive black coloration on its labellum with green and hairy lines [5]. The black orchid is an epiphytic sympodial orchid. Its pseudostems are either located closely or at a distance, each bearing two oblong and folded leaves. The flowers are arranged in clusters, with each cluster containing approximately 5-14 buds. The sepals and petals are lancet-shaped, pointed, and typically green. As an orchid, the seeds inside the fruit are microscopic in size due to the lack of endosperm [3]. Nowadays, the population of this orchid has greatly decreased in nature and is classified as a rare, endangered orchid [13]. The development of an efficient bioreactor system for micropropagation can contribute to the ex-situ conservation and propagation of plants, especially rare endemic orchid species in Indonesia.

Fig. 1. Morphology of the Black Orchid. A. A Plant with flower; B. Plant at the base of pseudobulbs; C. Close up a flower; D. Black labellum; E. Mature seeds; F. Fruit. Bars: 5 cm in A and B; 1 cm in C, D, and F; 0,5 mm in E [3].

In nature, black orchids thrive optimally at altitudes ranging from 1,000-15,000 meters above sea level, with humidity levels between 60-85%, CO_2 gas content at 500-1.200 ppm, light intensity at 40%-50%, and pH range of 6.4- 6.6 [7, 9]. Meanwhile, when cultivated *in vitro*, it is typically incubated at 21-23°C temperature and exposed to a TL lamp emitting around 1000 lux of light [10, 11].

Micropropagation using temporary immersion systems (TIS) is an alternative to reduce production costs, increase biological yield and quality, as well as reducing the culture time [4]. A TIS involves and combines the advantages of cultivating explants that are temporarily exposed to nutrient liquid culture and air in automated bioreactors systems for large-scale production plantlets under controlled conditions. The optimal system operation varies for each species due to many factors and variations in cultivation conditions. The bioreactor system makes it easy to monitor and control physical parameters including pH of media, temperature, and gas environment that supports plantlet growth. Furthermore, subculturing is unnecessary because the bioreactor system and the liquid medium allows for medium exchange. To date, many TIS variations have been developed, including the Temporary Immersion Bioreactor (TIB®), Temporary Immersion (RITA®), Ebb-and-Flow bioreactor, SETIS™, and Monobloc Advance TIS (MATIS®) [1].

Currently, technology developments are advancing rapidly. This can be seen from the widespread use of computers to facilitate the creation of automatic tools, harnessing technological advancements [12]. In the field of electronics control, we frequently encounter chips capable of storing and executing programmed manners, in which an electrical component known as microcontroller is employed to automate the operation of devices [14].

Recent technological advancements have unlocked innovations in designing tools with a system capable of automatically monitoring and controlling *in vitro* culture ecological conditions. The concept behind this tool is that it can perform momentary soaking as needed and adjust environmental variables, such as temperature, humidity, light intensity, pH, and CO_2 gas which can change rapidly based on external factors. This automated design, control, and monitoring can be used in the cultivation industry to ensure optimal plant growth. The creation of this tool involves an LCD monitor, which displays parameters within the culture room.

2 Materials and Discussion

2.1 Plant Materials and Culture Condition

Coelogyne pandurata Lindl. aged 8 months was used as plant material. Plantlets were obtained from the Biotechnology Laboratory at the Faculty of Biology, Gadjah Mada University. The liquid medium used was MS, supplemented with various combinations of BAP and 2,4-D hormones. Organic materials extracted from Ambon banana peels were also added as organic biocides. All cultures were maintained at a temperature range of 21-25°C, with 60-85% humidity, a medium pH range of 6.4-6.6, a CO_2 gas content range of 500-1.200 ppm, and a light intensity around 1000 lux using white LED lamps.

2.2 Design Modifications Bioreactors for Temporary Immersion Bioreactor Systems

The Temporary Immersion Bioreactor (TIB) system maximizes the gas exchange and nutrient absorption by alternating between plantlets immersion in the liquid medium and dry periods [2, 6]. Plantlets in the *in vitro* culture can absorb nutrients through both their roots and leaves. In the TIB, when plantlets are immersed in the liquid medium, their entire epidermis is exposed to it, thus maximizing the nutrient absorption [2, 6]. Therefore, the plantlets grown in the TIB system are expected to exhibit superior growth and reproduction rates compared to the plantlets grown in the conventional solid or liquid medium culture. Moreover, the TIB system enables the removal of any volatile compounds, such as ethylene, by promoting ventilation in the culture vessel through forced aeration.

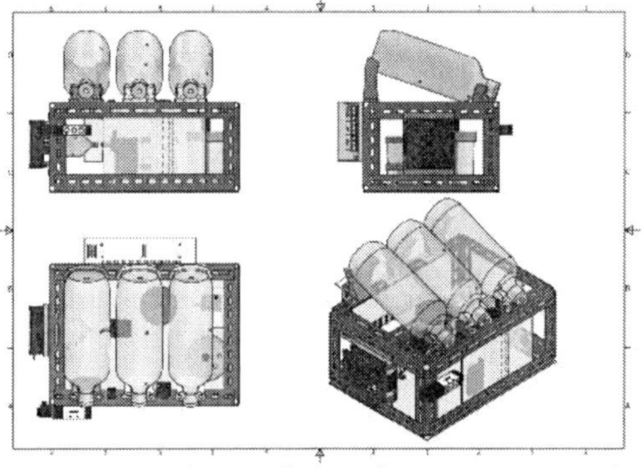

Fig. 2. 3D Design Hardware of Temporary Immersion Bioreactor Systems.

The modified temporary immersion bioreactor design in this paper consists of two chambers, the lower one for nutrient medium storage and the upper one for the culture room. The culture room shape is flexible to change depending on the needs. However, the culture room design demonstrated in this paper consists of three transparent glass bottles, each with 1 L volume and covered with a plastic port cap. The lower and upper containers are connected by silicone hose, and a 0.22 µm millipore filter is attached to each area where medium and gas exchange enters and exits to maintain the sterile condition (Fig. 2). The current optimal system is configured with a CO_2 gas injection volume of 0.1 vvm (aeration volume/medium volume/minute), and the timer is set to immerse the plantlet in the liquid medium for 1 minute every 20 minutes.

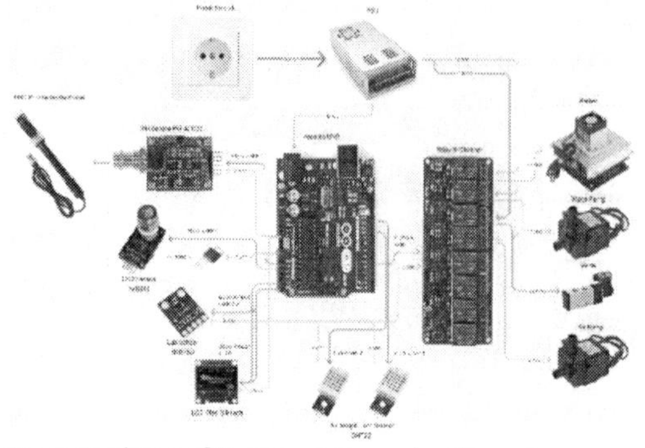

Fig. 3. Hardware of Temporary Immersion Bioreactor Systems.

The hardware components used in the temporary immersion bioreactor systems include a water pump for circulating the culture medium, an air pump for pushing the CO_2 gas and some fresh air into the system, a peltier device as the airflow cooler, a relay for connecting Arduino Uno to actuator systems, a 5 port pneumatic valve for adjusting airflow, a PSU for voltage supply, and sensor packages to monitor the live conditions inside the culture system. Several sensor systems installed in Arduino Uno include pH medium, light intensity, CO_2 gas, temperature, and humidity sensors (Fig. 3).

The installation of various sensors is crucial for monitoring and optimizing the TIB culture condition. Specifically, the temperature and humidity sensor (*DHT22*) is used to assess the environmental condition of the culture room. Additionally, a CO_2 sensor (*MG811*) was integrated to monitor CO_2 levels and gas exchange in the culture room. The effect of CO_2 supplementation was known to enhance the plantlets survivability during acclimatization phase. This improvement can be beneficial for micropropagation of commercial crops. A pH sensor (*PH4502C*) was also installed to assess the quality of the culture medium, as the provision of poor-quality culture medium adversely affects plantlet growth and development. Lastly, a light intensity sensor (*BH1750*) was added to measure the amount of light exposure in the culture room.

2.3 Automatic Software Development

The automatic controls in this system are regulated by signals generated from programmable digital timers. Synchronous schedules for both digital timers are manually assigned and must be created precisely to control the schedule. A well-scheduled assignment allows the temporary immersion bioreactor system to operate as expected. Failure to do so will result in unexpected output or even be able to damage the plantlet and equipment, for example if both valves are activated simultaneously, the compressed air could result in overpressure. Moreover, the excessive period of immersion could also cause hyperhydration and thus reduce productivity. Scheduling digital timers is a very complex task, which is especially important when multiple TIBs are activated simultaneously. A bad digital timer programming due to human error can occur and cause serious damage to plantlets and cultivation equipment.

Computer software for automatic dual sequence temporary immersion control on Arduino principle was developed. Operator can set the immersion period and forced ventilation of each TIB by uploading the settings on the Arduino uno with the times shown on the front panel of the GUI. The duration of soaking and ventilation (cleaning) can be determined by changing the numerical values on the GUI (image not shown). The SC module starts the immersion or forced ventilation process according to the specified value. The TIB status is monitored by the HI module, which sends signals to the targeted digital output devices. To control TIBs simultaneously, SC modules are arranged in series. Many global variables are specifically created for each TIB and are used to serve the parallel operation of GUI modules, SC modules, and HI modules.

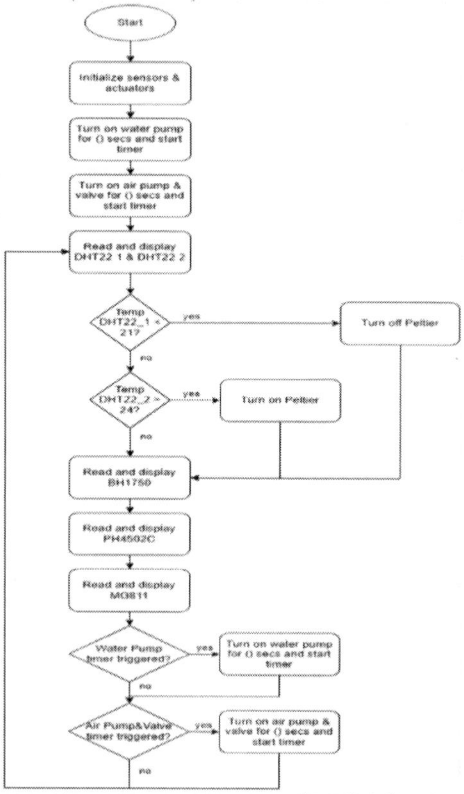

Fig. 4. Flow Chart of Temporary Immersion Bioreactor Systems.

The operational system flow starts from initializing sensors and actuators with multiple sequence control, then both pumps turn on for a few seconds and start the timer schedule. Media fluid transfer and gas exchange are controlled by Arduino Uno. The liquid medium culture is pumped into the culture bottle and remained there for twenty minutes, after which it flows back into the nutrient culture storage for reuse (Fig. 4).

Tool testing begins with the calibration process for each component used and testing is carried out on the entire series of microcontroller components that have been installed. Testing of DHT12 was successfully carried out, this sensor can control changes in temperature and humidity provided in TIB culture. The MG811 sensor succeeded in reading changes in CO2 levels in TIB culture. The PH4502C sensor succeeded in reading changes in the pH value of the solution injected in TIB culture. The BH1750 sensor succeeded in reading changes of light intensity when different watts of LED lights are used. Testing the 128x6 resolution LCD monitor confirms that it displays reading in accordance with the created system. The sensor reading for temperature, humidity, light intensity, pH value, and CO2 levels were successfully carried out as was the testing of the entire series of microcontroller components was successfully carried out according to the provided flow chart in (Fig. 4).

It was observed that using the developed computer software allows the operator to modify the immersion conditions with less effort and more precision when compared with using conventional digital timers. The programmed control system was able to control the sequence of multiple TIBs as given. However, it was observed that the liquid medium transfer rate between the storage tank and the culturing tank reduced when liquid medium hose was obstructed with small plantlets or when insufficient air pressure was supplied to TIB. Therefore, the filling and draining process times needed to be extended to complete the liquid medium transfer. Manual adjustment of the liquid medium transfer rate was required. Further study on the application of fluid flow and pressure measurements in TIBs together with appropriate computer algorithms is needed to improve capability of software to detect and solve faults in liquid medium transfer. The fault detection function of the software may prevent the damage of plantlets due to hyperhydricity caused by unintentional extended immersion duration. In this study, we observed black orchids grow well in the right duration of immersion because it was controlled by an arduino microcontroller. We also observed that the growth rate of plants on controlled TIB was faster when compared to plants on solid media (Fig. 5).

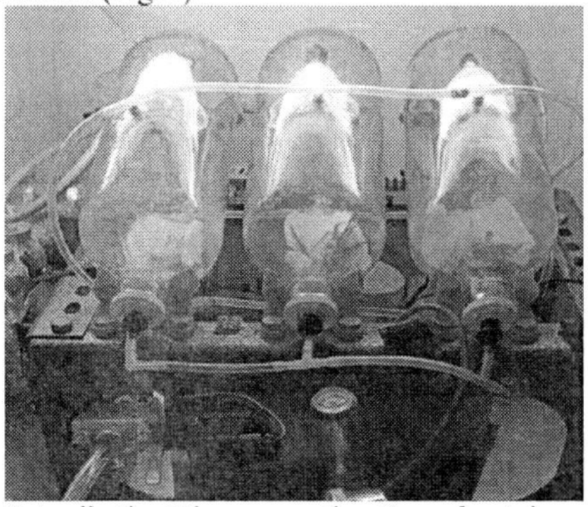

Fig. 5. Application Micropropagation *C. pandurata* in Temporary Immersion Bioreactor Systems.

2.4 Arduino Control System Algorithms

The output of these tools will be forwarded to the microcontroller as an entry point to be processed by the Arduino Uno Microcontroller system's brain. Arduino Uno Microcontroller receives data voltage and duration flow to compare it with the preceding value and makes a decision based on that input. Based on this input the decision is made by the Arduino Uno microcontroller. The main program module to support this project has been embedded in Arduino Uno microcontroller and is presented below.

```
//OLED - Display
#include <Adafruit_GFX.h>
#include <Adafruit_SSD1306.h>
#define SCREEN_WIDTH 128// OLED display width, in pixels
#define SCREEN_HEIGHT 64// OLED display height, in pixels
Adafruit_SSD1306 display(SCREEN_WIDTH, SCREEN_HEIGHT, &Wire, -1);

//DHT22 - Temperature & Humidity
#include "DHT.h"
#define DHTPIN_1 8    // DHT PIN 8
#define DHTPIN_2 9    // DHT PIN 9
#define DHTTYPE DHT22   // DHT 22 (AM2302), AM2321
DHT dht1(DHTPIN_1, DHTTYPE);
DHT dht2(DHTPIN_2, DHTTYPE);
float h1, t1, h2, t2 = 0;

//BH1750 - Lux
#include <BH1750.h>
BH1750 lightMeter

//PH-4502C - PH
#define PHPIN A0
#define PH_SAMPLE 5
#define ADC_RESOLUTION 1024.0

//MG811 - CO2
#define READ_SAMPLE_INTERVAL (50)
//define how many samples you are going to take in normal operation
#define READ_SAMPLE_TIMES (10)
//define the time interval between each samples in
#define DC_GAIN (8.5)
//define the DC gain of amplifier
#define ZERO_POINT_VOLTAGE (0.171)
//define the output of the sensor in volts when the concentration of CO2 is 400 PPM
#define REACTION_VOLTAGE (0.030)
//define the voltage drop of the sensor when move the sensor from air into 1000 ppm
#define MG_PIN (A1)
//define which analog input channel you are going to use
```

3 Conclusion

The Arduino program designed in this study is capable of controlling immersion bioreactor systems while working independently and integratedly. The program developed has been used in micropropagation of black orchids at the root and leaves elongation stages. TIB system controlled by the Arduino program was able to achieve a survival rate of 98%. Modification to immersion conditions for each temporary immersion system could be simply made on LCD screen or arduino software. The upper chamber design for *in vitro* culture can be flexibly adjusted as needed without the need to modify the control system. Additionally, the features of fluid transfer disturbance detection systems due to inhibition of small plantlets and insufficient air pressure should be developed to minimize risk of hyperhydricity explants.

Acknowledgements A.D.R., L.A.U., S.A.A.C and E.S. are funded by a 2023 Student Creativity Program grant organized by the Directorate of Learning and Student Affairs, Ministry of Education and Culture number 1686/E2/TU/2023.
Authors Contribution A.D.R., L.A.U., and S.A.A.C. wrote the manuscript, E.S. supervised all the process. All authors listed have made direct and intellectual contributions to the work, and approved it for publication.

References

1. A.H. Mirzabe, A. Hajiahmad, A. Fadavi, and S. Rafiee. Temporary immersion systems (TISs): A comprehensive review. Journal of Biotechnology. 357, 1 (2022).
2. E. Martínez-Estrada, B. Islas-Luna, J.A. Pérez-Sato, J.J. Bello-Bello. Temporary immersion improves *in vitro* multiplication and acclimatization of Anthurium andreanum Lind. Sci. Hortic. 249 (2019).
3. E. Semiarti, A. Indrianto, E. A. Suyono, R. L. Nurwulan, R. Restiani, Y. Machida, C. Machida. Genetic Transformation of the Indonesian Black Orchid (*Coelogyne pandurata* Lindley) through Agrobacterium tumefaciens for Micropropagation, in Proceedings of NIOC 2010, Nagoya Dome, Japan (2010).
4. H.N. Murthy, K-Y. Paek, and S-Y. Park. Micropropagation of orchids by using bioreactor technology. Greenhouses-Methods and Protocols. (2018).
5. J. D. Britto, Kamsinah, and L. Prayoga. Penambahan IAA dan BAP terhadap pertumbuhan kalus eksplan daun anggrek *Coelogyne pandurata* Lindl. Jurnal Ilmiah Biologi Unsoed. 3, 2 (2021).
6. N.Y. Kim, H.D. Hwang, J.H. Kim, B.M. Kwon, D. Kim, S.Y. Park. Efficient production of virus-free apple plantlets using the temporary immersion bioreactor system. Hortic. Environ. Biotechnol. 61 (2020).
7. R. Hasan, Pariyanto, and A. Sumantri. Keanekaragaman anggrek di sekitar kawasan Suban Kabupaten Rejang Lebon. Kependidikan. 2, 1 (2021).
8. R. Kartiman, D. Sukma, S.I. Aisyah, and Purwito, A. Multiplikasi in vitro anggrek hitam (*Coelogyne pandurata* Lindl.) pada perlakuan kombinasi NAA dan BAP. Jurnal Bioteknologi dan Biosains Indonesia. 5, 1 (2018).
9. R. Dwiyani, Y. Fitriani, and I.X. Mercuriani. The Alternative Media Supporting the Protocorm and Plantlet Growth of the Indonesian Black Orchid (*Coelogyne pandurata* Lindl.) Grown *In Vitro*. Caraka Tani: Journal of Sustainable Agriculture. 37, 1 (2022).
10. R. Restiani, E. Semiarti, and A. Indrianto. Konservasi anggrek hitam (*Coelogyne pandurata* Lindl.) melalui mikropropagasi pada berbagai media kultur. Symposium on Biology Education. (2016).
11. S-C. J. Luis and B-B. J. Jabín. CO2-enriched air in a temporary immersion system induces photomixotrophism during in vitro multiplication in vanilla. Plant Cell, Tissue and Organ Culture (PCTOC). (2023).
12. S. Samsugi, Z. Mardiyansyah, and A. Nurkholis. Sistem Pengontrol Irigasi Otomatis Menggunakan Mikrokontroler Arduino UNO. Jurnal Teknologi dan Sistem Tertanam. 1, 1 (2020).
13. T. S. Wahyudiningsih, Y. Jagau, and N. Ravenska. Konservasi *Coelogyne pandurata* Lindh. di Kalimantan Tengah: karakter morfologi, propagasi *in vitro*, dan pelestarian berbasis komunitas lokal. Jurnal Pengelolaan Lingkungan Berkelanjutan. 2, 2 (2018).
14. Y. Rahmanto, A. Rifaini, S. Samsugi, and S. D. Riskiono. Sistem Monitoring pH Air Pada Aquaponik Menggunakan Mikrokontroler Arduino Uno. Jurnal Teknologi dan Sistem Tertanam. 1, 1, (2020).

Identification of Maturity Stage of Cacao using Visible Near Infrared (Vis-NIR) and Shortwave Near Infrared (SW-NIR) Reflectance Spectroscopy

Riana Listanti[1], *Rudiati* Evi Masithoh[1*], *Arifin* Dwi Saputro[1], *Hanim* Zuhrotul Amanah[1]

[1] Department of Agricultural and Biosystems Engineering, Faculty of Agricultural Technology, Universitas Gadjah Mada, Jl. Flora No. 1, Bulaksumur, Yogyakarta 55281, Indonesia

Abstract. Choosing the cacao maturity stage is essential for producing high-quality cacao beans. Identifying indicators of the maturity level of cacao is a complex task because these fruits do not exhibit the characteristics of other fruits during the ripening period. Generally, cacao maturity is determined manually based on the estimated daily harvest date using sensory observation, which is marked by changes in the color of the cacao skin. This is certainly inaccurate because visual assessment is only performed subjectively. This is inaccurate because visual assessment is only performed subjectively, which is not in line with the demands of Industrial Revolution 4.0, which is a fast and accurate technology for sorting cacao. In this study, cacao maturity was identified using visible (350-1000 nm) and shortwave near-infrared spectra (SW-NIR) spectroscopy (1000-1600 nm). Chemometric analysis using principal component analysis-linear discriminant analysis (PCA-LDA) was used to classify cacao maturity. The results showed that SW-NIR spectroscopy yielded better performances with calibration and prediction accuracy of 92,50% and 85% using Savitzky–s 1st derivative (SGD1) spectra compared to Vis-NIR spectroscopy had calibration and prediction accuracies of 90% and 86% using raw spectra for PCA-LDA model.

Keywords: cacao, maturity, PCA-LDA, spectroscopy, SW-NIR, Vis-NIR

1 Introduction

Choosing the cacao maturity stage is essential for producing the highest quality cacao beans because, as the fruit ripens, compounds such as sugars, organic acids, methylxanthines, polyphenols, and proteins, all of which are crucial for the production of aroma precursors, are formed or altered [1]. Therefore, the correct level of maturity of cocoa pods and their homogeneity will contribute to good processing and, ultimately, to the quality of cocoa beans.

Ripe cacao is characterized by a change in the color of the cacao skin. In a comprehensive review article on chocolate manufacture [3], the authors stated that ripe cacao can be identified by its outer color, which can be yellow, red, or purple, depending on the variety. Furthermore, cacao should be harvested as it reaches optimum ripeness to prevent disease attacks and the possibility of germination of cacao beans [3].

Identifying indicators of the maturity level of cacao is a complex task because these fruits do not exhibit the characteristics of other fruits during the ripening period. Generally, cocoa fruit maturity is determined manually based on the estimated daily harvest date using sensory observation, which is marked by changes in the color of the cacao skin. This is certainly inaccurate because visual assessment is only performed subjectively. The demands of Industrial Revolution 4.0 require fast and accurate technology for sorting cacao. Therefore, a quick, non-destructive, and accurate evaluation is required to classify the maturity level of cocoa pods during the harvesting period. A comprehensive review [4] concluded that all reviewed applications of NIR spectroscopy in combination with chemometric analysis can play a valuable role in the food industry.

Applications of NIR spectroscopy for on-line and in-line food and beverage quality monitoring have been extensively reviewed by [5] and according to the authors, NIR spectroscopy is one of the most promising and widely accepted as one of on-line or in-line process control technology for non-destructive, reliable and accurate monitoring of chemical and physical parameters in food processing[6]. NIR spectroscopy provides analytical profiles of food products, which can be used to characterize and assess their authenticity [7]. In other studies, NIR spectroscopy has been widely used to detect the presence of fats from sources [8], identify bioactive compounds [9], [10], identify specific content for chocolate quality control or authentication [11], investigate chocolate quality parameters and predict cacao content in chocolate [12], and predict cacao solid content [13].

Accordingly, NIR spectroscopy offers a credible alternative for assessing the integrity and quality of cacao and their derivatives [14]. García-Muñoz et al. [15] recently developed a color guide to evaluate the maturity of cocoa clones using digital image processing methods. However, no studies have investigated the

* Corresponding author: evi@ugm.ac.id

application of NIR spectroscopy to nondestructively screen and classify the maturity levels of cacao. Therefore, the main objective of this research is to classify the maturity level of cacao using non-destructive methods, especially Vis-NIR and SW-NIR spectroscopy, combined with principal component analysis linear discriminant analysis (PCA-LDA). Principal component analysis (PCA), an unsupervised technique frequently used for dimension reduction and data visualization [16] and PCA combined with linear DA (PCA-LDA), a supervised technique used to classify data with a predetermined class [17] were the discrimination methods used in this study. The main objective of classifying the maturity level of cacao using PCA-LDA is to perform rapid automated sorting using NIR spectroscopy. With NIR analysis, information about the chemical attributes of cacao beans can be obtained quickly without damaging the beans, thus enabling a more efficient and accurate sorting process. When cocoa passes through an automatic sorting machine, it categorizes cocoa beans based on their maturity level. Immature cocoa beans may not be of good quality and can be directly processed into the final product. By separating immature cocoa beans, they can be fermented to improve the quality and flavor characteristics of the resulting chocolate. The long-term goal is that by using NIR spectroscopy technology and automated sorting machines, chocolate manufacturers and the cocoa industry can improve production efficiency, reduce wastage, and produce better quality and more consistent products.

2 Material and methods

2.1 Samples

The main material used for this research was 20 ripe lindak pods harvested from a garden in Gunung Kidul, Yogyakarta, Indonesia. The cacao used in this study had an average fruit length of 25 cm, average diameter of 9 cm, and average number of seeds per cacao of 40 cacao beans, with an average shell thickness of 1.40 cm. It is also known that the cacao beans produced from this cacao are generally included in the criteria for cacao beans with quality A to AA if measured by the number of beans per 100 g.

Figure 1. Cacao of different maturity levels used in this research (a) immature, (b) mature. The ripe cacao are greenish yellow or reddish orange and purple in color, with an average harvest age of around 5-6 months from the time the plants flower

2.2 Spectra Measurements

A modular spectrometer was used to measure the Vis-NIR and SW-NIR spectra of all cacao samples in reflectance mode. A visible/near-infrared spectrometer with a wavelength range of 400–1000 nm (Ocean Optics, Flame-T (Toshiba TCD1304AP)-Vis-NIR, USA) and a shortwave near-infrared spectrometer with a wavelength range of 1000–1700 nm (Ocean Optics, Flame-NIR+, USA) were used. Reflectance measurements were performed for both spectra. Fig. 2 shows the setup for combining SW-NIR and Vis-NIR spectroscopy acquisitions.

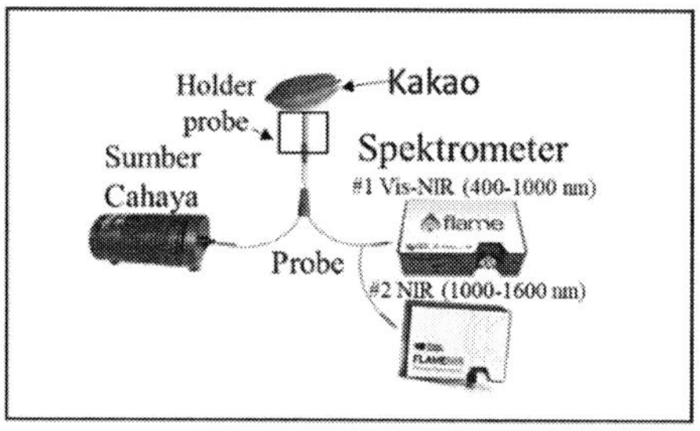

Figure 2. Spectra Measurement Schematic

A modular spectrometer system, reflection probes, tungsten halogen light, and computers (PCs/Laptops) with Ocean View 2.0.12 software were used in the experiment. The NIR instrument uses a semiconductor detector as a sensor and a halogen tungsten lamp as the radiation source, generating high-intensity radiation in the visible and IR spectra. Each cacao sample was placed on the optical probe. Ocean View 2.0.12 software was used to capture the reflectance spectra for the Vis-NIR and SW-NIR wavelengths, with integration times of 470 ms and 800 ms, respectively, scan to averages of 20 and 12, and a boxcar width of 1. Before collecting sample spectra, a calibration procedure was carried out, using a white reference using a ceramic diffuse reflectance standard (WS-1, Ocean Optics, USA) and a dark reference by switching off the light source.A total of 400 spectra were collected from 20 cacao samples, yielding 20 spectra from each sample. The same reflection probe could be used by both the Vis-NIR and SW-NIR spectrometers to read simultaneously, and the data collected could be automatically saved in different files.

 https://doi.org/10.1051/bioconf/20238006003

(a) (b)

Figure 3. (a) Flame-T-Vis-NIR Ocean Optics, (b) Flame-NIR+ Ocean Optics

2.3. Spectra Analysis

Ocean View 2.0.12, was used to obtain the Vis-NIR reflectance spectra (400-1000 nm) and SW-NIR reflectance spectra (1000-1600 nm), which were then separately compiled in Microsoft Excel®. Chemometric studies on 400 spectra were performed using the Unscrambler® X Camo software (version 10.4). To create and validate a multivariate model, the dataset was separated into calibration sets (75% samples) and prediction sets (25% samples). Six spectrum pre-processing techniques, including SNV, MSC, area normalization, Savitzky Smoothing, Savitsky-Golay 1st Derivative, and Savitsky-Golay 2nd Derivative, were utilized to reduce the sample-size effect. Pre-processing spectra are also employed to reduce problems that can impair the accuracy of chemical composition measurements, such as baseline variation, light scattering, and path length inconsistencies [18]. The pretreatment of spectra is an important stage in the interpretation of spectral data. When collecting spectral data, the data often noise, scattering spectra, and various other disturbances are also detected. Therefore, spectral pretreatment is needed to minimize the effect of disturbances, and the intention of interpreting spectral data can be enhanced [19].

Principal component analysis (PCA) is an unsupervised method used for dimensionality reduction by converting a number of correlated variables into fewer variables called principal components (PCs) [20]. In this study, PCA was used to cluster samples and detect the occurrence of outliers. A classification model to discriminate between cacao maturities was developed using principal component analysis linear discriminant analysis (PCA-LDA).

Principal component analysis (PCA), a supervised analysis method, and a method that combines PCA and linear discriminant analysis (PCA-LDA), both of which are chemometric tools, were used to evaluate the pre-processed spectra [21]. The PCA-LDA method is a qualitative analysis consisting of 2 steps [22]. First, the dimension of the spectral variable is reduced to become a new variable called a PC. Second, new variables are classified into two maturity levels: mature (M) and immature (T). PCA-LDA was performed using the spectrum as the independent variable (X) and maturity level as the dependent variable (Y).

The qualitative performance of the model was assessed using the resulting accuracy (Acc) value. Accuracy was calculated using Equation 1. The search equation for the model accuracy value was modified from [23], [24]. The accuracy value shows a comparison of the number of successful classifications of each class with the number of samples. A higher accuracy value (maximum 100%) indicated that the model had better predictions.

$$\text{Accuracy}=\frac{n^PT+n^PM}{nT+nM}x100\% \quad (1)$$

Descriptions:

n^PT = number of correct predictions of the immature sample
n^PM = number of correct predictions of the mature sample
nT = total sample of immature individuals
nM = total number of maturecells

3 Results and discussion

3.1.Visible near Infrared Spectra (Vis-NIR)

The Vis-NIR spectrometer recorded the spectral reflectance in the wavelength range of 345-1033 nm. However, the spectra used for the analysis were only in the range of 400–1000 nm because of the presence of severe noise at wavelengths below 400 nm and above 1000 nm. The number of variables in the wavelength range of 400-1000 nm was 3392 variables. This is due to interference at wavelengths below 400 nm and above 1000 nm measured from Vis-NIR Spectroscopy. This wavelength truncation avoids noise in the spectra, which could potentially interfere with the important information contained in the spectra. The reduction in resolution contributed minimally to the accuracy of the model. A reduction in the number of wavelength points can increase detection speed [25]. The truncation of the noise data at the beginning and end of the wavelength also aims to optimize the information. In addition to truncating the wavelength range, disturbances in the spectra due to light in addition to truncating the wavelength range, disturbances in the spectra due to light scattering can also be removed using preprocessing techniques that are applied before forming the calibration model so that it can be easier to analyzing the data [26].

The Vis-NIR spectral data obtained from the original spectra of the pure samples before spectral pre-processing are shown in Figure 4 (a). Spectral samples that react to light radiation determine these spectral characteristics. When light strikes cacao samples, it may be transmitted, absorbed, or reflected [27]. Using the original Vis–NIR spectra (Figure 4 (a)), the results showed that low reflectance or high absorbance at 450-500 nm wavelength is influenced by carotenoid pigments. According to [24] carotenoids are pigments that cause a bright yellow color in the fruit. Fruits with a dominant yellow color produce spectra with low reflectance values at wavelengths of 450-500 nm. Low reflectance or high absorbance at a wavelength of 500-550 nm contains some information related to the pigment content that gives rise to the red color of the fruit. The peak at 840 nm is correlated with oxalic acid. The acid content of cocoa beans is strongly related to the formation of chocolate flavor [28].

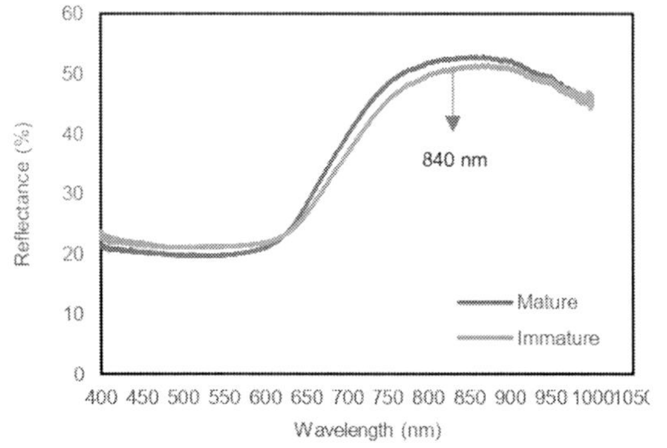

(a)

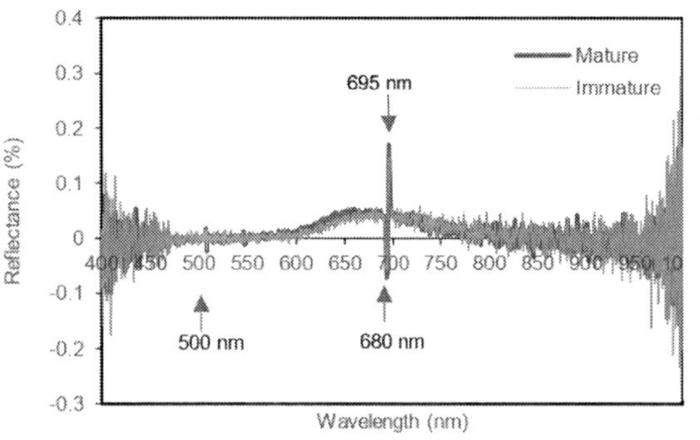

(b)

Fig. 4 **(a)** Original and **(b)** Savitzky–Golay 1st Derivative spectra mature and non mature of cacao observed with Vis-NIR Spectroscopy

According to [29], the absorbance at approximately 550 nm is influenced by anthocyanins. The absorption peak at 620 nm corresponds to respiratory pigments, which may be correlated with anthocyanin or red pigments [30]. Wavelength of 950-1000 nm indicated information related to the sugar groups and water molecules. At these wavelengths, the absorption bands of functional groups containing hydrogen atoms, such as O-H, C-H, and water molecules. such as O-H, C-H, and N-H bonds [31]. Fig. 4b. shows that some peaks and valleys are more distinct following pre-processing using the Savitsky Golay 1st derivative. The peak at 680 nm correlated with chlorophylls. Chlorophyll pigments contain C-H-N-O bonds that can be detected at Vis-NIR wavelengths [32]. The spectrum observed at a wavelength of 695 nm showed a high absorbance, which can be attributed to the presence of polyphenols in the sample. An increase in absorbance indicated an increased percentage of the total antioxidant activity of the polyphenolic fraction. [33].

3.2.Shortwave near-infrared spectra (SW-NIR)

The wavelength used in data analysis was 1000-1600 nm, which can be performed by SW-NIR Spectroscopy at 954-1700 nm. This is due to interference at wavelengths below 1000 nm and above 1600 nm measured by SW-NIR Spectroscopy. The SW-NIR spectral data obtained from the original spectra of the pure samples before spectral pre-processing are shown in Figure 5a. The original absorbance SW-NIR spectra of the cacao samples at various maturity stages are shown in Figure 5a. between 1000 nm and 1600 nm, respectively. Numerous factors, including the principal chemical concentrations (water, carbohydrates, etc.) and particle size, affect the shape of the spectra.

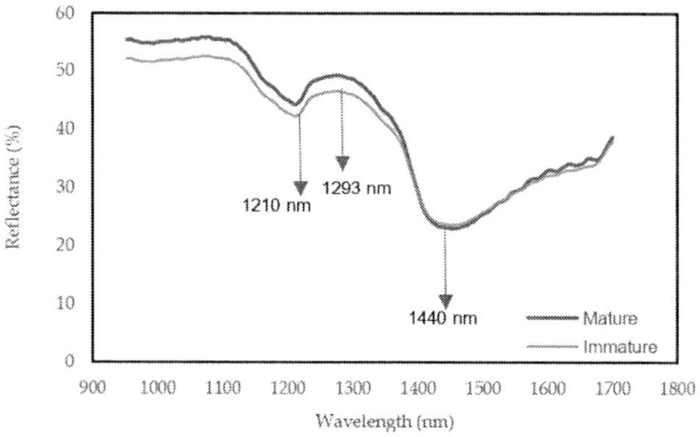

(a)

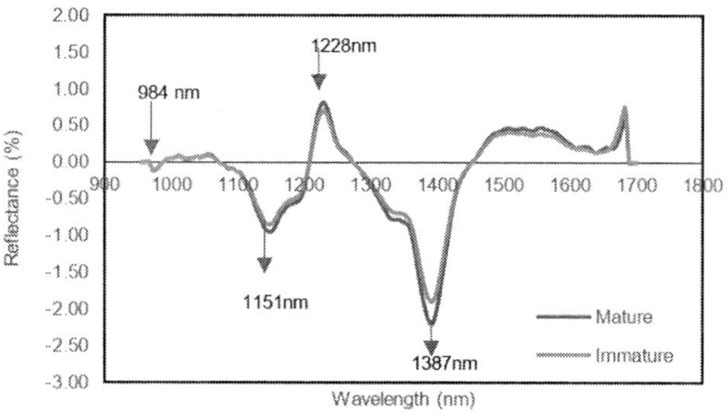

(b)

Fig. 5 **(a)** Original and **(b)** Savitzky–Golay 1st Derivative spectra mature and non mature of cacao observed with SW-NIR Spectroscopy

The lipid content and significant water band absorption at 1450 nm caused the spectra to peak at roughly 1200 nm. Peak absorption at wavelengths of 1210, 1161, 1188, 1212, and 1323 nm reveals the presence of lipids [34]. The bands in the 1150–1200 nm range are attributed to the second overtone of the C–H stretching modes, those in the 1400-1450 nm region are attributed to the first overtone of the O–H stretching modes, and the combination of the C–H stretching and C–H deformation modes is responsible for the bands in the region between 1450 and 1550 nm [34]. Another absorbance peak, a broad reflectance valley at around 1180 nm, which extends up to 1200 nm, is affected by proteins as well as the second overtone of C-H stretching in lipids [34]. Fig. 5b shows that some peaks and valleys

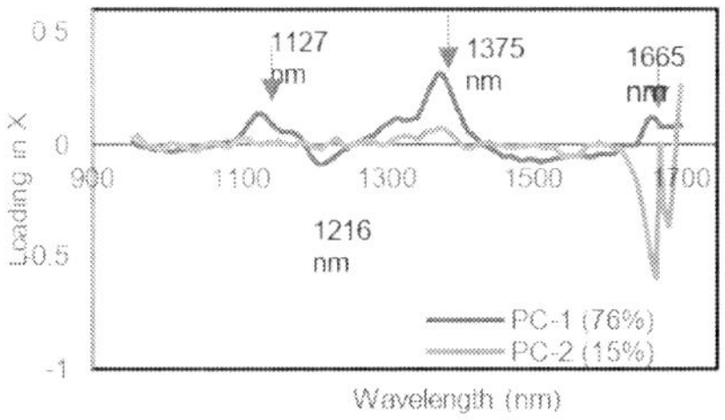

are more distinct following pre-processing using the Savitsky Golay 1st derivative. The absorbance peaks at 1380 nm and SW-NIR band in the 1355-1400 nm region represent the second overtone of the CH (carbohydrate, lipid) combination [35]. According to a previous study, the region in the window between 1200 and 1400 nm is related to several important chemical compounds such as lipids and proteins [31]. The NIR absorption coefficient increases with an increase in glucose concentration at wavelengths of 1630–1700 nm [36].

3.3. PCA Modeling

The spectroscopic data of mature and immature cacao were analyzed using a non-supervised multivariate Principal Component Analysis (PCA) approach. The main goal of this investigation was to evaluate the ability of samples to discriminate based on their quality [5]. PCA makes it possible to identify the key properties of spectra by reducing the dimensionality of the data in an unsupervised manner. Principal component analysis (PCA), a multivariate analysis, is used to produce new variables called principal components (PCs) and to minimize the spectral dimension without sacrificing significant data [33,34]. A typical approach involves applying PCA to reduce the dimensionality of the dataset while retaining most of the variance. This step helps remove redundant or irrelevant features and capture the major sources of variation in the data. In this investigation, spectra from 3648 variables for the Vis-NIR were used to construct 14 PCs with the Mahalanobis method and 128 variables for the SW-NIR were used to construct nine PCs using the Mahalanobis method. Principal component analysis (PCA) is an unsupervised technique.

3.3.1. PCA Modeling using Vis-NIR spectra

Fig. 6a displays a score plot of PC-1 and PC-2 constructed to detect principal groupings among observations using Savitzky-Golay 1st Derivative pre-processing spectra. The PCA model contained 96% of data variance, which explained variance of 90% from PC-1 and 6% in PC-2. According to Fig. 6b, the first principal component (PC1) explained 90% of the total explained variance (X-variance on the y-axis) in the data. Fourteen PCs accounted for more than 90% of the total explained variance in the calibration and validation processes (89,58% and 89,51%, respectively). Consequently, we decided to use 14 PCs as input variables for the PCA-LDA model. The portions of the spectra where the changes among the cacao samples are clearly evident are shown in Fig. 6c, which presents the PC1 loading values of the spectral data. When interpreting the interactions between variables and clusters in the score plot, the loading plot (Fig. 6c) was used to show the significance of each variable in the model.

(a)

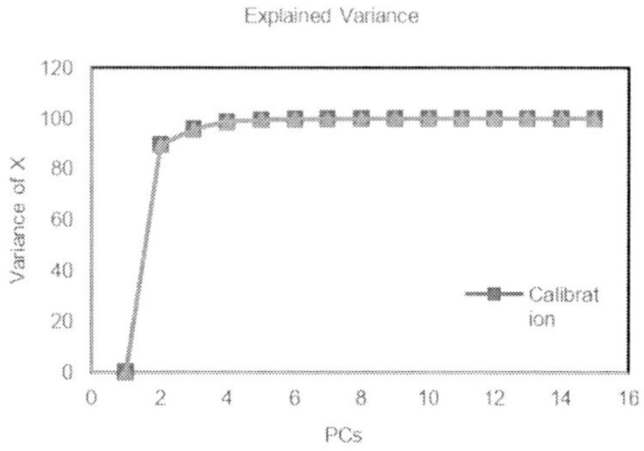

(b)

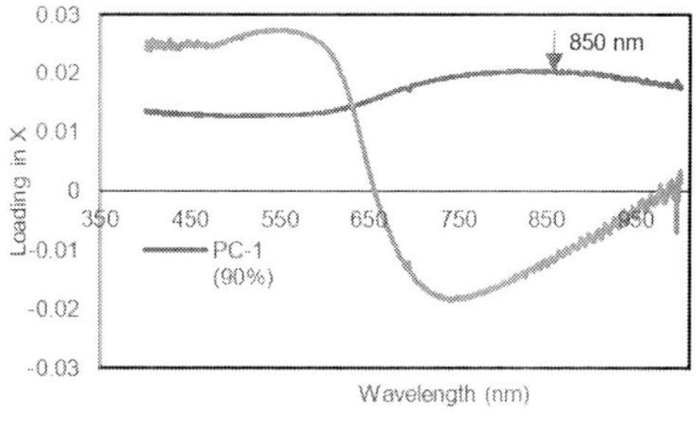

(c)

Fig. 6 Vis-NIR Analysis: **(a)** PC1/PC2 Score plot by PCA analysis on the calibration set : red (immature) ; blue (mature) ; **(b)** Plot of explained variance (%) of PCs used in the PCA model; **(c)** PC-1 and PC-2 loadings plot

3.3.2. PCA Modeling using SW-NIR spectra

Fig. 7a displays a score plot of PC-1 and PC-2 constructed to detect principal groupings among observations using Savitzky-Golay 1st Derivative pre-processing spectra. The PCA model contained 91% of data variance, which explained variance of 76% from PC-1 and 15% from PC-2. According to Fig. 7b, the first principal component (PC1) explained 76% of the total explained variance (X-variance on the y-axis) in the data. Nine PCs accounted for more than 76% of the total explained variance in the calibration and validation processes (76,40% and 76,06%, respectively). Consequently, we decided to use nine PCs as input variables for the PCA-LDA model. The portions of the spectra where the changes among the cacao samples are clearly evident are shown in Fig. 7c, which presents the PC1 loading values of the spectral data. When interpreting the interactions between variables and clusters seen in the score plot, a loading plot (Fig. 7c) was used to show the significance of each variable in the model.

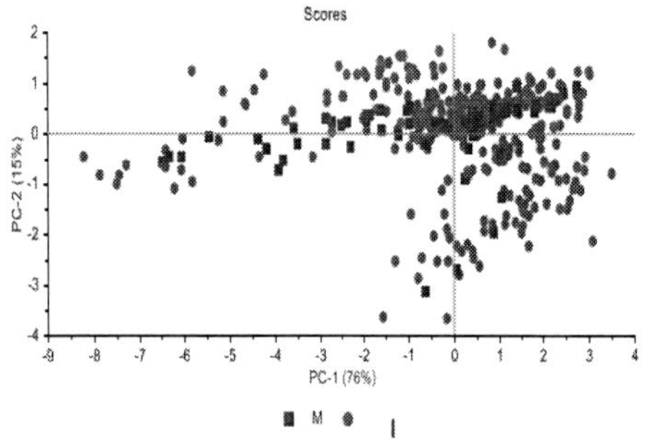

(a)

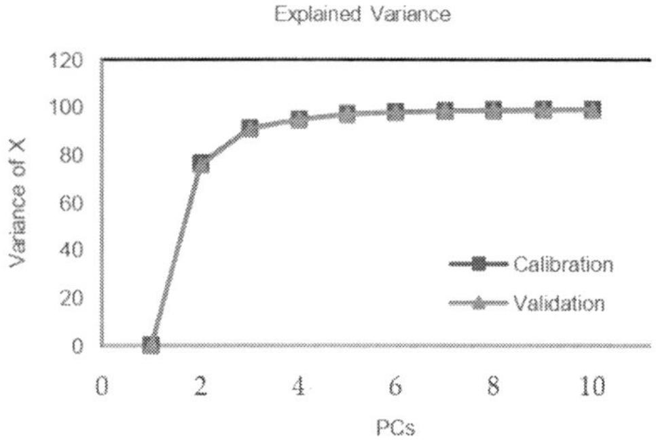

(b)

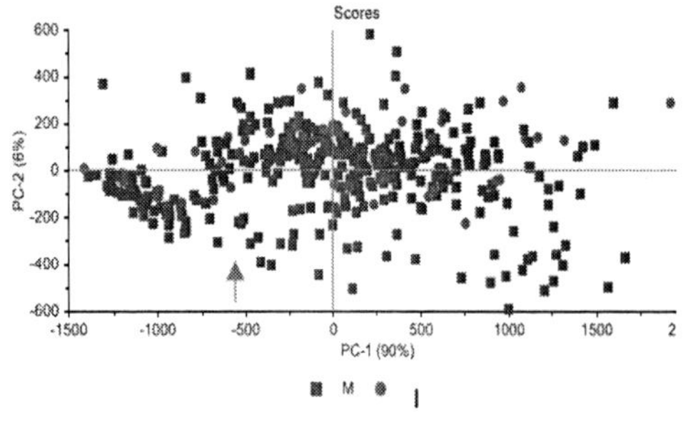

(c)

Fig. 7SW-NIR Analysis: **(a)** PC1/PC2 Score plot by PCA analysis on the calibration set : red (immature) ; blue (mature) ; **(b)** Plot of explained variance (%) of PCs used in the PCA model; **(c)** PC-1 and PC-2 loadings plot

3.4. PCA-LDA Modeling

After PCA, LDA was applied to the transformed data. LDA operates by finding a linear projection that maximizes between-class scatter while minimizing within-class scatter. It aims to project the data into a lower-dimensional subspace, where the classes are well separated and the data points within each class are tightly clustered. A supervised technique called linear discriminant analysis (LDA) was utilized to categorize category variables, such as the cacao maturity level.

3.4.1.PCA-LDA using Vis-NIR Spectra

The selection of the ideal number of principal components (PCs) to be used in the model is a critical step in the creation of the PCA-LDA framework. The total number of PCs is the same as the total number of initial variables. For many datasets, the top few PCs typically accounted for the majority of the variance, making it possible to ignore the remaining PCs with little information loss. The PCA-LDA model was developed using preprocessed spectra. The results showed that the cacao maturity level was perfectly classified using raw spectra for the cacao maturity classification Vis-NIR spectrometer using PCA-LDA methods.

The accuracy of a model can be determined by its sensitivity, specificity, and misclassification rate. Sensitivity allows the assessment of how well the model can identify samples that belong to a particular class, and specificity measures the capacity of the model to reject non-belonging samples. The misclassification rate was the ratio of false positives to the total number of samples. In this study, the misclassification rate was considered to evaluate model performance. The classification of mature and immature cacao using the PCA-LDA model performance with Vis-NIR spectroscopy is shown in Table 1. After spectral preprocessing using Area Normalization (AN) resulted that the method most inaccurate with the highest misclassification was 86%. The best model for Vis-NIR Spectroscopy was obtained using raw spectra, with an accuracy of 90% for calibration and 86% for prediction.

Table 1. PCA-LDA performance of Vis-NIR spectroscopy.

Performance	Calibration						
	Raw	Norm	SNV	MSC	SG	SGD1	SGD2
Acc(%)	**90**	86	88	89	88	85	85
Misclasified rate	10	**86**	11	11	11	11	11
Performance	**Prediction**						
	Raw	Norm	SNV	MSC	SG	SGD1	SGD2
Acc(%)	**86**	86	85	87	86	85	85
Misclasified rate	13	13	15	12	13	15	15

Note: Acc = accuracy; Raw = raw data; Norm = normalized data; MSC = multiple scatter correction; SNV = standard normal variate; SG= Savitzky Golay–s; SGD1 =Savitsky-Golay's 1st derivative, SGD2 = Savitsky-Golay's 2nd derivative.

The model's prediction performance also showed that it could accurately classify the cacao maturity level into each category (Table 1). Because the accuracy was greater than 90%, the model can be deemed to be a good one. In light of the overall amount of data employed, the accuracy demonstrates the model's capacity to properly categorize the cacao maturity level [37]. The outcomes of model generation using a set of samples for calibration (n=320). It seen that all samples were not correctly classified, resulting in an accuracy of 90 %. In Table 2, a confusion matrix is presented that exhibits the outcomes of the classification of prediction samples that utilized the Raw PCA-LDA model. Notably, the highest occurrence of false predictions (n=48) was observed for immature individuals, which were mistakenly classified as mature [34].

Table 2. The Result of a Classification of Cocoa Pods Samples using the PCA-LDA model from Vis NIR Spectra

True Class	Predicted Class	
	Calibration	Prediction
Non Mature	Non Mature (16)	Non Mature (1)
Non Mature	Mature (32)	Mature (11)
Mature	Non Mature (0)	Non Mature (0)
Mature	Mature (272)	Mature (68)
Accuracy (%)	90	86
Misclasification Rate (%)	10	14

3.4.2.PCA-LDA using SW-NIR spectra

The PCA-LDA model was developed using preprocessed spectra. The performance of the PCA-LDA model using the SW-NIR spectra is shown in Table 3. The results showed that the cacao maturity level was perfectly classified using SGD1 spectra for the cacao maturity classification SW-NIR spectrometer using PCA-LDA methods. The classification accuracies observed for differentiating the cacao samples through utilization the Vis-NIR original spectra were 91,56% and 86,25% during the calibration and prediction stages, respectively. After spectral preprocessing using Multiplicative Scatter Correction (MSC) resulted that the method most inaccurate with the highest misclassification was 9,69%. The best model for Vis-NIR Spectroscopy was obtained using Savitzky'-Golay 1st derivative pre-processing spectra, with an accuracy of 92,50% for calibration and 85% for prediction. The results show that PCA-LDA using SW-NIR spectra yielded a higher model than PCA-LDA using Vis-NIR spectra.

Table 3. PCA-LDA performance of SW-NIR spectroscopy.

Performance	Calibration						
	Raw	Norm	SNV	MSC	SG	**SGD1**	SGD2
Acc(%)	91,56	91,56	90,63	90,31	92,05	**92,50**	90,83
Misclasified rate	8.44	8.44	9.38	**9.69**	7.95	7.50	9.17
Performance	**Prediction**						
	Raw	Norm	SNV	MSC	SG	**SGD1**	SGD2
Acc(%)	86,25	85,00	85,00	85,00	86,25	**85,00**	85,00
Misclasified rate	13.75	15.00	15.00	15.00	13.75	15.00	15.00

Note: Acc = accuracy; Raw = raw data; Norm = normalized data; MSC = multiple scatter correction; SNV = standard normal variate; SG= Savitzky Golay–s; SGD1 =Savitsky-Golay's 1st derivative, SGD2 = Savitsky-Golay's 2nd derivative.

The model's prediction performance also showed that it could accurately classify the cacao maturity level into each category (Table 4). The ability of the model to accurately categorize cocoa pod maturity was further demonstrated by its strong prediction performance. The model's prediction performance also showed that it could accurately classify the cacao maturity level into each category (Table 3). Because the accuracy is greater than 92,50%, the model can be deemed to be a good one. In light of the overall amount of data employed, the accuracy demonstrates the model's capacity to properly categorize the cacao maturity level [34]. The outcomes of model generation using a set of samples for calibration (n=320). It seen that all samples were not correctly classified, resulting in an accuracy of 92,50 %. In Table 4, a confusion matrix is presented that exhibits the outcomes of the classification of prediction samples that utilized the Raw PCA-LDA model. Notably, the highest occurrence of false predictions (n=51) was observed for immature individuals, which were mistakenly classified as mature [27].

Table 4.The Result of a Classification of Cocoa Pods Samples using the PCA-LDA model from Vis NIR Spectra

True Class	Predicted Class	
	Calibration	Prediction
Non Mature	Non Mature (27)	Non Mature (0)
Non Mature	Mature (24)	Mature (12)
Mature	Non Mature (0)	Non Mature (0)
Mature	Mature (269)	Mature (68)
Accuracy (%)	92,50	85
Misclasification Rate (%)	7,50	15

4. Conclusion

The results show that the availability of Vis-NIR and SW-NIR spectroscopy for cocoa pod maturity can be distinguished using spectroscopy combined with chemometrics analysis. The results showed that SW-NIR spectroscopy yielded a better performance than SW-NIR spectroscopy. The best model for classifying cocoa pod maturity obtained by SW-NIR spectroscopy had calibration performance of 92% accuracy for the PCA-LDA model. The best model for classifying cocoa pod maturity obtained by Vis-NIR spectroscopy had calibration performance of 90% accuracy for the PCA-LDA model. As a result, it can be said that Vis-NIR and SW-NIR spectroscopy are promising techniques for identifying cocoa pods maturity.

Acknowledgment. This research was supported by the Center for Higher Education Funding (BPPT) and the Indonesian Endowment Fund for Education (LPDP).

References

[1] Y. K. T. Dang and H. V. H. Nguyen, "Effects of Maturity at Harvest and Fermentation Conditions on Bioactive Compounds of Cocoa Beans," *Plant Foods Hum. Nutr.*, vol. 74, no. 1, pp. 54–60, 2019, doi: 10.1007/s11130-018-0700-3.

[2] N. Elkhoshkhany, "UV–Vis-NIR spectroscopy, structural and thermal properties of novel oxyhalide tellurite glasses with composition TeO_2 -B_2O_3-$SrCl_2$-LiF-Bi_2O_3 for optical application," *Results Phys.*, vol. 13, 2019, doi: 10.1016/j.rinp.2019.102222.

[3] D. A. S. Saputri, M. Fahri Reza Pahlawan, B. M. A. Murti, and R. E. Masithoh, "Vis/NIR spectroscopy for non-destructive method in detecting soybean seeds viability," *IOP Conf. Ser. Earth Environ. Sci.*, vol. 1038, no. 1, 2022, doi: 10.1088/1755-1315/1038/1/012043.

[4] K. B. Beć, J. Grabska, and C. W. Huck, "In silico NIR spectroscopy – A review. Molecular fingerprint, interpretation of calibration models, understanding of matrix effects and instrumental difference," *Spectrochim. Acta - Part A Mol. Biomol. Spectrosc.*, vol. 279, no. March, 2022, doi: 10.1016/j.saa.2022.121438.

[5] W. Terouzi *et al.*, "Characterization and rapid detection of adulterations in sesame oil using FT-MIR and PCA-LDA," *Moroccan J. Chem.*, vol. 4, no. 4, pp. 4-4 (2016)1052-1060, 2016, [Online]. Available: https://revues.imist.ma/index.php/morjchem/article/view/5167

[6] J. Grabska and C. W. Huck, "Near-Infrared Spectroscopy in Bio-Applications," 2020.

[7] J. Van De Steene *et al.*, "Authenticity analysis of oregano: development, validation and fitness for use of several food fingerprinting techniques," *Food Res. Int.*, vol. 162, no. PA, p. 111962, 2022, doi: 10.1016/j.foodres.2022.111962.

[8] Y. B. Che Man, Z. A. Syahariza, M. E. S. Mirghani, S. Jinap, and J. Bakar, "Analysis of potential lard adulteration in chocolate and chocolate products using Fourier transform infrared spectroscopy," *Food Chem.*, vol. 90, no. 4, pp. 815–819, 2005, doi: 10.1016/j.foodchem.2004.05.029.

[9] N. N. Batista, D. P. de Andrade, C. L. Ramos, D. R. Dias, and R. F. Schwan, "Antioxidant capacity of cocoa beans and chocolate assessed by FTIR," *Food Res. Int.*, vol. 90, pp. 313–319, 2016, doi: 10.1016/j.foodres.2016.10.028.

[10] Y. Hu *et al.*, "Determination of antioxidant capacity and phenolic content of chocolate by attenuated total reflectance-Fourier transformed-infrared spectroscopy," *Food Chem.*, vol. 202, pp. 254–261, 2016, doi: 10.1016/j.foodchem.2016.01.130.

[11] I. A. Santos, D. G. Conceição, M. B. Viana, G. de J. Silva, L. S. Santos, and S. P. B. Ferrão, "NIR and MIR spectroscopy for quick detection of the adulteration of cocoa content in chocolates," *Food Chem.*, vol. 349, no. June 2020, pp. 1–6, 2021, doi: 10.1016/j.foodchem.2021.129095.

[12] J. Moros, F. A. Iñón, S. Garrigues, and M. de la Guardia, "Near-infrared diffuse reflectance spectroscopy and neural networks for measuring nutritional parameters in chocolate samples," *Anal. Chim. Acta*, vol. 584, no. 1, pp. 215–222, 2007, doi: 10.1016/j.aca.2006.11.020.

[13] D. N. de Oliveira, A. C. B. Camargo, C. F. O. R. Melo, and R. R. Catharino, "A fast semi-quantitative screening for cocoa content in chocolates using MALDI-MSI," *Food Res. Int.*, vol. 103, no. October 2017, pp. 8–11, 2018, doi: 10.1016/j.foodres.2017.10.035.

[14] E. K. Anyidoho, "Differentiation of Organic Cocoa Beans and Conventional Ones by Using Handheld NIR Spectroscopy and Multivariate Classification Techniques," *Int. J. Food Sci.*, vol. 2021, 2021, doi: 10.1155/2021/1844675.

[15] M. C. García-Muñoz, M. P. Tarazona-Díaz, N. A. Meneses-Marentes, G. González-Sarmiento, A. S. Pineda-Guerrero, and G. E. Gómez-Uribe, "Development of color guides to

evaluate the maturity of cacao clones by digital image processing," *Pesqui. Agropecu. Trop.*, vol. 51, no. November, 2021, doi: 10.1590/1983-40632021v5169621.
[16] R. E. Masithoh, S. Lohumi, W. S. Yoon, H. Z. Amanah, and B. K. Cho, "Development of multi-product calibration models of various root and tuber powders by fourier transform near infra-red (FT-NIR) spectroscopy for the quantification of polysaccharide contents," *Heliyon*, vol. 6, no. 10, p. e05099, 2020, doi: 10.1016/j.heliyon.2020.e05099.
[17] P. Khuwijitjaru, K. Boonyapisomparn, and C. W. Huck, "Near-infrared spectroscopy with linear discriminant analysis for green 'Robusta' coffee bean sorting," *Int. Food Res. J.*, vol. 27, no. 2, pp. 287–294, 2020.
[18] M. Ferreiro-González, E. Espada-Bellido, L. Guillén-Cueto, M. Palma, C. G. Barroso, and G. F. Barbero, "Rapid quantification of honey adulteration by visible-near infrared spectroscopy combined with chemometrics," *Talanta*, vol. 188, pp. 288–292, 2018, doi: 10.1016/j.talanta.2018.05.095.
[19] M. Caredda *et al.*, "Building of prediction models by using Mid-Infrared spectroscopy and fatty acid profile to discriminate the geographical origin of sheep milk," *Lwt*, vol. 75, pp. 131–136, 2017, doi: 10.1016/j.lwt.2016.08.053.
[20] D. Suhandy and M. Yulia, "Classification of lampung robusta Specialty coffee according to differences in cherry processing methods using UV spectroscopy and chemometrics," *Agric.*, vol. 11, no. 2, pp. 1–11, 2021, doi: 10.3390/agriculture11020109.
[21] M. E. Raypah, L. J. Zhi, L. Z. Loon, and A. F. Omar, "Near-infrared spectroscopy with chemometrics for identification and quantification of adulteration in high-quality stingless bee honey," *Chemom. Intell. Lab. Syst.*, vol. 224, no. March, p. 104540, 2022, doi: 10.1016/j.chemolab.2022.104540.
[22] P. Mishra *et al.*, "MBA-GUI: A chemometric graphical user interface for multi-block data visualisation, regression, classification, variable selection and automated pre-processing," *Chemom. Intell. Lab. Syst.*, vol. 205, no. August, p. 104139, 2020, doi: 10.1016/j.chemolab.2020.104139.
[23] E. Trullols, I. Ruisánchez, and F. X. Rius, "Validation of qualitative analytical methods," *TrAC - Trends Anal. Chem.*, vol. 23, no. 2, pp. 137–145, 2004, doi: 10.1016/S0165-9936(04)00201-8.
[24] L. S. Vieira, C. Assis, M. E. L. R. de Queiroz, A. A. Neves, and A. F. de Oliveira, "Building robust models for identification of adulteration in olive oil using FT-NIR, PLS-DA and variable selection," *Food Chem.*, vol. 345, no. August 2020, 2021, doi: 10.1016/j.foodchem.2020.128866.
[25] K. Ncama, "Application of Vis/NIR spectroscopy for predicting sweetness and flavour parameters of 'Valencia' orange (Citrus sinensis) and 'Star Ruby' grapefruit (Citrus x paradisi Macfad)," *J. Food Eng.*, vol. 193, pp. 86–94, 2017, doi: 10.1016/j.jfoodeng.2016.08.015.
[26] B. N. Peshlov, F. E. Dowelt, F. A. Drummond, and D. W. Donahue, "Comparison of three near infrared spectrophotometers for infestation detection in wild blueberries using multivariate calibration models," *J. Near Infrared Spectrosc.*, vol. 17, no. 4, pp. 203–212, 2009, doi: 10.1255/jnirs.842.
[27] X. Lin and D. W. Sun, "Recent developments in vibrational spectroscopic techniques for tea quality and safety analyses," *Trends Food Sci. Technol.*, vol. 104, no. June, pp. 163–176, 2020, doi: 10.1016/j.tifs.2020.06.009.
[28] A. F. Omar, H. Atan, and M. Z. MatJafri, "NIR spectroscopic properties of aqueous acids solutions," *Molecules*, vol. 17, no. 6, pp. 7440–7450, 2012, doi: 10.3390/molecules17067440.
[29] V. Cortés, C. Ortiz, N. Aleixos, J. Blasco, S. Cubero, and P. Talens, "A new internal quality index for mango and its prediction by external visible and near-infrared reflection spectroscopy," *Postharvest Biol. Technol.*, vol. 118, pp. 148–158, 2016, doi: 10.1016/j.postharvbio.2016.04.011.
[30] M. N. Merzlyak, A. E. Solovchenko, and A. A. Gitelson, "Reflectance spectral features and non-destructive estimation of chlorophyll, carotenoid and anthocyanin content in apple fruit," *Postharvest Biol. Technol.*, vol. 27, no. 2, pp. 197–211, 2003, doi: 10.1016/S0925-5214(02)00066-2.
[31] J. U. Porep, D. R. Kammerer, and R. Carle, "On-line application of near infrared (NIR) spectroscopy in food production," *Trends Food Sci. Technol.*, vol. 46, no. 2, pp. 211–230, 2015, doi: 10.1016/j.tifs.2015.10.002.
[32] J. H. Choi, "Portable, non-destructive tester integrating VIS/NIR reflectance spectroscopy for the detection of sugar content in Asian pears," *Sci. Hortic. (Amsterdam).*, vol. 220, pp. 147–153, 2017, doi: 10.1016/j.scienta.2017.03.050.
[33] R. Sahukari, S. Bhasha, H. Singamala, and V. Subbaiah, "Assessment of Potential Antioxidant Activity of Polyphenolic Fraction Separated from Acalypha Indica Leaves: An In vitro Approach Ethnopharmacology and Molecular Biology View project Transcriptome analysis View project," no. October 2017, 2015, [Online]. Available: https://www.researchgate.net/publication/320258303
[34] I. Novianty, K. B. Seminar, Irzaman, and I. W. Budiastra, "Improving the accuracy of near-infrared (NIR) spectroscopy method to predict the oil content of oil palm fresh fruits," *IOP*

Conf. Ser. Earth Environ. Sci., vol. 460, no. 1, 2020, doi: 10.1088/1755-1315/460/1/012004.

[35] T. J. Gutiérrez, "State-of-the-Art Chocolate Manufacture: A Review," *Compr. Rev. Food Sci. Food Saf.*, vol. 16, no. 6, pp. 1313–1344, 2017, doi: 10.1111/1541-4337.12301.

[36] J. Y. Tang, N. Y. Chen, M. K. Chen, M. H. Wang, and L. S. Jang, "Dual-wavelength optical fluidic glucose sensor using time series analysis of D(+)-glucose measurement," *Jpn. J. Appl. Phys.*, vol. 55, no. 10, 2016, doi: 10.7567/JJAP.55.106601.

[37] M. Rocha Baqueta, A. Coqueiro, P. Henrique Março, M. Mandrone, F. Poli, and P. Valderrama, "Integrated 1H NMR fingerprint with NIR spectroscopy, sensory properties, and quality parameters in a multi-block data analysis using ComDim to evaluate coffee blends," *Food Chem.*, vol. 355, no. March, p. 129618, 2021, doi: 10.1016/j.foodchem.2021.129618.

Improved image recognition via Synthetic Plants using 3D Modelling with Stochastic Variations

Chris C. Napier[1*], *David M.* Cook[1], *Leisa* Armstrong[1] and *Dean* Diepeveen[1,2]

[1]School of Science, Edith Cowan University, 270 Joondalup Drive, Joondalup, Australia
[2]Department of Primary Industry Resource Development, South Perth, Australia

Abstract. This research extends previous plant modelling using L-systems by means of a novel arrangement comprising synthetic plants and a refined global wheat dataset in combination with a synthetic inference application. The study demonstrates an application with direct recognition of real plant stereotypes, and augmentation via a plant-wide stochastic growth variation structure. The study showed that the automatic annotation and counting of wheat heads using the Global Wheat dataset images provides a time and cost saving over traditional manual approaches and neural networks. This study introduces a novel synthetic inference application using a plant-wide stochastic variation system, resulting in improved structural dataset hierarchy. The research demonstrates a significantly improved L-system that can more effectively and more accurately define and distinguish wheat crop characteristics. **Keywords:** Synthetic plants, Stochastic modelling, L-systems, Global Wheat, Inference

1 Introduction

In 2021 the Global Wheat Challenge (GWC) had a profound impact upon the expansion and extension of the Global Wheat Head Detection (GWHD) dataset, drawing renewed worldwide interest from the computer vision and agricultural science communities [1]. The global focus on improved quality and reduced cost of food production is driven by a rapidly increasing world population [2], [3], [4].

The ongoing pursuit of increased efficiencies in food production has resulted in the rapid growth of research of deep learning approaches towards new discoveries in crops adaptability to suit globally disparate agronomic and climatic conditions [5]. Whilst most of the work in this area has centred on using traditional deep learning approaches and large datasets to establish ground truth, this study demonstrates significant advantages in the application of alternative approaches to deep learning and crop modelling using the APSIM model [6], [7] and [8], or through L-systems [9], [10], and [11].

1.1 Background and Literature

An L-system is a set of rules and parameters which define plant growth, structure, and appearance [12]. This research proposes an extension to an existing L-system (L-NAP) with an improved wheat crop dataset; to add realistic stochastic growth variations; and to introduce a novel and important application in synthetic inference which will directly recognize wheat heads, and wheat characteristics under changing agronomic conditions.

L-systems typically evolve from a mathematical theory of plant and cellular growth [12]. They are based upon the Backus-Naur production rules and allow for extensive application of plant-based synthetic imagery [13], [14], [15].

L-systems are particularly useful in the ability to recognize and measure real wheat plants. Using Python as a language driver, L-system frameworks can be created to form algorithms to characterize and recognize features on wheat heads [16]. They are regularly described by means of small text-based explanations and imageries that require relatively small amounts of storage compared to real imagery [17]. The recognition success is dependent on pixels. An individual person's eyes respond to many light intensity levels. L-NAP must be able to "see" at low resolution and low scale. This increased level of recognition is made possible by refining the dataset to incorporate the range of scale and resolution contrasts.

The aim in synthetic inference is to achieve unabridged corollaries that create a higher level of reasoning for the interpretation of the observed images. The dataset will "carry" images at all growth scales, based on a wide range of grain-counts, which is not the same as image scaling. This is the necessary approach to see smaller images and still accurately recognize plant parts, since

* Corresponding author: d.cook@ecu.edu.au

smaller images are often caused by lower grain-count [18].

This is indeed what stochastic variation means, and it reflects real plant growth variations. This project will push the research envelope to use depth and stereo cameras of higher resolutions in line with improving miniature camera technology. In this sense the evolution of these stochastic augmentations grows in line with advances and upgrades in camera and image capture technologies.

2 Validation of Synthetic Images using the Global Wheat Head Detection (GWHD) dataset.

The Global Wheat Head Detection (GWHD) dataset of 2021 [1] provides renewed confidence in the expectation of improved recognition. In this study, a real wheat neural network was trained on a total of 677 images of the second domain of the GWHD. There were 40 field images in the second domain of around 25 wheat heads each. From these wheat heads two-thirds were used for training whilst one-third were used for validation (See Figure 1.).

This second domain was chosen for its higher quality, in combination with fewer images. It was used to validate the initial synthetic images and was designed to resemble that domain by means of accurate L-system parameter choices, and the stochastic variations of those parameters.

Some larger variations resulted in lower matching scores, while modest variations on the upper images gave higher validation. Here there is an advantage in the use of synthetic data, where this test indicated the best choice of L-system parameters for this domain. These choices in turn allow "domain adaptation" [19], [20] by simple parameter adjustments. These adjustments are guided in a similar test using real data, even if a limited amount, of that domain [21]. This typically demonstrates accurate results even in situations where the real data used to steer the recognition process is drawn from a limited number of real wheat head images.

2.1 Recognition of Awns

One example of this domain adaptation is in terms of elements such as the length of the awns on the wheat heads. These awns resemble long thin spikes and are predominant on the image set in Figure 2, which can be longer for other domains. Wheat heads represent a critical source of information about the quality of a wheat crop. The accurate recognition of awns is especially beneficial in determining the health of a crop since awns provide rich data about the health, maturity stage, and size of the associated wheat heads.

Fig. 1. Validated synthetic wheat with a neural network trained on the GWHD dataset, showing network object recognition errors on overlapping synthetic wheat head images. (Green box on right side of image).

Fig. 2. Validated synthetic wheat with a neural network trained on the GWHD dataset, showing clear recognition of Awns. (Blue box on left side of image).

3 L-NAP System Approach

The previous work of Napier et al (2023) described L-NAP as an application that accurately represented the hierarchy and growth of wheat heads by using synthetic models [16].

L-NAP fully separates an L-system algorithm, describing a synthetic plant, into two files. The first file defines the L-system rules relating to the structure and relation of the plant parts. The second file defines the L-system parameters, relating to sizes, angles, counts, textures, and plant growth. L-NAP communicates these varying parameters as necessary to the rules when the plant is grown, mimicking natural plant processes. The parameter calculations can be complex, but by this separation, they are contained within parameter modules.

Table 1. Steps in L-NAP towards dataset creation, application, recognition, and annotation of real plants.

L-NAP dataset steps		
Step A	Creation of Drawing Commands in an L-system algorithm	
Step B	Create 3D models from commands in Blender	
Step C	Create plant camera views of models	
Step D	Create synthetic dataset	
Step E	Apply dataset to recognise real images	

The L-NAP system has five steps, from A to E, which in turn create drawing commands from an L-system description; create 3D models from these commands; create plant views of the models; create a synthetic dataset from the views; and finally apply the dataset to recognize and annotate real plants (Table 1).

The method of taking this synthetic dataset approach means that the data processing complexity is handled during the dataset creation by the data pipelines, and by the dataset API access functions. The approach demonstrates its completion by the synthetic inference application which employs the synthetic dataset to accurately locate plant instances in real pasture images.

L_NAP has been extended to cover the L-system classes defined in the book "The Algorithmic Beauty of Plants" [12], and to introduce plant-wide stochastic variations, as show in the trees below. Figure 3 demonstrates the use of the L-NAP algorithm (below) which creates the trees by means of a python program.

Fig. 3. A full tree (on the left), and then the same tree re-grown (on the right) with only 80% branch growth, controlled by one L-NAP parameter.

The **complete** L-NAP rules which were used to grow the trees in Figure 3 are defined in this Python program:

```
from L_NAP import *
Tree = L_NAP()
Branch = Tree(Draw, [Turn_left, Branch],
              Draw, [Turn_right, Branch],
              Turn_left, Branch)
Axiom = Tree(Branch)
while Tree.next_stage(Axiom):
    Tree.grow()
```

The separate parameters file gives the number of growth stages (7), the branch length (1.0), width (0.1), and scaling per stage (0.5), the branch turning angle (20 deg.), and the branch growth probability (1.0 or 0.8), and these are applied to each grown branch (See Figure 3).

3.1 Modelling using Light Absorption

An important feature of the L-NAP framework is that since it is an L-system defined in a Python class, it may be extended with extra functionality, to create a new L-system framework of a new name, which for example, could have math functions, which depend on external light conditions, to guide the tree growth.

One such extension relates to the work of Renton et al, (2005) that considers modelling in relation to light [22]. This study concludes that new tree growth of (mountain birch) trees can be determined solely by absorbed light. Such a system can become an extension of the L-NAP framework.

In a similar fashion the L-PEACH framework [23], which has source-sink interaction functionality and is described as complex, could benefit from L-NAP. Such a framework could be further be extended to other tree types. The study on MAppleT [24] also constitutes a useful tool for simulating apple tree development in terms of the interaction with gravity. This tool demonstrates a strong alignment with the functionality of the L-NAP framework in terms of integration. The reflection, absorption, and transmission of light are considered to underpin the three principal steps that a light ray goes through upon arriving at the surface of a leaf [25].

4 The Synthetic Inference Approach

Synthetic Inference is a term introduced in this research to describe its novel plant recognition and measurement approach. It differs from traditional approaches in that synthetic inference does not require neural network training. Instead, a synthetic dataset is created as distinct parts preserving the structured knowledge of synthetic plant data and metadata in 2D, 2D + depth, and 3D representations.

Tu [26] (2020) and Li [27] discuss neural network image depth and stereo processing. This structured knowledge maintains the form and hierarchical structure of the plant parts and can be directly and efficiently accessed in synthetic inference applications. The dataset structure combines multiple views, representing

 https://doi.org/10.1051/bioconf/20238006004

positions and orientations of each 3D model, allowing the concept of the matching to individual plant models, and the user interface allows specified variations of these models in the recognition and measurement process. This contrasts with the "pooling" operations performed during neural network training, where individual details are assimilated and combined into a complex but problematic representation, as described by Hinton [28].

A dataset contains separate sections for different regions and different plant varieties, whereas there exist a common set of procedures to store and retrieve the data, and particular combinations of data and metadata, and orientations which match features in a real plant image. For example, to perform a grain-count measurement in synthetic inference, a set of wheat head images with the same orientation but with increasing grain-count can be requested from the dataset, to achieve a better recognition of a given real plant image. A test system was developed with twelve rotations of each view. This is effective in terms of improved image recognition.

5 A Synthetic Inference Application

An important synthetic Inference application was developed to harness the synthetic models, image data, and metadata and its hierarchal structure which is present on the created wheat head dataset.

Fig. 4. A Global Wheat dataset image, annotated by synthetic inference, indicating the best synthetic model numbers and matching IoU scores.

The dataset has 100, stochastically varied 3D models. Further, the dataset had 25 2D views of each model. The synthetic inference method directly locates wheat heads on Global Wheat pasture images, using this dataset with high visibility and clarity (See Figure 4).

It is important to discuss the significance of IoU scores using L-systems for synthetic images [19], [29], [30] and [31]. The Intersection over Union (IoU) is a standard object recognition metric. It is the intersection area divided by the union area of two objects, where area relates to 2D image pixels. Intersection is the common area of overlap, and the union area is the sum of both object areas less the common area, thus:

```
IoU = I / (A1 + A2 - I),
where the two objects have areas A1 and A2
and their intersection area is I.
```

The highest IoU score indicates the best matching synthetic wheat head at each real wheat head position. As a separate measure of overall success in matching a wheat field: Currently, 17 of the 23 wheat heads, are recognised on this one image, giving a score of 17/23 => 74%. This is preliminary since many refinements must be performed.

This synthetic inference application uses direct pixel comparisons between synthetic and real wheat image data. This is a novel approach to demonstrating an accurate method that saves time training neural networks.

6 The L-NAP Dataset

The L-NAP dataset gathers the plant data and makes it available to applications in a structured and relational manner. It follows the standard Detectron2 formalism [32], which uses a JSON file [33] describing the dataset contents. The data format is based on Microsoft COCO [34]. In the case of the Global wheat head recognition project, the dataset was stored in terms of its directory configuration (Table 2).

Table 2. Directory Contents

Meta	Metadata of synthetic wheat heads annotated in JSON files, referring to View and Model files.
Model	3D models, held in object and material files, referenced by JSON files
View	2D image view files, referenced by a JSON file

Each metadata JSON file describes a synthetic wheat head or several related wheat-heads, located on named 2D synthetic view files, as views of 3D models in a Wavefront format [35]. The wheat heads are described to be within given bounding box(es) on the view file, and within bit-mask segmentation(s), each defined by a list of Run-List-Encodings [36].

Each wheat head has categories defining its grain-count, orientation, scale, and partial-image class, and other possible classifications. When there is more than one wheat head defined on a JSON file, then the wheat heads are closely related, such as several orientations of a single view, or a range of wheat heads with increasing grain-count at the same orientation.

This JSON file is used in the synthetic inference application to access the synthetic wheat pixel data and to create relational information, which guides the application. The Model directory holds the original 2D wheat head models. It will be used in future applications to create digital representations of real pasture scenes. The View directory contains the pixel data named in the JSON file.

Note that the plant recognition system from Pl@ntNet.at [37] uses an image dataset that has been annotated by expert botanists. This system has achieved 88% plant recognition accuracy. This level is much higher compared to the lower scores for neural network-based mobile phone plant recognition systems, which generally have poorly annotated data. An algorithmic ranking of viable plant candidates was a key feature of their dataset.

7 Conclusion

This research demonstrates the advantages of using synthetic plants and 3D modelling to accurately recognize real plants by means of an L-systems approach. The study shows the efficacy in two key areas. Firstly, this paper shows the benefits of stochastic variations and the ability to gain a more accurate system of recognition. The second benefit is the ability to save time because this approach allows for synthetic mapping without the need to further train neural networks.

An innovative synthetic inference application has been successful in the accurate recognition of selected wheat head varieties of the Global Wheat dataset of 2021 [2]. This application is an alternative to the neural network application. Synthetic inference will be extended with higher accuracy to cover multiple wheat head varieties and multiple plant types. The stochastic variation method will be extended to allow the creation of increasingly realistic plants.

Research discovered that a smartphone plant recognition system from Pl@ntNet.at uses a dataset of images, categorised by botanists, to achieve 88% plant recognition [37] compared to much lower scores for neural network-based mobile phone plant recognition systems, which generally have poorly annotated data. An algorithmic ranking of possible plant candidates was a key feature. If we note that in the Global Wheat Competition of 2021 the best score was 84% [1], then we can confidently suggest that this type of L-systems approach will allow for image recognition training on a significantly smaller set of synthetic images. In future, the dataset must be refined, and experiments performed to create the best dataset interface, and the data for the best object recognition.

It is one task to find the best L-systems rules and parameters for a given plant, and this has been achieved. It is a separate task to introduce a suitable range of parameter variation within the dataset. To achieve this improvement requires that successful image matching is recorded and fed back to the L-NAP framework to allow it to adapt its L-system parameters and their variational ranges to a new domain or to a related plant type. Such a self-adaption would be a milestone in L-NAP development [38].

The use of stochastic variations is an important improvement in the ongoing research endeavour to create image recognition systems that can be trained on nature-based datasets without the need for enormously large numbers of images. They allow for a more robust set of validations that are helpful in terms of recognition in critical areas such as wheat heads. Stochastic variations represent important inclusions to the development of L-systems because they permit and develop greater flexibility and realism in modelling complex structures. The benefits of these inclusions are essential to food security in terms of resilience given the climatic challenges that are influencing global staples such as wheat.

The stochastic variations explored in this paper demonstrate the ability to save significant amounts of time. The 18 domains of wheat could easily be matched using this stochastic method. Having mapped one domain, this system would allow for the same application to map the other 17 global wheat domains. A further benefit is the significant value of recognising awns. Such an approach allows for a system of recognition using synthetic wheat and generating reliable mapping without the need to spend additional valuable time in individual domains.

7.1 Limitations and Future research

One of the limitations is the use of mono recognition of images. A future direction for this recognition approach will be to use a stereo image approach to determine a sharper image in terms of the quest for greater detail in wheat head characteristics. This is an important consideration given that synthetic data is highly applicable to stereo imagery, providing synthetic data that can be generated as 3D models.

This approach is transferrable to recognition in other plants. A new approach is being considered in terms of coffee to investigate the benefits of an L-systems approach for the classification and grading of coffee beans using synthetic modelling.

8 References

1. E. David, M. Serouart, D. Smith et al, *Global Wheat Head Detection 2021: An improved dataset for Benchmarking Wheat Head Detection Methods*, Plant Phenomics, Volume 2021. (2021).
2. Y. Kassem, H. Gokcekus, E. Alassi. *Identifying most influencing input parameters for predicting cereal production using an artificial neural*

network model. Modelling Earth systems and Environment, Volume 8. (2022).

3. T. Khan, J. Qiu, M. Qureshi, and M. Iqbal, *Agricultural fruit prediction using deep neural networks,* Procedia Computer Science, Volume 174, Science Direct. (2020).
4. H. Yalcion, and S Razavi, *Plant classification using convolutional neural networks,* 2016 Fifth International Conference on Agro-Geoinformatics (Agro-Geoinformatics), IEEE Explore. (2016).
5. J. Ubbens, M. Cieslak, P. Prusinkiewicz, and I. Stavness. *The use of plant models in deep learning: an application to leaf counting in rosette plants.* Plant Methods, Volume 14, 6. (2018).
6. D. Holzworth, N. Huth, J. Fainges, et al, *APSIM Next Generation: Overcoming challenges in modernising a farming systems model.* Environment Modelling and Software, Volume 103, pages 43 – 51. (2018).
7. J. Hyles, M. Bloomfield, J. Hunt, R.M. Trethowan. *Phenology and related traits for wheat adaptation.* Heredity, Volume 125, pp 417 – 430. (2020).
8. H. Brown, N. Huth, and D. Holzworth, Crop *Model Improvement in moderning and farming systems model,* European Journal of Agronomy, (2018).
9. J. Hanan, *Parametric L-systems and their application to the modelling and visualization of plants.* Dissertation University of Regina, Department of Computer Science, (1992).
10. R. Mech, *Modelling and simulation of the interaction of plants with the environment using L-systems and their extensions,* PhD Thesis, Computer Science Department, University of Calgary, (1997).
11. R. Rokhana, W. Herulambang, and R. Indraswari, *Machine Learning and Polynomial – L System Algorithm for Modelling and Simulation of Glycine Max* (L) Merrill Growth, International ElecLindentronics Symposium, (IES), pages 463 – 467, (2020).
12. P, Prusinkiewicz, A. Lindenmayer, and J. Hanah, *The Algorithmic beauty of Plants,* Springer Verlag, (2004).
13. N. Chomsky, *Syntactic Structures*, The Hague, Mouton and Co. (2002).
14. J.W.Backus, *The Syntax and semantics of the proposed international algebraic language of the Zurich ACM-GAMM Conference.* Proceedings of ICIP, UNESCO, pp. 125-132. (1959).
15. B.G. Fitch, P. Parslow, and K. Lunqvist. *Evolving complete L-systems: using genetic algorithms for the generation of realistic plants,* in Lewis, P.R. Headland C.J., Battle, S., and Ritsos, P.D. (Eds) Artificial Life and Intelligent Agents. Pp 16-23, Springer. (2018).
16. C. Napier, D.M. Cook, L. Armstrong, and D. Diepeveen, *A synthetic wheat L-system to accurately detect and visualise wheat head anomalies,* Proceedings of the 3rd International Conference on Smart and Innovative Agriculture (ICOSIA), (2023).
17. J. Bernard and I. McQuillan, *New techniques for inferring L-systems using genetic algorithm.* In Proceedings of the 8th International Conference on Bioinspired Optimization Methods and Applications. Lecture Notes in Computer Science, pp 13-25. Springer, (2018).
18. J. Guo, H. Jiang, B. Benes, O. Deussen, X. Zhang, D. Lischinski. And H. Huang. *Inverse Procedural Modeling of Branch Structures by Inferring L-Systems,* ACM Transactions on Graphics, Vol, 39, No. 5. (2020).
19. Z. Hartley, and A.P. French, *Domain adaptation of synthetic images for wheat head detection,* Plants, Volume 10, Issue 12. (2021)
20. A.Farahani, S.Voghoei, K. Rasheed. *A brief review of domain adaptation,* Advances in data science, Springer (2021)
21. D. Kuznichov, A Zvirin, Y. Honen, R.Kimmel. *Data augmentation for leaf segmentation and counting tasks in rosette plants,* Proceedings of the IEEE/CVF Conference on Computer Vision and Pattern Recognition (CVPR) Workshops, (2019).
22. M. Renton, P. Kaitaniemi, and J. Hanan, *Functional–structural plant modelling using a combination of architectural analysis, L-systems and a canonical model of function*, Ecological Modelling.Volume 184, Issues 2–4, (2005).
23. M.T., Allen, P. Prusinkiewicz, and T.M. DeJong, *Using L-systems for modelling source–sink interactions, architecture, and physiology of growing trees: the L-PEACH model.* New Phytologist, Volume 166: 869-880. (2005).
24. E. Costes, C. Smith, M. Renton, Y. Guédon, P. Prusinkiewicz, and C.Godin. *MAppleT: simulation of apple tree development using mixed stochastic and biomechanical models.* Functional Plant Biology Volume 35, 936-950. (2008).
25. F. Grisafi, T.M. DeJong, and S. Tombesi. *Fruit tree crop models: an update.* In Tree Physiology Volume 42, pp 441 – 457. (2021).
26. S. Tu, J. Ping, H. Liu, N. Zhuang, Y. Chen, C. Zheng, H, Wan, and Y. Xue. *Passion fruit detection and counting based upon multiple scale faster R-CNN using RGB-D images.* Precision Agriculture, Volume 21. Pp 1072 – 1091. Springer Link. (2020)
27. D. Li, L. Xu, C. Tan, E.D. Goodman, D. Fu, L. Xin. *Digitization and Visualization of Greenhouse Tomato Plants in Indoor Environments.* Sensors.; 15(2):4019-4051. (2015).
28. G. Hinton, S. Sabour, N. Frosst. *Matrix capsules with EM Routing.* Google Brain, Toronto, Canada, International Conference on Learning Representations (ICLR). (2018).

29. D. Li, L. Xu, C. Tan, E.D. Goodman, D. Fu, L. Xin. *Digitization and Visualization of Greenhouse Tomato Plants in Indoor Environments. Sensors.* 2015; 15(2):4019-4051. (2015).
30. E. Fiestas, O. Ramos, and S. Prado, *RPA and L-System based Synthetic Data Generator for Cost-Efficient Deep learning Model training.* 3rd IEEE Eurasia Conference on IOT, Communications and Engineering, (2021)
31. L. Du, R. Zhang, & X. Wang. Overview of two-stage object detection algorithms. In *Journal of Physics: Conference Series* (Vol. 1544, No. 1, p. 012033). IOP Publishing. (2020).
32. Y. Wu, A. Krillov, F. Massa, W.Lo, R.Girshick, *Detectron2.* https://github.com/facebookresearch/detectron2, (2019).
33. T. Bray. *The JavaScript object notation (JSON) data interchange format.* (2014).
34. TY. Lin, M. Maire, S. Belongie, L. Bourdev, R. Girshick, P. Perona, D. Ramanan, C.L. Zitnick, and P. Dollar. *Microsoft COCO: Common Objects in Context. CoRR, abs/1405.0312.* (2015).
35. W. Carlson. *History of Computer Graphics and Animation.* Ohio State University, (2017).
36. A. Apostolico, G. M. Landau and S. Skiena, "Matching for run-length encoded strings," *Proceedings. Compression and Complexity of SEQUENCES 1997 (Cat. No.97TB100171)*, Salerno, Italy, pp. 348-356, (1997).
37. Pl@nt.Net.at (PLoS One, doi.org/j479)
38. T. Bureš. *Self-Adaptation 2.0.* In 2021 International Symposium on Software Engineering for Adaptive and Self-Managing Systems (SEAMS) pp. 262-263. IEEE. (2021).

INNOVATION CHARACTERISTICS IN CONFIGURING FARMERS' DIGITAL LITERACY ON E-REPORTING POLYGON IN WUKIRSARI INDONESIA

Hisna Shania Alya[1*], *Widhiningsih* Diah Fitria[2], and *Subejo*[3]

[123]Department of Agricultural Socio-economics, Faculty of Agriculture, Universitas Gadjah Mada

Abstract. The e-reporting polygon is an effort of the Ministry of Agriculture to maximize agricultural cultivation activities, especially on the availability of subsidized fertilizers. However, not all agricultural extension agents and farmers are familiar with this innovation. It is possible that elder farmers will experience difficulties in accessing the innovation. Therefore, this study aims to: 1) Understand the characteristics of polygon e-reporting innovation and 2) Analyze the influence of the characteristics of polygon e-reporting innovation on farmers' digital literacy in Wukirsari. This research applied a quantitative approach and was conducted in Wukirsari Village, Cangkringan Sub-district, Sleman Regency. The 60 farmers were collected through simple random sampling The data were analyzed through simple regression. The results showed that the five categories of innovation characteristics presented good values in terms of relative advantage, compatibility, complexity, trialability and observability. The characteristics of innovation also positively and significantly influenced the digital literacy level of farmers in Wukirsari Village. Nevertheless, the digital literacy of farmers needed to be highly improved particularly on gadget operation, agricultural information access, networking, and information sharing to other farmers.
Keywords: digital innovation, e-reporting polygon, innovation characteristics, innovation dissemination, digital literacy

1 Introduction

The rapid growth of technology in the agricultural sector has already started a few years ago along with the emergence of the 4.0 industrial revolution. The Ministry of Agriculture gradually started to introduce cyber extension in the extension activities. A form of cyber extension that started to be developed in the agricultural sector is initiated from obtainable extension materials in online sources, followed by the existence of agricultural machinery that is remotely operated using gadgets, to the digitalisation of agriculture.

The agricultural digitalisation has received many responses from farmers be they positive or negative. Nonetheless, there are multiple benefits that can be gained from agricultural digitalisation such as a program of e-reporting polygon prompted by the Ministry of Agriculture. Previous study mentioned that when the literacy level of farmers is low, farmers are unable to optimally functionalize technology [1]. Therefore, to survive in the midst of technological onslaught, ones must balance that with the ability to respond to the technology itself. One of the ways is by improving the literacy level.

An example of digitalization starting to be implemented today is the E-reporting polygon which has started to be implemented for farmers in the Wukirsari Village since 2021. Through the implementation of this innovation, the characteristics of an innovation that can be accepted by the community are possible to be identified. Therefore, it is important to know how far the impact of the suitability of innovation characteristics with farmers' conditions is so that farmers' digital literacy levels can be measured.

2 Literature Review

Innovation

According to Rogers [2] in Ahmad [3] innovation is an idea, thought, practice, and product (goods and services) that has novelty, new characteristic, or different approach, but not new to study. The study stated that innovation is a diffusion process of ideas or new things in an effort to transform society continuously from a place to another, from a certain period to the next period, from a certain field to another into a group of members from the social system [3].

Agricultural innovation is understood as the implementation of thoughts, practices, or or new ideas on agriculture that becomes a new way which has commercial value but efficacious in improving the living standard of the community in an equitable way [4]. Thus, it can be concluded that agricultural innovation is an idea, thought, practice, and new thing

* Corresponding author: diah.fitria.w@ugm.ac.id

in the world of agriculture that has renewal and can be functional for society. Whether or not an innovation can be easily accepted by society can be indicated from the characteristics of the innovation.

Innovation Characteristics

According to [2], there are 5 characteristics of innovation, they are: relative advantage, compatibility, complexity, trialability, and observability. Relative advantage is related to the advantages of existing innovations both in terms of economy, performance, social and convenience. Compatibility is an assessment of an innovation, whether it is compatible with existing values and needs. Complexity refers to the complexity level of an innovation to be adopted. The easier an innovation to be learnt and understood, the easier it gets to be adopted. Trialability refers to whether the innovation can be put into a trial before it is implemented. Observability is a standard of how visible is the application of the innovation to other people. The easier people see the result of the innovation, the greater chance of adoption to occur.

Digital literacy

Digital literacy is the ability to create and share in different modes and forms; to create, collaborate and communicate more effectively, and to understand how and when to properly use digital technology to support those processes. The characteristics of digital literacy do not only refer to skills to operate and use various information and communication technology devices (hardware and software platforms), but also to the process of reading and understanding the contents of technological devices and the process of creating and writing new knowledge [5]. Digital literacy can also be interpreted as an individual's ability to apply functional skills to digital devices so that they can find and select information, think critically, create, collaborate with others, and most importantly, communicate effectively [6]. Digital literacy includes various types of literacy so that it becomes more complex. Similar to media literacy, digital literacy also requires the ability to critically analyze and evaluate so as to obtain quality understanding. In digital literacy, messages in the media are constructed in such a way thus they can function optimally in even more complex communication situations. Digital literacy has a broader scale and usually addresses important issues [7].

Innovation Dissemination

Dissemination of innovation is an important communication activity in encouraging the process of dissemination and application of technology in a rural social system. The act of dissemination is a prepared act according to careful planning, through discussions or other forums that are intentionally programmed, so agreement to carry out the innovation is reached [8]. Information dissemination in agricultural digitization is required in this era. This is intended to increase farmer literacy in receiving all available agricultural information.

3 Research Methods

The method used in this study is a descriptive method with a quantitative approach. Quantitative is a method of research using numerical data that serves a quantitative function to predict population conditions and or future trends [9]. The location of this research is the Wukirsari Village which is located in Cangkringan District, Sleman Regency. The hamlets selected in the research activities were Jambu Bangkok Hamlet, Glagahwero Hamlet, Dawung Hamlet and also Karangpakis Hamlet. Some of these locations were chosen because they have implemented agricultural digitization, namely the use of polygon e-reporting in the series of Definitive Plan for Group Activities (RDKK). The respondents in this study were 60 farmers from several hamlets in Cangkringan District, Sleman Regency, Yogyakarta Special Region who have received information about e-reporting polygons from relevant stakeholders. Data collection methods were interviews and questionnaires conducted on each farmer. Based on the number of respondents, 15 respondents were taken from each hamlet using simple random sampling techniques. The data analysis method used to answer the objectives of this study is descriptive statistical analysis and also a regression test.

3.1. Resources and Facilily Support in Cangkringan Sub-district

Human Resources

Wukirsari is located in the Cangkringan District. Cangkringan has 10,929 residents and is categorized as a very dense population based on Law no. 56/PRP/1960, as it reached 750.6 people/km2. The number of men and women has a sex ratio of 98.5% This figure means that the population of men and women has an almost equal ratio in which there are 5,161 male residents and 5,240 female residents. In terms of age, the majority of the population in Wukirsari belong to the productive age category, as there are 65.71% residents in the age range of 15-65 years old, while 22.66% and 11.63% of the population are aged 0-14 years old and over 65 years old respectively. The dependency ratio shows a figure of 52.2% that indicates medium category.

A crucial aspect used to measure digital literacy is education level. Most residents of Cangkringan are senior high school graduates. The following table shows the distribution of the population based on the education level.

Table 1. Education level of rural community in Cangkringan Sub-district

Education	Population	Percentage (%)
Not-school	1,736	15.88
Not graduated from elementary school	1,005	9.20
Graduated from elementary school	1,660	15.19
Graduated from junior high school	1,459	13.35
Graduated from senior high school	3,891	35.60

Diploma/ Undergraduate/ Graduate/ Postgraduate	1,178	10.78
Total	10,929	100.00

Source: Central Bureau of Statistics (2022)

The occupation data from Central Bureau of Statistics shows that Cangkringan consists of 13.89% of students, 5.92% of non-working group, 18.65% of households, 22.10% farmers in agrocomplex sector, and 39.44% non-agricultural workers. The farmers works in 429 hectare of upland area and 757.36 hectare of dry land area.

Farmers in Wukirsari cultivate superior commodities such as mangosteen, avocado and durian. According to the [10], the total production of mangosteen in Wukirsari Village reaches approximately 2,500 quintals. The total production of avocado plant commodities is approximately 6300 quintals. Meanwhile, the total durian production in Wukirsari Village reaches 1,500 quintals.

Internet Access for Agricultural Sector in Wukirsari Village

Most of the farmers in Wukirsari use smartphone-type devices. Smartphone is a communication device that is functioned not only for making calls and exchanging messages but also for other various activities such as information seeking, entertainment seeking, studying, business running, negotiating and so on. To support the use of a smartphone, it requires a network or connection to the internet. The available internet networks options are very diverse today, starting from cellular networks that use quota from a particular provider, wi-fi, personal hotspots and others. The farmers in Wukirsari Village use smartphones in their daily lives. They maximize their gadget's function with their personal cellular networks. The village authorities have not yet provided internet access to the community. Based on the interview with the farmer groups, internet access is also not provided by the farmer group. The following data shows the internet access that is available in Wukirsari.

Table 2. Internet access in Wukirsari Village

Internet Access	Physical Condition
Wi-fi	Owned by each person
Phone tower	6 towers in good condition
Operator for service	4 towers in good condition
Signal	Strong

Source: Central Bureau of Statistics (2022)

Based on [10] the Wukirsari sub-district has 4 cellular towers. This is the largest number of cellular towers among the number of towers in all of the sub-districts in Cangkringan. Similar to the cellular tower, the number of communication operator service providers in Wukirsari is also the largest in Cangkringan as it has 6 operators in total. While for the signal, it is also considerably strong.Therefore, it will be convenient for farmers in the Wukirsari sub-district to get internet access with such availability of many operators, cellular towers, and strong signal conditions. Therefore, the use of smartphones in Wukirsari can be optimized even though internet connection cost is unsubsidized by either village authorities or farmer groups.

3.2. Characteristics of e-Reporting Polygon Innovation

e-Reporting Polygon to support smart agriculture

Smart farming is an agricultural concept using digital and information technology that aims to support agricultural activities to be more effective and efficient. This is in line with the goal of e-reporting polygons implementation in which the e-reporting polygons play a significant role in arranging Definitive Plan for Group Activities (RDKK). The arrangement of the Definitive Plan for Group Activities (RDKK) is synced with the results of land calculations listed in the e-reporting polygon. When the implementation process of polygon e-reporting runs smoothly and optimally, it will affect the results given. The land area listed will be more accurate and simplify the RDKK process. The submission process of RDKK can be said to be simplified because when submitting the RDKK proposal to the relevant agency (agricultural ministry), the accuracy of land measurement determines the opportunity of the proposal to be approved. When the land area listed in the proposal complies with the actual land area, the proposal will be accepted and the fertilizer subsidy will be received by farmers in an optimal amount.

The Characteristics of e-Reporting Polygon

Table 3. Characteristics of e-Reporting Polygon innovation

No	Indicator	Category
	Relative Advantage	
1	Innovation support the networking among farmers	Average
2	Innovation enhance farmer's income	Low
3	Innovation effectively help farmers to achieve the group goals	High
4	Innovation efficiently support on-farm based on energy, time, and fund	High
	Average	Average
	Compatibility	
5	Innovation is compatible for farmers' condition (easy to apply)	Average
6	Innovation is compatible for social norms in farmers community	Average
7	Innovation compatible for for farmers' need	Average
	Average	Average
	Complexity	
8	Innovation is applicable based on farmers' skill	Low

No	Indicator	Category
9	Innovation is accessible	High
	Average	Average
	Triability	
10	Innovation is applicable in demonstration farm	Average
11	Innovation is applicable in group practice	High
12	Innovation can be implemented by limited labours	Average
	Average	Average
	Observability	
13	The results of innovation can be observed without additional tools	Average
14	The plant growth can be used for further research	Low
15	Innovation result can be observed by anyone in group or other farmers	Low
	Average	Low
	Total Average	Average

Source: Primary data analysis (2023)

On the characteristics of innovation, the relative advantage reported is 59.38%. This value indicates that the relative advantage is in the sufficient category. In the relative advantage indicator, it is found that digitalization (e-reporting polygon) can increase the effectiveness of agricultural activities, for instance by facilitating the farmer groups in achieving their goal. In this account, the achievements of farmer groups can be interpreted as the goal of farmer groups in maximizing farming activities through the fulfillment of fertilizer subsidies needed by the group. Optimizing the implementation of polygon e-reporting will facilitate the RDKK planning process and the need for fertilizer subsidies will be met to the fullest. In addition, it is also found that the digitalization of agriculture makes agricultural activities more efficient in terms of time, effort and cost. By the use of digitalization, it will make it easier to fulfill the need for fertilizer subsidies, which in turn can also lead to more efficient time, costs and manpower needed. The innovations carried out in the dissemination process to farmers in the Wukirsari Village are considered improving compared to previous innovations. The next innovation characteristic is compatibility of innovation (compatibility) that has a value of 52.22% which can be categorized as sufficient. Meanwhile, the complexity of innovation characteristics is in the range of 52.50%. This shows that the complexity level of this innovation is still in the sufficient category and can still be well adopted by farmers since the more complicated an innovation is the slower the innovation adoption process will be. Based on the trialability and observability characteristics of innovation, it has a value of 54.44% and 54.00% which means that the ease of the innovation to be observed and tested out is categorized as sufficient. The following table will present the characteristics of agricultural innovation more clearly and concisely.

e-Reporting Polygon Dissemination Process

Based on the results of interviews conducted with farmers in the four farmer groups in the Wukirsari Village, it is reported that each of them has a regular agenda in the form of group meetings. Each farmer group has a different schedule or meeting date according to the agreement of each group. For example, the farmer group in Dawung Hamlet holds a farmer group meeting every two months on the 25th day of that month. On the other hand, the farmer groups in Jambu Bangkok Hamlet hold group meetings every night of Legi Sunday (market day in the Javanese calendar). The farmer group in the Karangpakis Hamlet meets on the 2nd day of every month. As for the Glagahwero Hamlet farmer groups, they hold regular meetings on Saturday Pahing (Javanese calendar market day) which is held once every 35 days.

In these regular meetings various activities are carried out such as rotary savings (arisan), information exchange and also discussions on other topics. Farmer group meetings play a significant role in the innovation dissemination process. This is because the pattern of how innovations spread in the Wukirsari Village began in the Wukirsari Farmer group Association meeting at the end of each month which was attended by the chairmen or representatives of the farmer groups from every hamlet in the Wukirsari Village. The findings of the Farmer Groups Association meeting will be delivered by the head of the farmer group in the farmer group meeting and followed by a discussion on the delivered information. Based on the results of the interviews, it is also stated that the dissemination of agricultural information on polygon e-reporting, was first carried out in the Farmer Groups Association meetings. The Farmer Groups Association meeting was attended by agricultural extension agents in Wukirsari Village, the heads of farmer groups from each farmer group in the Wukirsari Village, and also resource persons with relevant expertise (in this case is the Directorate General of Agricultural Infrastructure and Facilities of the Central Ministry of Agriculture). In the meeting, information regarding polygon e-reporting is delivered, from the benefits and objectives of the program to the steps in utilizing e-reporting polygons. The follow-up of the Farmer Groups Association meeting was the transfer of information to the farmer groups which was carried out by the head of the farmer group from the Farmer Groups Association meeting report. From the presentation delivered by the researcher, it can be seen that the percentage value indicates that regular meeting agenda associated with the dissemination of innovations are often carried out in accordance to what is happening in the field. Although the dissemination activity has a low percentage value and is considered in the seldom category, regular meetings play a role in every agricultural innovation dissemination activity that takes place in the Wukirsari Village.

Table 4. e-Reporting Polygon Innovation Dissemination Process di Wukrisari Village

No	Indicator	Category
1	The innovation training from government	Rarely
2	The meeting for innovation dissemination	Often
4	Whatsapp group for sharing new information	Sometimes
5	Leaflet or poster informing the innovation is well disseminated (online or conventional)	Never
6	Discussion with extension agents (*door to door)* about the innovation	Never
	Average	Rarely

Sumber: Primary data analysis (2023)

Dissemination of innovation is an essential communication effort in promoting the process of dispersion and application of technology in a rural social system. The act of dissemination is an innovation effort that is prepared according to careful planning, through discussions or other **spreading forums** that are deliberately programmed, so that there is an agreement to carry out the innovation [11]. Agricultural innovation has an important role in increasing agricultural productivity. Wukirsari is one of the sub-districts in Cangkringan District which has implemented an agricultural innovation that is currently being promoted by the government as polygon e-reporting. E-reporting polygons is an innovation that has been promoted by the Ministry of Agriculture since 2021. This agricultural innovation has the aim of minimizing frauds in the agricultural sector related to false reports of land measurement. The e-reporting polygon is also a determinant of fertilizer subsidies that will be given by the government to farmers.

Wukirsari Sub-district has 32 farmer groups spread across each hamlet. Four out of 32 farmer groups from each hamlet are the focus of this research. The four groups include the Handayani farmer group in Glagahwero Hamlet, the Subur Makmur farmer group in Dawung Hamlet, the Tunas Fajar farmer group in Jambu Bangkong Hamlet and the Sido Makmur farmer group in Karang Pakis Hamlet. All farmer groups in the Wukirsari Village are members of the Makaryo Nyoto Farmer Group Association. The Makayro Nyoto Farmer Group has been established since 2009. All agricultural activities that take place in Wukirsari Village are inseparable from the existence of farmer groups and farmer group associations (Gapoktan). This is because Gapoktan is the first target for agricultural extension agents to deliver information before going directly to farmers. Hence, the presence of farmer groups becomes a bridge for information between farmers and agricultural extension agents. So it can be inferred that the information will reach farmers well after the previous stages are carried out well.

The process of disseminating agricultural innovations that took place in Wukirsari Sub-district began at a farmer groups association meeting which was held once a month. The meeting presented all the heads of farmer groups in Wukirsari Village, agricultural extension agents of Wukirsari Village and also stakeholder associated with agricultural innovation (e-reporting polygon) which at that time was attended by the Directorate General of Agricultural Infrastructure and Facilities of the Ministry of Agriculture. Submission of information related to agricultural innovation was delivered directly by the Directorate General of Agricultural Infrastructure and Facilities of the Ministry of Agriculture. As already mentioned, every Gapoktan meeting will be attended by all farmer group leaders, so the farmer group leaders are an important source of information in the dissemination of agricultural innovations. Furthermore, the follow-up that must be carried out by the head of the farmer group in disseminating this information is by conveying important matters or information at each routine meeting held by each farmer group.

The dissemination of agricultural innovations reached farmers when farmers attended farmer group meetings. This is because at the farmer group meeting, the information on agricultural innovation from the Farmer Group Association meeting was transferred. The results of the research in table 4 state that the process of disseminating agricultural innovations in the Wukirsari Village is dominated by regular farmer group meeting activities. Regular farmer group meeting activities are often carried out by each group. Each farmer group has a different schedule of regular meetings. For example, the Subur Makmur farmer group in Dawung Hamlet holds a farmer group meeting every two months on the 25th of that month. Unlike the Tunas Fajar farmer group in Jambu Hamlet, Bangkok, which holds group meetings every night of Legi Sunday (market day in the Javanese calendar). The Sido Makmur farmer group in Karangpakis Hamlet holds a meeting on the 2nd of every month. As for the Handayani farmer group in Glagahwero Hamlet, they hold regular meetings on Saturday Pahing (Javanese calendar market day) which is held once every 35 days. The high intensity of the meeting will boost the innovation dissemination process. This corresponds with a study by [1] which states that the high intensity of meetings will improve the existing innovation dissemination process. The higher the intensity of the meetings held, the higher the innovation dissemination process that occurs.

3.3. Farmers' Digital Literacy in succeeding the Adoption of e-Reporting Polygon Innovation

Digital literacy is the ability to create and share in different modes and forms; to create, collaborate and communicate more effectively, and to understand how and when to properly use digital technology to support those processes. The characteristics of digital literacy do not only refer to operating skills and using various information and communication technology devices (hardware and software platforms), but also to the process of reading and understanding the contents of technological devices and the process of creating and writing new knowledge [5].

Table 5. Digital Literacy of Farmers in Wukirsari Village

No	Indicator	Category
1	Ability to operate gadget	Low
2	Ability to use internet to access agriculture information	Low
3	Ability to find new agriculture information about in society	Average
4	Ability to analyze agriculture information	Average
5	Ability to evaluate the information	Average
6	Ability to communicate in digital space for networking	Low
7	Ability to use digital space to share agriculture information to other farmers	Low
	Average	Average

Sumber: Primary data analysis (2023)

The digital literacy of farmers in Wukirsari is evaluated from the ability to create and share in different modes and forms; to create, collaborate and communicate more effectively, and to understand how and when to properly use digital technology to support those processes. Researchers interpret the tabulation results using a number range with 0-20% (very low), 21-40% (low), 41-60% (average), 61-80% (high) and 81-100% (very high). Based on table 4 digital literacy has a value of 41.19% and belongs in the sufficient category. The ability of farmers to operate gadgets and also the internet to access information about agriculture is still relatively low. Even the ability of farmers to communicate through digital space and share agricultural information with members of other farmer groups is still relatively low.

Based on Table 5 digital literacy has a value of 41.19% and is in the sufficient category. The ability of farmers to operate gadgets and also the internet to access information about agriculture is still relatively low. Even the ability of farmers to communicate through digital space and share agricultural information with members of other farmer groups is still relatively low. Based on the results of interviews conducted by researchers, it is found that farmers are more comfortable communicating directly face to face.

The Influence of Innovation Characteristics towards Farmers' Digital Literacy

Innovation characteristics significantly influence farmers' digital literacy.

Table 6. Simple regression analysis (Backward method) of Innovation characteristics influence on farmers digital literacy

Variable	Regression Coef.	t-cal	Sig.	Result
Characteristics of -Reporting Polygon Innovation	0.253	4.508	0.000	**
Constant	2.221			
R squared	0.259			
Note: ** means significant at alpha 5%				

Source: Primary data analysis (2023)

The simple regression analysis shows that independent variable (X) or Innovation characteristics of e-reporting polygon gives significant influence on digital literacy of farmers in Wukirsari village. The result is based on the value of significance of 0.000 that is less than alpha of 5% or 0.05.

$$Y = 2.221 + 0.253 X$$

The regression equation describes a positive influence of the innovation characteristics towards farmers' digital literacy. The improvement of innovation characteristics will support the enhancement of farmers' digital literacy. When innovation characteristics value increase in a unit, the digital literacy will also raise in a 0.253 unit. The better the characteristics of an innovation, the more it will influence the level of digital literacy that farmers have. This is because farmers will more easily understand existing innovations if the innovations are easy to understand and suit the farmers' conditions

Farmers have strong understanding on how *e-reporting polygon* helps them to access the subsidized fertilizer. Thus, the implementation of this program is applicable because farmers have understood that the orientation of *e-reporting polygon* innovation is on the avalilability of input support from government such as fertilizers. The numbers of the fertilizers will be based on area report on *e-reporting polygon*.

4 Conclusion

The characteristics of e-reporting polygon innovation consisting of *relative advantage, compatibility, complexity, trialability, and observability* was sufficient (advantageous and easy to understand, try, and evaluate) for farmers. In addition, the characteristics of innovation has a significant influence of farmers' digital literacy. The enhancement of the characteristics of e-reporting polygon innovation will improve the digital literacy of farmers on how they understand and implement the digital system. It can help farmers to improve their digital literacy particularly on gadget operation, agricultural information access, networking, and information sharing to other farmers.

Since e-reporting polygon is beneficial for agricultural input support database, extension agents' role in introducing and understanding the innovation characteristics need to be improved. Farmers should have deep understanding about the benefit after finishing e-reporting polygon. Thus, they will be motivated and encouraged to learn more about the innovation. Thus, a training how to input the data on e-reporting polygon should be more intensive until their skills escalate. For the government, the innovation characteristics should be tested before the innovation is disseminated to the farmers as the obligation that will

https://doi.org/10.1051/bioconf/20238006005

affect the distribution of agricultural input such as fertilizer.

References

1. D. Apriliya, O. Anwarudin, Nazarudin. J., Inovasi 3 (2020)
2. E.M. Rogers, New York Free Press (1983)
3. Y. Ahmad, J. Agr. Sci **6**, 2 (2016)
4. Sujimin, L.M. Kolopaking, S.Sjaf, J. Sosiologi Pedesaan (2019)
5. I. Kurnianingsih, Rosini, N. Ismayati, J. Pengabdian Kepada Masyarakat **3**, 1 (2017)
6. A.H. Naufal, J.Perspektif (2021)
7. A. Restianti, Gunahuma **1**,1 (2018)
8. N.G. Muntaha, A.Amin, J. Pendidikan dan Konseling 5, 2 (2023)
9. A. Mukhid, Jakad Media Publishing (2021)
10. Central Bureau of Statistic of Sleman Regency (2022)
11. Muntaka, Alfauzan, (2023)

Multi-Agro: Design and Development of a Cartesian-based Self-Plantation Robot for Home-Scale Gardens with Mobile App

Endra Pitowarno[1*], *Adytia* Darmawan[2], *Fatih* Ali Fikri El Munba'its[3], and *Bagus* Pramuditya Affandy[4]

[1,2,3,4]Mechatronics Engineering, Department of Mechanical and Energy, Politeknik Elektronika Negeri Surabaya, Indonesia

Abstract. The increasing diversion of agricultural land in Indonesia has not been followed by a decrease in the need for food in society. This change in agricultural land use poses a serious threat to food security. To solve this problem, a home-scale farming tool called Multi-Agro has been developed, which can planting independently and automatically. This tool uses a cartesian method-based robot manipulator to determine the exact coordinate points during the planting, maintenance, and harvesting process of the plants. Plant watering time can be set on the app automatically so that plant health is maintained. In addition, a weed detection system using a camera can also improve plant health. With the automatic system and monitoring that can be accessed through the app, this independent farming becomes more efficient and does not require much time from agricultural actors. Based on the experiment conducted in a real-physical miniature model of home-scale farming it can be concluded that the proposed project has a significant improvement in smart urban farming. Hopefully, this research can help to overcome the increasing need for food even though agricultural land is increasingly limited.

1 Introduction

Currently more than 50% of the world's population lives in cities or downtown, and predicted by 2030 it will reach 80% [1]. The urbanization process leads to an increasing demand for food, especially in urban areas. However, the conversion of agricultural land in urban areas is increasingly prevalent today [2]. Agricultural land conversion is the process of shifting the function of agricultural land from agricultural use to other uses, on part or all of the land area which generally has a negative impact on the environment and on the potential of the land.

In general, the rate of land conversion is associated with the rate of population growth which results in increased fulfilment of needs based on land use, such as settlements and other public facilities. The process of land conversion is unavoidable in every developing region. Developing regions usually have a high population growth, followed by an increase in the need for land for settlements and other public facilities including for industry [3].

Agricultural land, especially paddy fields, is the land that has experienced the most conversion of functions. Changes in the use of agricultural land are a separate threat in achieving food security [4]. From the Badan Pusat Statistika (BPS), food crop production in Indonesia continues to decline every year [5].

One of the factors causing the decline in food crop production is the transfer of land from agricultural land to non-agricultural land. Reflecting on these problems, the idea of making Multi-Agro was born where one of the sectors that was focused on was plantations. It is called a self-planting robot with the aim that food crops can be planted independently by individuals by utilizing existing technology so that it does not require a lot of time and effort considering that the average population in Indonesia according to the Badan Pusat Statistika (BPS) works for 40.96 hours per week, so that if you grow your own traditionally it is not possible [5].

Self-plantation robot here is designed to be able to carry out planting with several basic treatments so that the yield can be maximized. With board media for planting, several coordinates will be set for the planting point by utilizing the use of a motor, the robot can have 3 axes to reach all parts of the media. Determination of the coordinates themselves uses a method commonly called the Cartesian method whose working principle is almost the same as the concept of a robot manipulator. Vacuum technology will be used here for the process of planting seeds at predetermined coordinates. Assisted by the Arduino and Raspberry microprocessors, this tool will be able to monitor the process using a simple mobile application. Where the data that has been obtained from sensor readings will enter the cloud and will be read by the application.

As more people move to cities, the agricultural area is being converted into urban areas. This has led to a reduction in land available for gardening and traditional farming. Robots for home-scale gardening offers a solution by enabling individuals to grow crops

* Corresponding author: epit@pens.ac.id

efficiently in limited spaces, such as balconies or rooftops, in urban environments. By integrating robotic technology into lawn care, it is hoped that the innovations that homeowners can gain several benefits, including time efficiency, better plant maintenance, and increased resource efficiency.

2 State of The Art

In this research, a combination of several idea in intelligent urban farming is carried out by regarding the successful of previous researches such as conducted by [6], [7] and [9]. These inspirations had been combined in this research and enhancing with a function of remote monitoring and automated farming via a mobile system such as in [8]. Once the physical system like the automation in urban farming carried out onto the web users can easily manage the farming everywhere and anytime.

3 Materials and Methods

This section describes the methodology used to implement the proposed robotic solution, as well as the various materials and inputs used.

3.1 Self-Plantation Robot Design

This robot uses linear movement in three position axes (X, Y, Z), with the length of each axis being X = 830 mm, Y = 1230 mm, and Z = 800 mm. The mechanical structure consists of several parts that need to be assembled and calibrated according to the desired design. The robot moves on an aluminium profile, which allows it to move smoothly and experience minimal friction. The components of the robot design can be seen in the Fig. 1. and explained as follows:

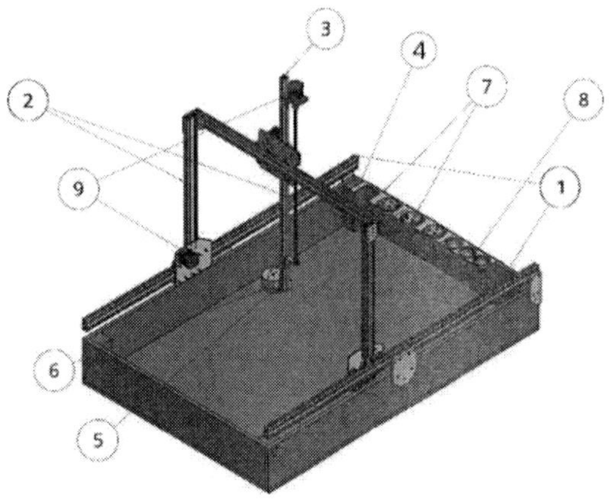

Fig. 1. Self Plantation Robot Design

Table. 1. Descriptions of Self Plantation Robot Design

No.	Descriptions
1	X-axis gantry track : Aluminum profile that supports the movement of the robot in the X-axis direction.
2	Y-axis path support : Consists of an aluminum profile that serves as a support for the robot's movement in the Y-axis direction. In geometric form, it can be likened to a lifting bridge connecting two X-axis profiles that in turn support the robot's Z-axis and electronics box. Movement on the path support allows reaching the Y coordinate in Cartesian Position (Z, Y, X).
3	Z-axis gantry track: Aluminum profile that supports the movement of the robot in the Z-axis direction.
4	Y-axis gantry track: Aluminum profile that supports the movement of the robot in the Y-axis direction.
5	Tool Holder: A tool holder for mounting tools to perform different jobs, located at the lower end of the Z-axis.
6	Camera: Camera for weed detection, located next to the tool holder.
7	Tools Bay: contains tools on the robot. Using magnets to be attached
8	Seeds Bay: a place that contains plant seeds that will be planted
9	Stepper motor: actuators to move the robot parts

The detailed size of the self-plantation robot can be seen in the Fig. 2 as follow.

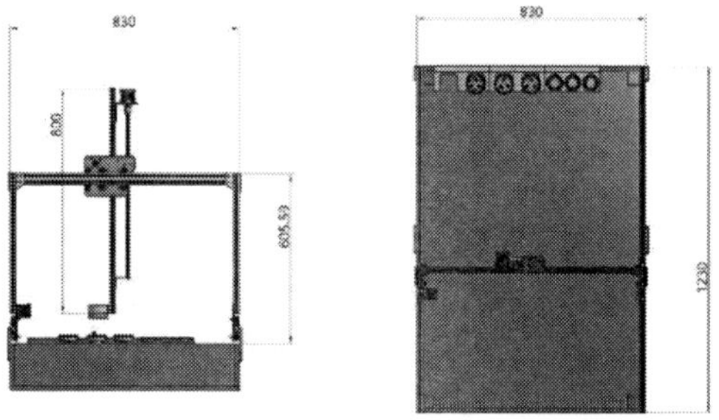

Fig. 2. Dimension of Self Plantation Robot Design

3.2 Hardware Design and Construction

The self-plantation is scheduled to be active for 24 hours and performs a series of processes. The cycle starts with seed planting to treatment until harvest time arrives, several systems in every 24 hours are:

a. Receiving and sending data from sensors to the cloud.
b. Processes the images before they are processed by one of the Convolutional Neural Network.
c. Generate and forward commands to actuators.

https://doi.org/10.1051/bioconf/20238006006

The purpose of data processing in the current system is to determine the threshold of moisture content in the soil, determine the irrigation dose for each individual plant so as not to overwater the plant, and periodically monitor the plants for maximum harvest.

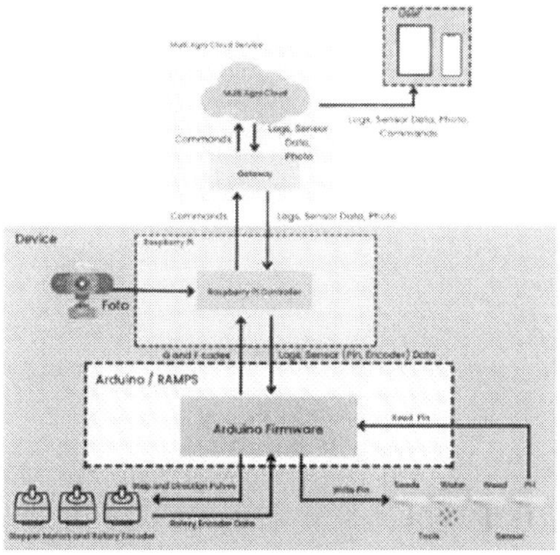

Fig. 3. The system design proposed

Details on the control part of this self-plantation robot can be seen in Fig.4. where Arduino uno as a microcontroller connected to CNC Shield. In CNC Shield, the DRV8833 motor driver is placed as the driver of each stepper motor that works in each axis. While the 2-channel relay module here will function as a state regulator of the tools on the robot, such as activating the vacuum when planting seeds, then activating the dc motor for plant watering needs. This 2-channel relay module will enter the digital pin CNC Shield which is connected to Arduino uno.

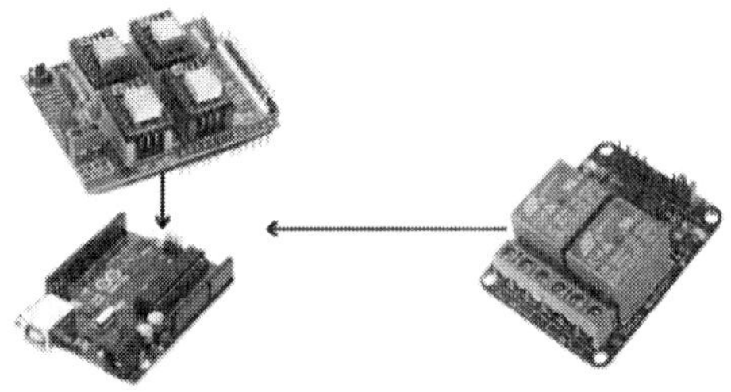

Fig. 4. Details on Self Plantation Robot Control

3.3 Robot Working System

In the self-plantation robot, the seed planting system is carried out automatically using a roller system with a motor for each axis located on the side of the planting medium and the aluminium frame of the robot. Each process carried out by this robot uses a different tool head. For the seed planting process, the tool head used is vacuum because to take seeds is done with a suction or vacuum system which then the robot will move to the planting coordinates and insert the seeds into the soil at those coordinates. In the watering process also has almost the same concept, it's just that there is a change of tool head because for watering there is its own so that in the design for this tool head the water can spread evenly and the mineral content of the plant can be said to be sufficient. For more details can be seen in the flowchart below.

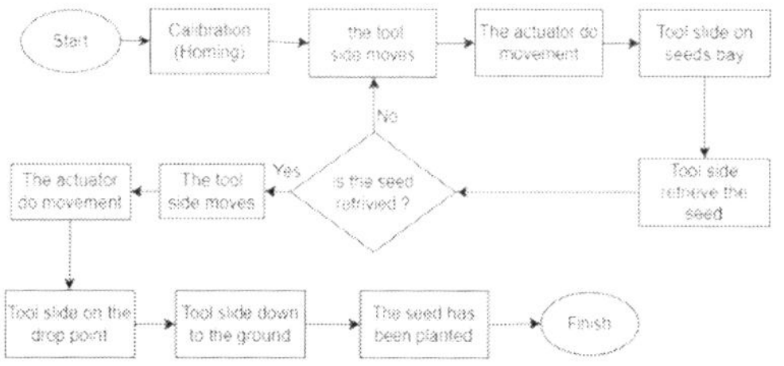

Fig. 5. Seed Planting Flowchart

The seed planting process will begin with the robot performing a calibration process (homing) to reach point 0 then the seed tool robot will aim above the seeds bay to take the seeds by sucking the seeds. When the seeds are successfully taken, the tool side will move to the planting point according to the existing coordinates and will go down to the ground to plant the seeds.

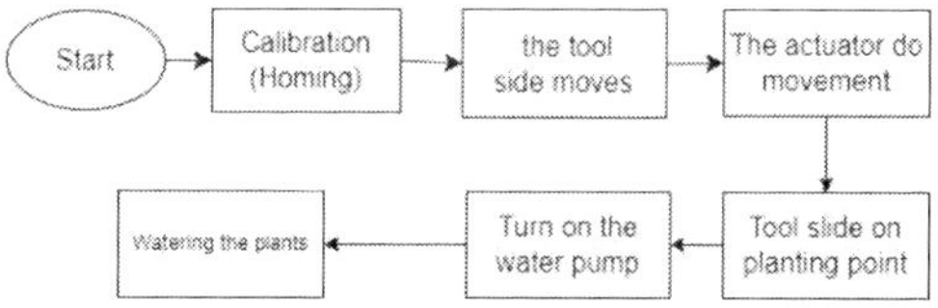

Fig. 6. Watering Flowchart

In addition to planting and watering, this tool will perform treatments such as periodically checking the soil moisture at the coordinates where the seeds have been planted. To maintain this moisture, if the coordinates that are checked are detected to be less moist, then proceed with the action of the plant watering process at that coordinate. In this robot, there will also be an app to support the effectiveness used as monitoring. Planters will be able to determine which points to plant so that they can have many options for planting layouts and seeds. When the seeds have been planted, the user app can monitor how the soil moisture was the last time the robot watered, then the health of the plants will also be displayed in the app.

3.4 Cartesian Method

Robot manipulators have basic structures such as cartesian, cylindrical, polar and angular configurations. Cartesian configuration states that the position of a point is done using cartesian coordinates (X, Y, Z), where the value taken by the end of the arm is a shared variable, corresponding to the coordinates. The cartesian method will be used as a robot mechanical system where the robot movement will be linear or prismatic in all three (X, Y, Z). Calculations using the cartesian method are used in determining the coordinates on the planting board, later each coordinate that has been determined on the planting board will become a seed planting point. Using the cartesian method or coordinates this robot will be able to determine automatically when planting. The use of the Cartesian method itself is inspired by its accuracy in the movements that can be done using this method and the ability to adapt to the size and app of planting board [6].

4 Apps

The system of this self-plantation robot has 3 stepper motors for the X, Y, Z axis with 4 sensors, each consisting of 3 limit switches. The sensors mentioned earlier will be connected and controlled through Arduino Uno which is then communicated with Raspberry Pi 4 using serial. Raspberry Pi 4 is used to detect weeds in the field where the detection results will decide the action of the destroyer for the weeds themselves. Data from sensors will be channelled to the cloud connected to the app so that it can be monitored in the app. As a database requirement in this research, utilizing Firebase as a data storage and communication from the microcontroller on the device with the smartphone app. Firebase itself is **mBaaS** (Mobile backend as a service) which means Firebase provides various services that will be provided in various backends for smartphone apps such as databases, notifications, machine learning, analytics, and so on. In this research, what is used is a database as a store of sensor values sent from the microcontroller [8].

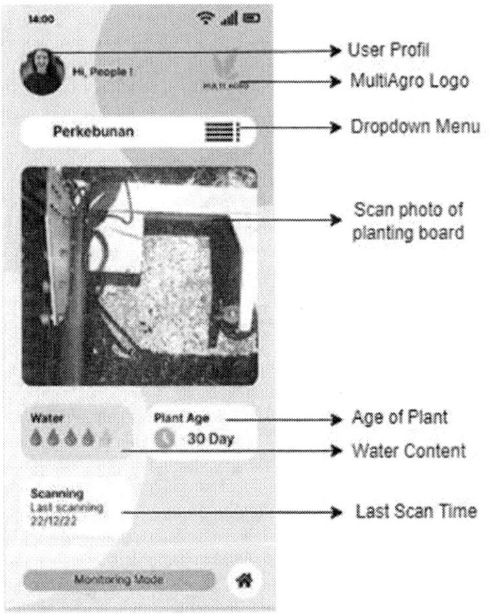

Fig. 7. Display of monitoring screen

The process of making apps begins with the selection of software and the usefulness of future apps. Because of the increasing number of iPhone users in Indonesia, the making of this app uses Flutter where one of the advantages of Android Studio is that when making apps it can be directly used for IOS, Android, and Web platforms [8]. The first step is to create a User Interface which is one of the important aspects in the app and can support user interest later. There are several menus offered for this self-plantation robot, as shown in Fig. 7.

On the monitoring screen there is some information that the user will get, such as soil moisture contained in the water menu indicated by several levels. In addition, there are the results of the last monitoring scan and when the monitoring was carried out. In addition to the monitoring menu in this app can also be used to plan seed planting, starting from selecting the row to be planted and which seeds to plant for that row as shown in Fig. 8.

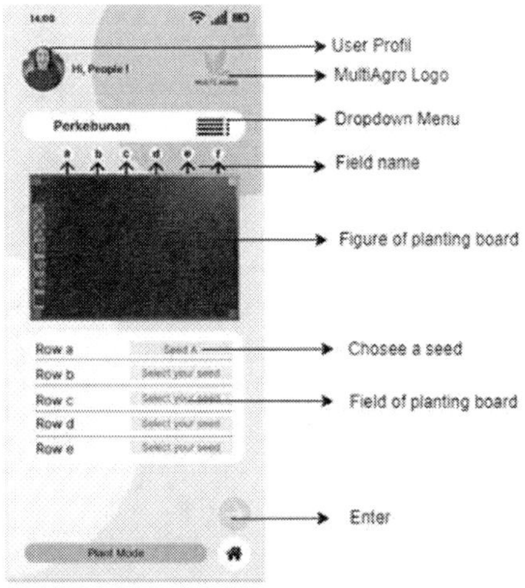

Fig. 8. Display of control screen

5 Results and Discussion

5.1 Realization of the Planting Board

In the realization process of the 3D design contained in the previous chapter, the self-plantation robot is made from plywood with a thickness of 1.5 cm. As for the track and gantry, it uses aluminium profile $20mm \times 40mm \times 1230mm$ for the track and $20mm \times 40mm \times 830mm$ for the Y-axis and $20mm \times 40mm \times 700mm$ for the Z-axis.

(a) (b)

Fig. 9. Media Planting Place

After the mechanical assembly of the self-plantation robot is carried out, it will continue with the stage of giving soil and electrical design. In the current case the installed electronics are not what they should be, because before they are completely refined the self-plantation robot will go through a trial process with several electronic components that have been proposed previously. The robot's mechanical transmission uses toothed belts and corrugated pulleys, which result in smooth movement on the axis. This transmission does not require ongoing maintenance and can work efficiently in humid environments with varying temperatures. The cable system is housed in a flexible cable carrier, which reduces wear and tear and protects it from sunlight and water.

5.2 Timing of Planting Robot's Movements

Robot's movements are carried out by giving commands from a python program that can be accessed through the MultiAgro app. In the planting test, the time is measured for each planting, while the distance of the vacuum tool to the seed is ± 1mm. In this test, the researcher made a determination where the vacuum tool will go down as far as - 324mm from point 0 of the Z axis to take the seeds contained in the seed bay, then will go up to point -200.32 for the Z axis and walk to the planting point, finally the vacuum tool will go down to plant seeds as far as -440mm from point 0 of the Z axis, which means the depth of the planting point is 25mm below the surface of the land, for the process can be seen in Fig. 10.

(a)

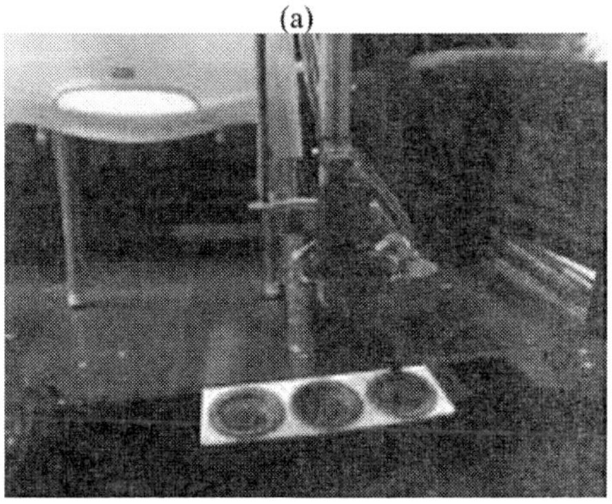

(b)

(c)

Fig. 10. (a) (b) (c) Processing of planting the seed

Table 2. is a planting test by measuring the time at each planting point and carried out at 9 planting points.

Table. 2. Measurement data of planting the seed

No.	Position	Time (sec)
1.	Vacuum tool picks up seeds	09.60
2.	Planting seeds at point A1	08.37
3.	Vacuum tool picks up the seed	08.52
4.	Planting seed at point A2	07.68
5.	Vacuum tool picks up the seed	08.73
6.	Planting seed at point A3	08.96
7.	Vacuum tool picks up the seed	08.86
8.	Planting seed at point B1	10.13
9.	Vacuum tool picks up the seed	09.07
10.	Planting seed at point B2	09.18
11.	Vacuum tool picks up seeds	08.98
12.	Planting seed at point B3	08.99
13.	Vacuum tool picks up the seed	08.48
14.	Planting seed at point C1	12.08
15.	Vacuum tool picks up seeds	10.31
16.	Planting seed at point C2	09.81
17.	Vacuum tool picks up seeds	10.13
18.	Planting seed at point C3	09.93
Total Timing		149.96
Average		08.33

The time required to complete the seed planting process is about 2 minutes 30 seconds for 9 planting points at a moderate speed. The time required is quite fast. But sometimes the robot is unstable at high speeds.

5.3 Timing of Watering Robot's Movements

The watering process is carried out based on a predetermined real time, twice a day. Watering tests are carried out by giving commands from the python program. In the watering test, the time is measured for each watering, while the distance of the vacuum tool to the flush point is ± 310mm. In this test, researchers made a determination where the vacuum tool would be at - 157.32mm from point 0 of the Z axis to carry out the watering process at all planting points.

Table 3. is a watering test by measuring the time at each watering point and carried out at 9 flush points.

Table. 3. Calculation Data of Watering

No.	Position	Time (sec)
1.	Water tool flushes point A1	03.22
2.	Water tool flushes point A2	05.19
3.	Water tool watering point A3	06.67
4.	Water tool watering point B1	05.98
5.	Water tool watering point B2	06.01
6.	Water tool watering point B3	05.75
7.	Water tool watering point C1	06.27
8.	Water tool watering point C2	06.03
9.	Water tool watering point C3	05.99
Total Timing		51.11
Average		05.67

The time taken to complete the watering process is less than 1 minute for 9 planting points at a moderate speed. The time required is very fast. But sometimes the robot is unstable at high speeds. The instability is due to the use of one motor on the X-axis makes the movement less stable at high speeds. movement is less stable at high speeds.

6 Conclusion and Recommendation

The real-physical miniature model of home scale farming with a fully automation features including web-based remote management system has been developed in this research and shows a significant improvement in the term of intelligent or robotic-based smart farming. Users can manage their own home-scale gardens from anywhere and anywhere. However, after finishing the (first step) research so far there is an important question, is it possible to expand and extend to a wider area as well as in the form of vertical farming. The using of fully autonomous robot in urban farming seems to be more challenging in the future.

Acknowledgement

We would like to express our deepest gratitude to Robotic Vehicle Mechatronics Engineering team of the Politeknik Elektronika Negeri Surabaya for supporting this research funding and process.

References

[1] N. Bakker, M. Dubbeling, S. Guendel, U. Sabel-Koschella and H. de Zeeuw, "Growing Cities, Growing Food: Urban Agriculture on the Policy Agenda: A Reader on Urban Agriculture," pp. 677-97, 2017.

[2] U. Nation, "Reconsidering Rural Development ", UN World Social Report, 2021

[3] R. Janah *et. al.*, "Alih Fungsi Lahan Pertanian dan dampaknya terhadap kehidupan Penduduk Kecamatan Sayung Kabupaten Demak," Agrisocionomics - Jurnal Sosial Ekonomi Pertanian, 2017.

[4] R. W. S. W. S. &. Y. E. Sari, "Identifikasi Dampak Alih Fungsi Lahan Pertanian Ke Non Pertanian Untuk Perumahan," *Kajian Ruang,* vol. 1, p. 255, 2021.

[5] B. P. Statistik, "Indikator Pertanian 2020," Badan Pusat Statistik, Indonesia, 2020.

[6] M. Moraitis, K. Vaiopoulos, and A. T. Balafoutis. "Design and Implementation of an Urban Farming Robot". Micromachines 2022, 13, 250. https://doi.org/10.3390/ mi13020

[7] C. Wang, Y. Wan and X. Liang, "Model establishment with Stribeck friction and fuzzy PID control for Cartesian picking robot system," Computer Science, Chinese Automation Congress (CAC), *IEEE,* 2017.

[8] A. A. &. Y. M. Rafiq, "Penerapan Firebase Realtime Database pada Pototype Aplikasi Pemesanan Makanan Berbasis Andoid," *Jurnal Teknik Informatika dan Sistem Informasi,* vol. 3, pp. 397-406, 2018.

[9] C. Choque, E. M. F. Sorogastúa and R. S. P. Gardini, "Efficient Implementation of a Cartesian Farmbot Robot for Agricultural Applications in the Region La Libertad-Peru," IEEE ANDESCON, 2018.

Performance of A Portable NIR Spectrometer to Distinguish Coffee Species Based on Qualitative Chemometric and Artificial Neural Network (ANN) Models

Agus Dharmawan[1], *Rudiati* Evi Masithoh[1] , and *Hanim* Zuhrotul Amanah[1]

[1] Department of Agricultural and Biosystems Engineering, Faculty of Agricultural Technology, Gadjah Mada University, Indonesia

Abstract. A wide range of genetic cultivars of coffee and their characteristics determine consumer preference and increase industrial actors' awareness of production and marketing. The primary objective of this study is to develop a method to distinguish coffee species based on spectral characteristics acquired from a portable near-infrared spectrometer. The performance of this spectrometer in addressing classification problems is evaluated by the classification accuracy obtained from qualitative chemometrics, such as PCA and LDA, and artificial neural networks (ANNs) models. In this study, the instrument was successfully used and gained moderate accuracy for discriminating two coffee species, Arabica and Robusta, from Temanggung and Toraja. The accuracy was fair and achieved greater than 75%. Therefore, the instrument can be implemented as it provides simple, real-time, and in-situ analyses and can reach reliable results.

1 Introduction

Coffee is one of the leading plantation commodities in Indonesia, which has become a vital role in economic development. In 2021, the production of coffee reached 786.2 thousand tons, with a majority of 99.32% produced from smallholder plantations (1.26 ha). The export value of coffee was 387 thousand tons, with foreign exchange earnings of US$ 859 million [1]. Mostly, 97.17% of the exported coffee was Arabica WIB/Robusta OIB, not roasted, and not decaffeinated. Two species of coffee are widely cultivated and marketed in Indonesia, including Robusta (*Coffea canephora*) and Arabica (*Coffea arabica*). Most exported coffees are not roasted, not decaffeinated, wet-processed Arabica coffee beans and dry-processed Robusta coffee beans. Arabica coffee grows well at an altitude of 1000–2100 m above sea level with an air temperature of 18–22^{0}C and an annual rainfall of at least 1500 mm, whereas Robusta coffee grows well in a hot and humid climate at an altitude of 100–1000 m above sea level with an air temperature of 22–26^{0}C and an annual rainfall of at least 2000 mm [2], [3].

The quality of coffee varies as a result of differences in cultivar genetics and locations where they are grown, affecting their taste and making the price and popularity of coffee differ considerably [4]. Coffee quality is regularly examined based on chemical, sensory, and physical attributes. The chemical assessments of coffee beans are determined from the chemical composition contained [5]. The organoleptic properties relate to coffee aroma, flavor, sweetness, acidity, and overall taste. The physical characteristics include shape, thickness, weight, and color [6]. Coffee characteristics and consumer preferences have become significant concerns in coffee industries, so industrial actors need to increase their awareness by confirming coffee species, varieties, and origins [4].

In distinguishing coffee species, the characteristics through visual aspects mostly play an important role [7]. The traditional recognition is carried out by experts/trained inspectors, is relatively high in cost, and the results may vary depending on the persons inspecting the food items. The physical assessments are difficult to identify and look similar because of their natural variability [8], [9]. Therefore, the accuracy of their inspections is frequently influenced by several factors, such as experts' moods, perceptions, and fatigue [10]. Spectroscopic methods in food analysis have now been developed and massively employed to address quantitative and qualitative issues. The quantitative issue is related to determining the chemical and compositional quality of foods. The qualitative issue is used to solve classification problems, including confirming geographic authenticity, distinguishing between species, the occurrence of adulteration, and the presence of defective beans [11]. The measurement procedure consists of (1) spectral data acquisitions, (2) data pre-processing to improve spectra, (3) building a classification model, and (4) evaluating the model's accuracy.

This study aimed to use a spectroscopic method, particularly in the NIR region, to distinguish coffee species. A portable pocket-sized near-infrared spectrometer (NIRS) was used to collect spectra from two species of coffee, Arabica and Robusta. Compared

https://doi.org/10.1051/bioconf/20238006007

to a benchtop NIR spectrometer, this portable spectrometer provides advantages such as non-destructive and in-situ analyses. The development has some critical factors, including measurement accuracy and high performance. But the implementation for industrial levels has increased interest, transforming large stationary analytical instruments into reliable-lightweight tools [12]. In this study, the discrimination between samples was carried out by qualitative chemometric models, including principle component analysis (PCA) and linear discriminant analysis (LDA), and a non-linear classification model, artificial neural network (ANN). Features in spectroscopic data is reduced using PCA and obtain lesser-dimensional data called principle components (PCs), which will become inputs to the classifiers, LDA and ANN.

Several studies have performed excellent classification results in distinguishing species of coffee using NIR spectroscopy. [13] used NIRS at a wavelength range of 1100–2500 nm to discriminate between arabica and robusta pure coffee varieties and blends of varied varietal composition. FTIRS was also used by [14] to discriminate coffee species by using PCA and classical discriminant analysis. [15] used FTIRS combined with KNN, SVM, PLS-DA, BP-ANN, and ELM to distinguish coffee species – Arabica and Robusta. The application of a portable spectrometer has only been found in a limited number of studies. [16] employed portable near-infrared spectroscopy to determine the quality of Arabica coffee by identifying and quantifying the adulterants with Robusta coffee (at different roasting levels) and adulterants with corn, peels, and sticks; the study combined two chemometric methods: partial least squares (PLS) and PCA. [17] employed handheld NIR spectroscopy combined with PLS-DA to evaluate directly cup profiles in roasted and ground coffee blends.

2 Methodology

2.1 Data Collection

The samples used in this study were green beans purchased from local markets in East Java (for Temanggung) and South Sulawesi (Toraja). The Arabica and Robusta samples were normal-shaped beans from full-wash and natural coffee processing, respectively, and had been cleaned from the remaining dry epidermis and endocarp/parchment. Spectral data were acquired using a pocket-sized connected micro-spectrometer (Consumer Physics, Inc. (SCiO), St. Cloud, Minnesota, USA). This spectrometer (Fig. 1) was integrated with user's mobile app to scan and real-time display and equipped with a back-end cloud to store the sample databases. This spectrometer worked in the NIR region, particularly 740 – 1070 nm, and generated 331 data points. A total of 2400 spectral data (2 species x 2 origin x 600 samples) were downloaded and stored in CSV format. The room temperature was maintained at 20±1 ^{0}C during spectral measurement. We only use raw spectra for further analyses.

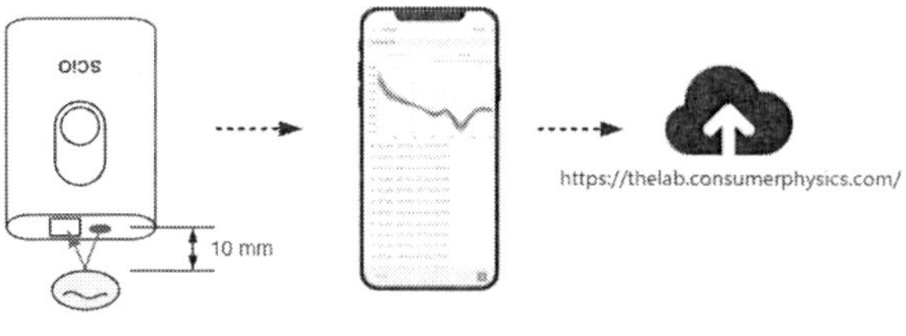

Fig 1. Scheme of data collection using a portable NIR spectrometer

2.2 Qualitative Chemometric Models

PCA was used to compress spectroscopic data and reduce its variable from a high dimensionality to a lesser dimensionality. The transformed data was called principle components (PCs), which were linear combinations of the original data. The number of PCs obtained for further analyses was obtained from those close to 100% of the explained variance. The new data generated (PC scores) was scaled using Z-score normalization.

In this study, LDA would provide more class separability between the sample classes. In contrast to PCA, LDA required matrix *Y*, consisting of two class labels for two coffee species, Arabica and Robusta. The LDA results were evaluated by plotting LDA values against label classes of samples to visualize the samples' distinguishment. The model effectiveness was examined using *k*-fold cross-validation and confusion matrix; see sections 3.4 and 3.5.

2.3 Artificial Neural Networks Model

The ANN model consisted of an input layer, two hidden layers, and an output layer. The structure of the ANN model is depicted in Fig 2. The number of input nodes varied depending on the number of components generated in PCA. The five nodes were defined for both hidden layers. The output layer would classify two label classes, label 0: Robusta and 1: Arabica. The one-hot encoding technique converted the class label into a new categorical column, assigning a binary value of 1 or 0. The model effectiveness was evaluated using *k*-fold cross-validation. The confusion matrix, ROC curves and AUC values were also performed to visualize and check the ANN classifier's performance.

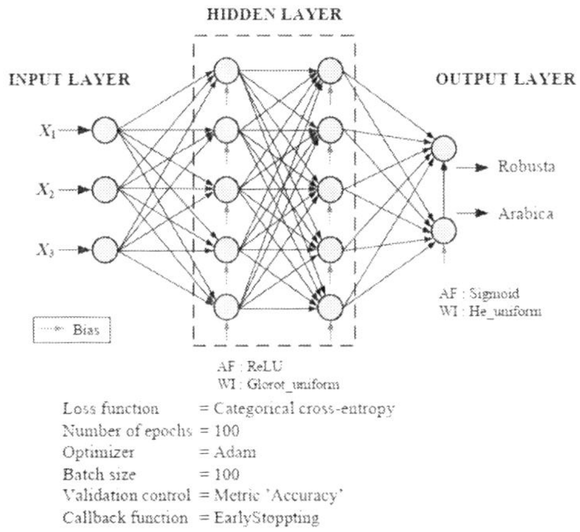

Fig 2. Structure of ANN classifier

2.4 Cross-Validation

The dataset was partitioned into a training set (randomly 2/3 of total samples) and a testing set (1/3 of random samples). *K*-fold cross-validation was used to evaluate the model's ability in a certain data to predict new data. This validation method used the split data from a training set. *k* referred to the number of folds (subsets) in the given data, the one-fold for testing and leaving the remaining $k - 1$ folds to train the model. The subsets were trained and validated in *k* iterations. The model's effectiveness was performed in mean+SD accuracy from all iterations.

2.5 Performance of Classification Models

A confusion matrix (Table 1) was used to evaluate the classification results of LDA and ANN. First, we must compute a set of predicted targets and compare them to the actual targets [18]. The predicted targets represented the values of the label classes as a result of the model, whereas the actual class represented the original values of the initial label classes [19]. The output TN (True Negative) showed the number of negative examples classified accurately, the TP (True Positive) indicated the number of positive examples classified accurately, the term FP (False Positive) value performed the number of actual negative examples classified as positive, and the FN (False Negative) was the number of actual positive examples classified as negative [20]. Positive samples were defined as classified Robusta beans, and negative samples represented classified Arabica beans.

Table 1. Scheme of two-class confusion matrix

		Predicted class	
		Positive	Negative
Actual class	Positive	TP	FN
	Negative	FP	TN

The evaluation of the model was calculated using performance metrics. Accuracy (AC) computed the percentage of correctly classified samples, see Eq 1, where *n* represented the total number of samples. Error (ER) was the percentage of wrongly assigned samples, and the formula is given in Eq 2. Recall (RE) was computed by the ratio of correctly classified samples to all positive samples, see Eq 3. Precision (PR) was the ratio of correctly classified samples as positive to all positively classified samples, as given in Eq 4. Specificity (SP) represented the ratio of incorrectly classified samples to all negative samples, see Eq 5. F1-score (Eq 6) was a weighted harmonic mean of PR and RE [21][22]. A good classifier was able to generate accuracy ideally close to 100% and error nearest to 0%. The specificity, precision, recall, and F1-Score were also declared 'good' if they were close to 100% or 1.00.

$$AC = \frac{TP + TN}{n} \quad (1)$$

$$ER = \frac{FP + FN}{n} \quad (2)$$

$$RE = \frac{TP}{TP + FN} \quad (3)$$

$$PR = \frac{TP}{TP + FP} \quad (4)$$

$$SP = \frac{TN}{TN + FP} \quad (5)$$

$$FS = \frac{2 \cdot R \cdot P}{R + P} \quad (6)$$

3 Experimental Results and Discussion

3.1 Spectral Profiles of Arabica and Robusta Coffee

A portable NIR spectrometer used in this study worked in the range of 740–1070 nm; the averaged spectra acquired are shown in Figure 3. In the NIR region, molecular vibration occurs when light strikes the samples and exhibits absorption bands in the molecules. Thus, in this region, we can also use it for chemical and compositional analyses of food products. The main absorption bands occurred at 900–1000 nm for the 3rd overtone of CH, CH_2, and CH_3 groups corresponding to phenolic compounds and caffeine content [17], [23][16]. The main difference between Arabica and Robusta coffee is their taste; Arabica coffee is more acidic, whereas caffeine carries a bitter taste in Robusta [24]. Robusta beans contain higher levels of caffeine than Arabica beans. Bitterness in Robusta is highly related to caffeine content [25]. Caffeine in the NIR Region corresponds to CH stretching [26].

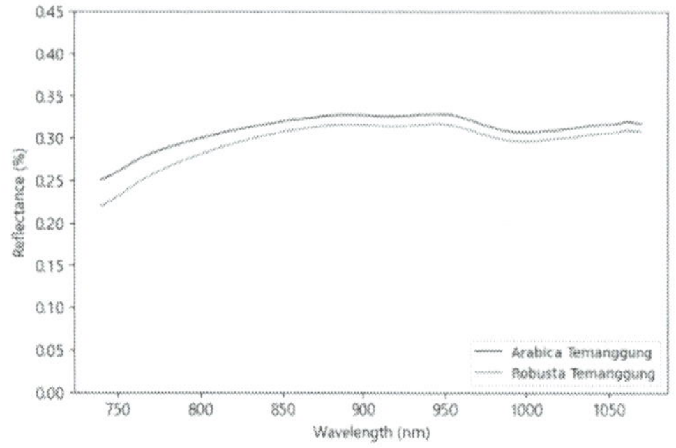

(a)

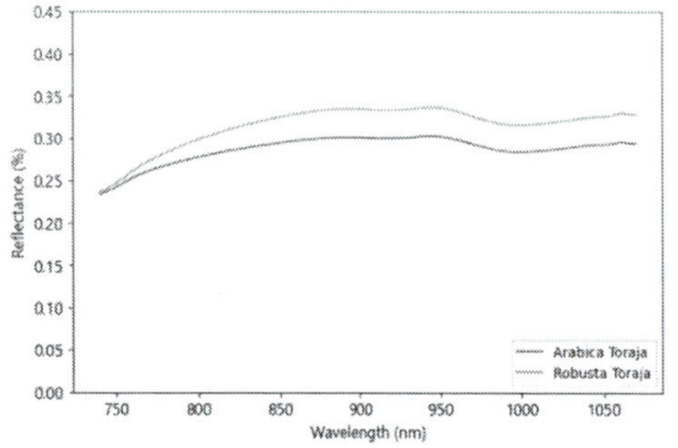

(a)

Fig 3. Original NIR Spectra of Arabica and Robusta coffee: (a) Temanggung and (b) Toraja

3.2 Results of PCA

PCA reduces variables of the spectroscopic data and generates new variables, called PCs, that are linear combinations of the original data and perform the maximum explained variance within the data [27]–[29]. The number of variables in the new data extracted by PCA may vary depending on the sorted eigenvalues. These eigenvalues are presented as percentages, also often called percent explained variance. We can also practically determine the number of PCs if the total population variance is over 80 to 90% [30]. The closer to 100 % of the total proportion of explained variance, the more faithful PCA represents the linear combinations of original data.

We determined the number of PCs used for further analyses was 3, which means three variables were obtained to perform LDA and ANN models. In Temanggung samples, the first three PCs accounted for 95.8% (PC1), 3.97% (PC2), and 0.2% (PC3) of the 99.97% explained total variance, while in Toraja samples, the PCA score accounted for 99.97% of the explained total variance from PC1: 96.88%, PC2: 2.95% and PC3: 0.14%. We also performed the separation between samples by plotting the two-dimensional PCA score plot (PC1 vs PC2), as depicted in Fig 4. Along PC1 and PC2, all samples seemed more dispersed and overlapped, but the distinguishment of Toraja samples showed satisfaction compared to Temanggung samples.

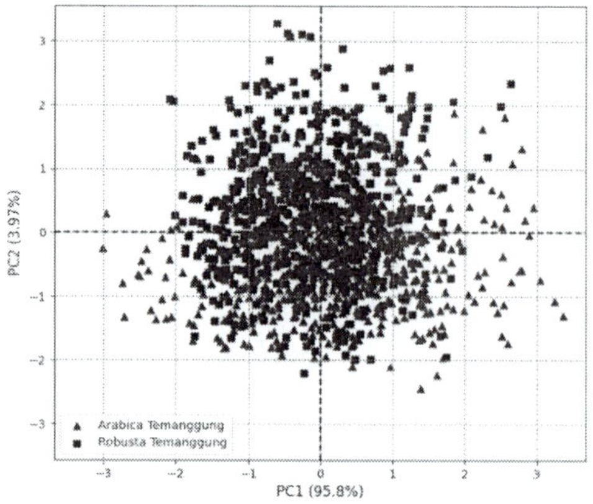

(a)

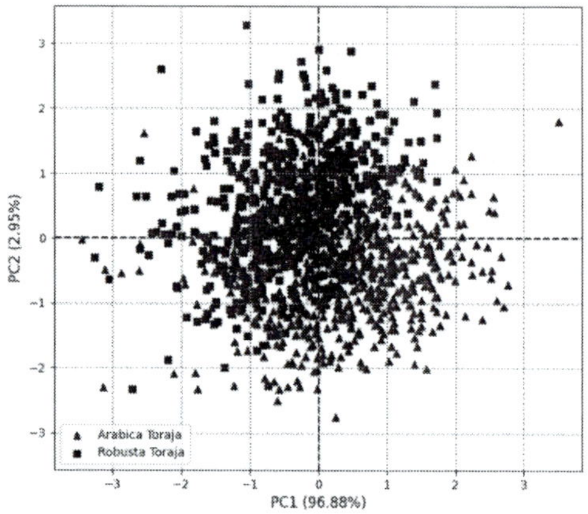

(a)

Fig 4. PCA score plots: (a) Temanggung and (b) Toraja

3.3 Classification Results of LDA

The new variables generated by PCA (the first three PCs) will be used as inputs to develop an LDA model. LDA provides more class separability and draws a decision boundary between the sample classes [31]. LDA computes an optimal data projection by minimizing within-class distance and maximizing between-class distance. Therefore, LDA is able to guarantee maximum class separation [32]–[34]. The projected data to visualize the discrimination between Arabica and Robusta coffee classes was performed in the LDA plot, see Fig 5. We still saw some Robusta samples were correctly predicted as Arabica and uncorrecly assigned as Robusta samples. Likewise, Arabica samples were predicted correctly and incorrectly.

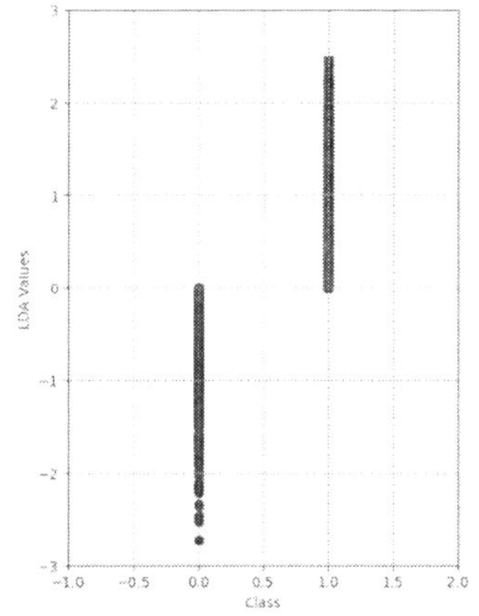

(a)

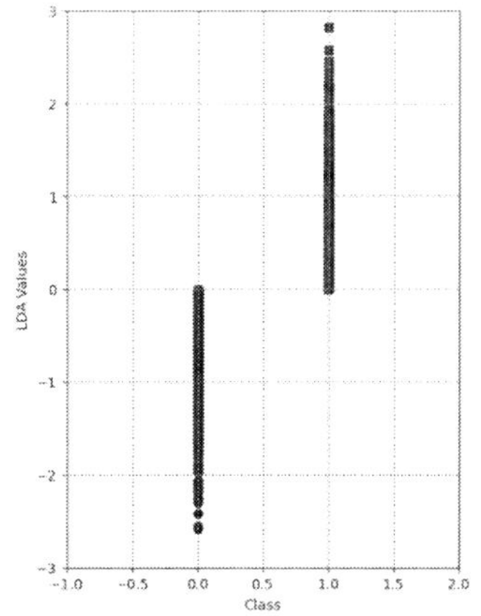

(b)

● Class 0 : Robusta
■ Class 1 : Arabica

Fig 5. LDA plots: (a) Temanggung, and (b) Toraja

The confusion matrix (Table 2) evaluated the LDA model and figured out the classification results. The LDA model produced an accuracy value of >75% for Temanggung and >80% for Toraja. The classification error value generated for Temanggung reached below 25% and 20% for Toraja. The specificity, precision, recall, and F1-score matrices also showed a value of >75% for Temanggung and >80% for Toraja.

3.4 Classification Results of ANN

This ANN classifier was categorized as supervised learning in which certain sample groups were determined based on feature data and its label [35], [36]. Fig 6 depicts the curves representing how this model works in the cross-validation process and to see the loss during training. The loss quantifies the discrepancy between the predicted and actual classes, while the accuracy measures the agreement. The ideal loss curve decreases and tends to flatten as it approaches 0.00 on the y-axis. Both loss curves did not closely approach the threshold value of 0.00 until the epochs ended, indicating some errors were generated during the training of the samples. The loss curve for Toraja obtained a lower error or closer to the threshold value of 0.00 than for Temanggung. A good model must achieve an accuracy value close to 1.00 in the accuracy curve. The accuracy curve of the Toraja represented more promising. It increased closer to a threshold value. All loss curves decreased to a stability point with a moderate gap between the two final curves. Continuing the training or more than 100 epochs to train the model may lead to 'overfit'.

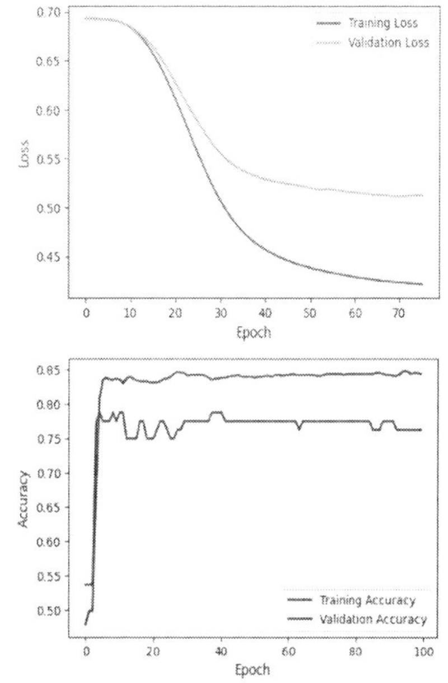

(a)

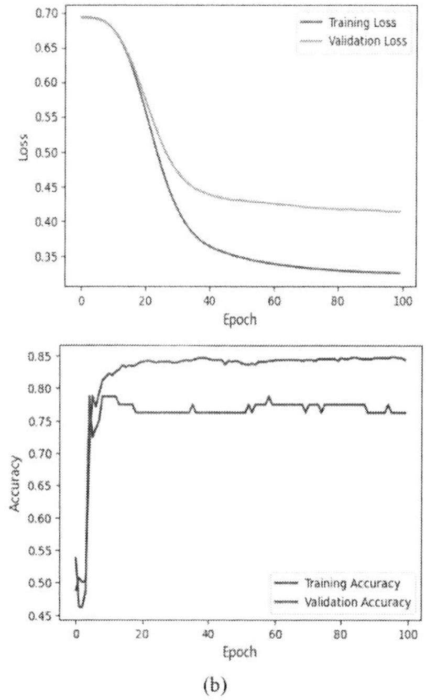

(b)

Fig 6. Loss and accuracy curves: (a) Temanggung and (b) Toraja

This study used three PCA-generated variables as inputs in the ANN model. The classification results were evaluated using a confusion matrix and performance metrics in Table 3. The results performed accuracy, specificity, precision, recall, F1-score, and error were not significantly different if we compared with the LDA results. Using a portable NIR spectrometer and ANN model, Toraja samples performed better in distinguishing coffee species compared to Temanggung samples.

The good ability of the ANN model to distinguish coffee species was also visualized in the ROC curve and AUC values (Fig 7). In Toraja samples, the ROC curves of Robusta and Arabica (Fig 7b) were closer to point (0,1) and consistently stayed away from above the ascending diagonal line. The AUC values showed closer to a value of 1, indicating that using this spectrometer and ANN model obtained better performance.

Table 2. Performance metrics of LDA

Samples			Confusion matrix		k-fold CV	AC*	SP*	PR*	RE*	FS*	E*
Toraja			R	A							
	Training	R	324	101	77.5±7.8	77.9	76.2	79.7	80.8	80.2	22.1
		A	77	302			79.7	76.2	74.9	75.6	
	Testing	R	164	47		79.3	77.7	81.1	82.4	81.7	20.7
		A	35	150			81.1	77.7	76.1	76.9	
Toraja			R	A							
	Training	R	336	67	83.3±6.1	83.3	83.4	83.2	83.4	83.3	16.7
		A	67	333			83.2	83.4	83.2	83.3	
	Testing	R	162	39		81.6	80.6	82.5	82.6	82.6	18.4
		A	34	161			82.5	80.6	80.5	80.5	

* All units in percentage

Table 3. Performance metrics of ANN

Samples			Confusion matrix		k-fold CV	AC*	SP*	PR*	RE*	FS*	E*
Toraja			R	A							
	Training	R	356	116	79.3 ± 5.6	80.4	70.9	75.4	89.9	82.0	19.6
		A	40	283			89.9	87.6	70.9	78.4	
	Testing	R	169	57		79.0	71.6	74.8	86.7	80.3	21.0
		A	26	144			86.7	84.7	71.6	77.6	
Toraja			R	A							
	Training	R	356	84	83.9 ± 3.1	83.6	78.9	80.9	88.1	84.4	16.4
		A	48	315			88.1	86.8	78.9	82.7	
	Testing	R	159	38		81.3	81.1	80.7	81.5	81.1	18.7
		A	36	163			81.5	81.9	81.1	81.5	

* All units in percentage

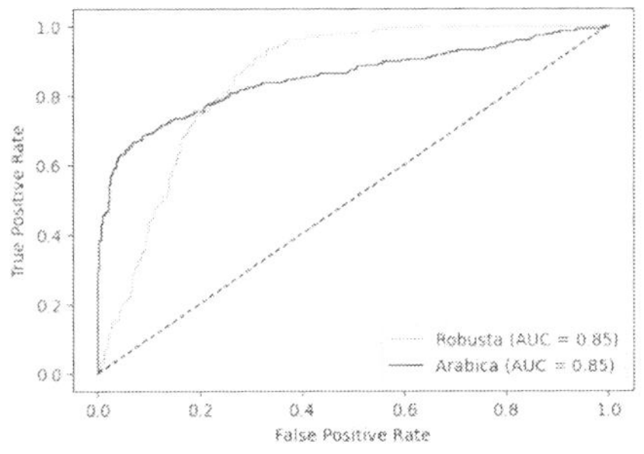

(a)

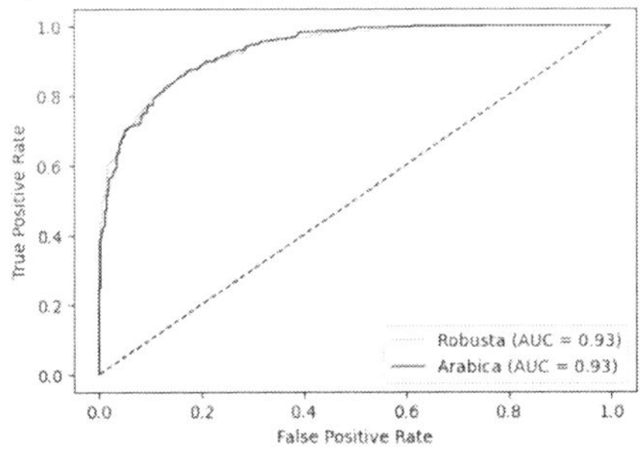

(b)

Fig 7. AUC-ROC curves of the ANN model: (a) Temanggung and (b) Toraja

4 Conclusions

The portable NIR Spectrometer associated with qualitative chemometric and artificial neural network models is moderately effective in distinguishing Arabica and Robusta coffee species. It is a promising technique that can be applied routinely for detecting and controlling coffee samples. It will reduce the time of analyses, be non-destructive without prior sample preparation, and provides in situ and real-time data acquisition. It can practically be applied at industrial levels to migrate from an analytical instrument to a reliable-lightweight tool. It may also become one of the promising methods for spectroscopy-based real-time food classification with a comparable result. Howefer, the accuracy of measurement is still one of the key challenges. Therefore, the implementation from different samples combined chemometric and machine learning models needs to be considered for further studies to guarantee consistency of generalization, not only to address classification problems but also to detect quality parameters of food.

References

[1] BPS, "Statistik Kopi Indonesia 2021," Jakarta, 2021.

[2] P. R. A. B. Toledo, L. Pezza, H. R. Pezza, and A. T. Toci, "Relationship Between the Different Aspects Related to Coffee Quality and Their Volatile Compounds," *Compr. Rev. Food Sci. Food Saf.*, vol. 15, no. 4, pp. 705–719, 2016, doi: 10.1111/1541-4337.12205.

[3] I. Hečimović, A. Belščak-Cvitanović, D. Horžić, and D. Komes, "Comparative study of polyphenols and caffeine in different coffee varieties affected by the degree of roasting," *Food Chem.*, vol. 129, no. 3, pp. 991–1000, 2011, doi: 10.1016/j.foodchem.2011.05.059.

[4] N. Liang, S. Sun, C. Zhang, Y. He, and Z. Qiu, "Advances in infrared spectroscopy combined with artificial neural network for the authentication and traceability of food," *Crit. Rev. Food Sci. Nutr.*, vol. 62, no. 11, pp. 2963–2984, 2022, doi: 10.1080/10408398.2020.1862045.

[5] M. R. Baqueta, A. Coqueiro, and V. Patricia, "Brazilian Coffee Blends: A Simple and Fast Method by Near-Infrared Spectroscopy for the Determination of the Sensory Attributes Elicited in Professional Coffee Cupping," *J. Food Sci.*, vol. 0, no. 0, 2019, doi: 10.1111/1750-3841.14617.

[6] H. T. M. Tran, C. A. C. Vargas, L. Slade Lee, A. Furtado, H. Smyth, and R. Henry, "Variation in bean morphology and biochemical composition measured in different genetic groups of arabica coffee (Coffea arabica L.)," *Tree Genet. Genomes*, vol. 13, no. 3, 2017, doi: 10.1007/s11295-017-1138-8.

[7] A. Khoshroo, A. Arefi, A. Masoumiasl, and G.-H. Jowkar, "Classification of Wheat Cultivars Using Image Processing and Artificial Neural Networks .," *Academia*, vol. 2, no. 1, pp. 17–22, 2014.

[8] A. Taner, Y. B. Oztekin, A. Tekgüler, H. Sauk, and H. Duran, "Classification of Varieties of Grain Species by Artificial Neural Networks," *Agronomy*, vol. 8, no. 7, 2018, doi: 10.3390/agronomy8070123.

[9] K. Kiliç, I. H. Boyaci, H. Köksel, and I. Küsmenoglu, "A classification system for beans using computer vision system and artificial neural networks," *J. Food Eng.*, vol. 78, no. 3, pp. 897–904, 2007, doi: 10.1016/j.jfoodeng.2005.11.030.

[10] M. Meenu, C. Kurade, B. C. Neelapu, S. Kalra, H. S. Ramaswamy, and Y. Yu, "A concise review on food quality assessment using digital image processing," *Trends Food Sci. Technol.*, vol. 118, no. PA, pp. 106–124, 2021, doi: 10.1016/j.tifs.2021.09.014.

[11] V. D. C. Martins *et al.*, "Fraud investigation in commercial coffee by chromatography," *Food Qual. Saf.*, vol. 2, no. 3, pp. 121–133, 2018, doi: 10.1093/fqsafe/fyy017.

[12] C. A. Teixeira Dos Santos, M. Lopo, R. N. M. J. Páscoa, and J. A. Lopes, "A review on the applications of portable near-infrared spectrometers in the agro-food industry," *Appl. Spectrosc.*, vol. 67, no. 11, pp. 1215–1233, 2013, doi: 10.1366/13-07228.

[13] I. Esteban-Díez, J. M. González-Sáiz, C. Sáenz-González, and C. Pizarro, "Coffee varietal differentiation based on near infrared spectroscopy," *Talanta*, vol. 71, no. 1, pp. 221–229, 2007, doi: 10.1016/j.talanta.2006.03.052.

[14] E. K. Kemsley, S. Ruault, and R. H. Wilson, "Discrimination between Coffea arabica and

Coffea canephora variant robusta beans using infrared spectroscopy," *Food Chem.*, vol. 54, no. 3, pp. 321–326, 1995, doi: https://doi.org/10.1016/0308-8146(95)00030-M.

[15] W. Zheng, X. Fu, and Y. Ying, "Spectroscopy-based food classification with extreme learning machine," *Chemom. Intell. Lab. Syst.*, vol. 139, pp. 42–47, 2014, doi: 10.1016/j.chemolab.2014.09.015.

[16] R. M. Correia *et al.*, "Portable near infrared spectroscopy applied to quality control of Brazilian coffee," *Talanta*, vol. 176, no. April 2017, pp. 59–68, 2018, doi: 10.1016/j.talanta.2017.08.009.

[17] M. R. Baqueta, A. Coqueiro, P. H. Março, and P. Valderrama, "Multivariate classification for the direct determination of cup profile in coffee blends via handheld near-infrared spectroscopy," *Talanta*, vol. 222, no. July 2020, p. 121526, 2021, doi: 10.1016/j.talanta.2020.121526.

[18] A. Géron, *Hands-on Machine Learning whith Scikit-Learing, Keras and Tensorfow*. 2019.

[19] I. Saputra and D. A. Kristiyanti, *Machine Learning untuk Pemula*. Bandung: Informatika, 2022.

[20] A. Kulkarni, D. Chong, and F. A. Batarseh, "5 - Foundations of data imbalance and solutions for a data democracy," F. A. Batarseh and R. B. T.-D. D. Yang, Eds. Academic Press, 2020, pp. 83–106.

[21] D. Ballabio and R. Todeschini, "Chapter 4 - Multivariate Classification for Qualitative Analysis," D.-W. B. T.-I. S. for F. Q. A. and C. Sun, Ed. San Diego: Academic Press, 2009, pp. 83–104.

[22] M. T. Ahad, Y. Li, B. Song, and T. Bhuiyan, "Comparison of CNN-based deep learning architectures for rice diseases classification," *Artif. Intell. Agric.*, vol. 9, pp. 22–35, 2023, doi: 10.1016/j.aiia.2023.07.001.

[23] F. Gordillo-Delgado, A. Bedoya, and E. Marín, "Study of the Pigments in Colombian Powdered Coffee Using Photoacoustic Spectroscopy," *Int. J. Thermophys.*, vol. 38, no. 1, pp. 1–8, 2017, doi: 10.1007/s10765-016-2144-z.

[24] N. Farhaty and Muchtaridi, "Tinjauan Kimia dan Aspek Farmakologi Senyawa Asam Klorogenat pada Biji Kopi: Review," *Farmaka Suplemen*, vol. 14, no. 1, pp. 214–227, 2019.

[25] J. S. Ribeiro, M. M. C. Ferreira, and T. J. G. Salva, "Chemometric models for the quantitative descriptive sensory analysis of Arabica coffee beverages using near infrared spectroscopy," *Talanta*, vol. 83, no. 5, pp. 1352–1358, 2011, doi: 10.1016/j.talanta.2010.11.001.

[26] S. S. N. Chakravartula, R. Moscetti, G. Bedini, M. Nardella, and R. Massantini, "Use of convolutional neural network (CNN) combined with FT-NIR spectroscopy to predict food adulteration: A case study on coffee," *Food Control*, vol. 135, no. January, p. 108816, 2022, doi: 10.1016/j.foodcont.2022.108816.

[27] H. Zhao, B. Guo, Y. Wei, and B. Zhang, "Near infrared reflectance spectroscopy for determination of the geographical origin of wheat," *Food Chem.*, vol. 138, no. 2–3, pp. 1902–1907, 2013, doi: 10.1016/j.foodchem.2012.11.037.

[28] Y. He, S. Feng, X. Deng, and X. Li, "Study on lossless discrimination of varieties of yogurt using the Visible/NIR-spectroscopy," *Food Res. Int.*, vol. 39, no. 6, pp. 645–650, Jul. 2006, doi: 10.1016/j.foodres.2005.12.008.

[29] J. N. Miller and J. C. Miller, *Statistics and Chemometrics for Analytical Chemistry*. Essex, England: Pearson Education Limited, 2010.

[30] R. A. Johnson and D. W. Wichern, *Applied Multivariate Statistical Analysis, 6th edition*. Essex, England: Pearson Education Limited, 2014.

[31] S. Balakrishnama and A. Ganapathiraju, "Linear discriminant analysis-a brief tutorial," *Inst. Signal Inf. Process.*, vol. 18, pp. 1–8, 1998.

[32] J. Ye and S. Ji, "Discriminant Analysis for Dimensionality Reduction: An Overview of Recent Developments," *Biometrics Theory, Methods, Appl.*, pp. 1–19, 2009, doi: 10.1002/9780470522356.ch1.

[33] A. Tharwat, T. Gaber, A. Ibrahim, and A. E. Hassanien, "Linear discriminant analysis: A detailed tutorial," *AI Commun.*, vol. 30, no. 2, pp. 169–190, 2017, doi: 10.3233/AIC-170729.

[34] C. H. Park and H. Park, "A comparison of generalized linear discriminant analysis algorithms," *Pattern Recognit.*, vol. 41, no. 3, pp. 1083–1097, 2008, doi: 10.1016/j.patcog.2007.07.022.

[35] R. Vadivambal and D. S. Jayas, *Bio-Imaging: Principles, Techniques, and Applications*. Florida: Taylor & Francis Group, LLC, 2016.

[36] H. O. Velesaca, P. L. Suárez, R. Mira, and A. D. Sappa, "Computer vision based food grain classification: A comprehensive survey," *Comput. Electron. Agric.*, vol. 187, no. September 2020, 2021, doi: 10.1016/j.compag.2021.106287.

Predicting physicochemical properties of melon (*Cucumis melo* L.) using ultrasonic technology and artificial neural network

Fahar Yazid Izdihar[1], *Nafis* Khuriyati[1*], and *Wagiman*[1]

[1]Department of Agroindustrial Technology, Faculty of Agricultural Technology, Universitas Gadjah Mada, 55281 DI Yogyakarta, Indonesia

Abstract. As one of the favorite fruits widely produced and consumed in Indonesia, quality testing for melon (*Cucumis melo* L.) fruit is mostly done using destructive testing. To overcome this problem, this study aims to predict physicochemical quality properties of melon fruit non-destructively using ultrasonic and artificial neural network (ANN). Fifty-nine Hami melons were tested to measure the attenuation value of ultrasonic wave emission as a non-destructive variable, along with density and age. Then, destructive testing was conducted to measure physicochemical properties consisting of fruit flesh firmness, total soluble solids (TSS), titratable acidity (TA), pH, and vitamin C. The test data obtained were processed using ANN to acquire prediction models, with non-destructive data as input variables and each destructive data as output variable. The results showed that the prediction values were still not accurate. Reliability analysis conducted on the test data set based on R^2 values (R^2_{test}) and Normalized Root Mean Squared Error (NRMSE) values showed that the predicted values were still unreliable with R^2_{test} values that were still very low, ranging from -0.776 to 0.485, although the NRMSE value was relatively good.

1 Introduction

Melon (*Cucumis melo* L.) is one of the favorite fruits in Indonesia [1]. Farmers prefer cultivating melons because they can fetch a good price in the market, ranging from IDR 10,000 - IDR 15,000 / kg, and the weight per fruit can reach 1 - 3 kg. On the other hand, melon is preferred by consumers due to its delightful flavor, refreshing aroma, and nutritional content that is not inferior to other fruits [1]. In addition, 129,147 tons of melons were produced in Indonesia in 2021 [2]. However, with this relatively high production and price, melon quality testing is still carried out destructively. Non-destructive fruit quality testing was developed to overcome the previous problems, one of which is using the attenuation value of ultrasonic wave emission.

Another study investigated ultrasonic technology usage for kiwifruit, avocado, and mango [3]. It found that ultrasonic was able to observe the development of changes in properties within the fruit tissue, from when the fruit is on the tree, harvesting, and during storage [3]. In melon fruit, an ultrasonic test has also been conducted on melon fruit slices, and it was found that there is a good correlation between the ultrasonic attenuation value and the quality properties of melon fruit, concluding that ultrasonic can be used to estimate the quality of melon fruit [4].

Ultrasonic technology has advantages over other non-destructive test technologies, of which ultrasonic waves can penetrate gases, liquids, and solid materials, which makes it utilizable for testing materials [5]. In addition, due to the large size, another non-destructive test technology, near-infrared spectroscopy, can only penetrate 2.1 cm into the melon fruit [6], making the widespread nature of ultrasonic waves an advantage of this method.

Previous research on melon fruit was limited to finding the relationship of attenuation value with a quality property, so the prediction model still relies on the measurement results of attenuation value as a single predictor. On the other hand, parameters such as fruit age and size used in similar studies have yet to be included in the resulting prediction model. Also, similar studies have not been conducted on whole melon fruits, but rather using cut pieces of melon fruit so that it cannot be classified as a non-destructive test. Thus, there are several variables that can be analyzed together to obtain better quality estimation on the whole melon fruit.

Currently, artificial neural network (ANN) has been developed to analyze relationships between variables, both linear and non-linear relationships [7]. ANN also does not require detailed knowledge of the system for modeling, does not require mathematical assumptions, is easy to implement and adaptive, and can handle large and complex systems with many variables [8]. Thus, ANN can be used to analyze the relationship between various parameters, such as attenuation value, fruit age, and fruit density to predict the physicochemical quality of melon fruit. Fruit age itself shows the changing trend for various fruit quality properties [9], potentially explaining variation better, while the density within a material, including most physical or chemical differences within, affects the attenuation value during propagation [10] and can be measured nondestructively.

* Corresponding author: nafis.khuriyati@ugm.ac.id

With the expensive cost of melon fruit per unit and the need for quality testing, it is necessary to develop a non-destructive test of the melon fruit as a whole. Then, in the absence of a model that considers a combination of variables for predicting quality properties, this study aims to predict and analyze the reliability of a non-destructive test result for the physicochemical quality properties of melon fruit.

2 Material and methods

2.1 Materials and tools

The melon fruits used in this study were Hami melons (*Cucumis melo* L. var. *reticulatus*), obtained from the Greenhouse Fablab Jogja at the Field Research Center of Gadjah Mada University, Wates, Kulon Progo. These melons were within the age range of 34 - 64 days after anthesis (DAA) and were categorized into 16 age groups, each with a 2-day interval. At each age group, three samples were collected for the ANN training dataset, resulting in a total of 48 training samples. Furthermore, an additional 11 samples of varying ages were selected for the testing dataset.

The tool used for acquiring ultrasonic attenuation data (as presented in Table 1) consists of a wave generator that utilizes a Teensy 3.6 Microcontroller, transmitter and receiver transducers operating at a frequency of 42 kHz, and an oscilloscope. Additional measurement tools included calipers, cloth meters, scales, an oven, a Fujiwara KM-5 fruit hardness tester, a blender, an Atago PAL-15S Digital Refractometer, a pH meter, and a titration tool set.

Table 1. Ultrasonic attenuation data acquisition tool

Component	Picture
Microcontroller Teensy 3.6	
Transducer 42 kHz	
Hantek DSO5102P Digital Oscilloscope	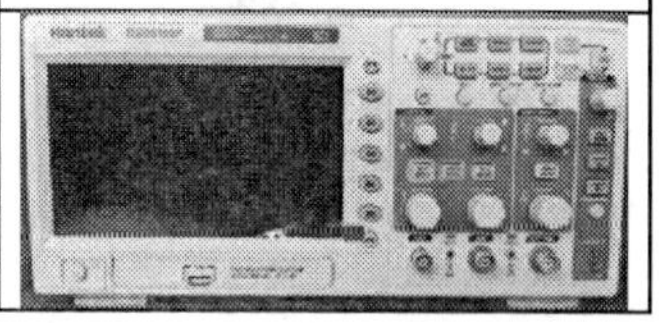

2.2 Non-destructive data acquisition

2.2.1 Ultrasonic attenuation measurement

Ultrasonic attenuation measurements were made at three placements setups of transmitter (T) and receiver transducer (R), as shown in Fig. 1.

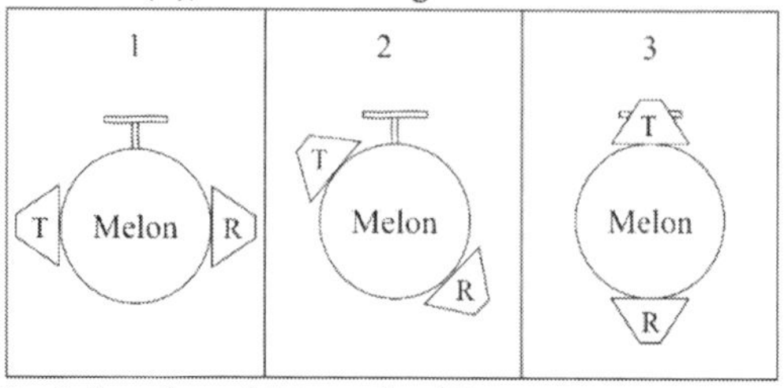

Fig. 1. Location of Attenuation Measurement

The data obtained are the initial amplitude value (A_0) transmitted by the transmitter transducer and the final amplitude value (A) received by the receiver transducer. The attenuation value is calculated with the following equation [11].

$$W = 20log(A_0/A) \tag{1}$$

In the equation 1, W is the attenuation in decibels (dB), A is the final amplitude value, A_0 is the initial amplitude value.

2.2.2 Density measurement

Density (mass/volume) measurements were carried out using scales and tools to measure fruit length and diameter. Calculation of fruit volume was done using the following formula [12].

$$Y = 0.984 * K * D^2 * L / 1000 \tag{2}$$
$$If\ D/L \leq 1,\ K = 0.1528 * (D/L) + 0.4152 \tag{3}$$
$$If\ D/L > 1,\ K = -0.2204 * (D/L) + 0.7872 \tag{4}$$

In the equation above, Y is the volume of the fruit (cm^3), K is the shape factor, D is the diameter of the fruit, and L is the length of the fruit.

2.3 Determination of physicochemical properties destructively

The part of the fruit used for the destructive data collection sample consists of 3 parts: the upper part (connected to the stalk), the middle part, and the lower part (Fig. 2).

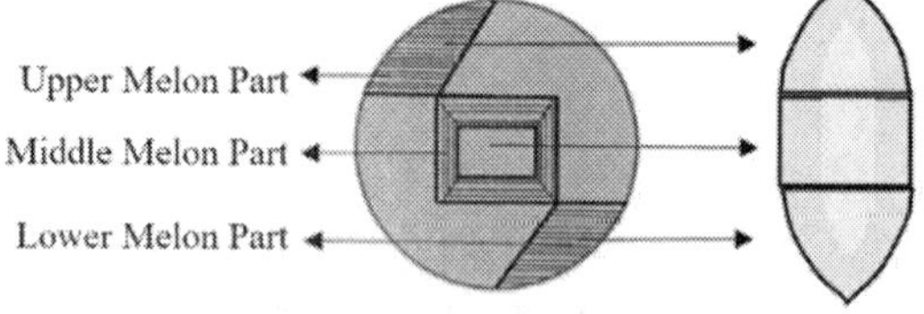

Fig. 2. Melon Parts Used for Sampling

2.3.1 Flesh firmness

Fruit flesh firmness measurements were measured using a Fruit Hardness Tester at 9 points to represent the overall fruit firmness level as shown in Fig. 3.

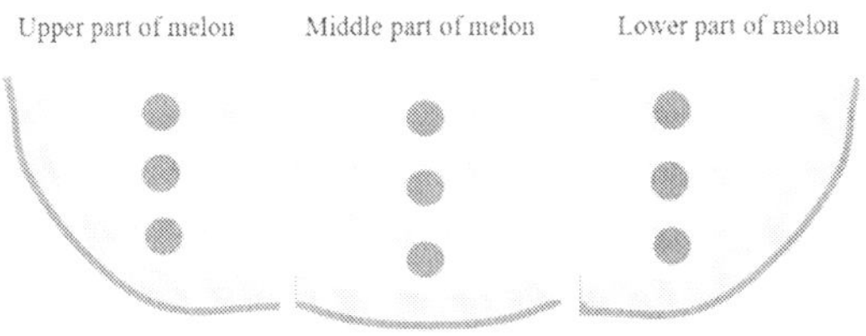

Fig. 3. Location of Firmness Measurement

2.3.2 Total soluble solids

Three sample parts of the fruit were taken and blended. The juice was taken and measured using ATAGO PAL-15S Refractometer to obtain the °Brix value.

2.3.3 Titratable acidity

Three sample parts of the fruit were blended. The juice was taken and dissolved in aquadest, and then the solution was titrated using a standardized 0.1 M NaOH solution. Phenolphthalein was used as an indicator. The total acid value was calculated using the following formula.

$$TA = mL\ NaOH * M\ NaOH * 0.064 * 100 / mL\ juice \quad (5)$$

2.3.4 Potential of hydrogen (pH)

Three sample parts were blended. The juice obtained was then measured using a calibrated pH meter.

2.3.5 Vitamin C

Determination of vitamin C content was done after ultrasonic attenuation data acquisition. Three parts of the fruit sample were collected and blended. The juice obtained was dissolved into aquadest to make a solution. Starch solution (1%) was added as an indicator. The solution was then titrated using a standardized 0.01 N Iodine solution. Vitamin C content (mg/100 g) was calculated by the following equation [13].

$$VC = (V * N) * EW * DF * 100 / mL\ juice \quad (6)$$

In the equation above, VC is mg vitamin C / 100 grams of sample, V is the volume of iodine solution, N is the concentration of iodine, EW is the equivalent weight of vitamin C (0.88), and DF is the dilution factor.

2.4 Data processing

The collected nondestructive and destructive data undergo processing through an artificial neural network. This processing serves as training for the neural network model, enabling it to predict output variables from input variable data. Throughout the training, the nondestructive data functions as the input, forecasting each physicochemical property that subsequently becomes an output variable.

The training was carried out as the procedure conducted by Amalia (2020) [7], which, sequentially, starts with determining the structure, then the activation function, and the learning rate, each step of which selects a result that provides the optimal SSE value. The artificial neural network model that produces the best SSE value is selected for prediction reliability analysis.

3 Results and discussions

3.1 Trends from the data obtained during fruit maturity

3.1.1 Ultrasonic attenuation

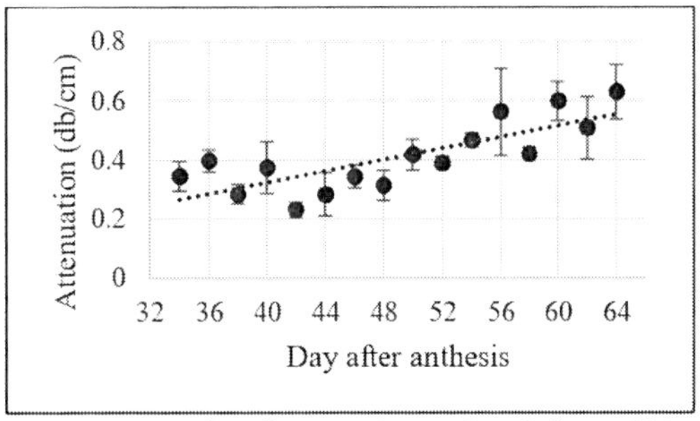

Fig. 4. Scatter plot of attenuation against fruit age (DAA)

Attenuation values increase with increasing fruit maturity (Fig. 4). Similar changes in attenuation values were also found in a study by [14], where the attenuation value negatively correlated with the flesh firmness of avocados, which softened over time. In this study, similar results were observed from the trend of melon flesh firmness values, which decreased as the fruit aged, while the trend of attenuation values increased.

3.1.2 Fruit density

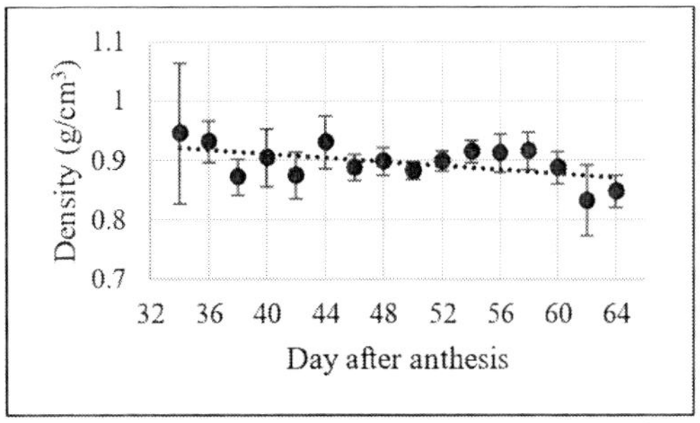

Fig. 5. Scatter plot of fruit density value against fruit age (DAA)

There is a trend of decreasing values for the density of melon fruits with increasing maturity (Fig. 5). The decrease occurs due to the ripening process, which causes the seed cavity to occupy a larger space [15].

3.1.3 Firmness

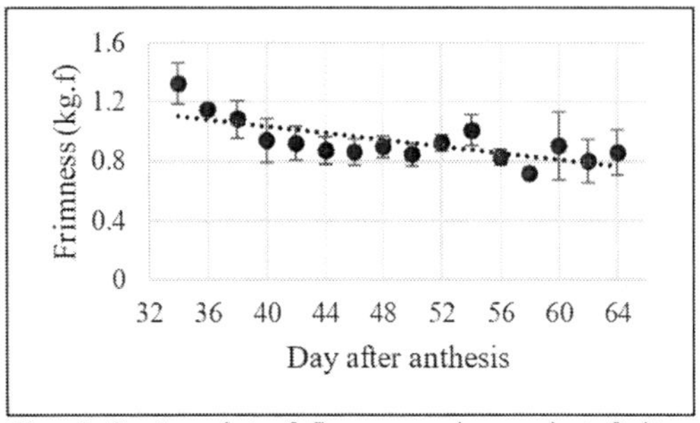

Fig. 6. Scatter plot of firmness value against fruit age (DAA)

The firmness value of melon fruit flesh had a decreasing trend (Fig. 6). Fruit softening occurs due to cell wall disassembly in the melon fruit ripening process, which is generally regulated by ethylene [16]. It involves various cell wall-modifying enzymes that modify and disassemble pectin.

3.1.4 Total soluble solids (TSS)

There is a trend of increasing Brix value as maturity increases (Fig. 7). Increase in °Brix value was caused by the formation of sucrose, glucose, and fructose, which were originally still in the form of starch (polysaccharides) [9]. In fully ripe melons, all carbohydrates have been converted into saccharides while at the same time increasing Brix value.

There are fluctuations in the °Brix value in the data obtained, whereas the brix value should have an increasing trend. This discrepancy arises as a consequence of manual watering practices in the greenhouse due to irrigation system malfunctions, where too much watering can cause lower Brix values [17]. In addition, the unstandardized sample cut can also be a factor in this discrepancy due to the non-uniform distribution of sugar in melon fruit, where the inner part of the melon fruit has higher sugar content [18].

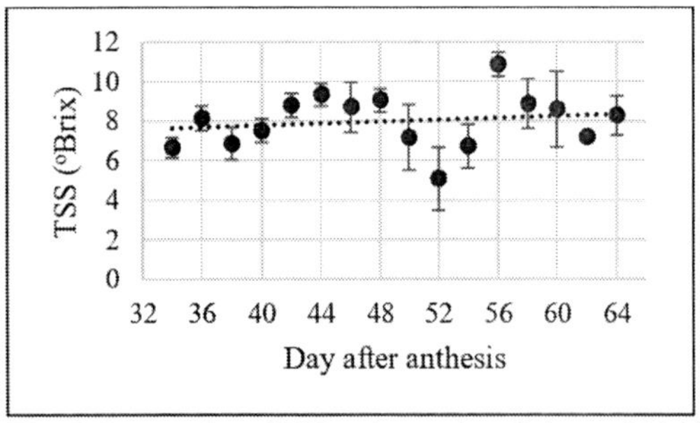

Fig. 7. Scatter plot of TSS value against fruit age (DAA)

3.1.5 Titratable acidity

The total acid value of the fruit had an upward trend during maturity (Fig. 8). This change in value was also observed in similar studies. Low-acid variety Reticulatus melon fruit increased up to a certain age point and then decreased [19-20].

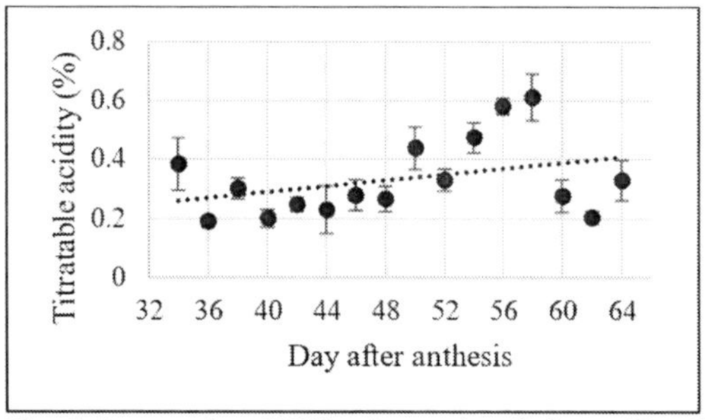

Fig. 8. Scatter plot of TSS value against fruit age (DAA)\

3.1.6 Potential of hydrogen (pH)

Hydrogen potential (pH) values tend to remain stable during growth (Fig. 9). This trend is similar to the study by Burger *et al.* (2003) [19] which showed that the pH values of low-acid melons were also relatively unchanged during fruit development.

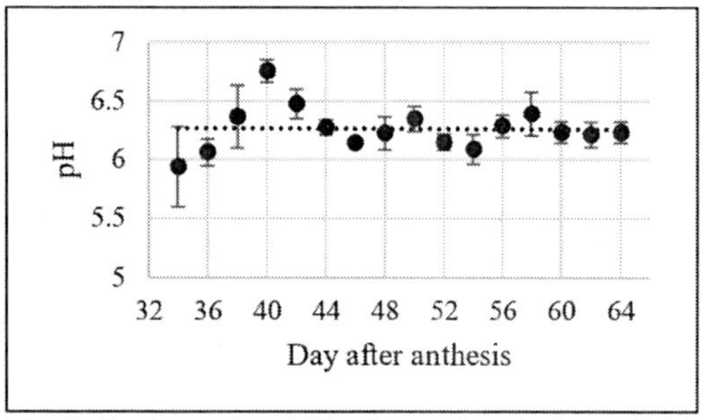

Fig. 9. Scatter plot of TSS value against fruit age (DAA)

3.1.7 Vitamin C

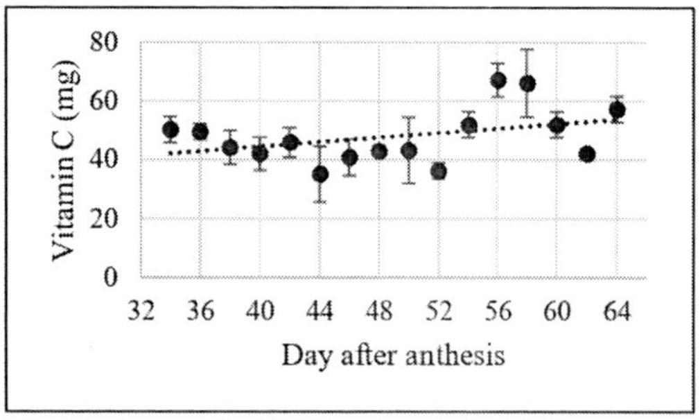

Fig. 10. Scatter plot of TSS value against fruit age (DAA)

Vitamin C value tends to increase with increasing maturity (Fig. 10). However, the vitamin C value reached its highest point at a certain age, namely at 56 - 58 days of age. This situation is in line with another study by Amzeri *et al.* (2022) [9], in which it is explained that at the beginning of growth, the fruit converts simple sugars, then the rate of the process increases due to the enzyme L-gluconolactone oxidase until finally reaching the peak, then decreases again due to oxidation of ascorbic acid into dehydroascorbic acid.

3.2 ANN training

A total of 48 melon samples from 16 age groups (3 samples per age) were used for training and validation with a ratio of 80:20, randomly selected with random numbers for the entire data analysis stage. Another 11 melon samples were used to test the model.

To train the neural network using SPSS software, standardization of the input data was performed as preprocessing. Standardization is a prerequisite step for machine learning that aims to transform numerical variables so that each variable has zero mean and variance so the backpropagation algorithm can work properly [21].

In the hidden layer training, 2 trainings were conducted for 1 and 2 hidden layers. For training 1 hidden layer, the network architecture used for this study are 3-1-1, 3-2-1, 3-3-1, 3-4-1, 3-5-1, 3-6-1, 3-7-1. Then, for training 2 hidden layers, the network architecture obtained with the help of random values are 3-1-3-1, 3-3-1-1, 3-1-1-1, 3-1-4-1, 3-2-2-1, 3-3-6-1, and 3-6-3-1. The architecture that produces the lowest error value for each physicochemical property is used for the second stage of training, which is activation function training.

In the SPSS software, activation function training is done with a combination of two functions for the input-hidden layer (ih) and hidden-output layer (ho) activation functions, which consist of sigmoid (s), hyperbolic tangent (ht) and identity (i). The identity function cannot be used as the input-hidden layer activation function, so, in total, there are 6 possible combinations (for ih-ho) comprising of hyperbolic tangent-sigmoid, hyperbolic tangent-hyperbolic tangent, hyperbolic tangent-identity, sigmoid-sigmoid, sigmoid-hyperbolic tangent, and sigmoid-identity. The activation function combination that produces the lowest error value for each physicochemical property is used for the third stage of training, which is learning rate training.

Learning rate training was conducted with 5 rate variations, namely 1; 0.1; 0.01; 0.001; and 0.0001, as conducted by Amalia (2020). The learning rate that produced the lowest error value for each physicochemical quality property became the most optimal model and was selected for further analysis. The training results that produced the lowest SSE values are compiled in Table 2.

Table 2. Neural network training

Properties	Structure	Activation function	Learning rate
Firmness	3-6-3-1	ht; s	0.01
TSS (°Brix)	3-5-1	ht; s	0.01
Titratable acidity	3-6-1	ht; s	0.001
pH	3-6-3-1	ht; s	0.001
Vitamin C	3-4-1	ht; s	0.0001

3.3 Prediction results and analysis

The predicted results are compared with the observed data as follows (Fig. 11 - Fig. 16).

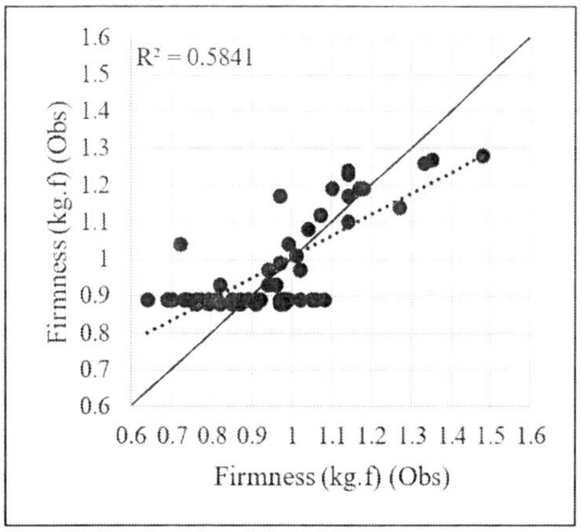

Fig. 11. Scatter plot of fruit firmness

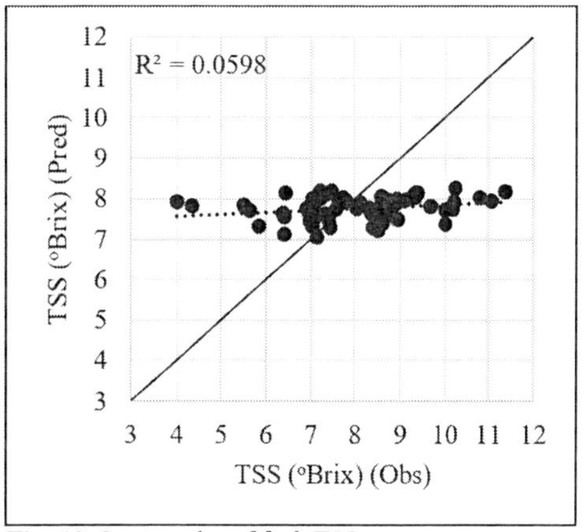

Fig. 12. Scatter plot of fruit TSS

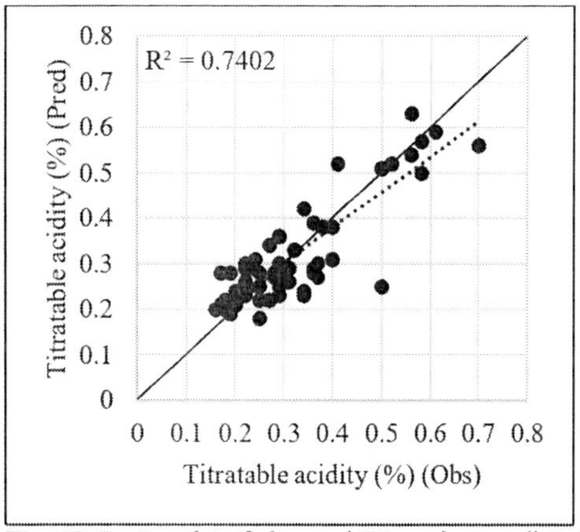

Fig. 13. Scatter plot of observed TA against predicted TA

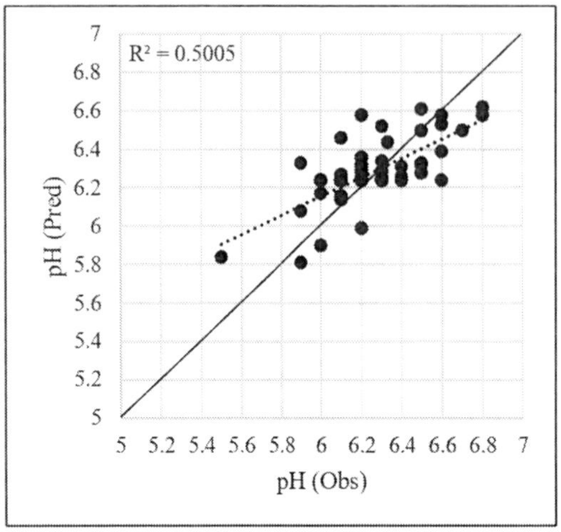

Fig. 14. Scatter plot of observed pH against predicted pH

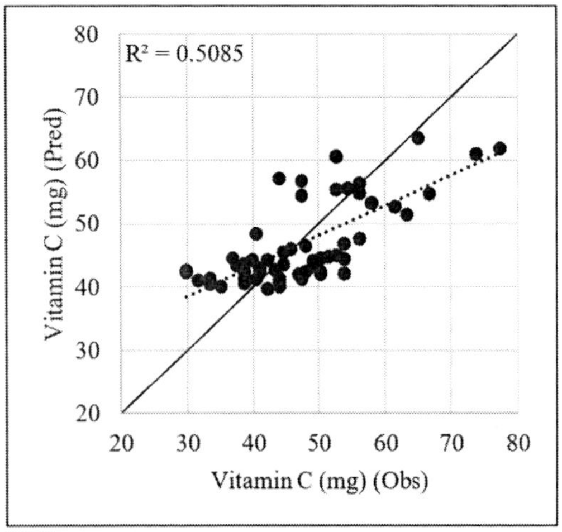

Fig. 15. Scatter plot of observed Vitamin C against predicted Vitamin C

The more the results of the graph plot form a diagonal line, the closer the prediction data is to the observation data. The prediction data had inaccurate values at low fruit flesh firmness levels for the flesh firmness (Fig. 11) properties but formed a diagonal line at higher fruit firmness levels. The scatter plot results for total soluble solids (Fig. 12) were scattered and did not form a diagonal, indicating that the prediction results for this physicochemical quality properties were inaccurate. The scatter plot graph shows that the prediction results for the quality properties of total acid (Fig. 13), pH (Fig. 14) and vitamin C (Fig. 15) formed a relatively diagonal pattern but were still scattered.

Reliability analysis is carried out by calculating the test data sets' R^2 value (R^2_{test}) and Normalized-RMSE value. The model can be good if the R^2 value is at least 0.7 [22]. Then, the NRMSE value is considered excellent if NRMSE <10%, good if 10-20%, fair if 20-30%, and bad >30% [23].

The results of R^2 values of the training dataset (R^2_{train}), R^2 of the test dataset (R^2_{test}), and NRMSE values of the test dataset are tabulated in Table 3.

Table 3. Prediction reliability

Properties	R^2_{train}	R^2_{test}	NRMSE
Firmness	0.596	0.485	11.89%
TSS (°Brix)	0.495	-0.419	13.99%
TA	0.761	-0.776	25.16%
pH	0.533	0.447	2.61%
Vitamin C	0.512	0.165	15.78%

The R^2_{test} values for the TSS and TA properties are negative. Such negative R^2_{test} values can be caused by predicting samples that were not used to build the model [24]. The negative R^2_{test} value also indicates a very poor fit of the prediction data [25].

The fruit firmness prediction result has an R^2_{test} value of 0.485, which is classified as unacceptable. Then, the NRMSE value of 11.89% can be categorized as good because it is in the range of 10-20%.

For the TSS prediction results that was measured in °Brix units, the R^2_{test} value obtained is (-0.419), meaning that the R^2_{test} value of the model cannot be accepted. Then, the NRMSE value of 13.99% can be categorized as good because it is in the range of 10-20%.

For the TA prediction results, the R^2_{test} value obtained is (-0.776), which is classified as unacceptable. Then, the NRMSE value of 25.16% can be categorized as sufficient because it falls within the range of 20% - 30%.

For pH prediction results, the R^2_{test} value obtained is 0.447, which is classified as unacceptable. Then, the NRMSE value of 2.61% can be categorized as very good because it is in the range of 0% - 10%.

For the vitamin C prediction results, the R^2_{test} value obtained is 0.165, which is classified as unacceptable. Then, the NRMSE value of 15.78% can be categorized as good because it falls within the range of 10% - 20%.

Of the five prediction results, all R^2_{test} values were not classified as acceptable, so all models could not be applied. Although some physicochemical properties prediction results have good NRMSE values, the low R^2_{test} values indicate that the prediction models from the artificial neural network cannot sufficiently explain the variations within the testing dataset.

Poor prediction results can be caused by the presence of cavities inside the melon fruit which can affect the propagation of ultrasonic waves which in turn affects the measured attenuation value. The presence of these cavities causes changes in the propagation of sound waves (including ultrasonic) and prevents transmission [26]. Another nondestructive testing study using ultrasonics on tomato fruit encountered similar problems due to the presence of pores in the tomato that cause scattering and interfere with wave transmission [27]. Therefore, further exploration on the behavior of ultrasonic wave propagation is needed for further exploration on ultrasonic nondestructive testing for melon fruit.

4 Conclusion

Prediction using artificial neural network through ANN training with input variables consisting of attenuation value, fruit age, and density for each output variable consisting of physicochemical quality properties of melon fruit, which are total acid, firmness, pH, vitamin C, total soluble solids (°Brix) resulted in R^2_{train} values of 0.761, 0.596, 0.533, 0.512, and 0.495, respectively.

Prediction reliability analysis that was carried out based on the calculation of R^2_{test} and NRMSE values from the test data set shows that the prediction value is still unreliable with unacceptable R^2_{test} values for each quality properties, although some properties have NRMSE values that are classified as good.

The presence of cavities inside the melon fruit may interfere with the wave propagation, affecting the acquired attenuation value, leading to inaccuracies. Therefore, further investigation into the propagation behavior of ultrasonic waves is necessary.

The authors would like to express their deepest gratitude to Universitas Gadjah Mada for funding this research through the *Final Project Recognition Grant* Universitas Gadjah Mada *Number* 5075/UN1.P.II/Dit-Lit/PT.01.01/2023. This funding is very important in enabling us to complete this research. The authors also greatly appreciate the facilities support of the Greenhouse Fablab Jogja, Field Research Center, Universitas Gadjah Mada.

References

[1] S. Paryadi, E. Hadiatna, *Budidaya Tanaman Melon* (2018)
[2] Badan Pusat Statistik (BPS), *Statistik Hortikultura 2021* (2022)
[3] O. P. Zhen, N. Hashim, B. Maringgal, JAFE, **1**, 0003 (2020)
[4] A. Mizrach, N. Galili, G. Rosenhouse, D. C. Teitel, ASAE, **34**, 5 (2001)
[5] Irwandy, *Ilmu Logam* (2013)
[6] L. Xu, J. Li, D. Zhang. JFPE, e12866
[7] R. Amalia, M. Ushada, A. P. Pamungkas, *Pengembangan Model Penentuan Waktu Istirahat Berbasis Lingkungan Kerja di UMKM Tahu Menggunakan Jaringan Saraf Tiruan*, Master Theses, Universitas Gadjah Mada (2020)
[8] L. H. Putro, *Biometana dari Air Limbah Pabrik Pengolahan Kelapa Sawit* (2021)
[9] A. Amzeri, K. Badami, A. Zuhri, G. Pawana, S. Suhartono, S. Khoiri, A. K. Umam, Y. Asmoro, S. A. Rohmatin, H. Sa'diyah, B. Badriyah, IJA, **2022**, 1-10 (2022)
[10] A. Mizrach, PBT, **48**, 3 (2008)
[11] S. N. Jha, *Nondestructive Evaluation of Food Quality* (2010)
[12] S. Jenni, K. A. Stewart, G. Bourgeois, D. C. Cloutier, HS, **31**, 7 (1996)
[13] R. Rompies, N. Mayulu, B. J. Kepel, H. Natanael, F. Faradila, FR, **5**, 5 (2021)
[14] U. Flitsanov, A. Mizrach, A. Liberzon, G. Zauberman, PBT, **20**, 3 (2000)
[15] F. Khoshnam, M. Namjoo, H. Golbakhshi, M. Dowlati, YYUJAS, **26**, 2 (2016)
[16] K. Nishiyama, M. Guis, J. K. C. Rose, Y. Kubo, K. A. Bennett, L. WangJin, K. Kato, K. Ushijima, R. Nakano, A. Inaba, M. Bouzayen, A. Latché, J. Pech, A. B. Bennett, JEB, **58**, 6 (2007)
[17] Y. Akhoundnejad, H. Daşgan, AEER, **17**, 4 (2019)
[18] J. Sugiyama, JAFC, **47**, 7 (1999)
[19] Y. Burger, U. Saar, A. Distelfeld, N. Katzir, Y. Yeselson, S. Shen, A. Schaffer, JASHS, **128**, 4 (2003)
[20] M. Tang, Z. Bie, W. Mingzhu, H. Yi, J. Feng, SH, **123**, 3 (2010)
[21] J. Loy, *Neural Network Projects with Python* (2013)
[22] B. T. II, *Introduction to Deep Learning using R* (2017)
[23] H. Zeng, L. Li, J. Hu, L. Liang, J. Li, B. Li, K. Zhang, TAC, **112**, 389-401 (2012)
[24] S. S. Du, J. Koushik, A. Singh, B. Póczos, ANIPS, 30 (2017)
[25] T. J. Cleophas, A. H. Zwinderman, *Kernel Ridge Regression in Clinical Research* (2019)
[26] T.S. Awad, H.A. Moharram, O.E. Shaltout, D. Asker, M.M. Youssef, Food Research International, **48**, 2, 410-427 (2012)
[27] F. Yildiz, A. T. Özdemir, S. Uluişik, Journal of Food Quality, **2019**, 6810865 (2019)

Space Based Data Usage for Smart Farming

Shin-ichi Sobue*

Japan Aerospace Exploration Agency 2-1-1 Sengen, Tsukuba, Ibaraki 305-8505, Japan

Abstract. Precision Agriculture and input optimization with knowledge sharing are key for smart farming. The use of new technologies such as satellites, drones, navigation, AI/ML, big data, IoT, cloud-computing makes farming and agriculture smarter and transparent. Use of such advanced technologies, governmental officers and farmers can create evidence-based prescription maps for variable rate application of inputs, such as fertilizers, pesticides, and irrigation. Also, smart farming improve efficiency, reduce costs, simplify forecasting, streamline recording and reporting, and boost the sustainability and environmentally friendly agriculture while addressing todays needs and helping future planning. This paper is a brief overview of space-based tools that are currently available for smart farming and also importance of earth observation for smart farming with some examples on rice crop in Asia.

1 Smart Farmimg

1.1 History of Smart Farming

Smart farming, also known as precision agriculture, involves the use of modern technologies and data-driven approaches to improve production efficiency, reduce production cost, simplify forecasting, streamline recording and reporting, and boost the sustainability and environmentally friendly agriculture. The current data and information driven agricultural technology has been evolved since the civilization of mankind. More than 10,000 years ago, agriculture with first commercial fields was deployed using manual or animal drawn operations. From 18th Century, industrial revolution had been occurred and the concept of smart farming has gained significant traction. The agricultural industry witnessed the advent of mechanization. The introduction of machines such as the seed drill, mechanical reaper, and threshing machines revolutionized farming practices and increased efficiency. In 1940 an American scientist started to experiment disease resistance high-yield varieties of wheat in Mexico and it was a great success. This success was soon spread worldwide in 1960s adopting many of these techniques to increase production through introduction of high yield seeds, soil fertilizer, pesticides. This new era was named as "Green Revolution". These innovations were combined with the establishment of agricultural research institutions and investments on agriculture research by governments. The Green Revolution had a significant impact on global food production. It helped increase the crop yields, particularly for staple food crops such as wheat, rice, and maize, which played a crucial role in alleviating food shortages and reducing famine in several countries. The introduction of high yield crop varieties and improved farming practices also led to increased farm incomes and improved living standards for many farmers. However, the Green Revolution also had negative impact on agricultural practices, and environmental and social environment too including human health issues. For instance, the heavy reliance on synthetic fertilizers and pesticides led to increased chemical pollution of soils and water bodies and the focus on a few high-yielding varieties led to a loss of agricultural biodiversity and the displacement of traditional crop varieties.

Starting with the launch of earth observation satellites in 1960s, and the development of remote sensing technologies, it became possible to gather valuable information about crops and soil conditions remotely (from a distance). Remote sensing techniques allowed farmers to monitor planting information and crop growth condition. From 1980s-1990s, by the integration of Global Positioning System (GPS) technology with GIS, agriculture entered a new era referred to as Smart Farming. GPS-enabled Agriculture machinery and farm devices provided accurate location data, allowing farmers to precisely navigate fields and target specific areas for proactive interventions such as irrigation, fertilization, and pesticide application. This led to the emergence of precision farming techniques, which aimed to improve yield by maximizing resource utilization and minimizing the wastage and addressing labour shortage.

From 2000s onwards, we entered the "Digital Agriculture" age using Internet of Things (IoT) and GIS as well as remote sensing data. The development of low-cost sensors and the rise of the Internet of Things (IoT) and Artificial Intelligence and Machine Learning further expanded the capabilities of smart farming. Sensors place in the field to collect real-time data on various parameters such as temperature, humidity, soil moisture, and crop condition. This data sends to cloud based or

* Corresponding author: sobue.shinichi@jaxa.jp

tailor-made systems for analysis, visualizations, sharing via IoT. Once they combine with advanced analytics and machine learning algorithms, farmers able to make data-driven decisions and implement appropriate and precise interventions. In other word, the increasing availability of large-scale and big volume of agricultural related data including remote sensing data, IoT base field data has enabled new avenues for smart farming. Farmers can now leverage big data analytics to gain insights of trends, predict yield, optimize resource allocation, and monitor and manage farm operations more efficiently. Overall, smart farming has evolved over time, incorporating various technologies such as earth observation, precise positing, IOT and artificial Intelligence and/or machine learning, and data-driven approaches to enhance agricultural practices. Its history reflects a continuous effort to leverage technology for increased productivity, sustainability, resource efficiency and eco-friendly agriculture.

1.2 Value of Earth Observation Data for Smart Farming

The use of Earth observation satellite data in smart agriculture brings several advantageous and benefits to the industry. Earth observation satellite data assists modern agriculture industry in number of ways through their remotely observed data in implementing precision agriculture techniques. By analysing multispectral or hyperspectral images captured by satellites, farmers can identify spatial variability of the status of crops and some phenological and soil properties of their fields. This information helps optimizing interventions, such as seeding, fertilizing, and pest/weed controlling, by targeting specific areas that require attention. Precision agriculture enhances the efficiency of the use of resource, reduces costs, and could minimizes environmental impacts.

Earth observation satellites also provide real-time and near-real-time weather and climate data. By integrating satellite-derived weather information with on-ground sensors, farmers can access accurate and localized weather forecasts. This helps them planning their agricultural activities, optimize irrigation scheduling, and mitigate risks associated with extreme weather events. Satellite data also aids in monitoring and mitigating natural disasters, such as droughts, floods, and wildfires, by providing timely information for early warning systems and emergency response planning.

Finally, integrating satellite data with advanced analytics and machine learning techniques allows the development of decision support systems for farmers. These systems provide actionable insights, predictive models, and recommendations based on satellite imagery and other relevant data sources. They help farmers make data-driven and evidence-based decisions, optimize resource allocation, and improve overall farm management practices.

Overall, the value and purpose of using Earth observation satellite data in smart agriculture lie in its ability to provide timely, accurate, and actionable information for farmers. By harnessing satellite technology, farmers can optimize their operations, increase productivity, minimize environmental impacts, and contribute to sustainable agricultural practices.

1.3 GIS with Earth Observation Data and IoT for Smart Farming

IoT (Internet of Thinking) and drones and earth observation data are integral components of smart farming, facilitating data collection, monitoring, and analysis on the geographic information system (GIS). These technologies enable precision agriculture, enhance resource management, improve crop health, and support decision-making processes. By harnessing the power of IoT and drones with earth observation data on GIS, farmers can achieve higher efficiency, sustainability, and profitability in their agricultural operations.

For data collection and monitoring, IoT devices, such as sensors and monitoring systems, are deployed throughout the farm to collect real-time data on various parameters like soil moisture, temperature, humidity, and crop health with satellite-based information. This data is crucial for understanding field conditions, identifying potential issues, and making informed decisions. Drones equipped with sensors and cameras can also provide high-resolution imagery and aerial data, enabling comprehensive field monitoring and analysis assisting sustainable agricultural management. IoT data with high temporal and hight resolution data are complementary to wide area earth observation satellite data.

The linkage between GIS, IoT and earth observation data is highly important for smart farming. It enables farmers to leverage spatial information and satellite imagery to gain valuable insights and make informed decisions in agricultural practices. Here are some key reasons why the linkage between GIS and earth observation data is crucial for smart farming. By overlaying satellite imagery onto GIS platforms, farmers can generate detailed maps that provide valuable information about land characteristics, crop distribution, soil types, and other spatial features. This spatial analysis helps in understanding the variability within fields, optimizing resource allocation, and planning precision agriculture practices. By integrating this data with GIS, farmers can also map and monitor crop performance, identify areas of concern, and take targeted management actions. This helps optimize irrigation, fertilization, pest control, and other agronomic practices.

Risk management is another important aspect that need to be evaluated for maintain agriculture sustainability as well as quantifying risks on the environment due to agricultural practices. Incorporating satellite imagery depicting land cover, vegetation indices, and environmental conditions into GIS models, farmers can evaluate the impact of farming practices on the environment. Also, it is possible to assess soil erosion risks, biodiversity hotspots, and areas requiring conservation efforts through GIS modelling and analysis. This information aids in sustainable land management and ensures compliance with environmental regulations. Since there are more than 40 years earth observation satellite data record as of now, by incorporating historical satellite imagery, weather data, and other spatial information into GIS models, farmers can analyse patterns and trends. This helps in predicting and mitigating risks associated with weather events, diseases, pests, and crop failure. Decision support systems utilizing GIS and earth observation data provide farmers with actionable insights and recommendations for effective risk management.

For precision Agriculture and input optimization, farmers can create prescription maps for variable rate application of inputs, such as fertilizers, pesticides, and water. This targeted approach optimizes resource usage, reduces costs, and minimizes environmental impacts.

2 JAXA Earth Observation for Agriculture

2.1 Overview of JAXA Earth observation mission for agriculture

JAXA has developed and launched several Earth observations satellites that contribute to assessment and monitoring agriculture activities. Some of the JAXA Earth observation satellites and their roles in supporting agriculture are shown in Figure 1.

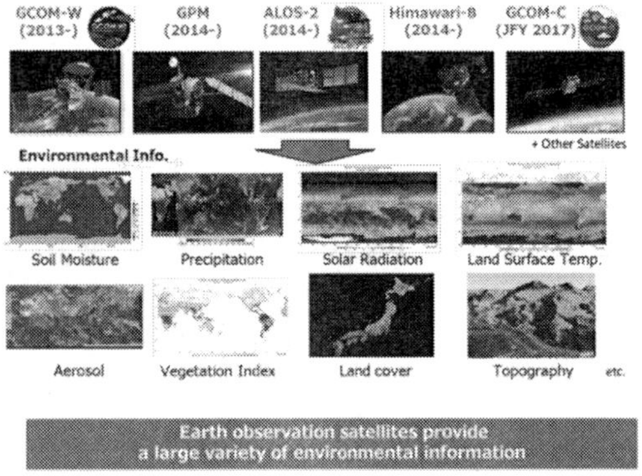

Figure1 Earth observation data for agriculture

GCOM-W (Global Change Observation Mission - Water) is part of the GCOM series focuses on observing water-related parameters. It provides data on soil moisture, snow cover, and vegetation health. This information is crucial for understanding water availability in agricultural areas, optimizing irrigation practices, and predicting crop yields. GCOM-C (Global Change Observation Mission – Climate) is a series of Earth observation satellites developed by JAXA. While GCOM-C is primarily designed for climate-related observations, its data can still be valuable in identifying status of land and assessing agricultural parameters such as land cover, vegetation health, and chlorophyll content. These measurements are crucial for monitoring crop growth, identifying areas of stress, and assessing the overall health of agricultural fields and in the end forecasting the crop yield.

ALOS (Advanced Land Observing Satellite): While not exclusively a satellite launched for agricultural use, ALOS plays a significant role in monitoring land and vegetation. Its data can be used to assess land cover changes, monitor crop health, and track deforestation and reforestation efforts, all of which are relevant to agriculture. The Japanese geostationary meteorological satellite named Himawari series, operated by the Japan Meteorological Agency (JMA), provide real-time imagery of weather patterns. This information is valuable for monitoring weather conditions such as rainfall, cloud cover, and temperature that are crucial for agricultural activities.

Soil moisture by GCOM-W and ALOS-2 are valuable for optimizing irrigation strategies, preventing waterlogging or drought stress, and improving water use efficiency in agriculture. Crop phenology by GCOM-C NDVI and ALOS-2 SAR backscattering aids in predicting planting and harvesting times, assessing crop development, and identifying anomalies that might affect yields. GSMaP has multi-satellite global precipitation map under the Global Precipitation Measurement (GPM) Mission, by using Dual-frequency Precipitation Radar (DPR) onboard GPM core satellites, other GPM constellation satellites including GCOM-W, and Geostationary satellites including Himawari. This precipitation information by GSMaP has been widely used for scientific purposes as well as for monitoring agricultural activities. On set of drought and flood as well as possible recovery time span derived from GSMaP are useful for agriculture. Timely agrometeorological information of such devastating events is crucial for implementing mitigation measures and protecting crops. In order to contribute to food security issue, Ministry of Agriculture, Forestry and Fisheries, Japan provides agrometeorological information for major commodities using satellite data with technical support from JAXA.

2.2 Land observing -ALOS

Rice is the main commodity crop in Asia. It is a staple food for a significant portion of the population in many Asian countries and holds immense cultural and economic importance. Asia is often referred to as the "Rice Bowl of the World" due to its dominant role in global rice production and consumption. Rice is not only a primary source of nutrition but also a key component of the agricultural economies of countries across the Asian continent. It is grown in a wide range of agroecological environments, from lowland paddies to upland fields and even in flood-prone areas. The methods of cultivation vary, but rice remains a central crop in the region, supporting livelihoods and economies. In many Asian countries, including major rice producers like India, China, Indonesia, and Thailand, the main rice crop season typically divided into two primary seasons: "wet/rainy" season and "dry" season. Wet/Rainy season is the main rice growing season that coincides with the monsoon rains. The availability of water from monsoon rains allows the cultivation of rice in flooded fields. This is the season when most of the rain-fed rice is planted and harvested.

Dry season follows the wet/rainy season, and this season is characterized by lower rainfall and may require irrigation for rice cultivation. In some regions, varieties of rice that are better adapted to drier conditions are grown during this season.

Given the dynamic of rice cultivation, extensive area coverage, dependency of weather patterns, satellite remote sensing can assist in many ways to monitor and assess status of a rice crop lands. Specifically, synthetic aperture data (SAR) is indispensable to monitor paddy field because SAR data can be used in all weather conditions given the persisting cloud cover during monsoon season.

ALOS-2 with the Phased Array type L-band SAR-2 (PALSAR-2), shown in Figure 2, was launched in 2014 and has been in operation until today. PALSAR-2 has a Spotlight mode (1 to 3 m) and a high resolution (Stripmap) mode (3 to 10 meter), while PALSAR has a 10-meter resolution. PALSAR-2 achieved the highest resolution of L-band radar by implementing 3-meter resolution using Stripmap mode while keeping wide observation swath.

Figure 2. ALOS-2

The value of L-band Synthetic Aperture Radar (SAR) lies in its unique characteristics that make it suitable for various applications, particularly in remote sensing and Earth observation. L-band SAR operates at a wavelength of around 1 to 2 GHz, which falls within the microwave range. Especially, L-band SAR can penetrate through vegetation and soil to a certain extent due to its longer wavelength compared to higher-frequency microwave bands like C-band or X-band. This penetration capability allows L-band SAR to provide information about the subsurface conditions, making it useful for applications like soil moisture estimation and mapping of root zone characteristics. It can also be used to estimate forest biomass, monitor deforestation, track agricultural crop growth, and assess soil moisture levels.

ALOS-4 is a satellite to observe the Earth's surface using its onboard phased array type L-band synthetic aperture radar (PALSAR-3). The L-band radar technology has continuously been developed in Japan. With further improved observation performance compared to the predecessor PALSAR-2 aboard the ALOS-2 are developing the satellite aiming at achieving both high resolution and a broader observation swath. Unlike observations by an optical sensor, radar images can be acquired day and night as it does not require sunlight. Moreover, since radio waves can penetrate cloud, the images can be obtained regardless of weather condition. The ALOS-4 will leverage these merits for observing and monitoring disaster-hit areas, forests, and sea ice. In addition, it will also challenge new areas such as monitoring infrastructure displacement.

3 Overview of Rice Crop Monitoring Service in Asia

3.1 Asia Rice in GEO GEOGLAM

GEOGLAM Asia Rice is a specific initiative that falls under the Global Agricultural Monitoring (GEOGLAM) under G20 declaration that aims to enhance the use of satellite and Earth observation data to monitor and provide information about agricultural production around the world. Especially, with the cooperation of FAO Agriculture Market Information System (AMIS), it contributes to food security and supports decision-making in agriculture.

The specific focus of GEOGLAM Asia Rice is to monitor rice production in the Asian region using satellite observations and remote sensing technologies. This initiative recognizes the importance of rice as a staple food in many Asian countries and aims to provide timely and accurate information to enhance rice production, manage risks, and improve food security. The main objectives of GEOGLAM Asia Rice are; rice crop monitoring, yield estimation, early warning, damage assessment, advanced research cooperation and knowledge sharing and outreach to satellite data usage for agriculture decision making support.

Under timely monitoring, rice crop outlook information is compiled to provide high accurate, transparent and up-to-date information on rice crop conditions, including planting, growth, and health, throughout the growing season. Yield estimation is to estimate rice yields in different regions using satellite data and modelling techniques. Rice crop outlook could further assist early warning and damage assessment to identify potential or ongoing issues such as droughts, floods, or pest outbreaks that could impact rice production, allowing early response and mitigation.

In addition, Asia Rice activity promotes not only data and derived information sharing but also knowledge sharing through workshop such as Asia Ocean GEO symposium and training in cooperation with space agencies and agriculture related government organizations in Asia.

3.2 ARSAF SAFE

The Asia-Pacific Regional Space Agency Forum (APRSAF) is an international forum that facilitates cooperation and collaboration among space agencies, public and private organizations, and experts from academia and research organizations in the Asia-Pacific region. Its goal is to promote peaceful and safe use of space technology and applications for socioeconomic development and environmental protection.

Space applications for the environment (SAFE) initiative launched in the Space Application Working Group of APRSAF addresses the use of space technology and satellite-based systems to monitor, understand, and manage various aspects of Earth's environment. These applications play a crucial role in gathering data, providing insights, and aiding decision-making processes related to environmental management, conservation, and sustainable development. As of March 2020, there have been 27 prototyping activities completed under the SAFE initiative. These prototyping efforts involve testing and developing various methodologies and technologies to address specific environmental or agricultural challenges. Since 2020, the implementing structure of SAFE was changed from a bilateral approach to a "multilateral" approach. This change allows more extensive collaboration with space agencies, research institutions, and international organizations within the Asia-Pacific region. Since then, three SAFE projects were launched including *agromet*, *rice crop monitoring* and *CH4Rice.*

The objective of Agromet project initiated in 2018 was to develop a mechanism and system for sharing agrometeorological information. This information is aimed at improving agriculture information services across the region, thereby enhancing agricultural planning and management. Second project is the rice crop monitoring project started in 2018. This project was designed to assist the region in monitoring food security status by using space-based data and technology to monitor rice crops. The newest project is CH4Rice project started in 2022. This CH4Rice project is designed to monitor water level in paddy field to reduce CH4 (Methane) emission from paddy field optimizing water level during growing season. Figure 3 shows a summary of all the above three projects conducted under APRSAF SAFE.

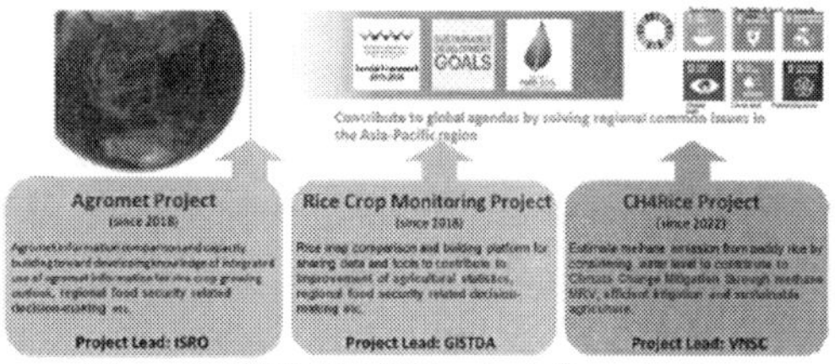

Figure 3 APRSAF SAFE projects for agriculture

3.3 INAHOR

A software package named International Asian Harvest mOnitoring system for Rice (INAHOR) was developed to estimate rice planted area using SAR data. This package uses data from L-band SAR of ALOS-2 acquired over an area for two or more time periods. During the monsoon when many Asian counties cultivate their major paddy crop, those areas are often covered with heavy cloud limiting the use of optical sensor data. However, the wavelength that uses on SAR sensors can penetrate the cloud allowing SAR sensors to monitor paddy fields regardless of the presence of cloud. Therefore, this software package was developed to address the needs in Asia with the collaboration of Asian space agencies and agriculture organizations under APRSAF SAFE. Also, this effort was supported by the Asian Development Bank and number of private companies. INAHOR was introduced and training assistance was provided in number of Asian countries to collect agricultural statistics data much effectively under APRSAF SAFE and GEO Asia Rice team activity for rice crop monitoring.

INAHOR uses ALOS-2 data and administrative boundary data provided in GIS format as inputs. The estimates of rice crop area for each administrative district is calculated and exports as a CSV file, and a Validation Sheet (jpg file format) with a Landsat image of the administrative district. In INAHOR, median filtering of ALOS-2 data is used for noise reduction and machine learning is used for classification. In order to promote easy and efficient usage of INAHOR, the software is installed and available for users on cloud-based systems such as Google Earth Engine. The concept of INAHOR is shown in Figure 4.

Figure 5 shows the result of Vietnam rice crop area estimation in state / provincial level. In most of provinces, INAHOR results are well matched with ground based statistical information of the country. Where there are discrepancies in ground based and satellite derived results, agriculture organizations were requested to explore the reasons for such disparity. Although there are some discrepancies, satellite estimations are very useful to improve national agricultural statistics. Further, INAHOR outcome is useful for smart farming such as advising planting period, fertilizing timing, water management, and pest management.

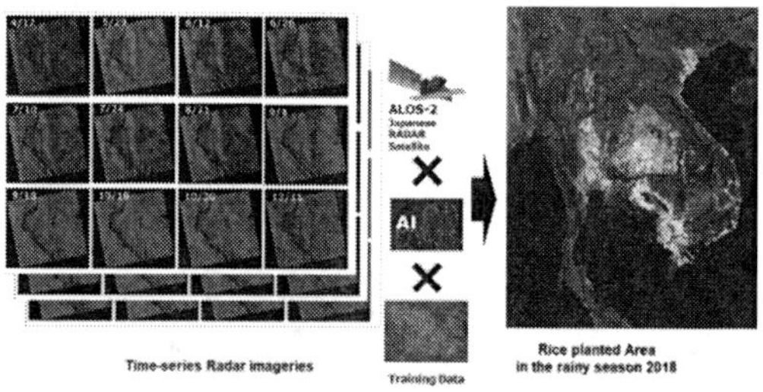

Figure 4 INAHOR concept

Vietnam, Hưng Yên 2022 wet season rice

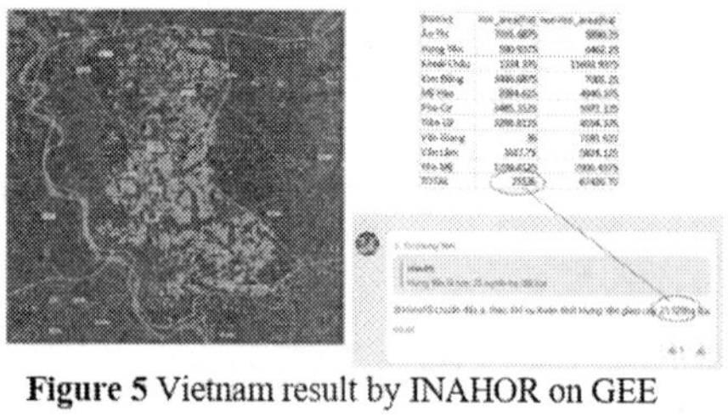

Figure 5 Vietnam result by INAHOR on GEE

INAHOR uses random forest algorithm as the machine learning technique. It is indispensable to collect training data on the ground for assist the train algorithm of the classification. During the ASEAN Food Security Information System (AFSIS) training activity using INAHOR on GEE platform, over 3000 points (rice and non-rice crop field) were collected as training data. This is a labour intensive and costly activity as field visits and other information such as optical satellite data and map information are required to identify paddy/non-paddy training points. Therefore, with the cooperation of supportive agriculture machinery companies, an automatic training data collection system was introduced to make the training data collection activity efficient.

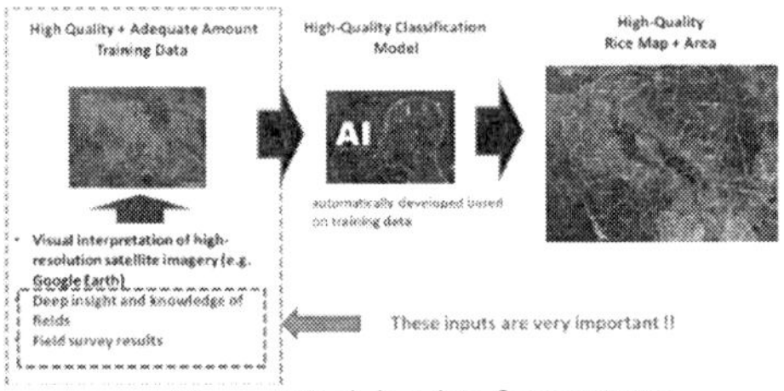

Figure 6 Training data for INAHOR

Overall, INAHOR with ALOS-2 showcases the power of space technology and remote sensing in addressing real-world challenges, such as accurate agricultural area monitoring. By using SAR data to estimate paddy rice planting areas, even in cloudy conditions, the system can provide valuable insights to support food security, resource allocation, and sustainable agricultural practices.

3.4 JASMAI

In 2021, JAXA and the Ministry of Agriculture, Forestry and Fisheries collaborated to promote the use of earth observation satellite data in the field of agriculture, forestry and fisheries by establishing the Japan Satellite Monitoring System for Agrometeorological Information (JASMAI). JASMAI utilizes JAXA's research and development results of a system that enables users to view agricultural weather data (temperature, precipitation, humidity, solar radiation, wind patterns, and soil moisture) and crop growth conditions (vegetation index) on the web in near-real time. These factors are crucial for understanding the climatic conditions that impact crop growth and the crop health. JASMAI utilizes satellite observation data from JAXA, Japan Meteorological Agency, NASA, and number of other countries to provide information on the growth of main production zones of major cereals and other crops, specifically in Asia in the form of maps and graphs.

JASMAI provides users with easy-to-understand visualizations and tools to interpret the data. Farmers, researchers, and policymakers can access these platforms to make use of the data to derive necessary information to make decisions about irrigation needs, planting strategies, harvesting time, and pest management plans etc. JASMAI can also offer data specific to different regions and areas, helping farmers make decisions tailored to their local conditions. In addition, JASMAI's historical data can help in identifying trends, making predictions about climate patterns, and adapting farming practices to changing conditions.

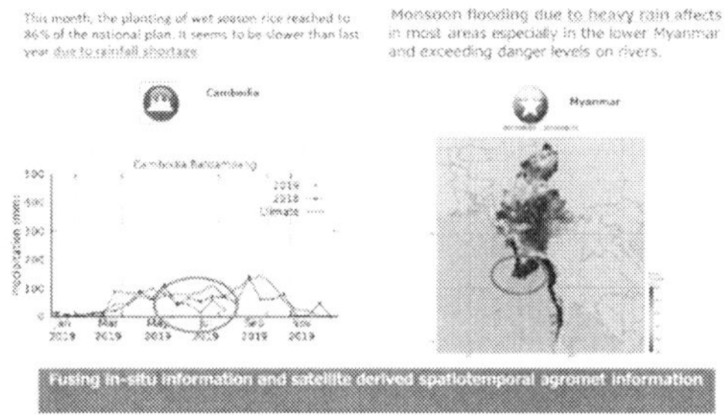

Figure 7 Agromet information for rice crop outlook using JASMAI/JASMIN in ASEAN

The rice crop outlook is a good example of the use of JASMAI. It can be used before planting seasons to guide farmers in making decisions about inputs, cultivation practices, and potential risks. Governments and organizations involved in agriculture often provide guidance based on these outlooks to support informed decision-making and mitigate potential challenges.

Also, JASMAI enhances agricultural resilience by providing timely and accurate information to manage risks associated with weather variability and climate change. It aids in optimizing resource use, increasing productivity, and contributing to sustainable agricultural practices.

4 Conclusion

Precision Agriculture and optimization of farmland in puts using earth observation data, IoT, GIS and AI/ML technologies, farmers can create strategic plans and evidence-based maps for variable rate application of inputs, such as fertilizers, pesticides, and water. This targeted approach optimizes resource usage, reduces costs, and minimizes environmental impacts. At present, earth observation satellite community provides enormous amount of time series data and derived information on certain agroclimatic parameters and land parameters for the possible use in agriculture. JAXA JASMAI and COPERUNICS by EU/ESA are good examples that demonstrates the provisions of space-based technologies. However, for individual farmer, the spatial resolution of those open and free data provided by the satellite communities are not sufficiently enough.

Therefore, for individual needs, farmers may have to turn to commercial providers or for ground-based observations including drone to monitor their individual fields. Furthermore, reasonable price of IT / GIS platform with data analysis system are also a hurdle in Asia because of cost and limited IT expert resources / skill to interpret earth observation information. With the current global commitments, it is expected that more and more high-resolution satellite data become available with improved spatial and temporal resolutions, and it is anticipated that the strengthened involvement of the commercial sector in reaching individual crop land for constructive and productive smart farming.

References

Mohamed, E., et al, 2021: Smart farming for improving agricultural management, The Egyptian Journal of Remote Sensing and Space Science, Volume 24, Issue 3, Part 2, December 2021, Pages 971-981 DOI: 10.1016/j.ejrs.2021.08.007

V. S. R, et al, 2022: Smart Farming: The IoT based Future Agriculture, 4th International Conference on Smart Systems and Inventive Technology (ICSSIT), pp. 150-155 DOI: 10.1109/ICSSIT53264.2022.9716331.

Inoue, Y, 2020: Satellite- and drone-based remote sensing of crops and soils for smart farming, pp. 798-810 , Soil Science and Plant Nutrition , Volume 66, 2020 - Issue 6 DOI: 10.1080/00380768.2020.1738899

Inoue, Y, Yokoyama, M, 2017: Drone-based Remote Sensing of Crops and Soils and Its Application to Smart Agriculture., Journal of the Remote Sensing Society of Japan 37: pp 224–235

Inoue, Y, et al, 2002: Season-long Daily Measurements of Multifrequency (Ka, Ku, X, C, and L) and Full-polarization Backscatter Signatures over Paddy Rice Field and Their Relationship with Biological Variables, Remote Sensing of Environment 81: pp 194–204

Hori, M., Murakami, H., Miyazaki, R., Honda, Y., Nasahara, K., Kajiwara, K., Nakajima, T. Y., Irie, H., Toratani, M., Hirawake, T., Aoki, T., 2018: GCOM-C Data Validation Plan for Land, Atmosphere, Ocean, and Cryosphere. Trans. JSASS Aerospace Tech. Japan, Vol. 16, No. 3, 218-223. DOI: 10.2322/tastj.16.218

Tanaka, K., Okamura, Y., Mokuno, M., Amano, T., Yoshida, J., 2018: First year on-orbit calibration activities of SGLI on GCOM-C satellite, Proc. SPIE 10781, Earth Observing Missions and Sensors: Development, Implementation, and Characterization V, 107810Q DOI: 10.1117/12.2324703

Kachi, M., et al., 2014: Recent improvements in the global satellite mapping of precipitation (GSMaP). IEEE Geoscience and Remote Sensing Symposium, 3762-3765, Doi.org/10.1109/IGARSS.2014.6947302.

Tadono, T., Ishida, F., Oda, S., Naito, K., Minakawa, H., 2014: Precise Global DEM Generation By ALOS PRISM. ISPRS Annals of the Photogrammetry, Remote Sensing and Spatial Information Sciences, Vol.II-4, 71-76.

Sobue, S., 2017: Status of the advanced land observing satellite-2 (ALOS-2) and its follow-on L-band SAR mission. IEEE International Geoscience and Remote Sensing Symposium (IGARSS), 2427-2429.

FAOSTAT, 2018: Food and Agriculture Organization Corporate Statistical Database (FAOSTAT) 29.

MAFF: Development of Smart Agriculture, https://www.maff.go.jp/e/policies/tech_res/smaagri/robot.html

JASMAI: https://jasmai.maff.go.jp/en/

JAXA INAHOR:

https://www.eorc.jaxa.jp/SAFE/project/ricemonitoring/04-20221114_INAHOR_on_GEE_Tutorial_FinalVer_221110.pdf

APRSAF SAFE: https://www.eorc.jaxa.jp/SAFE/

GEOGLAM Asia Rice : http://asia-rice.org/

Analysis of the Effect of Brand Equity on Purchasing Decision towards Organic Rice in Special Region of Yogyakarta

Nisa Puspita Mawadati[1], *Novita* Erma Kristanti[1*], and *Nafis* Khuriyati[1]

[1] Department of Agroindustrial Technology, Faculty of Agricultural Technology, Universitas Gadjah Mada, Indonesia

Abstract. The community's need for organic rice increases every year along with changes in people's lifestyles towards a healthy lifestyle. This has made many organic rice brands appear on the market. Each of these brands needs to implement the right marketing strategy to increase its competitive advantage. The study aims to analyze the influence of brand equity elements (brand awareness, brand association, perceived quality, and brand loyalty) on purchasing decisions for organic rice. The Structural Equation Modeling-Partial Least Square (SEM-PLS) is used to analyze the data, with a purposive sampling data collection technique. Respondents in this study were 105 consumers of organic rice "Lingkar Organik". The "Lingkar Organik" brand was chosen because it is the market leader in the branded organic rice market in Special Region of Yogyakarta. Based on the results of the study, it was found that the perceived quality and brand loyalty variables have a significant effect on the purchase decision towards organic rice. Brand association and brand awareness variables have no significant effect on purchasing decisions.

1. Introduction

The community's need for organic rice increases yearly along with changes in people's lifestyles towards a healthy lifestyle. The Covid-19 pandemic has made people more careful in choosing food that is safe for their health. This lifestyle toward a healthier society is one of the reasons for the existence of an organic farming system.

The organic farming system is farming that returns to nature without using chemical fertilizers or pesticides [1]. In 2010, the government issued a program called Go Organic through the Ministry of Agriculture. This program is a support for organic farming because it is considered more beneficial to health and the environment. The concern of the government to support organic farming programs can be seen in the policies written in the Nawacita Program until 2020. The Indonesian government launched the establishment of 100 Organic Villages, consisting of 600 Organic Food Villages, 250 Organic Horticulture Villages, and 150 Organic Plantation Villages to support the growth of the organic farm in Indonesia. [2].

The Special Region of Yogyakarta is a province starting to develop organic farming. The Kulon Progo Regency Government produce organic rice with a land area of 24 hectares in 2021 [3]. On the other hand, in 2022, Sleman Regency also cultivates organic rice with a total area of 20 hectares [4]. The increase in the area of land for organic rice has led to the emergence of more organic rice brands in the organic rice market in the Special Region of Yogyakarta. The branding of agricultural commodities, especially organic rice, is one of the efforts made by producers to provide added value to products [5]. However, the consumption of organic rice in DIY is still relatively low, only around 0.5% of total rice consumption [6].

One of the brands that have become a market leader in the organic rice market is organic rice with the Lingkar Organik brand from PT Lingkar Organik Indonesia. The company is engaged in the packaging and marketing industry for organic products, including organic rice. Lingkar Organik sells its products to several resellers and retailers across the Special Region of Yogyakarta. Therefore, researchers are interested in examining the effect of brand equity on purchasing decisions toward organic rice in the Special Region of Yogyakarta.

2. Theoretical Framework and Hypothesis Formulation

There are several studies that use the similar variables as the variables used in this study. Shafi & Madhaviaiah's research (2013) concerning the effect of brand equity on consumer behavior of organic products in India uses brand awareness, brand loyalty perceived quality, and brand association as variables of brand equity. Firmansyah et al's research (2021) concerning on marketing mix, brand equity, and purchase decision of packaged rice products, mentions purchasing decisions as one of the variables. Nonetheless, previous studies about the effect of brand equity on purchasing decision towards organic rice have not been discussed. This study aims to analyze the effect of brand equity on purchasing decisions for organic rice in the Special Region of Yogyakarta. This research is based on brand equity theory by Aaker (1991) which consists of several variables including brand awareness, brand association, perceived quality, brand loyalty, and other proprietary brands. However, this research has some limitations such as 1) The research was conducted on consumers of branded organic rice in the Special Region of Yogyakarta., 2) The organic rice brand was selected based on the market leader in the organic rice market in

*corresponding author: novita_erma@ugm.ac.id

DIY, namely "Lingkar Organic" organic rice, 3) The research consists of four elements of brand equity, namely brand awareness, brand association, perceived quality, and brand loyalty. The research framework will be figured out in Figure 1. The arrow shows the relationship between exogenous latent variables (brand awareness, brand association, perceived quality, and brand loyalty) and endogenous latent variables (purchasing decision). The hypothesis of this study are:

H1: Brand Awareness has a significant effect on purchasing decisions towards organic rice

H2: Brand Association has a significant effect on purchasing decisions towards organic rice

H3: Perceived Quality has a significant effect on purchasing decisions towards organic rice

H4: Brand Loyalty has a significant effect on purchasing decisions towards organic rice

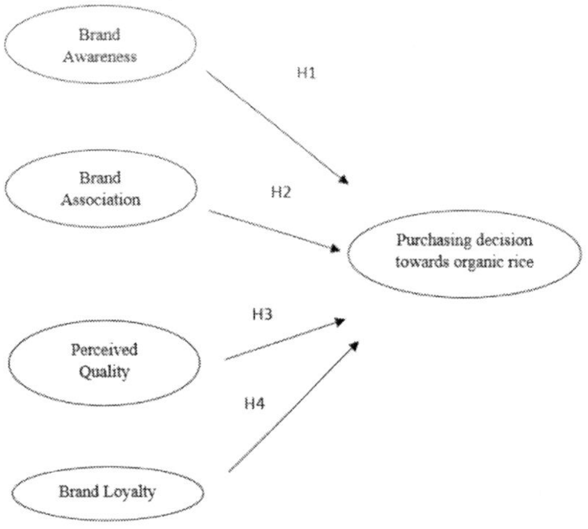

Fig 1. Research Framework

3. Research Method

The data used for this research paper was collected through an online questionnaire. The organic rice brand tested for brand equity is organic rice with the "Lingkar Organik" brand. The brand was chosen because the brand is the market leader in the branded organic rice market based on the results of market survey that have been conducted. It was found that "Lingkar Organik" has sales consistency in various retailers, even distribution of products, and is a popular and best-selling product in e-commerce. Data were obtained from 105 respondents who consumed organic rice branded "Lingkar Organic". The respondents criteria are (1) Domiciled in Special Region of Yogyakarta, (2) Male or Female aged 17 years and above, (3) Have consciously and intentionally purchased "Lingkar Organik" Organic Rice products in the last one year.

The data processing method uses the Partial Least Structural Equation Modeling method. The SEM-PLS method is used because this method can perform 3 activities together, such as checking the validity and reliability of the instrument which is a function of Comfirmatory Factor Analysis (CFA), testing the relationship between latent variables (path analysis), and obtaining a model that is useful for forecasting (regression analysis) [7].

This method was applied to examine the measurement model (outer model) and structural model (inner model). Evaluation of the measurement model (outer model) is carried out by conducting validity and reliability tests. The inner model is called the structural model that connects latent variables. At this stage, the evaluation of the structural model (inner model) is carried out by paying attention to the R-square (R^2) value.

4. Result

According to the respondents' profiles, the study found the majority (78%) were female, in the 25 – 34 year age group (42%), lives in Sleman (56%), education Undergraduate degree (51.4%), private employee (40%), salary Rp. 1.500.000 – Rp3.000.000 per month (39%), and purchase frequency between 1 – 3 times per month (85.7%). The demographic information of the respondents is shown in Table 1.

Table 1: Demographic information of the respondents

Characteristics	Category	Frequency	%
Gender	Male	82	78%
	Female	23	22%

Age	17 – 24 years old	43	41%
	25 – 34 years old	44	42%
	35 – 44 years old	8	8%
	45 – 55 years old	6	6%
	> 55 years old	4	4%
Domicile	Sleman Regency	59	56%
	Yogyakarta City	23	22%
	Bantul Regency	13	12%
	Gunungkidul Regency	9	9%
	Kulon Progo Regency	1	1%
Education	Senior High School	32	30.5%
	Diploma/Academy	4	3.8%
	Bachelor degree	54	51.4%
	Master degree	15	14.3%
Occupation	Private employee	42	40%
	College Student	32	30%
	Entrepreneur	11	10%
	Civil servant	6	6%
	House wife	6	6%
	Teacher/lecturer	4	4%
	Other	4	4%
Salary (per month)	<Rp.1.500.000	27	26%
	Rp1.500.000 - Rp3.000.000	41	39%
	Rp3.000.001 - Rp5.000.000	23	22%
	> Rp5.000.000	14	13%
Purchase frequency (per month)	1 - 3	90	85.7%
	4 – 6	10	9.5%
	>6	5	4.8%

In evaluating the measurement model (outer model) there are three tests, which are convergent validity, composite reliability, and discriminant validity. Convergent validity is calculated using the loading factor value and the average variance extracted value.Table 2 shows the value of the loading factors; Loading factors value is used to determine whether the statement items reflect the variables under study. Items are valid if the value > 0.7 [8], so all items are said to be valid. The expected Average Variance Extracted (AVE) value is > 0.5. AVE is the average squared loading of the construct indicators. An AVE value of 0.5 or greater indicates that on average a construct explains more than half of the variance of its indicators. Based on the calculation results, all latent variables already have a value > 0.5, so all the items are valid. The composite reliability value in Table 2 is said to be reliable if it has a value of more than 0.7 [8]. Based on the table 2, all variables have a value of more than 0.7, so it can be concluded that all variables are reliable or able to explain the variance of each indicator that measures them.

Table 2. Summary Result of the measurement model

Variable	Items	Loadings Factor	AVE	R - Square	Composite Reliability
Brand Awareness	AW1	0.885	0.729	-	0.915
	AW2	0.847			
	AW3	0.859			
	AW4	0.825			
Brand Association	AS1	0.887	0.809	-	0.895
	AS2	0.912			
Percieved Quality	PQ1	0.882	0.671	-	0.890
	PQ2	0.847			
	PQ3	0.788			
	PQ5	0.753			
Brand Loyalty	BL1	0.883	0.634	-	0.873
	BL2	0.717			
	BL3	0.833			
	BL4	0.743			

Purchase Decision	PD2	0.757	0.624	0.649	0.869
	PD4	0.789			
	PD6	0.737			
	PD8	0.760			

Discriminant validity is done to ensure that each concept of each latent model is different from other variables. Discriminant validity can be seen from the Fornell Larcker Criterion values in Table 3. All variables are valid based on the values in the Fornell Larcker table because the Fornell Larcker value in the target variable is greater than the Fornell Larcker value in the other variables. Based on this calculation, all items are valid.

Table 3. Discriminant validity

Latent Variable	AS	AW	PD	BL	PQ
Brand Associaton (AS)	**0.900**				
Brand Awareness (AW)	0.603	**0.854**			
Purchase Decision (PD)	0.600	0.597	**0.790**		
Brand Loyalty (BL)	0.569	0.681	0.701	**0.797**	
Perceived Quality (PQ)	0.642	0.563	0.735	0.614	**0.819**

Structural model evaluation (inner model) is calculated by calculating the R square value. Based on Table 2, the R-Square value of 0.649 means that brand awareness, brand associations, perceived quality, and brand loyalty can explain the purchasing decision variable of 64.9%. Other variables outside this research model explain the remaining 35.1%. Based on Fig.2, the path coefficient value can be seen from the arrows connecting exogenous latent variables to endogenous variables. The positive value obtained indicates that the related variables have a positive effect.

The bootstrapping analysis is used to test hypotheses and determine the significance of the effect of exogenous variables (Brand Awareness, Brand Association, Perceived Quality, Brand Loyalty) on endogenous variable (Purchase Decision). This test uses a significance level of 0.05, and the hypothesis is accepted if the result is greater than the t-table, which is 1.96. The results of the bootstrapping test are shown in Table 4 and Figure 2.

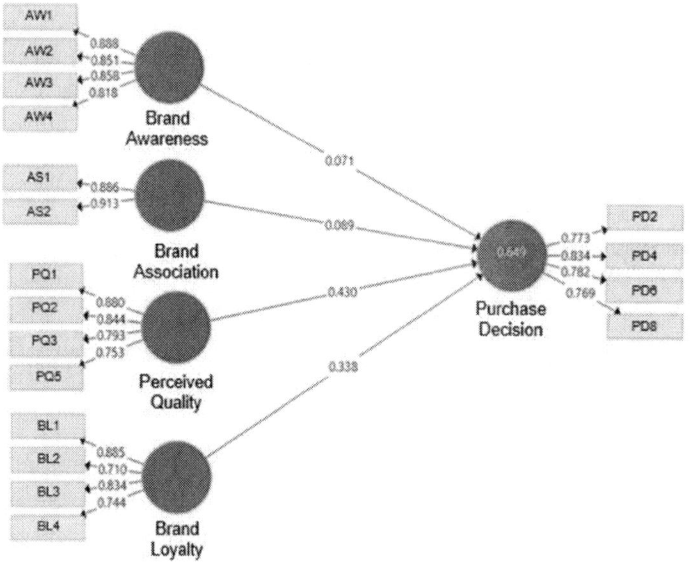

Fig 2. Path Analysis

Table 4. Summary of Findings

Latent Variable	*Original Sample* (O)	*Sample Mean (M)*	*t - table*	*t– statistics*	*Hypothesis*
Brand Awareness → Purchase decision (Hypothesis 1 (H1))	0.071	0.072	1,96	0.678	Not Significant
Brand Association → Purchase decision (Hypothesis 2 (H2))	0.089	0.081	1,96	0.945	Not Significant
Perceived quality → Purchase decision (Hypothesis 3 (H3))	0.430	0.428	1,96	4.067	Significant
Brand Loyalty → Purchase decision (Hypothesis 4 (H4))	0.338	0.348	1,96	3.066	Significant

The brand awareness variable has no significant effect on the purchase decision variable with a t-statistics value of 0.678 < t-table (1.96). This hypothesis does not follow the hypothesis that brand awareness influences purchasing decisions. These results can occur because brand awareness is not the main factor in consuming healthy foods such as organic products [9]. Limited promotions also makes brand awareness among costumers limited [10]. In the brand awareness variable, what is distinctive about organic products and can differentiate organic products from non-organic products is the presence of an organic label on the packaging. This organic label is a physical characteristic that consumers of an organic rice brand can quickly capture and realize quickly. This label can increase consumer confidence and increase purchasing power [11]. Building awareness of organic products can be done based on health awareness. Therefore, an approach can be taken based on consumer habits in consuming organic products [12].

The brand association variable has no significant effect on the purchasing decision variable with a t-statistics value of 0.945 <t-table (1.96). However, this is not in accordance with the hypothesis which states that brand association affects purchasing decisions. Building positive brand associations will have a direct effect on purchasing decisions. In the brand association variable, what is unique about organic products and can differentiate organic products from non-organic products is building strong brand associations for organic products. Brand association with organic products can be done by positioning products as natural, healthy, and environmentally friendly organic products that must be present in marketing communications [13].

The perceived quality variable has the highest path coefficient value of the other variables, with a value of 0.430. This value indicates that the perceived quality variable positively affects the purchasing decision variable. The perceived quality variable also significantly affects purchasing decisions with a t-statistics value of 4.067 > t-table (1.96). The most influential indicator in the perceived quality variable is the PK1 indicator (quality organic rice "Lingkar Organik"), with a loading factor value of 0.880. In the perceived quality variable, the characteristic of organic products that can be a differentiator between organic and non-organic products is the quality of the products obtained according to organic product processing procedures. Processing of organic products must follow the procedures for organic farming systems regulated in SNI 6729: 2016 concerning organic farming systems. Perceived quality in organic products is essential. Perceived quality is a consumer's assessment of the overall superiority of a product [14]. The market for organic products is consumers who desire to improve their health because quality organic products are perceived as healthier than other products [15]. This study's results follow previous studies showing that the quality of organic products has a significant effect on purchasing decisions. Organic products can provide more benefits than non-organic products, so consumers decide to buy products [16]

The brand loyalty variable has a path coefficient value of 0.430. This value indicates that the brand loyalty variable positively affects the purchasing decision variable. The brand loyalty variable also significantly affects purchasing decisions with a t-statistics value of 3.066 > t-table (1.96). The most influential indicator of the brand loyalty variable is BL1 (loyal consumers to organic rice brand "Organic Circle"), with a loading factor value of 0.885. In the brand loyalty variable, what is unique about organic products and can differentiate organic products from non-organic products is building brand loyalty based on consumer confidence in the quality of organic products. This result is in line with previous research on brand loyalty, which states that consumers who have confidence in the quality of a brand tend to be loyal [17]

5. Conclusion

Variables that have a positive and significant effect on purchasing decisions for organic rice are perceived quality and brand loyalty. Brand awareness and brand association variables have no significant effect on purchasing decisions for organic rice.

Acknowledgement

This paper is supported by "Rekognisi Tugas Akhir" Universitas Gadjah Mada program 2023 for students' thesis publication with contract number 5075/UN1.P.II/Dit-Lit/PT.01.01/2023

References

[1] T. B. Purwantini and N. Sunarsih,. Organic Agriculture: Concept, Performance, Prospects, and Constraints, Agro-Economy research forum. 37(2020)
[2] A. Charina, R. A. B. Kusumo, A. H. Sadeli, & Y. Deliana, Factors Influencing Farmers in Implementing Standard Operating Procedures (SOP) for Organic Farming Systems in West Bandung Regency. Extension Journal. 14 (2018) 68–78.
[3] Sutarmi. Kulon Progo organic rice production JOSS prevents short stature. Information on https://jogja.antaranews.com/berita/472596/kulon-progo-produksi-beras-organik-joss-cegah-tubuh-pendek (2021)
[4] V. S. Pranyoto, Sleman Regency Develops Organic Rice for Higher Selling Prices. Information on https://jogja.antaranews.com/berita/590409/sleman-kembangkan-padi-organik-untuk-harga-jual-lebih-tinggi (2022).
[5] M. A. J. Murti & W. Agung. Consumer Behavior In Purchasing Branded Rice In Jember Regency And The Factors That Influence It. 4 (2010)
[6] Sunartono. The percentage of organic rice consumption in DIY is still low. Information on https://jogjapolitan.harianjogja.com/read/2022/12/10/510/1120001/persentase-konsumsi-beras-organik-di-diy-masih-rendah (2022)
[7] M. R., Firmansyah, U. Sumarwan, & M. M. Ali, Marketing Mix, Brand Equity, and Purchase Decisions of Packaged Rice Products. Journal of Management and Agribusiness. (2021).
[8] Hair, J., Hult, T., Ringle, C., Sarstedt, M., Danks, N., & Ray, S, Partial Least Squares Structural Equation Modeling (PLS-SEM) Using R. (2014)
[9] S. Ali, S. Attiq, & N. Talb, Antecedents of Brand Hate: Mediating Role of Customer Dissatisfaction and Moderating Role of Narcissism. Pakistan Journal of Commerce and Social Sciences, 14 (2020) 603–628.
[10] R. Ekasari & I. M. L. M. Jaya, The Determinants of Consumer Purchasing Decisions of Health Food Products: An Empirical Study from Indonesia. Journal of Asian Finance, 8 (2021).
[11] N. Hidayati, M. Sarma & M. Syamsun, The Effect of Organic Vegetable Product Quality on Consumer Satisfaction in Forming Customer Loyalty. Journal of Applied Science, 4 (2014).
[12] C. A. Puspitasari, L. N. Yuliati, & F. Afendi, The Influence of Green Marketing, Environmental and Health Awareness on Purchase Decisions of Organic Food Products Through Attitudes. Journal of Business and Management Applications. (2021).
[13] V. Grigaliūnaitė & L. Pilelienė, How do brand associations affect purchase intentions? A case of organic products. Research For Rural Development, 2, (2017) 210–216.
[14] F. A. Konuk, The influence of perceived food quality, price fairness, perceived value and satisfaction on customers' revisit and word-of-mouth intentions towards organic food restaurants. Journal of Retailing and Consumer Services, 50 (2019) 103–110.
[15] S. Kamboj, M. Matharu, & M.Gupta, Examining consumer purchase intention towards organic food: An empirical study. Cleaner and Responsible Consumption, 9 (2023).
[16] P. Widyastuti, Quality and Price as the Most Important Variables in the Decision to Buy Organic Vegetables. Ekspektra: Journal of Business and Management, (2018) 17–28.
[17] S.Pattweekongka, Organic Food Brand Trust and Brand Loyalty: Evidence from Thailand. AgBioForum, 24(2022) 129–133.

Assessing the Viability of a Small Micro-Enterprise of an Urban Farming Group in Bausasran Village, Yogyakarta, in Developing a Vegetable-based Noodle Products

Habib Luthfi Ash Shiddiqie[1], *Ryan Anugrah* Putra*[,1], *Ardi* Wiranata[1], *Teguh* Ariyanto[2]

[1] Department of Mechanical & Industrial Engineering, Faculty of Engineering, Universitas Gadjah Mada, Jalan Grafika No. 2, 55281 Yogyakarta, Indonesia
[2] Department of Chemical Engineering, Faculty of Engineering, Universitas Gadjah Mada, Jalan Grafika No. 2, 55281 Yogyakarta, Indonesia

Abstract. Gemah Ripah Farmer Group, located in Bausasran Village, Danurejan District, Yogyakarta City, is one of the small farmer groups that have successfully practiced urban farming. The group grows vegetables and fruits. The harvest is then sold directly to consumers or distributors. To increase the economic value, the group plans to start a small micro-scale dry noodle product business by utilizing their cultivation results. In business planning, an economic feasibility analysis is required. This study aims to determine the business feasibility with the indicators of Net Present Value (NPV), Internal Rate of Return (IRR), Payback Period (PP), and Added value analysis with Hayami Method. This study simulates two business scenarios for dry noodle production, namely own production and third-party production with three variables of production capacity analyzed, namely 760, 950, and 1140 units/month. The results of this study show that self production scenario with a production capacity of 1140 pieces/month produces a higher level of feasibility with a value of NPV, IRR and payback period of IDR 42,038,344, 59% and 1.88 years respectively. Value-added analysis provides added value in the self-production scenario of IDR 55,527/kg at the production rate of 1140 units/month.

1 Introduction

According to data from the Horticulture Statistics Sub-directorate of the BPS, Indonesia experienced a 1.2 million hectare decrease in harvested area for vegetable crops between 2018-2022 [1]. Simultaneously, the country's population increased by 11 million individuals during this timeframe [2]. As the population grows, additional land is required for housing developments, with agricultural land being one of the proportions that must be sacrificed. In this way, the agricultural land area will decrease, disrupting food security. Therefore, innovation and creativity are necessary to utilize increasingly limited urban areas as micro-farming and urban farming land.

Urban farming involves cultivating crops within cities or urban areas to produce food locally in densely populated regions [3]. Urban farming can be accomplished through various methods, including vertical farming, rooftop gardening, urban parks operating as agricultural land, and utilizing modern technology, such as hydroponics or aquaponics [3]. Urban farming also offers a solution to improve food sustainability in urban areas, reduce dependence on imported food, and ensure access to fresh food for urban communities [4]. Furthermore, urban farming has substantial economic potential, including creating new business opportunities such as selling produce such as fruits, vegetables, and processed goods and reducing costs associated with food transportation and distribution [5].

One area that has successfully piloted the transformation of a narrow urban area into a vegetable cultivation village is Bausasran Village, located in Danurejan District of Yogyakarta City. Administratively, Bausasran Village spans an area of 470,000 square meters, of which 400 square meters can be utilized for vegetable cultivation [6]. Among the farmer groups operating in Bausasran Village, the Gemah Ripah Farmer Group is one of them. The farmer groups in Bausasran Village utilize the materials around them to build farms with soilless farming techniques such as hydroponics and aquaponics, as well as planting with soil in pots and verticulture. They garden either at the household scale (Home Farming) or at the community scale (Community Farming) to farm on narrow vacant lands. The cultivated plants are diverse, including vegetables, fruits, ornamental plants, and herbs. The crop yields are sold in the market, and the profits earned are reinvested into the agricultural business capital [12].

The agricultural crops of the Bausasran Village Farmers Group which are sold directly to consumers without undergoing processing into ready-to-eat food

* Corresponding author: ryan.putra@ugm.ac.id

have a relatively low economic value. To increase the economic value of their crops, a dry noodle production business is planned with one of the ingredients sourced from cultivated agricultural products. To determine whether a business is feasible to run, an economic feasibility analysis is needed so that the risk of loss and bankruptcy can be minimized.

2 Method

This research aims to investigate the economic analysis of expanding the dry noodle manufacturing business within the Gemah Ripah Farmers Group. Key indicators to consider include the Net Present Value (NPV), Internal Rate of Return (IRR), and Payback Period (PP). The NPV indicator calculates the ratio of present cash inflows to outflows over a set period, assessing the benefits of a production process [7]. The NPV can be calculated as follows:

$$NPV = \sum_{t=1}^{N} \frac{R_t}{(1+i)^t} \tag{1}$$

where N represents number of periods, i is rate of return and t is cash period.

The IRR indicator is utilized to calculate the interest rate at which the NPV equals 0. It serves as a representation of the percentage rate of return-on-investment capital, compared to the Minimum Acceptable Rate of Return (MARR) [10]. For a business to be feasible, the IRR value must exceed the MARR value [10]. The IRR value is derived using the following equation:

$$IRR = i_1 + \frac{NPV_1}{NPV_1 - NPV_2}(i_1 - i_2) \tag{2}$$

Where i_1 represents the interest rate that produces positive NPV values, and i_2 is interest rate that produces negative NPV values. NPV_1 on the other hand, represents a positive NPV, while NPV_2 represents a negative NPV.

The Payback Period indicator shows how long the business process takes to return investment capital from the annual cash inflow generated [11]. The equation for calculating PP is:

$$PP = t_i + \frac{M_t}{C_t} \tag{3}$$

t_i = the year right before the initial investment is fully returned

M_t = initial capital that is still not covered

C_t = cash flow in period t

In this study, the authors analyze the economic feasibility of two applicable scenarios. First, the business scenario for dry noodle production by procuring its own production equipment, starting from raw material processing, packaging, to sales. Second, the dry noodle business scenario involves dry noodle manufacturing services so that farmer groups only need to do packaging and sales. Figure 1 shows the schematic of scenario 1 (a) and scenario of scenario 2 (b).

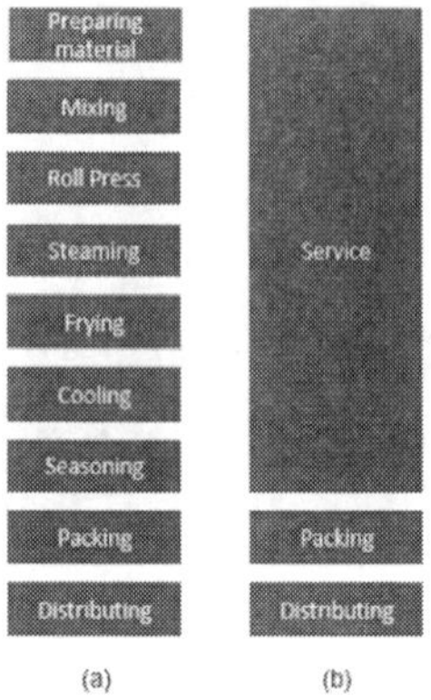

Fig. 1. Scenario 1 (a) and Scenario 2 (b) Scheme

To determine business feasibility in both scenarios, the authors create three production capacity variables in each scenario with NPV, IRR, and Payback Period indicators. The research flowchart is presented in Figure 2.

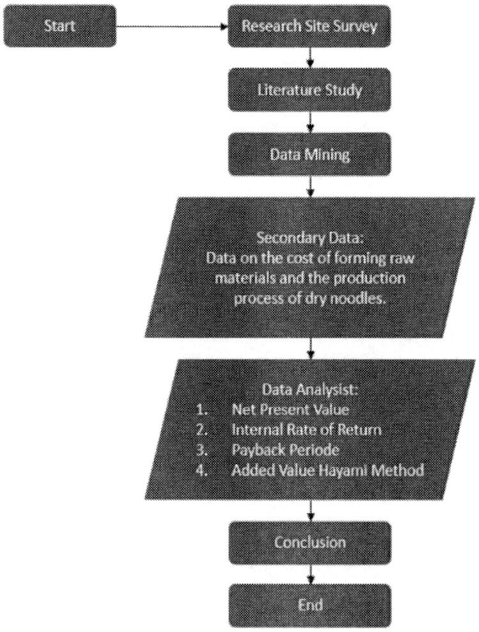

Fig. 2. Research methodology flowchart

3 Result and Discussion

3.1 Investment

Tool	Qty	Investment Cost (Rp)
Digital Balance	1	200,000
Horizontal Dough Mixer HMX-H02	1	2,700,000
Noddle Maker NOD1503	1	1,800,000

Food Dehydrator DHY-10H	1	3,200,000
FOMAC Electric Steamer STM-JEZ5	1	4,500,000
Gallon	1	32,000
Vaccum Sealer	1	300,000
Total		12,732,000

Table 1. Investment Cost of Scenario 1

The Bausasran Farmers Group plans to invest in processing the crops into processed dry noodle products. The Farmer Group plans to invest Rp 20,000,000 which will come from community service grants as well as profits from the sale of previous agricultural products. The amount of initial investment capital considers the price of production equipment and production costs for the next 3 months. The first scenario requires three workers working in production, packaging, and distribution. The production equipment investment cost component for scenario one is shown in table 1.

Cost Component	760 per month (Rp)	950 per month (Rp)	1140 per month (Rp)
Fixed Cost	2,148,224	2,148,224	2,148,224
Variable Cost	2,268,915	2,982,015	3,755,115
Semi-Variable Cost	780,000	900,000	1,000,000
Total Operational Cost	5,197,139	6,030,239	6,903,340
Sales Projection	5,737,388	7,177,994	8,522,011
BEP	6,740	6,142	5,797

Table 2. Cost Component per month of Scenario 1

The price per pack of 80 grams noodles is planned to be at a profit margin of 12-39% and is tested on production capacity variables of 760 per month, 950 per month and 1140 per month. The cost-forming components of scenario I are shown in table 2.

Then for the second scenario, namely the sale of dry noodles by involving the services of making dry noodles or in another sense it does not require production equipment as in the first scenario, except for a vacuum sealer. The workforce needed in the second scenario is only one person to communicate with service providers as well as carry out packaging and distribution. In this second scenario it is considered for analysis because of the large initial capital required by farmer groups in running a business. In this scenario, farmer groups only carry out packaging and sales activities without the production process. The amount of noodle making services is IDR 4,500/80-gram packaging. The cost-forming components of scenario II are shown in table 3. Table 4 presents the details of the fixed cost, variable cost, semi-variable cost components as a comparison between scenario I (self production) and scenario II (third party production) at a production rate of 760 per month.

Cost Component	760 per month (Rp)	950 per month (Rp)	1140 per month (Rp)
Fixed Cost	449,500	449,500	449,500
Variable Cost	3,724,000	4,730,000	5,756,000
Semi-Variable Cost	780,000	900,000	1,000,000
Total Operational Cost	4,953,500	6,079,500	7,025,500
Sales Projection	5,701,980	7,217,910	8,551,420
BEP	6,413	6,347	6,277

Table 3. Cost Component per month of Scenario 2

In scenarios I and II, the selling price is targeted at IDR 7,500/package. The projected sales target is expected to increase from the assumption of 60% capacity in the first year and then increase 5%/year until the 10th year.

Cost Component	Self Production of 760/month (Rp)	Third Party Production of 760/month (Rp)
Fixed Cost		
Maintenance Cost	539,280	42,000
Depreciation Cost	108,944	7,500
Salary Cost	1,200,000	400,000
Building Rent Cost	300,000	-
Variable Cost		
Materials/ Services Cost	1,689,400	3,420,000
Packaging Cost	304,000	304,000
Overhead Cost	275,515	-
Semi Variable Cost		
Marketing Cost	750,000	750,000
Administration Cost	30,000	30,000

Table 4. Detail of componen cost Self production Vs Third Party Production on 760 per month

3.2 Investment Valuation

To calculate the economic feasibility analysis, the authors assume a business age of 10 years, equipment maintenance costs of 4%/month, and a discount rate of 18%.

3.2.1 Scenario 1: Self Production

1. Net Present Value
   To calculate the NPV for each production capacity scenario per month, annual net cash flow data is needed.

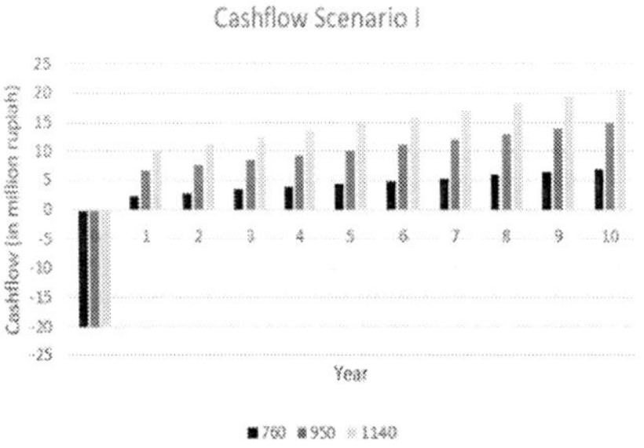

Fig. 3. Scenario I cash flow graph

The data in Figure 3 is entered into equation 1 so that the NPV value for each production capacity scenario is obtained and presented in table 5. Production of 760 per month produces a negative NPV which indicates business feasibility, whereas in the production capacity scenario 950 per month and 1140 per month generates a positive NPV value of IDR 22,745,376 and IDR 42,038,344 respectively. The greater the NPV, the higher and more profitable the return on capital.

	760 per month (Rp)	950 per month (Rp)	1140 per month (Rp)
NPV	-2,143,490	22,745,376	42,038,344

Table 5. NPV of Scenario 1

2. Internal Rate of Return
   The IRR calculation is an iterative calculation, in helping the author determine the IRR for each scenario, the help of Microsoft Excel software is used with the formula "=IRR (cash flow array along with initial capital, initial IRR guess)". From the calculations carried out, the IRR for each production capacity scenario is summarized in table 6. In the production capacity scenario of 760 per month, the IRR is 15% <18%, so this scenario is not feasible, while the production capacity scenario is 950 per month and 1140 per month produces an IRR value of 41% and 59% respectively. The greater the IRR, the greater and more profitable the percentage return on investment.

	760 per month	950 per month	1140 per month
IRR	15%	41%	59%

Table 6. IRR of Scenario 1

3. Payback Period
   To calculate the payback period, we can observe the cumulative cash flow from each scenario and see in what year the cumulative cash flow value is greater than the initial investment cost. Figure 4 shows the cumulative cash flow for each production capacity scenario.

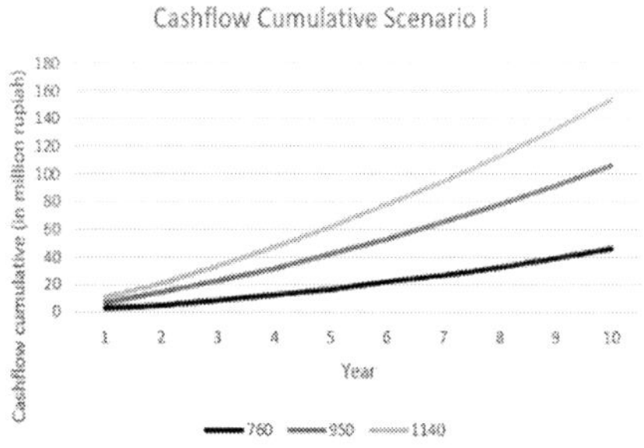

Fig. 4. Scenario 1 cash flow cumulative graph

At the production capacity of 760 per month, the amount of the initial investment will be returned in the year between the 5th and 6th year. Then at a production capacity of 950 per month, the amount of the initial investment will return between the 2nd and 3rd years. Then at a production capacity of 1140 per month, the amount of the initial investment will return between the 1st and 2nd years.

$$PP_{760} = 5 + \frac{20.000.000 - 16.750.410}{21.666.938 - 16.750.410} = 5{,}66\ years$$

$$PP_{950} = 2 + \frac{20.000.000 - 14.246.314}{22.699.450 - 14.246.314} = 2{,}68\ years$$

$$PP_{950} = 2 + \frac{20.000.000 - 10.070.436}{21.310.076 - 10.070.436} = 1{,}88\ years$$

3.2.2 Scenario 2: Third Party Production

1. Net Present Value
   As in the NPV calculation for scenario 1, Figure 5 presents the cash flow from scenario 2 for each production capacity. The NPV calculation results for scenario II are presented in table 7. Production of 760 per month produces a negative NPV which indicates business unfeasibility, while the production capacity scenario of 950 per month and 1140 per month produces a positive NPV value of IDR 10,071,883 and IDR 18,573,338 in succession.

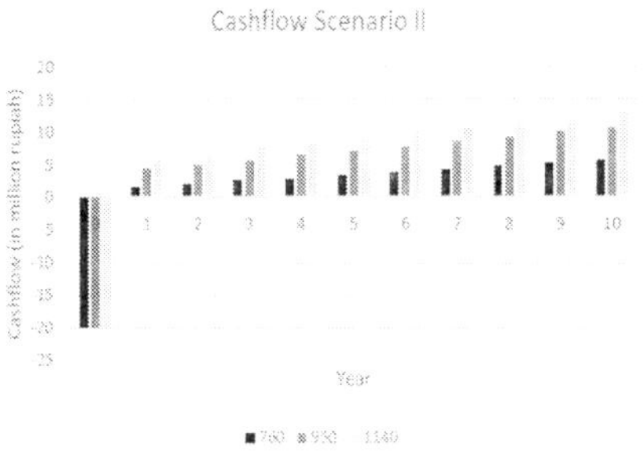

Fig. 5. Scenario 2 cash flow graph

	760 per month (Rp)	950 per month (Rp)	1140 per month (Rp)
NPV	-5.903.118	10.071.883	18.573.338

Table 7. NPV of Scenario 2

2. Internal Rate of Return
   From the calculations carried out, the IRR for each production capacity scenario is summarized in table 8. In the production capacity scenario of 760 per month, the IRR is 11% <18%, so this scenario is not feasible, while the production capacity scenario is 950 per month and 1140 per month produces an IRR value of 29% and 37% respectively.

	760 per month	950 per month	1140 per month
IRR	11%	29%	37%

Table 8. IRR of Scenario 2

3. Payback Period
   In Figure 6, the cumulative cash flow from scenario II for each production capacity scenario is attached.

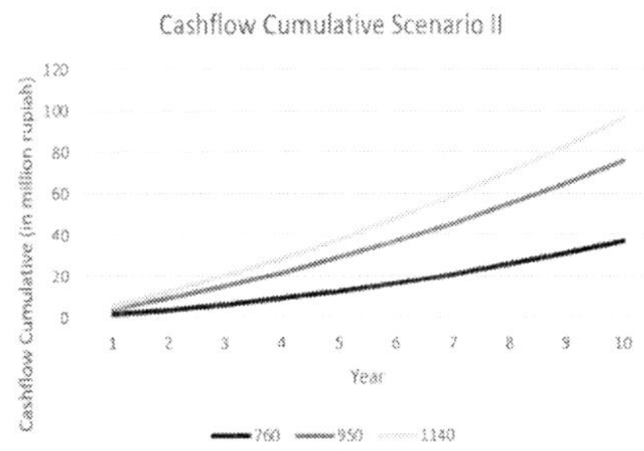

Fig. 6. *Scenario 2 cash flow cumulative graph*

As an example, consider the production capacity of 760 per month, the initial investment amount will be returned in the year between the 6th and 7th year. Then at a production capacity of 950 per month, the amount of the initial investment will return between the 3rd and 4th years. Then at a production capacity of 1140 per month, the amount of the initial investment will return between the 2nd and 3rd years.

$$PP_{760} = 6 + \frac{20.000.000 - 16.876.656}{21.324.240 - 16.876.656} = 6{,}70\ years$$

$$PP_{950} = 2 + \frac{20.000.000 - 15.460.794}{22.016.484 - 15.460.794} = 3{,}69\ years$$

$$PP_{1140} = 3 + \frac{20.000.000 - 12.718.800}{20.316.528 - 12.718.800} = 2{,}96\ years$$

3.3 Added Value Analysis

In addition to the business feasibility analysis, it is important to conduct a value-added analysis for the assessment of services obtained by business actors and measure the employment opportunities created. The higher the added value of a product will trigger tighter competition in the acquisition of raw materials and product marketing because it is more profitable [9]. Research on added value using the Hayami method. The Hayami method is presented in table 9.

No	Output, Input, and Price	Score
1	Output (Unit/production)	(1)
2	Input Materials (Kg/production process)	(2)
3	Input Workers (working day/production process)	(3)
4	Conversion Factors	(4) = (1)/(2)
5	Labor Coefficient	(5) = (3)/(2)
6	Output Price (IDR/unit)	(6)
7	Average wage for workers (IDR/working day)	(7)
	Revenue and Income	
8	Price of Raw materials (IDR/Kg)	(8)
9	Other input contributions (IDR/Kg)	(9)
10	Output price (IDR/unit)	(10) = (4) x (6)
11 a	Added Value (IDR/Kg)	(11a) = (10) - (9) - (8)
b	Additional Value Ratio (%)	(11b) = (11a)/(10) x 100%
12 a	Workers' allowance (IDR/working day)	(12a) = (5) x (7)
b	Workers' share (%)	(12b) = (12a)/(11a) x 100%
13 a	Income (IDR/Unit)	(13a) = (11a) - (12a)
b	Income level (%)	(13b) = (13a)/(10) x 100%
	Remuneration for the Owner of production factors	
14	Margin	(14) = (10) - (8)
a	Labor income (%)	(14a) = (12a)/(14) x 100%
b	Other input contributions (%)	(14b) = (9)/(14) x 100%
c	Company income (%)	(14c) = (13a)/(14) x 100%

Table 9. Variables and Score in Added Value Analysis using Hayami Method

Table 10 presents a table comparing the value of each sub-variable used in the Hayami Method calculation in the self-production vs third party production scenario. The added value of self-production is Rp 55,527 per kg. This implies that 1 kg of vegetable noodle material can produce an added value of Rp 55,527. This value is greater than the third-party production at the same production rate, which is Rp 32,800.

Variable	No.	Sub Variable	Self Production		Third Party Production	
Output, Input, and Price	1	Output (Unit/production)		1140		1140
	2	Input Materials (Kg/production process)		90		90
	3	Input Workers (working day/production process)		20		20
	4	Conversion Factors		12.67		12.67
	5	Labor Coefficient		0.22		0.22
	6	Output Price (IDR/unit)	Rp	7,500	Rp	7,500
	7	Average wage for workers per working day (IDR/production hour)	Rp	85,500	Rp	28,500
Revenue and Income	8	Price of Raw materials (IDR/Kg)	Rp	27,473	Rp	57,000
	9	Other input contributions (IDR/Kg)	Rp	12,000	Rp	5,200
	10	Output price (IDR/unit)	Rp	95,000	Rp	95,000
	11 a	Added Value (IDR/Kg)	Rp	55,527	Rp	32,800
	b	Additional Value Ratio (%)		58.45		34.53
	12 a	Workers' allowance (IDR/working day)	Rp	19,000	Rp	6,333
	b	Workers' share (%)		34.22		19.31
	13 a	Income (IDR/Unit)	Rp	36,527	Rp	26,467
	b	Income level (%)		38.45		27.86
Remuneration for the Owner of production factors	14	Margin	Rp	67,527	Rp	38,000
	a	Labor income (%)		28.14		16.67
	b	Other input contributions (%)		17.77		13.68
	c	Company income (%)		54.09		69.65

Table 10. Added Value Analysis of Self production vs Third party production in production rate of 1140 per month

4 Conclusion

The results of this study can be concluded that building a dry noodle production business is more feasible by implementing scenario I, namely by buying production equipment and then carrying out the production process by members of farmer groups. This is based on the indicators NPV, IRR, Payback Period which show better results in scenario I. Then with the production capacity variable it is found that the production capacity of 760 packages per month is not feasible to run because the NPV value is negative, and the IRR value is smaller than the discount factor in both scenarios. A larger increase in production capacity provides greater profit because it can reduce production margin costs as seen in the production capacity scenario of 950 per month and 1140 per month in both scenarios which results in greater profits in the production capacity scenario of 1140 per month. The NPV, IRR, and PP at a production capacity of 1140 per month in scenario I is IDR 42,038,344; 59%; and 1.88 years respectively, while the NPV, IRR, and PP at a production capacity of 1140 per month in scenario II is IDR 18,573,338; 37%; and 2.96 years respectively. In scenario I, farmer groups require large initial investment capital to provide production equipment for the production, packaging, and distribution process, while in scenario II, farmer groups only need to carry out packaging and distribution. Then from the results of the added value analysis using the Hayami method, the added value obtained in self-production is Rp 55,527 which is greater than third party production of Rp 32,800 at the same production level of 1140 units/month.

The authors would like to thank for Mrs. Winarsih as chairman of the Gemah Ripah farmer group for his cooperation in providing information for this research.

References

1. BPS-Statistics Indonesia, Agricultural Statistic for Horticulture SPH, Harvested Area of Vegetable Crops by Province and Crop Type 2018-2022. Accessed on https://www.bps.go.id/indikator/indikator/view_data_pub/0000/api_pub/bXNVb1pmZndqUDhKWElUSjhZRitidz09/da_05/1 (2023)
2. BPS-Statistics Indonesia, Mid-Year Population 2018-2022. Accessed on https://www.bps.go.id/indicator/12/1975/1/jumlah-penduduk-pertengahan-tahun.html (2023)
3. Abdurrohman, A. Arkasala, F.F., Nurhidayah, N. Penerapan Konsep Urban Farming-Based Resilient City dalam Pengembangan Kota yang Berketahanan Pangan di Kota Surakarta. Desa-Kota: Jurnal Perencanaan Wilayah, Kota, dan Permukiman, **Vol. 3**, Nomor 2, hal 162-170 e-ISSN: 2656-5528 (2021)
4. Wijaya, K., Permana, A. Y., Hidayat, S., & Wibowo, H. Pemanfaatan Urban farming Melalui Konsep Eco-Village Di Kampung Paralon Bojongsoang Kabupaten Bandung. Jurnal Arsitektur ARCADE, **4**(1), 16. Accessed on https://doi.org/10.31848/arcade.v4i1.354 (2020)
5. Suwarlan, S. A. Perancangan Urban farming Pada Pesisir Kampung Kelembak Kepulauan Riau. Jurnal Linears, **3**(1), 20–25. Accessed on https://doi.org/10.26618/jlinears. v3i1.3134 (2020)
6. Aditiyarini, D. Pemberdayaan Kelompok Tani dan Pengembangan Kampung Sayur Bausasran di Masa Pandemi Covid 19. Prosiding Sendimas VI Tahun 2021, hal 160-165, e-ISSN: 2451-3805 (2021)
7. Abuk, G.M., Rumbino, Y. Analisis Kelayakan Ekonomi Menggunakan Metode Net Present Value (NPV), Metode Internal Rate of Return (IRR) Payback Period (PBP) Pada Unit Stone Crusher di CV. X Kab. Kupang Prov. NTT. Jurnal Ilmiah Teknologi FST Undana **Vol. 14**, No. 2, hal 68-75 ISSN: 1693-9522 (2020)
8. Hadi, W. Menggali Potensi Kampung Wisata di Kota Yogyakarta Sebagai Daya Tarik Wisatawan. Journal of Tourism and Economic **Vol.2**, No.2, 2019, Page 129-139 ISSN: 2622-4631 (2019)
9. Nuzuliyah, L. Analisis Nilai Tambah Produk Olahan Tanaman Rimpang. Jurnal Teknologi dan Manajemen Agroindustri. **Vol.7**, No.1, hal 31-38, ISSN: 2548-3582 (2018). Accessed on https://doi.org/10.21776/ub.industria.2018.007.01.4 (2018)
10. Afiq, M., Eliza, Kurnia, D. Analisis Usaha Agroindustri Tahu (Studi Kasus Pada Usaha Agroindustri Tahu Bapak Henry di Desa Pandan Wangi Kecamatan Peranap Kabupaten Indragiri Hulu). Indonesian Journal of Agricultural Economics (IJAE). **Vol. 10**, No. 2, hal 24-38, ISSN 2087-409X (2019)
11. Dewi, I.A., Effendi, U., Wijana, S., Sari, D.N. Analisis Kelayakan Finansial Produksi Setup buah Nipah Pada Skala Industri Kecil Menengah (IKM). Jurnal Teknologi Pertanian. **Vol. 20**, No. 1, hal 25-32. Accessed on https://doi.org/10.21776/ub.jtp.2019.020.01.3 (2019)
12. Kusumawati, E. et al. Pemberdayaan Masyarakat Kampung Sayur Perkotaan Menjadi Kampung Wisata Berbasis Edukasi Melalui Implementasi Urban Farming. Seminar Nasional Pemberdayaan Masyarakat.**Vol. 3**, hal 406-411, ISSN 2685-9017. (2021)

Effect of Varying Color LED Lights on *Porang* (*Amorphophallus muelleri*) Bulbil Seed Germination and *Porang* Plant Vegetative Growth Phase

Aryanis Mutia Zahra[1,2,*], Esty Indrayanti[1], Bayu Dwi Apri Nugroho[1], and Rudiati Evi Masithoh[1]

[1] Department of Agricultural and Biosystems Engineering, Faculty of Agricultural Technology, Universitas Gadjah Mada, Yogyakarta, Indonesia

[2] Division of Postharvest Technology, School of Bioresources and Technology, King Mongkut's University of Technology Thonburi, Bangkok, Thailand

* aryanismutiazahra@ugm.ac.id

Abstract. Intensive cultivation is one of the strategies for developing *Porang* commodities to increase production by using LED lights as optimum lighting for *Porang* cultivation. *Porang* bulbil was germinated and planted in a growth chamber with 100% red, 100% blue, 100% white, and 67%:20%:13% red:blue:white lighting. Compared to other LED treatments and sunlight, red:blue:white LED had the best effect on *Porang* plant germination parameters in the first 30 days, with 100% of germination parameters (GP %), 17.2 days of mean germination time (MGT), 6.33%/day of germination rate index (GRI), 5.81% of germination velocity coefficient (CVG), and 138.0 of germination index (GI). For the next 30 days, red:blue:white LED had the best effect on *Porang* plant vegetative growth, with 29.64 cm plant height, 11.43 cm crown diameter, 21.96 cm root length, 17.342 g fresh weight, 1.368 g dry weight, and 57.53 leaf green index. The plant height prediction model is used to predict crop yield, with the coefficient of determination (R^2), root means square deviation (RMSE), mean absolute error (MAE), and average absolute percentage error (MAPE) values up to 0.9854, 0,6894, 1.2623, and 3,9413, respectively, under model validation of the combination of LED.

1 Introduction

Amorphophallus muelleri Blume, also known locally as *Porang*, is a plant species that is being developed as a food source that could be one of the solutions in response to the rising demand for food brought about by a rise in the human population [1-3]. Despite widespread application in food, as a source of 91–95% glucomannan, *Porang* is utilized as a material in the pharmaceutical, biological, and cosmetic industries [2, 4]. Along with many potential applications and purposes of *Porang*, the demand for its industrial raw materials as an export commodity of sliced and dried chips is also increasing significantly [3-5]. However, supply only met the demand for 20% of the available dry chips [6]. The dearth of demand for *Porang* is because it is not cultivated intensively in Indonesia. In addition, the growing demand in export markets will necessitate extensive management of *Porang* cultivation in areas like land preparation for nurseries and planting, plant and land maintenance, and the promotion of *Porang* cultivation as an agro-forestry product and export commodity [1, 6, 7].

Utilizing leaf tubers or bulbils, which are found at the intersections of the leaf stalks, it is possible to achieve intensive *Porang* production through seed propagation [1, 8]. When the petiole spontaneously detaches, indicating complete dormancy, the *Porang* bulbil is harvested as seedlings. Bulbil quality is determined by dormancy and the absence of decaying, peeling, or fungal infections in the seeds [9, 10]. Among the three alternatives, bulbil propagation is the most advantageous in ease, cost, and environmental impact. In every plant's period, the bulbil generated decreases in size as the distance between the leaves and the main branch increases. Production of corms and bulbils varied during the beginning, second, and tertiary crop periods [1, 8].

Using the highest quality seeds is essential to obtain high-yielding, high-quality tubers and seedlings. Numerous factors, such as genotype seeds and their progenitor habitat, landraces, seed pureness, seed shape, seed dimensions, and seed maturity, have impacted seed quality [1, 11]. Because a larger bulbil develops more rapidly than a smaller one, seed size is essential for determining the total reserve amount. A lengthier petiole is indicative of greater viability for heavier seeds. The larger the buds, the taller the plantlets, the drier the branches and roots, the wider the tubers, and the heavier the corms and bulbils during harvest [12, 13]. The success of *Porang* cultivation depends on the availability of seedlings with the proper physiological quality and sufficient quantity [6].

In addition to being genetically determined, seed dormancy, germination, and vegetative growth are also significantly influenced by environmental factors. Light has become one of the most influential environmental factors in determining the extent of dormancy, timing of germination, and growth [11]. The responsiveness of seedlings to light varies significantly between species, with some seeds germinating more readily only in darkness or light, whereas some are unaffected by this variable. Although light is not essential for germination in all species, it mitigates the adverse effects of germination when the incubation temperature exceeds the optimal range [14]. In recent years, light-emitting diodes (LEDs) have been utilized in innovative methods, allowing for plant-targeted development and enhancing quality and productivity concerning environmental control due to LED light's narrow spectral bands and controllable intensity and duration. Success in the LED lighting industry enables using monochromatic wavelengths or their combinations in the red and blue light regions to optimize plant growth and development by providing an efficient light spectrum at the absorption peak of photosynthetic pigments [11, 15-20].

To optimize plant growth, extensive research has been conducted on applying red, blue, and their combination in various tubers [15-20]. In response to light, as an external cue recognized by plants and translated into internal cues, endogenous phytohormones regulate the physiology of seed metabolism [11]. This research was focused on investigating the effect of red, blue, and white LED lights, as well as the combination of red:blue:white, on the germination and vegetative growth phases of the *Porang* plant. This research will contribute to the preparation procedure for planting *Porang* under artificial lighting.

2 Materials and methods

2.1 Sample preparation

Porang seeds (*A. muelleri*) were manually and visually sorted under normal conditions based on their uniform size, approximately 5.00 grams, and absence of infection or damage (Fig. 1). During the third and fourth vegetative phases, *Porang* bulbil samples were collected between June and August from Karangmojo, Kalasan, Yogyakarta, Indonesia.

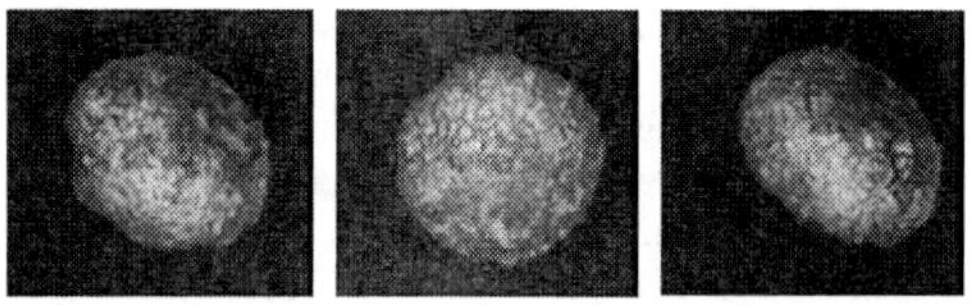

Fig. 1. The *Porang* bulbil with the normal condition and uniform size.

2.2 Experimental design and growth condition

Porang seeds were germinated and cultivated in a mixture of soil, compost, and husk charcoal as planting medium (approximately 150 g; pH 7.0) in an iron-framed growth chamber with 95% para net coverage. Four different LEDs with varying lux, PPFD, and wavelength, i.e., 100 lux, 44.26 mol·m^{-2}·s^{-1} of PPFD, and 470 nm of wavelength for 100% blue; 600 lux, 20.32 mol·m^{-2}·s^{-1}, and 660 nm of wavelength for 100% red; 1000 lux, 95.92 mol·m^{-2}·s^{-1}, and 550 nm of wavelength for 100% white; 350 lux, 37.33 mol·m^{-2}·s^{-1}, with 660, 470, and 550 nm of wavelengths for the combination of 67%:20%:13% red:blue:white, respectively, were used as artificial light, with the photoperiod was 12 h per day (Fig. 2). For the control treatment, bulbil *Porang* was germinated and cultivated in laboratory conditions of indirect sunlight and an ambient environment. PPFD, lux, and wavelength were measured using a PAR meter and quantum sensor (LICOR LI-250Q PAR comprising LI-250 Light Meter and LI-190R Quantum Sensor), a light meter (Lutron LX-107), and VIS-NIR Ocean Optics spectrometer using a set of fiber-optic probes and a tungsten halogen lamp (HL-2000-HP-FHSA Ocean Optics).

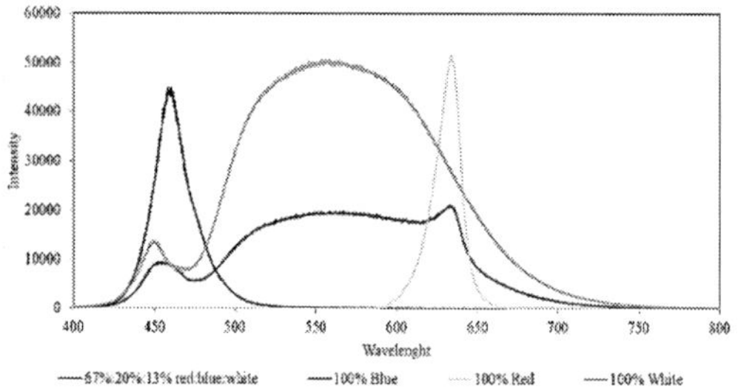

Fig. 2. The light spectrum of red, blue, white, and red:blue:white LED that used in this study.

Porang seedlings were germinated in a 60 cm x 30 cm x 3.5 cm tray with 5 ml of water and no additional nutrients for 30 days. The seeds were then transferred into polybags 15 cm x 15 cm for the next 30 days and then were cultivated under varied LED lights and control with a depth of approximately 5 cm and daily watering of 10 ml without additional nutrition. The environment's temperature is daily recorded in the range of 25-35°C and relative humidity of 50-80%.

2.3 Determination of germination parameters

After thirty days of seedling growth and development, as indicated by the first apparition of a hypocotyl longer than 0.2 centimeters, the parameters of *Porang* seed germination were measured. Germination Percentage (GP), Mean Germination Time (MGT), Germination Rate Index (GRI), Coefficient of Velocity of Germination (CVG), and Germination Index (GI) were calculated from the observed germination parameters using the equation (1) - (5) [14, 21].

$$GP = \frac{Number\ of\ gGerminated\ seeds}{Number\ of\ tested\ seed} \times 100\% \quad (1)$$

$$MGT = \sum_{i=1}^{n} N_i G_i / \sum_{i=1}^{n} G_i \quad (2)$$

$$GRI = \frac{GP1}{d1} + \frac{GP2}{d2} + \ldots \frac{GPx}{dx} \quad (3)$$

$$CVG = \left(\sum_{i=1}^{k} f_i / \sum_{i=1}^{k} f_i x_i\right) 100 \quad (4)$$

$$GI = (10 \times n1) + (9 \times n2) + \ldots + (1 \times n10) \quad (5)$$

where Ni is the number of days from the day of sowing to the end of the observation; Gi is the number of sprouts on the day of observation; GP1, GP2, and GPx represent the germination percentage on dx days after planting; d1, d2, dx is the day of germination; fi is the number of newly germinated seeds; xi is the number of days from seeding related to fi; k is the final day of germination; n1, n2,... n10 represent the number of seeds that germinated on the first, second, and following day; 10, 9,... 1 represents the weight assigned to the number of seeds that germinate on the first, second, and following day.

2.4 Determination of vegetative growth parameters

Plant height was measured at 7-day intervals, while planting yield was measured 30 days after planting using crown diameter (cm), root length (cm), plant height (cm), fresh weight (gram), and dried weight (gram) of plants. The Computer Vision System (CVS) method determined leaf color by capturing images from both leaf surfaces using a black background and sufficient illumination. Red, green, and blue color values are extracted from the image results using an image processing application.

2.5 Statistical analysis

For each germination and vegetative growth parameter, a one-sample T-test with a significance level of 0.05 was used to compare the effects of the light treatment to the control. The evaluated value represented the statistical difference between the change value and the value at zero. The homogeneity test, Kolmogorov-Smirnov test for normality, and Kurskal Walis's test were used as rank-based nonparametric tests with a significance level of 0.05 to determine the effect of LED light illumination differences. With a significance level between 0.05 and 0.01, Spearman correlation analysis was used to measure the intensity and direction of the linear relationship between the two variables of vegetative growth parameters.

The plant height of 50 samples grown under various color LED lights for 5 weeks was modeled using regression polynomials, beginning on days zero, seven, fourteen, twenty-one, and twenty-eight. Polynomial Regression multiplies independent variables by a polynomial degree in linear Regression. The analysis structures of multiple linear and polynomial regression models are identical. Each rank or order in a polynomial model modifies the initial variable, transforming a free variable into a non-free variable [22].

This investigation used a calibration set of seven samples to build the model, while a validation set of three samples was used to evaluate the model. The predictive error values of the calibration and prediction sets were examined to evaluate the accuracy of the model's performance. The model was evaluated using root mean square error (RMSE), mean absolute error (MAE), mean absolute percentage error (MAPE), and coefficient of determination (R^2). The polynomial equation and model evaluation are listed in equations (6) - (10); n is the number of samples, ý is the predicted value of the plant height at the i-point, and y represents the actual plant height at the i-point [22].

$$y = \beta_0 \sum_{j=i}^{d} \beta_j x_i^j + \varepsilon_{i,j} = 1,2,3,n \quad (6)$$

$$RMSE = \sqrt{\frac{1}{n}\sum_{i=1}^{n}(y_i - \acute{y}_i)^2} \quad (7)$$

$$MAE = \frac{1}{n}\sum_{i=1}^{n} | y_i - \acute{y}_i | \quad (8)$$

$$MAPE = \frac{1}{n}\sum_{i=1}^{n} \frac{| y_i - \acute{y}_i |}{\acute{y}_i} \quad (9)$$

$$R^2 = \frac{\sum_{i=1}^{n}(y_i - \acute{y}_i)(\acute{y}_i - \acute{y}_i)^2}{\sum_{i=1}^{n}(y_i - \acute{y}_i)^2 \sum_{i=1}^{n}(\acute{y}_i - \acute{y}_i)^2} \quad (10)$$

3 Results and discussion

3.1 Relationship between the Growth Chamber's Light Intensity, Temperature, and Humidity

Fig. 3 shows the effect of various lighting colors on the microclimate condition in the growth chamber. In this investigation, light intensity measurements showed that each LED color had a different intensity. White LEDs emit the lightest with an average irradiance of 12000 lux, followed by combination LEDs at 1300 lux, red LEDs at 700 lux, and open space controls at 300 lux, while blue LEDs gave the lowest light intensity at 240 lux. LED lights maintained a more consistent temperature than sunlight. White LEDs had the highest temperature (31.6°C), mixture LEDs had the lowest (30.2°C), and the control treatment had a temperature of 30.6°C. With LED lighting, the growth chamber humidity increased, while the control treatment humidity fluctuated during each observation. White LED treatment had the lowest humidity at 61.8%. This LED treatment makes the growing chamber dry and hotter than others. Red and blue LED lights produce 66% and 63.6% humidity. The 71% combination LED treatment is closest to the control treatment's 79% humidity. According to other reports, this study's microclimate temperature and humidity are appropriate for the optimal condition; porang plants thrive at 25-35°C and 50-80% RH [22].

3.2 Germination parameters of Porang bulbils under varying LED lights

Table 1 shows the effect of various lighting colors on the evaluated seed germination parameters, with the best GP, MGT, GRI, CVG, and GI values obtained by applying the LED combination of red:blue:white, with respective values of 100%; 17.2 days; 6.33% per day; 5.81%; and 138.00.

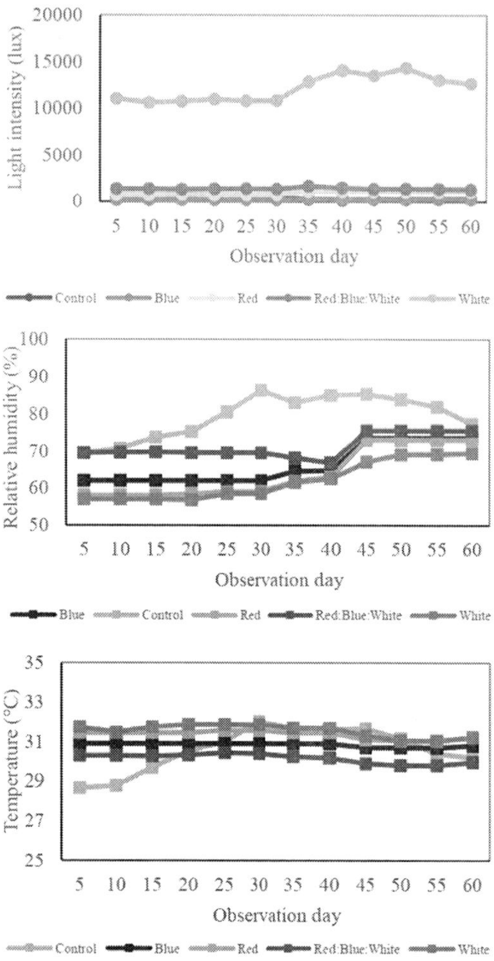

Fig. 3. Effect of differences in light treatment on microclimate conditions observed during the growth of bulbil porang.

GP (%) shows the percentage of final germination achieved by *Porang* bulbils under various irradiation color LED lights. Within 30 days, cultivation under LED combination of red: blue: white produced the best germination rate (100%), followed by control treatment (90%), blue LED (80%), and the lowest was red LED and LED white (70%). The development of the number of seeds that germinated as germination capacity (%) for each treatment is shown in Fig. 4. The seeds in each treatment showed an increase in germination. Seeds can respond differently to various types of light, not only as a source of energy to produce organic compounds but also as a supply of information to plants that regulate their growth and development in response to environmental changes [11].

The MGT (days) is calculated over 30 days, revealing that the average yield of seed germ indicating more rapidly occurs between days 15 and 20 (Fig. 4). The shorter the MGT (days), the more rapidly a seed population germinates [14, 21]. It is known that red:blue:white LEDs generated the maximum MGT of 17.2 days for *Porang* bulbil, whereas red LEDs resulted in the lowest MGT of 20.7 days.

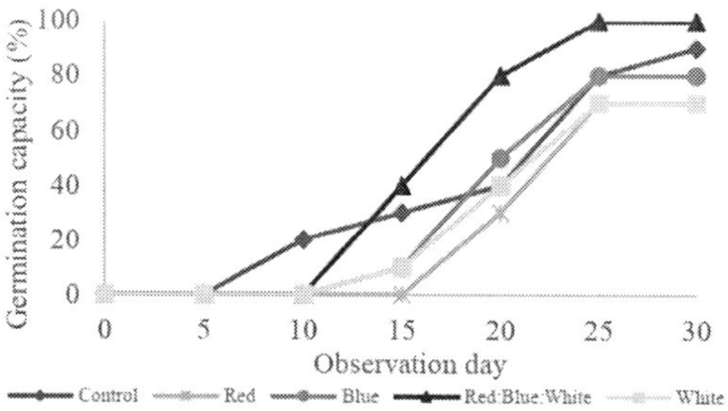

Fig. 4. The germination capacity of Porang bulbil increases by 30 days.

GRI (percent per day) represents the daily germination percentage of *Porang* bulbil seeds; the more significant the GRI value, the higher and more rapid the germination process [21]. *Porang* bulbil germinated under red: blue: white LED light had the highest GRI value of 6.33%/day, followed by the control treatment with 5.79%/day and the red LED treatment with 3.45%/day. Like GRI, GI focuses on the germination rate and rate of germination. In GI calculations, however, the highest value is assigned to seeds germinating on the first day, and the lowest is germinating on the following day [21]. In germinated *Porang* bulbils, the red:blue:white LED combination treatment provided the highest GI value of 138.00, just above the control treatment of 113.00. In contrast, the red LED treatment provided the lowest GI value of 72.00.

Table 1. Effect of differences in light treatment on bulbil *Porang* germination parameters.

Treatment	GP (%)	MGT (days)	GRI (%/days)	CVG (%)	GI
Control	90	18.4	5.79	5.42	113.00
Red	70	20.7	3.45	4.83	72.00
Blue	80	19.0	4.43	5.26	96.00
Red:Blue:White	**100**	**17.2**	**6.33**	**5.81**	**138.00**
White	70	19.0	3.92	5.26	84.00
Kruskal Walis's test	ns	ns	ns	ns	ns
One sample T-test	*	*	*	*	*

Note: * are significantly different at $\alpha = 0.05$.

CVG signifies the rate of germination, whose value rises as the proportion of seeds that germinate and the time needed for germination decrease [21]. The obtained CVG (%) indicates a slow pace germination process, which may be due to the length of time bulbil *Porang* seeds require to germinate. *Porang* bulbil germinated under a red:blue:white LED combination had the maximum CVG (%) at 5.81%, while the red LED treatment had the lowest CVG (%) at 4.8%.

Red and blue LED light has been regarded as the essential spectrum for plant growth, as it is not only the primary source of light for photosynthesis but also controls numerous morphogenetic responses in plants via photoreceptors [11].

3.3 Vegetative growth parameters of *Porang* bulbils under varying LED lights

Porang plant growth and development respond differently to each LED exposure. Concerning plant height, *Porang* grown under red light grew taller than *Porang* grown under blue light (Table 2), but both grew shorter than *Porang* grown under a combination of red and blue light. The monochromatic red light induces a shade avoidance syndrome, which results in excessive stem elongation and a less refined plant body. In contrast, blue light inhibits plant organ elongation [23]. In addition, blue light has a lower capacity for energy absorption than red light [24].

Table 2. Effect of differences in light treatment on vegetative growth parameters of *Porang* plant.

Treatment	Plant height (cm)	Crown diameter (cm)	Root length (cm)	Fresh weight (g)	Dry weight (g)	Leaves color		
						R	G	B
Control	26.20 ± 24.86	8.20 ± 9.41	14.22 ± 8.57	15.615 ± 8.948	1.347 ± 0.694	29.61	38.50	11.91
Red	22.02 ± 17.40	4.80 ± 6.45	19.64 ± 9.74	15.237 ± 7.939	1.272 ± 0.627	39.12	43.27	14.81
Blue	18.59 ± 17.06	7.45 ± 9.36	13.77 ± 12.41	14.153 ± 7.035	1.216 ± 0.360	32.31	40.84	8.79
Red:Blue:White	**29.64 ± 16.55**	**11.43 ± 9.26**	**21.96 ± 6.02**	**17.342 ± 8.891**	**1.368 ± 0.331**	**47.40**	**57.53**	13.56
White	13.30 ± 9.92	5.35 ± 6.64	21.47 ± 9.27	12.864 ± 5.633	1.259 ± 0.438	30.15	36.06	8.18
Kruskal Walis's test	ns	ns	ns	ns	ns	ns	ns	ns
One sample T-test	*	*	*	*	*	*	*	*

Note: * are significantly different at α = 0.05.

In this study, *Porang* plants grown under a red:blue:white LED combination had the highest values for plant height, crown diameter, root length, plant weight, and color parameters (Table 2). The red, blue, and white LED combination has a ratio of 67%, 20%, and 13%, respectively, which is close to the specifications for a suitable LED color combination for plant growth. When applied at the right intensity, red light combines with blue light to govern plant responses, and an optimal R:B ratio enhances photosynthetic capability, growth, and yield [25].

The use of various wavelengths on different LED light colors caused a difference in the height of the *Porang* plant, with the highest value being 29.64 cm for the red:blue:white LED combination and 13.30 cm for the white LED, as a result of the auxin hormone, plant height increases in low light intensities. Auxin, a plant hormone that affects cell elongation, is susceptible to injury or degradation when exposed to intense light. However, some reported that white light is conducive to developing various horticultural products and advantageous for photosynthesis because more light can penetrate the canopy to lower leaves than red and blue lights [15].

In addition to plant height, the use of different wavelengths in various LED light colors affects the diameter of the *Porang* crown, with the highest diameter obtained from the LED combination of red:blue:white at 11.43 cm and the lowest value obtained from white LED at 5.35 cm. The leaves, branches, and twigs contribute to the crown diameter. Since crown morphology can affect plant light absorption, crown diameter is an essential parameter in *Porang* cultivation. In the vegetative growth phase of plants, root growth is crucial because long roots can maximize the absorption of water and minerals from the medium, resulting in increased plant canopy growth [26, 27]. The use of spectra of various LED light colors affects the root length of *Porang* plants, with the LED combination of red:blue:white producing the longest roots at 21.96 cm and the blue LED producing the shortest roots at 13.77 cm.

The red:blue:white LED spectrum produced the highest fresh weight yield of *Porang* plants (17,342 g) compared to the control treatment (15,615 g). The fresh weight under the red LED light treatment was remarkably comparable to the control treatment (15,237 g), followed by the blue LED treatment (14,153 g) and the white LED treatment (12,864 g). The LED combination of red:blue:white can increase the overall yield of *Porang* plants by 10% over the control condition and by 34.8% over the white LED condition. Red and blue LEDs can increase plants' fresh and dried weight [28-30]. In comparison, the lowest weight of plants grown under white LED lighting is due to slow plant growth, which can reduce total plant weight [31]. The use of different wavelengths on different LED light colors caused differences in the weight of the *Porang* plant, with the highest fresh and dry weight obtained from the LED combination of red:blue:white by 17,342 g and 1.368 g, respectively, the lowest fresh weight obtained from white LEDs by 12.864 g and the lowest dry weight obtained from blue LEDs by 1.216 g.

The plants absorb red light more efficiently than blue light [24]. Applying LEDs between 380 and 660 nm induced multiple phytochromes in seed potato tubers to produce green color intensity in response to irradiance [20]. However, by stimulating chlorophyll production in plants, blue light affects plant compaction and leaf discoloration [32]. In this study, plants exposed to monochromatic red light performed better than those exposed to blue LED light. Compared to the LED combination of red, blue, and white treatment, the monochrome blue or red-light treatment resulted in subtler leaf colors and a lighter shade of green. The wavelengths of distinct LED light colors affect the leaf color of the *Porang* plant; the LED combination of red:blue:white produced the best results with a green (G) color index value of 57.33, a red (R) color index value of 47.40, and a blue color index value of 13.56. Other reports mentioned that the LED combination of red:blue: and white produced the highest accumulated photosynthetic pigment and induced green color and was claimed as appropriate light for potato pre-basic seed tuber cultivated under plant factory conditions [19].

The Spearman correlation between the vegetative growth characteristics of the *Porang* plant, such as plant height, root length, crown diameter, fresh and dry weight, and leaf color parameters, was measured and shown in Table 3. Positive and highly significant correlations exist between plant height and crown diameter (rho = 0.726), fresh weight (rho = 0.887), and leaf color parameters of R, G, and B index (rho = 0.720, 0.798, and 0.700, respectively). The crown diameter was positively and highly correlated with fresh weight (rho 0.816) and G index (rho 0.814), yielding a high correlation coefficient.

3.4 The vegetative growth model of the *Porang* plant based on plant height

Plant height in the vegetative growth phase of *Porang* plants cultivated under LED lighting with varying colors is shown in Fig. 5. Based on the graph, the final height of *Porang* plants exposed to the LED combination of red:blue:white light was the greatest (27.96 cm) when compared to other LED treatments. Based on the average weekly height during the five-week cultivation of *Porang* plants, a model of the vegetative growth of *Porang* plants under blue LED, red LED, combination LED, and white LED treatments was developed.

Table 3. Correlation between *Porang* plant vegetative growth parameters.

Vegetative growth parameter		Plant height	Root length	Crown diameter	Fresh weight	Dry weight	R	G	B
Plant height	P	1.000	.407**	**.726****	**.887****	-.366**	**.720****	**.798****	**.700****
	Sig.	.	.003	.000	.000	.009	.000	.000	.000
Root length	P		1.000	.515**	.454**	-.373**	.435**	.450**	.485**
	Sig.		.	.000	.001	.008	.002	.001	.000
Crown diameter	P			1.000	**.816****	-.496**	**.709****	**.814****	.620**
	Sig.			.	.000	.000	.000	.000	.000
Fresh weight	P				1.000	-.407**	**.788****	**.855****	**.710****
	Sig.				.	.003	.000	.000	.000
Dry weight	P					1.000	-.514**	-.501**	-.617**
	Sig.						.000	.000	.000
R	P						1.000	**.933****	**.820****
	Sig.							.000	.000
G	P							1.000	**.734****
	Sig.								.000
B	P								1.000
	Sig.								.

Note: * are significantly different at α = 0.05, and ** are significantly different at α = 0.01.

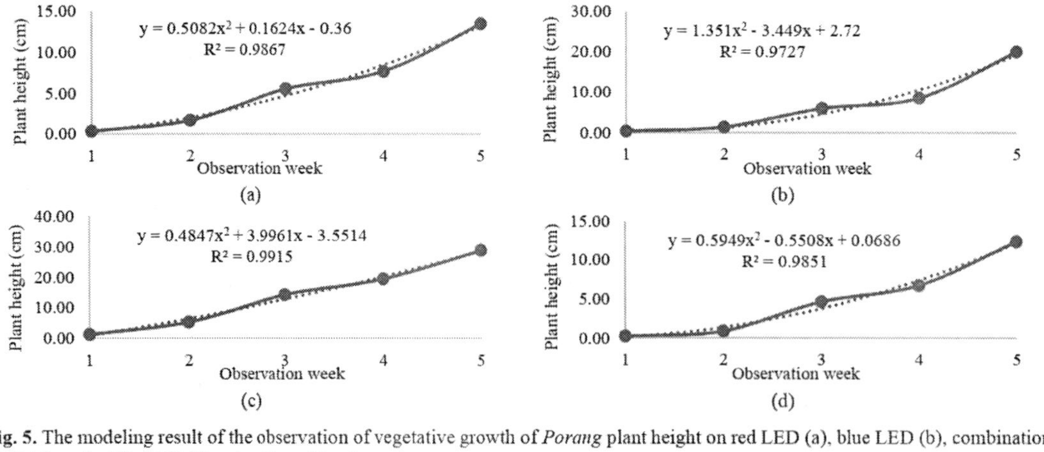

Fig. 5. The modeling result of the observation of vegetative growth of *Porang* plant height on red LED (a), blue LED (b), combination LED (c), and white LED (d) using the calibration set.

The various treatments of 100% red LEDs, 100% blue LEDs, a combination of 67%:20%:13% red:blue:white LEDs, and 100% white LEDs have coefficients of determination of 0.9760, 0.9867, 0.9727, 0.9915; 0.9851, respectively. These results indicate a strong correlation between the height development of *Porang* plants and planting time. In addition, the five models can be used to predict the height of *Porang* plants under varying LED light color variations.

Based on the results of the evaluation of the plant height prediction model on the calibration and validation sample set, the most appropriate performance model was obtained from the LED combination of red:blue:white, with an R^2 of 0.9915; RMSE of 0.4052; MAE of 0.7699; and the MAPE of 2.4472, in the calibration set, and with an R^2 of 0.9854; RMSE of 0.6894; MAE of 1.2623; and the MAPE of 3.9413, in the validation set (Table 4). In general, the four models have a deficient error between the predicted value and the actual value; based on the reliable performance model, the four models can be used as an appropriate measure for estimating the growth of *Porang* plant height during the vegetative phase of growth under varying color LED lighting as artificial lighting.

Table 4. Performance model comparisons of *Porang* plant height cultivated under varying LED lighting.

Treatment	R^2	RMSE	MAE	MAPE
Calibration set				
Blue	0.9867	0.2403	0.4414	2.2683
Red	0.9727	0.5222	0.9337	6.3973
Red:Blue:White	**0.9915**	**0.4052**	**0.7699**	**2.4472**
White	0.9851	0.2408	0.4578	5.3712
Validation set				
Blue	0.9729	2.9442	4.1959	6.3792
Red	0.9845	2.1986	1.7081	5.9990
Red:Blue:White	**0.9854**	**0.6894**	**1.2623**	**3.9413**
White	0.9606	0.5354	0.7150	5.5202

4 Conclusions

The use of varying color light, 100% red, 100% blue, 100% white, and the combination of 67%:20%:13% red:blue:white, affects *Porang* bulbil germination and *Porang* plant vegetative growth, with all parameter values better than the control treatment. Percent germination (GP) by LED combination of 67%:20%:13% red:blue:white achieved 100% as the highest value, with the shortest average germination time (MGT) of 17.2 days, and other parameters, such as germination rate index (GRI), germination rate coefficient (CVG), and germination index (GI), 6.33%/day; 5.81%; and 138.00, respectively. In addition, the application of LED combination can provide more optimal light requirements for the vegetative growth of *Porang* plants, with a higher plant height of 29.64 cm, crown diameter of 11.43 cm; root length of 21.96 cm, fresh weight of 17.342, and dry weight of 1.368, with 57.53 leaf green index. The polynomial model's performance analysis was precise and reliable, with a high R^2 value for varying color LED lights, with the obtained RMSE, MAE, and MAPE values also relatively modest. Based on plant height, the *Porang* vegetative growth model allows for predicting the height of *Porang* plants grown under artificial lighting.

Acknowledgments

The authors would like to thank the *Porang* Farmer in Kalasan, Yogyakarta, Indonesia, for their assistance with the bulbil of *Porang*.

References

[1] Handayani T, Effect of Growing Media on Seed Germination and Seedling Growth of *Porang* (Amorphophallus muelleri Blume), In: *The SATREPS Conference,* pp 119-28 (2019)

[2] Anggela A, Setyaningsih W, Wichienchot S and Harmayani E, *Indonesian Food and Nutrition Progress* **17** (2021)

[3] Nurlela N, Ariesta N, Santosa E and Mulandri T, *Food Research* **6** 345-53 (2022)

[4] Yanuriati A, Marseno D W, Rochmadi and Harmayani E, *Carbohydr Polym* **156** 56-63 (2017)

[5] Sakaroni R, Suharjono S, and Azrianingsih R, Identification of potential pathogen fungi which cause rotten on *Porang* (*Amorphophallus muelleri* Blume) tubers. In: *International Conference on Biology and Applied Science (Icobas),* (2019)

[6] Riptanti E W, Irianto H and Mujiyo, *Open Agriculture* **7** 566-80 (2022)

[7] Dermoredjo S K, Azis, M. Saputra, Y H, Susilowati, G and Sayaka, B, Sustaining *Porang* (*Amorphophallus muelleri* Blume) production for improving farmers' income. In: *1st International Conference on Sustainable Tropical Land Management*: IOP Publishing) p 012032 (2021)

[8] Nugrahaeni N, Hapsari, R T, Indriani, F C, Amanah, A, Yusnawan, E, Mutmaidah, S, Baliadi, Y and Utomo, J S, Morphological characteristics of Madiun 1, the First *Porang* (*Amorphophallus muelleri* Blume) released cultivar in Indonesia. In: *2nd ICFST 2021*: IOP Publishing) (2021)

[9] Hidayah N, Suhartanto, M R and Santosa, E, *Buletin Agrohorti* **6** 405-11 (2018)

[10] Sari M, Santosa, E, Pieter Lontoh, A, Kurniawati, A, *Jurnal Ilmu Pertanian Indonesia* **24** 144–50 (2019)

[11] Yan A and Chen Z, *The Botanical Review* **86** 39-75 (2020)

[12] Ibrahim M S D, Sulistiyorini, I and Tresniawati, C, Effect of 6-benzyl amino purine on the multiplication ability of shoots of various sizes of

Porang (*Amorphophallus muelleri* Blume) bulbils. In: *The 2nd International Conference on Sustainable Plantation*: IOP Publishing) (2022)

[13] Soedarjo M, Sasmita, P, In: *2nd ICFST 2021,* pp 0-8 (2021)

[14] Guo C, Shen, Y and Shi, F, *Forests* **11** 300 (2020)

[15] Li R, Long J, Yan Y, Luo J, Xu Z and Liu X, *HortScience* **55** 71-7 (2020)

[16] Chen L, Zhang K, Gong X, Wang H, Gao Y, Wang X, Zeng Z and Hu Y, *Journal of Integrative Agriculture* **19** 108-19 (2020)

[17] He J and Qin L, *J Plant Physiol* **252** 153239 (2020)

[18] Upadhyaya C P, Pundir R K, Pathak A, Joshi N and Bagri D S, *Plant Science Today* **7** 406-16 (2020)

[19] Rahman M H, Azad M O K, Islam M J, Rana M S, Li K H and Lim Y S, *Plants* **10** (2021)

[20] Mølmann J A B and Johansen T J, *Potato Research* **63** 199-215 (2019)

[21] Calone R, Sanoubar, R., Noli, E. and Barbanti, L., *Agriculture* **10** 29 (2020)

[22] Nurrahmah N A, Zahra A M, Rahayoe S, Masithoh R E, Rahmawati L and Pahlawan M, Mathematical Model of Vegetative Growth of *Porang* (*Amorphophallus muelleri*) with Different Seed Quality. In: *Proceedings of the International Conference on Sustainable Environment, Agriculture and Tourism (ICOSEAT 2022)*: Atlantis Press) pp 245-53 (2022)

[23] Li C X, Xu Z G, Dong R Q, Chang S X, Wang L Z, Khalil-Ur-Rehman M and Tao J M, *Front Plant Sci* **8** 78 (2017)

[24] Liu J and van Iersel M W, *Front Plant Sci* **12** 619987 (2021)

[25] Paradiso R and Proietti S, *Journal of Plant Growth Regulation* **41** 742-80 (2021)

[26] A'yun Q, Harijati, N, Mastuti, R, *Journal of Environmental Engineering and Sustainable Technology* **6** 30-5 (2019)

[27] Farooq M, Hussain, M, Ul-Allah, S and Siddique, K H, *Agricultural Water Management* **219** (2019)

[28] Piovene C, Orsini F, Bosi S, Sanoubar R, Bregola V, Dinelli G and Gianquinto G, *Scientia Horticulturae* **193** 202-8 (2015)

[29] Naznin M T, Lefsrud M, Gravel V and Azad M O K, Blue Light added with Red LEDs Enhance *Plants* **8** (2019)

[30] Kamal K Y, Khodaeiaminjan, M, El-Tantawy, A A, Moneim, D A, Salam, A A, Ash-shormillesy, S M, Attia, A, Ali, M A, Herranz, R, El-Esawi, M A and Nassrallah, A A, *Physiologia plantarum* **169** 625-38 (2020)

[31] Lobiuc A, Vasilache V, Pintilie O, Stoleru T, Burducea M, Oroian M and Zamfirache M M, *Molecules* **22** (2017)

[32] Yudina L, Sukhova, E., Mudrilov, M, Nerush, V, Pecherina, A, Smirnov, A A, Dorokhov, A S, Chilingaryan, N O, Vodeneev, V, and Sukhov, V, *Biology* **11** 60 (2022)

Identification of Sustainable Supply Chain Performance in Primary Cocoa Processing (Case Study in Patuk, Yogyakarta)

Tasyarafa Shelia Khairani[1], Muhammad Prasetya Kurniawan[1], Kuncoro Harto Widodo[1], Tri Marwati[2], Titiek Farianti Djaafar[2]

[1]Department of Agro-industrial Technology, Faculty of Agricultural Technology, Universitas Gadjah Mada, Indonesia 55281, tasyarafashelia@mail.ugm.ac.id (T.S.K); kurniawan_prasetya@ugm.ac.id (M.P.K); kuncorohw@ugm.ac.id (K.H.W)
[2]National Research and Innovation Agency, Indonesia 55281, trim011@brin.go.id (T.M); titi033@brin.go.id (T.F.J)

Corresponding author: Muhammad Prasetya Kurniawan (kurniawan_prasetya@ugm.ac.id)

Abstract

This study identified the handling of dry fermented cocoa bean production, analyzed the level of cocoa supply chain (SC) performance in Patuk, and developed recommendations for poor performance to improve SC performance. In this study, environmentally friendly handling was carried out by observing the SC and processing process, and SC performance was measured based on the Supply Chain Operations Reference (SCOR) model in the scope of reliability, responsiveness, and agility supported by the 2.2 Analytical Hierarchy Process (AHP) weighting method using pairwise comparison. The result of this study indicates that sustainable practices from the seeding, nurturing, harvesting, and shipping stages are easier to implement due to the small scale and proximity to residential areas and processing units. Cocoa SC performance in Patuk is divided into 3 parts, namely for farmer performance (79.79), collective traders (70.23), and processing industry (83.29). Improvements that can be made to improve the performance of the SC are by increasing plant maintenance, recording production, and making contracts. The benefits of this study are to encourage the adoption of environmentally friendly cocoa cultivation and processing practices, provide information on SC factors on SC performance, and provide recommendations for performance improvement.

Key words: *Analytical hierarchy process, Cocoa, Performance, Processing,* Supply chain operations reference

1. Introduction

The aspect of environmental sustainability involves the wise management of natural resources. In cocoa farming, the use of water, soil, and energy must be carefully managed so as not to over-drain these resources or damage natural ecosystems. Sustainable farming practices should support biodiversity, including protecting plants, insects, and microorganisms that contribute to the balance of agricultural ecosystems. This biodiversity can help control natural pests and increase agricultural productivity, which in turn affects the income of supply chain actors, especially farmers [1]. In cocoa bean processing, waste management and processing must be done properly. The selection of environmentally friendly methods to dispose of waste and the use of renewable energy can reduce negative impacts on the environment. This proper management can contribute greatly to sustainable development [2].

Cocoa is one of the plantation commodities in Indonesia that plays an important in the development of agro-industry [3]. One of the regions in Indonesia that has many cocoa plantations is the Special Region of Yogyakarta. 13.77% of cocoa plantations in Yogyakarta are located in Gunungkidul Regency [4], covering an area of 1,373.5 ha with an average production of 0.75 tonnes/ha and the plantations in Gunung Kidul involve 8,752 heads of farming families [5]. One of the sub-districts in Gunungkidul that has the most cocoa land area is Patuk Sub-district with a cocoa land area of 710 ha.

Cocoa supply chain activities in Patuk Sub-district involve several levels ranging from farmers, collective traders, and processing industries. Farmers ini this supply chain are the cocoa plantation owners. They carry out activities to plant and cultivate cocoa land and produce wet cocoa beans as their end products. The collective trader's role in this supply chain is to process the wet cocoa beans delivered by the farmers into dry cocoa beans which will be handed over to the processing industries. Processing industries play a role in producing various processed cocoa products as final products such as pure cocoa powder which can be used to produce various types of processed cocoa products.

Cocoa fruit has a harvesting period of about 4-5 months which causes production at each level to adjust to the availability of cocoa fruit. Based on preliminary observations in the field, cocoa beans ordered by collective traders often do not match the demand where collective traders request wet cocoa beans in the amount of 25 kg, but the order of wet cocoa beans coming from farmers is only 21.8 kg. This shows that sometimes the number of orders coming is below the demand. Some cocoa farmers in the Patuk Subdistrict also do not routinely deposit wet cocoa beans to collective traders following the routine schedule, so sometimes the number of cocoa fruit suppliers is different every month. The inconsistent procurement of raw materials causes production to have no fixed schedule and fluctuating production results to fulfill the demands of processing industries that have certain specifications for the cocoa beans ordered.

Based on the description of the cocoa supply chain condition in Patuk Sub-district, several things need to be analyzed related to the supply chain activities of each supply chain actor because each activity affects the other. This study aims to analyze the level of performance of the cocoa supply chain in the Patuk Sub-district and develop development recommendations for poor performance. Performance metrics in the SCOR model are used as indicators of supply chain performance measurement because it uses a vertical approach and the weighting of performance indicators is determined using AHP to determine the level of importance of performance indicators. In this study, SCOR and AHP can be used together or separately in the cultivation and processing of fresh cocoa pods into fermented dry beans at the farm level to ensure that these practices support environmental sustainability in the long term.

2. Methods

2.1 SCOR Model

Key Performance Indicators (KPIs) are determined based on business processes at farmers, collective traders, and processing industries using the SCOR Model. The SCOR model is based on 6 main processes, namely Plan, Source, Make, Deliver, Return, and Enable [6]. This research focuses on customers so it only uses 3 performance attributes, namely reliability, responsiveness, and agility.

2.2 Analytical Hierarchy Process

Determination of the level of importance of processes, performance attributes, and KPIs in this study was carried out using weighting with the Analytical Hierarchy Process (AHP) method to determine which KPIs had poor performance so that recommendations for improvement could be given.

The weighting is carried out on the core process of SCOR which consists of 6 processes, namely plan, source, make, deliver, return and enable. The weighting is also carried out on SCOR performance attributes which focus on customer aspects, namely reliability, responsiveness and agility. These consumer focused attributes are used in this research since the research focuses on the performance of the process. weighting is also carried out on KPIs identified from SCOR performance metrics based on existing business processes at each tier to determine the performance of each KPI.

Data for weighting was obtained based on the results of a questionnaire using pairwise comparisons filled out by 23 respondents consisting of 17 farmers, 4 collective traders, and 2 processing industries. The data from the questionnaire was processed using Microsoft Excel to weigh the processes, performance attributes, and KPIs for each tier.

2.3 Snorm de Boer

The final performance value is calculated by multiplying the AHP weighting results by the normalization results, namely using Snorm de Boer which functions to equalize the matrix values used as indicators measured using the SCOR model [7]. In Snorm de boer normalization, the normalization categories used are higher is better and lower is better [8]. The following is the normalization formula for the categories of each indicator, namely:

1. Higher is better

$$Snorm = \frac{Si - Smin}{Smax - Smin} x\ 100 \quad (1)$$

2. Lower is better

$$Snorm = \frac{Smax - Si}{Smax - Smin} x\ 100 \quad (2)$$

3. Results and Discussions

Cocoa supply chain actors in the Patuk Sub-district consist of farmers, collective traders, and processors. Figure 1 shows the cocoa supply chain in the Patuk Sub-district. Cocoa cultivation in Nglanggeran is a transition from the initial activity of processing cassava into cassava chips. The traditional cocoa supply chain starts with farmers planting cocoa beans, tending them, and finally harvesting the beans. During the maintenance period, farmers often use pesticides and herbicides to eradicate common cocoa diseases such as cocoa pod borer (CPD) and vascular disease (VSD). The characteristics of cocoa farmers can be divided into three, namely owner farmers, tenant farmers, and field extension workers from the local agricultural office. The harvest can be processed directly by farmers or purchased by processing units and supplied to local and regional companies. The fermentation and shelling process can improve the quality of dried cocoa beans, so these additional activities can increase the income of farmers, collectors, and processing units.

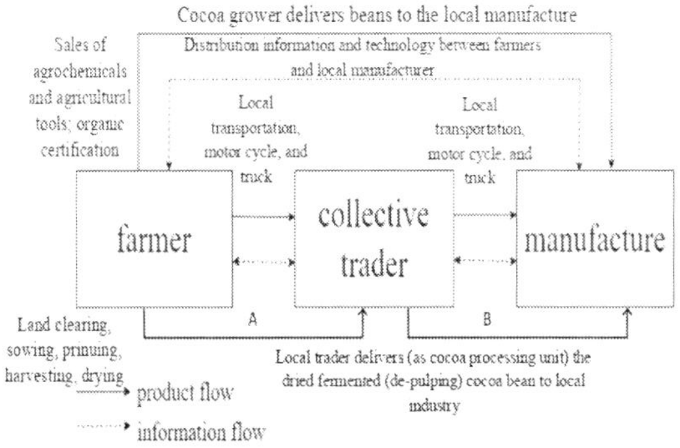

Fig 1. Local supply chain cocoa in the Patuk area

In each tier, farmers, collective traders, and processors consist of plan, source, make, deliver, return, and enable processes. KPIs were identified based on the business processes carried out by each tier which were then validated by all supply chain actors using an assessment questionnaire. KPI validation was processed using Aiken's V validation. The KPI validation results show that there are 15 KPIs from the reliability, responsiveness, and agility performance attributes that will be measured in the research. Figure 2 shows the AHP hierarchy model for this research.

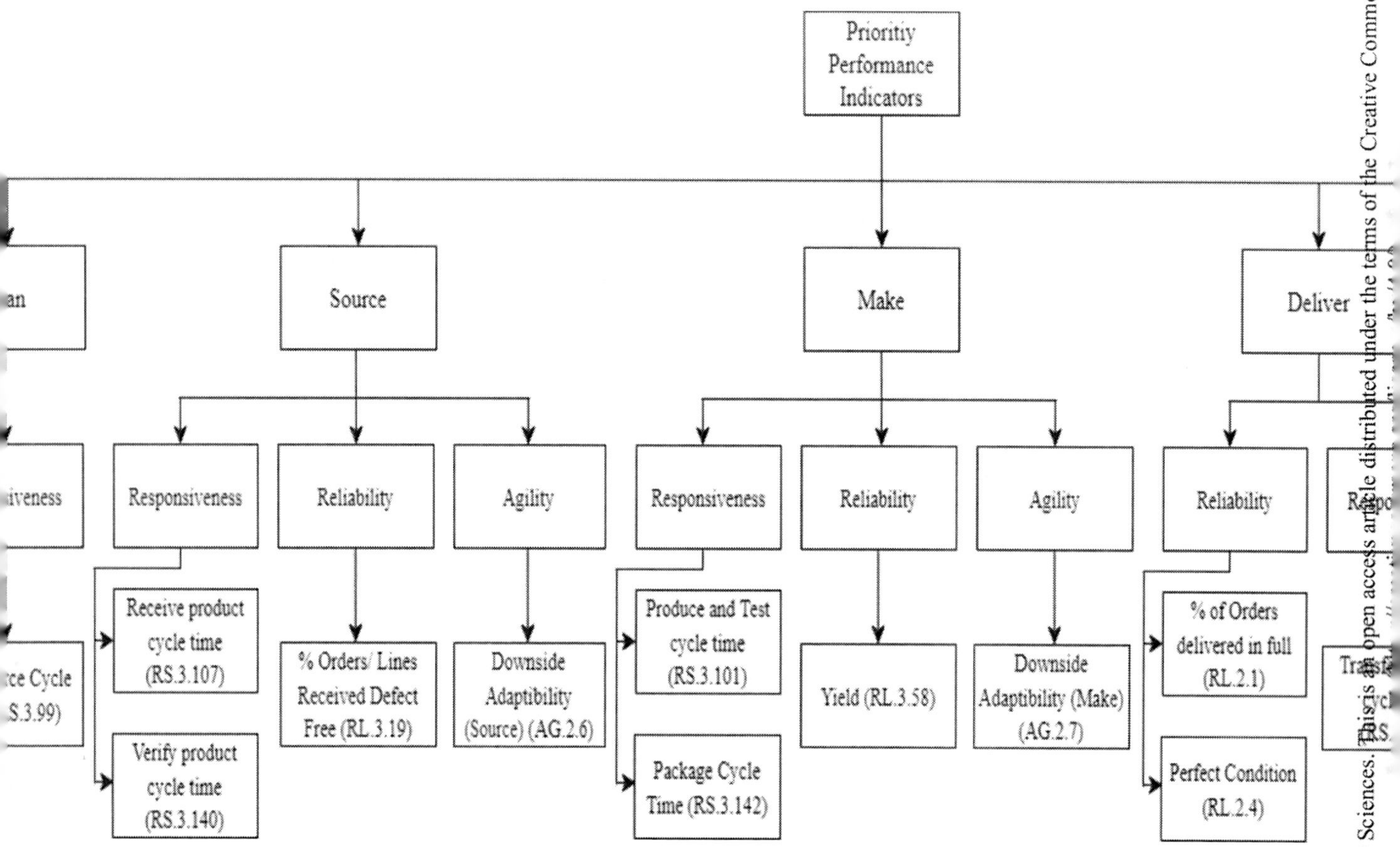

Fig 2. AHP Hierarchy

The final performance value is obtained using Snorm de Boer normalization multiplied by the AHP weighting results. Figure 3 is a table of performance calculation results for farmers and logistics scenarios for collective traders, and Figure 5 is a table of performance calculation results for the processing industry.

Table 1. Performance of Farmers

KPI	Weight	Snorm
Plan Source Cycle Time (RS.3.99)	0.09	92.61
Receive product cycle time (RS.3.107)	0.1	100
Verify product cycle time (RS.3.140)	0.04	100
% Orders/ Lines Received Defect Free (RL.3.19)	0.17	95.77
Downside Adaptability (Source) (AG.2.6)	0.05	84.33
Produce and Test cycle time (RS.3.101)	0.09	68.57
Package Cycle Time (RS.3.142)	0.03	75
Yield (RL.3.58)	0.18	44.94
Downside Adaptability (Make) (AG.2.7)	0.08	72.51
% of Orders delivered in full (RL.2.1)	0.03	84.69
Perfect Condition (RL.2.4)	0.02	66.46
Transfer product cycle time (RS.3.139)	0.03	75
Current Delivery Return Volume (AG.3.31)	0.02	100
Current source return volume (AG.3.41)	0.01	100
Manage MAKE Information Cycle Time (RS.3.68)	0.06	92.61

Table 2. Performance of Collective Traders

KPI	Weight	Snorm
Plan Source Cycle Time (RS.3.99)	0.06	73.33
Receive product cycle time (RS.3.107)	0.03	81.25
Verify product cycle time (RS.3.140)	0.01	83.33
% Orders/ Lines Received Defect Free (RL.3.19)	0.2	58.45
Downside Adaptability (Source) (AG.2.6)	0.08	52.63
Produce and Test cycle time (RS.3.101)	0.13	80
Package Cycle Time (RS.3.142)	0.02	100
Yield (RL.3.58)	0.18	69.71
Downside Adaptability (Make) (AG.2.7)	0.09	50
% of Orders delivered in full (RL.2.1)	0.01	73.91
Perfect Condition (RL.2.4)	0.08	67.39
Transfer product cycle time (RS.3.139)	0.05	70
Current Delivery Return Volume (AG.3.31)	0.02	100
Current source return volume (AG.3.41)	0.01	100
Manage MAKE Information Cycle Time (RS.3.68)	0.04	73.33

Table 3. Performance of Processing Industries

KPI	Weight	Snorm
Plan Source Cycle Time (RS.3.99)	0.1	64
Receive product cycle time (RS.3.107)	0.03	50

https://doi.org/10.1051/bioconf/20238007004

Verify product cycle time (RS.3.140)	0.01	76.68
% Orders/Lines Received Defect Free (RL.3.19)	0.19	92.59
Downside Adaptability (Source) (AG.2.6)	0.08	76.92
Produce and Test cycle time (RS.3.101)	0.09	100
Package Cycle Time (RS.3.142)	0.03	62.5
Yield (RL.3.58)	0.27	85.19
Downside Adaptability (Make) (AG.2.7)	0.05	68.75
% of Orders delivered in full (RL.2.1)	0.01	100
Perfect Condition (RL.2.4)	0.01	80
Transfer product cycle time (RS.3.139)	0.01	65.63
Current Delivery Return Volume (AG.3.31)	0.02	100
Current source return volume (AG.3.41)	0.02	100
Manage MAKE Information Cycle Time (RS.3.68)	0.08	74

Based on the results of the supply chain performance calculation for each tier, it is known that the performance levels of farmers, collective traders, and processors based on the performance indicator monitoring system are good (79.79), average (70.23), good (83.29), respectively. In each tier, the KPI that has the lowest value for farmers is yield, for collective traders is downside adaptability (make), and the processing industries is receive product cycle time.

The problem in the farmer tier is poor yield performance so that several proposals are needed to improve their yield performance. Farmers need to design SOPs that discuss the sorting process of wet cocoa beans and the implementation of crop maintenance. Farmers also need to improve crop maintenance to avoid fruit rot disease attacking plants. Improving plant care can be done by creating a more planned schedule so that plant care can be carried out regularly and by increasing the intensity of wrapping cocoa pods in plastic to avoid disease. Apart from that, the local government also needs to conduct training related to cocoa plant care so that farmers can understand and implement the knowledge gained.

At the collective trader tier, the performance of concern is downside adaptability (make). The proposed improvement strategy that can be given for this problem is that the local government needs to hold training related to maintenance for collectors because one of the causes of this problem is limited and damaged equipment. The training aimed to give insights to this tier so that there is SOP can be designed regarding equipment maintenance and used as a guidance for them. Collective traders also need to conduct regular meetings on a more scheduled basis to avoid miscommunication between them and the farmers because some farmers did not deliver the materials at the specified time and quantity. They can also pick up materials directly from farmers to avoid miscommunication so that the production process can be carried out immediately without having to wait for late materials. Erratic seasons are also a problem at the collector tier, the proposed strategy is for the local government to provide assistance in the form of additional drying equipment so that the dried cocoa beans are not exposed to rain even though the drying process is not maximized and requires a longer time.

The lowest performance in the processing industry tier is receive product cycle time. The proposed improvement strategy that can be given for this problem is that the industry should pick up dry cocoa beans directly to collective traders without having to wait for them to send materials if the stock of dry cocoa beans has run out. This action is given based on the reason that the delivery from collective traders depend on a certain amount of products, while if the production of collective tarders is decreasing, it will extend the time for receiving materials by the industry. In addition, this direct pick-up of materials is also because some collective traders are also constrained by transportation that is not available to make deliveries. Miscommunication also often occurs between the processing industry and collective traders due to the absence of contracts related to delivery schedules and delivery conditions so it is necessary to have a contract that regulates these matters so that the process of delivering and receiving materials from suppliers to the processing industry can be more scheduled and clear. In addition, the processing industry also needs to record material stocks to avoid material shortages due to increased waiting time for materials to arrive from suppliers. By recording material stock, the processing industry can estimate the time of the next order to the supplier.

There are potential challenges in implementing the proposed improvements. The main implementation challenge is the government's limited access to provide training to all farmers and collective traders because they are located in different locations and not all tier parties feel that the training is urgent to improve the performance of their activities.

The source cycle time plan of farmers and collective traders enables monitoring and measurement of supply chain performance in terms of received product cycle time, adaptability, and product transfer, each of which contributes to the environmental impact of inputs and outputs. Monitoring of resource use efficiency, estimation of waste, and estimation of emissions, should be disseminated to farmers according to their adaptability. Furthermore, AHP is for the selection of nutrients and pesticides at the cultivation stage. Such monitoring can lead to sustainable cocoa farming [9] Processing techniques (drying and fermentation), energy use for transport, and waste utilization can be used to assess environmental risks. Farmers need to improve their ability to assess high environmental risks and take steps to minimize their impact in the future because one of the pillars of a sustainable cocoa supply chain is economic sustainability [9] and the stability of cocoa bean prices is one of the factors that need to be improved as an effort to improve sustainability in the cocoa supply chain in terms of socioeconomics [10].

4 Conclusion

Cocoa supply chain performance in Patuk Sub-district is divided into 3 parts, namely for farmer performance of 79.79, collective trader performance of 70.23, and processing industry performance of 83.29. Capacity building related to environmentally friendly handling and improvement recommendations given to improve farmer performance is to increase plant care, for collective traders is to make contracts with farmers and record production, for processing industries is to make contracts related to delivery schedules.

Acknowledgment

The authors gratefully acknowledge support from the Directorate of Research & Development Universitas Gadjah Mada, National Research and Innovation Agency, Indonesia, and Unit Pengolah Hasil Kelompok Tani Ngudi Raharjo II, Patuk, Gunung Kidul, Yogyakarta for the support and contribution to conduct these projects.

References

[2] Alizadeh, S., Vali, F., Vatani, Z., & Avami, A. (2023). Sustainable analysis of Waste-to-Energy systems in cities by eco-efficiency assessment using DEA approach: A case study of Iran's municipalities. *Sustainable Cities and Society*, 104825.

[4] Badan Pusat Statistik. (2016). Luas Tanaman Perkebunan menurut Jenisnya dan Kabupaten/Kota di D.I. Yogyakarta. *Bps.go.id*. Diakses pada 26 februari 2023 melalui https://yogyakarta.bps.go.id/indicator/54/63/1/luas-tanaman-perkebunan-menurut-jenisnya-dan-kabupaten-kota-di-d-i-yogyakarta-.html.

[5] Badan Pusat Statistik. (2022). Statisik Kakao 2021. *Bps.go.id*. Diakses pada 3 Januari 2023 melaluihttps://www.bps.go.id/publication/2022/11/30/be404f7a76a56887462b5187/statistik-kakao-indonesia-2021.html.

[6] Gonzalez-Pascual, E., Nosedal-Sanchez, J., & Garcia-Gutierrez, J. (2021). Performance evaluation of a road freight transportation company through SCOR metrics. *Case Studies on Transport Policy*, *9*(4), 1431-1439. DOI: 10.1016/j.cstp.2021.07.001.

[1] Ingram, V., Van Rijn, F., Waarts, Y., & Gilhuis, H. (2018). The impacts of cocoa sustainability initiatives in West Africa. *Sustainability*, *10*(11), 4249.

[9] Iskandar, E., Amanah, S., Hubeis, A. V. S., Sadono, D., & Ginting, L. N. (2022). Improving the sustainability of cocoa smallholders farming in Aceh, Indonesia. In *IOP Conference Series: Earth and Environmental Science* (Vol. 951, No. 1, p. 012036). IOP Publishing.

[3] Ministry of Industry. (2007). *Gambaran Sekilas Industri Kakao*. Jakarta: Sekertariat Jenderal Kementrian Perindustrian.

[7] Prasetyo, D. S., Emaputra, A., & Parwati, C. I. (2021). Pengukuran Kinerja Supply Chain Management Menggunakan Pendekatan Model Supply Chain Operations Reference (SCOR) pada IKM Kerupuk Subur. *Jurnal Penelitian dan Aplikasi Sistem & Teknik Industri (PASTI)*, Vol. 15 (1): 80-92. DOI: 10.22441/pasti.2021.v15i1.008.

[8] Purnomo, H., Kisanjani, A., Kurnia, W. I., & Suwarto, S. (2019). Pengukuran Kinerja Green Supply Chain Management Pada Industri Penyamakan Kulit Yogyakarta. *Jurnal Ilmiah Teknik Industri*, *18*(2): 161-169. DOI:10.23917/jiti.v18i2.8535.

[10] Rahim, A., Antara, M., Rauf, R. A., Lamusa, A., Safitri, D., & Mulyo, J. H. (2020). Sustainability of cocoa production in Indonesia. *Australian Journal of Crop Science*, *14*(6), 997-1003.

 https://doi.org/10.1051/bioconf/20238007005

Nuisance Insects of Rearing mass Asian Citrus Psyllid (ACP) *Diaphorina citri* Under Controlled Conditions

Haris Setyaningrum[1,2*], *Edhi* Martono[2], *Alan* Soffan[2], *Jianhua* Mo[3], and *Siti* Subandiyah[2]

[1] Department of Agrotechnology, Faculty of Science and Technology, University of Darussalam Gontor, Ponorogo 63471, East Java, Indonesia
[2]Department of Plant Protection, Faculty of Agriculture, University of Gadjah Mada, Sleman, Yogyakarta 55281, Indonesia
[3]Department of Primary Industries, New South Wales (NSW) Government, 2198 Irrigation Way, Yanco New South Wales 2703, Australia

Abstract. Rearing insects for pest experiment it's very critical. The condition of rearing such insects impacted the smooth running of research. This condition also happens in the rearing of Asian Citrus Psyllid (ACP), *Diaphorina citri*. ACP is one of the most dangerous pests of citrus since responsible for the huanglongbing disease vector in citrus. Many researchers, either laboratory or field-based use a *D.citri* for object experiments. A clear and recent description of the types of insects or arthropods that disturb the rearing ACP is needed, especially in Indonesia. The research was conducted in the greenhouse of Faculty of Agriculture, Gadjah Mada University, from January 2021 to September 2022. The rearing of ACP uses orange jasmine (*Muraya paniculata*) and citrus (*Citrus nobilis*). The collected insect and spider preserved in alcohol or pined in insect tray and identify them based on related literature. This study proposed to assess the nuisance insect that infested on rearing of D.citri based on greenhouse conditions. Our results showing that the many nuisance insects that disturb the growth of rearing in *D.citri*. A total of nine insect and spider attacking *D citri,* they are black ants (*Dolichoderus* sp), aphid (*Aphis* sp), leaf roller caterpillar (*Phyllocnistis citrella*), cockroach (*Blattella* sp), ladybug (*Exochomus nigromaculatus*), mite (*Tetranychus sp*), Mealy bug (*Pseudoccocus* sp), Scale insect (*Aonidella aurantia*) and Spider. Those nuisance insects and spiders attack *D.citri* colonies in different ways and stages. The attacking nuisance insect affected to affected the size of the ACP culture.

* Corresponding author: haris.setyaningrum@unida.gontor.ac.id

1 Introduction

Rearing of any insects is important for providing insects in any level of generation for specific research interests. Rearing main purpose is to provide reliable and affordable high-quality insects (Leppla, 2009), for sure it's dependent on researcher needs. Rearing of insect including ACP, either mass or laboratory scale is key roles for successful of any insect treatment. Researchers or users need to pay attention to insect rearing since some effects on insects from laboratory adaptation, inbreeding depression, inadvertent selection, or direct rearing effects (Sorensen et al., 2012). Since rearing insects is critical to any scope of insect research, the rearing condition becomes very important.

Diaphorina citri (DC) or known as Asian Citrus Psyllids (ACP) are one of the most responsible vectors, transmitting Huanglongbing disease in almost every country around the world (Berk, 2016; Urbaneja et al., 2020). Rearing *D.citri* has been recently very important for studies of the transmission of bacteria causing hualongbing, *Candidatus Liberibacter asiaticus* (Las) (Li et al., 2006; Paris et al., 2013). Providing good and healthy D.citri during the experiment are challenging since any nuisance insect or other organisms bothering *D.citri* growth. The nuisance insect recorded in disturbing *D.citri* growth and colony such as *Tamarixia radiata* (Irvin et al., 2021; Kalile et al., 2022), coccinellids (Coleoptera: Coccinellidae), lacewings (Neuroptera: Chrysopidae), syrphids (Diptera: Syrphidae) and spiders (Arachnida: Araneae), and *Diaphorencyrtus aligarhensis* Shafee, Alam & Argarwal (Corallo et al., 2021; Michaud, 2004). The extensive method of rearing ACP described by (Skelley & Hoy, 2004) includes prerequisites of their parasitoids.

2 Materials and method

2.1 Rearing set

The rearing of ACP uses two common plants, orange jasmine (*Muraya paniculata*) and Sweet orange (*Citrus nobilis*). The rearing is set in a 1x1x1 meter square cage covered with net cloth, the cage maintained under a greenhouse with 27±1 °C and 60% Humidity. Every cage contains six to eight individual plants depending on their canopy. The plant was pruned two weeks before use as a feed for ACP. Every month plants are fertilized with common NPK fertilizer through diluted two spoons (equal to 2 grams) of fertilizer in 10 liter fresh water and poured to plant media as needed. The designs of the greenhouse and rearing set are described in Fig 1. Greenhouses typically consist of two separate doors (no.1 and 2) to prevent insects and other unwanted organisms from entering the rearing area. Inside of green house consisted two benches separately (no. 3) and four cages (1X1X1m) (no. 4) on the bench. The greenhouse is also provided with a fan (no.7) and humidifier apparatus (no. 5) to maintain temperature and humidity under optimal conditions for ACP rearing. Meanwhile, for watering, a water tap equipped with a hose is provided (no. 6)

2.2 *D.citri* infestation and nuisance insect observation

Ten to fifteen of matured adults ACP laid on new shoots of *M.paniculata* and *C.nobilis* using an aspirator. The colony was covered with a 10 cm diameter of a plastic tube covered with net cloth. The colony was left inside the tube for six days and opened immediately after the egg was laid. Eggs are allowed to hatch naturally and are monitored regularly. During the development of egg *D.citri*, observations are conducted at least every week to see if nuisance insects are coming and disturbing. The individuals of nuisance insects are collected and noted if available. The collected insect and arthropod are preserved in alcohol or pinned depend on their size and type of body. All the collected insect and and arthropod are identify based on relevant literature and online catalogue such as Encyclopedia of Life (https://eol.org/).

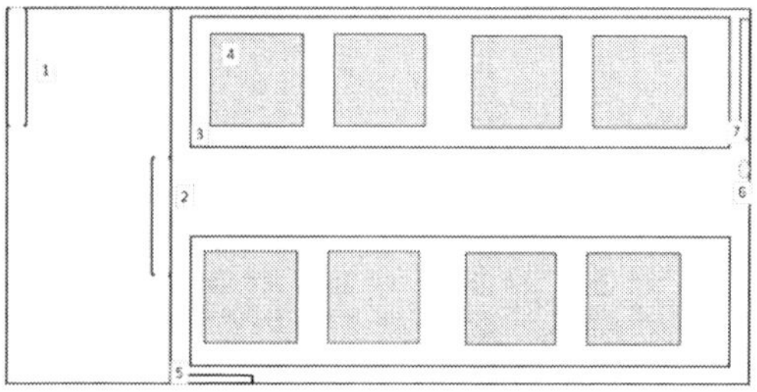

Note: 1) First door; 2) Second door; 3) Bench; 4) cages (1x1x1m); 5) Humidifier; 6) Water tap; 7) Fan

Fig. 1. Design of Greenhouse and rearing set during observations of nuisance insect.

3 Results and discussion

The results during observation rearing facilities of ACP are provided in Table 1. Every insect and arthropod have different type of disturbing even predator, parasitoid, or competitor. Those roles are describing as possible their type of disturbing and stadia of ACP.

Table 1. Kind of nuisance insect and arthropod in rearing facilities of ACP.

No	Organism	Possible role	Symptom appearance	Stadia disturbed
1	Black ants (*Dolichoderus* sp)	Predator	Unknown	Egg and nymph
2	Aphid (*Aphis* sp)	Competitor	Leaf curling, especially	Egg and nymph

			in new flush	
3	Leaf roller caterpillar (*Phyllocnistis citrella*)	Competito r	The leaves roll up and stick to each other	Egg and nymph
4	Cockroach (*Blattella* sp)	Predator	Unknown	Egg, Nymph, and imago
5	Lady bug (*Exochomus nigromaculatu s*)	Competito r, Predator	pale spots on leaves	Nymph and imago
6	Mite (*Tetranychus* sp),	Competito r, Predator	solid white spots on the leaf surface	Egg, Nymph, and imago
7	Mealy bug (*Pseudoccocus* sp),	Competito r	yellow spot and adherent white colonies on leave and stem	Egg, Nymph, and imago
8	Scale insect (*Aonidella aurantia*)	Competito r	The leaves turn white and the stems dry white	Egg, Nymph, and Imago
9	Spider	Predator	marked by the emergence of the web	Egg, Nymph, and imago

Some insects are commonly pests in the rearing facilities of ACP in every country. Skelley & Hoy (2004) described them very well of Scale insects and Aphid as quarantine pests who mostly attack ACP rearing mass. In our greenhouse and rearing facility's Scale insects, *Aonidella aurantia,* and Aphid, *Aphis sp* are two common pests who disturb the rearing colony in almost every season. The possibilities of Scale insects and aphids coming from the unclear plant, *Muraya*, and *Citrus*, when they were used as prey for rearing. These problems are common in ACP rearing as (Skelley & Hoy, 2004) mentions in their notes. *Muraya* and *Citrus* are favoured host plants for ACP (Hall, 2014; Setyaningrum et al., 2023) which also naturally favor other insects such as Thrips, *Frankliniella spp* ((Hall, 2014), scale insects (Poudel et al., 2022), and any other insect from various orders (Zuhran et al., 2021).

Spider, especially from the Aranea order is the mostly diverse group, consisting of over 42,000 species in 109 families, which are all predators and primarily feed on insects (Humenik et al., 2011). This situation makes Aranae very easy to find in every place where insects settle. In tropical rearing ACP facilities such as, it's almost found in every season both wet and dry season. It's also happened in nature or citrus orchards such as West Kalimantan (Zuhran et al., 2021). Mites are citrus pests, that have a symptom bronzing or silvering effect on the leaves where they attacked on(KENNETT et al., 1999). Differences from the spider, mites mostly attacked plants rather than *D.citri* colony. In these situations, mites become competitors or specific predators, especially on eggs of D.citri.

The black ant, *Dolichoderus sp* is commonly associated with seasonal fruit trees, including Sapodilla and Citrus (Jutsum et al., 1981; Van Mele & Cuc, 2001). Black ants in some cases came to the citrus trees associated with another insect, such as Scale insect (Jutsum et al., 1981; Michael Githae et al., 2020). These situations may happen in the greenhouse, since we found Scale insects *Aonidella aurantia* and *Dolichoderus* sp, in citrus and Muraya. Controlling Scale insects will impact directly or indirectly to *Dolichoderus* infestation inside rearing facilities. Mostly, the infestation of scale insects comes from the previous stock of plant rearing. As we know, scale insects are one of the serious insects that are difficult to control even when applying pesticides. It could happen also in the Leaf roller caterpillar (*Phyllocnistis citrella*) which is a common pest in citrus (Moore & Duncan, 2017), especially in the dry season (Foda et al., 2021). The leaf roller disturbing rearing since young leaves, which *D.citri* are the suitable places for growth. By infestation leaf roller, *D.citri* is unable to lay eggs. The possibilities of pathway's nuisance insect and arthropod-infested *D.citri* rearing are described in Fig 2.

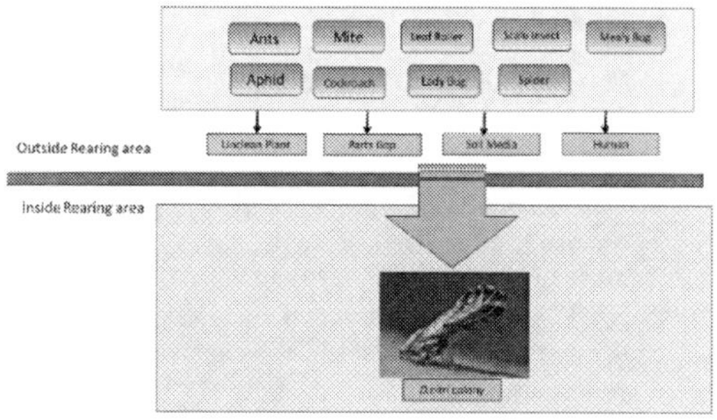

Fig. 2. Nuisance Insect and Arthropod Pathways Disturbance of ACP rearing

4 Conclusion

Nuisance insects of rearing ACP are black ants (*Dolichoderus* sp), aphid (*Aphis* sp), leaf roller caterpillar (Phyllocnistis citrella), cockroach (*Blattella sp*), ladybug (*Exochomus nigromaculatus*), mite

(*Tetranychus sp*), Mealy bug (*Pseudococcus sp*), Scale insect (*Aonidella aurantia*) and Spider. That nuisance insect played a different role in disturbing the growth of ACP colony. The possible role could be competitor or predator which affected to any stages of ACP growth. The potential pathways that allow nuisance insects and arthropods to pass through unclean plant host sources, gaps in rearing facilities, or through human movement during the rearing process.

5 Acknowledgment

I would like to express my gratitude to DRPM Ministry of Research and Higher Education for publication funding of this research through PDD Scheme (Number 0217/E5/PG.02.00/2023; 018/E5/PG.02.00.PL/2023. I would also thank the Australian Center for International Agricultural Research (ACIAR) for providing field research funding(HORT/2019/164) and accessing facilities. This research is a part of the PhD dissertation research of the corresponding author.

6 Selected references

1. Leppla, N. C. Rearing of Insects. Encycl. Insects 866–869 (2009) doi:10.1016/B978-0-12-374144-8.00227-7.

2. Sørensen, J. G., Addison, M. F. & Terblanche, J. S. Mass-rearing of insects for pest management: Challenges, synergies and advances from evolutionary physiology. Crop Prot. 38, 87–94 (2012).

3. Urbaneja, A. et al. Citrus pests in a global world. The Genus Citrus 333–348 (2020) doi:10.1016/B978-0-12-812163-4.00016-4.

4. Berk, Z. Diseases and pests. Citrus Fruit Process. 83–93 (2016) doi:10.1016/B978-0-12-803133-9.00005-9.

5. Li, W., Hartung, J. S. & Levy, L. Quantitative real-time PCR for detection and identification of Candidatus Liberibacter species associated with citrus huanglongbing. J. Microbiol. Methods 66, 104–115 (2006).

6. Paris, T. M., Rohde, B. B., Allan, S. A., Mankin, R. W. & Stansly, P. A. SYNCHRONIZED REARING OF MATED AND UNMATED DIAPHORINA CITRI (HEMIPTERA: LIVIIDAE) OF KNOWN AGE. Florida Entomol. 96, 1631–1634 (2013).

7. Irvin, N. A., Pierce, C. & Hoddle, M. S. Evaluating the potential of flowering plants for enhancing predatory hoverflies (Syrphidae) for biological control of Diaphorina citri (Liviidae) in California. Biol. Control 157, 104574 (2021).

8. Kalile, M. O. et al. UV light attracts Diaphorina citri and its parasitoid. Biol. Control 170, 104928 (2022).

9. Corallo, A. B., Pechi, E., Bettucci, L. & Tiscornia, S. Biological control of the Asian citrus psyllid, Diaphorina citri Kuwayama (Hemiptera: Liviidae) by Entomopathogenic fungi and their side effects on natural enemies. Egypt. J. Biol. Pest Control 31, 1–9 (2021).

10. Michaud, J. P. Natural mortality of Asian citrus psyllid (Homoptera: Psyllidae) in central Florida. Biol. Control 29, 260–269 (2004).

11. Skelley, L. H. & Hoy, M. A. A synchronous rearing method for the Asian citrus psyllid and its parasitoids in quarantine. Biol. Control 29, 14–23 (2004).

12. Hall, D. G. Interference by Western Flower Thrips in rearing Asian citrus psyllid: Damage to host plants and facultative predation. Crop Prot. 60, 66–69 (2014).

13. Setyaningrum, H., Martono, E., Soffan, A. & Mo, J. Best Practices Intercropping Citrus Controlling Asian Citrus Psyllids (Diaphorina citri) in Indonesia. Adv. Biol. Sci. Res. 29, 591–596 (2023).

14. Poudel, A., Sapkota, S., Pandey, N., Oli, D. & Regmi, R. Causes of citrus decline and its management practices adopted in Myagdi district, Nepal. Heliyon 8, e09906 (2022).

15. Zuhran, M., Mudjiono, G. & Dyah Puspitarini, R. Pengaruh pengelolaan agroekosistem terhadap kelimpahan kutu loncat jeruk Diaphorina citri Kuwayama (Hemiptera: Liviidae). J. Entomol. Indones. 18, 102–102 (2021).

16. Humenik, M., Scheibel, T. & Smith, A. Spider silk: Understanding the structure-function relationship of a natural fiber. Prog. Mol. Biol. Transl. Sci. 103, 131–185 (2011).

17. KENNETT, C. E., MCMURTRY, J. A. & BEARDSLEY, J. W. Biological Control in Subtropical and Tropical Crops. Handb. Biol. Control 713–742 (1999) doi:10.1016/B978-012257305-7/50074-6.

18. Van Mele, P. & Cuc, N. T. T. Farmers' Perceptions and Practices in Use of Dolichoderus thoracicus (Smith) (Hymenoptera: Formicidae) for Biological Control of Pests of Sapodilla. Biol. Control 20, 23–29 (2001).

19. Jutsum, A. R., Cherrett, J. M. & Fisher, M. Interactions Between the Fauna of Citrus Trees in Trinidad and the Ants Atta cephalotes and Azteca Sp. J. Appl. Ecol. 18, 187 (1981).

20. Michael Githae, George O. Ong'amo, John Nderitu, Gillian W. Watson & Wanja Kinuthia. Ants (Hymenoptera: Formicidae) associated with scale insects (Hemiptera: Coccomorpha) on citrus trees in Coastal and Lower Eastern Counties, Kenya. journal of Agricultural Science and Practice 245–249 https://www.researchgate.net/publication/348381949_Ants_Hymenoptera_Formicidae_associated_with_scale_insects_Hemiptera_Coccomorpha_on_citrus_trees_in_Coastal_and_Lower_Eastern_Counties_Kenya (2020).

21. Moore, S. D. & Duncan, L. W. Microbial Control of Insect and Mite Pests of Citrus. Microb. Control Insect Mite Pests From Theory to Pract. 283–298 (2017) doi:10.1016/B978-0-12-803527-6.00019-6.

22. Foda, Y. L., Wibowo, L., Lestari, P. & Hasibuan, R. INVENTARISASI DAN INTENSITAS SERANGAN HAMA TANAMAN JERUK (Citrus sinensis L.) DI KECAMATAN SEKAMPUNG UDIK KABUPATEN LAMPUNG TIMUR. J. Agrotek Trop. 9, 367 (2021).

https://doi.org/10.1051/bioconf/20238007006

The quality analysis of "JOSS" (Jatisarono Organik Sehat Sejahtera) organic rice

Hasna Difa Salisa[1], *Novita* Erma Kristanti[1*], and *Anggoro* Cahyo Sukartiko[1]

[1] Department of Agro-industrial Technology, Faculty of Agricultural Technology, Universitas Gadjah Mada, Indonesia

Abstract. Consumption of organic rice continues to increase accompanied by the increasing number of organic rice products in the market. This results in increasingly tight competition for organic rice product. One of the efforts for a new product to be accepted by consumers is by paying attention to its quality. In addition, it is also necessary to know the level of consumer acceptance on the product. The main subject of this research is JOSS Organic Rice, a local product from Yogyakarta Special Region that is currently in the phase of introduction to the market. The purpose of this study is to conduct an analysis of the quality profile of JOSS Organic Rice and its level of consumer acceptance. The rice samples used in this study are white organic aromatic rice from JOSS Organic Rice (Menor), Brand X (Mentik Susu), and Brand Y (Pandan Wangi). The samples have the same limitations in terms of location, price, and production period. The physical quality testing includes attributes such as rice color, rice dimensions, and rice hardness. Meanwhile, the chemical quality testing includes moisture content, whiteness, and cooked rice shelf life. Rice dimensions are calculated using ImageJ, hardness is tested using a Universal Testing Machine, moisture content is measured using an moisture analyzer, rice color is assessed using a chromameter, rice stickiness is measured using a spektrofotometer. JOSS Organic Rice has the characteristic of having a dull white color, medium rice grain shape, but it tends to break easily. The texture of the cooked rice from JOSS Organic Rice is not too sticky nor too loose. Based on consumer acceptance tests, JOSS Organic Rice scored highest in rice stickiness, while the attribute related to rice appearance had the lowest consumer acceptance score. These results proved that the most disliked of JOSS Organic Rice's appearance surprisingly resulted in the most preferred cooked rice by consumers.

1 Introduction

The consumer interest in a product becomes more focused and diverse over a time. This has a positive impact, leading to the emergence of various brands of organic rice products in the market. The abundance of organic rice products in the market creates intense competition for new products to be accepted by consumers. One of the efforts for a new product to be accepted by consumers is by paying attention to its quality attributes to align with consumer expectations. Hence, there is a need to conduct an quality analysis of JOSS Organic Rice based on its quality attributes. In addition, it is also necessary to anlyze the level of consumer acceptance on the product.

Jatisarono Organik Sehat Sejahtera Organic Rice, referred as JOSS Organic Rice, is an organic rice productfrom the Rice Farmers Group in Kulon Progo Regency, specifically located in Jati Ngarang Lor Village. JOSS Organic Rice first introduced its rice products to the market in 2013 and began promoting organic rice in 2020. They produce two types of rice: white rice and red rice, using various rice varieties. The rice varieties that commonly used are Ciherang, Menor, Rojolele, Nutrizinc, and Pandan Wangi. JOSS Organic Rice also has a division known as the Internal Control System (ICS) to monitor and oversee theproduct quality throughout the production process, both inthe fields and post-harvest. For the JOSS Organic Rice company, product quality is a top priority.

One of the existing problems based on the researcher's interview with the company is related to the wholeness of the rice grains. The company was dissatisfied with the final shape especially rice wholeness after the post-harvest handling process. Based on manual visual observation, JOSS organic rice grains have prominent characteristics. Rice grains tend not to be intact or cracked. These problems can undoubtedly impact sales, where consumers wonder and assume the product is not fresh and has poor quality.

According to the expert, Heizer & Rander (2001), quality is defined as the entirety of the form and characteristics of goods or services that demonstrate theirability to satisfy both evident and hidden needs. The quality of rice is regulated by the SNI 6128-2020, which applies to both organic and non-organic rice, consisting general requirements and quality specifications. For organic rice is also requiring to adhere to SNI 6729:2016 (Organic Farming System). [6]

2 Research Method

In this study, the quality of rice will be assessed through few physicochemical testing. The physical quality testing includes attributes such as rice color, rice dimensions, and rice hardness. Meanwhile, the chemical quality testing includes moisture content and rice stickiness. Rice color testing is conducted using a

*corresponding author: novita_erma@ugm.ac.id

chromameter and employs color parameters such as L*, a*, b* values. Rice shape measurement is identified through dimention test and hardness test. Rice dimension measurement involves measuring the length and width of the rice. Length and width are calculated using the ImageJ software with ten repetitions for each organic rice sample. Hardness testing is conducted using texture analysis with a Universal Testing Machine (UTM) and repeated twice. To determine the stickiness level of cooked organic rice can be found from amylose and amylopectin content and tested using spectrophotometer an instrument for measuring sample absorbance at specific wavelengths. The moisture content of rice is measured using the Ohaus Moisture Analyzer through the thermogravimetric method. Consumer acceptance is determined through hedonic testing. The sensory testing method that used in this research is consumer acceptance test that refers to SNI 01-2346- 2006.

Tabel 1. Data Collecting and Processing Methods

Data	Data Collecting	Data Processing
Rice Dimension	Length and width measurements using ImageJ Software	- Normality Test - Kruskall-Wallis Test - Mann-Whitney Test
Rice Hardness	Hardness test using Universal Testing Machine (UTM)	
Rice Color	Color test using chromameter	
Rice Moisture Content	Moisture test using Ohaus Moisture Analyzer	
Rice Stickiness	Amylose content test	
Consumer Acceptance	Hedonic Test	
Quality Attributes	Literature	Descriptive

3 Sample

The sample is determined based on the market survey by selecting 1 sample at different market positions. Based on market surveys that have been conducted, Brand Y (Pandan Wangi) rice samples are included as the market leader, Brand X (Mentik Susu) is included as the market challenger, and JOSS organic rice is a market follower. Based on a preliminary survey, the classification is determined by looking at market data, sales, and product distribution in the Yogyakarta Special Region. Market surveys were conducted before the start of the study, and data was generated in the form of the availability of organic rice products in traditional retail in each city/district in Yogyakarta.

3.1 Brand X (Mentik Susu)

Mentik Susu variety is one of the main varieties used in selling Brand X aromatic white organic rice products. Brand X organic rice produces organic rice together with partner farmer groups based on mutual trust. Brand X organic rice is from rice grown in the DIY and Magelang regions. The Mentik Susu variety is selected because Mentik Susu is the mainstay local variety of organic rice farming in the area where the farmer group is located. Mentik Susu variety is included in aromatic rice, whose shape is closer to the shape of white sticky rice. The Mentik Susu variety has a fluffier texture and whiter color (Handayani et al., 2018).

3.2 Brand Y (Pandan Wangi)

Brand Y Organic Rice uses the pandan wangi variety, also included in the aromatic rice type. The pandan wangi variety was chosen as the main variety in selling Brand Y organic white rice because it is one of the local superior varieties. Similar to Brand X, Brand Y organic rice is taken from rice grown in the DIY and Magelang areas. The production system is also the same as Brand X, namely by partnering with several farmer groups. According to Syamsiah et al. (2020), the Pandan Wangi rice variety has a characteristic according to its name, namely producing pandanus-scented rice with a fluffy texture.

3.3 JOSS (Menor)

Organic is Unlike the other two samples, JOSS Organic Rice is produced by the farmers themselves. JOSS Organic Rice is produced from rice grown in the Nanggulan area using the Menor variety, a superior local rice variety from the Menoreh Mountains region. Menor rice is a local rice that is developing in Kulon Progo Regency.

4 Result and Discussion

4.1 Rice Shape

4.1.1 Rice Dimension

The measurement of rice dimensions was conducted with ten repetitions for each sample of organic rice. Rice dimensions include length and width, from which the elongation value (length/width) was derived. The steps involved in measuring the dimensions included capturing images of rice grains placed on a millimeter-block base next to a ruler. These images were then processed using ImageJ software to calculate the length and width of the rice grains. The results of the length and width measurements using ImageJ software can be seen in Table 2.

Table 2. Dimensions of organic rice

Sample	Length (cm)	Widht (cm)	Ratio (L/W)	Length category	Shape
Brand X (Mentik Susu)	0,560^{a}	0,230^{a}	2,433	sedang	medium
Brand Y (Pandan Wangi)	0,579^{a}	0,264^{b}	2,197	sedang	medium
"JOSS" (Menor)	0,625^{b}	0,234^{a}	2,674	sedang	medium

*different letter notations in the same column indicate significant differences with a significance level of 5%.

https://doi.org/10.1051/bioconf/20238007006

Based on the measurements of length and width from the three samples of organic rice, it can be observed that there are significant differences among some samples. In terms of length measurements, JOSS Organic Rice exhibits a significant difference from the other two samples. The length of JOSS Organic Rice using the Menor variety is the longest and tends to be longer, measuring 0.625 cm. According to the Department of Agriculture (2003), based on length measurements, rice can be classified into several groups, including Short (<5.51 mm), Medium (5.51-6.60 mm), Long (6.61-7.50 mm), and Very Long (>7.50 mm). Consequently, the length measurements of these three samples fall within the Medium rice category, which is 5.51-6.6 mm in length.

In terms of width measurements, JOSS Organic Rice doesn't show a significant difference compared to Organic Rice Brand X (Mentik Susu), while Rice Brand Y (Pandan Wangi) exhibits a significant difference, tending to be wider with a value of 0.264. The elongation values of the three samples don't present significant differences; all of them fall into the classification of rice with a medium shape size, which corresponds to a ratio value between 2.1-3.0.

4.1.2 Rice Hardness

In a package of organic rice, the initial visual perception for consumers is rice dimensions and also rice wholeness. To assess the level of rice wholeness in each organic rice sample, this can be done by calculating the percentage of broken rice or by conducting a hardness test on the rice grains. The hardness test is carried out using a Universal Testing Machine (UTM) by measuring Fmax, which represents the maximum force required to compress the product until it breaks for the first time. The results of the rice hardness test can be seen in Table 3.

Table 3. Hardness test of organic rice

Sample	Unit	Average
Brand X (Mentik Susu)	Fmax (N)	405,43 ± 1,60[a]
Brand Y (Pandan Wangi)	Fmax (N)	405,55 ± 1,15[a]
Rice "JOSS" (Menor)	Fmax (N)	403,83 ± 1,90[a]

*different letter notations in the same column indicate significant differences with a significance level of 5%.

From the results of the hardness test, it can be observed that the three samples do not exhibit significant differences. The Fmax values for Rice Brand X (Mentik Susu) are not substantially different from the Fmax values for Rice Brand Y (Pandan Wangi), measuring 405.43 N and 405.55 N respectively. JOSS Organic Rice has the lowest Fmax value among the three samples, measuring 403.83 N. The relatively lower value can potentially contribute to the susceptibility of JOSS Organic Rice grains to breakage. According to research conducted by Widiatmoko (Hernawan & Meylani, 2016), the hardness value of rice is influenced by factors such as moisture content and the duration of rice storage. Broken rice can also result from the milling process. The use of a new milling stone can lead to more broken rice, while a worn milling stone tends to produce fewer broken rice grains (Handayani et al., 2018). [4,5]

The hardness value of rice affects the level of rice wholeness in each sample. The greater the hardness value of the rice, the more resistant it is to breaking or fracturing. This aligns with the calculation of broken and chalky rice in each sample. JOSS Organic Rice is a sample with the highest percentage of chalky and broken rice in a single package. The percentage of chalky and broken rice in a package of JOSS Organic Rice reaches 60%, surpassing the percentage of intact rice. This can be attributed to overly harsh milling, resulting in a higher amount of broken and chalky grains (BSN, 2015).

4.2 Rice Moisture Content

Rice moisture content is the amount of water content present in rice, expressed as a percentage. The hardness level of rice is greatly influenced by its moisture content. Moisture content is also a crucial quality attribute as it can impact the appearance, texture, and taste of food products. The rice moisture content measurements for the three samples can be seen in Table 4.

Table 4. Moisture test of organic rice

Sample	Unit	Average
Brand X (Mentik Susu)	(% Wb)	10,40 ± 0,54[a]
Brand Y (Pandan Wangi)	(% Wb)	11,18 ± 0,040[a]
Rice "JOSS" (Menor)	(% Wb)	13,49 ± 0,060[b]

*different letter notations in the same column indicate significant differences with a significance level of 5%.

The analysis results indicate that only the moisture content of JOSS Organic Rice shows a significant difference compared to the other two samples. JOSS Organic Rice has the highest moisture content at 13.49%. The moisture content of Organic Rice Brand X (Mentik Susu) is at 10.40%, while Organic Rice Brand Y (Pandan Wangi) has the lowest moisture content at 11.18%. According to the quality requirements for organic and non-organic rice, all three samples still meet the SNI criteria, which state a maximum moisture content value of 14%.

Stated by Arsyad & Saud (2020), excessively high moisture content (over 14%) can affect the rice milling process; the rice grains become too soft or mushy, leading to broken rice. Excessively low moisture content can also make the rice too dry and prone to breakage. Aside from moisture content, rice breakage can also be caused by inadequate drying processes. According to research by Prasetyo et al. (2008), using excessively high drying temperatures can lead to rapid drying rates, inducing stress within the material, creating moisture content variations, resulting in cracks, and lowering rice quality, characterized by a decrease in the percentage of head rice.

4.3 Rice Color

Rice color is greatly influenced by the rice variety used. There are rice varieties with bright white color, while others have a dull white appearance. Based on the color test conducted using a chromameter, the L*, a*, and b* values of the three rice samples can be seen in Table 5.

Table 5. The value of brightness, redness and yellowness of organic rice color

Sample	Lightness (L*)	Redness (a*)	Yellowness (b*)
Brand X (Mentik Susu)	67,81^{b}	-1,42^{a}	7,71^{a}
Brand Y (Pandan Wangi)	57,20^{a}	-1,84^{a}	8,13^{a}
Rice "JOSS" (Menor)	59,82^{a}	-1,68^{a}	7,51^{a}

*different letter notations in the same column indicate significant differences with a significance level of 5%.

The rice types in the three samples are white rice, so the color test results can be observed in the L* parameter, which represents the brightness level of color, with a value of 0 for black and 100 for white. The a* value indicates the level of redness for positive values, while the b* value indicates the level of yellowness for positive values. Based on the testing results using the chromameter, JOSS Organic Rice has a brightness level (L*) of 59.82. JOSS Organic Rice and Rice Brand Y (Pandan Wangi) have similar values, while Rice Brand X (Mentik Susu) has a higher value, measuring 67.81.

Based on the color test results, it is evident that Rice Brand X (Mentik Susu) has the highest level of brightness, measuring 67.81. This aligns with the theory that variety of Mentik Susu tends to have a whiter color, resembling the color of milk. Mentik Susu rice is a variety with a milky white color due to genetic factors that result in a higher calcium content in the organic rice grains of the Mentik Susu variety (Handayani et al., 2018). The color test results for JOSS Organic Rice and Rice Brand Y (Pandan Wangi) are not significantly different. Both samples have a similar color, falling within the category of dull white.

4.4 Rice Stickiness

The testing of rice stickiness level is conducted by examining the amylose content of the rice using a spectrophotometer, which is a testing instrument used to measure the absorbance of a sample at a specific wavelength. The amylose content results for the three samples can be observed in Table 6.

Table 6. The amylose levels of organic rice

Sample	Unit	Average
Brand X (Mentik Susu)	(% Wb)	14,48 ± 0,05^{a}
Brand Y (Pandan Wangi)	(% Wb)	17,99 ± 0,06^{b}
Rice "JOSS" (Menor)	(% Wb)	15,46 ± 0,06^{c}

*different letter notations in the same column indicate significant differences with a significance level of 5%.

In terms of statistical analysis, the amylose content values of the three samples do not show significant differences, but there are notable differences in the values. Among the three samples, the amylose content of JOSS Organic Rice falls in the middle range when compared to the other two samples. The amylose content value of JOSS Organic Rice is 15.46%, which classifies it as medium amylose rice, resulting in rice that is neither overly sticky nor too loose. Organic Rice Brand Y has the highest amylose content at 17.99%, while Organic Rice Brand X has the lowest amylose content at 14.48%.

According to the classification by Kumar and Kush as cited in Sari et al. (2020), JOSS Organic Rice and Organic Rice Brand Y (Pandan Wangi) fall under the medium amylose rice classification, while Organic Rice Brand X (Mentik Susu) is categorized as low amylose rice. Based on the Indonesian National Standard (SNI) Number 6128 of 2015 regarding rice, one of the factors determining the rice texture, whether sticky or loose, is the amylose content. A content of amylose > 25% results in a sticky texture, amylose content between 20-25% leads to a loose texture, amylose content between 15-20% results in a very loose texture, and amylose content < 15% is associated with a sticky (glutinous) texture (BSN, 2015). [7]

4.5 Consumer Acceptance

The Consumer acceptance test determines consumer assessment of a product's properties or quality. The acceptance test used in this study is the hedonic test, a sensory (organoleptic) test used to determine the preference level for a food or beverage product. A hedonic test assessment was conducted on 47 panelists with different age ranges and domiciles. In the consumer acceptance test, mathematical data processing was carried out using the Kruskal-Wallis method to determine whether the results of the three samples were significantly different. The results showed that rice color, wholeness, and texture attributes were significantly different. There is no significant difference in the rice aroma attribute and overall results.

Table 7. Average score of hedonic test

Attributes	Average Score of Hedonic Test		
	Brand X (Mentik Susu)	Brand Y (Pandan Wangi)	JOSS (Menor)
Rice color	5,85 ± 0,932^{a}	4,85 ± 1,489^{b}	4,66 ± 1,109^{b}
Rice wholeness	5,19 ± 1,345^{a}	5,77 ± 1,047^{b}	3,91 ± 1,265^{c}
Cooked rice stickiness	4,83 ± 1,372^{a}	4,57 ± 1,379^{a}	5,43 ± 1,137^{b}
Cooked rice aroma	4,72 ± 1,136^{a}	5,28 ± 1,246^{a}	5,00 ± 1,319^{a}
Overall	4,74 ± 1,276^{a}	5,30 ± 1,30^{a}	5,19 ± 0,947^{a}

*different letter notations in the same column indicate significant differences with a significance level of 5%.
1 = Dislike very much 2 = Dislike moderately 3 = Dislike slightly 4 = Neutral 5 = Like slightly 6 = Like moderately 7 = Like very much

The acceptance value of Brand X (Mentik Susu) in rice color attributes has a significant difference with the two samples, while Brand Y (Pandan Wangi) and JOSS Organic Rice (Menor) have similar or not significantly different values. In the rice wholeness attribute, each sample has a significantly different value. Based on general assessments, consumers will give a low acceptance value to rice with a dull color (Mardiah et al., 2016) At the rice texture attribute, Brand X (Mentik Susu) has a consumer acceptance value that is not significantly different from Brand Y (Pandan Wangi), while JOSS Organic Rice has a significant difference compared to the other two samples. The rice aroma attribute has no significant difference in the results of all samples, and this can be caused because all organic rice samples used are included in fragrant rice varieties. In addition, the aroma of rice that is easily lost or faded can also be a factor causing the value of consumer acceptance of the rice aroma attribute is not significantly different.

The overall attribute does not have a significant difference in value, and this can be due to consumer preferences that balance the quality of rice and rice and are not too affected by the differences in each sample as a whole. Overall, JOSS Organic Rice is only limited to a score of 5 or Somewhat Like. This result can be due to consideration of the physical quality of rice, which has a low level of wholeness.

Based on the table, it can be seen that JOSS Organic Rice received a relatively low score on the rice wholeness attribute. This proves that the results of the quality characteristics show that JOSS Organic Rice breaks easily have a significant impact on consumer acceptance. Consumers tend to like rice with a high wholeness level and do not break easily. Based on the level of liking, JOSS Organic Rice is most significant in the attribute of rice stickiness, with an average score of 5,4. That score proves that the test results for the amylose content of JOSS Organic Rice, which is in the middle range, produce rice with the stickiness/fluffiness most liked by consumers.

5 Conclusion

JOSS Organic Rice, currently positioned as a market follower, exhibits quality characteristics that enable it to compete with the other two samples. JOSS Organic Rice has a brightness value (L*) of 59.82, with a characteristic dull white color. Its rice shape is medium and tends to be breakable. The resulting moisture content value is 13.49%, and the rice hardness is measured at 403.83 N. The rice texture from this rice is neither too sticky nor too loose, with an amylose content of 15.46%.

Based on the consumer acceptance test, JOSS Organic Rice has the highest acceptance value on the rice texture attribute, while the quality attribute with the lowest acceptance value is the rice wholeness attribute. These results prove that the least favorable level of wholeness of JOSS Organic Rice produces the most favorable rice texture for consumers.

Acknowledgments. The author would like to express the gratitude toward Gadjah Mada University for providing the funding to this research through RTA Program 2023 contract number : 5075/UN1.P.II/Dit-Lit/PT.01.01/2023

References

[1] Badan Standarisasi Nasional (BSN). Rice SNI 6128 2015. (2015)

[2] Badan Standarisasi Nasional (BSN). Organic Farming System SNI 6729:2016. (2016).

[3] Departement of Agriculture, B. P. and P. P. K. N. P. Nutfah. Guide to Paddy Characterization and Evaluation System. Secretariat of the National Germplasm Commission. (2003).

[4] E. Hernawan, V. Meylani. Analysis of physicochemical characteristics of white rice, brown rice, and black rice (Oryza sativa L., Oryza nivara and Oryza sativa L. indica). Journal of Health: Journal of Nursing Sciences, Health Analyst and Pharmacy, 15(1), 79–91, (2016).

[5] S. Handayani, M.I. Affandi, & S. Susanti. Analysis of quality characteristics of organic rice varieties of Mentik Susu and Sintanur. Journal of Food Systems and Agribusiness, 75–82, (2018).

[6] J. Heizer, B. Rander. Operations Management Principles. Salemba Four. (2001).

[7] A. R. Sari, Y. Martono, and F. S. Rondonuwu. Quality Identification of White Rice (Oryza sativa L.) Based on the Content of Amylose and Amylopectin in Traditional and "Selepan" Markets in Salatiga City. Science Titian: Scientific Journal of Multi Sciences, 24–30 (2020)

[8] T. Prasetyo, A. Kamaruddin, K. D. Made, H.T. Armansyah & N. Leopold. The effect of drying time and tempering on the quality of rice in thin layer grain drying. Semesta Teknika, 11(1), 29–37 (2008).

Using Superior Variety and Application of CI200 to Support Food Security in South Sumatra Swamp Land

Waluyo[1], *Agus* Suprihatin[2*], *Suparwoto*[1] and *Jumakir*[1]

[1]Research Center for Food Crops, National Research and Innovation Agency (BRIN), CWS CWS Palembang, Kancil Putih street, Rusa alley, Palembang city, South Sumatra 30137
[2]Research Center for Food Crops, National Research and Innovation Agency (BRIN), CWS KSE Achmad Baiquni, Babarsari Street, Tambak Bayan, Caturtunggal, Depok district, Sleman regency, DI Yogyakarta. 55281

Abstract. Swamp paddy fields with their various problems have enormous potential to support the achievement of national food security. Using superior varieties and determining the exact planting season in the swamp rice cultivation is expected to increase rice productivity and farmers' income per unit area. The research aims to increase rice production and assess the feasibility of CI200 in swamps. The study used an oversite design with 5 replications in a farmer's swamp land with an area of 1 ha, in the Gelebek Dalam village, Rambutan District, Banyuasin Regency, South Sumatra Province. Factors tested: 1) superior varieties, there are 4 varieties namely Inpari 32, Inpara 2, Mekongga, and Ciherang, 2) seasons, application of CI200 so that planting is carried out throughout the year, namely the DS (April-August 2021) and RS (October 2021-February 2022). Inpari 32 has the highest productivity compared to other varieties, reaching 6.98 t HDG ha^{-1}. The value of the B/C ratio reached 1.51-2.44 in DS and 1.09-1.98 in RS. Therefore, the application of the CI200 cropping pattern by using NSV which has been adapted can be widely developed to support food security in swamps land.

1 Introduction

Rice is the main food crop as a staple food in various parts of Indonesia. Rice farming in Indonesia is still the backbone of the rural economy. The availability of domestic rice stocks is very important in the framework of the sustainability of national food security, which is expected to be self-sufficient in food (rice) can be achieved [1]. The Ministry of Agriculture has targeted that Indonesia will be able to be self-sufficient in sustainable food since 2010. Increased production and productivity of food crops is one of the performance parameters of the Ministry of Agriculture.

In general, swamp land has not been exploited optimally for agricultural business, even though with the application of land management and management technology and the application of integrated agricultural commodities, swamp land can become one of the mainstay sources of agribusiness growth and national food security [2]. Swamps in South Sumatra Province have two types of swamps that have been used for rice, namely swamp lands covering an area of 285,941 ha and tidal lands covering an area of 273,919 ha. Banyuasin Regency is one of the regencies that have the widest swamp after Ogan Komering Ilir Regency and Ogan Ilir Regency [3].

The demonstration was an extension method that directly contacted primary and business farmers. Demonstration farming was a technique used to share research outcomes and practices to hasten the adoption of technology among users. It involved organizing various displays to showcase effective methods for promoting agricultural development. Demfarm could accelerate the adoption of agriculturally beneficial technology constructed by end-users according to the potential of the commodity This method was utilized to speed up adoption [4].

One of the main technological components that has been proven to increase rice production is the use of new superior varieties (NSV). The use of superior varieties has a very important role in efforts to increase rice productivity [5]. The use of NSV can increase farmers' productivity and income [6]. The varieties used must be certified superior varieties and follow the land to be used [7].

Superior rice cultivars are crucial to enhancing productivity, managing pests and diseases, improving fertilization efficiency, and adapting to adverse environmental conditions. Compared to other production technologies, superior varieties are accepted by farmers more quickly because they are more effective and efficient [8]. High-yielding varieties accounted for 56% of the production components, with the main focus being on increasing yield through the development and enhancement of new high-yielding varieties [9].

Inundation and drought that cannot be predicted with accuracy are the main obstacles to the development of lowland swamps which depend on hydrographic conditions, rainfall, and river water levels [10]. In general, rice productivity in lowland swamps is

* Corresponding author: at33ns@gmail.com

relatively low. The opportunity to increase production is quite large considering the area is quite large, the technology for its utilization is already available, and the agricultural cultivation applied by farmers is generally not as recommended.

Efforts to increase the productivity of food crops in swamp lands can be carried out by utilizing the land by intensifying the application of technology and expanding the planting area by increasing the cropping index (CI) [2]. The farmers' existing rice cultivation in swampy lands still implements CI100, a condition caused by high water stagnation during the rainy season and drought during the dry season [11].

Local farmers in swamp areas in Indonesia tend to avoid new superior varieties despite the fact that local rice varieties have lower productivity and longer lifespans [12]. The yield of rice in a swamp paddy field was below 3 t ha^{-1} during the second planting season. The rice sector's productivity may increase by increasing inputs or implementing innovative technology [13].

Three superior varieties were planted in Sungai Dua Village, Rambutan District, Banyuasin Regency have smaller B/C ratio values, namely IPB 3S (0.80), Inpari 32 (0.79), and Ciherang (0.74), were obtained [14]. Therefore, this study aims to increase rice production by implementing superior varieties that implement CI200 and to assess the feasibility of CI200 rice farming in swamp lands.

2 Method

The research was conducted in a 1-hectare shallow swamp farmer's paddy field, in Gelebek Dalam Village (-3.0633° S, 104.8522° E) Rambutan District, Banyuasin Regency, South Sumatra Province (Figure 1). Geographically, Banyuasin Regency is located at 02°43'48" - 03°09'00" South Latitude, and 104°10'48" - 105°07'12" East Longitude.

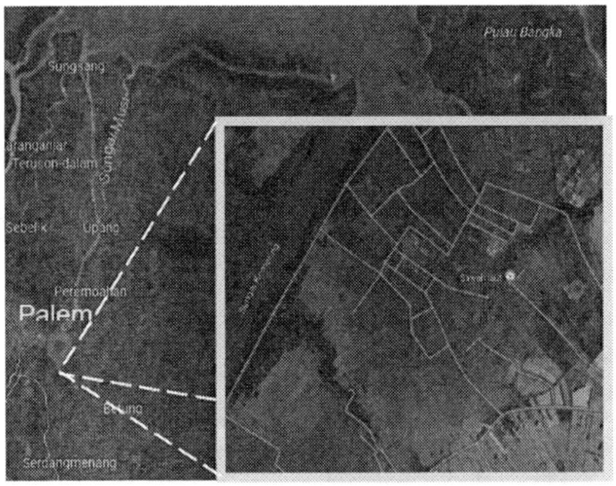

Fig. 1. Research location of swamp paddy field, Gelebak Dalam Village, Rambutan District, Banyuasin Regency, South Sumatra Province.

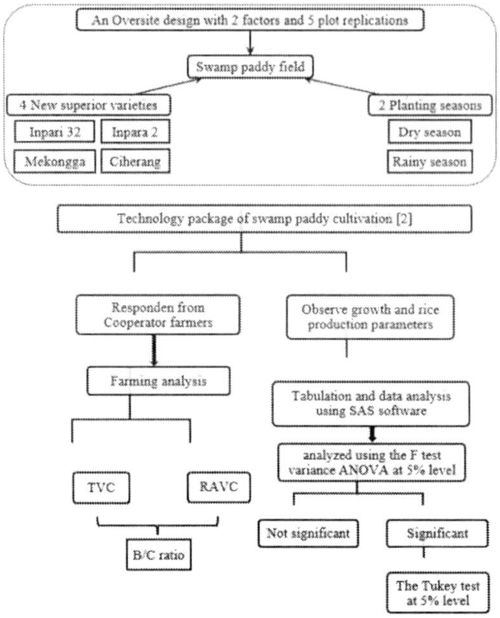

Fig. 2. Research location of swamp paddy field, Gelebak Dalam Village, Rambutan District, Banyuasin Regency, South Sumatra Province.

This study used an oversite design with 5 replications. There are 2 factors tested, namely 1) Varieties, there are 4 superior varieties, namely Inpari 32, Inpara 2, Mekongga, and Ciherang. 2) planting season, there are 2 planting times, namely dry season (April-August 2021) and rainy season (October 2021-February 2022). The technology package utilized for cultivating swamp rice is available for reference in Table 1.

Table 1. Components of shallow swamp rice cultivation technology [15]

Technology component	Research Activities	
Planting Season	April-August (DS)	October-February (RS)
Seedling Age (DAP)	25-30	25-30
Number of seeds per clump	2-3	2-3
Seed needs (kg ha^{-1})	25-30	25-30
Planting distance (cm)	25 x 25	25 x 25
How to plant	transplanting	transplanting
Variety	superior variety	superior variety
Water management	naturally	naturally
Fertilization (kg ha^{-1})	*PUTR*, LCC	*PUTR*, LCC
Manure	1 t ha^{-1}	1 t ha^{-1}

Soil processing	tillage	tillage
Pests and diseases	integrated	integrated
Harvest	90 % flowering	90 % flowering

DS: dry season; RS: rainy season; DAP: day after planting; PUTR: swamp soil test device; LCC: leaf color chart.

Parameters observed included: 1) characteristics of the study area, 2) growth parameters (plant height and number of productive tillers), and crop production (number of filled grains, number of empty grains, thousand grains weight, and harvest dry grains (HDG), and 3) farm analysis (costs, revenues, profits, and the value of the B/C ratio). Plant samples were taken randomly, each parameter consisting of 10 plants. Grain production data was taken with 5m x 5m tiles 3 times. The rice growth and production parameters' data were analyzed using SAS software through an Oversite design. Meanwhile, Analysis of the financial feasibility of rice farming includes net income and the value of the B/C ratio using the input-output analysis method [16].

$$B/C\ ratio = \frac{(RAVC)}{TVC} \quad (1)$$

Description:
B/C ratio = Revenue to cost ratio
P = Rice selling price (IDR kg^{-1})
TVC = Total cost (IDR ha^{-1} $season^{-1}$)

$$RAVC = (Q \times P) - TVC \quad (2)$$

Q = Total rice production (kg ha^{-1} $season^{-1}$)
Meaning:
B/C ratio > 1, farming is feasible
B/C ratio = 1, The farm is at the break-even point
B/C ratio < 1, farming is not feasible

3 Result and discussion

3.1 The research site characteristics

The area of swamps in Rambutan District reaches 28,158 ha, 46.70% of the total area of the sub-district, consisting of 12,134 ha (20.12%) of shallow swamp lands, 7,504 ha (12.44%) of middle swamps and 7,504 ha (12.44%) of deep swamps 8,520 ha (14.13%). Shallow swamp land has the potential to be developed into lowland rice fields [3]. Gelebak Dalam village has a land area of 3.77% (16,98 km^2) of the Rambutan District, Banyuasin Regency [18].

The development of swamp land for agricultural activities, especially rice plants, has various obstacles, both physical and socio-economic constraints. To increase rice production in lowland swamps, it is necessary to adapt water management technology for the cultivation of food crops as a result of the effects of climate change by controlling the height of groundwater and rivers to reduce flooding or inundation of lowland swamps [18].

The land at the study site has a flat to undulating relief so that the higher parts of the swamp land are planted with rice first, followed by the lower parts of the swamp. The condition of the research land has a low fertility rate with an acid soil pH (4.2 – 5.4) so the application of fertilizer and liming will be very beneficial. Irrigation systems in swamps are almost completely natural, flooded during the rainy season and dry during the dry season. The main obstacle faced in increasing production is inundation or drought which cannot be predicted precisely.

The average yield of swamp rice in Gelebak Village, Rambutan District during the first planting season amounted to 4.11 t ha^{-1}. In contrast, the productivity of swamp rice during the second planting season was recorded at 2.64 t ha^{-1}. Rice plants contain aerenchymal tissues that permit air passage to organs (roots) that are buried in wet soil, so rice plants can adapt to paddy fields that are always or occasionally flooded with water [19]. Sustainable land management in sub-optimal wetlands can be implemented through three approaches, namely water, soil, and plant management [20].

Applied CI200 cropping pattern on swamp land of Ogan Ilir Regency, South Sumatra namely rice planting in the dry season is carried out in February-May, the age of the seedlings is 30-40 DAP, while during the rainy season it is carried out in October-February, seedling age 25-30 DAP. The number of seeds is 2-3 per hole and the number of seeds is 40 kg ha^{-1} [21].

Several factors that need to be considered in developing CI200 cropping patterns are a) sufficient water availability including time and duration of availability, b) soil conditions including physical, chemical, and soil surface characteristics, c) altitude above sea level, especially its relation to air temperature, soil, and water, d) existence of chronic and potential plant pests and diseases, e) availability and accessibility of rice types and varieties, f) accessibility and smoothness of production results with infrastructure support, g) capital capacity of farmers to develop these cropping patterns, and h) socio-cultural characteristics of farmers related to technology adoption and development [21].

3.2 Growth and productivity of rice in swamp land

Growth parameters such as plant height and number of productive tillers were not significantly different between superior varieties planted both in the rainy and dry seasons. Meanwhile, the production parameters such as the number of empty and filled grains and HDG productivity were significantly different between superior varieties and planting seasons. Planting superior varieties of rice during the rainy and dry seasons provides a significant interaction on the thousand grains weight (Table 2).

The height of the superior varieties of rice tested has moderate postures, with a plant height of 90-125 cm [22]. The Inpara 2 variety has a lower posture compared to the other varieties tested, while Ciherang has a taller plant posture (Table 3). The shorter rice posture is generally liked by farmers because the plants will fall

Table 2. Anova table for plant height, number of productive tillers, number of empty grains, number of filled grains, thousand grains weight, and harvest dry grains as influenced by superior varieties and planting season

Dependent variables	plant height		number of productive tillers		number of empty grains		number of filled grains		thousand grains weight		Harvest Dry Grains	
	MS	F value	MS	F value	MS	F value	MS	F value	MS	F value	MS	F value
Varieties (V)	315.67	1.77^{ns}	154.67	0.58^{ns}	12.09	2.35^{ns}	6747.29	316.03**	0.09	43.00**	4.44	29.09**
Season (S)	422.50	2.36^{ns}	102.40	0.38^{ns}	24.03	4.68*	172.23	8.07*	0.04	18.00**	1.40	9.19*
V x S	196.97	1.1^{ns}	238.67	0.89^{ns}	1.76	0.34^{ns}	10.56	0.49^{ns}	0.01	3.67*	0.02	0.11^{ns}

*,** Significant difference at 0.05 and 0.01 probability levels, respectively; ns: nonsignificant.

due to the wind blowing, but at the time of the research, none of the varieties collapsed. The difference in plant height of the four varieties tested was due to genetic factors and environmental influences, such as soil fertility, spacing, water conditions, and temperature [23]. Plant height is one of the selection criteria for rice plants, but high posture does not guarantee high production [24] and productivity [25].

Table 3. Rice growth and productivity parameters of several superior varieties were planted during the dry and rainy seasons.

Factor	Parameter			
	plant height (cm)	number of productive tillers	number of empty grains	number of filled grains
Varieties:				
Inpari 32	99.6 a	16.8 a	13.8 a	192.7 a
Inpara 2	95.5 a	16.4 a	12.8 a	151.1 b
Mekongga	103.4 a	14.0 a	12.1 a	141.3 c
Ciherang	108.7 a	23.2 a	11.2 a	135.0 c
Season:				
Dry	98.55 a	16.0 a	11.7 b	157.1 a
Rainy	105.05 a	19.2 a	13.25 a	152.95 b

Different letters in the same column indicate significant differences according to the F test (P ≤ 0.05).

The Ciherang variety has a higher number of productive tillers per clump compared to other varieties, while the Mekongga has the lowest number of productive tillers (Table 3). According to Ramadhani et al [26] research, the Ciherang varieties display more productive tillers per clump and demonstrate resistance to pests and adaptability. The difference in the number of productive tillers is caused not only by the type of variety but also by environmental factors. As stated by Waluyo and Suparwoto [25] which stated that the height of waterlogging in swamps greatly affects the number of tillers. Factors that influence tiller formation are spacing, fertilizer, varieties, and planting season [27]. Genetic and environmental factors determine the growth and development of plants. The number of productive tillers is a factor that greatly influences rice productivity [28].

The number of filled grains of the superior varieties tested was independently influenced by genetic (variety) and environmental (season) factors, while the number of empty grains was strongly influenced by the environment (Table 2). Lack of water at the flower formation stage will cause a decrease in the number of grains per panicle, but if it occurs at the pollination/fertilization stage it will cause an increase in the number of empty grains per panicle [29]. The percentage of the number of empty grains to the number of filled grains of the superior varieties tested has a value of less than 10%. The Inpara 32 variety had the highest number of filled grains compared to other varieties, and the highest number of empty grains occurred during the rainy season (Table 3). The high or low percentage of filled grains per panicle is caused by differences in response and resistance of each variety to unfavorable environmental conditions, especially during the reproductive phase.

Table 4. The thousand grains weight (g) of several superior varieties planted during the dry and rainy seasons

Season	Varieties			
	Inpari 32	Inpara 2	Mekongga	Ciherang
Dry	27.2 ab	27.0 c	27.28 ab	27.2 ab
Rainy	27.24 ab	27.14 bc	27.3 a	27.24 ab

Different letters in the same column indicate significant differences according to the F test (P ≤ 0.05).

The use of high-yielding rice varieties and the determination of the planting season provide a very real interaction with the parameter weight of a thousand grains. The highest of a thousand grains weight was achieved by the Mekongga variety which was planted in the rainy season. Meanwhile, the Inpara 2 variety planted during the dry season had the lowest thousand grains weight (Table 4). The high and low rice grains per panicle are caused by differences in response and resistance of each rice variety to unfavorable environmental conditions, especially during the reproductive and ripening phases. The higher the percentage of rice grains, the lower the percentage of empty grains.

The Inpari 32 variety had the highest productivity of 6.98 t HDG ha^{-1}, and the Ciherang variety had the lowest productivity of 5.03 t HDG ha^{-1} (Table 5). Paddy cultivation in swamps during the rainy season can produce 5.0-6.0 t HDG ha^{-1} [12]. Ciherang planted during the rainy season in swamp rice fields in Kotadaro II Village, Rantau Panjang Regency, Ogan Ilir District, South Sumatra also produced a productivity of 5 t ha^{-1} [30]. Inpari is more suitable in swampy rice fields which are inundated most of the year [12]. Swamp rice cultivation by using new superior varieties can reach productivity up to 5.0-7.0 t HDG ha^{-1} [31] and 10 t ha^{-1} [32]. The capacity of plants to relocate assimilates during seed filling and accumulate dry matter before heading affects the variation in rice yield [33].

Table 5. The harvest of dry grains from several superior varieties planted during the dry and rainy seasons

Season	Varieties				mean
	Inpari 32	Inpara 2	Mekongga	Ciherang	
Dry	7.27	5.90	5.50	5.30	5.99 a
Rainy	6.70	5.40	5.17	4.77	5.51 b
mean	6.98 a	5.65 b	5.33 b	5.03 b	(-)

Different letters in the same column indicate significant differences according to the F test ($P \leq 0.05$).

Table 5. showed that planting superior varieties of rice in swamps can be carried out throughout the year because they are not affected by the choice of planting season, whether the dry or rainy season. This is as conveyed by Waluyo and Suparwoto [34] and Paiman *et al.* [12] stated that one of the efforts to improve and maintain national food security is through the use of suitable and adaptive varieties, both location and season-specific.

Superior rice varieties reliably boost output because they yield more, are more resistant to disease attacks and pests, have shorter lives [12,35,36], and have fluffier tastes [35]. Rice varieties with high yields must have a yield index of 0.45–0.50, medium plant age (100–120 days), and a yield potential of 10 t ha^{-1} in the dry season and 7–8 t h^{-1} in the rainy season [37]. The yield potential of a rice variety is determined by four components, namely panicle length, number of grains per panicle, percentage of filled grains, empty grains, and thousand grains weight [38].

Based on the results of the correlation analysis, show that the plant height parameter has a positive correlation with a thousand grains weight, with a correlation coefficient of 0.72. The parameter of the number of filled grains is very significantly positively correlated with harvested dry grains, with a correlation coefficient value of 0.97 (Table 6). This shows that a higher number of filled grains will further increase the productivity of harvested dry grains per unit area. Therefore, efforts to increase rice productivity in lowland swamps can be made by increasing the number of filled grains per panicle, not per plant cluster. This can be seen in Table 6. that the number of productive tillers per clump is not significantly correlated with rice productivity, namely the number of filled grains and harvested dry grains.

Table 6. Correlation analysis between parameters observed in swamp rice cultivation using CI200

Parameter	NPT	TGW	HDG
PH	0.46999	0.72079*	-0.50352
NG	-0.21435	0.63453	0.97453**

*,** Significant difference at 0.05 and 0.01 probability levels, respectively; NPT: number of productive tillers: TGW: thousand grains weight; HDG: harvest dray grains, PH: plant height; NFG: number of filled grains.

The Ciherang variety with a higher number of productive tillers and a taller plant posture, but with the lowest number of filled grains, and lowest productivity compared to other superior varieties. Therefore, the selection of rice varieties greatly determines the productivity of rice in swamps. Compared to higher posture plants, the moderate posture is typically more resilient to falling over brought on by the wind or other cultivation variables [39].

The introduction of NSV is expected to be able to increase rice yield compared to varieties that have been commonly planted in previous seasons. The results of the study by Sirappa et al. [40] showed that the introduction of new superior varieties of rice supported by appropriate cultivation technology could increase yields by 21-54%. Therefore, the achievement of a variety must be supported by technology and an optimal growing environment. Lowland rice farming on shallow land has been found to significantly improve production. The rice production in microclimates within the South Sumatra Province contributes to national rice production.

3.3 Feasibility of CI200 Rice Farming in Swamp land

The results of the analysis of the feasibility of rice farming from the four varieties tested and planted at the time DS and RS provided quite diverse income, net revenue, and B/C ratio values. Rice farming of the Inpari 32 variety that is planted during the dry and rainy seasons provides the highest profit value in terms of cash and total costs compared to other superior varieties. Meanwhile, the Menkongga variety planted during the dry and rainy seasons provides the lowest profit value in terms of cash or total costs. Therefore, the value of the B/C ratio of rice farming that planted the Inpari 32 variety throughout the year had the highest values, namely 2.45 (Table 7) and 1.98 respectively (Table 8). This shows that the CI200 cropping pattern in swamp paddy fields using superior varieties is very feasible to develop on a large scale.

Table 7. Farming analysis of swamp rice cultivation using several superior varieties in the dry season (Maret-July)

No	Production facilities	Unit	Unit price (IDR)	Amount (IDR)
1.	Seeds (kg)	30	10,000	300,000
2.	Fertilizer			
	Urea (kg)	150	2,250	337,500
	Phonska (kg)	150	2,300	345,000
	Manure (kg)	1,000	500	500,000
3.	Pesticides			
	Reagent (250 ml)	1,500	71,500	572,000
	Antracol 70 WP (500 g)	3,500	155,000	1,085,000
	Sidamethrin 50 EC (400 ml)	3,200	42,000	336,000
	Spontam 400 SL (500 ml)	3,500	60,000	420,000
	Amount (IDR)			**3,550,500**
4.	Labor (man.days)			
	Land processing	35	50,000	1,750,000
	Nursery	5	50,000	250,000
	Planting	20	50,000	1,000,000
	Weeding	20	50,000	1,000,000
	Fertilization	5	50,000	250,000
	Spraying	5	50,000	250,000
	Harvest and post-harvest	50	50,000	2,500,000
	Amount (IDR)			7,000,000
	Total cost (IDR)			**10,550,500**
5	Income (IDR)			
	Inpari 32 (kg)	7,250	5,000	36,250,000
	Inpara 2 (kg)	5,900	5,000	29,500,000
	Mekongga (kg)	5,500	5,000	27,500,000
	Ciherang (kg)	5,300	5,000	26,500,000
6	Net Revenue (IDR)			
	Inpari 32			25,699,500
	Inpara 2			18,949,500
	Mekongga			16,949,500
	Ciherang			15,949,500
7	B/C ratio			
	Inpari 32			2.44
	Inpara 2			1.80
	Mekongga			1.61
	Ciherang			1.51

Table 8. Farming analysis of swamp rice cultivation using several superior varieties in the rainy season (October-February)

No	Production facilities	Unit	Unit price (IDR)	Amount (IDR)
1.	Seeds (kg)	30	10,000	300,000
2.	Fertilizer			
	Urea (kg)	150	2,250	337,500
	Phonska (kg)	150	2,300	345,000
	Manure (kg)	1,000	500	500,000
3.	Pesticides			
	Reagent (250 ml)	1,500	71,500	572,000
	Antracol 70 WP (500 g)	3,500	155,000	1,085,000
	Sidamethrin 50 EC (400 ml)	3,200	42,000	336,000
	Spontam 400 SL (500 ml)	3,500	60,000	420,000
	Klerat/rat poison (kg)	10	69,950	699,500
	Amount (IDR)			**4,250,000**
4.	Labor (man.days)			
	Land processing	35	50,000	1,750,000
	Nursery	5	50,000	250,000
	Planting	20	50,000	1,000,000
	Weeding	20	50,000	1,000,000

	Fertilization	5	50,000	250,000
	Spraying	5	50,000	250,000
	Harvest and post-harvest	50	50,000	2,500,000
	Amount (IDR)			7,000,000
	Total cost (IDR)			**11,250,000**
5	Income (IDR)			
	Inpari 32 (kg)	6,700	5,000	33,500,000
	Inpara 2 (kg)	5,400	5,000	27,000,000
	Mekongga (kg)	5,200	5,000	26,000,000
	Ciherang (kg)	4,700	5,000	23,500,000
6	Net Revenue (IDR)			
	Inpari 32			22,250,000
	Inpara 2			15,750,000
	Mekongga			14,750,000
	Ciherang			12,250,000
7	B/C ratio			
	Inpari 32			1.98
	Inpara 2			1.40
	Mekongga			1.31
	Ciherang			1.09

Shallow and middle swamps could apply CI100 and CI200 planting patterns throughout the year. However, the deep swamps are only implemented CI100 during the dry season [41]. Rotation of superior varieties planted is necessary to prevent the explosion of pests and diseases. However, this will cause a loss of information when the highest productivity is achieved by one superior variety in the swamp paddy field. At the beginning of the rainy season, NSV should be selected which is resistant to planthoppers and resistant to several diseases. The selection of NSV must pay attention to the existence of endemic pests and diseases. In brown planthopper endemic areas, the use of Mekongga, Ciherang, Cibogo, Inpari 2, and Inpari 3 varieties is highly recommended, while those that are endemic to tungro should be planted with Inpari 7, 8, and 9. Planthopper pest populations are more numerous during RS compared to DS, while borer attacks are not looking at the season [21].

4 CONCLUSION

The Inpari 32 variety had the highest productivity of 6.98 t HDG ha^{-1}. The value of the B/C ratio reached 1.55-2.45 in the DS and 1.13-1.98 in the RS. This shows that the application of the Rice Cultivation Technology Package in Lebak rice fields using superior rice varieties effectively supports the implementation of CI200. Therefore, the application of the CI200 planting pattern using the adapted NSV can be widely developed to support food security in swamp lands.

References

1. A. Suryana. Orasi Pengukuhan Profesor Riset Bidang Sosial Ekonomi Pertanian, Badan Litbang pertanian. Bogor (2007)
2. Waluyo, Fuadi Irsan, dan Yanter Hutapea, *Inovasi teknologi pengelolaan sumberdaya lahan rawa lebak di Sumatera Selatan*, dalam Buku Bunga Rampai Teknologi Unggulan Pengelolaan Lahan Sub Optimal di Sumatera Selatan, Unsri Press, Palembang (2021)
3. Badan Pusat Statistik Sumsel, *Luas lahan menurut penggunaan di Sumatera Selatan*, dalam Badan Pusat Statistik Provinsi Sumatera Selatan, Palembang (2022)
4. T. Kartinaty, R. Massinai, D. P. Widiastuti, J. TABARO, **5**(2): 20-29 (2022)
5. Kiswanto dan F. Y. Adriyani, *Uji adaptasi varietas unggul baru padi sawah di Kecamatan Pubiana Lampung Tengah*, dalam Prosiding Seminar Nasional Pengkajian dan Diseminasi Inovasi Pertanian Menduking Program Strategi Kementrian Pertanian Buku 2, Balai Besar Pengkajian dan Pengembangan Teknologi Pertanian, Cisarua 9-11 Desember 2010, Bogor (2011)
6. F. D. Arianti, R. Nurlaily, M. N. Setiapermas, *Peningkatan produktivitas padi melalui penggunaan varietas unggul baru dan pemupukan di lahan sawah tadah hujan*, dalamProsiding Seminar Nasional Teknologi Pertanian Kesiapan Sumber Daya Pertanian dan Inovasi Spesifik Lokasi Memasuki Era Industri 4.0, Balai Besar Pengkajian dan Pengembangan Teknologi Pertanian, pp 191-196, Bogor (2020)
7. I. N. Arnama, Perbal: Jurnal Pertanian Berkelanjutan **8** (2): 166-175 (2020)
8. Y. Rina and K. Koesrini, J. Researcher. Question. Plant. Food, **2**(2): 85-94 (2018) doi: 10.21082/jpptp.v2n2
9. N. Chairuman, J. Pertan. Trop., **1**(1): 47–54 (2013) doi: 10.32734/jpt.v1i1.2867
10. Helmi, Jurnal Pertanian Tropik, **2**(2):78- 92 (2015)
11. E. Saleh, C. Irsan, M. U. Harun, *Evaluasi pola tanam pada lahan rawa lebak dangkal Ogan Keramasan*, dalam Laporan Penelitian, Universitas Sriwijaya, Palembang (2013)

12. Paiman, Ardiyanta, M. Ansar, I. Effendy, and B. T. Sumbodo, Agricultural Science, **8**: 300–309 (2020)
13. D. Abubakar, L. Anggraeni, and A. Fariyanti, J. Ekon. And Wise Developer, **8**(2): 120–144 (2019) doi: 10.29244/jekp.8.2.2019.120-144.
14. Y. Nearti, B. Fachrudin and N. Jayanti, Sriwijaya J. of Environment, **7**(2): 118-128 (2022) http://dx.doi.org/10.22135/sje.2022.7.2.118-128
15. Waluyo, F. Irsan and Y. Hutapea. *Inovasi teknologi pengelolaan sumberdaya lahan rawa lebak di Sumatera Selatan,* dalam Bunga Rampai Teknologi Unggulan Pengelolaan Lahan Sub Optimal di Sumatera Selatan, hal: 108-127, UNSRI Press, Palembang (2021)
16. R. Hendayana, *Analisis data sosial ekonomi: Cerdas dan cermat menggunakan alat analisis data untuk karya tulis ilmiah*, IAARD PRESS, Bogor (2015)
17. BPS. *Kecamatan Rambutan dalam angka.* Penerbit: BPS Kabupaten Banyuasin, Pangkalan Balai, 137 p (2022). https://banyuasinkab.bps.go.id/publication/download.html?nrbvfeve=MjI0ZDU0NTc3YzUwY2UyZWIwYWYwOWNh&xzmn=aHR0cHM6Ly9iYW55dWFzaW5rYWIuYnBzLmdvLmlkL3B1YmxpY2F0aW9uLzIwMjIvMDkvMjYvMjI0ZDU0NTc3YzUwY2UyZWIwYWYwOWNhL2tlY2FtYXRhbi1yYW1idXRhbi1kYWxhbS1hbmdrYS0yMDIyLmh0bWw%3D&twoadfnoarfeauf=MjAyMy0wOS0yMyAyMTo0OTo1Mw%3D%3D
18. A. N. Triana, *Kajian pengelolaan tanah dan air lahan rawa lebak*, dalam Prosiding Seminar Nasional Hari Air Dunia, 21 Maret 2019, Palembang (2019)
19. R. Velásquez, F. García, D. Diamont, M. Perdomo, G. Torrealba, Agronomía Trop. **64** 39–47 (2014)
20. M. E. Armanto, R. H. Susanto, and E. Wildayana, Sriwijaya Journal of Environment. **2**(1): 1-7 (2017)
21. Balibangtan, Agroinovasi, dalam Sinar Tani, Edisi 8-14 Februari No. 3443 XLII, Jakarta (2012)
22. IRRI, Standard Evaluation System for Rice (SES), International Rice Research Institute, Philippines (2002)
23. Waluyo, Suparwoto, dan Sudaryanto, J. Hidrosfir Indonesia **3**(2): 57–66 (2008)
24. Suprapto dan A. Dradjat. Buletin Plasma Nutfah Vol. **11** No. 1 (2005)
25. Waluyo dan Suparwoto, *Pemanfaatan lahan rawa lebak dalam mendukung ketahanan pangan melalui teknologi spesifik lokasi di Sumatera Selatan*, dalam Prosiding Seminar Nasional Membangun Good Govermance Menuju Desa Mandiri Pangan dan Energi Menuju ERA MEA, FakultasPertanian Sebelas Maret, 27-28 April 2016, Surakarta (2016)
26. S. Ramadhani, N. B. Astuti, and M. Mahdi, JOSETA J. Socio-economics Trop. Agric., **1**(3): 68–77 (2019) doi: 10.25077/joseta.v1i2.147
27. H. R. Sugeng, *Bercocok Tanam Padi*, CV Aneka Ilmu, Semarang (2001)
28. O. S. Lesmana, H. M. Toha, I. Las, dan B. Suprihanto, *Varietas unggul baru padi*, Balai Penelitian Tanaman Padi, Sukamandi (2004)
29. Sujinah dan A. Jamil. Iptek Tanaman Pangan, **11**(1): 1-7 (2016)
30. Irmawati, H. Ehara, R. A. Suwignyo, and J. I. Sakagami, Trop. Agr. Develop. **59**(1): 35-39 (2015)
31. Suparwoto, Waluyo, dan Syahri, Laporan akhir: PTT lahan rawa lebak Kabupaten Ogan Ilir Sumatera Selatan. Balai Pengkajian Teknologi Pertanian Sumatera Selatan (tidak dipublikasi), Palembang (2014)
32. M. D. Hapsah, *Potensi, peluang, dan strategi pencapaian swasembada beras dan kemandirian pangan nasional.* dalam Inovasi Teknologi Padi Menuju Swasembada Beras Berkelanjutan. Buku Satu. Balitbangtan, Badan Litbang Pertanian, Hal. 55-70, Jakarta (2005)
33. A. Hambali dan I. Lubis, Bul.Agrohorti **3**(2): 137–14 (2015)
34. Waluyo dan Suparwoto. *Pengkajian beberapa varietas unggul baru (VUB) padi di lahan rawa lebak kabupaten Ogan Komering Ilir Sumatera Selatan*, dalam Prosiding Seminar Nasional Lahan Suboptimal. Universitas Sriwijaya, 08-09 Oktober Palembang (2015)
35. M. N. H. Miah, T. Yoshida, Y. Yamamoto, and Y. Nitta, J.Crop Sci. **65**: 672–685 (1996)
36. Arinta dan I. Lubis, Buletin Agrohorti, **6**(2): 270-280 (2018)
37. A. Dobermann and T. H. Fairhurst, Better Crops International Special Suplement, **16**: 7-11 (2002)
38. S. Yoshida, *Fundamentals of rice crop science*, International Rice Research Institute, Los Banos, Laguna, Philippines (1981)
39. Sobrizal and M. Ismachin, Plant Mut. Rep. **1**(1) 18–21 (2006)
40. M. P. Sirappa, A. N. Susanto, dan Y. Tolla, Jurnal Pengkajian dan Pengembangan Teknologi Pertanian, Vol. **9**(1): 18-28 (2006)
41. Y. Pujiharti, Jurnal Litbang Pertanian, **36**(1):13-2 (2017)

AUTHOR INDEX

Abdilah, Moch Reza Wahyu 200
Abdusshamad, Bimo Rizki 346
Affandi, Fahrizal Yusuf 292
Affandy, Bagus Pramuditya 375
Agussabti, ... 114
Ahmadi, Tyasto Prima 9
Ahmed, Rezwan 49
Aji, Wicaksono Satrio 27
Akter, Tanjima 1
Alam, Alfindy Juliansari 22
Aline, Umuhoza 1
Alya, Hisna Shania 368
Amanah, Hanim Zuhrotul 351, 381
Amri, Ikhwan 42
Andarwati, Siti 66
Andhini, Wahyu Putri 128
Andika, Yan 42
Anggraini, Syifa Ayulia 33
Ardiansyah, Fiqri 98
Arfiandi, Ahmad 207
Ariyanto, Teguh 409
Armstrong, Leisa 361
Asrofi, Muhammad Zayyin 22, 27, 33
Astuti, Putri Kusuma 323
Ayu, Alifia Dyah 136
Ayuningsih, Ira 318
Azani, Muhammad Raidan 22
Bagi, Zoltán 323
Baihaqi, Akhmad 114
Bintoro, Nursigit 276
Bodrogi, Lilla 323
Budianta, Wawan 207
Budiman, Kresna 346
Bunyakanchana, Thasorn 284
Cen, Shiddharta Arya Anggoro 346
Cho, Byoung-Kwan 1
Cook, David M. 361
Darmawan, Adytia 375
Dewi, Astarina Eka 79, 141
Dharmawan, Agus 381
Dian, Arianti Ade Maya 27
Diepeveen, Dean 361
Djaafar, Titiek Farianti 423
Dwinata, Vincentius Ferry Surya 307
Edward, ... 284
Ekasari, Indriani 176
Endriani, ... 72, 121
Fahmi, Ahmad Arief 141
Fajardo, Roland 323
Fajri, Muhammad 66
Fajri, ... 114
Faqeerzada, Mohammad Akbar 1
Fauzi, Raka Achmad 221
Fauziyah, Eva 151
Fitria, Widhiningsih Diah 368
Fitriawan, Muhammad Bagus 17
Ginting, Litna Nurjannah 114
Guntoro, Budi 66
Hairani, Anna 264
Hamid, Ahmad Humam 114
Handini, Esti 207
Hani, Aditya 151
Harianti, Mimien 160
Harini, Rika 221
Herviyanti, ... 160
Hidayati, Iswari Nur 243
Hoque, Md. Ahasanul 49
Idwar, ... 227
Imron, Muhammad Ali 141
Indrajaya, Yonky 79
Indrayanti, Esty 415
Iqbal, Muhammad 42
Irham, ... 59
Ishigaki, Genki 335
Izdihar, Fahar Yazid 389
Izzatillah, Asalia Raudhati 42
Jihad, Aqmal Nur 98
Jumakir, ... 439
Juwita, Pitri 160
Kamal, Mustafa 331
Kamphayae, Sukanya 335
Karyadi, Joko Nugroho Wahyu 307
Khairani, Tasyarafa Shelia 423
Khoirotunnisa, Safira Defiani 146
Khuriyati, Nafis 389, 403
Khurnianto, Rezha Tri 160
Kinanti, Kartika Ayu 233
Kistanti, Anisa 292
Kristanti, Novita Erma 403, 434
Kurnia, Rifaangga 176
Kurniawan, Labriyantoko 346
Kurniawan, M. Prasetya 128
Kurniawan, Muhammad Prasetya 423
Kurniawati, Dyah Ayu 313
Kurniawati, Musyrifah 292
Kusumandari, Ambar 169

Kusza, Szilvia 323
Leksono, Bagas Oktavianto 339
Lestari, Sri 79, 141
Listanti, Riana 351
Listyarini, Diah 72, 121
Lita, Arestha Leo 160
Lubis, Nursiani 227
Mahanani, Riski Sri 307
Martono, Edhi 430
Marwati, Tri 423
Marzuki, Ahmad Fatih 9
Masithoh, Rudiati Evi 351, 381, 415
Masrianti, Satwika Indri 169
Maulana, Amsar 160
Mawadati, Nisa Puspita 403
Mo, Jianhua 430
Muamar, Muhammad 339
Muflikhun, Muhammad Akhsin 346
Muguerza, Melody 335
Mulyawan, Arif 233
Munba'Its, Fatih Ali Fikri El 375
Murdiati, Agnes 303
Mustika, I. Wayan 346
Muyyasir, ... 114
Nafisa, Gina Isna 146
Napier, Chris C. 361
Nelvia, ... 227
Ngadisih, ... 146
Ningrum, Andriati 303
Nitthaisong, Pattama 335
Nu'Man, Muchammad 146
Nugraha, Bayu 292, 318
Nugraha, Favian Daniswara 33
Nugroho, Agus 114
Nugroho, Bayu Dwi Apri 415
Nugroho, Darmawan Ari 9, 128
Nurkhalifah, Azizah 22
Nurudin, Makruf 108
Ohran, Husein 323
Pahlawan, Muhammad Fahri Reza 1
Pattaraprachyakul, Wasin 284
Pintér, Tímea 323
Pitowarno, Endra 339, 375
Pramudyani, Lelya 264
Prasetyo, Teguh Budi 160
Prasojo, Yogi Sidik 331, 335
Pratiwi, Dian 79, 141
Primatama, Satria 292
Priyanti, Devi 318
Pulungan, Nur Ainun H. J. 108, 193
Purwanto, Benito Heru 183
Putra, Eka T. S. 264
Putra, Ryan Anugrah 409
Putranti, Khoirida Dian 108
Putri, Febriana Rias 270
Putri, Melta Meliano 37
Putri, Ratih Fitria 17, 22, 27, 33, 37
Putri, Vincencia Septaviani Issera Sulistya 243
Raharjo, Sri 303
Rahayoe, Sri 318
Rahayu, Anisa Dewi 346
Rahmah, An-Nidaa' Fatkhur 318
Rahmawati, Dian Kharisma 307
Ramadhani, Dehafiyyan 200
Rekso, Gatot Trimulyadi 103
Ridyo, Arum Mutiarra 59
Rijanta, R. 221
Risada, Nabila Husna 17
Rizqi, Fathi Alfinur 193
Rizqiyah, Elia Laila 183
Rosmindari, Shiammita 276
Rusdi, Muhammad 114
Ryswaldi, Ridho 160
Salisa, Hasna Difa 434
Samodra, Ailsa Saraswati 307
Sanudin, ... 151
Saputra, Ferdi 227
Saputro, Arifin Dwi 276, 351
Saroyo, Pujo 313
Satriagasa, Muhammad Chrisna 136
Sean, Leonard 346
Semiarti, Endang 346
Senawi, ... 98
Setyaningrum, Haris 430
Setyawan, Chandra 146, 200
Shiddiqie, Habib Luthfi Ash 409
Sholihah, Umi Mar Atus 193
Sinuhaji, Prasetya Febriyansyech Personanta 307
Siripitakpong, Panattida 284
Siswoko, Bowo Dwi 331
Skoda, Gabriella 323
Sobue, Shin-Ichi 396
Soeherman, Yudistira 141
Soffan, Alan 430
Soraya, Emma 98
Sosiawan, Hendri 233
Subandiyah, Siti 430
Subejo, ... 368
Sukartiko, Anggoro Cahyo 270, 434
Sukmana, Arkan Roja 331
Supartono, Wahyu 270, 313
Suparwoto, ... 439
Suppavorasatit, Inthawoot 284
Suprihatin, Agus 439

Suryatmojo, Hatma 136, 169
Susanti, Devi Yuni 307
Susanto, Sahid 200
Suyantohadi, Atris 9
Syahda, Dyfany Aurariel 37
Tafrichan, Muhammad 141
Taryono, ... 264
Tongdeenok, Piyapong 252
Umami, Lathief Al 346
Umami, Nafiatul 331, 335
Umarhadi, Deha Agus 98
Utama, Zaki 303
Utami, Sri N. H. 264
Utami, Sri Nuryani Hidayah 183, 233
Utomo, Eko Budi 339
Wagiman, ... 270, 389
Wahyu, Utami Arini 59
Waluyo, ... 439
Wanjala, George 323
Wardhana, Wahyu 98
Widada, Jaka 233
Widiyatno, ... 136
Widodo, Kuncoro Harto 423
Widura, Elsa 42
Wijayanti, Lies Rahayu 243
Wiranata, Ardi 409
Wongprasert, Thanakorn 284
Wulan, Putri Yuko Nawang 37
Yaseva, Yosi 17, 37
Yasin, Syafrimen 160
Yusuf, Wahina Anisa 233
Yuwati, Tri Wira 79, 141
Zahra, Aryanis Mutia 415
Zaki, Muhamad Khoiru 200
Zulfatri, ... 227